7/10/90

Nuclei and Particles
Second Edition

Nuclei and Particles

An Introduction to Nuclear and Subnuclear Physics

SECOND EDITION
Completely revised, reset, enlarged

EMILIO SEGRÈ

Professor of Physics (Emeritus)
University of California, Berkeley

Addison-Wesley Publishing Company, Inc.
The Advanced Book Program
Redwood City, California • Menlo Park, California • Reading, Massachusetts
New York • Amsterdam • Don Mills, Ontario • Sydney • Bonn • Madrid
Singapore • Tokyo • San Juan • Wokingham, United Kingdom

NUCLEI AND PARTICLES
First edition, 1964
Second edition, completely revised, reset, enlarged, 1977

First printing, 1977
Second printing, with revisions, 1980
Third printing, 1982

Library of Congress Cataloging in Publication Data

Segrè, Emilio.
  Nuclei and particles.

  Bibliography: p.
  includes indexes.
  1. Nuclear physics. 2. Particles (Nuclear physics)
I. Title.
QC776.S4 1977   539.7   77-10845
ISBN: 0-8053-8601-7

Foreign-language versions of the first edition:
  Italian: N. Zanichelli, Bologna
               1966
  Japanese: Yoshioka Shoten, Kyoto
               1975
  Spanish: Editorial Reverté, Barcelona
               1972

Manufactured in the United States of America

GHIJ-HA-898

Contents

Preface to the Second Edition   xiii

Preface to the First Edition   xv

Selected Bibliography   xvii

CHAPTER 1	**HISTORY AND INTRODUCTION**	1
1-1	Units	10
1-2	Terminology and Definitions	12
	Bibliography	15
	Problems	15

Part I	**Tools**	
CHAPTER 2	**THE PASSAGE OF RADIATIONS THROUGH MATTER**	19
2-1	Introduction	19
2-2	Rutherford Scattering	22
2-3	Energy Loss due to Ionization	28
2-4	Energy Loss of Electrons	36
2-5	Polarization Effects—Cerenkov Radiation	41
2-6	Ionization in Gases and Semiconductors	42
2-7	Multiple Scattering	44
2-8	Straggling	48
2-9	Passage of Gamma Rays through Matter	54
2-10	Radiation Loss by Fast Electrons	62

2-11   Radiation Length; Showers   71
2-12   Positron Annihilation   74
2-13   Polarization Phenomena   78
        Bibliography   82
        Problems   83

CHAPTER 3   DETECTION METHODS FOR NUCLEAR RADIATIONS   86

3-1   Classification   86
3-2   Ionization Chambers and Solid-State Detectors   88
3-3   Proportional and Geiger-Müller Counters   94
3-4   Scintillation Counters   97
3-5   Cerenkov Counters   101
3-6   Photographic Emulsions   104
3-7   Nuclear Tracks in Solids   107
3-8   Cloud Chambers   109
3-9   Bubble Chambers   112
3-10   Spark Chamber and Streamer Chamber   114
3-11   Electronics   117
3-12   Complex Nuclear Instruments   120
3-13   Charged-Particle Spectrometers   123
3-14   Gamma-Ray Spectrometers   126
        Bibliography   128
        Problems   128

CHAPTER 4   PARTICLE ACCELERATORS   130

4-1   Introduction and Classification   130
4-2   Potential-Drop Accelerators   135
4-3   The Betatron   136
4-4   The Cyclotron   141
4-5   Phase Oscillations and Stability   143
4-6   The Synchrotron and Proton Synchrotron   149
4-7   Strong Focusing   149
4-8   Linear Accelerators   153
4-9   Colliding Beams   156
4-10   Beam-Transport Apparatus   161
        Bibliography   166
        Problems   167

CHAPTER 5   RADIOACTIVE DECAY   168

5-1   Continuum Theory—One Substance   169
5-2   Continuum Theory—More Than One Substance   170
5-3   Branching   175
5-4   Some Units Used in Radioactivity; Dosimetry   176
5-5   Fluctuations in Radioactive Decay—General Theory   180

5-6	Fluctuations in Radioactive Decay—Applications	189
5-7	Method of Maximum Likelihood	193
5-8	Methods of Measuring Decay Constants	195
5-9	Chronological and Geological Applications	200
	Bibliography	202
	Problems	202

Part II   The Nucleus

CHAPTER 6   ELEMENTS OF THE NUCLEAR STRUCTURE AND SYSTEMATICS   207

6-1	Charge	208
6-2	Mass	211
6-3	Nuclear Radii	219
6-4	Mesic and Exotic Atoms	227
6-5	Nuclear Statistics	232
6-6	The Nucleus as a Fermi Gas	236
6-7	The Liquid Drop Model	240
6-8	Electric Moments of Nuclei	249
6-9	Spin and Magnetic Moments I	254
6-10	Spin and Magnetic Moments (Measurements) II	265
6-11	Nuclear Polarization	272
6-12	Values of Spin, Magnetic Moments, and Quadrupole Moments	274
6-13	Shell Model	279
6-14	The Pairing Interaction	287
6-15	Collective Nuclear Model	293
6-16	Rotational Levels	300
6-17	Concluding Remarks	311
	Bibliography	313
	Problems	314

CHAPTER 7   ALPHA EMISSION   319

7-1	Introduction: Barrier Penetration	319
7-2	Fine Structure of Alpha Spectra	328
7-3	Systematics of Alpha Decay	330
7-4	Virtual Binding	334
	Bibliography	339
	Problems	340

CHAPTER 8   GAMMA EMISSION   341

8-1	Introduction	341
8-2	Selection Rules	351
8-3	Transition Probabilities	355

8-4	Internal Conversion	362
8-5	Nuclear Isomerism	369
8-6	Angular Correlations in Gamma Emission	373
8-7	Coulomb Excitation	375
8-8	Nuclear Fluorescence	377
	Bibliography	388
	Problems	388

CHAPTER 9   BETA DECAY   391

9-1	Introduction	391
9-2	Experiments on the Neutrino	395
9-3	Energetics of Beta Decay	402
9-4	Classification of Interactions. Parity	404
9-5	Fermi's Theory of Beta Decay	410
9-6	Matrix Element	421
9-7	Further Experiments on the Weak Interaction	426
9-8	Theory of the Beta Interaction	431
9-9	Quantitative Study of Some Matrix Elements	436
9-10	Conservation of Leptons	440
9-11	Universal Fermi Interaction	443
	Bibliography	444
	Problems	445

CHAPTER 10   THE TWO-BODY SYSTEMS AND NUCLEAR FORCES   446

10-1	The deuteron	447
10-2	Low-Energy Neutron-Proton Scattering	450
10-3	Proton-Proton System and Scattering	455
10-4	Charge Independence of Nuclear Forces—Isotopic Spin	458
10-5	Spin-Dependent and Tensor Forces	469
10-6	Nucleon-Nucleon Forces; Exchange Forces	471
10-7	Polarization; High-Energy Nuclear Scattering	481
10-8	Capture of Slow Neutrons by Hydrogen	491
10-9	Photodisintegration of the Deuteron	496
	Bibliography	498
	Problems	498

CHAPTER 11   NUCLEAR REACTIONS   501

11-1	Introduction	501
11-2	General Features of Cross Sections	505
11-3	Inverse Reaction—Detailed Balance	510
11-4	Reaction Mechanisms. The Compound Nucleus (Qualitative)	512
11-5	Formal Developments—Scattering Matrix	526
11-6	Resonances	535
11-7	Optical Model	547

11-8   Compound Nucleus—Level Density   552
11-9   Direct Reactions   560
11-10   The Fission Process   572
11-11   Heavy-Ion Nuclear Reactions   595
11-12   Photonuclear Reactions   601
11-13   "Intermediate Energy" Reactions   608
        Bibliography   612
        Problems   613

CHAPTER 12   NEUTRONS   616

12-1   Neutron Sources   618
12-2   Slowing Down of Neutrons   621
12-3   Energy Distributions of Neutrons from a
        Monoenergetic Source   626
12-4   Mean Distance from a Point Source vs. Energy   628
12-5   Diffusion Theory—Introduction   631
12-6   The Age Equation   632
12-7   Diffusion of Thermal Neutrons   636
12-8   Chain-Reacting Pile   638
12-9   Pile Kinetics   649
12-10   Breeding and Converting   652
12-11   Fusion Reactions   652
12-12   Effect of Chemical Binding of Hydrogen Scatterer   657
12-13   Low-Energy Scattering from Complex Nuclei   659
12-14   Determination of Scattering Lengths   660
12-15   Scattering in Ortho- and Parahydrogen   664
12-16   Interference Phenomena in Crystals   665
12-17   Index of Refraction   670
12-18   Polarization of Slow-Neutron Beams   673
        Bibliography   675
        Problems   675

Part III   Particles

CHAPTER 13   INTRODUCTION TO PARTICLE PHYSICS   681

13-1   General Ideas, Nomenclature, and Catalogue of Particles 681
13-2   Associated Production; Strangeness   696
13-3   Interactions; Conservation Laws   704
13-4   Some New Symmetries and Selection Rules   707
13-5   CPT Theorem   712
13-6   Crossing Relations   713
13-7   Experiments on Mass, Life, and Other Particle Properties 716
        Bibliography   725
        Problems   725

CHAPTER 14   LEPTONS   730

  14-1   Neutrinos   731
  14-2   Muon Production and Decay   732
  14-3   Muon Capture   737
  14-4   Spin and Magnetic Moment of Muons   740
         Bibliography   747
         Problems   747

CHAPTER 15   PIONS AND OTHER BOSONS   748

  15-1   The Yukawa Interaction   748
  15-2   Spin of the Pions   754
  15-3   Intrinsic Parity   757
  15-4   Isotopic Spin of Pions   763
  15-5   Pion-Nucleon Scattering and Resonances   767
  15-6   Nuclear-Collision Production and Photoproduction
         of Pions   777
  15-7   The ρ, ω, and Other Strongly Decaying Bosons   779
  15-8   Dalitz Plots   784
  15-9   The η and K Mesons   790
  15-10 Peripheral Collisions   795
         Bibliography   798
         Problems   799

CHAPTER 16   BARYONS   801

  16-1   Baryon Generation   801
  16-2   Baryon Spin Measurements   809
  16-3   Hyperfragments   810
         Bibliography   814
         Problems   814

CHAPTER 17   CLASSIFICATION OF HADRONS, QUARKS, AND SU (3)   816

  17-1   Sakata's Model; Quarks   816
  17-2   Combination of Quarks   819
  17-3   Mass Formulas   824
  17-4   Cross-Section Predictions by SU (3) and Quarks   829
  17-5   Regge Recurrences   829
  17-6   Charm   832
  17-7   Color   836
         Bibliography   839
         Problems   840

CHAPTER 18   FORM FACTORS AND e^+e^- COLLISIONS   841

  18-1   Form Factors for Nucleons   841
  18-2   Electron-Proton Inelastic Scattering   846

18-3   Electron-Positron Collisions   851
18-4   The ψ-Particles   854
        Bibliography   860
        Problems   860

CHAPTER 19   WEAK INTERACTIONS REVISITED   861

19-1   Conserved Current   862
19-2   Selection Rules in Strange Decays; Cabibbo's Theory   866
19-3   Neutral Currents; Unification of Electromagnetism and
        Weak Interactions   871
19-4   Intermediate Bosons   874
19-5   Some Further Examples of Weak Decays   875
19-6   The K^0-K^0 Doublet   880
19-7   CP Violation in K Decay   887
        Bibliography   889
        Problems   889

CHAPTER 20   HIGH-ENERGY COLLISIONS OF HADRONS   891

20-1   Introduction   891
20-2   Statistical Theory of High-Energy Collisions   895
20-3   Main Features of High-Energy Collisions   897
20-4   Diffraction Scattering   902
20-5   Exchange Collisions; Regge Poles   907
20-6   Inclusive Reactions; Scaling   910
        Bibliography   918
        Problems   919

APPENDIX A   Scattering from a Fixed Center of Force   920
APPENDIX B   Effective Range   926
APPENDIX C   Description of Polarized Beams (Spin $\frac{1}{2}$) 929
              Problem   935
APPENDIX D   Kinematics of Binary Collisions   936
              Bibliography   938
              Problems   939
APPENDIX E   Composition of Angular Momenta   942

Author Index   949

Subject Index   959

Preface to the Second Edition

The second edition preserves the goals, level, and spirit of the first. In the last decade nuclear and particle physics have become increasingly technical, and both theory and experiment have grown more complicated. This is an unavoidable trend that tends to increase the gap between textbooks and original research literature in the journals, which often is intelligible to only a very restricted circle of initiates. However, the student should somehow be given a glimpse of what the specialists are doing before joining them and should acquire an idea of the forest before concentrating on the tree. One of the aims of this book is to convey such a general, but not superficial, view of the subject.

It gave me great pleasure to find out that colleagues often consulted the first edition for orientation in areas remote from their specialty, and to discover that in several physics libraries the book looks soiled and worn from the use it has had.

I fully realize that one cannot do justice to many important ideas sketched in the text without a heavier theoretical equipment than is used here, but I decided not to change the theoretical level because doing so would have completely altered the character of the book.

I have revised the whole text, updating numerical data and improving derivations and writing style. I have added or replaced many figures, and a study of the illustrations should be rewarding for the inquisitive student. The bibliographies have been modernized, especially by the addition of easily accessible review articles that are known to me for their quality and clarity. Problems have been added and changed. I have tried not to increase the length of the book and, whenever possible, subjects that are common knowledge or are usually taught in other courses have been omitted.

Occasionally I have been able to shorten some chapters, especially in Part 1 ("Tools").

In Part 2 ("The Nucleus"), Chapter 6 ("Elements of the Nuclear Structure and Systematics") attempts to give an idea of modern trends in nuclear models without becoming too technical. Chapter 9 ("Beta Decay") has been correlated with Chapter 19 ("Weak Interactions Revisited"). The parts of Chapter 11 ("Nuclear Reactions") that deal with fission and heavy-ion reactions have been rewritten. I have entirely reorganized Part 3 ("Particles") to take into account the substantial progress, including the recent developments on quarks, that has taken place in this area.

Many friends and colleagues have helped me materially by explaining certain subjects to me and by reading or improving sections of the book. In particular I am indebted to W. Chinowsky, E. Commins, D. Jackson, D. Judd, L. Moretto, D. Prosperi, A. Rindi, G. Shapiro, H. Shugart, M. Suzuki, and G. Temmer. CERN, LBL, SLAC, FNAL, and other laboratories are among those that graciously supplied illustrations. I especially thank Gary Lum for the questions, suggestions, and comments he made while valiantly helping to ferret out errors and check the proofs.

E. SEGRÈ

Preface to the First Edition

This book is addressed to physics students, chemists, and engineers who want to acquire enough knowledge of nuclear and subnuclear physics to be able to work in this field. The book is definitely an introduction. The coverage is rather broad, but the treatment has been kept as simple as possible, compatible with a professional understanding of the subject.

While I have tried to convey as much as possible an intuitive understanding of the phenomena encountered, it is nonetheless impossible to discuss nuclear physics without a moderate use of quantum mechanics; and it is assumed that the person who wishes to become acquainted with nuclear physics is also willing to acquire the background in physics necessary to an understanding of this specialized subject. In my opinion, the presentation of nuclear physics at the introductory level should be the same for both the future theoretical physicist and the future experimental physicist. This account should be useful to a beginner regardless of his intended specialization.

The book aims at the same level as Rasetti's *Elements of Nuclear Physics* and Fermi's notes collected by Orear, Rosenfeld, and Schluter. I have drawn liberally from the latter. In the selection of materials I have followed my personal preferences. This may have resulted in an emphasis on some parts of nuclear physics better known to me, at the expense of others equally important. Furthermore most of my own work has been experimental, and this necessarily affects my outlook. The permissible length of such a book and my own knowledge of the subject are some of the limiting factors. However, I hope that some loose ends will provide food for thought to interested students and stimulate them to further reading.

The bibliography appended to every chapter lists for the most part review articles and monographs suited to more detailed study. Little of the original literature is included. Although many important and exciting papers are thus omitted, the references selected are probably the most profitable for a beginner. The Selected Bibliography gives the catalog of a small, basic, personal library for a research worker in the area covered by this book.

Problems vary considerably in difficulty. Some are simple numerical applications, and others contain interesting supplements to a discussion in the text.

In teaching an undergraduate course in nuclear physics, I have found it possible to cover in two semesters a large portion of the material in this book. For a shorter course one might omit completely Part I (which is of interest mainly to prospective experimental nuclear physicists and which contains fewer fundamentals) and reduce drastically the content of some of the other chapters. In general, I have tried to make the individual chapters relatively independent of each other, even at the risk of repetition. Sections beginning and ending with a heavy black dot may be omitted in the first reading; these often require a deeper knowledge of quantum mechanics on the part of the student than does the rest of the book.

I want to acknowledge gratefully the help of numerous colleagues who have read sections of this book and often have given me valuable advice. In particular, I want to thank Drs. Chamberlain, Fano, Frauenfelder, Goldberger, Judd, Rasmussen, Rosenfeld, Shafer, Steiner, Telegdi, Trilling, and Wu. The responsibility for any errors in this text is, of course, entirely mine. My thanks also go to Mrs. Patricia Brown, who patiently typed the manuscript, and to my wife Elfriede Segrè, who compiled the indexes.

E. Segrè

Berkeley, California
July 1964

Selected Bibliography

The following selected bibliography lists books covering a wider range of subjects than the works cited at chapter ends. In referring to this bibliography we shall use an abbreviated form of the last name(s) of the author(s) and the year of publication of the book. For example, (**Ra 56**) means Ramsey, N. F., *Molecular Beams*, Oxford University Press, New York, 1956.

INTRODUCTORY

BM 56 Bethe, H. A., and Philip Morrison, *Elementary Nuclear Theory*, 2nd ed., Wiley, New York, 1956.

Co 71 Cohen, B. L., *Concepts of Nuclear Physics*, McGraw-Hill, New York, 1971.

El 66 Elton, L. R. B., *Introductory Nuclear Theory*, Saunders, Philadelphia, 1966.

Fe 50 Fermi, Enrico, *Nuclear Physics*, University of Chicago Press, Chicago, 1950.

Fr 66 Frazer, W. R., *Elementary Particles*, Prentice-Hall, Englewood Cliffs, N.J., 1966.

FKM 64 Friedländer, G., J. W. Kennedy, and J. Miller, *Nuclear and Radiochemistry*, Wiley, New York, 1964.

Le 73 Leon, M., *Particle Physics: An Introduction*, Academic Press, New York, 1973.

Meyerhof, W. E., *Elements of Nuclear Physics*, McGraw-Hill, New York, 1967.

Pe 72 Perkins, D., *Introduction to High Energy Physics*, Addison-Wesley, Reading, Mass., 1972.

MORE ADVANCED TEXTS

BW 52 Blatt, J. M., and V. F. Weisskopf, *Theoretical Nuclear Physics*, Wiley, New York, 1952.

BM 69 Bohr, A., and B. R. Mottelson, *Nuclear Structure, Volumes I and II*, Benjamin, New York, 1969–75.

De Benedetti, S., *Nuclear Interactions*, Wiley, New York, 1964.

Feld, B. T., *Models of Elementary Particles*, Blaisdell, Waltham, Mass., 1969.

Gasiorowicz, S., *Elementary Particle Physics*, Wiley, New York, 1966.

Ka 64 Källén, G., *Elementary Particle Physics*, Addison-Wesley, Reading, Mass., 1964.

Nishijima, K., *Fundamental Particles*, Benjamin, New York, 1965.

O 71 Omnès, R., *Introduction to Particle Physics*, Wiley-Interscience, New York, 1971.

P 62 Preston, M. A., *Physics of the Nucleus*, Addison-Wesley, Reading, Mass., 1962.

Se 59 Segrè, Emilio, (ed.), *Experimental Nuclear Physics*, Wiley, New York, 1953–1959.

W 71 Williams, W. S. C., *An Introduction to Elementary Particles*, 2nd ed., Academic Press, New York, 1971.

IMPORTANT MONOGRAPHS

AS 60 Ajzenberg-Selove, Fay, *Nuclear Spectroscopy*, Parts A and B, Academic Press, New York, 1960.

BH 55 Bethe, H. A., and F. de Hoffmann, *Mesons and Fields*, Row, Peterson, Evanston, Ill., 1955.

Com 73 Commins, E. D., *Weak Interactions*, McGraw-Hill, New York, 1973.

Dewitt, C., and M. Jacob (ed.), *High Energy Physics*, Gordon & Breach, New York, 1965.

En 62 Endt, P. M., and P. B. Smith (eds.), *Nuclear Reactions*, North-Holland, Amsterdam, 1962.

Fe 51 Fermi, Enrico, *Elementary Particles*, Yale University Press, New Haven, Conn., 1951.

Fey 61 Feynman, R., *The Theory of Fundamental Processes*, Benjamin, New York, 1961.

GE 52 Glasstone, Samuel, and M. C. Edlund, *The Elements of Nuclear Reactor Theory*, Van Nostrand, Princeton, N.J., 1952.

GW 64 Goldberger, M. L., and K. M. Watson, *Collision Theory*, Wiley, New York, 1964.

H 63 Hagedorn, R., *Relativistic Kinematics*, Benjamin, New York, 1963.

Haïssinsky, M., *Nuclear Chemistry and Its Applications*, Addison-Wesley, Reading, Mass., 1964.

Hev 48   Hevesy, G., *Radioactive Indicators*, Wiley-Interscience, New York, 1948.

Hu 53   Hughes, D. J., *Pile Neutron Research*, Addison-Wesley, Reading, Mass., 1953.

Ko 58   Kopfermann, Hans, *Nuclear Moments*, Academic Press, New York, 1958.

La 66   Lamarsch, J. R., *Introduction to Nuclear Reactor Theory*, Addison-Wesley, Reading, Mass., 1966.

Livingston, M. S., and J. P. Blewett, *Particle Accelerators*, McGraw-Hill, New York, 1962.

MJ 55   Mayer, M. G., and J. H. D. Jensen, *Elementary Theory of Nuclear Shell Structure*, Wiley, New York, 1955.

Persico, E., E. Ferrari, and S. E. Segre, *Principles of Particle Accelerators*, Benjamin, New York, 1968.

Ra 56   Ramsey, N. F., *Molecular Beams*, Oxford University Press, New York, 1956.

Ro 53   Rossi, Bruno, *High Energy Particles*, Prentice-Hall, New York, 1952.

Si 65   Siegbahn, Kai, *Beta- and Gamma-Ray Spectroscopy*, North-Holland, Amsterdam, 2nd ed., 1965.

Sm 46   Smyth, H. D., *Atomic Energy for Military Purposes*, Princeton University Press, Princeton, N.J., 1946.

WW 58   Weinberg, A. M., and E. P. Wigner, *The Physical Theory of Neutron Chain Reactors*, University of Chicago Press, Chicago, 1958.

W 69   Wilkinson, D. H. (ed.), *Isospin in Nuclear Physics*, North-Holland, Amsterdam, 1969.

WM 66   Wu, C. S., and S. A. Moszkowski, *Beta Decay*, Wiley-Interscience, New York, 1966.

ATLASES

General Electric Co., *Nuclear Chart*, Schenectady, N.Y., 1972.

GMLB 54   W. Gentner, H. Maier-Leibnitz, and W. Bothe, *An Atlas of Typical Expansion Chamber Photographs,* Pergamon Press, London, 1954.

LHP 67   Lederer, C. M., J. M. Hollander, and I. Perlman, *Table of Isotopes*, 6th ed., Wiley, New York, 1967.

PFP 59   Powell, C. F., P. H. Fowler, and D. H. Perkins, *The Study of Elementary Particles by the Photographic Method*, Pergamon Press, London, 1959.

RW 52   Rochester, G. D., and J. G. Wilson, *Cloud Chamber Photographs of the Cosmic Radiation*, Pergamon Press, London, 1952.

RPP   Particle Data Group, "*Review of Particle Properties.*" Published annually in *Rev. Mod. Phys.* or *Phys. Letters*.

In addition to these books, there are excellent review articles in
Annual International Conference on High Energy Physics (RoC).
Annual Reviews of Nuclear Science (*Ann. Rev. Nucl. Sci.*).
Progress in Elementary Particle and Cosmic Ray Physics (*Progr. Elem. Particle Cosmic Ray Phys.*).
Progress in Nuclear Physics (*Progr. Nucl. Phys.*).
Reviews of Modern Physics (*Rev. Mod. Phys.*).
and in many summer schools, such as Les Houches, or Varenna.

Almost none of the publications listed above are completely up to date, and the year of publication must always be kept in mind in consulting them. The list is by no means complete but should be sufficient as a starting point for finding more detailed bibliographical data.

A set of notes on quantum mechanics to be used for reference is
Fe 61   Fermi, Enrico, *Notes on Quantum Mechanics*, University of Chicago Press, Chicago, 1961.

Standard books on quantum mechanics are, e.g.,
Kae 65   Kämpffer, F. A., *Concepts in Quantum Mechanics*, Academic Press, New York, 1965.
Ma 57   Mandl, F., *Quantum Mechanics*, 2nd ed., Butterworth, London, 1957.
Me 61   Messiah, Albert, *Quantum Mechanics*, 2 vols., Wiley, New York, 1961–1962.
Sc 68   Schiff, L. I., *Quantum Mechanics*, 3rd ed., McGraw-Hill, New York, 1968.

As a reference for electromagnetism, we suggest
Ja 75   Jackson, J. D., *Classical Electrodynamics*, 2nd ed., Wiley, New York, 1975.

General references on many subjects are
CO 67   Condon, E. U., and H. Odishaw, *Handbook of Physics*, McGraw-Hill, New York, 1967.
Fl E   Fluegge, S., *Encyclopedia of Physics*, Springer, Berlin, 1955– .

The "constants" of physics, including masses of particles, etc., are in a continuous process of revision, as measurement precision improves. A recent list of physical constants is given in *Rev. Mod. Phys.* **41**, 477 (1969), and in *RPP*.

Nuclei and Particles
Second Edition

CHAPTER

1

History and Introduction

Although speculation on the nature of matter appears at the very dawn of Greek philosophy, a scientific study of this subject, in the modern sense, was not initiated until the sixteenth century, when experiment and mathematical analysis, which together constitute what we today call the "scientific method," were first used in conjunction with each other. However, simpler and easier problems had to be solved before the structure of matter could be investigated scientifically. It is true that the original steps in the kinetic theory of gases, employing strictly atomic models, were taken by Daniel Bernoulli in the eighteenth century (1738). But the branch of science in which atomic concepts first assumed a fundamental importance was chemistry. The tremendous success of the atomic hypothesis (Dalton, 1803) in explaining both qualitatively and quantitatively the innumerable facts of chemistry, the construction of tables of atomic weights, the discovery of Avogadro's law (1811) and of Faraday's laws of electrolysis (1833) are all major achievements of the first part of the nineteenth century. They made the atomic hypothesis highly credible, and it is surprising, perhaps, that the very existence of atoms should have remained the subject of a deep skepticism lasting into the early years of the twentieth century.

It must be pointed out, however, that an explanation of all the facts of chemistry then known required only a very general hypothesis, one almost

Emilio Segrè, Nuclei and Particles: An Introduction to Nuclear and Subnuclear Physics, Second Edition

ISBN 0-8053-8061-7

completely lacking any details of the specific properties of atoms, such as their mass, size, and shape. Detailed knowledge of atomic structure was acquired only after 1910. In the early history of this last development, chemistry, kinetic theory, and the study of electrical discharges in gases played a very important role. With the advent of quantum theory, spectroscopy became the main tool for the exploration of the "outer layers" of the atom. In recent years, the nucleus has come under intense study, and the burgeoning of a subnuclear physics is already apparent.

The first experimental discoveries that made possible an attack on the structure of the atom followed each other in rapid succession. In 1895 Röntgen discovered X rays; early in 1896 Becquerel discovered radioactivity; and a little later Sir J. J. Thomson, Wiechert, and Kaufmann gave proof of the independent existence of the negative electron. These were soon followed by the introduction into physics of the idea of quanta of energy, a concept that developed in a rather roundabout way (Planck, 1900). Quantum concepts, which originated in thermodynamics, were destined to dominate the entire field of the physics of small objects. Together with Einstein's special theory of relativity (1905), they form the foundation on which modern physics rests. While it is impossible to recount here the fascinating history of the interrelations of all these lines of inquiry, a single example will not be out of place.

Becquerel discovered radioactivity while (at the suggestion of Poincaré) investigating in uranium salts a hypothetical relation between optical fluorescence and the then recently discovered X rays. This relation proved to be illusory, but the pertinent studies opened the gate to momentous developments. Marie Curie observed that, although the radioactivity of uranium compounds taken from pure chemicals was proportional to the uranium content, the ores from which they were extracted showed much more radioactivity than could be accounted for by the uranium content alone. She then performed chemical analyses of the ores and measured the radioactivity of the different fractions she had isolated (Fig. 1-1). This method, which was, and still is, fundamental to radiochemistry, led to the discovery of polonium and radium (1898). Surprise followed surprise when it was found that the radioactive atoms changed their chemical identity with time. Intense study of the phenomenon led to the theory of radioactive decay (Rutherford and Soddy, 1903; von Schweidler, 1905). According to this theory, radioactive atoms of a certain species will disintegrate spontaneously, the number that disintegrate per unit time being, on the average, proportional to the total number present but showing fluctuations characteristic of random phenomena. The law is expressed by the differential equation for the average number of atoms or nuclei, N,

$$- dN = \lambda N \ dt$$

or by its integral

$$N(t) = N(0)e^{-\lambda t}$$

ISBN 0-8053-8601-7

ISBN 0-8053-8601-7

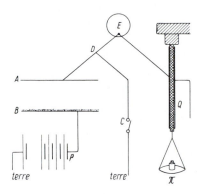

Figure 1-1 Apparatus used by Mme M. Curie to measure the conductivity of air under the influence of radioactive radiations. *AB*, plate capacitor; the radioactive substance is in *B*; *CD*, ground connection; *E*, electrometer; *p*, battery; *Q*, piezoelectric quartz; *π*, weight. [Mme Curie, 1899.]

where the "decay constant" λ is a characteristic of the nucleus. Investigation of the rays emitted during disintegration led to their classification into three types (Fig. 1-2).

Alpha rays are strongly ionizing particles and are absorbed by a few centimeters of air. The deflection of alpha particles in electric and magnetic fields identified them as helium atoms with a double positive charge, or He^{++}. This conclusion was confirmed by a direct experiment in which alpha particles were introduced into an evacuated glass tube. After a sufficient number had accumulated, an electric discharge in the tube showed the helium spectral lines (Rutherford and Royds, 1908) (see Fig. 1-3).

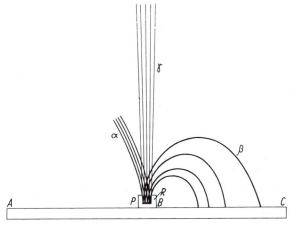

Figure 1-2 Deflection of alpha, beta, and gamma rays in a magnetic field. The nomenclature is due to Rutherford (1899). [Mme Curie, Thesis, 1904.]

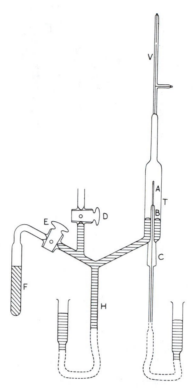

Figure 1-3 Radon contained in the thin-walled capillary tube *AB* expels alpha particles through the walls. Helium accumulates in the evacuated space *T* and when compressed in the capillary *V* shows, in an electric discharge, the characteristic spectrum. [E. Rutherford and T. Royds, 1908.]

Beta rays penetrate aluminum a few tenths of a millimeter thick and are identical with atomic electrons. They were identified by their deflection in electric and magnetic fields.

Gamma rays penetrate several centimeters of lead, are undeflected by electric or magnetic fields, and are high-energy electromagnetic radiation identical in nature to X rays. Other types of radiation or particles of great importance to nuclear physics, such as the neutron, were discovered much later and were not considered in the early days following the discovery of radioactivity.

The changes in chemical identity resulting from the emission of particles by radioactive substances were summarized in the displacement law enunciated by Russell, Soddy, and Fajans (1913): A substance is transformed (1) by emission of an alpha particle, into the substance that precedes it by two places in the periodic system; (2) by emission of a beta particle, into the substance that follows it by one place in the periodic system. The atomic weight diminishes by four units in the first case, and remains unchanged in the second. An alpha disintegration followed by two beta disintegrations

ISBN 0-8053-8601-7

should, therefore, result in a substance having a different atomic weight but "occupying the same place" (isotope) in the periodic table as the "parent" substance. The concept of isotope thus developed for radioactive elements was extended by J. J. Thomson and Aston to the ordinary stable elements (1913–1919).

The discovery by von Laue, Friedrich, and Knipping (1912) of the diffraction of X rays opened up a new field of spectroscopy. In the hands of Moseley it led to the concept of atomic number Z and to the clarification of the concept of a chemical element. All atoms having the same atomic number belong to the same element. A little earlier, in 1911, Rutherford in studying the scattering of alpha particles by foils of different materials had found that he could account for the experimental results by a planetary model of the atom. The atomic number Z was then interpreted as the charge of the nucleus in units of the same magnitude as the charge of the electron. Applying to this simple model the idea of the quantum, Bohr (1913) accounted for the hydrogen spectrum with admirable precision. This discovery was the starting point for the tumultuous development of atomic physics that culminated in the late 1920s in the establishment of quantum mechanics (de Broglie, Heisenberg, Born, Schrödinger, Dirac, Pauli, and others). The application of quantum mechanics to atomic phenomena was particularly fertile because, although the frame of the theory was new and different, the Coulomb law of electricity was all that was needed to explain innumerable phenomena of spectroscopy, chemistry, and solid-state physics and, in general, to give a complete account of the atomic and molecular properties as distinguished from the nuclear ones. In all these studies it was usually sufficient to schematize the nucleus as a point charge with a certain mass. For a while the theoretical study of the nucleus profited only slightly from quantum mechanics, for the interaction law governing the nuclear constituents, and even their identity, was unknown. Indeed present knowledge of the interaction law is still far from complete.

In 1919 Rutherford succeeded in breaking up nitrogen nuclei by alpha-particle bombardment and in showing that hydrogen nuclei were emitted (Fig. 1-4). For these the word *proton* ($\pi\rho\tilde{\omega}\tau o\nu$ = first) was coined, because it was thought that they were a sort of primordial universal substance, reviving an old hypothesis of Prout (1815).

Because protons often appeared in nuclear disintegrations, and some nuclei were known to emit electrons, the most natural hypothesis about the nuclear construction of a complex nucleus was that it consisted of A protons and $A - Z$ electrons, where A is an integer called the *mass number*. This accounted for the fact that the single isotopes have atomic masses approximately equal to integral multiples of the atomic mass of hydrogen. Departures from the exact integral numbers were taken to represent the binding energy of the nuclei according to Einstein's relation (1905)

$$E = \Delta mc^2$$

where E is the binding energy and Δm the mass defect, or difference between

ISBN 0-8053-8601-7

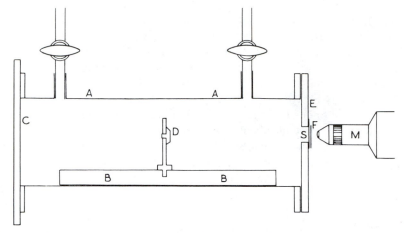

Figure 1-4 Apparatus used by Rutherford for disintegrating the nitrogen nucleus. The alpha particles emitted by the source *D* were absorbed in the gas used to fill the box. When the gas through which the particles passed was nitrogen, scintillations were observed on the screen *F* even when the absorbing matter between the source and the screen was sufficient to stop the primary alpha particles. Rutherford concluded that the scintillations were due to protons ejected with great speed from the nitrogen nucleus by the impact of an alpha particle. [E. Rutherford, 1919.]

the sum of the masses of the constituents and the mass of the complex nucleus.

However, the hypothesis of a nucleus composed of electrons and protons proved untenable, for reasons that will be discussed later, and was replaced by a model, now universally accepted, according to which the nucleus comprises Z protons and $A - Z$ neutrons. The neutron, discovered in a dramatic succession of events in which Bothe, Joliot, and Chadwick played a vital part, is neutral and has a mass approximately equal to that of the proton (1932).

In 1934 I. Curie and F. Joliot discovered that many stable elements under the bombardment of alpha particles became radioactive isotopes of other commonly stable elements (artificial radioactivity). Soon thereafter Fermi, Amaldi, Pontecorvo, Rasetti, and Segrè showed that neutrons could be slowed down to thermal energies and that at low velocities they were particularly effective in disintegrating other nuclei. This discovery was followed by the fission of uranium (Hahn and Strassmann, 1938), a particular reaction in which neutrons split the uranium nucleus into two large fragments, with the emission of several additional neutrons. This opened the way to the liberation of nuclear energy on a large scale (Fermi, 1942) and to its practical application.

Quantum mechanics in the hands of Dirac had predicted the existence of a positive electron. The positive electron, or positron, was discovered by C. D. Anderson in 1932 (Fig. 1-5), and the similar antiparticle on the nuclear level,

ISBN 0-8053-8601-7

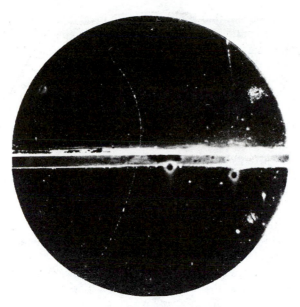

Figure 1-5 Positron. A particle travels from the lower to the upper part of a cloud chamber, as indicated by the fact that it loses energy in crossing the central lead plate. Its charge is positive, as indicated by the curvature in the magnetic field applied to the chamber. The curvature and specific ionization indicate that it has electronic mass. [C. D. Anderson, 1932.]

the antiproton, was discovered by Chamberlain, Segrè, Wiegand, and Ypsilantis in 1955.

In the early 1930s the first particle accelerators (Cockcroft and Walton, 1931; Lawrence, 1931) were invented and built. Artificial acceleration of particles gave a new dimension to nuclear and particle physics, making possible intensities and energies many orders of magnitude greater than those attainable with natural sources; it also made possible the use of projectiles that were not available naturally. It freed particle physics from its dependence on cosmic rays, which, although still unsurpassed as energy, are much less adaptable to systematic experimentation. Accelerator science is now a chapter of physics in itself, one in which ingenious new ideas, such as phase stability (Veksler, McMillan, 1945) and strong focusing (Christofilos, 1950; Courant, Livingston, Snyder, 1952), implemented by very advanced technology, make possible the attainment of beam energies and intensities that are limited only by money.

The phenomenology of beta decay presented great puzzles, which were in part overcome by Fermi (1933) with the help of the neutrino hypothesis of Pauli (1930). Fermi's theory introduced a new type of force in Nature, the so-called weak interaction. Later this force was found to be important in many phenomena of particle physics. It reserved a great surprise when Lee

ISBN 0-8053-8601-7

and Yang (1957) surmised that it might not be invariant with respect to parity transformation, a hypothesis soon confirmed by experiment (Wu, Ambler, Hayward, Hoppes, and Hudson, 1958). This failure of parity conservation was one of the major discoveries of the 1950s. It stimulated many experiments and led to the subsequent surprising result that the decay of neutral K's (a particle) violates not only the conservation of parity but also the combined parity and charge conjugation transformation (Christenson, Cronin, Fitch, and Turlay, 1964) which is obeyed by beta decay. The origin of this violation is still a major puzzle.

Fermi's theory also furnished the model that inspired H. Yukawa's theory of nuclear forces (1935). In his theory Yukawa postulated the existence of a particle (the meson or pion) having a mass intermediate between the mass of the electron and that of the proton.

Particles of intermediate mass between that of the proton and of the electron were detected in the thirties by several investigators, including Anderson, Neddermeyer, Rossi, Stevenson, and Street. However, they proved to be, not the expected ones (Conversi, Pancini, and Piccioni, 1947), but another unsuspected type. Nevertheless, the Yukawa meson was ultimately found (Lattes, Occhialini, and Powell, 1947; Fig. 1-6). There promptly followed the discovery of several other particles that are still only slightly understood.

With the improvement of technical means of production and detection, such as the giant accelerators and the bubble chamber, the subnuclear world has shown a great complexity. Many "particles" have been discovered, each accompanied by its antiparticle. To order this complexity it has become necessary to introduce a new quantum number, *strangeness* (Gell-Mann and Nishijima, 1953), and we see now the beginning of a classification based on group theory [Gell-Mann (1962) and Ne'eman (1961)].

Experiments at Brookhaven National Laboratory and at CERN have demonstrated the existence of two different kinds of neutrinos (1962–63). In 1974, some new particles, different from all known ones because of their relatively long life and high mass have emerged from different reactions at BNL and at SLAC. From the theoretical side we see some speculations that indicate deep relations between weak, electromagnetic, and perhaps even strong interactions.

Particle physics is still in a state of flux and many fundamental problems remain unsolved. For instance, one cannot relate the quantization of the electric charge to other phenomena, nor does one know why the proton is stable or what determines the mass difference between muon and electron.

This extremely sketchy historical outline has touched only on the milestones in the development of nuclear physics, without regard to applications or techniques. It cannot be emphasized enough that the whole development bears a resemblance to the evolution of a living being. All parts are deeply and vitally interrelated, and the development of highly abstract theories is as necessary to progress as the construction of gigantic accelerating machines.

ISBN 0-8053-8061-7

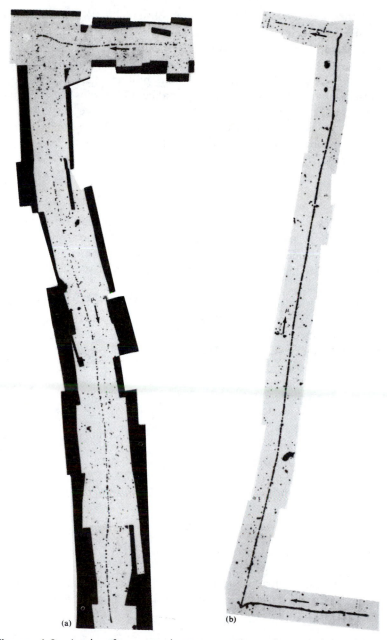

(a) (b)

Figure 1-6 A pion from cosmic rays seen in a photograph emulsion. (a) First observation of the decay of a pion. [Lattes, Muirhead, Occhialini, and Powell, 1947.] (b) An early observation of the $\pi-\mu-e$ decay. The particles travel in the direction of increasing ionization. The range of the μ is 600 microns. [Courtesy Prof. C. F. Powell.]

The influence exerted on other branches of science by the unfolding of nuclear physics has been deep and powerful: In chemistry artificial elements have been created (Segrè, 1937), and the periodic table has been extended considerably (McMillan, Seaborg, and others). The use of isotopes (Hevesy and Paneth, 1913) as tracers in chemistry and biology has added a tool to these sciences probably comparable in importance with the use of the balance and the microscope. Similarly, the medical sciences have been benefited greatly, and even geology and archaeology have felt the effect of nuclear methods. And, of course, the tremendous technological possibilities, in both peace and war, that nuclear physics has opened up are obvious.

1-1   UNITS

Nuclear physics employs several systems of units particularly suited to its problems. In order to be able to use them freely as the occasion arises, we shall mention them briefly. The widely used cgs system comes first, but since it is often numerically inconvenient, different branches of physics have developed their own systems.

The quantity $e^2/\hbar c = \alpha = 1/\text{"}137\text{"} = 1/137 \cdot 0360$, being dimensionless, has the same numerical value in all of them and is called the *Sommerfeld fine-structure constant*.

In atomic physics Hartree provided a system of "atomic units" in which the charge of the electron,

$$e = 4.8032 \times 10^{-10} \text{ esu} = 1.6022 \times 10^{-19} \text{ coulomb} \qquad (1\text{-}1.1)$$

is the unit of charge; the rest mass of the electron,

$$m = 9.1095 \times 10^{-28} \text{ g} = 0.51100 \text{ MeV} \qquad (1\text{-}1.2)$$

is the unit of mass; and Planck's constant divided by 2π,

$$\hbar = h/2\pi = 1.0546 \times 10^{-27} \text{ erg sec}$$

$$= 6.5822 \times 10^{-22} \text{ MeV sec} \qquad (1\text{-}1.3)$$

is the unit of angular momentum or action. In this system the unit of length is the radius of the first Bohr orbit in hydrogen, if the mass of the proton were infinite

$$a_0 = \hbar^2/me^2 = 0.52918 \times 10^{-8} \text{ cm} \qquad (1\text{-}1.4)$$

The unit of velocity is the electron velocity in this orbit,

$$v_0 = e^2/\hbar \qquad (1\text{-}1.5)$$

ISBN 0-8053-8601-7

and the unit of energy is twice the ionization potential of hydrogen,

$$me^4/\hbar^2 = 4.3598 \times 10^{-11} \text{ erg} = 27.2116 \text{ eV} \tag{1-1.6}$$

In practical nuclear physics lengths are sometimes expressed in fermis (or femtometers, F), 10^{-13} cm, and nuclear cross sections are generally measured in barns, 10^{-24} cm^2. Energies are expressed in eV or MeV (electron volts, million electron volts),

$$1 \text{ eV} = 1.60219 \times 10^{-12} \text{ erg} \tag{1-1.7}$$

Masses may also be measured in MeV by using mc^2 in place of m. For instance, the rest mass m of the electron is 0.51100 MeV, because numerically mc^2 expressed in MeV takes that value. Similarly, momenta may be measured in MeV/c. To do this, multiply the momentum by the velocity of light. The result has the dimensions of energy and can be expressed in MeV. This number then gives the measure of the momentum in MeV/c.

The relativistic relation between total energy E (including rest energy) and momentum p is

$$c^2p^2 + m^2c^4 = E^2 \tag{1-1.8}$$

With the two limiting cases

$$E = mc^2 + (p^2/2m) \qquad \text{nonrelativistic, or n.r.} \tag{1-1.9}$$

$$E = cp \qquad \text{extreme relativistic, or e.r.} \tag{1-1.10}$$

In the e.r. case, the energy in MeV is numerically equal to the momentum in MeV/c. In the n.r. case, calling E_{kin} the kinetic energy, we have

$$p = (2mE_{\text{kin}})^{1/2} \tag{1-1.11}$$

and the momentum in MeV/c is given numerically by this relation if the energy and the mass are expressed in MeV, as indicated above.

Velocities are often measured by taking the velocity of light

$$c = 2.997925 \times 10^{10} \text{ cm/sec} \tag{1-1.12}$$

as the unit. They are then indicated by the dimensionless quantity

$$\beta = v/c \tag{1-1.13}$$

ISBN 0-8053-8061-7

The quantity

$$\gamma = \frac{1}{(1-\beta^2)^{1/2}} = \frac{E}{mc^2} = \frac{p}{mc\beta} \tag{1-1.14}$$

is also important.

In theoretical work a system of units in which

$$\hbar = c = 1 \tag{1-1.15}$$

is frequently used.

If the mass m of some important particle, such as the electron or the proton, is chosen as unit of mass, the length $\hbar/mc$, called the Compton wavelength, has the numerical value 1; it is thus the unit of length. Similarly, the time $\hbar/mc^2$ has the numerical value 1 and is the unit of time. The energy mc^2 has the numerical value 1 and is the energy unit, etc. In Table 1-1 we give the cgs value of some units for $\hbar = c = 1$ and $m_e = 1$ or $m_p = 1$.

TABLE 1-1   UNITS FOR $\hbar = c = 1$

		$m_e = 1$	$m_p = 1$
Length	$\hbar/mc$	3.862×10^{-11} cm	2.103×10^{-14} cm
Time	$\hbar/mc^2$	1.288×10^{-21} sec	7.015×10^{-25} sec
Mass	m	9.109×10^{-28} g	1.6726×10^{-24} g
Energy	mc^2	8.187×10^{-7} erg	1.503×10^{-3} erg
		0.5110 MeV	938.26 MeV

In the same units $\hbar = c = 1$ the charge of the electron is $(1/\text{"}137\text{"})^{1/2} = 8.542 \times 10^{-2}$ and the Fermi constant of beta decay (see Chap. 9) is $1.025 \times 10^{-5}/m_p^2$.

Another unit of length often used is the classical radius of the electron,

$$e^2/mc^2 = r_0 = 2.81794 \times 10^{-13} \text{ cm} \tag{1-1.16}$$

The classical radius of the electron, its Compton wavelength, and the first Bohr radius are related according to the equations

$$a_0 = \lambda_c/\alpha = r_0/\alpha^2 \tag{1-1.17}$$

In mass-spectrographic work the unit of mass was one-sixteenth the mass of ^{16}O. In 1961 it was changed to one-twelfth the mass of ^{12}C (see Chap. 5).

We shall use these different units according to convenience and, when necessary, shall specify them in detail.

For utmost precision in numerical values of physical constants and errors, see the current RPP.

1-2   TERMINOLOGY AND DEFINITIONS

A few terms that may not be familiar from elementary physics are defined briefly below.

A *nucleon* is a neutron or a proton. A *nuclide* is a certain species of

ISBN 0-8053-8061-7

nucleus characterized by the atomic number Z and the mass number A. The terms nucleus and nuclide can often be interchanged without confusion. All nuclides with the same Z are *isotopes*; all nuclides with the same A are *isobars*; all nuclides with the same $A - Z$ are *isotones*. Nuclides that have the same A and Z but different states of excitation are called *isomers*.

One of the most important concepts in nuclear physics is that of cross section. Consider a beam of intensity I, a beam of protons, for example, for which I protons cm^{-2} sec^{-1} cross a region containing target nuclei. The target contains N nuclei cm^{-3} of material. A thickness dx of target will contain $N\,dx$ nuclei cm^{-2}, and the beam crossing it will be attenuated by collisions. If dI is the change in the intensity, we expect dI to be proportional to I and to $N\,dx$.

$$dI = -I\sigma N\,dx \qquad (1\text{-}2.1)$$

The proportionality factor σ, which has the dimensions of area, is called the *nuclear cross section* of the target for the particles of the beam. The reason for this name is that if we think of the target as extremely magnified and look at it from the direction of the impinging beam, we see a picture like that in Fig. 1-7. The irregularly disposed spots represent the nuclei of the target. Their number is $N\,dx$ per square centimeter. The fractional area they occupy is $\sigma N\,dx$, and if the particles hitting them are removed from the beam, the variation of the intensity is indeed given by Eq. (1-2.1).

Integration of Eq. (1-2.1) gives, for σ constant,

$$I(x) = I(0)e^{-N\sigma x} \qquad (1\text{-}2.2)$$

where $I(0)$ is the incident intensity and x is the thickness traversed. The quantity

$$N\sigma = \mu \qquad (1\text{-}2.3)$$

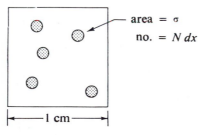

Figure   1-7   Five nuclei, each of cross section σ, are shown in a target of unit area. The probability that a particle crossing the unit area will hit a nucleus is $N^{*}\sigma$, where N^{*} is the number of nuclei per unit area and σ is the nuclear cross section; $N^{*} = $ (number of nuclei per unit volume) $\times$ (thickness of target).

ISBN 0-8053-8061-7

is sometimes used and is called the *absorption coefficient*; its reciprocal is the mean free path λ. The number

$$I(x)/I(0) = e^{-N\sigma x} \tag{1-2.4}$$

is often called *attenuation*.

Not all radiation is attenuated according to Eq. (1-2.2), because when the beam changes its properties (e.g., energy) in crossing the target, σ may depend on x. In such cases we cannot speak of an absorption coefficient.

We may also make finer distinctions in the cross section. For instance, we may consider whether the collision is elastic or inelastic, designating the corresponding cross sections σ_{el} and σ_{inel}. By *elastic collision* is meant a collision in which projectile and target before and after the collision are the same and have the same energy in the center-of-mass (c.m.) system. To illustrate, a proton and an aluminum nucleus upon collision are deflected from their trajectories in the c.m. system without change of energy and spin orientation; the collision is then elastic. If the proton produces a transmutation or leaves the aluminum in an excited state, the collision is inelastic.

The number of particles per unit area unit time scattered elastically is given by

$$dI_{el} = I\sigma_{el}N \; dx \tag{1-2.5}$$

The corresponding number undergoing inelastic processes is given by

$$dI_{inel} = I\sigma_{inel}N \; dx \tag{1-2.6}$$

The total attenuation will be

$$-dI = dI_{el} + dI_{inel} = I(\sigma_{el} + \sigma_{inel})N \; dx \tag{1-2.7}$$

and we write

$$\sigma_{tot} = \sigma_{inel} + \sigma_{el} \tag{1-2.8}$$

We may also consider in which direction the particles are scattered and may define a differential scattering cross section in the direction (θ, φ) by considering the number of particles scattered in the infinitesimal solid angle $d\omega$ in the direction (θ, φ),

$$dI_{sc} \text{ into } d\omega \text{ in direction } (\theta, \varphi) = \frac{d\sigma(\theta, \varphi)}{d\omega} IN \; dx \; d\omega \tag{1-2.9}$$

In the case of inelastic scattering, we may consider the energy of the scattered particle and may define a differential scattering cross section in the energy

ISBN 0-8053-8061-7

interval between E and $E + dE$,

$$dI_{sc} \text{ in energy interval } dE = \frac{d\sigma}{dE} \, IN \, dx \, dE \qquad (1\text{-}2.10)$$

From these examples it is clear how to define $d^2\sigma/dE \, d\omega$ for particles scattered in a certain direction and at a certain energy. It must be pointed out that occasionally differential cross sections are written, not as $d\sigma/d\omega$, but simply as $\sigma(\theta, \varphi)$.

Cross sections are often measured in barns (10^{-24} cm^2) and differential cross sections correspondingly in barns per steradian.

BIBLIOGRAPHY

Beyer, R. T. (ed.), *Foundations of Nuclear Physics*, Dover, New York, 1949.

Birks, J. B., *Rutherford at Manchester*, Benjamin, New York, 1963.

Boorse, H. A., and L. Motz (ed.), *The World of the Atom*, Basic Books, New York, 1966.

Curie, Ève, *Madame Curie*, Doubleday, New York, 1937.

Curie, Marie, *Oeuvres de Marie Skłodowska Curie*, Państwowe Wydawnictwo Naukowe, Warsaw, 1954.

Eve, A. S., *Rutherford*, Macmillan, London, 1939.

Fermi, E., *Collected Papers of Enrico Fermi*, University of Chicago Press, Chicago, 1962–65.

Fierz, M., and V. Weisskopf (eds.), *Theoretical Physics in the Twentieth Century*, Wiley-Interscience, New York, 1960.

Frisch, O. R. (ed.), *Trends in Atomic Physics*, Wiley-Interscience, New York, 1959.

Joliot-Curie, Frédéric et Irène, *Oeuvres scientifiques complètes*, Presses Universitaires de France, 1961.

Les Prix Nobel, Imprimerie Royale, P. A. Norstedt and Söner Publishers, Stockholm, 1902–present.

Millikan, R. A., *The Autobiography of R. A. Millikan*, Prentice-Hall, Englewood Cliffs, N. J., 1950.

Planck, M., *Physikalische Abhandlungen und Vorträge*, F. Vieweg, Braunschweig, 1958.

Rozental, S. (ed.), *Niels Bohr*, Wiley, New York, 1967.

Rutherford, E., *The Collected Papers of Lord Rutherford of Nelson*, Vols. I–III, Wiley-Interscience, New York, 1962–65.

Segrè, E. *Enrico Fermi, Physicist*, University of Chicago Press, Chicago, 1970.

Thomson, J. J., *Recollections and Reflections*, G. Bell, London, 1936.

Whittaker, E. T., *A History of Theories of Aether and Electricity*, 2 vols., Harper Torchbooks, New York, 1960.

Yukawa, H., and K. Chihiro, "Birth of the Meson Theory," *Am. J. Phys.*, **18**, 154 (1950).

PROBLEMS

1-1 In the system of units $\hbar = c = 1$ we may further assume an arbitrary unit of mass. For instance, the mass of the electron is often taken as a unit. Other derived units are then given by

Notation	Meaning	Customary notation	Value
m	Mass of electron	m	
m	Energy	mc^2	
m	Momentum	mc	

ISBN 0-8053-8061-7

Notation	Meaning	Customary notation	Value
m	Frequency	$mc^2/\hbar$	
$1/m$	Time	$\hbar/mc^2$	
$1/m$	Length	$\hbar/mc$	
e^2	Fine-structure constant	$e^2/\hbar c$	$1/137.04$
e^2/m	Classical radius of electron	e^2/mc^2	
$1/me^2$	Bohr radius	$a_0 = \hbar^2/me^2$	

Fill in the last column, using the units you prefer. For instance, you may use MeV for energy, MeV/c for momentum, fermis for length, etc.

1-2   Make universal curves of p/m, E/mc^2, and E_{kin}/mc^2, using a table of hyperbolic functions with $\beta = v/c$ as the independent variable. The argument of the hyperbolic functions λ is connected to β by $\tanh \lambda = \beta$. Show that the relativistic addition of velocities corresponds to the formula for $\tanh(\lambda_1 + \lambda_2)$ as a function of $\tanh \lambda_1$, $\tanh \lambda_2$.

1-3   Calculate the mean free path corresponding to the cross section of 1 b per nucleon in liquid hydrogen, air at STP, Al, Cu, and Pb.

1-4   Plot a graph of energy in MeV versus momentum in MeV/c for electrons, pions, protons from $E = 10^{-3}$ to 10^3 MeV.

1-5   A beam of protons of 1 MeV energy containing 10^8 protons per second falls on a silver foil 0.05 mg cm^{-2} thick and is scattered at 45 deg. How many protons per second fall on a detector 10 cm^2 in area located 1 m from the foil. For the scattering cross section use Eq. (2-2.17).

1-6   A beam of neutrons has a momentum equal to 10 GeV/c. How far do the neutrons have to travel before 50% have decayed? (Mean life of the neutron, 1013 sec.)

ISBN 0-8053-8601-7

PART

1

TOOLS

CHAPTER 2

The Passage of Radiations through Matter

Although this subject is really a part of atomic physics rather than of nuclear physics, the effects of the passage of "radiations" through matter are of paramount importance to all nuclear experiments; in fact, a thorough knowledge of these effects is absolutely indispensable to the experimental nuclear physicist. Many arguments treated in this chapter are primarily applications of electromagnetism; for these (Ja 75) is an excellent reference.

2-1 INTRODUCTION

The radiations we consider can be divided into uncharged particles (neutrons, neutrinos) and those subject to electromagnetic interaction. For neutrons see Chap. 12. For neutrinos of energy up to several MeV the weak interaction, the only one to which they respond, gives mean cross sections of the order of 10^{-48} cm^2; thus even the whole earth is transparent to them. The radiations for which electromagnetic phenomena are important are of three main types: charged heavy particles of mass comparable with the nuclear mass, electrons, and light quanta. For all of these we must consider the energy loss produced by the electromagnetic interaction, the only appreciable one for electrons and muons. In addition to the electromagnetic energy loss,

Emilio Segrè, Nuclei and Particles: An Introduction to Nuclear and Subnuclear Physics, Second Edition

19

which is the most important one up to energies of the order of 100 MeV, there is for protons and other particles subject to nuclear forces an energy loss due to nuclear collisions. Roughly speaking, this loss gives rise to an exponential absorption with a mass absorption coefficient of 100 cm^2 g^{-1} for both protons and neutrons.

A striking difference in the absorption of the three types of radiation is that only heavy charged particles have a range. That is, a monoenergetic beam of heavy charged particles, in traversing a certain amount of matter, will lose energy without changing the number of particles in the beam. Ultimately they will all be stopped after having crossed practically the same thickness of absorber. This minimum amount of absorber that stops a particle is its range: e.g., the range of polonium alpha particles, of energy 5.30 MeV, is 3.84 cm of air at STP (15°C and 760 mm Hg). For electromagnetic radiation, on the other hand, the absorption is exponential. Energy is removed from the beam and degraded; i.e., the intensity decreases in such a way that

$$-\frac{dI}{I} = \mu \, dx \tag{2-1.1}$$

where I is the intensity of the primary radiation, μ is the absorption coefficient, and dx is the thickness traversed. Electrons exhibit a more complicated behavior. They radiate electromagnetic energy easily because they have a large value of e/m and hence are subject to violent accelerations under the action of electric forces. Moreover, they undergo scattering to such an extent that they follow irregular trajectories.

We shall now define a few terms which recur frequently in this chapter. Consider a parallel beam of monoenergetic particles (e.g., protons) moving through an absorber. As they travel, they lose energy. The energy lost per unit path length is the *specific energy loss* and its average value is the *stopping power* of the absorbing substance. The *specific ionization* is the number of ion pairs produced per unit path length. The specific energy loss and the specific ionization are subject to fluctuations; hence we define a mean specific energy loss, a mean specific ionization, etc. The fluctuations in energy loss also produce fluctuations in range (*straggling*). A plot of the number of particles in the beam penetrating to a certain depth gives the curve of Fig. 2-1. The abscissa R_0 of the point passed by half the particles is called the *mean range*. The abscissa R_1, the intersection of the x axis with the tangent at the point of steepest descent, is called the extrapolated range. The difference between the extrapolated and mean range is sometimes called the *straggling parameter*.

The curve showing the specific ionization as a function of the residual range is known as a *Bragg curve*. It is necessary to distinguish between the Bragg curve of an individual particle (Fig. 2-2) and the average Bragg curve for a beam of particles (Fig. 2-3).

The thickness is often measured in grams per square centimeter (g cm^{-2}) of absorber. One then speaks of a mass absorption coefficient, mass stopping

ISBN 0-8053-8061-7

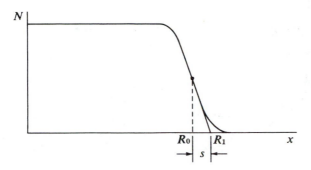

Figure 2-1 Range curve showing the number of particles in a beam penetrating to a given depth.

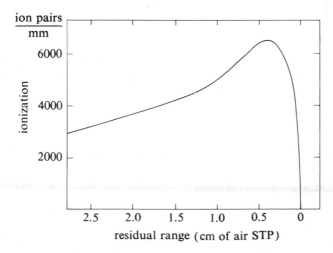

Figure 2-2 Bragg curve of an individual alpha particle. Ionization of an alpha particle, in ion pairs per millimeter, as a function of its residual range, according to experiments by M. G. Holloway and M. S. Livingston. [*Phys. Rev.*, **54**, 29 (1938).] In experiment $\rho_{air} = 1.184$ mg cm^{-3} (15°C, 760 mm Hg).

power, etc. The relation between the absorption coefficient μ and the mass absorption coefficient μ' is found by noting that the thickness x (in cm) is related to the thickness t (in g cm^{-2}) by

$$\rho x = t \tag{2-1.2}$$

where ρ is the density of the medium. Consequently,

$$\mu x = \mu t / \rho = \mu' t \tag{2-1.3}$$

and hence the mass absorption coefficient is

$$\mu' = \mu / \rho \tag{2-1.4}$$

ISBN 0-8053-8061-7

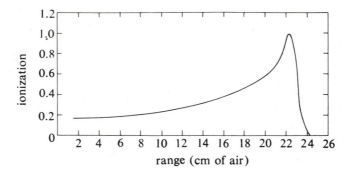

Figure 2-3 Specific ionization for a beam of protons as a function of range. In experiment $\rho_{air} = 1.166$ mg cm^{-3}. Ordinate scale arbitrary. [R. R. Wilson, Cornell University.]

The atomic absorption coefficient μ_a is sometimes employed when thicknesses are measured in atoms per square centimeter; we have then, by an argument similar to the previous one,

$$\mu_a = \mu / N \tag{2-1.5}$$

where N is the number of atoms per cubic centimeter.

2-2   RUTHERFORD SCATTERING

Consider a particle of charge ze traversing matter of atomic number Z, for instance, a proton traversing a piece of aluminum. Occasionally the proton will collide elastically with an aluminum nucleus and will undergo "Rutherford scattering"; i.e., the electrostatic repulsion from the nucleus will deflect it.

Elastic nuclear collisions give rise to large changes in the direction of the impinging particle but not, on the average, to significant energy losses. In a cloud-chamber picture (Fig. 2-4) a nuclear collision is easily distinguishable. In addition there are collisions with the extranuclear electrons. These constitute the main cause of energy loss at energies below several hundred MeV, although they produce only an extremely small scattering of heavy particles. Inelastic nuclear collisions are treated in Chap. 11.

The effect of a nuclear collision can be calculated classically as follows: Assume that the scattering center has an infinite mass (in other words, is fixed) and that it exerts a repulsive electrostatic force on the impinging proton, given by Ze^2/r^2. This force, which has a potential

$$V(r) = \frac{Ze^2}{r} \tag{2-2.1}$$

ISBN 0-8053-8601-7

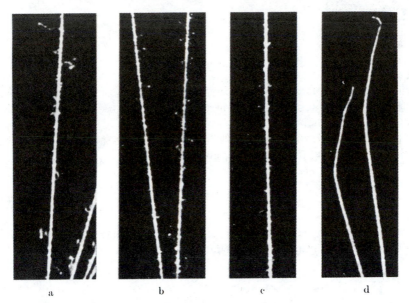

Figure 2-4 Cloud-chamber tracks of alpha rays showing delta rays. The first picture is in air, the last three in helium; the gas pressure in the chamber is such that the tracks cross about 10^{-5} g cm^{-2} of air equivalent. Note nuclear collisions in the section on the right. [T. Alper, *Z. Physik*, **67**, 172 (1932).]

produces a motion whose orbit lies in the plane of the fixed center and the initial velocity vector. If **r** is the radius vector from the force center (located at the origin) to the proton and $\mathbf{p} = m\dot{\mathbf{r}}$, the proton's momentum, Newton's second law of motion gives

$$\dot{\mathbf{p}} = \frac{Ze^2\mathbf{r}}{r^3} \tag{2-2.2}$$

Multiplying both sides vectorially by **r**, we have

$$\mathbf{r} \times \dot{\mathbf{p}} = 0 \tag{2-2.3}$$

Thus angular momentum

$$\mathbf{L} = \mathbf{r} \times \mathbf{p} \tag{2-2.4}$$

is a constant of the motion, since its time derivative is zero.

The total energy

$$\frac{p^2}{2m} + \frac{Ze^2}{r} = E \tag{2-2.5}$$

is another constant of the motion.

ISBN 0-8053-8601-7

The "Lenz" vector

$$\boldsymbol{\epsilon} = \frac{-1}{Ze^2m} \mathbf{L} \times \mathbf{p} + \frac{\mathbf{r}}{r} \tag{2-2.6}$$

which lies in the plane of motion, is a third constant of the motion as can be verified by calculating $\dot{\boldsymbol{\epsilon}}$ according to Eq. (2-2.2), and the formula for the vector triple product.

Scalar multiplication of Eq. (2-2.6) on both sides by $\mathbf{r}$, by using the formula for the mixed triple product, gives

$$\boldsymbol{\epsilon} \cdot \mathbf{r} = \frac{L^2}{me^2Z} + r \tag{2-2.7}$$

This equation can be easily interpreted by using polar coordinates with the polar axis in the direction of $\boldsymbol{\epsilon}$. Equation (2-2.7) then reads

$$\epsilon r \cos \varphi = \frac{L^2}{me^2Z} + r \tag{2-2.8}$$

or

$$r = \frac{-L^2/me^2Z}{1 - \epsilon \cos \varphi}$$

which for $\epsilon > 1$ is the equation of a hyperbola of eccentricity ϵ.

The angle between the asymptotes not containing the hyperbola (Fig. 2-5) defines the deflection of the particle θ. It is found by determining the difference between the values of $\varphi = \pm \varphi_1$ for which the denominator is zero and taking the supplementary angle to this difference. One finds

$$\cos \varphi_1 = 1/\epsilon = \sin (\theta/2) \tag{2-2.9}$$

or

$$Ze^2/2Eb = \tan (\theta/2) \tag{2-2.10}$$

where $b = L/(2mE)^{1/2}$ is the impact parameter, defined as the distance between the center of force and the limiting line of flight of the particle for large values of r.

We can now calculate the probability of a deflection θ for protons crossing a foil of a substance of atomic number Z. We assume that the deflection is the consequence of a single nuclear collision. This is the case for large deflections. Small deflections are generally the result of the combined action

ISBN 0-8053-8061-7

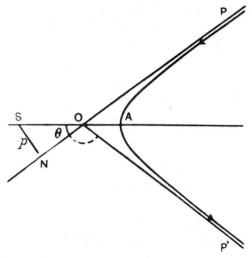

Figure 2-5 Orbit of a particle undergoing Rutherford scattering. [Original from E. Rutherford, *Phil. Mag.*, **21**, 672 (1911).] Note that θ and p of this figure are φ_1 and b of our text.

of many collisions, as will be seen later. We shall thus evaluate the nuclear-scattering cross section $d\sigma/d\omega$ and the probability of scattering through an angle between θ and $\theta + d\theta$ in crossing a foil of thickness x of a material containing N nuclei per unit volume. The probability $P(\theta)\,d\omega$ for scattering through angle θ into an element of solid angle $d\omega$ is given by

$$P(\theta)\,d\omega = \frac{d\sigma}{d\omega}\,Nx\,d\omega \qquad (2\text{-}2.11)$$

Consider one nucleus of the scatterer and an incident beam containing one proton per unit surface area. If a proton has an impact parameter b with respect to the scatterer, the deflection is given by (2-2.10).

The number of protons dn having an impact parameter between b and $b + db$ is $2\pi b\,db = dn$, where, from Eq. (2-2.10),

$$db = -\frac{Ze^2}{4E}\,\frac{d\theta}{\sin^2(\theta/2)} \qquad (2\text{-}2.12)$$

Hence

$$|dn| = \pi\left(\frac{Ze^2}{2E}\right)^2\frac{\cos(\theta/2)}{\sin^3(\theta/2)}\,|d\theta| \qquad (2\text{-}2.13)$$

This is the number of particles deflected through an angle between θ and $\theta + d\theta$. They pass with uniform density between two cones of aperture θ and

ISBN 0-8053-8601-7

$\theta + d\theta$. The solid angle included between these cones is

$$d\omega = 2\pi \sin \theta \; d\theta \tag{2-2.14}$$

and

$$\frac{dn}{d\omega} = \frac{1}{4} \left(\frac{Ze^2}{2E} \right)^2 \frac{1}{\sin^4(\theta/2)} \tag{2-2.15}$$

The quantity $dn/d\omega$ is dimensionally an area, and comparison with Eq. (2-2.11) shows that the differential-scattering cross section is

$$\frac{d\sigma}{d\omega} = \frac{1}{4} \left(\frac{Ze^2}{mv^2} \right)^2 \frac{1}{\sin^4(\theta/2)} \tag{2-2.16}$$

This is the famous Rutherford scattering formula (Rutherford, 1912). Put in numerical form it is

$$\frac{d\sigma}{d\omega} = \frac{0.1295 Z^2}{E^2 \, (\text{MeV}) \sin^4(\theta/2)} \times 10^{-26} \; \text{cm}^2 \; \text{per nucleus} \tag{2-2.17}$$

The experimental verification was carried out in detail by Rutherford, Geiger, Marsden, and others and led to the formulation of the planetary model of the atom.

Formula (2-2.16) can be extended to other particles besides the proton by replacing Z with Zz, where z is the atomic number of the projectile. Equation (2-2.16) is not relativistic and refers to a fixed center. Moreover, it considers only Coulomb forces, neglects both the finite size of the nucleus and specific nuclear forces, and is calculated classically without regard to quantum mechanics. In spite of all these approximations, the equation gives excellent results in many practical cases, notably for the scattering of particles having a computed minimum distance of approach to a target larger than approximately $1.2 \times 10^{-13} A^{1/3}$ cm, where A is the mass number of the target. The failure of Eq. (2-2.16) for cases where this distance becomes smaller is evidence that specific nuclear forces become operative. In fact, it was just such failure that provided the first indication of the "nuclear radius."

The generalization of Eq. (2-2.16) that takes into account the finite mass of the target is

$$\frac{d\sigma}{d\omega} = \left(\frac{e^2 zZ}{mv^2} \right)^2 \frac{1}{\sin^4 \theta} \frac{\left\{ \cos \theta \pm \left[1 - (m/M)^2 \sin^2 \theta \right]^{1/2} \right\}^2}{\left[1 - (m/M)^2 \sin^2 \theta \right]^{1/2}} \tag{2-2.18}$$

ISBN 0-8053-8601-7

where M is the mass of the target and m is the mass of the projectile. For $m < M$ the positive sign only should be used before the square root. For $m > M$ the expression should be calculated for positive and negative signs and the results added to obtain $d\sigma/d\omega$. The angle θ is the laboratory angle. If the colliding particles are identical, important quantum-mechanical corrections are necessary, and Eq. (2-2.18) is no longer applicable (Mott, 1930) (see Chap. 10).

An important limiting case of Eq. (2-2.16), also valid relativistically, is obtained when the deflection angle θ is small compared with 1 rad,

$$\frac{d\sigma}{d\omega} = \left(\frac{2Zze^2}{p\beta c} \right)^2 \frac{1}{\theta^4} \qquad \beta = \frac{v}{c} \tag{2-2.19}$$

For extremely small angles, which correspond to large impact parameters, the nuclear charge is screened by the atomic electrons, and Eq. (2-2.19) is invalid. It is this screening effect that prevents the equation from diverging for $\theta \to 0$. An important practical application of Eq. (2-2.19), to the problem of multiple scattering, will be discussed later.

The Rutherford scattering formula can also be obtained by means of quantum mechanics; in this case Born's approximation gives the correct result. The simple derivation is given here as an example of the application of Born's approximation. The fundamental formula (see Appendix A) is

$$\frac{d\sigma}{d\omega} = \frac{1}{4\pi^2\hbar^4} \frac{p^2}{v^2} |U_{pp'}|^2 \tag{2-2.20}$$

Use of Eq. (2-2.20) requires the calculation of the matrix element $U_{pp'}$ for the potential Ze^2/r. We have

$$U_{pp'} = Ze^2 \int \frac{\exp{(i/\hbar)(\mathbf{p} - \mathbf{p}')\cdot\mathbf{r}}}{r} \, d\tau \tag{2-2.21}$$

This integral is best calculated by transforming it to polar coordinates with a polar axis in the direction $\mathbf{p} - \mathbf{p}'$. Designate by θ the scattering angle and by μ the cosine of the angle between $\mathbf{r}$ and $\mathbf{p} - \mathbf{p}'$, and observe that $|\mathbf{p}| = |\mathbf{p}'|$ and $|\mathbf{p} - \mathbf{p}'| = 2p \sin(\theta/2) \equiv k\hbar$. In the integral (2-2.21) the volume element becomes $2\pi \, d\mu \, r^2 \, dr$, and we have

$$U_{pp'} = Ze^2 \int_{-1}^{1} \int_{0}^{\infty} \frac{e^{ik\mu r}}{r} \, 2\pi \, d\mu \, r^2 \, dr \tag{2-2.22}$$

Integrating with respect to μ gives

$$U_{pp'} = Ze^2 4\pi \int_{0}^{\infty} \frac{\sin kr}{kr^2} r^2 \, dr \tag{2-2.23}$$

ISBN 0-8053-8061-7

This last integral oscillates in value when the upper limit is considered, but it is easy to prove, for instance by replacing Ze^2/r by $(Ze^2/r)e^{-\alpha r}$ and after integration going to the limit $\alpha \to 0$, that we must take 0 as the value at the upper limit. We thus obtain

$$U_{\text{pp}'} = \frac{4\pi Ze^2}{k^2} = \frac{\pi \hbar^2 Ze^2}{p^2 \sin^2(\theta/2)} \tag{2-2.24}$$

and by using Eq. (2-2.20) we find

$$\frac{d\sigma}{d\omega} = \frac{1}{4\pi^2 \hbar^4} \frac{p^2}{v^2} \frac{\pi^2 \hbar^4 Z^2 e^4}{p^4 \sin^4(\theta/2)} = \frac{Z^2 e^4}{4p^2 v^2} \frac{1}{\sin^4(\theta/2)} \tag{2-2.25}$$

which is identical to Eq. (2-2.16).

2-3 ENERGY LOSS DUE TO IONIZATION

In addition to the nuclear collisions mentioned above, a heavy charged particle moving through matter also collides with atomic electrons. The greatest part of the energy loss occurs in these collisions. Sometimes electrons are detached from atoms and are clearly visible in cloud-chamber pictures (Fig. 2-4, delta rays). Sometimes the atom is excited but not ionized. In any case, the energy for these processes comes from the kinetic energy of the incident particle, which is thereby slowed down. Figure 2-3 gives a plot of specific ionization versus range. Since the energy spent in forming an ion pair in a gas happens to be approximately independent of the energy of the particle forming the ions, this curve approximates the curve of specific energy loss.

To calculate the rate of energy loss by a particle of charge ze as it progresses through a medium containing $\mathfrak{N}$ electrons cm^{-3}, we first consider the electrons as free and at rest. The force between the heavy particle and the electron is ze^2/r^2, where r is the distance between them. The trajectory of the heavy particle is not appreciably affected by the light electron, and we can consider the collision as lasting such a short time that the electron acquires an impulse without changing its position during the collision. By this hypothesis the impulse acquired by the electron must be perpendicular to the trajectory of the heavy particle and can be calculated by

$$\Delta p_\perp = \int_{-\infty}^{\infty} e\mathcal{E}_\perp \, dt = \int e\mathcal{E}_\perp \frac{dx}{v} = ze^2 \int_{-\infty}^{\infty} \frac{1}{r^2} \cos\theta \, \frac{dx}{v} \tag{2-3.1}$$

where $\mathcal{E}_\perp$ is the component of the electric field at the position of the electron normal to the trajectory of the particle (Fig. 2-6) and v is the velocity of the heavy particle, which is taken to be constant during the collision. The integral is easily evaluated by applying Gauss's theorem to a cylinder of radius b

ISBN 0-8053-8601-7

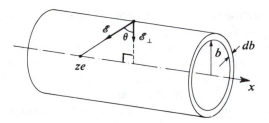

Figure 2-6 Transfer of momentum to an electron by a moving heavy charge.

having the trajectory as its axis. Note that the flux of $\mathcal{E}$ through this cylinder is given by

$$\phi = \int \mathcal{E}_\perp \, 2\pi b \, dx = 4\pi \, ze \tag{2-3.2}$$

Replacing the second integral of Eq. (2-3.1) by its value obtained from Eq. (2-3.2),

$$\Delta p = 2ze^2/bv \tag{2-3.3}$$

The energy transferred to the electron is then

$$\frac{(\Delta p)^2}{2m} = \frac{2}{m}\left(\frac{ze^2}{bv}\right)^2 \tag{2-3.4}$$

and since there are $2\pi \mathcal{N} b \, db \, dx$ electrons per length dx that have a distance between b and $b + db$ from the heavy ion, the energy loss per path length dx is

$$-\frac{dE}{dx} = 2\pi \, \mathcal{N} \int b \, db \, \frac{(\Delta p)^2}{2m} = 4\pi \, \mathcal{N} \, \frac{z^2 e^4}{mv^2} \int_{b_{min}}^{b_{max}} \frac{db}{b}$$

$$= 4\pi \, \mathcal{N} \, \frac{z^2 e^4}{mv^2} \, \log \frac{b_{max}}{b_{min}} \tag{2-3.5}$$

This is the stopping power of the absorbing medium. At first, one might be tempted to extend the integral from zero to infinity, obtaining a divergent result. To do so, however, would be incompatible with the hypotheses under which Eq. (2-3.4) was derived; for instance, distant collisions last a long time, and the corresponding energy transfer is not given by this equation.

Equation (2-3.5) also shows why the energy loss due to collisions with nuclei is negligible compared with the energy loss to electrons. In nuclear collisions we would find a factor Z^2 in the numerator, and the nuclear mass

ISBN 0-8053-8601-7

instead of the electron mass in the denominator. The increase in the denominator is the dominating factor.

We shall now discuss the values of b_{max} and b_{min} which are suitable to the problem. For b_{max} we consider that the electrons are not free but are bound in atomic orbits. The adiabatic principle of quantum mechanics states that one cannot induce transitions from one quantum state to another by a time-dependent perturbation if the variation of the perturbation is small during the periods τ of the system. In our case it can be assumed that the duration of the perturbation is the time b/v during which the heavy particle is near the electron and that in order to produce transitions the condition $b/v < \tau = 1/\nu$ must be fulfilled. This determines b_{max} as $< v/\langle \nu \rangle$, where $\langle \nu \rangle$ is an appropriate average of the frequencies of the atom. Taking relativistic corrections into account, the duration of the perturbation is shortened by a factor $(1 - \beta^2)^{-1/2} = \gamma$. The limit for b_{max} given by the adiabatic condition then becomes

$$b_{max} = v\gamma/\langle \nu \rangle \tag{2-3.6}$$

The limits for b_{min} are several: first, in an elastic collision it is impossible to change the momentum of an electron by an amount greater than $2mv$, as can easily be seen if we consider the heavy particle at rest and the electron impinging on it. This implies, according to Eq. (2-3.3), a minimum classical impact parameter

$$b_{min\ cl} = \frac{ze^2}{mv^2} > \frac{ze^2}{mc^2} = zr_0 \tag{2-3.7}$$

where r_0 is the classical radius of the electron, 2.8×10^{-13} cm.

Quantum mechanics gives another limit to b_{min} inasmuch as the electron can be localized with respect to the heavy ion only to the accuracy of its de Broglie wavelength; that is,

$$b_{min\ qm} > \frac{\hbar}{p} = \frac{\hbar(1 - \beta^2)^{1/2}}{mv} \tag{2-3.8}$$

We must now introduce in Eq. (2-3.5) the smallest value of b_{max} and the largest value of b_{min}. Over a large velocity interval this gives

$$-\frac{dE}{dx} = \frac{4\pi z^2 e^4}{mv^2} \mathfrak{N} \log \frac{mv^2}{\hbar\langle \nu \rangle (1 - \beta^2)} \tag{2-3.9}$$

The quantity $\hbar\langle \nu \rangle$ is a special average of the excitation and ionization potentials in the atom of the stopping material. It can be calculated by using the Thomas–Fermi model of the atom. Bloch (1933) found that it is approxi-

ISBN 0-8053-8061-7

mately proportional to Z,

$$\hbar\langle\nu\rangle = I = BZ \tag{2-3.10}$$

A better semiempirical formula is $I/Z = 9.1(1 + 1.9Z^{-2/3})$ eV for $Z \geqslant 4$. For liquid hydrogen $I = 20.4 \pm 1$ eV. A more precise calculation of the stopping power, performed by Bethe, gives

$$-\frac{dE}{dx} = \frac{4\pi z^2 e^4}{mv^2}\, \mathfrak{N}\left[\log\frac{2mv^2}{I(1-\beta^2)} - \beta^2\right] \tag{2-3.11}$$

A sample of the stopping-power curve is given in Fig. 2-7.

At very low velocities, i.e., when v is comparable with the velocity of the atomic electrons around the heavy particle (in the case of hydrogen, $v = c/137$), the moving ion neutralizes itself by capturing electrons for part of the time. This results in a rapid falloff of ionization at the very end of the range. On the other hand, at extremely high energies, with $\gamma > 5$, ionization increases, for several reasons. The relativistic contraction of the Coulomb field of the ion increases b_{max} according to Eq. (2-3.6) and decreases b_{min} according to Eq. (2-3.8). Furthermore, part of the energy is carried away as light (Cerenkov radiation). This last effect will be discussed in Sec. 2-5. The result is that the stopping power for charged particles by ionization increases slowly and reaches a plateau for $\gamma = 100$ at an energy loss of about 1.2–1.4 minimum ionization.

The general form of Eq. (2-3.11) indicates some conclusions of considerable practical importance. We can write Eq. (2-3.11) as

$$-\frac{dE}{dx} = z^2\lambda(v)$$

or, remembering that the kinetic energy of a particle of mass M is $E = M\epsilon(v)$, where ϵ is a function of the velocity only, we have, using E as a variable,

$$-\frac{dE}{dx}(E) = z^2\lambda_E(E/M) \tag{2-3.12}$$

or, using v as a variable,

$$-\frac{dv}{dx}(v) = \frac{z^2}{M}\lambda_v(v) \tag{2-3.13}$$

Relations (2-3.12) and (2-3.13) allow us to write the energy loss as a function of energy for any particle, once the energy loss as a function of energy is known for protons. In particular, protons, deuterons, and tritons of the same velocity have the same specific energy loss.

ISBN 0-8053-8601-7

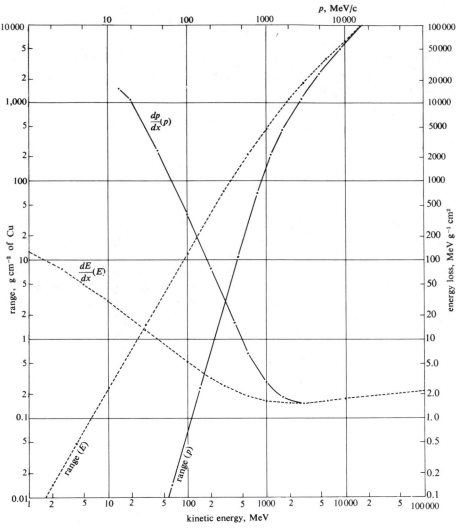

Figure   2-7   Graph of stopping power versus energy and of specific momentum loss versus momentum for heavy particles in copper. Range–energy and range–momentum relations for protons in copper are shown. All scales are logarithmic. The figure may be used for other particles. Remember $-dE/dx = z^2\lambda(v)$ and the scaling laws of Sec. 2-3. For the dp/dx curve use the left ordinate scale reading 1 MeV/c gr^{-1} cm^2 at ordinate 1.

ISBN 0-8053-8061-7

Similar scaling relations obtain for the range. Using the velocity as a variable, one has

$$R_v(v) = \int_0^v \left(\frac{dv}{dx} \right)^{-1} dv = \frac{M}{z^2} \int_0^v [\lambda_v(v)]^{-1} dv = \frac{M}{z^2} \rho_v(v) \quad (2\text{-}3.14)$$

or, using energy as a variable,

$$R_E(E/M) = (M/z^2)\rho_E(E/M) \quad (2\text{-}3.15)$$

For clarity we have indicated explicitly the independent variable to be used in the functions.

Equation (2-3.14) is not exact, for the neutralization phenomena occurring at the end of the range and other corrections are neglected; but it is sufficiently accurate for most cases, excluding very low energies. As an example of the application of Eq. (2-3.15) we can verify that a deuteron of energy E has twice the range of a proton of energy $E/2$.

A semiempirical power law valid from a few MeV to 200 MeV for the proton range–energy relations is

$$R(E) = (E/9.3)^{1.8}$$

where E is in MeV and R is in meters of air.

Sample numerical data on range–energy relations are provided in Fig. 2-7 and in Fig. 2-8, which presents a nomogram useful for approximate estimates.

The mass stopping power is used more often than $-dE/dx$, the (linear) stopping power. The mass stopping power depends on the factors $\mathfrak{N}/\rho$ and I of the stopping substances. The number of electrons per cubic centimeter, $\mathfrak{N}$, is roughly proportional to the density ρ. If this proportionality were exact, only the dependence of I on Z would influence the mass stopping power. Actually, $\mathfrak{N}/\rho$, and hence the mass stopping power, decreases with Z of the medium.

Heavier ions, such as ^{12}C, ^{16}O, and ^{40}A, are slowed down by ionization loss in much the same way as alpha particles. In particular, the energy loss to electrons is subject to an approximate law similar to Eq. (2-3.12). The main difference is that z is replaced by $z_{eff} = \gamma(v)z$ where $\gamma(v)$ is an increasing function of the velocity reaching its limiting value of 1 for $v/c \simeq 2(z/137)$.

When the velocity is decreased, the stopping power first increases, reaching a maximum that for ^{12}C and ^{40}A occurs at approximately $v/c = 0.037$ and 0.059, corresponding to energies of 8 and 65 MeV, respectively. At lower energies the decrease in the effective nuclear charge overcompensates the effect of the diminishing velocity and the stopping-power decreases with energy; the behavior is the same, on an exaggerated scale, as that observed at the end of the Bragg curve for protons and alpha particles (Fig. 2-2).

ISBN 0-8053-8601-7

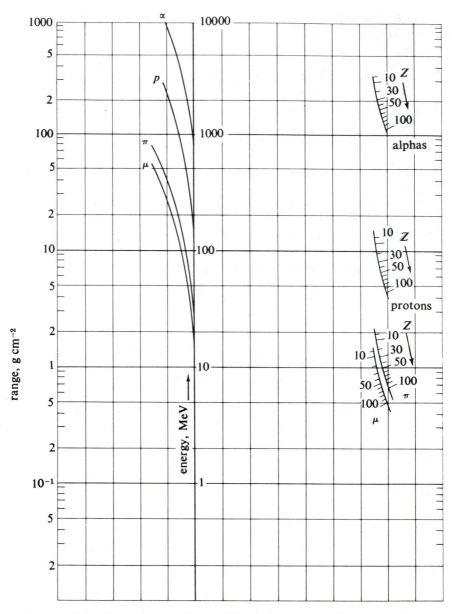

Figure 2-8 Nomogram by R. R. Wilson for range–energy relation. Left scale, range in grams per square centimeter. Middle scale, kinetic energy in MeV. Right scale, atomic number Z of stopping material and mass of particle. To use, connect range, energy, and Z by a straight line. [(Se 59).]

ISBN 0-8053-8061-7

Figure 2-9 emphasizes the scaling law by presenting a plot of the energy in MeV per atomic mass unit ($\mathcal{E} = E/m$) versus the thickness in $\mathcal{X}$ of aluminum, mg cm^{-2} times z^2/m. The graph gives the energy left in certain heavy ions, all of which started with $\mathcal{E} = 10$, after having crossed a certain thickness $\mathcal{X}$ of aluminum.

A particularly interesting case is that of fission fragments. Their Z_{eff} is large, reaching about 20 at the beginning of the range; and nuclear collisions are an important source of energy loss. If a fragment of atomic number Z_1 crosses a medium of atomic number Z_2 and nuclear mass M_2, the specific energy loss to nuclei is proportional to

$$Z_1^2 \frac{Z_2^2}{M_2} \tag{2-3.16}$$

whereas the loss to electrons is proportional to

$$Z_{1\,\text{eff}}^2 \frac{Z_2}{m} \tag{2-3.17}$$

The first equation, (2-3.16), applies to close nuclear collisions where the entire

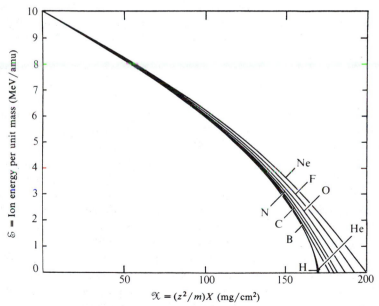

Figure 2-9 Energy loss curves for various heavy ions in aluminum. The near-universality in shape is emphasized by plotting $\mathcal{E} = E/m$ versus $(z^2/m)X = \mathcal{X}$. [L. C. Northcliffe, *Phys. Rev.*, **120**, 1744 (1960).] Ions of species indicated enter an aluminum foil with $\mathcal{E} = 10$ MeV/amu; at depth $\mathcal{X} = (z^2/m)X$ mg/cm^2 they have the energy indicated in graph.

ISBN 0-8053-8061-7

charge of the fragment and the entire charge of the target are effective. In the case of electronic collisions only the net charge $Z_{1\,eff}$ of the fission fragment, with whatever electrons it carries along, is effective, and the target electrons have unit charge. The factor Z_2 of Eq. (2-3.17) arises from the presence of Z_2 electrons per nucleus. The approximate value of $Z_{1\,eff}$ is obtained by assuming that the fragment will lose all the electrons whose orbital velocity in the atom is smaller than the velocity of the fragment itself.

The two causes of energy loss considered above may be comparable, but the energy loss due to nuclear collisions is concentrated in few events, whereas the electronic collisions are much more uniformly distributed along the range. The nuclear collisions give rise to the peculiar branches observable in cloud-chamber pictures of fission fragments. The concentration of the nuclear energy loss in a few events is the cause of the great value of the straggling shown by fission fragments.

Up to now we have paid no attention to the crystalline structure of solids, but when heavy ionizing particles such as Kr ions pass through single crystals, the energy loss depends on the orientation of the trajectory with respect to the crystalline axes. The intuitive reason is immediately realized by reflecting on how the appearance of a field of regularly planted trees differs, depending on the direction in which we look. Thus in a face-centered cubic lattice the direction perpendicular to a $\langle 101 \rangle$ face is the most transparent. As an example, 40-keV ^{85}Kr ions were found to penetrate Al crystals for about 4000 Å in the direction perpendicular to a $\langle 101 \rangle$ face, but for only 1500 Å in the direction perpendicular to a $\langle 111 \rangle$ face.

2-4 ENERGY LOSS OF ELECTRONS

The energy loss of electrons is a much more complicated phenomenon than the energy loss by ionization of heavy ions, for it involves an energy loss due to the electromagnetic radiation (*bremsstrahlung*) emitted in the violent accelerations that occur during collisions (Fig. 2-10). We shall consider the two effects separately, confining ourselves here to the energy loss due to ionization. The combination of radiation and ionization energy loss will be treated in Sec. 2-11. At low energies ($\ll 2mc^2$) the loss by ionization is much greater than that by radiation. For the derivation of the equations and a bibliography see the article of Bethe and Ashkin in (Se 59).

The energy loss by ionization may be treated in a manner similar to that used for heavy ions, but there are several important differences. It is necessary to take into account the identity of the particles involved in the collision and their reduced mass. The formula for nonrelativistic electrons is

$$-\frac{dE}{dx} = \frac{4\pi e^4 \mathfrak{N}}{mv^2} \left(\log \frac{mv^2}{2I} - \frac{1}{2} \log 2 + \frac{1}{2} \right) \qquad (2\text{-}4.1)$$

Except for small factors in the logarithmic term, this formula is the same as

ISBN 0-8053-8061-7

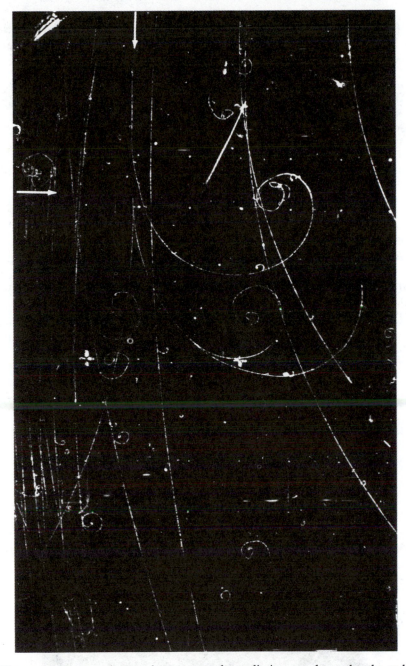

Figure 2-10 An electron loses energy by radiation, as shown by the sudden increase in curvature of its trajectory. The emitted quantum makes an electron–positron pair. [Propane bubble chamber, courtesy Lawrence Radiation Laboratory.]

Eq. (2-3.11); hence, electrons and protons of the same nonrelativistic velocity will lose energy at the same rate. For highly relativistic velocities the energy loss of electrons is

$$-\frac{dE}{dx} = \frac{4\pi e^4}{mc^2} \, \mathfrak{N} \left[\log \frac{2mc^2}{I} - \frac{3}{2} \, \log(1-\beta^2)^{1/2} - \frac{1}{2} \, \log 8 + \frac{1}{16} \right] \quad (2\text{-}4.2)$$

whereas for protons it is

$$-\frac{dE}{dx} = \frac{4\pi e^4}{mc^2} \, \mathfrak{N} \left[\log \frac{2mc^2}{I} + 2 \log \frac{1}{(1-\beta^2)^{1/2}} - 1 \right] \quad (2\text{-}4.3)$$

At equal values of β, the two expressions differ by less than 10% up to proton energies of 10^{10} eV. The difference between the average energy loss of electrons and positrons is even smaller.

An important practical difference between the behavior of heavy particles and that of electrons arises from the fact that the trajectories of electrons in matter at low energies ($E \ll mc^2$) are not straight lines. For this reason the actual path length of an electron passing through two points may be appreciably longer than the distance between these points measured on a straight line, as can be seen in Fig. 2-11. Thus, electrons of the same energy are not all

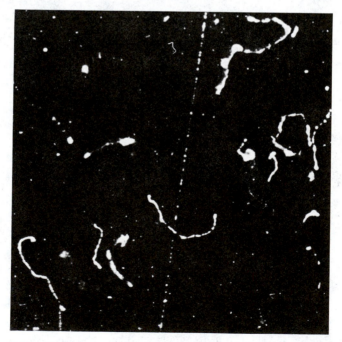

Figure   2-11   Slow electrons showing a curved path due to scattering. A fast electron goes straight. [Original from C. T. R. Wilson, 1923.]

ISBN 0-8053-8601-7

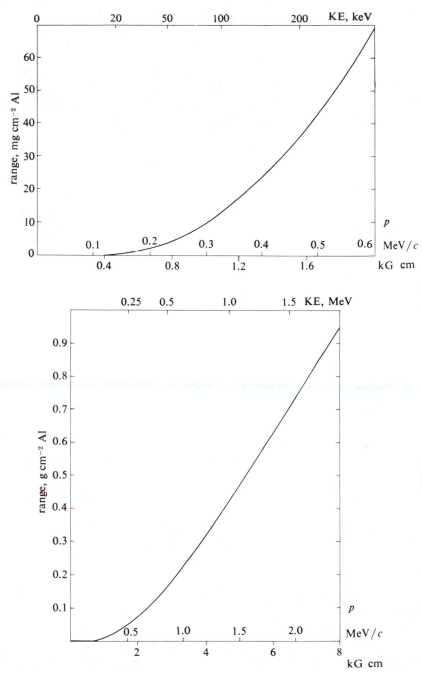

Figure 2-12 Range of electrons in aluminum. Abscissa on lower scale, momentum in MeV/c and in $Br = cp/e$ (3327 G cm = 1 MeV/c). Abscissa on upper scale, KE in MeV. [(Se 59).]

ISBN 0-8053-8601-7

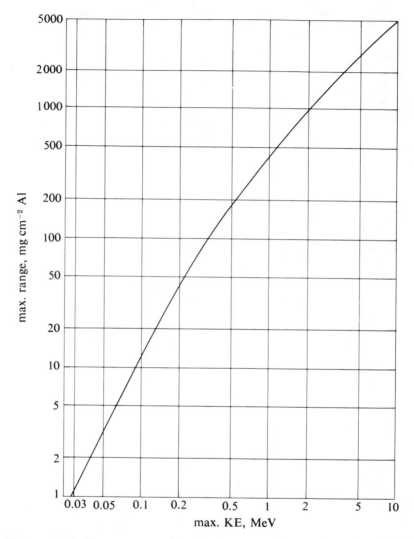

Figure 2-13   Range–energy plot of some common beta emitters (logarithmic scales). [(Si 55).]

stopped by the same thickness of material, and the concept of range has a limited validity.

For practical measurements of electron energy we can use the extrapolated range. It is important to note, however, that the geometry of the apparatus influences the result. Thus, in order to use the data found in the literature, one must reproduce the experimental arrangement used to obtain them (Fig. 2-12).

In the case of beta rays the electrons have a continuous energy spectrum,

ISBN 0-8053-8061-7

but it is still possible to find a relation between the upper limit of the energy of the spectrum E and the maximum range R (in g cm^{-2} of aluminum) of the electrons (Feather, 1938). A relation frequently used for a rapid determination of E (MeV) is

$$R = \begin{cases} 0.542E - 0.133 & 0.8 < E < 3 \\ 0.407E^{1.38} & 0.15 < E < 0.8 \end{cases} \tag{2-4.4}$$

See Fig. 2-13 for a range–energy plot usable for beta emitters.

2-5 POLARIZATION EFFECTS—CERENKOV RADIATION

The derivation of Eq. (2-3.11) did not take into account the electrical polarization of the medium in which the heavy ion moves. The dielectric constant of the medium weakens the electric field acting at a distance from the ion, causing a decrease of the energy transfer to atoms located far from the ion, and hence a decrease in the mass stopping power. Thus, in the case of a medium in two phases of different density, such as water and vapor, the lower density phase has a higher mass stopping power. This effect is appreciable, however, only for relativistic velocities and seldom amounts to more than a few percent.

Another important effect of the dielectric constant is the production of Cerenkov radiation (Cerenkov; Frank and Tamm, 1937). If a charge moves with a velocity βc in a medium of refractive index n, its electric field propagates with velocity c/n; and if $\beta c > c/n$, a phenomenon similar to the production of a bow wave results. Figure 2-14 gives the Huyghens' construction for the electromagnetic waves emitted by the particle along its path. At time $t = 0$, the particle is at O. One second later it is at P after traveling a distance $OP = \beta c$. The front of the electromagnetic wave is on the surface of the cone of aperture $\sin^{-1}(1/n\beta)$, which means that the rays of the corresponding light make an angle $\theta = \cos^{-1}(1/n\beta)$ with the trajectory of the particle. The intensity of the Cerenkov light can be calculated semiclassically [see for instance (Ja 75)]: for the number of quanta radiated per unit length with frequency

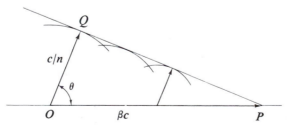

Figure 2-14 Huyghens' construction for electromagnetic waves emitted by a moving charged particle. Origin of Cerenkov radiation.

ISBN 0-8053-8061-7

between ν and $\nu + d\nu$ one has

$$dN = \frac{2\pi e^2}{\hbar c}\left(1 - \frac{1}{n^2\beta^2}\right)\frac{d\nu}{c} = \frac{2\pi e^2}{\hbar c}\sin^2\theta\,\frac{d\nu}{c} \qquad (2\text{-}5.1)$$

In the spectral region between 3000 and 6000 Å Eq. (2-5.1) thus gives approximately $750\sin^2\theta$ photons per centimeter. The spectrum is a continuum.

The light is polarized, with its electric vector pointing in the PQ direction. The measure of the angle θ of Cerenkov light may be used to determine the value of β for the particle. The density effect and the Cerenkov light are interrelated, both being functions of the dielectric constant of the medium.

At extremely relativistic energies another effect is useful for particle detection: transition radiation. A charged particle crossing an interface between a material having an electron plasma frequency $\omega_p^2 = 4\pi\mathfrak{N}e^2/m$ (where $\mathfrak{N}$ is the electron density and m the electron mass) and a vacuum radiates X rays in a narrow cone in the direction of motion. The energy radiated per surface crossed is $W = \frac{2}{3}(e^2/c)\omega_p\gamma$; it is thus proportional to the particle energy $M\gamma c^2$. In favorable circumstances this radiation may be used for measuring γ.

2-6   IONIZATION IN GASES AND SEMICONDUCTORS

A charged particle passing through a gas ionizes it. However, only part of the energy goes into ionizing the gas and into imparting kinetic energy to the electrons. A sizable fraction is spent in exciting the atoms below the ionization limit, and some of it is then transformed into detectable scintillation light. The average amount of energy required per ion formed is remarkably independent of the charge, mass, and velocity of the particle producing the ionization, but depends on the gas in which the ions are formed. There is no simple physical explanation of this fact, which is the consequence of the relative importance of a series of processes. Table 2-1 gives the average energy spent per ion pair formed in some of the most important gases.

TABLE 2-1 $w = $ eV **PER ION PAIR FOR DIFFERENT GASES**[a]

He	Ne	Ar	Xe	H_2	O_2	N_2	CO_2	Air	
42.7	36.8	26.4	21.9	36.3	32.5	36.5	34.3	35.0	Polonium α particle
42.3	36.6	26.4	22.0	36.3	30.9	34.9	32.9	33.8	Tritium β rays

[a]Note that extremely small amounts of impurities can affect w noticeably, especially in helium.

The primary act of ionization results in electrons, some of which (delta rays) have sufficient energy to produce secondary ions. If the ions are formed in a gas subject to an electric field, they move under the action of the field. The average velocity v in the direction of the field, called the *drift velocity*, is proportional to the field. The constant of proportionality μ is called the *mobil-*

ISBN 0-8053-8061-7

ity. We can see the essential factors involved in the drift velocity if we assume that the ions have a certain mean free path λ in the gas and that their average random velocity is u. The velocity u may be equal to the velocity of thermal agitation, but it is not necessarily so. If we think of an electron bouncing between heavy atoms, we see that it will not transfer much of its energy to the atoms unless it can excite them by inelastic collision. In the case of noble gases the required energies are of the order of 10 eV. Under the action of the field the electron will acquire a large velocity u, which is randomized in direction by the collisions until an inelastic collision causes a considerable decrease in velocity. If the electric field $\mathcal{E}$ lies in the z direction, an electron will drift in this direction in time τ between collisions by an amount

$$\frac{1}{2}\frac{e}{m}\mathcal{E}\tau^2 = \frac{1}{2}\frac{e\mathcal{E}}{m}\left(\frac{\lambda}{u}\right)^2 \tag{2-6.1}$$

because it is subject to the force $e\mathcal{E}$. The number of collisions per second is u/λ; hence the drift velocity is

$$v = \frac{1}{2}\frac{e\mathcal{E}}{m}\frac{\lambda}{u} = \mu\mathcal{E} \tag{2-6.2}$$

Clearly a low value of the random velocity u increases the mobility μ. For heavy ions the mobility is of the order of 1 cm sec^{-1} per V cm^{-1} at STP; for electrons it is about 1000 times larger. In very pure noble gases the mobility of electrons may be increased by as much as a factor of 10 by adding a small amount of a polyatomic impurity. The electrons then lose their energy by exciting rotational or vibrational states of the impurity molecules. This keeps u low and increases λ, for the Ramsauer effect produces an increase in λ on decreasing u, and both effects tend to increase v.

If, between collisions, the electron acquires enough kinetic energy to ionize the gas, the conditions for a multiplicative discharge are satisfied. Multiplicative processes and discharges are of fundamental importance in gas counters, proportional counters, and Geiger–Müller counters. For discharge mechanisms in counters, see Korff in (Fl E).

The energy spent per ion pair formed in solid semiconductors is about ten times smaller (3 eV) than in gases. A typical mobility in a semiconductor is 10^3 cm sec^{-1} per V cm^{-1}.

In practice one often uses the measurement of the ionization produced by a stopping particle to infer its energy. The total charge released eN is on the average related to the energy spent by $e\langle N\rangle = E/w$ where E is the energy of the particle and w the average energy spent to form an ion pair. This relation, however, obtains only for the average values; assuming E constant, N fluctuates around the average. The magnitude of this fluctuation depends on the probability of each elementary process giving rise to a specific energy loss and ionization. Clearly if all processes were identical and each gave rise to the same energy loss per ion pair produced, N would not fluctuate at all. Calling σ_N the standard deviation from the average number of ions, we have

$$\sigma_N^2 = \langle N^2 \rangle - \langle N \rangle^2 = \langle N \rangle F \tag{2-6.3}$$

where the number F, called the *Fano factor*, depends on the probability of energy losses of different magnitude per ion pair formed. The Fano factor for semiconductors is about 0.1; for gases it is about 0.4.

2-7   MULTIPLE SCATTERING

Section 2-2 treated the scattering due to the Coulomb force of a single nucleus acting on a charged particle. In addition to this single scattering, we must consider the cumulative effect of many small nuclear deflections that produce a deviation θ from the original direction of a particle. Consider a particle impinging normally on a foil and emerging with a given deviation θ. This deviation may result from a single scattering event or from an accumulation of many small scatterings. One criterion for deciding whether a particular deflection is most likely to be the result of one collision or of multiple scattering is obtained by considering an angle θ_1 such that in crossing the foil the particle is likely to have only one scatter as large as θ_1. Then deflections larger than θ_1 are very likely to be the result of a single scattering; deflections smaller than θ_1, the result of multiple scattering.

Clearly the distinction between deflections due to single scattering or to the multiple scattering cannot be made solely on the basis of the deflection angle θ. Large deflections may be caused by a few relatively wide-angle scatterings (plural scattering). The complete theory (Molière, 1948; Snyder and Scott, 1949; Nigam, Sundaresan, and Wu, 1959), however, is beyond the scope of this book. Figure 2-15 illustrates the relation between single, plural, and multiple scattering.

● In order to give an elementary theory of multiple scattering, we shall neglect the effects of plural scattering and limit the discussion to deflections $\theta \ll \theta_1$ due to many collisions.

Using an argument similar to the one preceding Eq. (2-3.3), we consider the collision of a rapidly moving incident particle of charge ze against a nucleus of charge Ze and find for the transverse impulse given to the incident particle

$$\Delta p / p = 2zZe^2 / bvp \tag{2-7.1}$$

This expression remains correct for relativistic momenta p. Furthermore, for small deflections we have in a single collision

$$\Delta p / p = \theta \tag{2-7.2}$$

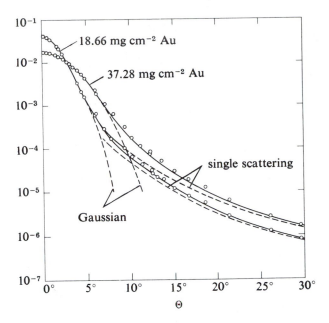

Figure 2-15 Angular distribution of electrons scattered from Au at 15.7 MeV. Solid lines indicate the distribution expected from the Molière theory for small- and large-angle multiple scattering, with an extrapolation in the transition region; dashed lines, the distributions according to the Gaussian and single-scattering theories. The ordinate scale gives the logarithm of the fraction of the beam scattered within 9.696×10^{-3} sr. [R. D. Birkhoff in (Fl E).]

If a particle traverses a foil, there will be many collisions, each deflecting the particle by a small angle θ_i, and the resulting deviation will be the sum of all the θ_i. Naturally in this sum one must take into account the fact that the θ_i occur in different directions, incoherently. We shall designate by Θ the total angular deviation of each particle and by Θ_x, Θ_y its projections in two fixed directions perpendicular to each other and to the direction of motion.

Designating φ_i by the azimuth of the plane of any single collision giving rise to the single deflection θ_i, we have, summing over single collisions,

$$\Theta_x = \sum_i \theta_i \cos \varphi_i \tag{2-7.3}$$

$$\Theta_y = \sum_i \theta_i \sin \varphi_i \tag{2-7.4}$$

from which we derive

$$\Theta^2 = \Theta_x^2 + \Theta_y^2 = \left(\sum_i \theta_i \cos \varphi_i \right)^2 + \left(\sum_i \theta_i \sin \varphi_i \right)^2 \tag{2-7.5}$$

Taking into account that the individual scatterings are incoherent and hence that

$$\overline{\sin \varphi_i \, \sin \varphi_j} = \overline{\cos \varphi_i \, \cos \varphi_j} = \tfrac{1}{2}\delta_{ij} \tag{2-7.6}$$

(indicating by the bar the average over collisions), we have

$$\langle \Theta^2 \rangle = \left\langle \sum_i \theta_i^2 \right\rangle = \sum_i \overline{\theta_i^2} \tag{2-7.7}$$

(indicating by $\langle \ \rangle$ the average over particles). Note that the most probable value of Θ and of Θ_x or Θ_y is zero but that $\langle \Theta \rangle$ and $\langle \Theta^2 \rangle$ are necessarily positive, whereas $\langle \Theta_x \rangle$ is zero. There are also the relations

$$\langle \Theta^2 \rangle = 2\langle \Theta_x^2 \rangle = 2\langle \Theta_y^2 \rangle \tag{2-7.8}$$

The study of the distribution of Θ_x for many particles passing the foil gives a distribution which for small Θ_x is approximately Gaussian. This approximation will not suffice for the wings of the distribution. Here a very few large deflections are much more likely to cause large deviations than an accumulation of many small deflections. Confining ourselves to the Gaussian region, we shall assume that the probability of finding a deviation between Θ_x and $\Theta_x + d\Theta_x$ is

$$P(\Theta_x)\, d\Theta_x = \frac{e^{-\Theta_x^2/\langle \Theta^2 \rangle}}{(\pi \langle \Theta^2 \rangle)^{1/2}}\, d\Theta_x \tag{2-7.9}$$

To evaluate $\langle \Theta^2 \rangle$, start from Eq. (2-7.7) and replace $\overline{\theta_i^2}$ by the value obtained from squaring Eq. (2-7.2), taking into account Eq. (2-7.1). By integrating over the range of permissible values of the impact parameter b, we obtain

$$\langle \Theta^2 \rangle = 2\pi Nx \int_{b_{\min}}^{b_{\max}} \left(\frac{2zZe^2}{bvp} \right)^2 b \, db \tag{2-7.10}$$

where N is the number of nuclei per cubic centimeter and x the thickness of the absorber. It is implicitly assumed that the thickness x is practically equal to the path length of the particle in the absorber. Because of the screening of the nucleus by the atomic electrons, the effective Z depends on b, but we shall take Z out of the integral and shall take into account the screening effect by a proper choice of $b_{\max}$. Then, neglecting the change in v and p in passing through the foil, we have

$$\langle \Theta^2 \rangle = \frac{8\pi NxZ^2 z^2 e^4}{v^2 p^2} \log \frac{b_{\max}}{b_{\min}} \tag{2-7.11}$$

As pointed out previously, the effective charge is a function of b. At $b_{\max}$ the charge should be completely screened. In a crude way this can be approximated by choosing

$$b_{max} = a_0/Z^{1/3} \tag{2-7.12}$$

where a_0 is the Bohr radius. This value may be justified by the Fermi–Thomas statistical model of the atom. On the other hand, we require that b_{min} give a deflection angle (in a single collision) small compared with 1 rad. This gives

$$b_{min} = 2zZe^2/vp \tag{2-7.13}$$

Other considerations, such as the nuclear radius or the de Broglie wavelength of the particle, may govern the choice of b_{min}. The result is not very sensitive to the choice of b_{min} and b_{max}, which appear only logarithmically in Eq. (2-7.11). ●

Figure 2-16 shows a graph of Θ versus the fraction of the range traversed for different particles.

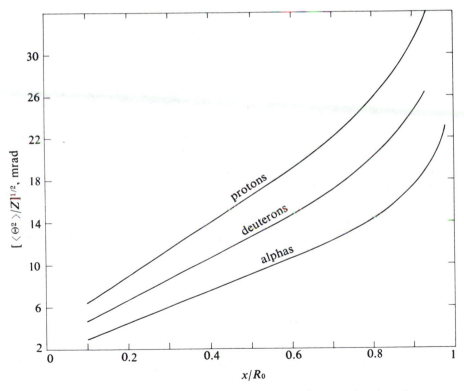

Figure 2-16 Multiple-scattering angle (unprojected) versus fraction of range traversed by protons, deuterons, and alpha particles. [R. H. Milburn and L. Schecter, UCRL 2234 (rev. ed.).]

As a very crude approximation of Eq. (2-7.11) pv is, passing from the nonrelativistic to the extreme-relativistic case, always between twice the kinetic energy and the kinetic energy of the particle. Neglecting all other functions of the velocity in Eq. (2-7.11), we can write, remembering Eq. (2-7.7),

$$\langle \Theta^2 \rangle \sim Z^2 x / (KE)^2 \qquad (2\text{-}7.14)$$

and as a numerical example for electrons

$$\langle \Theta^2 \rangle = \frac{600x}{[E(\text{Me}V)]^2} \qquad x = \text{cm of lead}$$

$$\langle \Theta^2 \rangle = \frac{7000x}{[E(\text{keV})]^2} \qquad x = \text{cm of air} \qquad (2\text{-}7.15)$$

Another formula for $\langle \Theta^2 \rangle$ that is practically useful is due to Rossi and Greisen. They measured the thickness t' in radiation lengths (see Sec. 2-11) and found (Θ in radians)

$$\langle \Theta^2 \rangle = z^2 (E_s / pv)^2 t' \qquad (2\text{-}7.16)$$

where $E_s = (4\pi \times 137)^{1/2} \times mc^2 = 21.2$ MeV, p is in MeV/c, and v in units of c. For the projected angle Θ_{proj} a semiempirical formula is

$$\Theta_{\text{proj}} = z(15/pv)(t')^{1/2} \qquad (2\text{-}7.17)$$

The distribution of Θ is, however, not truly Gaussian. The corresponding projected displacement y on traversing an absorber of thickness L is $y = 3^{-1/2} L \Theta_{\text{proj}}$.

An important and elegant application of multiple scattering is its use in measuring the pv of a particle traversing a photographic emulsion (see Sec. 3-6).

2-8   STRAGGLING

The energy loss calculated in Sec. 2-3 is an average value. For each particle the actual value fluctuates around the average value, with two consequences. For a given path length the energy loss and ionization fluctuate. For a given energy loss the path length fluctuates. The latter phenomenon is called *straggling*.

● Assume that in crossing a thickness x the average energy loss is E_0 and that in a specific case the energy loss is E. We want to calculate $\langle (E - E_0)^2 \rangle$.

Remembering that $\langle E \rangle = E_0$ by definition and developing the square, we have immediately

$$\langle (E - E_0)^2 \rangle = \langle E^2 \rangle - \langle 2E_0 E \rangle + \langle E_0^2 \rangle = \langle E^2 \rangle - E_0^2 \qquad (2\text{-}8.1)$$

Now suppose that many particles, all of the same initial energy, are sent through the absorber. In crossing the thickness x there will be collisions for which the energy loss is E_r. Their number is ν_r for a specific particle; averaged over all the particles it is $\langle \nu_r \rangle$. The numbers ν_r and $\langle \nu_r \rangle$ are related by (see Chap. 5)

$$\langle (\nu_r - \langle \nu_r \rangle)^2 \rangle = \langle \nu_r \rangle \qquad (2\text{-}8.2)$$

which is valid if the numbers ν_r found in many trials are statistically independent. Considering a large number P of particles, we then have

$$\langle E^2 \rangle - E_0^2 = \frac{1}{P} \sum_{i=1}^{P} \sum_r \left(\nu_r^{(i)} E_r - \langle \nu_r \rangle E_r \right)^2 = \sum_r \langle \nu_r \rangle E_r^2 \qquad (2\text{-}8.3)$$

Now

$$\langle \nu_r \rangle = N \int \sigma_r \, dx \qquad (2\text{-}8.4)$$

where σ_r is the differential cross section for energy loss E_r and N is the number of atoms per cubic centimeter in the absorber. Replacing ν_r in Eq. (2-8.3) gives

$$\langle E^2 \rangle - E_0^2 = N \sum_r \int \sigma_r E_r^2 \, dx \qquad (2\text{-}8.5)$$

Using the results of Sec. 2-3, we can evaluate the sum over r by observing that

$$-\frac{dE}{dx} = N \sum_r \sigma_r E_r = \frac{4\pi z^2 e^4}{m v^2} NZ \int \frac{db}{b} \qquad (2\text{-}8.6)$$

Now E_r and b are related by

$$E_r = 2z^2 e^4 / m v^2 b^2 \qquad (2\text{-}8.7)$$

which on logarithmic differentiation gives

$$\frac{dE_r}{E_r} = -\frac{2 \, db}{b} \qquad (2\text{-}8.8)$$

Using Eq. (2-8.7), we have approximately

$$N \sum_r \sigma_r E_r^2 = \frac{2\pi z^2 e^4 NZ}{mv^2} \int_{E_{min}}^{E_{max}} dE = \frac{2\pi z^2 e^4 NZ}{mv^2} (E_{max} - E_{min}) \quad (2\text{-}8.9)$$

In this formula we can take $E_{max} = 2mv^2$, which is the classical limit for heavy particles, and $E_{min} = 0$. We thus obtain from Eqs. (2-8.9) and (2-8.5)

$$\frac{d}{dx} (\langle E^2 \rangle - E_0^2) = 4\pi z^2 e^4 NZ \quad (2\text{-}8.10)$$

This formula was given by Bohr in 1915. Although only approximate, it gives results in reasonable agreement with experiment. For a finite thickness Δx Eq. (2-8.10) gives

$$\langle E^2 \rangle - E_0^2 = \langle (E - E_0)^2 \rangle = 4\pi z^2 e^4 NZ \, \Delta x \quad (2\text{-}8.11)$$

If it is assumed that the actual energy loss has a Gaussian distribution around the average value E_0, it follows from Eq. (2-8.10) that this distribution takes the form

$$P(E) \, dE = \frac{dE}{(8\pi^2 z^2 e^4 NZx)^{1/2}} \exp\left[-\frac{(E - E_0)^2}{8\pi z^2 e^4 NZx} \right] \quad (2\text{-}8.12)$$

where x is the thickness traversed.

This discussion has considered only collisions with electrons giving an energy transfer very small compared with the total energy and even with the energy straggling of the heavy particle. Larger energy losses such as occur in nuclear collisions are in most cases rare and do not contribute much to the average loss, but they influence the fluctuations appreciably, giving to the Gaussian distribution a tail on the side of the high energy losses (Fig. 2-17), just as large single-scattering deflections influence the angular distribution. This fact sometimes makes the distinction between average and most probable energy loss important.

The straggling effects are much more important for electrons than for heavy particles, because an electron may lose half its energy by simple elastic collision, whereas a heavy particle may lose only a fraction of its energy (of the order of m/M). Radiation losses add further to electron straggling. Thus electron straggling reaches values of the order of 0.20 of the total energy loss. Moreover, the most probable energy loss for an electron is much less than the average energy loss (Fig. 2-18). For example, for high-energy electrons ($E \gg mc^2$), Goldwasser, Mills, and Hanson (1953) have found the semiempirical

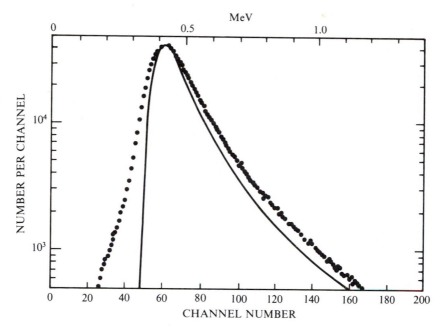

Figure 2-17   Pulse-height distribution from a silicon detector 1.09 mm thick
when 730-MeV protons pass through it. Semilog plot. The solid curve is the
theoretical curve according to Lindhard *et al.* (1963). [R. Lander *et al., Nucl.
Instrum. and Methods*, **42**, 261 (1966).]

formula for the most probable energy loss,

$$\Delta E_p = 2\pi N e^4 \, \frac{Zx}{mc^2} \left(\log \frac{x}{a_0} - 0.37 \right)$$

where x is the thickness of the material traversed and a_0 is the Bohr radius;
$\hbar^2/me^2 = 0.53 \; 10^{-8}$ cm. The average energy loss according to Eq. (2-4.2) is

$$\Delta E_{av} = 2\pi e^4 \, \frac{NZx}{mc^2} \left(\log \frac{E^3}{2mc^2 I^2} + \frac{1}{8} \right)$$

(compare Fig. 2-18).

Turning now to the second problem, that of the fluctuations in range for a
given initial energy, we also postulate here an approximate Gaussian distribution of ranges around the average range R_0,

$$P(R) \, dR = (1/2s) \exp \left[-(\pi/4s^2)(R - R_0)^2 \right] dR \qquad (2\text{-}8.13)$$

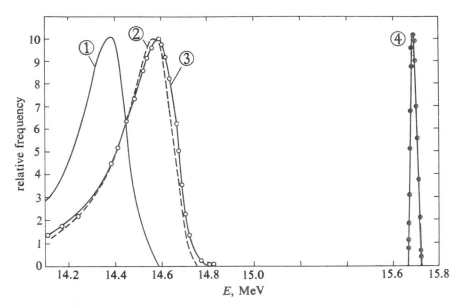

Figure 2-18 Energy distribution of an "unobstructed" electron beam and the calculated and experimental distributions of electrons that have passed through 0.86 g cm^{-2} of aluminum. (1) Landau theory without density correction; (2) Landau theory with Fermi density correction; (3) experiment; (4) incident beam. [E. L. Goldwasser, F. E. Mills, and A. O. Hanson, *Phys. Rev.*, **88**, 1137 (1952).]

The parameter s^2 has the value

$$s^2 = (\pi/2)\langle (R - R_0)^2 \rangle \tag{2-8.14}$$

as can be seen by calculating $\langle (R - R_0)^2 \rangle$ from Eq. (2-8.13). The quantity s, called straggling (Fig. 2-19), is the difference between the average range R_0 and the extrapolated range R_1 in the case of a Gaussian distribution of ranges (see Fig. 2-1).

The fluctuation in range and in energy loss are related. For a small thickness dx and a small energy loss dE, the fluctuations in residual range and the fluctuations in energy loss are related by

$$(\Delta E^2)_{dx} = \left(\frac{dE}{dx} \right)^2 (\Delta x^2)_{dE} \tag{2-8.15}$$

where

$$(\Delta E^2)_{dx} = \langle (E - E_0)^2 \rangle \tag{2-8.16}$$

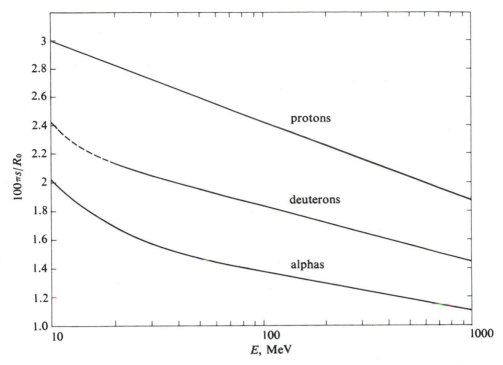

Figure 2-19 Range straggling as a function of energy. Abscissa, logarithmic scale in MeV. Ordinate, $100\pi s/R_0 = 100\pi^{3/2}2^{-1/2}\langle(R - R_0)^2\rangle^{1/2}R_0^{-1}$. [R. H. Milburn and L. Schecter, UCRL 2234 (rev. ed.).]

relative to the traversal of thickness dx and

$$(\Delta x^2)_{dE} = \langle(x - x_0)^2\rangle \tag{2-8.17}$$

is the standard deviation squared for the residual range after an energy loss dE. For a finite thickness we have, summing the $(\Delta x^2)_{dE}$, and recalling Eq. (2-8.11), which shows that $(\Delta E^2)_{dx}$ is proportional to dx,

$$\langle(x - x_0)^2\rangle = \int_0^{x_0} \frac{(\Delta E^2)_{dx}}{dx} \left(\frac{dE}{dx}\right)^{-2} dx$$

$$= \int_0^{E_0} \frac{(\Delta E^2)_{dx}}{dx} \left(\frac{dE}{dx}\right)^{-3} dE$$

$$= 4\pi z^2 e^4 NZ \int_0^{E_0} \left(\frac{dE}{dx}\right)^{-3} dE \tag{2-8.18}$$

This relation is only approximate. For numerical values, see Fig. 2-18. The effects of straggling are much greater for electrons than for heavy ions, but they cannot be treated in a simple fashion. In the passage through matter of very heavy ions—fission fragments, for example—the phenomena of scattering and straggling acquire great importance, because there are occasional single collisions of anomalous importance, as discussed in Sec. 2-3. ●

2-9 PASSAGE OF GAMMA RAYS THROUGH MATTER

The interaction of electromagnetic radiation with matter produces three main types of phenomena:

1. Photoelectric effect
2. Scattering on free electrons (Thomson, Rayleigh, and Compton)
3. Electron–positron pair production

Each of these processes is itself fairly complex, being accompanied by secondary effects such as the emission of Auger electrons and fluorescent radiation in (1), the emission of recoil electrons in (2), and the subsequent annihilation of positrons in (3). The three processes have different relative importance in different spectral regions, depending on the atomic number of the absorber, as shown in Fig. 2-20.

Photoelectric Absorption

In photoelectric absorption the energy of one photon is used to remove one of the electrons from an inner shell of an atom of the absorber element. Clearly this process can occur only if the incoming gamma ray has an energy higher than the binding energy of the electron to be removed. We have thus a series of jumps in the curve of the absorption coefficient, corresponding to the binding energy of the different orbits (see Fig. 2-21). These energies are given approximately by Moseley's law,

$$E = Rhc \frac{(Z - \sigma)^2}{n^2} \tag{2-9.1}$$

where $Rhc = 13.605$ eV, Z is the atomic number, σ the screening constant, and n the principal quantum number of the different electronic orbits. In the K series $n = 1$, in the L series $n = 2$, etc. The screening constant σ has the approximate value 3 for the K shell and 5 for the L shell, but for precise values one must consider the different screening for the S, P, and other orbits and the spin–orbit interaction. Tables of X-ray levels are found in (LHP67). Figure 2-21 shows the mass absorption coefficient for platinum, separating the contributions of different levels. Figure 2-22 shows the atomic absorption coefficient of substances with different Z for X rays of 1 Å wavelength ($\hbar\omega$ = 12 398 eV). There are several semiempirical expressions and extensive

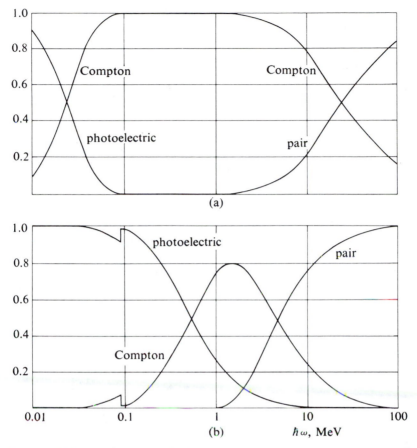

Figure 2-20 Relative contributions of various photon interactions to the total attenuation coefficient for (a) carbon and (b) lead. Abscissa, logarithmic scale in MeV. Ordinate, fraction of total attenuation due to phenomena indicated on curve. [White, private communication; (Fl E).]

tables for the mass absorption coefficient. The order of magnitude of the photoelectric atomic absorption cross section is $r_0^2 Z^5 / 137^4$.

The jumps in absorption coefficients provide a method for bracketing X-ray wavelengths in small intervals by measuring the absorption in elements with adjacent values of Z, one having the K or L absorption edge above and one having it below the wavelength to be measured.

The vacancy created by the ejection of an electron from the inner shells is filled by outer electrons falling into it, and this process may be accompanied by the emission of *fluorescent radiation*. Alternatively it is possible that no fluorescent radiation is emitted but that an electron from an outer shell is ejected. For instance, a vacancy in a K shell may be filled by an L electron

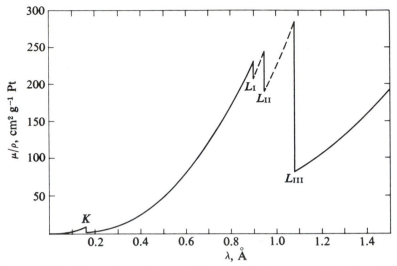

Figure   2-21   Mass absorption coefficient of platinum as a function of the wavelength of the incident X rays. [W. Bothe, in Geiger Scheel, *Handbuch der Physik*, Springer, Berlin, 1933.]

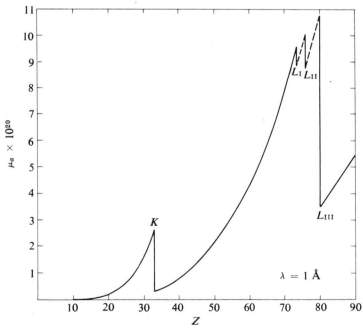

Figure   2-22   Atomic absorption coefficient of substances of different Z for X rays having $\lambda = 1$ Å. The atomic absorption coefficient $\mu_a = \mu A / \rho N_0$, where A is the atomic weight of the absorber and N_0 is Avogadro's number. [W. Bothe, in Geiger Scheel, *Handbuch der Physik*, Springer, Berlin, 1933.]

and another L electron emitted with energy $E_K - 2E_L$. Similarly, if the K vacancy is filled by an L electron and an M electron is emitted, the ejected electron will have the kinetic energy $E_K - E_L - E_M$. Here E_K and E_L are the binding energies of the K and L electrons. These and similar processes are called *Auger processes*, after their discoverer.

The emission of fluorescent radiation and the emission of Auger electrons are competing processes. The number of X rays emitted per vacancy in a given shell is called its *fluorescent yield*, indicated by β. One can analyze the phenomenon in more detail and distinguish the fluorescent yields for the different upper and lower levels. One has obviously

$$\beta_K = \beta_{KL} + \beta_{KM} \cdots \tag{2-9.2}$$

where β_{KL} indicates the partial fluorescent yields for filling the K level from the other levels. Figure 2-23 shows the fluorescent yield for the K level as a function of Z. Fluorescent yields are important in the study of orbital electron capture processes.

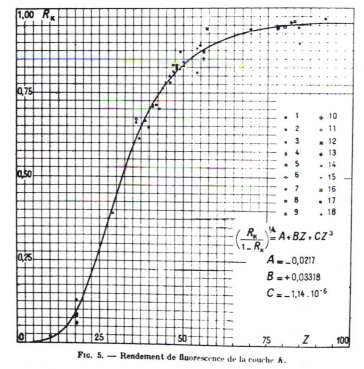

Fɪɢ. 5. — Rendement de fluorescence de la couche K.

Figure 2-23 Fluorescent yield for K level as a function of Z.

Thomson and Compton Scatterings

In the photoelectric effect it is essential that the electron be bound. However, even a free electron in the variable electromagnetic field of a beam of X rays is set in oscillatory motion and radiates as an oscillator. The radiation appears as scattered X rays. A classical theory on this effect was given by J. J. Thomson as follows: Consider a plane sine wave traveling in the x direction, with the electric vector polarized in the z direction and having intensity I_0 (Fig. 2-24). From electromagnetic theory we have the following relations:

$$I_0 = \frac{\langle \mathbf{E}^2 \rangle}{4\pi} c = \frac{\langle \mathbf{H}^2 \rangle}{4\pi} c \qquad (2\text{-}9.3)$$

where $\mathbf{E}$ and $\mathbf{H}$ are the instantaneous amplitudes of the electric and magnetic fields. An electron in this field will be subject to the force

$$e\mathbf{E} = e\mathbf{E}_0 \sin \omega t \qquad (2\text{-}9.4)$$

where ω is the frequency of the wave, and it acquires the acceleration

$$\mathbf{a} = (e\mathbf{E}_0/m) \sin \omega t \qquad (2\text{-}9.5)$$

Electromagnetic theory shows that a charge e subject to an acceleration $\mathbf{a}$ radiates energy at the average rate

$$\langle W \rangle = \frac{2}{3} \frac{e^2}{c^3} \langle a^2 \rangle \qquad (2\text{-}9.6)$$

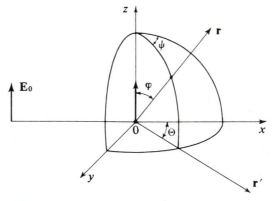

Figure 2-24 Notation used in treating Thomson scattering of an electromagnetic wave.

and in our case the average power radiated, if we recall that $\langle \sin^2 \omega t \rangle = \frac{1}{2}$, is

$$\langle W \rangle = \frac{1}{2} \frac{2}{3} \frac{e^4}{c^3} \frac{E_0^2}{m^2} = \frac{8\pi}{3} \left(\frac{e^2}{mc^2} \right)^2 I_0 \tag{2-9.7}$$

This power is subtracted from the primary beam, and we can equate it to the intensity falling on the scattering cross section of the electron σ_T. There results from Eq. (2-9.7)

$$\sigma_T = \frac{8\pi}{3} \left(\frac{e^2}{mc^2} \right)^2 = \frac{8\pi}{3} r_0^2 = 0.665 \ 10^{-24} \ \text{cm}^2 \tag{2-9.8}$$

where

$$r_0 = e^2 / mc^2 = 2.82 \times 10^{-13} \ \text{cm} \tag{2-9.9}$$

and is called the *classical radius* of the electron.

The radiation emitted in the case of polarized incident radiation has the same polarization and angular distribution as that emitted by a dipole oriented in the direction of the incident field. The intensity must obviously be axially symmetric around the direction of the field (z axis); and if φ is the angle between **E** and the direction of observation **r**, we know from the classical theory of electricity that the intensity at distance r decreases as r^2 and has an angular distribution

$$I = \frac{I_0}{r^2} \left(\frac{e^2}{mc^2} \right)^2 \sin^2 \varphi \tag{2-9.10}$$

The scattered rays are polarized with the electric vector in the rz plane. If the primary light is unpolarized, the intensity of scattered light must be axially symmetric around the direction of propagation x. The intensity can be calculated by superimposing incoherently the effects of primary light polarized in the z and y directions each of intensity $I_0/2$. Because of the axial symmetry around x, let us consider only the xy plane and in it the direction **r'** making an angle Θ with x. The beam polarized in the z direction gives an intensity

$$I_z = \frac{I_0}{2r^2} r_0^2 \tag{2-9.11}$$

because for this beam $\varphi = 90$ deg. The beam polarized in the y direction gives an intensity

$$I_y = \frac{I_0}{2r^2} r_0^2 \cos^2 \Theta \tag{2-9.12}$$

because the angle between E_y and $\mathbf{r}'$ is $(\pi/2) - \Theta$. The sum of the intensities is thus

$$I = \frac{I_0}{2r^2} r_0^2 (1 + \cos^2 \Theta) \qquad \text{or} \qquad \frac{d\sigma_T}{d\omega} = \tfrac{1}{2} r_0^2 (1 + \cos^2 \Theta) \quad (2\text{-}9.13)$$

The polarization of this radiation is

$$\frac{I_z - I_y}{I_z + I_y} = \frac{1 - \cos^2 \Theta}{1 + \cos^2 \Theta} \qquad\qquad (2\text{-}9.14)$$

It will be noted that Eq. (2-9.13) for angular dependence does not contain the frequency of the radiation.

Thomson's theory gives an atomic-scattering cross section proportional to Z; in fact, even before the present atomic model was developed, the atomic number of the lightest elements was measured by Barkla and others while observing the scattering of X rays. However, if the momentum transferred by the photon is small compared with the momentum of the electron in the atom, one must sum the amplitudes of the X rays scattered by all electrons in an atom. This scattering is called *Rayleigh scattering*, and the corresponding atomic-scattering cross section is of the order of

$$r_0^2 Z^2$$

Thomson's scattering does not take into account the quantum aspects of light and for energies comparable with mc^2 or larger gives results in disagreement with experiment. For one thing, it has been found experimentally that the frequency of the scattered light differs from that of the incident light. That this must be the case follows from conservation of energy and momentum if we interpret scattering as the elastic collision of light quanta with electrons. Assume that the quanta have an energy $\hbar\omega$ and a momentum $\hbar\omega/c$ directed in the direction of propagation of light. Conservation of energy and of momentum give the following relation between the wavelength of the incident and scattered light (λ and λ', respectively) and the direction of scattering:

$$\lambda' - \lambda = \frac{2\pi\hbar}{mc}(1 - \cos\Theta) \qquad \frac{2\pi\hbar}{mc} = 24.262 \times 10^{-11}\,\text{cm} = \lambda_c \quad (2\text{-}9.15)$$

The change of wavelength or of frequency on scattering is the Compton effect, named for its discoverer, A. H. Compton (1922).

The recoil electron goes in a direction Φ such that

$$\tan \Phi = \frac{\cot(\Theta/2)}{1 + \alpha} \qquad \alpha = \frac{\hbar\omega}{mc^2} = \frac{\lambda_c}{\lambda} \qquad (2\text{-}9.16)$$

and its kinetic energy is

$$E_{\text{kin}} = \hbar\omega \frac{2\alpha \cos^2 \Phi}{(1 + \alpha)^2 - \alpha^2 \cos^2 \Phi} \qquad (2\text{-}9.17)$$

as shown in Fig. 2-25. The problems of intensity and polarization cannot be treated by elementary means. We must confine ourselves to stating the result for the Compton scattering cross section for an electron (formula of Klein and Nishina), averaging over polarizations

$$\frac{d\sigma_c}{d\omega} = \frac{r_0^2}{2} \frac{k^2}{k_0^2} \left(\frac{k_0}{k} + \frac{k}{k_0} - \sin^2 \Theta \right) \qquad (2\text{-}9.18)$$

with $k_0 = 2\pi\hbar/\lambda$ and $k = 2\pi\hbar/\lambda' = k_0[1 + \alpha(1 - \cos\Theta)]^{-1}$. Figures 2-26 and 2-27 show a diagram of the angular distribution of Compton-scattered X rays for different values of α. (See Ja 75).

Pair Production

The last mechanism of absorption to be considered is the transformation of a gamma ray into an electron–positron pair, also called *materialization*. The principle of conservation of energy and momentum prevents this from occurring in free space. There must be a nucleus or an electron in order to balance energy and momentum in the transformation. The threshold energy

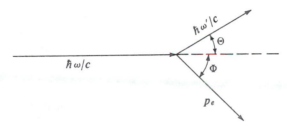

Figure 2-25 Compton scattering. Notation and diagram showing the conservation of energy and momentum.

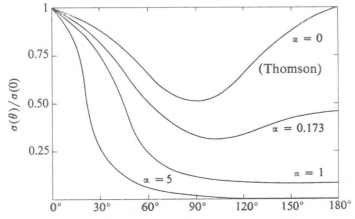

Figure 2-26 Angular distribution of Compton-scattered X rays as a function of energy ($\alpha = \hbar\omega/mc^2$).

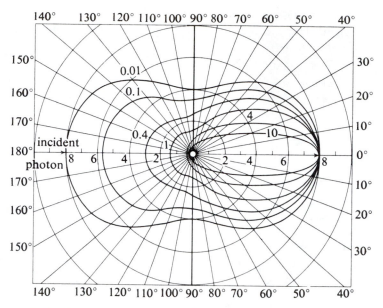

Figure 2-27 Polar diagram of the differential cross section per electron for Compton scattering. Curves labeled according to $\alpha = \hbar\omega/mc^2$ of incident photon. Units of $d\sigma/d\omega = 10^{-26}$ cm^2 sr^{-1}.

in the c.m. system for the materialization process is $2mc^2 = 1.022$ MeV.

This energy is very close to the threshold in the laboratory system for materialization near a nucleus, which by its recoil ensures the conservation of momentum. When the recoil is absorbed by an electron, the threshold required by the conservation of energy and momentum in the laboratory system is $4mc^2$, and there are two electrons and a positron acquiring appreciable momentum. In a cloud or bubble chamber they form a *triplet* (Fig. 2-28). Pairs can also be produced by heavy-particle collision, by electron–electron collisions, in mesonic decay, and by internal conversion in some gamma transitions. Some of these phenomena will be described later.

The atomic-pair-production cross section is of the order of $r_0^2 Z^2/137$ for relativistic energies.

Figure 2-29 summarizes the main results on the absorption of gamma rays and gives the absorption coefficient as a function of energy and Z in typical examples.

2-10 RADIATION LOSS BY FAST ELECTRONS

As mentioned before, the main cause of energy loss for a very fast electron ($E \gg mc^2$) traversing matter is the electromagnetic radiation it emits because of the accelerations to which it is subject (see Fig. 2-10). At low

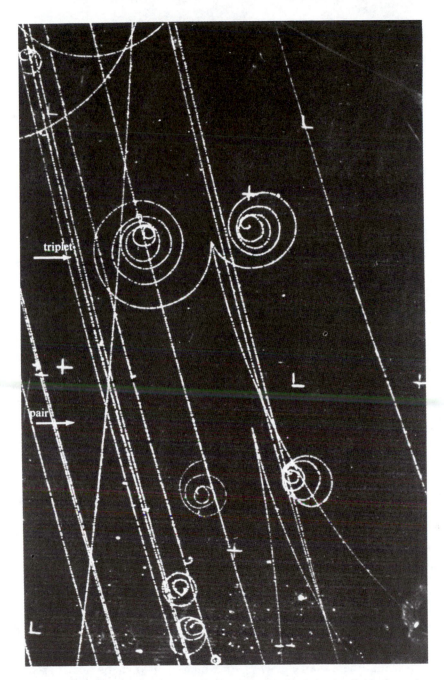

Figure   2-28   Formation of an electron–positron pair in the field of an electron (triplet). Formation of a pair in the field of a proton (pair). (Hydrogen bubble chamber.) [Courtesy Lawrence Radiation Laboratory.]

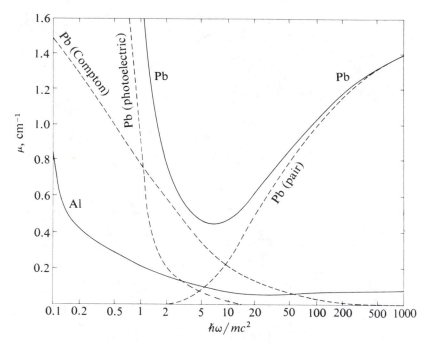

Figure   2-29   Total absorption coefficients of X rays by lead and aluminum as a function of energy (solid lines). Photoelectric absorption of aluminum is negligible at the energies considered here. Dashed lines show separately the contributions of photoelectric effect, Compton scattering, and pair production for Pb. Abscissa, logarithmic energy scale; $\hbar\omega/mc^2 = 1$ corresponds to 511 keV.

energy ($E \ll mc^2$) this radiation loss is unimportant compared with the ionization loss, but at high energy it becomes preponderant. The energy loss by radiation is proportional to Z^2 of the material and increases linearly with the electron's energy. The energy loss from ionization and excitation is proportional to Z and increases only logarithmically with energy. Thus, radiation loss predominates at the higher energies. More quantitatively, the approximate ratio of the two losses is (E in MeV)

$$\frac{(dE/dx)_{\text{rad}}}{(dE/dx)_{\text{ion}}} = \frac{EZ}{1600mc^2} = \frac{EZ}{800} \tag{2-10.1}$$

It is clear that there is an energy E_c for which the two energy losses are equal.

Table 2-2 summarizes some properties of materials of interest in experimental work (RPP).

TABLE 2-2 ATOMIC AND NUCLEAR PROPERTIES OF MATERIALS[a]

Material	Z	A	Nuclear cross section σ[b] (b)	Nuclear collision length L_{coll}[c] (g cm^{-2})	(cm)	Absorption length λ[c] (cm)	Minimum dE/dx[d] (MeV g^{-1} cm^2)	(MeV cm^{-1})	Radiation length L_{rad}[e] (g cm^{-2})	(cm)	Density[f] (g cm^{-3}) ( ) is for gas (g liter^{-1})	Refractive index n[f] ( ) is $(n-1) \times 10^6$ for gas
H_2	1	1.01	0.039	43.0	607.4	789.5	4.12	0.292	62.8	887	0.0708 (0.090)	1.112 (140)
D_2	1	2.01	0.074	45.1	273.3	341.7	2.07	0.342	126	764	1.65	1.128
He	2	4.00	0.134	49.6	396.5	477.8	1.94	0.243	93.1	745	0.125 (0.178)	1.024 (35)
Li	3	6.94	0.215	53.6	100.4	120.6	1.65	0.902	83.3	156	0.534	—
Be	4	9.01	0.270	55.4	30.0	36.7	1.61	2.97	66.0	35.7	1.848	—
C	6	12.01	0.340	58.7	≈37.8	49.9	1.78	≈2.76	43.3	≈67.0	≈1.55[g]	—
N_2	7	14.01	0.390	59.7	73.8	99.4	1.82	1.47	38.6	47.8	0.808 (1.25)	1.205 (300)
Ne	10	20.18	0.520	64.4	53.7	74.9	1.73	2.08	29.1	24.3	1.207 (0.90)	1.092 (67)
Al	13	26.98	0.650	68.9	25.5	37.2	1.62	4.37	24.3	9.0	2.70	—
A	18	39.95	0.890	74.5	53.2	80.9	1.51	2.11	19.7	14.1	1.40 (1.78)	1.233 (283)
Fe	26	55.85	1.160	79.9	10.2	17.1	1.48	11.6	13.9	1.77	7.87	—
Cu	29	63.54	1.270	83.1	9.3	14.8	1.44	12.9	13.0	1.45	8.96	—
Sn	50	118.69	2.040	96.6	13.2	22.8	1.28	9.4	8.9	1.22	7.31	—
W	74	183.85	2.810	108.6	5.6	10.3	1.17	22.6	6.8	0.35	19.3	—
Pb	82	207.19	3.080	111.7	9.8	18.5	1.13	12.8	6.4	0.56	11.35	—
U	92	238.03	3.380	116.9	≈6.2	12.0	1.09	≈20.7	6.1	≈0.32	≈18.95	—

TABLE 2-2 (CONTINUED)

66

Material	Nuclear collision length L_{coll} [c] (g cm^{-2})	(cm)	Absorption length λ [c] (cm)	Minimum dE/dx [d] (MeV g^{-1} cm^2)	(MeV cm^{-1})	Radiation length L_{rad} [e] (g cm^{-2})	(cm)	Density (g cm^{-3}) ( ) is for gas (g liter^{-1})	Refractive index n [f] ( ) is $(n-1) \times 10^6$ for gas
Air (gas at 20°C)	60.2	50 000[h]	67500	1.82	0.0022[h]	37.2	30 870[h]	0.001205[h] (1.29)	— (293)
H$_2$O	58.3	58.3	78.8	2.03	2.03	36.4	36.4	1.00	1.33
H$_2$ (bubble chamber 26°K)[i]	43.0	683	887	4.12	0.26	62.8	990	≈0.063[i]	1.112
D$_2$ (bubble chamber 31°K)[i]	45.1	322	403	2.07	0.29	126	900	≈0.140[i]	1.110
H–Ne mixture (50 mol %)[y]	62.9	154.5	215	1.84	0.75	29.9	73.5	0.407	1.092
Propane (C$_3$H$_8$)[k]	55.0	134	176	2.28	0.98	45.9	110	0.41[k] (2.0)	1.25[k] (1005)
Freon 13B1 (CF$_3$Br)[k] (See other Freons below.)	74.3	49.5	73.5	1.52	≈2.3	16.7	≈11	≈1.50[k] (8.71)	1.238[k] (750)
Ilford emulsion	79.5	20.8	39.1	1.44	5.49	11.2	2.94	3.815	—
NaI	91.9	25.0	41.3	1.32	4.84	9.5	2.59	3.67	1.775
LiF	61.1	23.1	30.7	1.69	4.46	39.3	15.1	2.64	1.394
Polyethylene (CH$_2$)	55.7	59.6	78.4	2.09	≈1.95	45.3	≈49	0.92–0.95	—
Mylar (C$_5$H$_4$O$_2$)	58.5	42.1	56.1	1.91	2.65	40.5	29.2	1.39	—
Polystyrene, scintillator (CH)[l]	57.0	55.2	68.5	2.03	1.97	44.3	43	1.032	1.581

Plexiglas ($C_5H_8O_2$) (also called Lucite)	57.7	48.9	65.0	1.95	1.65	40.5	34.5	1.16–1.20	≈1.49
Spark or proportional chamber[m]	0.5%		0.3%		0.073	2.7%		0.046	—
Shielding concrete[n]	64.9	26.0	32.2	1.70	4.25	26.6	10.6	2.5	—
Cerenkov counter materials									
CO_2	60.4	33800	46000	1.82	0.0033	36.2	20210	(1.79)[o]	(420)[o]
Freon 12 (CCl_2F_2)	68.1	13800	20200	1.64	0.0081	23.7	4810	(4.93)[o]	(1080)[o]
Freon 13 ($CClF_3$)	66.0	15000	21400	1.70	0.0072	27.15	6380	(4.26)[o]	(720)[o]

[a] From Particle Data Group, "Review of Particle Properties," *Phys. Letters*, **50B**, No. 1 (1974), p. 30. Table revised January 1974 by J. Engler and F. Mönnig. For details and references, see CERN NP Internal Rept. 74-1.

[b] σ of neutrons ($\approx \sigma$ protons) at 20 GeV from *Landolt-Bornstein*, New Ser. I, 5. Energy dependence for all nuclei $\approx \frac{1}{2}\%$ GeV^{-1} (from 5 to 25 GeV).

[c] $L_{coll} = A/(N \cdot \sigma)$. In the absorption length the elastic scattering is subtracted.

[d] From W. H. Barkas and M. J. Berger, *Tables of Energy Losses and Ranges of Heavy Charged Particles*, NASA SP-3013 (1964).

[e] Mainly from O. I. Dovzhenko and A. A. Pomanskii, *Soviet Phys. JETP*, **18**, 187 (1964); for some recent to-be-published results, see Y. Tsai, SLAC-PUB 1365 (1974).

[f] Values for solids, or the liquid phase at boiling point, except where noted. Values in parentheses for gaseous phase STP (0°C, 1 atm), except where noted.

[g] Density variable.

[h] Gas at 20°C.

[i] Density may vary about ±3%, depending on operation conditions.

[j] Values for typical working condition with H_2 target; 50 mol %, 29°K, 7 atm.

[k] Values for typical chamber working conditions: Propane ~57°C, 8–10 atm. Freon 13B1 ~28°C, 8–10 atm.

[l] Typical scintillator; e.g., PILOT B and NE 102A have an atomic ratio H/C = 1.10.

[m] Values for typical construction: two layers 50-μm Cu-Be wires, 8-mm gap, 60% argon, 40% isobuthane or CO_2; two layers 50-μm Mylar–Aclar foils.

[n] Standard shielding blocks, typical composition O_2 52%, Si 32.5%, Ca 6%, Na 1.5%, Fe 2%, Al 4% plus reinforcing iron bar. Attenuation length $l = 115 \pm 5$ g cm^{-2}; also valid for earth (typical $\rho = 2.15$) from CERN–LRL–RHEL Shielding exp. UCRL-17941 (1968).

[o] At 26°C and 1 atm. Indices of refraction from E. R. Hayes, R. A. Schluter, and A. Tamosaitis, ANL-6916 (1964).

The radiated energy appears as X rays and forms the bremsstrahlung, or continuous X-ray spectrum observed in ordinary X-ray tubes. At high energy the phenomena accompanying the bremsstrahlung are quite complicated, and they reach colossal proportions in the formation of showers containing millions of particles in the case of high-energy cosmic rays. For the theory of bremsstrahlung, we shall confine ourselves to a semiquantitative treatment of a special case, following Fermi. We shall also, however, give the results for cases where Fermi's hypotheses do not apply.

The main idea of Fermi's treatment (Fermi, 1924; Weizsäcker, 1937; E. J. Williams, 1933) is to consider the electromagnetic field produced by a charge in motion, to Fourier-analyze it, and to calculate the effect on other charges of the single Fourier components. This is simple if the cross section for the electromagnetic interaction of X rays on the charges is known.

To take a specific example, consider an electron of velocity $v \cong c$ passing by a nucleus of charge Ze. However, first use a system of reference in which the electron is at rest as the nucleus passes by it, and calculate the effect of the nuclear electromagnetic field by performing a Fourier analysis and treating the single components according to Thomson's formula (Sec. 2-9). The result will be the energy radiated by the electron in its rest system, which will ultimately be transformed back to the laboratory system.

In the rest system of the electron the nucleus moves with velocity $-v$. The electric field of the nucleus, $\mathbf{E}$, is contracted in the direction of motion, and a magnetic field $\mathbf{H}$ appears perpendicular to $\mathbf{E}$ and to the direction of motion. Its magnitude is almost equal to $\mathbf{E}$. These two fields are indistinguishable from a plane electromagnetic wave when seen from the electron. This wave undergoes scattering on the electron, and the scattered quanta, when viewed from the laboratory system, appear as the bremsstrahlung.

The calculation is developed in detail in (Fe 50) as well as in (Ro 52) and (Ja 75). As final result one finds that the spectrum extends from $v = 0$ to $2\pi\hbar v_{max} = mc^2(\gamma - 1)$ and a plot of the energy radiated versus v has an approximately rectangular form. Figure 2-30 shows σv versus v/v_{max}. Equation (2-10.2) gives an approximate expression for the average radiation loss of an electron crossing a substance a nucleus of atomic number Z and atomic density N:

$$-\left\langle \frac{dE}{dx} \right\rangle_{rad} = \int_0^{v_{max}} h v N \Sigma(v)\, dv$$

$$= 4Z^2 \frac{N}{137} r_0^2 h v_{max} \log \frac{183}{Z^{1/3}} \tag{2-10.2}$$

This expression is approximate and is valid for $E_0 \gg 137 mc^2 Z^{-1/3}$. The formulas valid under different hypotheses differ only in having included in the logarithm a function of energy.

Further, since an electron may lose an appreciable fraction of its energy to a single photon, the actual energy loss fluctuates widely from the average value.

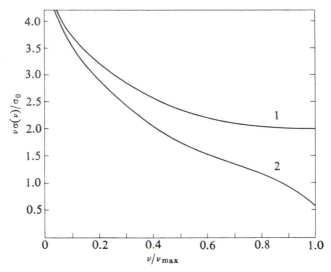

Figure 2-30 Integrated bremsstrahlung cross section versus frequency. Curve 1 refers to an energy of about $27Z^2$ eV, curve 2 to an energy of $4300Z^2$ eV. Frequency is in units of the upper frequency limit; $\sigma_0 = \frac{16}{3} Z^2 \alpha^3 a_0^2 (\text{Ry}/E_0)$ $= (Z^2 \text{Ry}/E_0) \times 5.8 \times 10^{-23}$ cm^2. [(Fl E).]

There are many detailed questions of great theoretical and practical importance concerning the bremsstrahlung. We shall mention only the polarization and angular distribution of the radiation. At low energy ($E \ll mc^2$), as in ordinary X-ray tubes, most of the X rays near the upper frequency limit are polarized with their electric vector parallel to the direction of motion of the electron, and the intensity has a maximum in the direction perpendicular to the direction of motion. At very high energy ($E \gg mc^2$) the average angle of emission of a quantum is

$$\theta = mc^2/E \tag{2-10.3}$$

independent of the energy of the emitted quantum. This distribution gives the characteristic narrow pencils of electromagnetic radiation beams observed in electron accelerators.

The electric vector of high-energy bremsstrahlung is prevalently normal to the plane of the incident electron and the X ray.

If the electron beam generating the bremsstrahlung is polarized, the polarization of the bremsstrahlung itself is affected. Notably, if the electron spin is in the direction of motion, the radiation tends to be circularly polarized with the angular momentum parallel to the spin of the electron.

Pair production is closely related to the bremsstrahlung, as can be seen by considering pair production as the absorption of a gamma quantum (in the presence of a nucleus) by an electron in a negative energy state, which is thereby excited to a positive energy state, and the bremsstrahlung as the transition of an ordinary electron from one positive energy state to another in the presence of a nucleus (accompanied by the emission of a gamma ray).

Calculation of the pair-production cross section gives for the case in which $h\nu \gg mc^2$

$$\sigma_{\text{pair}} = \frac{e^2}{\hbar c} Z^2 r_0^2 \left(\frac{28}{9} \log \frac{183}{Z^{1/3}} - \frac{2}{27} \right) \tag{2-10.4}$$

which is similar to Eq. (2-10.2).

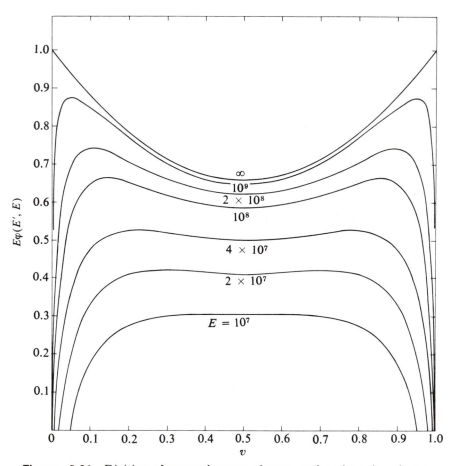

Figure   2-31   Division of energy between electron and positron in pair production. Differential probability of pair production per radiation length of lead for photons of various energies. Abscissa, $v = $ (total energy of one electron divided by energy of primary photon) $= E'/E$. Ordinate, $E \times \varphi(E', E)$. $\varphi(E', E)$ is the probability per radiation length of producing electron in a unit of energy interval at E. The curves are labeled according to E in eV. [B. Rossi and K. Greisen, *Rev. Mod. Phys.*, **13**, 259 (1941).]

The energy of the gamma ray is apportioned between the electron and positron as shown in Fig. 2-31, and both particles travel in approximately the same direction as the gamma ray for $E \gg mc^2$. The average angle between the direction of motion of the created electron and the gamma ray is

$$\theta = mc^2 / E \qquad (2\text{-}10.5)$$

in striking analogy to Eq. (2-10.3).

2-11 RADIATION LENGTH; SHOWERS

Bremsstrahlung and pair production combine to produce the spectacular phenomenon of showers. Starting with a single high-energy electron or gamma quantum, a multiplicative process (see Fig. 2-32) produces a number of electrons and gamma rays and forms the *shower*. A precise mathematical

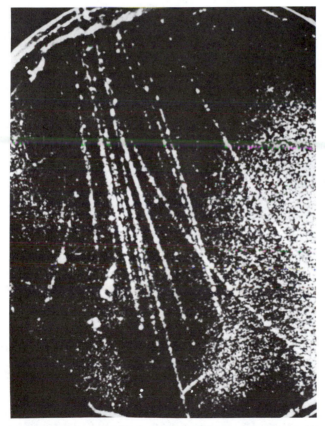

Figure 2-32 First shower observed by Blackett and Occhialini in a cloud chamber triggered by counters in coincidence above and below the chamber. [*Proc. Roy. Soc. (London)*, **139**, 699 (1933).]

theory of showers is very complicated, even when restricted to average behavior; furthermore, the fluctuations from the average are usually large. Monte Carlo calculations of showers have been made and are useful in practical cases. Some basic concepts, such as radiation length, are of great importance in shower theory as well as in other phenomena; we shall now develop them very briefly, together with an extremely simplified shower model.

Equation (2-10.2) shows that, on the average, an electron loses its energy by radiation according to

$$-\frac{dE}{E} = \frac{dx}{X_0} \tag{2-11.1}$$

with

$$\frac{1}{X_0} = \frac{4Z^2N}{137} r_0^2 \log \frac{183}{Z^{1/3}}$$

Equation (2-11.1) gives, on immediate integration,

$$\langle E \rangle = E_0 e^{-x/X_0} \tag{2-11.2}$$

This relation is valid when the radiation loss predominates over ionization loss, as is the case for energies larger than a certain critical energy E_c. The quantity X_0 is called *radiation length* and depends on Z and ρ, the density of the medium; it is roughly proportional to $1/Z\rho$. The critical energy E_c is crudely represented according to Eq. (2-10.1) by

$$E_c(\text{MeV}) = 800/Z \tag{2-11.3}$$

(See also Table 2-2.)

Similarly, gamma rays have a mean free path for pair production X_p, which is obtained from Eq. (2-10.4) and is found to be

$$X_p = (9/7)X_0 \tag{2-11.4}$$

Let us now consider the oversimplified shower model. The conservation of momentum gives to the whole shower an axis in the direction of the momentum of the first electron. The shower spreads laterally much less than it propagates longitudinally, as is clear from Eqs. (2-10.3) and (2-10.5). Thus the following discussion will be limited to the shower propagation in the direction of the initial particle.

An electron of energy E gives rise in one radiation length to about three gamma rays, which, in turn, give rise in a mean free path to about three pairs. The number of particles in the shower, N, starts to grow exponentially with the progress of the shower,

$$N(x) = e^{\gamma x} \tag{2-11.5}$$

Gamma is such that for

$$x_1 = X_0 + X_p \qquad N = e^{\gamma x_1} = 6 \tag{2-11.6}$$

but

$$X_0 + X_p = (16/7)X_0 \tag{2-11.7}$$

and hence

$$\gamma = \frac{\log 6}{(16/7)X_0} = \frac{0.78}{X_0} \tag{2-11.8}$$

In an actual shower, electrons, positrons, and photons are simultaneously present. Furthermore, the number of photons, especially of low energy, ends by being approximately twice as large as the number of particles. The number of particles in the shower increases until energy starts to dissipate, primarily by ionization and the Compton effect rather than radiation and pair production. At this point the particles have reached the energy E_c, and if we assume that the average energy of electrons, positrons, and gamma rays is the same, namely, E_c, their number is $E_0/3E_c$. This occurs at a distance

$$X_c = \frac{X_0}{0.78} \log \frac{E_0}{3E_c} \tag{2-11.9}$$

from the origin; and

$$N_{max} = E_0/3E_c = e^{0.78X_c/X_0} \tag{2-11.10}$$

This estimate neglects energy losses by ionization, the energy going into the rest mass of the pairs, and other important factors that depress N_{max} to a value nearer $\frac{1}{6}(E_0/E_c)$. More detailed expressions for high energy and low Z are

$$N_{max} = 0.31 \frac{E_0}{E_c} \left(\log \frac{E_0}{E_c} - 0.37 \right)^{-1/2} \tag{2-11.11}$$

and

$$X_c = 1.01 X_0 [\log(E_0/E_c) - 1] \tag{2-11.12}$$

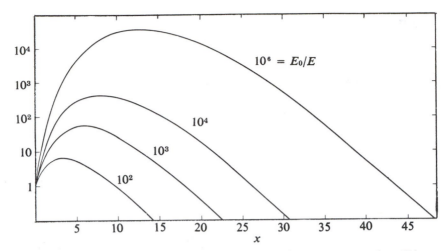

Figure 2-33 The average number of electrons of energy greater than E in a shower initiated by an electron of energy E_0 as a function of depth x (in radiation lengths). Ordinate in logarithmic scale. Each curve is labeled according to a value of E_0/E. [(Ro 53) should be consulted for approximations involved.]

The results of more precise evaluations are shown in Fig. 2-33. At low energy, 300 MeV for example, the general treatment becomes very complicated, in part because the cross sections depend on the energies. Useful results are obtainable, however, by the Monte Carlo method (Fig. 2-34). The lateral distribution of a shower has also been studied, but for more details we must refer to (Ro 53).

2-12 POSITRON ANNIHILATION

The annihilation process of the positron and electron is of great theoretical and practical importance. It is impossible to conserve energy and momentum if only one gamma quantum is emitted in the annihilation of a pair. The simplest possible process, starting from a pair *with small relative velocity v*, is then the annihilation into two gamma rays, each with an energy $\hbar\omega = mc^2$ and traveling in opposite directions. This is the most important mode of annihilation; only when this mode is forbidden by some special selection rule do more complicated processes occur, for example, three-gamma annihilation. The cross section for the annihilation process can be presumed to be proportional to r_0^2. Moreover, the cross section should be proportional to v^{-1}. This important result is supported by the following argument: A positron crossing a foil of electron density $\mathfrak{N}$ and thickness x, with velocity v, has a probability of annihilation of

$$\sigma(v)\,\mathfrak{N}x = P \qquad (2\text{-}12.1)$$

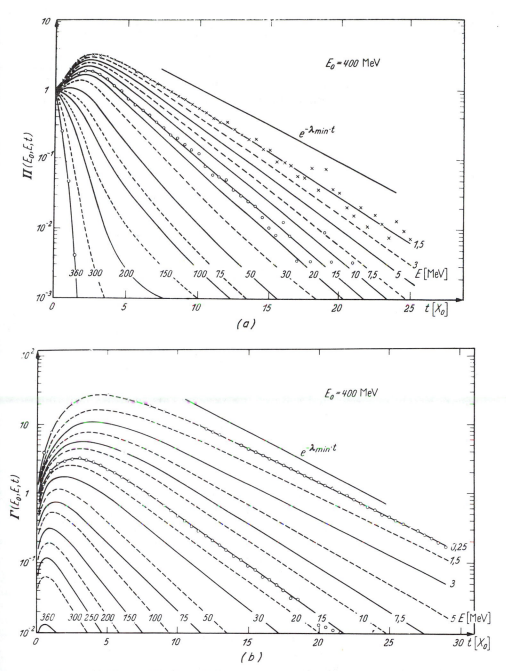

Figure   2-34   (a) Number of electrons $\pi(E_0, E, t)$ with energy larger than E at depth t for a primary electron of energy E_0, by Monte Carlo calculation. (b) Same for gamma rays $\Gamma(E_0, E, t)$. (Depth t in radiation lengths.) [H. H. Nagel, *Z. Phys.*, **186**, 319 (1965).]

As long as the velocity of the positron is small compared with the atomic velocities of the electrons, the probability of annihilation will be proportional to the time $t = x/v$ spent by the positron in the foil, because the relative velocity of the positron and the electrons is little affected by the small velocity of the positron. On changing velocity the only way in which P in Eq. (2-12.1) can become proportional to t is to make $\sigma = \alpha/v$, with α constant. Then

$$P = (\alpha/v)\,\mathfrak{N}x = \alpha\,\mathfrak{N}t \qquad (2\text{-}12.2)$$

as required. Indeed, a detailed (and advanced) calculation gives

$$\langle \sigma \rangle = \pi r_0^2 c/v \qquad (2\text{-}12.3)$$

where the average is over the two possible mutual directions of the spin; this is valid if

$$(e^2/\hbar c)mc^2 \ll T \ll mc^2 \qquad (2\text{-}12.4)$$

where T is the relative kinetic energy of the positrons.

● A positron–electron system bound together as a sort of hydrogen atom is called *positronium*. Positronium is unstable because of the electron–positron pair annihilation, which gives rise to gamma rays. The annihilation rate depends on the overlapping of the electron–positron clouds; annihilation thus occurs mainly in the ground state of the system. This state has no orbital angular momentum; the spins of the two particles can be parallel ($^3S_1$) or antiparallel ($^1S_0$). An argument given in Sec. 13-4 shows that only positronium in the $^1S_0$ state can annihilate into two quanta. These are correlated as follows: The gamma rays, in order to conserve momentum, must be collinear; moreover, if a gamma ray is analyzed with an apparatus which detects linear polarization and is found to be linearly polarized in a given direction, its conjugate will be polarized in a direction perpendicular to the first. Similarly, if one is analyzed with an apparatus which detects circular polarization and is found to be, e.g., left circular, its conjugate will be found to be right circular. Positronium in $^3S_1$ state annihilates into three quanta.

For an unbound S state, Eq. (2-12.3) is made more specific by considering separate annihilation cross sections for singlet or triplet states. The cross sections for two-quanta annihilation are

$$^3\sigma_{2\gamma} = 0 \qquad ^1\sigma_{2\gamma} = 4\pi r_0^2 c/v \qquad (2\text{-}12.5)$$

The probability of annihilation per unit time, or the reciprocal of the mean life, can be written as

$$1/\tau = \sigma_{2\gamma}\,\mathfrak{N}v \qquad (2\text{-}12.6)$$

where $\mathfrak{N}$ is the density of the electrons and v the relative velocity of the positron and electron. In the bound case $\mathfrak{N}$ is the density of the electron at the positron position, which is given by the absolute square of the wave function ψ at the origin. For the singlet state

$$\frac{1}{\tau_{2\gamma}} = {}^1\sigma_{2\gamma}v|\psi(0)|^2 = 4\pi r_0^2 \frac{c}{v} v \frac{1}{\pi}\left(\frac{1}{2na_0}\right)^3$$

$$= \frac{1}{2}\left(\frac{e^2}{\hbar c}\right)^5 \frac{mc^2}{\hbar} \frac{1}{n^3} = \frac{1}{1.24 \times 10^{-10} n^3} \text{ sec}^{-1} \qquad (2\text{-}12.7)$$

where a_0 is the Bohr radius and n is the total quantum number of the orbit.

For triplet states the mean life is about $1115 \approx 8\hbar c/e^2$ times longer, and in the case of free collisions the ratio between three- and two-quanta annihilation is

$$\frac{\frac{3}{4}{}^3\sigma_{3\gamma}}{\frac{1}{4}{}^1\sigma_{2\gamma}} = \frac{3}{1115} = \frac{1}{372} \qquad (2\text{-}12.8)$$

The positronium atom itself was detected by Deutsch (1949) by slowing down positrons in freon (CCl_2F_2) and measuring the mean life of the triplet state that is formed in three-fourths of the captures. Starting observation 10^{-7} sec after the arrival of a positron in the gas, singlet positronium, which is formed in one-fourth of the captures, has already decayed by the time observation begins and is thus not found. What is seen is only the decay of

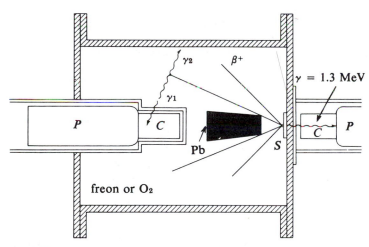

Figure 2-35 Apparatus used by Deutsch to measure the lifetime of positrons in gases. *P*, photomultipliers; *C*, sensitive crystals of NaI (Tl activated); *S*, source of ^{22}Na; γ_1, γ_2, 0.51 MeV annihilation gamma rays. [M. Deutsch, *Phys. Rev.*, **83**, 866 (1951).]

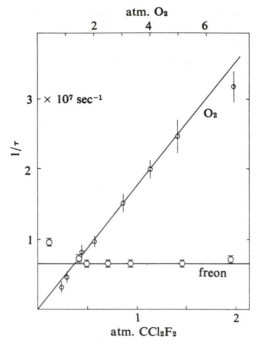

Figure 2-36 Pressure dependence of the decay rates of positrons in O_2 and in freon. [M. Deutsch, *Phys. Rev.*, **83**, 866 (1951).]

triplet positronium, provided that this is not converted by collisions to the singlet state. When the conversion does not occur, the mean life of positronium is independent of the pressure of the gas in which the positrons are stopped.

When capture occurs in the presence of molecules with a finite magnetic moment, such as O_2, the triplet states are converted to singlet states by collision and the singlet states decay in about 10^{-10} sec. The apparent mean life is due to free positrons decaying in collisions. The apparatus used and the dependence on pressure of the decay constant of positrons in freon and in O_2 are shown in Figs. 2-35 and 2-36. The interpretation is confirmed by the observation of three gamma rays in coincidence, corresponding to the annihilation of orthopositronium (triplet states), and of the two gamma rays, corresponding to the annihilation of parapositronium (singlet states). ●

2-13 POLARIZATION PHENOMENA

The study of polarization phenomena in electron and X-ray beams has become of great importance in recent years. As a consequence, experimental techniques for polarization and analysis have been developed.

The electron has spin $\frac{1}{2}$ and presents the simplest polarization phenomena. The polarization of a beam is defined as a vector **P** in the direction of the expectation vector of the spin and of length $2\langle s\rangle$. Operationally **P** is measured by observing the number of electrons in the beam that have a spin parallel (N_+) or antiparallel (N_-) to a given direction, giving the quantity

$$P = \frac{N_+ - N_-}{N_+ + N_-} \qquad (2\text{-}13.1)$$

The direction in which P is maximum is the direction of **P**, and $|\mathbf{P}|$ is given by Eq. (2-13.1) measured in that direction. Clearly $|\mathbf{P}|$ varies between 0 and 1, the former case corresponding to an unpolarized beam, the latter to that of a completely polarized beam. The electrons of a beam have a momentum **k**. If **P** and **k** are parallel, the beam is longitudinally polarized; if **k** and **P** are perpendicular, the beam is transversally polarized. The quantity

$$\mathcal{H} = \frac{\mathbf{P}\cdot\mathbf{k}}{|P|\,|k|} \qquad (2\text{-}13.2)$$

is often called the *helicity* of the electron. If the spin points in the direction of **k**, the helicity is $+1$; if it points in the opposite direction, the helicity is -1. An ordinary right-handed screw has positive helicity; i.e., in turning, the relation between the linear momentum with which it advances and its angular momentum corresponds to a positive helicity (see Chap. 9).

Beta decay often gives rise to longitudinally polarized electrons, with P near 1; however, for experimental reasons it is generally desirable to have beams transversally polarized. The rotation of the spin can be accomplished by deflecting the beam in an electric field. In the nonrelativistic case (assuming the value $g = 2$ for the ratio of the magnetic moment in Bohr magnetons and the spin in $\hbar$), the spin remains parallel to itself, while the vector **k** may rotate (Fig. 2-37). Note that in a magnetic field the angle between **P** and **k**

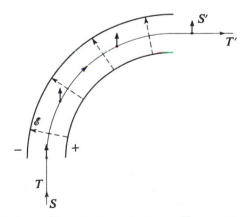

Figure 2-37 Spin rotation. Deflection of an electron beam in an electric field; nonrelativistic approximation. Dashed lines, electric-field lines; short arrows, spin orientation (S and S' are initial and final orientation); T, T', trajectory of electron; the polarization changes from longitudinal to transverse.

does not vary; therefore if the polarization is initially longitudinal, it remains longitudinal. For these simple conclusions the hypothesis $g = 2$ and the non-relativistic approximation are essential.

Electrons can be polarized by several methods and correspondingly analyzed for polarization. We shall mention only the methods that have found practical application. The oldest is Coulomb scattering by nuclei, which was originally proposed by Mott and is known as *Mott scattering*. This method may be used for both polarization and analysis. Figure 2-38 shows schematically a polarization analysis by scattering. Figure 2-39 gives the asymmetry to be expected. In Mott scattering we obtain transverse polarization and observe only transverse polarization.

Scattering of electrons or positrons on electrons depends on the relative spin orientation (Møller and Bhabha scattering). In practice it is possible to partially polarize the spins of the scatterer by using a magnetized ferromagnetic target. The change in scattering cross section is observed as a variation of intensity of the scattered electrons. Measurements are always made on a differential basis, i.e., by magnetizing the target first in one direction and then in the opposite direction. This is especially necessary because the best electron polarization obtainable in iron involves only 2 out of 26 electrons, and the resulting effects are very small.

The annihilation cross section for an electron–positron pair also depends on the relative spin orientation (see Sec. 2-12). Consequently, the polarization

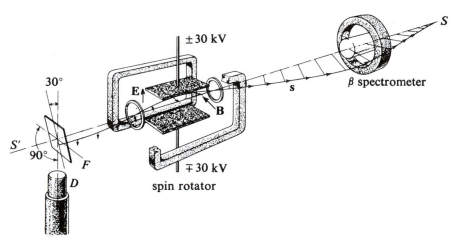

Figure   2-38 Crossed electric and magnetic fields to change electron polarization from longitudinal to transverse. A beta spectrometer is used to select the energy of the electrons from source S. The spin rotator changes the spin orientation, indicated by s, but has no action on the trajectory ($E = Bv/c$). The gold foil F scatters by 90 deg the electrons which are detected by D. The detector and foil assembly can rotate around SS' to determine asymmetry. [P. E. Cavanagh *et al.*, *Phil. Mag.*, **2**, 1105 (1957).]

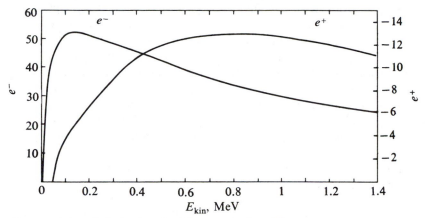

Figure 2-39 Polarization in Mott scattering. The asymmetry percentage $(200 \, |a|)$ in a single-scattering experiment of a totally polarized beam as a function of the energy for electrons and positrons $(Z = 80)$. Scattering angle 90 deg. [H. A. Tolhoek, *Rev. Mod. Phys.*, **28**, 277 (1956).]

of a slow positron beam can be detected by passing it through a magnetized ferromagnetic foil.

X rays also show polarization phenomena. For their description we may use the conventional optical nomenclature and distinguish between linear and circular polarization. Compton and Thomson scattering depend on the polarization of the incident beam. Thus for plane-polarized X rays the complete Klein–Nishina formula is:

$$\frac{d\sigma}{d\Omega} = \tfrac{1}{4} r_0^2 \frac{k^2}{k_0^2} \left[\frac{k_0}{k} + \frac{k}{k_0} - 2 + 4 \cos^2 \mu \right] \qquad (2\text{-}13.3)$$

where μ is the angle between the directions of polarization of the incident and scattered quantum. The quantities k_0 and k have the same meaning as in Eq. 2-9.18 and thus contain the scattering angle. Equation 2-13.3 coincides with Thomson's formula for $\hbar\omega \ll mc^2$; for $\hbar\omega \gg mc^2$ it has interesting limiting cases for small or large angle of scattering.

The variation of the scattered intensity as a function of the azimuth around the incident beam keeping the scattering angle constant indicates the polarization of the incident beam.

In order to analyze circularly polarized X rays, we need a scatterer with polarized electrons, such as a magnetized ferromagnet. The scattering cross section for circularly polarized X-rays having their direction of rotation parallel or antiparallel to that of the electron spin is different. Therefore, circularly polarized X rays are attenuated differently according to whether they cross a ferromagnet magnetized parallel or antiparallel to their direction of propagation.

The photodisintegration of the deuteron affords another method of detecting the linear polarization of X rays, because the proton is ordinarily ejected in the direction of polarization of the incident beam.

The polarization of electrons and X rays also gives rise to very interesting and important second-order phenomena in which they are mutually interrelated. For example, the bremsstrahlung produced by a longitudinally polarized electron is usually circularly polarized in the same direction as the electron, giving rise to an apparent multiplication of the angular momentum. It is thus sometimes possible to detect the polarization of primary radiation by transferring it to a secondary one and then observing the latter.

High-energy X rays may be polarized by producing bremsstrahlung shooting electrons on single crystals, for instance of diamond. For suitable orientation of the incident beam with respect to the crystalline lattice, the bremsstrahlung shows intensity peaks where the radiation is highly polarized.

Finally, a powerful method of obtaining any desired polarization and any energy of X rays uses the scattering of polarized laser light on a beam of fast electrons moving in the opposite direction. Seen from the electron frame of reference this is a simple Compton effect; in the lab system it generates high-energy X rays. (See Sec. 4-9.)

Annihilation in flight of positrons may be used as a source of monochromatic gamma rays and if the positrons and electrons are polarized, one may even obtain monochromatic polarized gamma rays.

Protons may also be polarized by nuclear scattering, especially at high energies (i.e., above 100 MeV). Detection of polarization is achieved most frequently by a double-scattering experiment. Energy loss by ionization occurs without appreciably affecting the polarization of protons. Low-energy polarized protons are formed in certain nuclear reactions.

BIBLIOGRAPHY

The proof of several formulas contained in this chapter requires elaborate calculations. The student who is particularly interested should consult Bethe and Ashkin in (Se 59); (Ro 53); (He 54); and Fano, Spencer, and Berger in (Fl E).

Barkas, W. H., and M. J. Berger, "Tables of Energy Losses and Ranges of Heavy Charged Particles," NASA SP-3013 (1964).
Bethe, H. A., and J. Ashkin, "Passage of Radiations through Matter," in (Se 59), Vol. I.
Betz, H-D. "Charge States and Charge-Changing Cross Sections of Fast Heavy Ions Penetrating through Gaseous and Solid Media," *Rev. Mod. Phys.*, **44**, 465 (1972).
Bohr, N., "The Penetration of Atomic Particles through Matter," *Kgl. Danske Videnskab. Selskab, Mat.-Fys. Medd.*, **18**, 8 (1948).
Burhop, E. H., *The Auger Effect and Other Radiationless Transitions*, Cambridge University Press, New York, 1952.
Crispin, A., and G. N. Fowler, "Density Effect in the Ionization Energy Loss of Fast Charged Particles in Matter," *Rev. Mod. Phys.*, **42**, 290 (1970).
Datz, S., C. Erginsoy, G. Leibfried, and H. O. Lutz, "Motion of Energetic Particles in Crystals," *Ann. Rev. Nucl. Sci.*, **17**, 129 (1967).
DeBenedetti, S., and H. Corben, "Positronium," *Ann. Rev. Nucl. Sci.*, **4**, 191 (1954).

Diambrini Palazzi G., "High Energy Bremsstrahlung and Electron Pair Production in Thin Crystals" *Rev. Mod. Phys.*, **40**, 611 (1968).

Fagg, L. W., and S. S. Hanna, "Polarization Measurements on Nuclear Gamma Rays," *Rev. Mod. Phys.*, **31**, 711 (1959).

Fano, U., "Penetration of Protons, Alpha Particles, and Mesons," *Ann. Rev. Nucl. Sci.*, **13**, 1 (1963).

Fano, U., L. V. Spencer, and M. J. Berger, "Penetration and Diffusion of X-rays," in (Fl E), Vol. 38.2.

Gentner, W., H. Maier-Leibnitz, and W. Bothe (GMLB 54).

Jackson, J. D. (Ja 75).

Koch, H. W., and J. W. Motz, "Bremsstrahlung Cross-Section Formulas and Related Data," *Rev. Mod. Phys.*, **31**, 720 (1959).

Northcliffe, L. C., "Passage of Heavy Ions through Matter," *Ann. Rev. Nucl. Sci.*, **13**, 67 (1963).

Page, L. A., "The Polarization Measurements on Beta and Gamma Rays," *Rev. Mod. Phys.*, **31**, 759 (1959) and *Ann. Rev. Nucl. Sci.*, **12**, 43 (1962).

Rossi, B. (Ro 53).

Rutherford, E., *Radiations from Radioactive Substances*, Cambridge University Press, London, 1930.

RPP published annually (see bibliography).

Scott, W. T., "The Theory of Small Angle Multiple Scattering of Fast Charged Particles," *Rev. Mod. Phys.*, **35**, 231 (1963).

Tolhoek, H. A., "Electron Polarization, Theory and Experiment," *Rev. Mod. Phys.*, **28**, 277 (1956).

Willis, B. H., and J. H. Atkinson, Jr., *High-Energy Particle Data*, UCRL 2426, June 1957.

Yuan, L., and C. S. Wu, *Methods of Experimental Physics—Nuclear Physics*, Academic Press, New York, 1962.

PROBLEMS

2-1 Derive Eq. (2-2.10) for a small value of θ in the form $\theta = Ze^2/Eb$. Do the same for any value of θ by considering first the momentum transfer in any central force and specializing to the particular case of an inverse-square force.

2-2 What is the probability that an alpha particle of 5-MeV energy crossing a gold leaf of 1 mg cm^{-2} will be deflected through an angle between 10 and 11 deg?

2-3 If the alpha particle mentioned in Prob. 2-2 is deflected by 20 deg, what is its impact parameter and distance of maximum approach?

2-4 Prove Eq. (2-2.18) by introducing the relative coordinates of the colliding bodies and their reduced mass.

2-5 Find the maximum angle of scattering in the laboratory system for alpha particles impinging on protons.

2-6 Use Fig. 2-7 to find the range in copper of an alpha particle having an energy of 30 MeV and of a muon having an energy of 75 MeV. What is their specific energy loss in copper? Estimate the ranges of the same particles in aluminum, lead, and hydrogen, by considering the effect of the change in I. Note that the number of electrons per gram of absorber is approximately twice as large for hydrogen as for other substances and hence its mass stopping power is also twice as large.

2-7 Show by a dimensional argument based on the Thomas-Fermi atomic model that the constant I of Eq. (2-3.10) is proportional to Z [F. Bloch, *Z. Physik*, **81**, 363 (1933)].

2-8 Justify the following formula used in photographic emulsion studies. The number n of delta rays having an energy between W_1 and W_2 per unit length of the track is

$$n = \frac{2\pi \mathfrak{N} Z^2}{\beta^2} \left(\frac{e^2}{mc^2} \right)^2 \left(\frac{mc^2}{W_1} - \frac{mc^2}{W_2} \right)$$

where $\mathcal{N}$ is the number of electrons per unit volume of emulsion and β and Z refer to the particle making the track.

2-9   The following equation has been given for the energy loss per centimeter by fission fragments,

$$\frac{1}{N}\frac{dE}{dx} = \frac{4\pi e^4}{mv^2}(Z_1^{\text{eff}})^2 Z_2 \log\frac{1.123 mv^3\hbar}{Ie^2 Z_1^{\text{eff}}}$$

$$+ \frac{4\pi e^4}{M_2 v^2} Z_1^2 Z_2^2 \log\left(\frac{M_1 M_2}{M_1 + M_2}\frac{v^2 a_{\text{scr}}}{Z_1 Z_2 e^2}\right)$$

where 1 refers to the fission fragment and 2 to the absorber material. The quantity a_{scr} is the impact parameter beyond which the energy loss resulting in nuclear collisions is zero because of screening. The first term refers to electron collisions, the second to nuclear collisions. Justify and discuss the equation.

2-10   Show that Rutherford's scattering formula gives, for the probability that a particle of charge z and energy E transfers to an electron the energy E' in crossing a thickness of dx g cm^{-2} of a substance of atomic mass A and atomic number Z, the expression

$$\phi(E, E')dE' \, dx = 2\pi N \frac{Zz^2}{A\beta^2} r_e^2 m_e c^2 \frac{dE'}{E'^2} dx$$

where N is Avogadro's number, $\beta = v/c$ of the initial particle, and r_e is the classical radius of the electron.

2-11   An absorption curve of a sample emitting beta and gamma rays was taken, using a Lauritsen electroscope, with aluminum absorbers. The data obtained were:

Absorber thickness, g cm^{-2}	Activity, divisions min^{-1}	Absorber thickness, g cm^{-2}	Activity, divisions min^{-1}
0	5.8	0.700	0.11
0.070	3.5	0.800	0.10
0.130	2.2	1.00	0.10
0.200	1.3	2.00	0.092
0.300	0.60	4.00	0.080
0.400	0.28	7.00	0.065
0.500	0.12	10.00	0.065
0.600	0.11	14.00	0.040

(a) Find the maximum energy of the beta spectrum (in MeV). (b) Find the energy of the gamma ray. (c) What would be the absorption coefficient of this gamma ray in lead?

2-12   Evaluate the angle θ_1 of Sec. 2-7 and find

$$\theta_1^2 = \frac{4\pi NZ(Z+1)z^2 e^4}{p^2 v^2} t$$

where t is in grams per square centimeter. See (Se 59), Vol. I, p. 285.

2-13   Calculate $\langle\Theta^2\rangle$ due to multiple scattering for 100-MeV protons crossing 2 mm of copper.

2-14   The discontinuities in absorption coefficients give a method for bracketing X-ray wavelengths in small intervals by measuring the absorption in elements with adjacent values of Z. Consider two absorbers: one with the K absorption edge above, one with the edge below, the wavelength to be measured. Plan experiments to identify by this method the K radiation of Tc and At.

2-15   A gamma ray is Compton-scattered backward ($\theta = 180$ deg). Calculate the energy of the scattered quantum for a primary quantum having $\hbar\omega = 0.01, 0.1, 1.0, 10, 100, 1000$ MeV.

2-16   Calculate the total Compton cross section

$$\sigma_c = \pi r_e^2 \frac{1}{\alpha} \left\{ \left[1 - \frac{2(\alpha + 1)}{\alpha^2} \right] \log(2\alpha + 1) + \frac{1}{2} + \frac{4}{\alpha} - \frac{1}{2(2\alpha + 1)^2} \right\}$$

and the limiting cases for $\alpha = \hbar\omega/mc^2 \ll 1$ or $\gg 1$. [Result: For $\alpha \ll 1$, $\sigma_c = (8\pi/3)r_e^2(1 - 2\alpha)$ and for $\alpha \gg 1$, $\sigma_c = \pi r_e^2 \alpha^{-1}(\frac{1}{2} + \log 2\alpha)$.]

2-17   A gamma ray has the absorption coefficient of 0.6 cm^{-1} in lead. Give its mass absorption coefficient and the absorption cross section per electron. What can you say about the energy of the gamma ray? If you measured the absorption coefficient in aluminum, what would you expect to find?

2-18   Electrons of 1-GeV energy pass through a lead plate 1 cm thick. Calculate the mean square of the linear deviation of the emergent points from the geometrical shadow point on the exit surface of the lead plate. What is the probability that the electron will produce a quantum? Repeat the calculation for a proton.

2-19   In the annihilation of positronium in the $^1S_0$ state (see Sec. 14-3), one finds two quanta with the eigenfunctions

$$LL - RR$$

Show that the linear polarizations of the two quanta are perpendicular to each other. Repeat the analysis for the case $LL + RR$. See (BH 59).

2-20   To reduce the energy of a proton beam by using a copper moderator, what are the required thickness of copper, the fraction of the primary beam passing through, and the value of $\langle\Theta\rangle$ due to multiple scattering in the following cases:

E initial	E final
1 GeV	100 MeV
100 MeV	30 MeV
3 MeV	0.3 MeV

Discuss the effect of nuclear absorption in all three cases, and estimate the final momentum spread, $\Delta p/p$, in each instance.

2-21   Calculate the Fano factor if two kinds of collision are equally probable, one with energy loss $w + \epsilon$ and the other $w - \epsilon$. Both produce on average one ion pair.

CHAPTER

Detection Methods for Nuclear Radiations

Experiments in nuclear physics depend upon the detection of nuclear radiations. This detection is made possible by the interaction of nuclear radiations with atomic electrons, which was treated in Chap. 2. Here we shall describe the main instruments which are used in practice.

3-1 CLASSIFICATION

For reasons of convenience, we shall classify these instruments as shown in Table 3-1. The same detector can be used in studies of various radiations, for the interaction of the radiations with the sensitive part of a detector may depend on different phenomena. Thus, a gas ionization chamber may detect heavy charged particles through their direct ionization effect; gamma rays through the photoelectrons, Compton electrons, or electron–positron pairs they produce in the gas or at the walls; high-energy neutrons through the ionization produced by recoil protons; and slow neutrons through the alpha particles produced by nuclear capture in boron or by the fission of ^{235}U. Similarly, a Cerenkov counter can be used to detect charged particles and gamma rays if the latter have enough energy to produce detectable electrons.

Instruments also differ widely in the information they yield. Some provide

Emilio Segrè, Nuclei and Particles: An Introduction to Nuclear and Subnuclear Physics, Second Edition

ISBN 0-8053-8061-7

TABLE 3-1 SUMMARY OF DETECTION METHODS

Type	Detection of single events	Detection of many particles
Electric	Ionization chamber	Ionization chamber
	Proportional counter	
	Geiger–Müller counter	
	Semiconductor counter	
	Spark chamber	
Optical	Scintillation counter	Photographic blackening
	Cerenkov counter	Materials damage
	Photographic emulsion	
	Cloud chamber	Chemical detectors
	Bubble chamber	
	Spark chamber	

accurate determinations of the time at which an event occurs (say, within 1 nsec) but give little spatial resolution; a large Cerenkov counter is an example of this type. Other instruments provide excellent spatial resolution but no information about the time at which the event took place. Photographic emulsions used to record tracks of particles are a case in point. Instruments that make possible very precise measurements of the total energy released in an event may provide only a moderately accurate time resolution. An example would be a solid-state counter used to detect the energy of alpha particles.

The efficiency of an instrument, i.e., the probability of its detecting a radiation crossing it, is another important characteristic. This may vary greatly. For most of the directly ionizing radiations, the efficiency is 1. In the detection of neutrons the efficiency depends strongly on their energy, ranging from a few percent at high energy to nearly 1 for slow neutrons. There is a similar dependence of efficiency on energy in the case of gamma rays.

Almost all instruments, after detecting an event, lose their sensitivity for a certain period known as *dead time*. This period sets a limitation on the number of events per unit time that an instrument can count. It is clear that in order to count with high efficiency, the fraction of the total time the counter is paralyzed must be small compared to 1, $rt_d \ll 1$ where t_d is the dead time and r the counting rate.

According to Hofstadter, "a perfect detector might have the following characteristics: (i) 100 percent detection efficiency; (ii) high-speed counting and timing ability; (iii) good energy resolution; (iv) linearity of response; (v) application to virtually all types of particles and radiations; (vi) large dynamic range; (vii) virtually no limit to the highest energy detectable; (viii) reasonably large solid angles of acceptance; (ix) discrimination between types of particles; (x) directional information; (xi) low background; and (xii) picturization of the event." To which we would add a reasonable cost. Obviously perfection is not of this world, and in each experiment the specific requirements often conflict, making compromises necessary in the choice and design of detectors.

ISBN 0-8053-8061-7

Typical numbers for parameters of different detectors are given in Table 3-2. Most of the radiation detectors listed in the table cannot function usefully without auxiliary electronic equipment, such as amplifiers, discriminators, coincidence circuits, and scalers. For example, the light emitted by a single particle passing through a Cerenkov counter can be detected only by a sensitive photomultiplier tube; the electrical impulse of the photomultiplier must then be amplified so that it can be used in a coincidence circuit or recorded on a scaler. In complex experiments, a computer, used to record and analyze the data from a system of many counters or to interpret the data obtained from a bubble chamber, becomes part of the instrument.

TABLE 3-2   DETECTOR PARAMETERS

Detector	Time resolution, sec	Dead time, sec	Space resolution, cm	Volume, cm^3
Ionization chamber	10^{-3}	10^{-2}	a	$1 - 10^5$
Geiger–Müller counter	10^{-6}	10^{-4}	a	$1 - 10^4$
Semiconductor counter	10^{-8}	10^{-6}	0.5	10
Scintillation counter	10^{-8}	10^{-6}	a	$1-10^5$
Cerenkov counter	10^{-9}	10^{-8}	a	$10-10^5$
Photographic emulsion			10^{-4}	10^4
Cloud chamber	10^{-2}	100	0.05	10^5
Bubble chamber	10^{-3}	1	5×10^{-3}	10^7
Spark chamber	10^{-6}	10^{-2}	0.05	5×10^6
Plastics				10^7

^{a}Depends on the size of the instrument.

3-2   IONIZATION CHAMBERS AND SOLID-STATE DETECTORS

Ionization chambers are one of the oldest types of instruments used in nuclear physics and are extremely versatile. An ionization chamber is essentially a vessel containing some substance, usually gas, which becomes ionized when charged particles pass through it. The ions are collected with the aid of an electric field, and the ionization current is measured. We recall from Chap. 2 that it takes about 35 eV to form an ion pair in air. If the primary radiation loses 1 MeV in the chamber, it will form approximately 2.86×10^4 ion pairs— a quantity of electric charge equal to 4.6×10^{-15}C for each sign. A minimum-ionizing, singly charged particle produces about 30 ion pairs in traversing 1 cm of air at atmospheric pressure. In general, ionization chambers are designed to collect all the ions produced in them, and a chamber is said to reach *saturation* when such total collection is made. At the same time, however, ion multiplication by secondary phenomena must be avoided. If it occurs, the apparatus works differently and is not normally called an ionization chamber

ISBN 0-8053-8061-7

(see Sec. 3-3). A schematic design of a typical ionization chamber is given in Fig. 3-1.

The rate of collection of the ions is determined by their mobility, which is defined as the ratio of the ion velocity to the electric field in the chamber. In air at STP, the ion mobility is of the order of 1 cm sec^{-1} per V cm^{-1}.

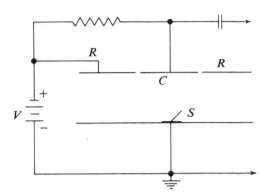

Figure 3-1 Typical ionization chamber. Parallel-plate ionization chamber with guard ring. R, guard ring; S, sample emitting an alpha particle; V, battery; C, collecting electrode.

In cases where one is interested in integrating the current over a period of time long compared with the time for collecting the ions, the mobility is unimportant. However, when one is interested in measuring the current produced by a single primary particle crossing the chamber, a short collecting time becomes desirable. A short collecting time is achieved by collecting only electrons whose velocity is about 1000 times greater than that of the positive ions produced. (See Sec. 2-6.) For this purpose one uses pure noble gases or, better, mixtures such as Ar–CO_2 or Ar–CH_4, in which electrons remain free and hence available for rapid collection. The comparatively high velocity allows the electrons to be collected in a time during which the positive ions barely move. As a consequence, with the usual type of apparatus (Fig. 3-2a) the output voltage depends on the site at which the ion is formed. In Fig. 3-2a, if only the negative charge q moves, the output voltage ΔV of the detector is given by

$$\Delta V = \frac{q}{C} \frac{x}{d} \qquad (3\text{-}2.1)$$

where C is the electrical capacity of one electrode with respect to the other. The dependence of ΔV upon the site of ion formation is often undesirable and can be obviated by introducing an auxiliary grid, as shown in Fig. 3-2b.

ISBN 0-8053-8601-7

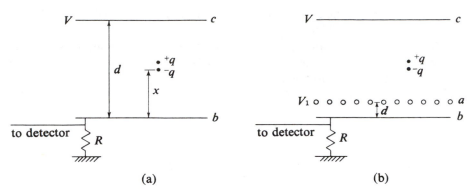

Figure 3-2   (a) Schematic drawing of an ion pair in a parallel-plate ionization chamber; (b) same as (a) but with an auxiliary grid. V, V_1 are the potentials of electrodes c and a.

Since the electrons produce a significant output voltage only when they travel
between grid a and the collecting electrode b, the same signal is obtained
regardless of where the electron was liberated between a and c.

Ionization chambers and other detectors depending on ionization by
charged particles can also be used to detect uncharged particles, for instance,
gamma rays and neutrons. Fast neutrons traversing a chamber filled with
hydrogen gas can be detected through the ionization produced by the recoil
protons that arise from neutron–proton collisions. For slow neutrons one can
use a gas, such as BF_3, containing boron nuclei that disintegrate under slow
neutron bombardment, emitting detectable alpha particles. Sometimes the
chambers contain substances having special nuclear properties that make
them sensitive only to particular radiations. For example, if the electrodes of
a multielectrode chamber are coated with ^{235}U, the chamber becomes sensitive to slow neutrons, which cause the fission of the ^{235}U nuclei. If the electrodes are coated with Bi, the chamber becomes sensitive to neutrons with
energy greater than 50 MeV—the threshold for producing fission in bismuth
by means of neutrons.

There are many possible geometric arrangements for ionization chambers,
ranging from extremely small ones used for dosimetry to very large ones
sometimes employed in cosmic-ray studies. Figure 3-3 shows an example of
the first case, in which a small tube the size of a fountain pen contains the
chamber and the electrode. Ionization chambers have also been made by
substituting solids or liquids in place of gases.

Semiconductors have found an important application in the detection of
nuclear events. In a $p–n$ or $p–i–n$ type semiconductor (e.g. Si or Ge) subjected
to an electric field, an ionizing particle crossing the $p–n$ junction produces a
current pulse and an associated voltage pulse. A fast pulse ($\sim 10^{-8}$ sec) results when the ionizing particle passes through the thin junction region. The
pulse amplitude is proportional to the energy lost in the junction layer over a

ISBN 0-8053-8601-7

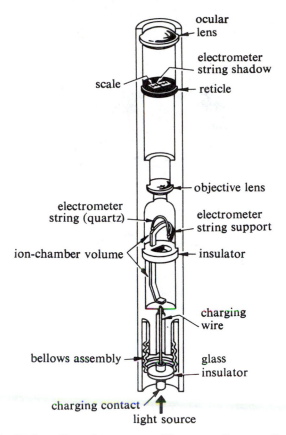

ocular
lens

electrometer
string shadow

scale

reticle

objective lens

electrometer
string (quartz)

electrometer
string support

ion-chamber volume

insulator

charging
wire

bellows assembly

glass
insulator

charging contact

light source

Figure   3-3   Pocket fiber electrometer. Victoreen direct-reading dosimeter. The charging wire contacts the electrometer string support for charging only. The electrometer string is of platinized quartz and moves when the electrometer is discharged. The optical system provides for measuring the displacement of the quartz fiber. [Courtesy Victoreen Company.]

very wide range of ionization density. However, the semiconductor must be operated at liquid air temperature. Figure 3-4 shows the assembly of such a counter; the field effect transistor (FET) in the first stage of the amplifier is the best electronic device as far as noise is concerned and it limits the resolution on its own account to approximately 60 eV.

The smallness of w and of the Fano factor help to give semiconductor detectors excellent energy resolution; furthermore, the relatively high absorption by solids of X rays helps to give good efficiency. This is illustrated in Figs. 3-5, 3-6, 3-7 and 3-8. Very thin silicon transmission detectors give a signal proportional to dE/dx, while total absorption of the particle gives E. The combination of the two signals makes it possible to identify the particle generating them; this is illustrated in Fig. 3.6

ISBN 0-8053-8601-7

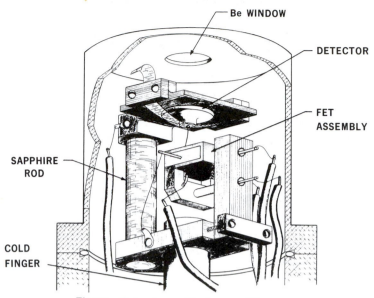

Figure 3-4 Assembly for a solid-state counter.

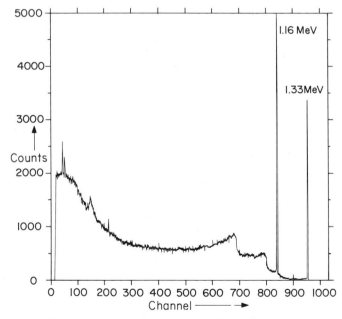

Figure 3-5 The gamma-ray spectrum from ^{60}Co observed by a germanium detector. Note the two full-energy peaks due to the 1.16-MeV and 1.33-MeV gamma rays, and the Compton distribution at energies less than that of the peaks. The sharp drops in the curve at channels No. 680 and No. 790 are the so-called Compton edges corresponding to the maximum collision energy given to electrons by 1.16-MeV and 1.33-MeV gamma rays. [F. S. Goulding and Y. Stone, *Science*, **170**, 280, (1970)].

ISBN 0-8053-8601-7

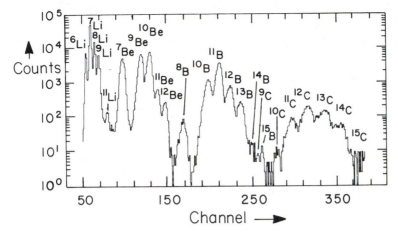

Figure 3-6 The spectrum of isotopes produced in fragmentation of uranium nuclei by 5-GeV protons as observed by a detector telescope and particle identifier. Each product particle is passed through a very thin silicon detector that measures the specific ionization, and is stopped in a second detector that measures the total energy. The signals are combined permitting the identification of the particles. This spectrum represented the first observation of the short-lived isotopes ^{11}Li, ^{14}B, and ^{15}B. [F. S. Goulding and Y. Stone, *Science*, **170**, 280 (1970).]

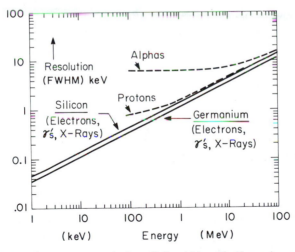

Figure 3-7 The energy resolution (full width at half maximum) as a function of energy for silicon and germanium detectors. These curves take into account the basic statistical limitations involved in the charge-production process and in atomic collisions. The Fano factor F is assumed to be 0.08 for germanium, and 0.1 for silicon. No account is taken of the amplifier-noise contributions to resolution. [F. S. Goulding and Y. Stone, **170**, 280 (1970).]

ISBN 0-8053-8061-7

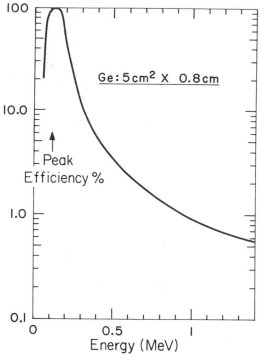

Figure 3-8 Gamma-ray efficiency as a function of energy for a 5-cm²
germanium detector 0.8 cm thick. Peak efficiency is defined as the ratio of
counts appearing in the peak to the total number of photons entering the
detector. [F. S. Goulding and Y. Stone, *Science*, **170**, 280 (1970).]

3-3 PROPORTIONAL AND GEIGER–MÜLLER COUNTERS

A schematic sketch of a gas-filled cylinder containing an isolated wire
along its axis is shown in Fig. 3-9. When ionizing radiation enters such a
cylinder, the current produced is a function of the impressed voltage V. At
very low voltages, the current is proportional to V; then, over a certain voltage interval, the current is independent of V. At this point the apparatus
performs as an ionization chamber, and the current is said to be saturated;
i.e., only the primary ions are being collected.

If the voltage is now increased beyond the saturation interval, the primary
ions begin to produce secondaries by collisions with the gas; consequently,
the primary ionization is multiplied by a factor that depends on the geometry
of the apparatus and on the applied voltage V, and the device functions as a
proportional counter. In this way it is possible to obtain multiplication factors
up to several millions using appropriate gas mixtures. The effective time of
collection, which also depends on the geometry and V, can be reduced to
microseconds. We shall not discuss here the mechanism of the multiplication

ISBN 0-8053-8061-7

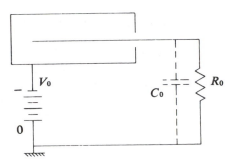

Figure 3-9 Geiger counter with external quenching resistance.

process. Proportional counters can be used to investigate soft beta rays and X rays through the photoelectrons they produce. When traversed by energetic particles, they also provide information on the ionization density produced by the particles.

Returning to the apparatus of Fig. 3-9, if we increase the voltage still further, a point is reached (the threshold) at which the presence of ions in the cylinder triggers a self-sustaining discharge that stops only if some special device lowers the potential difference across the counter. This procedure is called *external quenching*. The current in the discharge is independent of the primary ionization and the apparatus functions as a Geiger–Müller (G–M) counter. The ultraviolet-light quanta emitted in the discharge play an essential role in the working of a G–M counter. However, we shall not describe here the gas-discharge phenomena occurring in such counters.

Geiger–Müller counters are generally filled with argon–alcohol mixtures at pressures of about 10 cm Hg argon and 2 cm Hg ethyl alcohol. Occasionally other mixtures, consisting largely of a noble gas and an organic vapor, are used. The addition of the organic vapor renders the counter self-quenching; i.e., the discharge stops spontaneously without the necessity of lowering the voltage across the counter. A constant flux of ionizing particles through a G–M counter gives a counting rate independent of V over a certain voltage interval i.e., the counter has a "plateau." An example is shown in Fig. 3-10. Note that the threshold indicated for the G–M counter is above the voltage region where the tube works as a proportional counter. After a discharge, a G–M counter has a certain dead time, of the order of 100 μsec, during which it cannot register a new pulse. One must therefore be careful not to send too high a flux through the counter; otherwise the successive ions arrive during the dead time, when the counter is *paralyzed*.

A G–M counter will respond to a single ion formed in the gas; thus any charged particle crossing the tube is certain to be detected. X rays, on the other hand, rarely form an ion pair, and their detection depends on the material of the counter walls from which electrons are ejected by the X rays. Generally walls made of materials having high Z, such as lead, are the most favorable, but even then the efficiency for X rays or gamma rays is generally

ISBN 0-8053-8601-7

of the order of 1%. Typical constructions of G–M counters are shown in Figs. 3-11 and 3-12.

A G–M counter can be rendered sensitive to neutrons by filling it with a boron-containing gas, commonly BF_3. It is also possible to embed such a counter in paraffin (Fig. 3-13) in such a way that its sensitivity to neutrons is more or less independent of energy in the range from a few thousand to a few million electron volts.

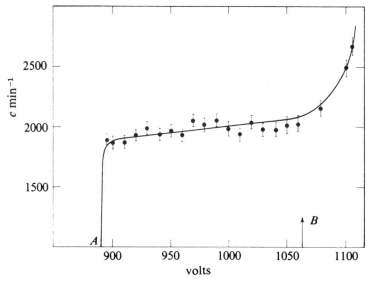

Figure   3-10   Dependence of counting rate of a Geiger counter on applied voltage. *A*, threshold voltage; *B*, limit of plateau. Plateau is 150 V long.

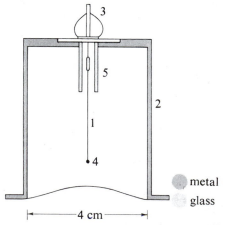

Figure   3-11   Thin mica-window counter. Window thickness 3 mg cm^{-2}. Diameter approximately 4 cm. 1, wolfram anode wire; 2, metal shell; 3, Kovar-glass seal; 4, glass bead; 5, glass sleeve. [Courtesy Radiation Counter Laboratories, Inc., Skokie, Ill.]

ISBN 0-8053-8601-7

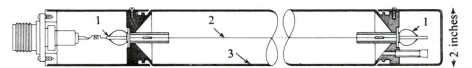

Figure 3-12 Cosmic-ray Geiger–Müller counter. Actual counter 30 in. long. 1, Kovar-glass seals; 2, anode wire; 3, metal cathode. [Courtesy Radiation Counter Laboratories, Inc., Skokie, Ill.]

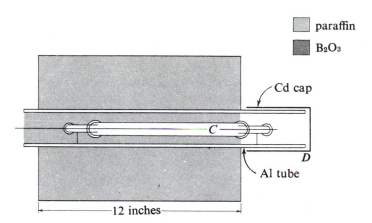

Figure 3-13 A neutron counter designed by Hanson and McKibben with uniform sensitivity (to within $\sim 10\%$) from 10 keV to 3 MeV. C is a BF_3 counter; D, cadmium cap and shield. [A. O. Hanson and J. L. McKibben, *Phys. Rev.*, **72**, 673 (1947).]

3-4 SCINTILLATION COUNTERS

At the beginning of this century, Crookes introduced a device for counting single alpha particles. He called the apparatus a *spinthariscope*. This instrument worked on the principle that an alpha particle impinging on a screen covered with zinc sulfide containing additional activating impurities produces scintillations that can be observed with a low-power microscope (Crookes; Elster and Geitel, 1903). The dials of luminous watches owe their luminosity to such scintillations; they are made by adding a small amount of a radioactive substance to a paste containing activated zinc sulfide.

The spinthariscope was an extremely important instrument in the hands of Rutherford and his pupils, who employed it extensively in establishing, for instance, Rutherford's scattering law. However, because the method was extremely laborious, it fell into disuse with the development of G–M counters. Recently, modern versions have come to the forefront through the use of photomultipliers to detect scintillations, thus dispensing with dependence on the human eye and the accompanying labor and uncertainty of efficiency in detection. Many substances scintillate when bombarded by nuclear radia-

ISBN 0-8053-8061-7

tions, but those commonly used fall into three classes: plastics, liquids, and organic and inorganic crystals. Noble gases (Kr, Xe) and liquid helium have been used for special purposes such as the detection of fission fragments or alpha particles. Table 3-3 gives typical characteristics of a few scintillating substances. The scintillation light is emitted for the most part over a continuous spectrum with a maximum between 3800 and 4500 Å. When the scintillations occur in the extreme ultraviolet, one uses a light shifter, i.e., a substance that, when excited in the extreme ultraviolet, fluoresces in a more convenient spectral region.

TABLE 3-3   CHARACTERISTICS OF SCINTILLATORS

Substance	Density, g cm^{-3}	Refractive index	Decay constant, nsec	Light (anthracene = 100)	Wavelength of max Å
Anthracene	1.25		32	100	4470
Stilbene	1.16	1.62	4	60	4100
Plastic^a	1.04	1.58	3	36	4500
Liquid^b	0.88	1.50	3	40	3820
NaI(Tl)	3.67	1.77	250	230	4130
ZnS(Ag)	4.09	2.36	200	300	4500

^{a}Polystyrene + 16 g liter^{-1} p-terphenyl.
^{b}Toluene + 3 g liter^{-1} 2, 5-diphenyl oxazole; scintillators with a mineral oil base have similar properties.

It is possible with modern photomultipliers to detect a particle that loses only 4 keV in a scintillator. This is a practical figure and takes into account the problem of the collection of light. A phototube gives one electron at the cathode for about 20 light quanta, and the tube itself multiplies the electron by a factor of 10^8 in a period of a few picoseconds. The electron avalanche striking the anode may last $\sim 3 \times 10^{-9}$ sec. The fluctuations in this period depend on the structure of the photomultiplier tube.

The density of ionization in some scintillating media influences the time dependence of light emission. If the ionization density is low, the light is emitted faster than if the ionization density is high. This effect has important applications in distinguishing different types of particle, such as protons and alpha particles, entering a scintillator. A very striking case is that of neutrons and gamma rays. The first act through recoil protons, the second mainly through Compton electrons. By measuring the total amount of light emitted in a time interval of the order of 0.5 μsec and the maximum instantaneous intensity (at about 0.1 μsec after the excitation), it is possible to distinguish between scintillations produced by X rays and those produced by neutrons (Fig. 3-14).

Sodium iodide crystals activated with thallium are especially valuable for detecting X rays. They can be made nearly 100% efficient, and the amount of

ISBN 0-8053-8601-7

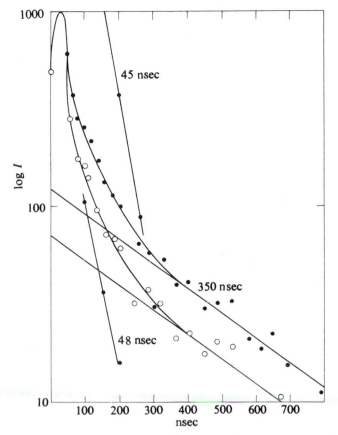

Figure 3-14 Light emission as a function of time for a scintillator (quaterphenyl) excited by neutrons or gamma rays. Ordinate is on a logarithmic scale. Open circles, gamma rays; closed circles, neutrons. [R. B. Owen, *IRE Trans.*, **NS-5**, 198 (1958).]

light emitted can be related to the energy of the impinging radiation. This, however, requires a pulse-height analysis like the one shown in Fig. 3-15 and an interpretation of the various peaks.

Large liquid scintillators, loaded with cadmium or boron and having a volume of hundreds of liters, are among the most efficient neutron detectors. Efficiencies greater than 95% are possible.

Plastic and crystal scintillators can be made in all shapes and sizes. They are sometimes put in direct optical contact with a photomultiplier. However, for convenience they may be connected to the photomultiplier by *light pipes*. These are highly polished pieces of plastic (Plexiglas) in which the light travels by multiple reflections from the walls. The attenuation in light pipes is generally small (in the visible spectrum), but it must be remembered that it is

ISBN 0-8053-8061-7

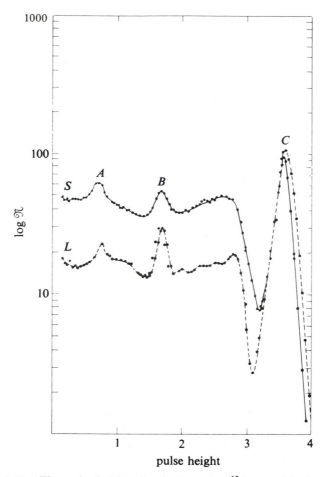

Figure 3-15 The pulse-height distributions for ^{65}Zn on $1\frac{1}{2}$- by 1-in. (S curve) and 3- by 3-in. crystals (L curve) of activated sodium iodide. Ordinate is number of pulses on a logarithmic scale; abscissa, pulse height. The peaks observed are A, back-scattered radiation; B, annihilation radiation (0.51 MeV); C, primary radiation (1.12 MeV). [P. R. Bell, in (Si 55).]

not possible to funnel the light from a large scintillator to a small photocathode surface without an attenuation of light that is at least in the ratio of the surface areas at each end of the light pipe.

At very high energies the phenomena by which the energy of the incoming particles is dissipated are complicated: for electrons and gammas they reduce to shower production; for strongly interacting particles one has a nuclear cascade. One way to measure the energy of the primary is to absorb in the detector all the secondaries. This is sometimes possible using large scintillation counters, e.g., of NaI or other media preferably of high Z. For examples of the resolving power obtainable see Fig. 3-16.

ISBN 0-8053-8601-7

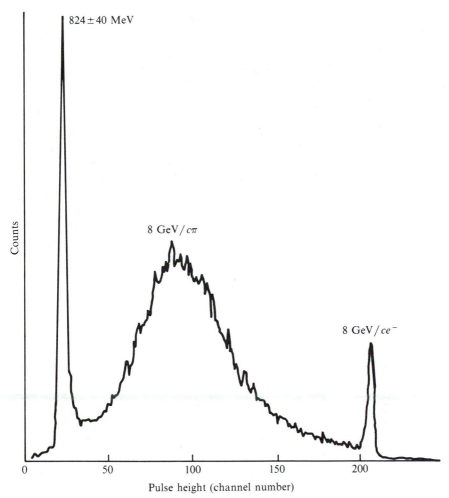

Figure 3-16 Spectrum obtained by bombarding a NaI(Tl) crystal 11.2 in. in diameter and 56 in. long with 8 GeV/c pions. The peak at the right is a calibration peak obtained with electrons. The shower is completely absorbed and the peak is narrow. The width of the peak is due to incomplete absorption of the nuclear cascade. The peak at the left is due to muons and the few pions which lose energy only by electromagnetic processes. [R. Hofstadter, *Science*, **164**, 1471 (1969).]

3-5 CERENKOV COUNTERS

The Cerenkov counter works on the principle that radiation is emitted by charged particles whose velocity βc exceeds the velocity of light in a given medium, i.e., when $\beta > 1/n$, where n is the refractive index of the medium. The intensity of Cerenkov light in the visible spectrum is about 100 times weaker than typical scintillation light in a plastic. A particle moving with

ISBN 0-8053-8061-7

velocity c in a medium having $n = 1.5$ emits about 250 light quanta per centimeter of path (see Sec. 2-5).

Even though Cerenkov light is comparatively weak, Cerenkov counters are very useful in high-energy nuclear physics because of the directional property of the light. That is, Cerenkov radiation is emitted in a cone of half angle θ, the axis of the cone being the incident-particle direction, and $\cos \theta = 1/n\beta$.

Thus transparent Cerenkov radiators, coupled with optical devices and photomultipliers, make useful velocity selectors for high-energy particles. Figures 3-17 and 3-18 show examples of such detectors. The radiator may be a solid, in which case one measures the angle between the beam direction and the Cerenkov light. Often the radiator is a gas, and one varies its refractive index by changing its pressure. For example, in carbon dioxide, by changing the pressure between 1 and 200 atm and the temperature between 25°C and 50°C, we cause the refractive index to take all values between 1.0004 and 1.21. The corresponding minimum β detected then varies from 0.992 to 0.825.

Cerenkov counters can also be used to detect high-energy X rays. In this application a large block of heavy glass is employed as the Cerenkov radiator.

THRESHOLD:

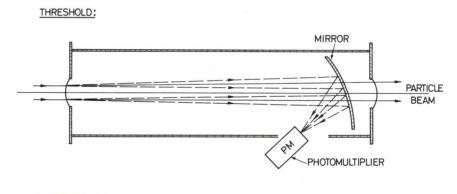

DIFFERENTIAL:

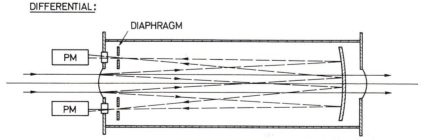

Figure 3-17 (a) A threshold Cerenkov counter and (b) a differential Cerenkov counter. In (a) the Cerenkov light is focused on a single photomultiplier. In (b) the light is focused on an annular diaphragm and detected by a crown of photomultipliers behind the diaphragm. The radiator is a gas, e.g., He, N_2, CO_2 [J. Litt and R. Meunier, *Ann. Rev. Nucl. Sci.*, **23**, 1 (1973).]

ISBN 0-8053-8601-7

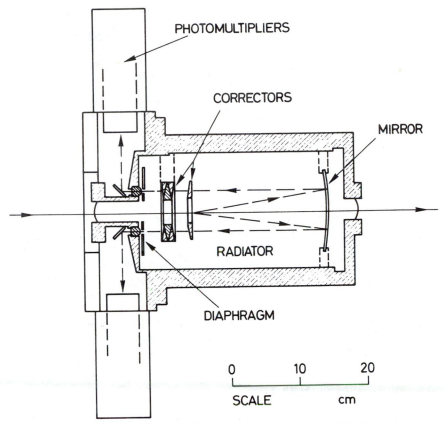

PHOTOMULTIPLIERS

CORRECTORS

MIRROR

RADIATOR

DIAPHRAGM

0   10   20

SCALE   cm

Figure 3-18 An achromatic Cerenkov counter used at CERN in the 20-GeV beam. Its resolving power is about 5×10^{-5} and it is used in the $0.996 < \beta < 1$ region. The radiator is gaseous. [J. Litt and R. Meunier, *Ann. Rev. Nucl. Sci.*, **23**, 1 (1973).]

The X rays produce a shower in the glass, and one can gain some idea of the energy of the X ray from the total amount of light emitted by the shower. Glass especially suited for this purpose can be obtained in large blocks with good transparency. A typical glass has the composition shown in Table 3-4. This particular glass has a density of 4.49 g cm^{-3}, $n = 1.72$, and radiation length $X_0 = 2$ cm. An example of such a glass Cerenkov counter is shown in Fig. 3-19.

ISBN 0-8053-8061-7

TABLE 3-4 GLASS COMPOSITION

Compound	Na_2O	K_2O	PbO	SiO_2	As_2O_3
Percentage by weight	1.0	3.4	61.7	33.1	0.8

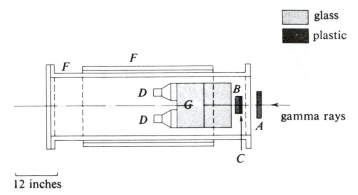

glass

plastic

gamma rays

12 inches

Figure 3-19 A large glass block used as a Cerenkov counter for high-energy gamma rays. The instrument can be used as a spectrometer. The schematic arrangement of the spectrometer shows the glass, phototubes, and magnetic shield, as well as the anticoincidence counter, lead converter, and coincidence counter. These two scintillation counters ensure that the electron showers, which are pulse-height-analyzed, start in the 6-mm lead converter *C*, and thus all begin at the same place in the system. *F*, iron magnetic shields; *A*, plastic scintillator in anticoincidence with *B* and *D*: *B*, plastic scintillator in coincidence with *D*; *C*, Pb converter 6 mm thick; *G*, glass Cerenkov radiator; *D*, detectors (photomultipliers). [J. M. Brabant, B. J. Moyer, and R. Wallace, *Rev. Sci. Instr.*, **28**, 421 (1957).]

3-6   PHOTOGRAPHIC EMULSIONS

The blackening of photographic plates was the first effect of nuclear radiation ever observed (Becquerel, 1896), and the blackening of plates or films is still a very important detection method. It is interesting to note that the reciprocity law according to which the blackening of a plate by X rays depends only on the product $I \cdot t$ (I = intensity, t = time of exposure) is valid in most cases. Special film for X rays is being manufactured currently and is used with or without reinforcing screens, which increase the photographic action of the primary radiation by emitting secondary light or electrons.

Single particles are detectable in special photographic emulsions rich in silver bromide. This old technique (Kinoshita, 1912) has been perfected and is now very useful. A typical modern emulsion (G-5 of Ilford) is described in Table 3-5.

A minimum-ionizing, singly charged particle produces about 30 developable grains per 100 μm of path length. The grain density is proportional to $-dE/dx$ and can thus be used to measure the velocity of a particle, at least so long as $v \ll c$ and the grain density does not reach saturation. Saturation occurs at $-dE/dx \cong 10^3$ MeV g^{-1}cm^2. For such a high specific energy loss and the consequent high specific ionization, all grains on the track are developable: a further increase of the specific ionization cannot increase the grain density. The determination of the specific ionization is often accomplished in

ISBN 0-8053-8601-7

TABLE 3-5   DATA ON ILFORD G-5 EMULSION[a]

Density, g cm^{-3}	3.907	Ag	1.85 g cm^{-3}
Atoms, cm^{-3}	8.12×10^{22}	Br	1.36 g cm^{-3}
Mean A	28.98	I	0.024 g cm^{-3}
Mean Z	13.17	C	0.27 g cm^{-3}
Mean Z^2	456	H	0.056 g cm^{-3}
Radiation length, cm	2.93	O	0.27 g cm^{-3}
		S	0.010 g cm^{-3}
		N	0.067 g cm^{-3}

[a]The emulsion considered in the table is assumed to be in equilibrium with air of 50% relative humidity.

practice by measuring the length of the gaps, i.e., of the regions free of developed grains along a track (Fig. 3-20). The range of a particle traversing a photographic emulsion is also a function of the particle energy and can be found in range–energy tables (Fig. 3-21).

For velocities near c, it is still possible to measure the quantity $p\beta$ (where p is momentum) from the multiple scattering. One has approximately

$$\langle \alpha \rangle = K \left(\frac{l}{100 \ \mu m} \right)^{1/2} \frac{z}{p\beta} \qquad (3\text{-}6.1)$$

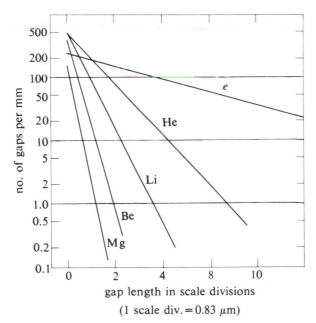

y-axis: no. of gaps per mm

x-axis: gap length in scale divisions

(1 scale div. = 0.83 μm)

Figure   3-20 Graph for determining the specific ionization by measuring "gaps." Variation with gap length l of the number of gaps of length equal to or greater than l, for particles of different specific ionization. [(PFP 59).]

ISBN 0-8053-8061-7

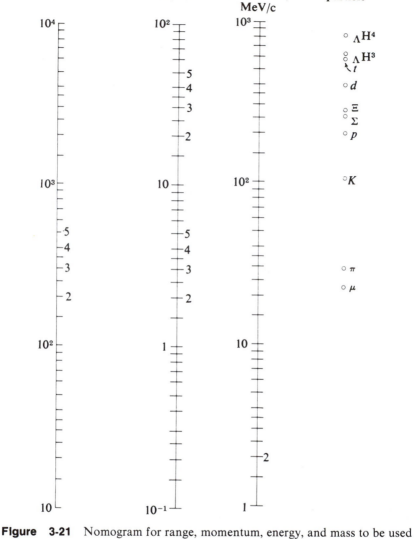

Figure 3-21 Nomogram for range, momentum, energy, and mass to be used in emulsion work. It relates to four variables by laying a straight edge across the scales. [Courtesy W. Barkas.]

where $\langle \alpha \rangle$ is the projected angle between chords of successive track segments of length l (in microns). For $l = 100$ μm the constant K is 25.8 deg for single charged particles and for $p\beta$ in MeV/c. The density of delta rays along a heavily ionizing track provides another means of measuring the energy of the particles.

Emulsions can be loaded with elements such as B, Bi, and U for special purposes; they can also be assembled in relatively large "stacks" (e.g., 30 by 20 by 10 cm thick, with single layers 600 μm thick) and subsequently dis-

ISBN 0-8053-8601-7

assembled for development. Each emulsion is then mounted on a glass plate and observed separately. With suitable markings, it is possible to follow the individual tracks through the successive emulsion layers. Figure 1-6 shows the historic photograph in which the pi meson was discovered.

3-7 NUCLEAR TRACKS IN SOLIDS

The study of tracks produced by ionizing particles in solid insulators is another powerful detection method. When an ion passes through a crystalline solid it shatters the lattice along its track and leaves a microscopic hole. Amorphous solids such as plastics are also damaged because the polymer chains are broken and new reactive molecules are formed. Where the hole intersects a free surface it can be enlarged by etching with a suitable agent. For instance, mica is etched with hydrofluoric acid, zircon with hot phosphoric acid, and most plastics with oxidizing solutions of hypochlorites. After etching, the entrance hole takes on a conical shape depending on the ratio of the etching rates of the damaged and undamaged material. The damage density in turn depends on the charge and velocity of the impinging ions (See Fig. 3-22). Study of the tracks combined with proper calibrations permits the identification of the atomic number of the ion causing them. Recoils from

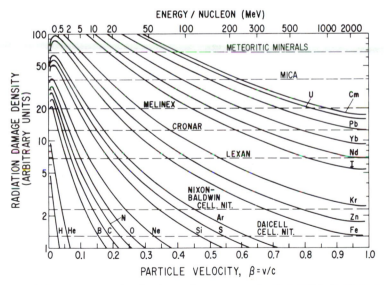

Figure 3-22 Damage density in various nonconducting solids is given as a function of velocity and of energy per nucleon for a number of nuclei. The damage density increases with increasing charge, or atomic number, from hydrogen (H) to curium (Cm); it also increases as the particle slows down (until it is going so slowly that it becomes less ionized). Broken horizontal lines represent the thresholds for track recording in materials ranging from sensitive plastics (bottom) to typical constituents of stony meteorites (top).

ISBN 0-8053-8061-7

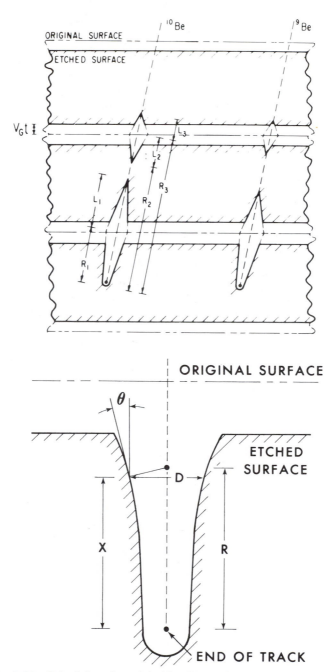

Figure 3-23 Principles of particle identification by etch rate measurements in solid track detectors. Ionization rates at various residual ranges R can be determined by measurements of either (a) the etched cone length L or (b) the taper angle θ or diameter D. [P. B. Price and R. L. Fleischer, *Ann Rev. Nuc. Sci.*, **21**, 298 (1971).]

ISBN 0-8053-8061-7

alpha decay produce detectable damage even when the alpha is undetectable. The track study technique has found increasing application to many problems, including those of geochronology and space physics. (See Figs. 3-23 and 3-24.)

3-8 CLOUD CHAMBERS

C. T. R. Wilson observed that a supersaturated vapor tends to condense on ions and that the resulting droplets become visible under bright illumination. In 1912, he developed the first cloud chamber with which he obtained the beautiful pictures in Fig. 3-25.

A cloud chamber is essentially a box, filled with a gas and a vapor (e.g., air and H_2O), that can be expanded rapidly by a piston or a moving wall.

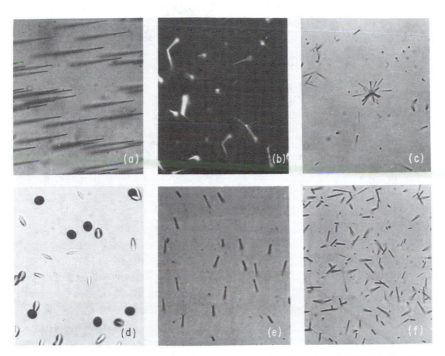

Figure 3-24 (a) Lexan polycarbonate plastic irradiated with 330-MeV sulfur ions. (b) Thorium silicate crystal irradiated with 410-MeV argon ions, which induce binary and ternary fission. Only the fission fragments (not the argon ions) leave tracks. (c) Fossil tracks from spontaneous fission of uranium impurities in 300-million-year-old mica. (d) Fossil tracks from spontaneous fission of uranium impurities in 50-million-year-old obsidian. (e) Cellulose nitrate plastic irradiated with 4.5-MeV alpha particles. (f) Fossil tracks from spontaneous fission of ^{244}Pu (now extinct) in a diopside crystal from a 4.5-billion-year-old meteorite. [Courtesy B. Price.]

ISBN 0-8053-8061-7

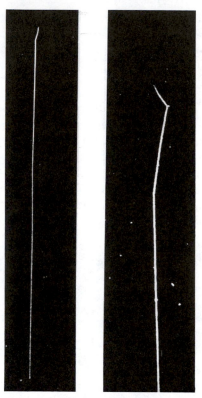

Figure 3-25 Original pictures of alpha particles obtained by C. T. R. Wilson in 1912.

Adiabatic expansion cools the gas and makes the vapor supersaturated. The saturated vapor pressure on a small charged drop is smaller than the saturated vapor pressure on a plane surface. The vapor will therefore condense around charges, forming droplets that grow rapidly in size. The expansion ratio necessary to obtain suitable condensation conditions varies with the gas and vapor: for example, for air–alcohol mixtures at 20°C, a suitable expansion ratio is 1.31.

The cloud chamber remains sensitive for about 0.1 sec after expansion, and the ions must be swept away by an electric field before a new compression–expansion cycle begins. Cloud chambers are usually illuminated by electric flashes in xenon, and the tracks are photographed stereoscopically in order to be able to reconstruct the event in three dimensions.

Magnetic fields are usually applied to the chamber to measure the momentum of the particles producing the tracks. The accuracy of the measurement of curvature is limited by turbulence in the chamber and by the

ISBN 0-8053-8601-7

unavoidable multiple scattering in the gas. The first effect can be avoided by careful construction and operation.

The expansion mechanism of a cloud chamber may be triggered at regular time intervals or in synchrony with the output of an artificial source such as an accelerator. It is also possible to arrange around a chamber counters or other devices that will trigger the chamber only when they detect a particular event. This is feasible because the chamber retains the ability of forming the tracks for long enough to allow the motion of the parts determining the expansion (Blackett and Occhialini, 1933; see Fig. 2-32).

It is also possible to build continuously operating cloud chambers which do not need any expansion mechanism. These are boxes having a vertical temperature gradient, with the top warm and the bottom cold, and containing a mixture of gas and vapor. A layer is formed in which the vapor is supersaturated and in which tracks can be produced. This type of chamber can be operated with different mixtures of gas and vapor and at pressures up to several atmospheres. Demonstration chambers can be built very easily (Fig. 3-26).

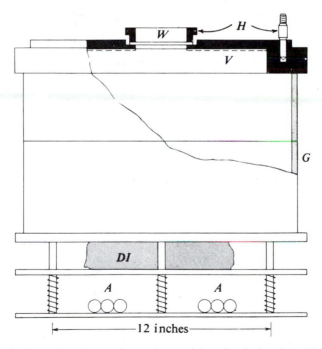

Figure 3-26 A simple, continuously sensitive cloud chamber. *W*, window; *V*, velvet pad; *G*, glass rings; *A*, counters; *DI*, dry ice; *H*, heaters. [E. W. Cowan, *Rev. Sci. Instr.*, **21**, 991 (1950).]

ISBN 0-8053-8061-7

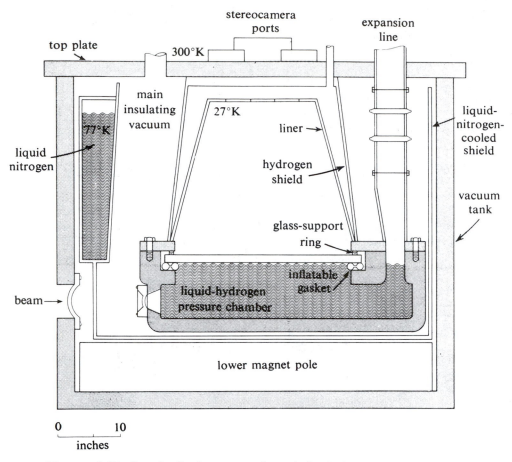

Figure 3-27 Longitudinal cross section of the 72-in. hydrogen bubble chamber. [Lawrence Radiation Laboratory; Alvarez group.]

3-9   BUBBLE CHAMBERS

An important development (Glaser, 1952) in the art of detecting nuclear radiation is the bubble chamber. This is a chamber full of a liquid at a pressure and temperature such that it is below the boiling point. When the pressure is suddenly lowered, thereby lowering the boiling point, the liquid starts to develop bubbles, which form preferentially around ions. Like the droplets in a cloud chamber, the bubbles are visible under strong illumination. Bubble chambers are sensitive for times of the order of 10 msec. The decompression typically takes 10 msec; at its end the beam is injected and 1 msec later the lights illuminating the bubbles are flashed, after which the chamber is recompressed. The cycle is repeated about every second. Bubble chambers have been operated with a variety of liquids, including hydrogen,

ISBN 0-8053-8061-7

deuterium, Ne solution in hydrogen, helium, propane, and xenon. Their great advantage lies in the fact that the detecting material is much denser than the gas in cloud chambers and of simpler composition than photographic emulsions. For instance, in a hydrogen bubble chamber, hydrogen has a density of 0.07 g cm^{-3} at the boiling point under 1 atm pressure, whereas at 0°C and 1 atm the density of the gas is 0.0899×10^{-3} g cm^{-3}. (See Table 2-2).

The large hydrogen bubble chamber developed by Alvarez and his coworkers had the capacity to hold 500 liters of liquid hydrogen (Fig. 3-27). The bubble chamber is especially suited to the study of the complex phenomena that occur in the interactions of high-energy particles with hydrogen. Several of the illustrations in this book show the power of this method.

Huge bubble chambers, such as the 38-m^3 hydrogen or deuterium chamber (BEBC) at CERN, the 32-m^3 hydrogen chamber at the Fermi National Accelerator Laboratory in Batavia, Illinois, (Figs. 3-28, 3-29) and the 10-m^3 freon or propane bubble chamber (Gargamelle) at CERN, are extremely complex and costly devices. They also require magnetic fields of the order of

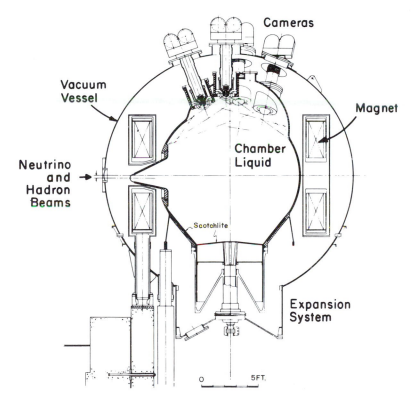

Figure 3-28 Diagram showing major components in FNAL's liquid hydrogen bubble chamber. It permits observation of 15 ft of track.

ISBN 0-8053-8061-7

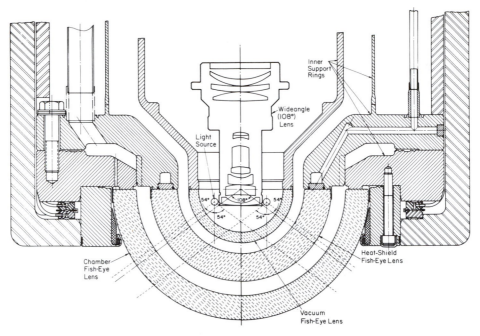

Figure 3-29 Detail of a camera port of FNAL's 15-ft bubble chamber, showing the fisheye lenses. There are six such ports, arranged in two triangular arrays.

30 kG on their volume. However, they are indispensable for the exploitation of the big accelerators.

The cloud chamber and bubble chamber give a picture of the event being studied that must be analyzed to extract the pertinent information. The event is first reconstructed in space by using two or more photographs taken from different angles (i.e., a stereoscopic projection). The momenta of the particles involved are measured either from their range or from the curvature of their trajectories in a known magnetic field. The analysis is laborious and is greatly assisted by computing machines. The reader is referred to the literature for this special subject (see Rosenfeld and Humphrey, 1963).

3-10   SPARK CHAMBER AND STREAMER CHAMBER

The spark chamber, a device based on old principles but developed only in recent years (Conversi and Gozzini, 1955; Cranshaw and DeBeer, 1957; Fukui and Miyamoto, 1959), combines some of the features of counter arrays and of the bubble chamber. In its simplest form a spark chamber is an array of thin conducting plates. The even-numbered ones are grounded, and the odd-numbered can be brought to a high voltage. The gaps between plates are

ISBN 0-8053-8061-7

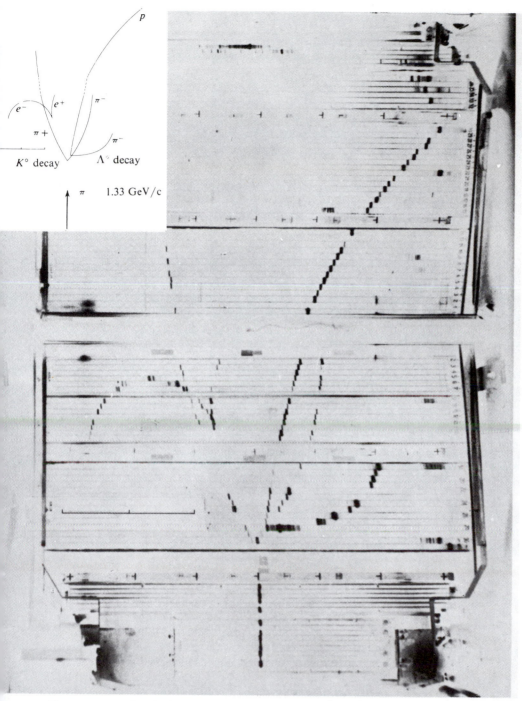

Figure 3-30 Spark-chamber picture of $\pi^- + p \rightarrow \Sigma^0 + K^0$ reaction and subsequent reactions $\Sigma^0 \rightarrow \Lambda^0 + \gamma$; $\Lambda^0 \rightarrow p^+ + \pi^-$; $\gamma \rightarrow e^+ + e^-$. The spark chamber is in a field of approximately 13.5 kG. [Courtesy A. Roberts.]

115

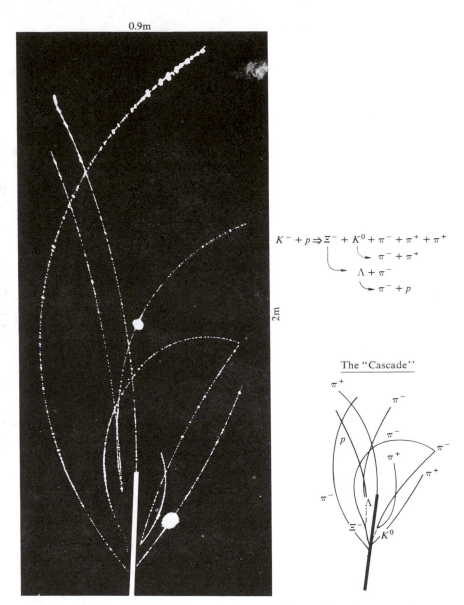

Figure 3-31 Streamer-chamber photograph of a 7.1 GeV/c K^- interaction in a liquid H_2 target. A clear example of the "cascade" configuration ($\Xi^- \to \Lambda\pi^-$) can be seen. Neither the incoming beam nor its point of primary interaction with a proton is visible because the beam is directed into a small, tubular, opaque target (1.3 cm in diameter, 40 cm long), filled with liquid hydrogen, that has been inserted in the chamber. The remainder of the chamber, visible to the cameras above it, is filled with neon gas.

In the analysis of this event all of the outgoing particles were visible or had visible decay products; there was no "missing mass" left over to indicate missing neutral particles. The interaction is $K^- + p \Rightarrow \Xi^- + K^0 + \pi^- + \pi^+ + \pi^+$; $\Xi^- \to \Lambda + \pi^-$; $K^0 \to \pi^- + \pi^+$; $\Lambda \to \pi^- + p$.

Courtesy of A. Barbaro-Galtieri.

ISBN 0-8053-8061-7

filled with a noble gas at atmospheric pressure. The gaps are about 8 mm wide, and a voltage of about 10 kV is applied across them for a short time (~ 0.2 μsec) at a chosen moment. The chamber is sensitive for about 0.5 μsec preceding the application of the voltage. Any charged particle passing through the chamber during this period produces along its path a visible and audible discharge which can be located by photographing and by acoustical or electrical methods. A clearing field is applied between the plates to sweep away unwanted electrons and ions.

From this description it is apparent that the chamber can be triggered with counters by applying the sensitizing field. The short interval during which the chamber is sensitive allows it to photograph the desired events without a serious background even when the flux of particles is several million per second. Ordinarily two cameras at right angles to one another are used to photograph an event in order to reconstruct the tracks in space (Fig. 3-30).

The photographing of spark chambers entails a lengthy analysis. The modern tendency is to replace the photographing by other spark detection methods: acoustical and a variety of electrical techniques are in use. The sparks are never seen, but their location is fed directly into a computer, which stores the data. The spark chamber has been further developed by replacing the flat electrodes with arrays of thin parallel wires. The charge collected by the single wires is detected. Furthermore, it is possible to choose voltages and gas pressure in such a way that the single wires act as proportional counters. These proportional multiwire chambers do not need auxiliary trigger counters and have better time and space resolution than ordinary spark chambers.

The streamer chamber derives from the spark chamber, but works on a different principle: the electrodes are far apart and only two in number. They establish for a very short time, typically 10 nsec, a field of about 15 kV cm^{-1} in a large volume of gas (1 m^3). If at that moment there is an ionizing track in the gas, streamers develop along the track in a direction parallel to the field, but their length is short because it is limited by the duration of the voltage pulse; the typical length is a few millimeters. These streamers can be photographed. Streamer chambers operate with noble gases. Hydrogen may be inserted in a separate container. The advantage of the streamer chamber is that it can be triggered by external counters and thus be made as highly selective as the spark chamber. Figure 3-31 shows a picture obtained with a streamer chamber.

3-11 ELECTRONICS

In addition to actual detecting devices, a wide variety of electronic apparatus and techniques form an integral part of nuclear instrumentation. Their detailed description falls outside the scope of this book. We shall confine ourselves to a partial catalogue including some performance data. Many of the devices are now commercially available.

ISBN 0-8053-8061-7

Almost all detecting devices include amplifiers, i.e., apparatus that multiply the voltage applied by the primary detector until it becomes easily measurable. The ratio between the output and the input voltage of the amplifier is called *voltage gain*, or simply *gain*. It is often important that the gain be constant over a wide range of the input voltage. An apparatus with constant gain is called a *linear amplifier*. The constancy of the gain in time is another important characteristic: a gain drift of less than 1% in 24 h is a usual requirement.

Direct-current amplifiers are required for ionization chambers in a steady state. They are essentially different from pulse amplifiers, and their significant parameter is current sensitivity: 10^{-15} A is measurable to an accuracy of a few percent.

For pulse-detecting instruments the output depends not only on the maximum voltage applied but also on the time dependence of the input voltage. A common way to characterize the time response of an amplifier is to give its *rise time*. If we suddenly apply a voltage v to the input terminal, the output will rise from 0.1 to 0.9 of its maximum value within the rise time. The decay time of the amplifier is defined as the time taken to return to $1/e$ of the maximum value of the input voltage (Fig. 3-32). Rise times of 10^{-9} to 10^{-6} sec are used in practice. Decay times are usually 1 to 100 times as long as rise times. Generally speaking, speed of the amplifier and linearity of the output are mutually exclusive for high-gain devices.

Another important characteristic of an amplifier is its noise. The voltage at the output fluctuates, even in the absence of a signal, in an irregular fashion. Thermal agitation (Johnson noise) and the fluctuation in current due to the finite charge of the electron (shot effect) are unavoidable causes of noise.

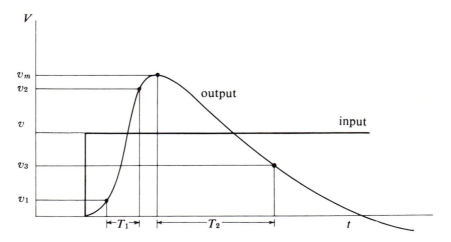

Figure 3-32 Input and output voltages in an amplifier v_m = maximum output voltage; $v_1 = 0.1v_m$; $v_2 = 0.9v_m$; $v_3 = v_m/e$; T_1 = rise time; T_2 = decay time.

ISBN 0-8053-8061-7

In a detector the signal-to-noise ratio must be kept as high as possible. For this reason, amplification of the primary signal is often performed in two stages: a first stage comprises a preamplifier of modest gain (5–10) but of very low noise. Its output is then fed through a transmission line, if necessary over a distance of many meters, to the amplifier proper. Another reason for using a preamplifier is that one can adapt it to any special purpose desired and feed its output into a main amplifier of standard design. The output voltages of common amplifiers are 1 10 V.

In many applications it is also important to have automatic apparatus that will record the number of pulses applied to them. Such apparatus are called scalers. Their primary component is a flip-flop circuit. This is a circuit with two stable states of current. An input pulse changes the current from one state to the other.

The most important characteristic of a flip-flop circuit is its speed, i.e., how many times per second it can change from one state to the other; 10^6 times per second is a representative figure. It must be remembered, however, that this refers to equally spaced pulses. Random pulses occurring at the average rate of n per second are counted with an approximate fractional counting loss $n\tau$, where τ is the minimum time required for a change from state a to state b, or vice versa. Generally, the voltage pulses (output of an amplifier) need be counted only if they are above a certain threshold. This is accomplished by a *discriminator* circuit. Discriminators are triggered only by pulses above their threshold, and once triggered, most of them give a pulse the amplitude and duration of which are independent of the triggering input.

One of the most powerful electronic tools used in nuclear physics is the coincidence circuit (Rossi, 1930), an apparatus that responds only to two simultaneous input pulses. By "simultaneous" we mean occurring within a very short time interval (resolving time), for example, 10^{-8} sec. After a pulse the circuit remains dead for a time of the same order of magnitude as the resolving time. The output for pulses occurring separately is appreciably smaller than the output for simultaneous pulses and will not trigger a discriminator. Thus the whole apparatus responds only to coincident pulses. Conversely, it is possible to form a circuit that responds only to separate pulses, but not to coincident pulses (anticoincidence). In all coincidence and anticoincidence circuits the input pulses must be similar in shape and amplitude in the two channels.

The fundamental circuits may be combined into extremely versatile and complicated instruments, such as multichannel pulse analyzers. In such an instrument the input pulses are sorted according to their magnitude and counted in different channels (up to 4000 channels are commercially available). A semiconductor detector connected to a multichannel pulse analyzer gives at once the whole gamma-ray spectrum emitted by a source.

Besides the fundamental apparatus mentioned above, a number of indispensable auxiliary apparatus are used in nuclear measurements; regulated

ISBN 0-8053-8061-7

power supplies of high and low voltage, electrostatic voltmeters, pulse generators for calibration purposes, and gating circuits that make a detector sensitive only for an assigned time after a given type of event are examples of the
many circuits available. These elaborate electronic techniques are, of course,
in a state of continuous development.

3-12 COMPLEX NUCLEAR INSTRUMENTS

The detectors described thus far are the fundamental components of the
more elaborate instruments used in nuclear physics. It is obviously impossible
to describe these in detail. In fact every experiment requires its own combination, and the design of such combinations is an important part of the experimental art. We shall confine ourselves to a few simple examples.

A powerful auxiliary in all counting methods is the use of coincidence
counters (Rutherford, 1912; Bothe, 1928). A single example will illustrate this
method. Suppose that a radioactive substance emits a beta radiation and a
gamma radiation and that we want to know whether the gamma radiation is
emitted within a very short time of the beta disintegration, as would be the
case if the nucleus were left in an excited state by the beta decay, or whether
the gamma ray is uncorrelated in time with the beta decay, as would in fact
happen if it preceded the beta decay. We could solve this problem by arranging a beta-ray counter (e.g., a thin-window G–M counter) and a gamma-ray
counter (e.g., a NaI scintillator) as in Fig. 3-33. We count the single counts in
the two counters and the coincidences between them, i.e., the times when
both counters register a pulse within a very short time interval τ. We assume
now that the beta emission precedes the gamma emission by a time, on the
average much smaller than τ and that the emission of the beta and gamma
rays is isotropic and uncorrelated in space. The number of counts per second
registered by the beta counter is

$$\nu_\beta = \nu\eta_\beta \tag{3-11.1}$$

where η_β is the efficiency of the counter, i.e., the probability that a beta ray
emitted by the source is counted by the detector, and ν is the rate of disintegrations in the source. The number of counts per second registered by the
gamma counter is

$$\nu_\gamma = \nu\eta_\gamma \tag{3-11.2}$$

and the rate of coincidences if the β and γ emission are practically simultaneous, as happens when the beta ray precedes the gamma ray, is

$$\nu_c = \nu\eta_\beta\eta_\gamma \tag{3-11.3}$$

From these three equations it is possible to find ν, η_β, and η_γ. This example is
extremely simple, but it illustrates the principle of the coincidence method.

ISBN 0-8053-8061-7

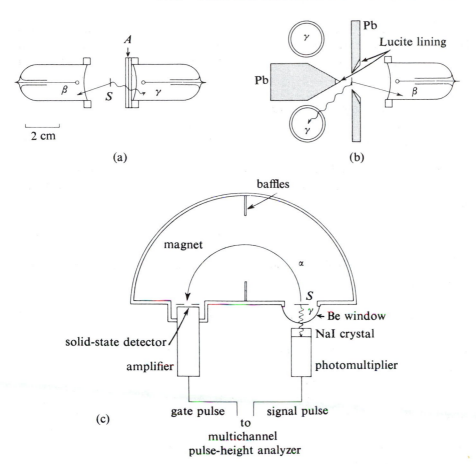

Figure 3-33 Types of coincidence measurements. (a) Simple coincidence arrangement, not suitable for gamma–gamma coincidences, because scattered quanta may produce spurious coincidences. S, source; A, absorber. (b) Coincidence arrangement for beta–gamma and gamma–gamma coincidences. The probability of scattering of electrons or soft gamma rays from one counter to another is reduced by the lead shields (Pb). (c) Schematic diagram of an alpha spectrograph arranged for alpha–gamma coincidence studies.

The time τ plays an essential part. In particular, when we use the terms "uncorrelated," "coincident," and so on, it is always with reference to τ.

Let us now consider what happens if the gamma emission precedes the beta emission. We may assume that the time between the two is on the average much larger than τ. The rate of coincidences is then the rate of "random," or "accidental," coincidences, which we evaluate as follows. Let us consider the number of times per second in which the beta counter is activated. This is given by Eq. (3-11.1). Similarly, for the gamma counter we

ISBN 0-8053-8061-7

have Eq. (3-11.2). Every time that the beta counter is activated, the apparatus will register a coincidence if the gamma counter is also activated within a time τ_β. The time τ_β is the time during which the "gate" opened by the beta count stays open. During this time we shall have on the average $\nu\eta_\gamma\tau_\beta$ gamma impulses, and thus we have a rate of random coincidences, in which the beta ray precedes the gamma ray, given by

$$\nu_{\beta r} = \nu^2 \eta_\gamma \eta_\beta \tau_\beta \qquad (3\text{-}11.4)$$

where the index βr indicates a random coincidence initiated by a beta ray. In the same way we compute the rate of accidental coincidences, in which the gamma ray precedes the beta ray,

$$\nu_{\gamma r} = \nu^2 \eta_\gamma \eta_\beta \tau_\gamma \qquad (3\text{-}11.5)$$

Their sum gives the total rate of accidental coincidences. If, as often happens, $\tau_\beta = \tau_\gamma = \tau$, then

$$\nu_{\beta r} + \nu_{\gamma r} = \nu_r = 2\nu^2 \eta_\gamma \eta_\beta \tau_\beta = 2\nu_\beta \nu_\gamma \tau \qquad (3\text{-}11.6)$$

Note that the rate of accidental coincidences is proportional to ν^2, whereas that of true coincidences is proportional to ν.

In measuring coincidences the rate of accidental coincidences must be considered as a background and subtracted. A good way to check the rate of accidental coincidences is to use two separate sources, one for the beta counter, the other for the gamma counter, giving the same rate ν_β, ν_γ as the single common source. The coincidence rate obtained from the two separate sources is the rate of accidental coincidences.

Coincidence can also be used to locate events in space. Thus the coincidence between two counters can be used to trigger a cloud chamber located between them, which will show the particles that did the triggering. A system of this kind allowed Blackett and Occhialini (1933) to detect showers in a cloud chamber (Fig. 2-32).

The coincidence system can be refined to show the time of flight between two detectors, a system that was used for the identification of the antiproton and one that is frequently used in neutron spectroscopy. Correlation between the direction of emission of two particles can also be detected by coincidence methods. For instance, two-quanta annihilation of positrons with electrons at rest produces coincidences between two gamma-ray detectors only when these lie on a straight line with the source in between them.

The effective operation of a coincidence system depends mainly on its electronics. The important Rossi circuit mentioned previously has undergone many variations, but its principle is still the basis of most coincidence circuits. Anticoincidence circuits also have important uses. It is required of them that

ISBN 0-8053-8061-7

only one counter, not both of a pair, register a pulse. For example, if we want to be certain that a particle has stopped in a certain absorber, we can insert two counters in anticoincidence, as in Fig. 3-34.

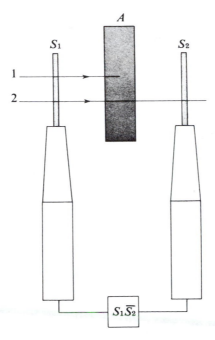

Figure 3-34 Method for selecting particles of a given range, i.e., stopping in A. S_1, S_2, scintillators; A, absorber; particle 1 triggers scintillator 1 but not 2; particle 2 triggers scintillators 1 and 2. Particle 1 of the desired kind is registered as $S_1 \bar{S}_2$; particle 2 as $S_1 S_2$. The bar on S_2 means that scintillator 2 does not give the signal.

3-13 CHARGED-PARTICLE SPECTROMETERS

A large group of instruments is devoted to the measurement of the energy of alpha and beta particles. These instruments almost always involve a magnetic field, which deflects the trajectory of the particle; the degree of curvature gives the momentum according to the equation

$$Br = (c/e)\, p \qquad (3\text{-}12.1)$$

For electrons

$$Br = (3.335 \times 10^3)\, p \qquad (3\text{-}12.2)$$

where Br is in G cm and p in MeV/c.

ISBN 0-8053-8061-7

The spectrometers are mainly of two kinds, according to whether high resolving power or high luminosity is the most important characteristic. One of the simplest of the first type is used in measurements of electron energies and is shown in Fig. 3-35. It focuses the electrons at 180 deg from the source, as indicated. The actual detection can be made by a photographic plate or by counters. The preparation of the source is extremely important, because its thinness, absence of back scattering from the support, and small dimensions are primary determinants of the quality of the spectrum (Fig. 8-7). For electrons, spectrometers of high luminosity are often of the type illustrated in Figs. 3-36 and 3-37. The current in the magnet is varied and the counting rate in the detector is given as a function of the current in Fig. 3-38. Apertures up to 1% are obtainable with $\Delta p / p \cong 0.005$.

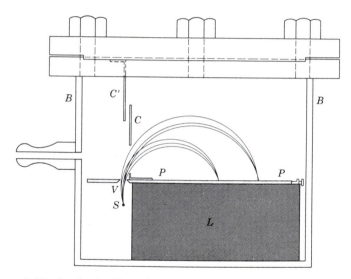

Figure   3-35 Rutherford's and Robinson's design for a beta spectrometer. Electrons from the source S, passing through the entrance slit V, are focused on the photographic plate P; L is a lead shield; C are baffles to stop scattered electrons.

Alpha spectrometers generally follow the lines of beta spectrometers, except that much higher magnetic fields are ordinarily required. Figures 3-37 and 3-38 show a spectrometer and a spectrum. In the most favorable circumstances the absolute energy of alpha particles can be measured to a few parts in 10^4. The planning of spectrometers is a problem in electron optics, and we shall not explore it further.

ISBN 0-8053-8601-7

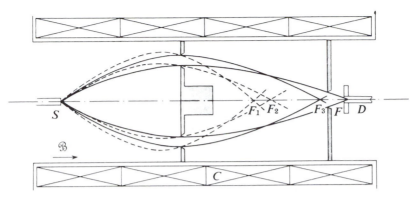

Figure 3-36 The principle of helical spectrometers. Coils *C* produce a uniform magnetic field *B*. Electrons emitted by source *S*, having different momenta, are focused in F_1, F_2, etc. When the focus coincides with the entrance of counter *D*, one has a maximum in the counting rate. By changing the current in *C* one obtains the peaks corresponding to different momenta.

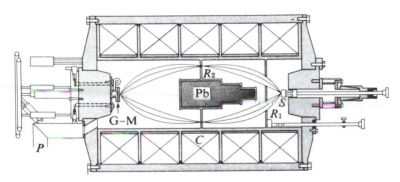

Figure 3-37 Axial section of the first intermediate-image beta-ray spectrometer. *S*, source; *C*, coils; *P*, pump; *R*, shutter; G–M, detecting counter. A first ring image is formed at R_2. Remember that here and in Fig. 3-36 trajectories of electrons are not plane curves. [K. Siegbahn and H. Slätis.]

Alpha-particle spectrometry can be carried out by magnetic spectrometers and by a refined use of semiconductor counters.

Gamma rays and X rays are often measured through their conversion electrons; thus beta spectrometers also provide information on electromagnetic radiation. If internal conversion is used, extremely precise results (1 keV or better) can be obtained. External conversion is much less precise and generally can be used only for qualitative work.

ISBN 0-8053-8601-7

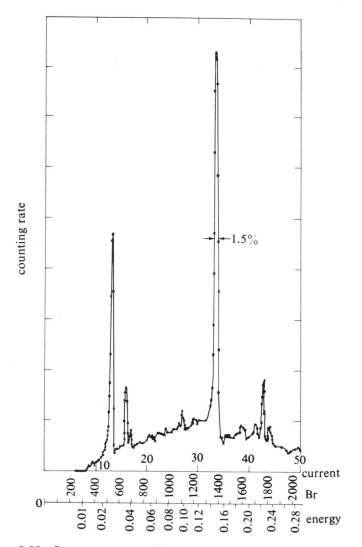

Figure 3-38 Beta spectrum of ThB taken with a spectrometer of the type illustrated in Fig. 3-37. On the abscissas: current in the coils is in amperes; Br in G cm; energy in MeV.

3-14 GAMMA-RAY SPECTROMETERS

Spectrometers that measure the wavelength of gamma rays directly are of the crystal type, based on the Bragg equation

$$n\lambda = 2\,d\sin\theta \qquad (3\text{-}13.1)$$

where λ is the wavelength to be measured, n is the order of the spectrum, d is the lattice constant of the crystal, and θ is the angle between the beam and the reflecting surface.

ISBN 0-8053-8061-7

The main difficulty arises from the smallness of θ: for instance, for annihilation radiation (511 keV) $\lambda = 0.0243$ Å, and for a lattice constant of 3.02 Å (calcite), we have in the first order $\theta = 4.02 \times 10^{-3}$ rad, which means that special spectrometers have to be built. Conspicuous examples of precision instruments based on bent quartz crystal have been built by DuMond. The X rays are detected by counters. Semiconductor counters are also precision detectors, especially at low and medium energies (0.5–100 keV).

At high energies ($h\nu \geqslant mc^2$) pair production makes it possible to measure the energy of the primary X ray. The kinetic energy of the electron and positron in the pair is at the relativistic limit $cp^{\pm}$. The corresponding radii of curvature of the orbit of the electron and positron in a magnetic field are

$$r^{\pm} = (c/e)(p^{\pm}/B) \tag{3-13.2}$$

and the sum

$$r^{+} + r^{-} = h\nu/eB \tag{3-13.3}$$

where p^{+}, r^{+} are the momentum and radius of curvature of the positron orbit and p^{-}, r^{-} are the same quantities for the electron orbit. For high energies the electron and the positron move initially in the line of flight of the X ray, as required by the principle of conservation of momentum, and a pair spectrometer measures $r^{+} + r^{-}$ (Fig. 3-39).

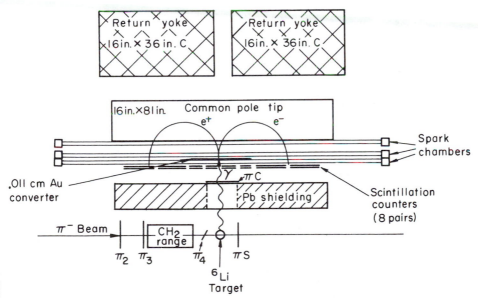

Figure 3-39 Electron–positron pair spectrometer for gamma rays up to 200 MeV. This instrument was used to measure the energetic gamma-ray spectrum produced by π^{-} mesons stopping in various nuclei. [J. A. Bistirlich et al., *Phys. Rev.*, **5C**, 1867 (1972).]

ISBN 0-8053-8061-7

BIBLIOGRAPHY

Barkas, W., *Nuclear Research Emulsions*, Academic Press, New York, 1963, 1973.

Bradner, H., "Bubble Chambers," *Ann. Rev. Nucl. Sci.*, **10**, 109 (1960).

Charpak, G., "Evolution of the Automatic Spark Chambers," *Ann. Rev. Nucl. Sci.*, **20**, 195 (1970).

Deutsch, M., and O. Kofoed-Hansen, "Gamma Rays" and "Beta-Rays," in (Se 59).

Fairstein, E., and J. Hahn "Nuclear Pulse Amplifiers, Fundamentals and Design Practice," *Nucleonics*, **23**, 7, 9, 11 (1965); **24**, 1, 3 (1966).

Fleischer, R. L., Price, P. B., and R. M. Walker, *Nuclear Tracks in Solids: Principles and Applications*, U.C. Press, Berkeley, 1975.

Fluegge, S., *Encyclopedia of Physics*, Vol. 45, Springer, Berlin, 1955 (Fl E).

Gentner, W. H., Maier-Leibnitz, H., and W. Bothe, *An Atlas of Typical Expansion Chamber Photographs*, Pergamon Press, London, 1954 (GMLB 54).

Glaser, D., "The Bubble Chamber," in (Fl E), Vol. 45.

Korff, S. A., *Electron and Nuclear Counters*, 2nd ed., Van Nostrand, Princeton, N. J. 1955.

Malmstadt, H. V., Enke, C. G., and S. R. Grouch, *Electronic Measurements for Scientists* Benjamin, Reading, Mass., 1974.

Powell, C. F., Fowler, P. H., and D. H. Perkins, *The Study of Elementary Particles by the Photographic Method*, Pergamon Press, London, 1959 (PFP 59).

Ritson, D. M., *Techniques of High Energy Physics*, Wiley-Interscience, New York, 1961.

Rosenfeld, A. H., and W. E., Humphrey, "Analysis of Bubble Chamber Data," *Ann. Rev. Nucl. Sci.*, **13**, 103 (1963).

Sayres, E., and C. S. Wu, "Gas Scintillators," *Rev. Sci. Instr.*, **28**, 758 (1957).

Snell, A. H. (ed.), *Nuclear Instruments and Their Uses*, Wiley, New York, 1963.

Tavendale, A. J., "Semiconductor Nuclear Radiation Detectors," *Ann. Rev. Nucl. Sci.*, **17**, 73 (1967).

Wenzel, W. A., "Spark Chambers," *Ann. Rev. Nucl. Sci.*, **14**, 205 (1964).

Yuan, L. C., and C. S. Wu, *Methods of Experimental Physics–Nuclear Physics*, Academic Press, New York, 1961 and 1963.

PROBLEMS

3-1 How much current would one obtain in an ionization chamber full of argon at atmospheric pressure traversed by 10^9 protons sec^{-1}? The energy of the protons is 300 MeV, the depth of the chamber 10 cm.

3-2 With the chamber of Prob. 3-1 what would be the current produced by 10^9 gammas sec^{-1} if the energy of the gammas is 5000; 10 000; 10^5; and 10^7 eV? Neglect the effect of the walls. What would you do to make the wall effect negligible?

3-3 How would you measure the energy and intensity of a beam of protons in the energy domain from 100 keV to 10 GeV? Indicate how to identify the particles, and devise appropriate systems for each energy region considered.

3-4 The same as Prob. 3-3 for electrons or positrons.

3-5 The same as Prob. 3-3 for gamma rays.

3-6 The same as Prob. 3-3 for neutrons from 0.01 eV to 10 GeV. Pay special attention to the differentiation from gamma rays.

3-7 Plan an experiment to measure the $p - p$ scattering cross section at 100 MeV. Plan counters, target, beam geometry, etc.

3-8 Plan an experiment to detect the rare modes of decay

$$\pi^+ \to e^+ + \nu \quad \text{and} \quad \pi^+ \to \pi^0 + e^+ + \nu$$

having the branching ratios 10^{-4} and 10^{-8}, respectively, compared with the decay $\pi^+ \to \mu^+ + \nu$.

ISBN 0-8053-8061-7

3-9   Plan an experiment to measure the spontaneous fission of ^{238}U. Consider all possible causes of background, the thickness of your sample, the time of observation, etc.

3-10   Plan an experiment for the precision measurement of the energy of the electron–positron annihilation radiation.

3-11   Calculate the energy of gamma rays produced by the annihilation in flight of positrons against electrons. Consider the X rays in the direction of the line of flight.

$$\text{Extreme-relativistic answer: } \hbar\omega = \frac{E + mc^2 \pm pc}{2}$$

3-12   A laser beam of $\lambda = 6943$ Å scatters on free electrons with $\gamma = 30$ moving in the direction opposite to that of the light. Calculate the energy of the Compton-scattered gamma rays in the lab as a function of the lab scattering angle.

ISBN 0-8053-8061-7

CHAPTER

4

Particle Accelerators

The subject of this chapter is a branch of electromagnetism and mechanics. However, accelerators are an indispensable tool of nuclear experimentation, and a brief report on them is in order.

4-1 INTRODUCTION AND CLASSIFICATION

The development of particle accelerators specifically designed for use in nuclear experiments began in the late 1920s. Several of the early accelerators involved the production of high voltages applied to an evacuated tube (Breit, Tuve, Lauritsen, Van de Graaff). The first nuclear disintegrations with artificially accelerated protons were achieved in 1930 by Cockcroft and Walton at the Cavendish Laboratory. Their apparatus, which accelerated protons to 300 keV, produced $^7\text{Li}(p, 2\alpha)$ reactions.

The great difficulties connected with high voltages (arcs and corona discharges among others) prompted the invention of machines that do not require high electric fields and in which the high energy of the projectile is achieved by multiple accelerations or by electromagnetic induction. The first multiple accelerators were built by Wideroe in 1928 and, on the same principle, by Lawrence and Sloan in 1930. These machines, however, had no important practical applications at the time.

Emilio Segrè, Nuclei and Particles: An Introduction to Nuclear and Subnuclear Physics, Second Edition

ISBN 0-8053-8061-7

All the accelerators mentioned so far used only electric fields. The cyclotron (Lawrence, 1929) was the first accelerator to employ a magnetic field, which by bending the orbit of the particle in a spiral, forces it to pass many times through an accelerating electric field. It is thus possible to achieve great energies with a relatively compact apparatus. The idea of the cyclotron occurred independently to several physicists (Thibaud, Lawrence, Szilard), but the development of an effective machine from small models into large cyclotrons was the work of E. O. Lawrence and his associates at Berkeley. Accelerators have had a major effect on the evolution of nuclear physics.

The cyclotron has inherent limitations that prevent it, in its simplest form at least, from accelerating particles to relativistic energies. These difficulties have been circumvented by the ingenious schemes of Veksler (1945) and McMillan (1946), which involve "phase stability."

The combination of phase stability with the application of the alternating-gradient principle (strong focusing), invented by Christophilos (1950) and independently by Courant, Livingston, and Snyder (1952), has no apparent energy limit. At present the cost of the accelerators is the main obstacle to the achievement of ever higher energies; 5×10^{11} eV are about to be reached. However, one has to bear in mind that in the extreme-relativistic case the center-of-mass energy, which accounts for physical effects, grows with the square root of the laboratory energy. Thus in a proton–proton collision $E_{c.m.} = (2m_p E_{lab})^{1/2}$.

The cyclotron and its derivative apparatus accelerate light nuclei. Electromagnetic induction was used by Kerst (1940) to accelerate electrons to many million electron volts. Kerst's apparatus is called the *betatron*. In later developments the betatron principle was combined with other acceleration methods (*synchrotron*).

All accelerators require an ion source. For electrons a hot filament is universally used; for nuclei many types of discharge tubes have been developed. As an example, we mention an ion source in which electrons coming from a hot filament oscillate in a volume occupied by hydrogen and ionize it by collision. The ions are "pulled out" by an electric field generating ion currents of the order of milliamperes (Fig. 4-1). There are also sources that give polarized ions, that is, ions whose spins are directed mainly in one direction. The spin orientation can be preserved during acceleration and thus one obtains polarized beams.

We shall limit our discussion to certain types of accelerators that are widely used and practically important, and for these we shall give only a summary of the principles, without technical details. To classify accelerators we shall divide them according to Table 4-1, partially illustrated in Fig. 4-2.

In addition to energy and intensity, the *duty cycle*, defined as the ratio of the time during which the beam emerges to the total time, is important. Generally a large duty cycle is desirable and the experimental use of an accelerator is influenced by the duty cycle and by the intensity as a function of time.

ISBN 0-8053-8061-7

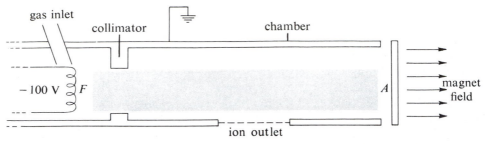

Figure 4-1 A typical positive-ion source. Electrons emitted by the filament *F* at − 100 V oscillate between it and the anode *A*, while they curl around the lines of force of the containing magnetic field. The electrons form positive ions by impact in the gas present in the chamber. The mixture of positive ions and electrons, called plasma, fills the shaded region. Positive ions are extracted from the plasma through the aperture. [R. S. Livingston and R. J. Jones.]

TABLE 4-1   CLASSIFICATION OF ACCELERATORS

Accelerator	*Particle accelerated*	**E**	**H**	*Orbit*	*Typical energy,* MeV
Electrostatic, or Van de Graaff	$e, p, d, \alpha,$ or other	Constant	None	Straight	12
Tandem Van de Graaff	$p, \alpha,$ or other	Constant	None	Straight	21
Multiplier circuit, or Cockcroft–Walton	e, p, d, α	Constant	None	Straight	4
Betatron	e	None	Variable	Circular	300
Cyclotron	p, d, α	Fixed ω	Constant	Spiral	25 for p
Sector-focused cyclotrons	$p, d, \alpha,$ or other	Fixed ω	Variable with θ	Sectored spiral	75- for p
Synchrocyclotron	p	Variable ω	Constant	Spiral	700
Synchrotron	e	Fixed ω	Variable	Circular	10^4
Proton synchrotron	p	Variable ω	Variable	Circular	10^4
Strong-focusing	p	Variable ω	Variable	Circular	5×10^5
Linear (rf)	p, d	$\omega \sim 200$–800 MHz	None	Straight	800
Linear (conventional)	e	$\omega \sim 3000$ MHz	None	Straight	2×10^4
Heavy-ions (Linac)	$^{12}C, {}^{16}O,$ etc.	$\omega \sim 70$ MHz	None	Straight	$10 \times A$ of ion

ISBN 0-8053-8061-1

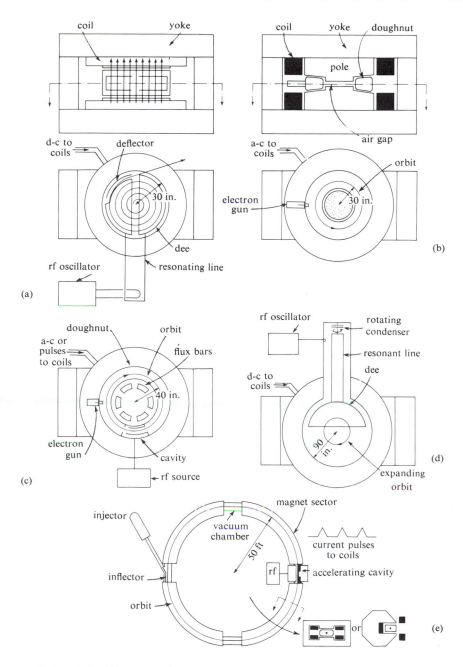

Figure 4-2 Diagrams of some of the most important circular (magnetic) accelerators, with typical dimensions: (a) cyclotron, (b) betatron, (c) electron synchrotron, (d) synchrocyclotron, (e) proton synchrotron (or electron race track). [*Nuclear Engineering Handbook*, McGraw-Hill, New York, 1958.]

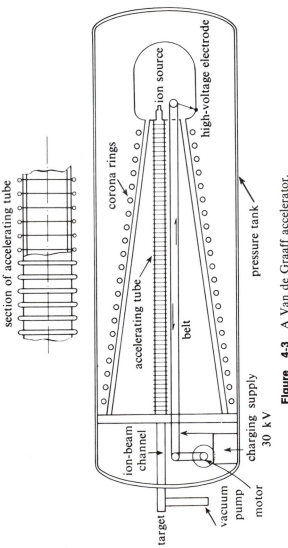

Figure 4-3 A Van de Graaff accelerator.

ISBN 0-8053-8061-7

4-2 POTENTIAL-DROP ACCELERATORS

The simplest accelerator, in principle, is an electrostatic machine connected to a discharge tube. The Van de Graaff accelerator (Fig. 4-3) is a modern realization of such an apparatus. A moving belt, charged by corona discharge, transports the charge into the interior of a large conducting sphere. There another discharge removes the electricity from the belt, transferring it to the sphere. The sphere is the high-voltage electrode connected to the accelerating tube. The whole system is usually enclosed in a pressurized tank containing nitrogen, to which a few percent of freon (CCl_2F_2) has been added. The purpose of this choice of gases is to prevent discharges and destructive fires.

The maximum attainable voltage is about 10 MV and may be measured by special voltmeters. Fixed points on the voltage scale are given by certain sharp nuclear thresholds or resonances; for example, $^7Li(p, n)^7Be$ at 1880.7 $\pm$ 0.4 keV or $^7Li(p, \gamma)^8Be^*$ at 441.2 $\pm$ 0.3 keV. The voltage produced is quite constant (within 1000 V or less), and currents of the order of 100 μA are

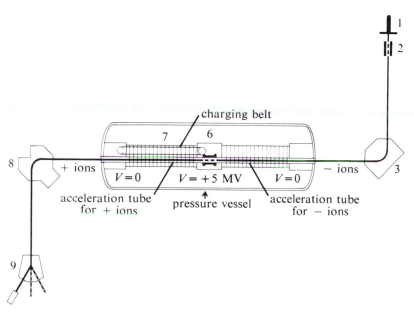

Figure 4-4 Two-stage tandem accelerator. 1, source of positive ions. 2, electron-adding canal. The ions are passed through gas at low pressure where they capture electrons and become negatively charged. 3, the negative ions are preaccelerated to 80 keV and injected into the Van de Graaff accelerator, where they acquire an energy of 5 MeV. 6, the ions are stripped of electrons and charged positively by passing through gas at low pressure. 7, the positive ions are accelerated to 10 MeV. 8, deflecting and analyzing magnet. 9, switching magnet. [Courtesy High Voltage Engineering Corp., Burlington, Mass.]

ISBN 0-8053-8061-7

obtainable. An interesting device to double the energy attainable by an electrostatic accelerator is used in tandem accelerators, in which the charge of a
negative ion is changed from negative to positive by electron stripping (Fig.
4-4). The ion is thus accelerated to twice the potential generated by the
machine.

Another method of generating energies of about 10^6 eV involves voltage-
multiplier circuits (Greinacher, Cockcroft, Walton), in which a transformer
feeds alternating current of a certain voltage into a rectifying and multiplying
apparatus (Fig. 4-5). The voltages obtainable are given in Table 4-1. Apparatus of the type described here are often used as injectors for larger accelerators.

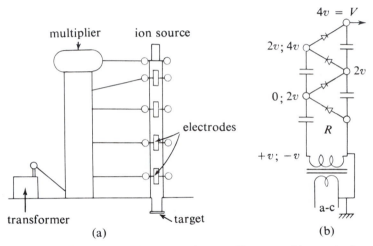

Figure   4-5 A Cockcroft–Walton accelerator. The general layout is shown in
(a). The multiplication principle is indicated in (b). Alternating current travels
up the line of capacitors to the left and is distributed to all the rectifiers R
returning to ground through the capacitors to the right. The dc flows through
the rectifiers in series. When no current is drawn at V the potentials on the
capacitors to the right are constant and have the values indicated in the figure.
The potentials on the capacitors on the left oscillate between the limits indicated in the figure. The voltage V is four times as large as the peak input
voltage v.

4-3   THE BETATRON

As its name indicates, the betatron accelerates electrons. The particles
circulate in an evacuated doughnut and the accelerating force is produced by
electromagnetic induction rather than by an applied electric field (Fig. 4-2b).

To keep a particle in its orbit in the doughnut, we need a magnetic field
B_0 (at the orbit) such that

$$(e/c)B_0R = p \tag{4-3.1}$$

ISBN 0-8053-8601-1

where R is the radius of the orbit and e and p are the charge and momentum magnitude, respectively, of the particle. This equation is relativistically correct. In right hand cylindrical coordinates we chose $\dot{\theta} > 0$. Then $B_0 = B_z(R)$ has sign opposite to e and is positive for electrons.

When the radius of an orbit changes slowly, that is, when $\Delta R/R$ per revolution $\ll 1$, we shall speak of an *instantaneous orbit* and an *instantaneous radius*, meaning by this the orbit corresponding to the radius of curvature at time t.

If we increase the flux φ of **B** linked with the orbit, the particle experiences an electric field such that

$$\oint E \, ds = 2\pi R E = \frac{-1}{c} \frac{d\varphi}{dt} \tag{4-3.2}$$

and hence its momentum increases according to

$$\frac{dp}{dt} = eE = -\frac{1}{c} \frac{e}{2\pi R} \frac{d\varphi}{dt} \tag{4-3.3}$$

Introducing $\langle B \rangle$, the average value of B over the area enclosed by the orbit, we have by definition

$$\varphi = \langle B \rangle \pi R^2 \tag{4-3.4}$$

and, keeping the radius of the orbit constant,

$$\frac{d\varphi}{dt} = \pi R^2 \frac{d\langle B \rangle}{dt} \tag{4-3.5}$$

which combined with Eq. (4-3.3) gives

$$\frac{dp}{dt} = -\frac{e}{c} \frac{R}{2} \frac{d\langle B \rangle}{dt} \tag{4-3.6}$$

On the other hand, if we want to keep the radius of the orbit constant, Eq. (4-3.1) requires, with proper sign, that

$$-R \frac{e}{c} \frac{dB_0}{dt} = \frac{dp}{dt} \tag{4-3.7}$$

and comparison with Eq. (4-3.6) gives

$$2 \frac{dB_0}{dt} = \frac{d\langle B \rangle}{dt} \tag{4-3.8}$$

which is the fundamental condition for the betatron. The final momentum

ISBN 0-8053-8061-7

obtained by the electron is given by integrating Eq. (4-3.3),

$$p - p_0 = \frac{-1}{2\pi R} \frac{e}{c} (\varphi - \varphi_0) \tag{4-3.9}$$

where φ_0 and p_0 indicate, respectively, the initial momentum and initial flux associated with the orbit. Note that it is possible to choose independently the flux φ_0 and the initial guiding field B_0. For the *biased* betatron in particular the sign of φ_0 is the opposite of that of φ, and such a machine may give, for the same magnet, about twice the momentum obtainable with a machine in which φ_0 is nearly 0.

The electron gains energy as it travels in its orbit. In the case of most betatrons the energy gained per turn is of the order of a few hundred electron volts. Hence, a large number of turns is required, and the beam must be focused. Otherwise it would spread out and dissipate by hitting the walls of the doughnut.

The focusing in the betatron, which is similar in many respects to that of other accelerators, will now be discussed briefly. Let us call the circle of radius R the ideal orbit. This orbit has a constant radius in spite of the change of momentum as long as the betatron condition of Eq. (4-3.8) is satisfied. The field **B** at the orbit has two components: a z component, in the direction of the axis of symmetry of the machine, and a radial component B_r, in the radial direction. The plane of the orbit has $z = 0$. Suppose that a particle (electron) for some reason (e.g., a collision with the residual gas) is displaced from its position in the ideal orbit. The displacement can have three components: in the z, r, and s direction, the last being the arc along the orbit. For the first two there will be restoring forces that will force the particle to oscillate in the z or r direction with respect to the ideal orbit. The s direction is of no importance for the betatron, but in other accelerators there are "phase oscillations" in which the particle will be alternately ahead of or behind the ideal particle in the orbit.

We shall first treat the z and r oscillations. Introduce for this two dimensionless variables $\zeta = z/R \ll 1$ and $\rho = (r - R)/R \ll 1$ where R is the radius of the ideal orbit. The equations for ρ and ζ can then be linearized.

We write the equations of motion of a particle of charge e (negative for the electron) subject to the Lorentz force (see Ja 75)

$$\mathbf{F} = e\left(\mathbf{E} + \frac{1}{c} \mathbf{v} \times \mathbf{B}\right)$$

in cylindrical coordinates as

$$\frac{d}{dt} p_r - mr\dot{\theta}^2 = e \frac{r}{c} \dot{\theta} B_z \tag{4-3.10}$$

$$\frac{d}{dt} p_z = - \frac{e}{c} r\dot{\theta} B_r \tag{4-3.11}$$

ISBN 0-8053-8061-7

where by m we indicate the relativistic mass and by p_r and p_z the momentum components in the r and z directions.

The equation for the component of the *canonical* momentum p_θ (θ is the azimuthal angle in our cylindrical coordinate system) is

$$p_\theta \equiv mr^2\dot{\theta} + (e/c)rA_\theta = \text{constant} \qquad (4\text{-}3.12)$$

A_θ is the vector potential component in the θ direction. $\mathbf{A}$ also depends on the time and $-(1/c)(\partial \mathbf{A}/\partial t)$ gives the accelerating electric field. However, in our consideration of the focusing action we do not need to consider the time variations of $\mathbf{A}$.

The z component of the magnetic field in the vicinity of the orbit is described by

$$B_z = B_0(r/R)^{-n} \qquad (4\text{-}3.13)$$

where n is the field index and is positive. Equation (4-3.13) can be expanded to give in a first approximation

$$B_z = B_0\left(1 - n\,\frac{r - R}{R} + \cdots\right) = B_0(1 - n\rho) \qquad (4\text{-}3.14)$$

B_r can now be determined from Maxwell's equation $\nabla \times \mathbf{B} = 0$, which in our case gives

$$\frac{\partial B_r}{\partial z} = \frac{\partial B_z}{\partial r} \qquad (4\text{-}3.15)$$

Hence, remembering that $B_r = 0$ for $z = 0$, we have

$$B_r = z\,\frac{\partial B_r}{\partial z} = -\frac{nz}{R}B_0 + \cdots \qquad (4\text{-}3.16)$$

Introducing the dimensionless variables ρ and ζ, we substitute the expressions for B_z and B_r of Eqs. (4-3.14) and (4-3.16) into Eqs. (4-3.10) and (4-3.11). Neglecting the change in mass during the motion considered here and quantities of the second order in ρ and ζ, we obtain after some calculation

$$\ddot{\rho} - (eB_0/mc)(1 - n)\rho\dot{\theta} = (1 + \rho)\dot{\theta}^2 + (eB_0/mc)\dot{\theta} \qquad (4\text{-}3.17)$$

$$\ddot{\zeta} - (eB_0/mc)\dot{\theta}\,n\zeta = 0 \qquad (4\text{-}3.18)$$

To complete the linearization of these equations we must find $\dot{\theta}$. For the ideal orbit $\dot{\theta} = \omega_0 = -eB_0/mc$, and this determines the constant of Eq. (4-3.12).

ISBN 0-8053-8061-7

Developing this equation with sufficient approximation, we find $\dot{\theta} = \omega_0(1 - \rho)$, which when inserted in Eqs. (4-3.17 and 18) yields

$$\ddot{\rho} + \omega_0^2(1 - n)\rho = 0 \tag{4-3.19}$$

$$\ddot{\zeta} + \omega_0^2 n\zeta = 0 \tag{4-3.20}$$

These are the linearized equations for ρ and ζ.

For Eqs. (4-3.19) and (4-3.20) to describe oscillations, the coefficients of ρ and ζ must be positive, that is, $1 > n > 0$. This means that the field must decrease from the center to the periphery; or in other words that the lines of force must be concave toward the axis of the machine. This decrease, however, must not be too rapid ($n < 1$).

The oscillations described by Eqs. (4-3.19) and (4-3.20) are called *betatron oscillations*, and they are naturally damped as the field grows. The frequency of the radial oscillations is

$$\omega_r = (1 - n)^{1/2}\omega_0 \tag{4-3.21}$$

That of the vertical oscillations is

$$\omega_z = n^{1/2}\omega_0 \tag{4-3.22}$$

hence both frequencies are always less than ω_0. In the cyclotron radial oscillations are relatively unimportant. In the betatron and in other accelerators in which the beam is contained in a doughnut, the amplitude of both radial and vertical oscillations must be small, because if the oscillations exceed the dimensions of the cavity, the beam obviously strikes the walls and is lost.

In a betatron the energy necessary to energize the magnet travels between the magnet and a large bank of capacitors. The exchange occurs with a frequency of about 60 Hz.

Electrons are injected with a special gun at energies of the order of 50 keV. After acceleration the orbit is either expanded or displaced and the electrons strike a target, producing a beam of bremsstrahlung (see Chap. 2). If the target is thin, this beam has a small angular aperture of the order of mc^2/E and can be very intense. For example, a 300-MeV betatron may give 15 000 roentgen (see Sect. 5-4) min^{-1} at 1 m from the target.

The practical limit of betatron energies is about 300 MeV. The electrons traveling in a circular orbit radiate electromagnetic energy. The resulting energy loss per turn is given by

$$L = \frac{4\pi}{3} \frac{e^2}{R} \left(\frac{E}{m_0 c^2} \right)^4 \beta^3 \tag{4-3.23}$$

Inserting numbers for $R = 100$ cm, one obtains 8.8 eV per turn at 100 MeV. These losses, increasing with E^4, very quickly become prohibitive for

ISBN 0-8053-8601-7

electron accelerators, whereas they are negligible in all proton machines. The energy is radiated in a continuous spectrum that has a maximum for frequencies near

$$\nu_{max} = \frac{3}{2} \left(\frac{E}{m_0 c^2} \right)^3 \frac{\omega_0}{2\pi} \qquad (4\text{-}3.24)$$

For a 100-MeV betatron having $R = 100$ cm, the maximum occurs in the visible region and the corresponding light has been observed.

The radiation loss forces the electrons to spiral toward the center of the betatron where they can be caught by a target that "scrapes" the inner side of the beam.

4-4   THE CYCLOTRON

In the Lawrence cyclotron a constant magnetic field guides the ions (nuclei) in a spiral path. The acceleration is imparted by an electric field that has the correct direction any time that the particle is subject to it (Fig. 4-2a).

For the apparatus to function, it is essential that the particle arrive at the place where the electric field is applied at the right time. The two dees oscillate with a period of $2\pi/\omega$. A particle between the dees at time 0 will again cross the slit between the dees at time

$$\pi r / v = t \qquad (4\text{-}4.1)$$

where r is the radius of its orbit. At that time the electric field must be reversed from what it was at time 0. That is, t must be π/ω. We thus have

$$r/v = 1/\omega \qquad (4\text{-}4.2)$$

or, recalling (4-3.1),

$$eB/mc = \omega \qquad (4\text{-}4.3)$$

This is the fundamental equation of the cyclotron. In any case m is the total *relativistic* mass. Note that r does not appear in the equation; hence the orbit may spiral (variation of r) while B and ω are kept constant in time. This is at least true as long as m is constant (nonrelativistic energy).

From Eq. (4-4.3) it is apparent that, for a given frequency and field, a cyclotron can accelerate different particles having nearly identical values of e/m, such as deuterons and helium ions. In fact, by tuning the apparatus for different masses it has been possible to accelerate a number of ions up to carbon and nitrogen.

In the cyclotron the ions must be focused, and the magnetic field of the cyclotron provides a weak focusing action, as described for the betatron (Fig.

ISBN 0-8053-8601-7

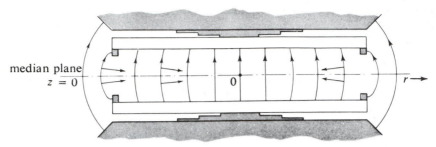

median plane
$z = 0$

$r \longrightarrow$

Figure 4-6 Restoring forces for particles in orbits above or below the median plane in the radially decreasing magnetic field of a cyclotron. The flat pyramidal shims shown in the shimming gaps are arranged to provide a small radial decrease in the field in the center of the gap between the poles. Ring shims at the periphery increase the region of useful field.

4-6). In addition, on crossing the gap between the dees there are vertical components of the electric force. Their main effect, albeit small, is focusing if the particle crosses the dee while the electric field is decreasing. Radial oscillations are less important for the cyclotron than for the betatron. After the particles have spiraled out toward the edge of the magnetic field, they are usually deflected electrically into a channel and may be brought out of the machine.

A cyclotron of this type can generate a circulating current of many milliamperes and a deflected beam of the same order of magnitude. Its energy is limited by relativistic considerations; for since the mass of the particle increases as it moves toward the edge of the magnetic field, we should also increase B in order to satisfy Eq. (4-4.3), with ω constant. This, however, conflicts with the focusing conditions, requiring that the magnetic lines of force be concave toward the symmetry axis of the machine. Relativity thus sets a limit to the maximum energy of a conventional cyclotron. For protons this limit is, in practice, about 30 MeV.

To reach higher energies, it is possible to introduce a magnetic field that increases with r (which, as we have seen, defocuses the ions), but to overcompensate for this defocusing action by making the magnetic field dependent on θ in such a way as to have a total focusing action. This expedient was indicated by L. H. Thomas in 1939, and there are now machines based on this method of focusing, which can be enhanced by spiraling the boundaries between the hills and valleys on the pole surfaces.

Another way out of the difficulty is to vary the electrical frequency applied to the dees in order that a particle always be in step with the accelerating field. This is accomplished by using the principle of phase stability (Veksler, 1945; McMillan, 1946).

ISBN 0-8053-8601-7

4-5 PHASE OSCILLATIONS AND STABILITY

In any accelerator operating on the multiple-acceleration principle, it is essential that the successive impulses be imparted at the proper time. When an extremely large number of accelerations (many thousands) is required, the problem of keeping the circulating particles in step with the accelerating field appears to be formidable. Actually there is a mechanism that makes this possible. With suitable arrangements of fields the particles tend automatically to cross the accelerating gap at a time at which they receive the amount of energy necessary to keep them in resonance with the electric field. This is the principle of phase stability.

To understand the phenomenon qualitatively, let us consider first a particle circulating in an orbit through a spatially constant magnetic field and passing a gap with an applied alternating electric field. Assume that the velocity of the particle is such as to cross the gap when the electric field is 0, and neglect radiation and other losses. The particle will circulate indefinitely in this orbit, at a constant velocity. The energy, frequency, and radius of this orbit are denoted by the adjective *synchronous*.

Suppose now that another particle arrives at the gap at time t_1, a little earlier than a synchronous particle (Fig. 4-7). It will then gain energy in crossing the gap. Having gained energy, its angular velocity (which is related to the energy by

$$\omega = eB/mc = eBc/E \qquad (4\text{-}5.1)$$

where E is the total energy) will be lowered, and the particle will next cross the gap at time t_2, a little later in phase than it did the first time. This process continues until the gap is crossed at the time when there is no electric field. By now, however, the energy will be higher than that required to reach the

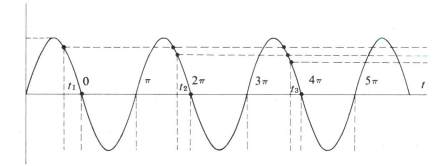

Figure 4-7 Variation of accelerating potential with time, showing origin of phase oscillations.

ISBN 0-8053-8601-7

gap at zero field, and after more turns the ion will cross the gap in the decelerating part of the cycle. Thus the particle will lose energy and hence its frequency will increase. This process will continue in the decelerating part of the cycle. The situation is now reversed, and the particle gains in frequency and loses energy until it is returned to the zero-phase position. A particle thus performs oscillations in s (the arc along the synchronous orbit) with respect to a synchronous particle.

Thus the phase, frequency, energy, and radius of the orbit of a particle oscillate around the equilibrium situation. A group of particles forms a "bunch" in the orbit, and each particle migrates back and forth within the bunch, requiring many thousands of revolutions to complete the cycle.

We may now take the advantage of this property of the orbits in order to accelerate the particles by increasing the magnetic field very slowly, keeping the frequency constant. In the case of a relativistic particle (electron), the momentum increases as the radius of the orbit oscillates slightly around the equilibrium value. This is the principle of the synchrotron. Conversely, we may keep the magnetic field constant and slowly decrease the frequency of the accelerating field in a cyclotron. The radius of the orbit then increases, along with the energy of the particle. This is the principle of the synchro-cyclotron.

We may also keep the radius constant and slowly change both the frequency and the magnetic field (proton synchrotron). In every case the energy oscillates around the synchronous energy, which slowly increases with time.

We shall now consider this subject more quantitatively in a special example. Let us put the line $\theta = 0$ at the gap between the dees of a cyclotron. Call the potential difference at this gap

$$V \sin \omega_e t \qquad (4\text{-}5.2)$$

and define the phase φ of a particle as the angle

$$\varphi = \theta - \omega_e t \qquad (4\text{-}5.3)$$

where θ is the azimuthal angular position of the particle.

The particle has an instantaneous angular velocity $\omega = 2\pi/T$. If ω remains equal to ω_e, the phase is constant and the particle is synchronous. On each revolution, with two gap crossings, its energy will increase by

$$\Delta E = 2eV \sin \varphi \qquad (4\text{-}5.4)$$

Equation (4-5.4) shows the importance of the phase concept: the phase determines the increase of energy per revolution. For a synchronous particle having $\varphi = 0$ or π, ΔE is 0. For a synchronous particle the radius r_s and the

ISBN 0-8053-8601-7

energy are determined by the relations

$$\omega_s = \omega_e = eB_s c / E_s \qquad (4\text{-}5.5)$$

$$r_s = \frac{c}{\omega_e} \qquad \beta = \frac{c}{\omega_e} \left[1 - \left(\frac{m_0 c^2}{E_s} \right)^2 \right]^{1/2} \qquad (4\text{-}5.6)$$

and the energy increase per revolution [Eq. (4-5.4)] may be rewritten by using Eq. (4-5.5) as

$$2eV \sin \varphi_s = \frac{dE_s}{dt} T_s = \frac{2\pi ce}{\omega_e} \frac{d(B_s/\omega_e)}{dt} \qquad (4\text{-}5.7)$$

This equation allows two synchronous phases φ_s and $\pi - \varphi_s$. One is stable and the other unstable.

● We shall now prove that, for a particle having the parameters r, ω, etc., near synchronism, the phase oscillates around a fixed value. This fact is of the greatest importance, because it tells us that it is not necessary to adjust all parameters in strict synchronism (a practically impossible task), but that the motion itself takes care of keeping the particle in phase (phase stability). Also by slowly changing B (increasing) or ω_e (decreasing), we cause the synchronous energy to increase, and hence the particles that are locked in the synchronous orbit are accelerated.

Let us indicate by

$$E - E_s = \delta E \qquad T - T_s = \delta T \qquad \text{etc.} \qquad (4\text{-}5.8)$$

the difference between the energy, period, etc., of a particle and the corresponding quantity for a synchronous particle, and introduce the number $(\delta p/p)/(\delta s/s) = \alpha$, called the *momentum compaction*. It indicates the relative change in length of the trajectory between passages in the acceleration gaps per relative change of momentum. For nearly circular orbits, obviously $\alpha = (\delta p/p)/(\delta R/R)$ and starting from

$$T = 2\pi r / \beta c \qquad (4\text{-}5.9)$$

we obtain by logarithmic differentiation

$$\frac{\delta T}{T} = \frac{\delta r}{r} - \frac{\delta \beta}{\beta} = \frac{1}{\alpha} \frac{\delta p}{p} - \frac{\delta \beta}{\beta} \qquad (4\text{-}5.10)$$

But because $cp = \beta E$ and $E = mc^2 \gamma$ we also have

$$\frac{\delta \beta}{\beta} = (1 - \beta^2) \frac{\delta p}{p} = \frac{1}{\gamma^2} \frac{\delta p}{p}$$

ISBN 0-8053-8601-7

and hence

$$\frac{\delta T}{T} = \left(\frac{1}{\alpha} - \frac{1}{\gamma^2} \right) \frac{\delta p}{p} \qquad (4\text{-}5.11)$$

This equation connects the change of period to the change of momentum. The sign of the coefficient is important because a change in sign produces an instability at the energy at which the sign change occurs.

To further specify a realistic case, assume that the particles move in a magnetic field as described by Eq. (4-3.13). We then have

$$cp = eBr = eB_0 r^{1-n} R^n$$

and $\delta p / p = (1 - n)(\delta r / r)$ or $\alpha = 1 - n$ and by Eq. (4-5.11)

$$\frac{\delta T}{T} = \left(\frac{1}{1-n} - \frac{1}{\gamma^2} \right) \frac{\delta p}{p} \qquad (4\text{-}5.12)$$

or by using $\delta p / p = (1/\beta^2)(\delta E / E)$ we obtain

$$\frac{\delta T}{T} = \left[1 - \frac{n}{\beta^2 (n-1)} \right] \frac{\delta E}{E} = K \frac{\delta E}{E} \qquad (4\text{-}5.13)$$

The quantity in brackets reduces to $(1 - n)^{-1}$ for $\beta = 1$. Moreover, if $0 < n < 1$, as is required for vertical and radial stability, $K > 1$. Taking the time derivative of Eq. (4-5.3), using Eq. (4-5.13) and the relation $\delta E = \omega_s r_s \delta p$, we find

$$\dot{\varphi} = \omega - \omega_e = \delta \omega = - \omega_s \frac{\delta T}{T} = - K r_s \omega_s^2 \frac{\delta p}{E} \qquad (4\text{-}5.14)$$

We return now to the problem of establishing the phase variation in time. The canonical equation gives

$$\dot{p}_\theta = \text{external torque} = (2eV/2\pi)\sin \varphi$$

in which we have replaced by a continuous torque the impulsive torque which the particle undergoes on crossing the slit between the dees.

Writing the expression for p_θ, we have

$$\frac{d}{dt} \left(mr^2 \dot{\theta} + \frac{e}{c} rA_\theta \right) = \frac{e}{\pi} V \sin \varphi \qquad (4\text{-}5.15)$$

This equation, for A_θ constant in time, states that the gain in energy per unit time is equal to the work done by the external electric torque. If A_θ contains

ISBN 0-8053-8601-7

the time explicitly, the gain in energy is determined by the sum of the external electric torque and the induced emf (betatron) acceleration. The equation neglects energy loss by radiation, which is unimportant for protons; it could be taken into account by subtracting $L/2\pi$ from the right side, where L is the energy lost, per turn, by radiation [Eq. (4-3.23)].

For a synchronous particle Eq. (4-5.15) is valid provided that we use the synchronous quantities p_s, r_s, etc. In order to see how a particle oscillates in a synchronous orbit, we "linearize" the problem by taking the difference between Eq. (4-5.15) and the corresponding equation for a synchronous particle. Recalling the definition of δ [Eq. (4-5.8)], we have, to the first order,

$$\frac{d}{dt}\left[\delta(mr^2\dot{\theta}) + \frac{e}{c}\delta(rA_\theta)\right] = \frac{e}{\pi}V(\sin\varphi - \sin\varphi_s) \qquad (4\text{-}5.16)$$

Now, if A_θ does not contain the time explicitly, the quantity in brackets is

$$p\,\delta r + r\,\delta p + (e/c)\,\delta(rA_\theta) \qquad (4\text{-}5.17)$$

This can be simplified by observing that $p = -(e/c)Br = -(e/c)[\partial(rA_\theta)/\partial r]$, from which $p\,\delta r = -(e/c)\delta(rA_\theta)$. We thus obtain

$$\frac{d}{dt}(r_s\,\delta p) = \frac{e}{\pi}V(\sin\varphi - \sin\varphi_s) \qquad (4\text{-}5.18)$$

Taking $r_s\,\delta p$ from Eq. (4-5.14) and substituting it in Eq. (4-5.18), we have

$$\frac{d}{dt}\left(\frac{\dot{\varphi}E_s}{\omega_s^2 K}\right) + \frac{e}{\pi}V(\sin\varphi - \sin\varphi_s) = 0 \qquad (4\text{-}5.19)$$

Here φ is the unknown function of t and E_s, ω_s^2, K, and V are to be regarded as constant to the first approximations. Equation (4-5.19) then represents an oscillatory motion. If $\varphi - \varphi_s = \epsilon$ is small, Eq. (4-5.19) can be approximated by

$$\frac{E_s}{\omega_s^2 K}\ddot{\epsilon} + \frac{eV}{\pi}\epsilon\cos\varphi_s = 0 \qquad (4\text{-}5.20)$$

which gives the frequency of the phase oscillation as

$$\omega_\varphi = \omega_s\left(\frac{eVK}{\pi E_s}\cos\varphi_s\right)^{1/2} \qquad (4\text{-}5.21)$$

For either sign of eVK one of the two synchronous phases, φ_s or $\pi - \varphi_s$ will give a stable oscillation with real ω_φ. The root is always $\ll 1$, because $eV \ll E_s$, and hence the phase oscillations are slow compared with the synchronous frequency. ●

ISBN 0-8053-8601-7

Even if phase oscillations are not small, Eq. (4-5.19) may easily be discussed on the basis of the mechanical model shown in Fig. 4-8. The correspondence between the parameters of the model and the slowly varying quantities in Eq. (4-5.19) is given in the figure caption. It is clear that if $\varphi_{max} > \pi$, the motion loses its oscillatory character. The phase oscillations are accompanied by oscillations of radius, energy, etc. Equation (4-5.19) may be generalized to take into account radiation losses, and it may be transformed in various ways suitable for the synchrotron, linear accelerators, etc.

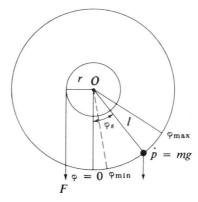

Figure 4-8 The weight mg gives a torque $mgl \sin \varphi$ with respect to O, which is balanced by the constant torque Fr in the equilibrium position $\varphi = \varphi_s$. The equation of motion is

$$l\ddot{\varphi} = -g(\sin \varphi - \sin \varphi_s)$$

and it is identical to Eq. (4-5.19) if

$$g/l = +(eV/\pi)(K/E_s)\omega_s^2$$

φ_{min} and φ_{max} represent the minimum and maximum values of φ in an oscillation.

The main differences between the operation of the ordinary cyclotron and the synchrocyclotron are that:

1. In the latter the particles may turn around 10^5 times before reaching the maximum energy, instead of 10^2 times, as in the ordinary cyclotron. Thus the potential applied to the dees needs to be 10^{-5} of the maximum obtainable energy.

2. The frequency is modulated.

3. The ions form "slugs" that circulate in the machine and come out in spurts lasting about 50 μsec and repeating at the rate of 100 sec^{-1}.

The variable frequency is obtained by inserting in the radio-frequency oscillator a capacitor that periodically varies its capacitance by either a rotary mechanism or a vibrating reed.

ISBN 0-8053-8601-7

4-6   THE SYNCHROTRON AND PROTON SYNCHROTRON

The synchrocyclotron requires magnetic-pole pieces of the diameter of the orbit. The maximum momentum obtainable is

$$(e/c)BR = p \quad \text{or} \quad T^2 + 2m_0c^2T = e^2B^2R^2 \quad (4\text{-}6.1)$$

where T is the kinetic energy. Since B is limited in practice to about 22×10^3 G, the radius and hence the cost of the machine becomes prohibitive for protons above about 700 MeV.

The advantage of a ring-shaped machine is obvious, and the synchrotron is just such a machine. Its characteristic is a constant radius of orbit. The ions move in a doughnut-shaped evacuated channel, to which is applied the magnetic field necessary to maintain the ions in an orbit of the proper radius. The accelerating electric field is provided in one or more gaps. For electrons, which are extremely relativistic, it suffices to increase the magnetic field and keep the frequency of the electric field constant. For protons, which are not extremely relativistic, it is necessary to increase the frequency as the magnetic field increases. In large machines, to avoid the problems of an extremely weak magnetic field at injection, one injects the ions at an energy of many MeV. The injection is accomplished by auxiliary accelerators. The orbit of a large machine always has some field-free regions through which the ions move in a straight line. These regions are necessary for injection, acceleration, and deflection.

4-7   STRONG FOCUSING

In order to reach higher and higher energies, it is necessary to increase the radius of the accelerator, since B is limited by practical considerations, even assuming the possibility of superconducting magnets. It is clear that it is then imperative to keep the ions in their trajectory with great precision, because in their extremely long path they could easily strike the wall of the doughnut. Moreover, the smaller the cross section of the tank, the cheaper it is to build and supply power to the magnet. These considerations put a premium on keeping the amplitudes of the radial and vertical oscillations small.

This requirement is related to keeping the oscillation frequencies high. For instance, if at injection the ions start with $z = 0$, $\dot{z}(0)$ different from 0, the vertical oscillation will have the amplitude

$$z_{\max} = \dot{z}(0)/\omega_z \quad (4\text{-}7.1)$$

as can be seen immediately. Now $\omega_z = \omega_0(n)^{1/2}$, according to Eq. (4-3.22), and a large n is required to keep the value of $z_{\max}$ low. However, we know that radial stability requires $n < 1$, and it seems that the situation is hopeless.

Christofilos (1950) and, independently, Courant, Livingston, and Snyder (1952) found a way out of the difficulty. The magnet is built of successive

ISBN 0-8053-8601-7

segments having, alternately, n large and positive and n large in magnitude
but negative. The first segment focuses vertically but defocuses horizontally.
The opposite happens with the second segment. However, the sum total for
both vertical and radial motion is focusing. This fact, unexpected at first
sight, can be qualitatively understood by considering two optical lenses, one
convergent and the other divergent. We have indicated in Fig. 4-19 the rays
through a system of two magnetic lenses showing the resultant focusing action. Alternating-gradient focusing also has the major advantage of markedly
reducing the radial excursions due to momentum differences.

In an actual accelerator the vacuum tank is surrounded by a succession of
magnets having alternately large n positive and negative. The magnets can be
so arranged as to have a net focusing action. In Fig. 4-9 we show the trajectory in a section of the magnet. The whole ring is made by the periodic
repetition of the elementary section. The precise arrangement must be calculated by taking the stability of the orbit into account. The combination of

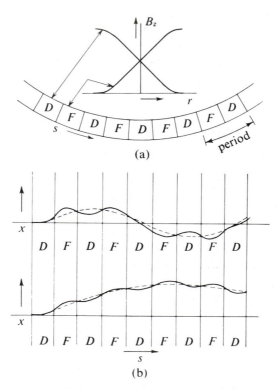

Figure   4-9   (a) A succession of focusing and defocusing lenses as arranged
in a strong-focusing accelerator. (b) Two cases of the displacement of the
beam along s showing focusing and defocusing actions. [Green and Courant,
(Fl E).]

ISBN 0-8053-8601-7

Figure 4-10   View of the alternating-gradient proton synchrotron of Brookhaven National Laboratory, showing a few of the 240 magnet sections. [Courtesy Brookhaven National Laboratory.]

ISBN 0-8053-8601-7

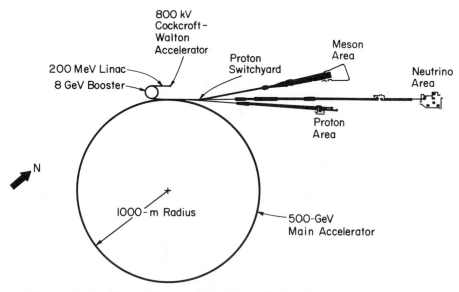

Figure 4-11 General layout of the Fermi National Accelerator Laboratory at Batavia, Illinois. Four stages of acceleration; final energy 500 GeV. For data see Table 4-2. [Courtesy FNAL, Batavia, Illinois.]

TABLE 4-2   ALTERNATING-GRADIENT SYNCHROTRON CHARACTERISTICS

	Accelerator	
	BNL[a]	FNL[b]
Maximum proton energy (GeV)	33	500
Number of protons per second (internal beam)	8×10^{12}	6×10^{12}
Repetition rate (pulse sec^{-1})	0.5	1/12
Ring diameter (m)	257	2000
Steel weight (tons)	4000	9000
Copper coil weight (tons)	400	850
Magnetic field		
At injection (G)	250	396
At 33 GeV and 300 GeV (kG)	13.1	13.5
Rise time (sec)	0.45	2.4
Flat top time (sec)	1	1
Mean power input in magnet (MW)	2.4	36
Radio-frequency range (MHz)	2.5–4.61	53.08–53.10
Radio-frequency input peak (kW)	1000	1800
Cost (millions)	$31[c]	$230[d]

[a] Brookhaven National Laboratory proton synchrotron.
[b] Fermi National Accelerator Laboratory.
[c] Construction period 1953–1960.
[d] Construction period 1969–1972.

ISBN 0-8053-8601-7

Figure 4-12 Aerial view of CERN looking north over Laboratory I to the site of Laboratory II. In the right foreground can be seen the administration buildings; in the center, the 200-m diameter ring of the 28-GeV proton synchrotron; and beyond, the 300-m diameter ring of the ISR. To the right of the main road is part of the site of CERN's Laboratory II, where a 46-m shaft descends to the SPS tunnel, now. [Courtesy of CERN, Geneva.]

strong focusing and the principle of the proton synchrotron has made it possible to accelerate protons to energies of approximately 5×10^{11} eV. Figure 4-10 shows a section of the Brookhaven alternating-gradient machine. Figure 4-11 shows the general layout of the accelerator at the Fermi National Accelerator Laboratory in Batavia, Illinois. This is the largest accelerator in existence. A similar one is under construction at CERN (Geneva, Switzerland; Fig. 4-12 is a view of the CERN grounds). Table 4-2 gives some of the vital data on the Brookhaven and Batavia accelerators.

4-8 LINEAR ACCELERATORS

The principle of multiple acceleration is also used in linear accelerators. The trajectory of the ions is approximately straight, however. There are many types of linear accelerators, for electrons, protons, and heavy ions such as ^{12}C. We shall describe briefly only two types, one for electrons of very high energy, and the other for heavy ions.

ISBN 0-8053-8601-7

As we mentioned earlier, above a certain energy radiation losses become prohibitive for circular electron accelerators. In this respect linear accelerators have an obvious advantage. For heavy ions the specific charge Q/A is often low, because the atoms forming the ions are only partially ionized. (The usual unit for Q is the protonic charge and for A one twelfth the mass of ^{12}C.) A magnetic field would bend the trajectories of these ions only slightly, and it would be more costly to use a circular machine than a linear accelerator. At present, energies of the order of 10 MeV per nucleon are secured for heavy ions by means of linear accelerators. However, heavy ions (up to Fe) have also been accelerated with the Bevatron combined with a special injector, and energies of several billion electron volts per nucleon are attainable.

A linear accelerator of historical interest is shown in Fig. (4-13). The even and odd electrodes are connected to opposite poles of an oscillator. In the gaps between the drift tubes located on the axis of the cavity there is an electric-potential difference

$$V = V_0 \cos \omega t \tag{4-8.1}$$

An ion in the gap is subject to the related field. An ion traveling inside a drift tube does not feel any field. If an ion crosses the gaps at the appropriate times, for example, at 0, $T = 2\pi/\omega$, $2T$, etc., it receives multiple accelerating impulses. The distance between gaps must increase if the ion is to cross the gap at the right time.

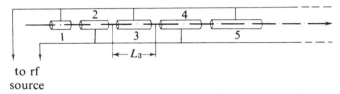

to rf
source

Figure 4-13 Lawrence–Sloan type linear accelerator. Drift tube 3 has a length $L_3 = [3(2e/m)V_0]^{1/2}(T/2)$.

The distance L between gaps must be such that an ion enters a drift tube when the tube is negative and leaves it when it is positive; hence

$$L = vT/2 \tag{4-8.2}$$

where v is the velocity in the drift tube. Nonrelativistically, after crossing j gaps, v is

$$[(2e/m)jV_0]^{1/2} \tag{4-8.3}$$

Hence

$$L_j = [j(2e/m)V_0]^{1/2}(T/2)$$

ISBN 0-8053-8601-1

Relativistically, for electrons of high energy, $v \to c$, and the distance between gaps is constant,

$$L = cT/2 \qquad (4\text{-}8.4)$$

The type of accelerator illustrated in Fig. 4-13 was actually built and used by Lawrence and Sloan (1931) but has had little use in nuclear physics. Its principle, however, is important.

Modern accelerators make use of waveguides to establish the electric field. High frequencies of up to 3000 MHz are currently employed. A waveguide is a pipe of conducting material in which an oscillating electromagnetic field is established. The electromagnetic field can form a standing wave in the cavity, which then acts as a resonator, or it can form a traveling wave. The standing wave may of course be considered as the superposition of two traveling waves progressing in opposite directions. An ion that moves with the same velocity as the traveling wave is subject to a constant accelerating force.

In the case of high-energy electrons moving at a velocity very near c, the electromagnetic wave must move with a phase velocity c. Such modes of oscillation are achieved by inserting partitions in the cavity and exciting it at the proper frequency. An example of such a cavity is shown schematically in Fig. 4-14. The conducting disks give to the line the desired characteristic phase velocity. A 2-mile-long linear accelerator at Stanford reaches 22 GeV with a current of 30 μA.

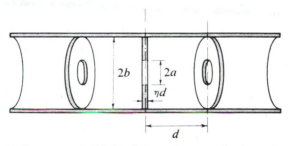

Figure 4-14 Structure of disk-loaded accelerator, showing important design dimensions.

As another example we mention the heavy-ion linear accelerators used to accelerate heavy ions (e.g., C, N, Ne) to energies of about 9 MeV per nucleon (Fig. 4-15). Positive ions of small charge are first accelerated in a Cockcroft-Walton type of accelerator having a potential drop of about 800 kV. On emerging from this accelerator, the ions are "bunched" so that they can be injected into the next section, a linear accelerator of the standing-wave type. In the second section the ions acquire an energy of about 1 MeV, per nucleon. It is then possible to ionize them further by making them pass through a thin carbon foil and thus to obtain a higher specific charge, which is desirable for

ISBN 0-8053-8061-7

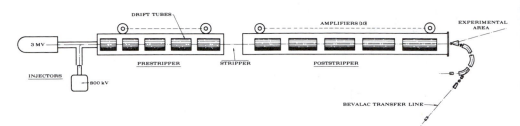

Figure 4-15 The heavy-ion linear accelerator accelerates ions up to Xe (ultimately U) to an energy of 8.5 MeV/A (A is the ion mass in units of $^{12}C = 12$). During acceleration, the ionic charge Q (and hence the specific charge Q/A) varies, but because the frequencies and the geometry of the accelerator are constant, $QE/A = a$ (E, electric field; a, acceleration), the acceleration of the ions must be independent of Q (otherwise the ions are lost). The ions are accelerated by one of the Cockcroft and Walton accelerators (one for light, the other for heavy ions). They enter the prestripper with $Q/A \geqslant 0.05$ and $\beta = 0.015$. At the prestripper's exit they have $\beta = 0.05$. The stripper carbon foil then increases Q/A to ~ 0.33 for Ar (to 0.17 for U). The poststripper accelerator brings β to 0.13 (energy to about 8.5 MeV/A). At its exit, particles may be used directly or injected into a synchrotron (Bevatron), where they reach up to 2.5 GeV/A.

the next acceleration. The part of the machine where this process takes place is called the *stripper*. The ions now enter a new section, also of the standing-wave type, where they are further accelerated to an energy of about 9 MeV per nucleon ($v/c = 0.13$).

Linear accelerators, like circular accelerators, have phase stability, the most important difference being that phase stability in the former is achieved during the part of the cycle in which the potential increases, rather than during the decreasing potential part of the cycle used in ordinary weak-focusing circular accelerators.

The ions are focused by magnetic-quadrupole lenses (see Sec. 4-10) or other devices to prevent spreading of the beam.

4-9 COLLIDING BEAMS

The construction and operating costs of accelerators are at best proportional to the energy in the laboratory system and in many cases grow faster than this energy. On the other hand, the important energy to the physicist is the energy in the center-of-mass system. This quantity for a particle of mass m impinging on a particle of the same mass is given in the extreme-relativistic

ISBN 0-8053-8601-7

case, $E \gg mc^2$, by $E_{\text{c.m.}} = (2mc^2 E_l)^{1/2}$. This formula shows how hard the race to high energy is. A proton accelerator of 500-GeV laboratory energy gives only 30.6 GeV in the c.m. system. To overcome this problem, beams traveling in opposite directions have been made to collide. Here the c.m. energy of two beams, each with energy E_l and the same mass, is $2E_l$. Two colliding proton beams of 15 GeV each are equivalent, as far as energy is concerned, to a 500-GeV beam on a stationary target.

The problem, however, is the intensity. The rate at which a reaction takes place can be written as

$$\dot{n} = L\sigma \tag{4-9.1}$$

where σ is the cross section at the relative velocity, $\dot{n}$ is the number of processes per second, and L is the *luminosity*. The luminosity thus defined has the dimensions $l^{-2}t^{-1}$. For instance, for a beam of 10^{12} particles sec^{-1}, such as is obtained from a large multi-GeV accelerator, impinging on a liquid hydrogen target 10 cm thick, $L = 4.3 \times 10^{36}$ cm^{-2}sec^{-1}. With the best colliding beams available, L at present is about 10^6 times smaller. The formula for the luminosity of a beam hitting a stationary target is $L = n_1 v_1 A N$ where n_1 is the number of particles per unit volume in the beam, v_1 its velocity with respect to the target containing N nuclei cm^{-2}, and A the area of the target. For colliding beams containing n_1 and n_2 particles per unit volume and moving with respect to each other with velocity v_1, v_2, the luminosity *per unit volume* is

$$\frac{L}{V} = n_1 n_2 \left[(\mathbf{v}_1 - \mathbf{v}_2)^2 - \frac{(\mathbf{v}_1 \times \mathbf{v}_2)^2}{c^2} \right]^{1/2} \tag{4-9.2}$$

or for the case of $\mathbf{v}_1 = -\mathbf{v}_2$

$$L/V = n_1 n_2 2v_1 \tag{4-9.3}$$

Equation (4-9.2) may be rewritten for $v = c$ as

$$L = \frac{I_1 I_2}{e^2 c \tan(\theta/2)} \frac{1}{h_{\text{eff}}}$$

where θ is the intersection angle of the beams and I_1, I_2 are the beam currents. The parameter h_{eff} describes the overlap of the beams. Assume that the beams are in a horizontal plane and extend vertically for a certain distance. The current density as a function of the vertical coordinate is $\rho_1(z)$, $\rho_2(z)$ for each beam. Then

$$h_{\text{eff}} = \int \rho_1(z)\, dz \int \rho_2(z)\, dz \Big/ \int \rho_1(z)\rho_2(z)\, dz$$

ISBN 0-8053-8601-7

Thus L is a characteristic of the machine and its working conditions. It can be determined either by calibration with a known σ (Coulomb scattering at an extremely small angle), or by direct measurement of ρ_1, ρ_2, or by other means.

In any case, it is clear that it is essential to obtain as high a current density as possible. Colliding beams from an ordinary accelerator give too low a luminosity to be of practical interest. To increase the currents available, one loads a "storage ring" by means of an ordinary accelerator. It is then possible to obtain very high circulating currents, much higher than from the injector. There are limits to the current density obtainable. Theoretically, Liouville's theorem sets an absolute limit, unless dominated by radiation damping and quantum fluctuations, as will be the case in electron and positron storage rings.

The volume of phase space occupied by a "bunch" is $\Delta\Omega = \Delta x\,\Delta y\,\Delta z\,\Delta p_x\,\Delta p_y\,\Delta p_z$, where Δx, etc., and Δp_x, etc., are the spreads of coordinates and momenta of the particles, and the volume is a constant of the motion. New particles injected in the ring must go in regions of phase space not previously occupied and must be stacked in a new region of phase space available in the ring. Thus the number of particles storable is at most

$$N = n(\Omega_R/\Omega_i) \tag{4-9.4}$$

where n is the number of particles in an injection burst "bucketful," Ω_R the phase space available in the ring determined by its volume and the interval of momentum for which the motion is stable, and Ω_i is the phase space of one bucketful determined by its physical dimensions and momentum spreads at injection. In practice, the particles are grouped in bunches rotating in storage rings, like beads on a rosary.

Radiation losses in electron storage rings are significant [see Eq. (4-3.23)] and must be compensated by a radio-frequency source. At the same time, such storage rings are important sources of light for research in the extreme ultraviolet or soft X-ray region. Figure 4-16 shows the general layout of the intersecting storage rings for protons at CERN. (An aerial view of CERN appears in Fig. 4-12.) There may be one storage ring for both beams, or a separate ring for each. The luminosity for an effective beam area A, a frequency of revolution f, and k bunches per ring is

$$L = f(kN_1N_2/A)(c^2p/vE) \tag{4-9.5}$$

where N_1N_2 is the number of particles per bunch and the last parenthesis for $v \to c$ is 1. The importance of keeping A small is obvious and this puts severe requirements on the vacuum, precision of focusing, and adjustment of the rings.

Other theoretical and practical limits are important. The magnetic field must be very accurately determined and the vacuum must be extremely good

ISBN 0-8053-8601-7

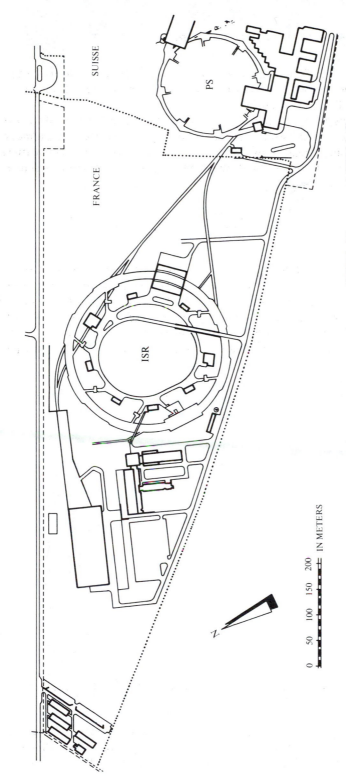

Figure 4-16 General layout of the intersecting proton storage rings at CERN. [Courtesy of CERN, Geneva.]

(better than 10^{-9} mm Hg). The current can then circulate for hours or days without loss (except by radiation) and reach intensities of amperes (for protons) or 0.02 A for electrons.

Colliding beams of light and electrons have also been used for experimental purposes. The visible light produced by a ruby laser of momentum k_i is made to collide head-on with high-energy electrons. The light scattered backward at a small angle θ with the direction of the incoming electron has the momentum k_f ($c = 1$)

$$k_f = \frac{4E_e^2 k_i}{m_e^2 + 4k_i E_e} \left[1 + \left(\theta^2 k_{f_{max}}/4k_i \right) \right]^{-1} \qquad (4\text{-}9.6)$$

where $k_{f_{max}}$ is the first factor on the right-hand side. The scattering may be considered in the electron rest frame as ordinary Compton scattering. In that frame the incident gamma quantum has an energy $k_0 = 2\gamma k_i$ with $\gamma = E_e/mc^2$. For ruby laser photons (1.78 eV) and electrons of 16 GeV, $k_{f_{max}}$ is about 4.8 GeV and the effective scattering cross section about 0.02 b. If the incident beam is polarized, the polarization is preserved in the scattered beam (Fig. 4-17).

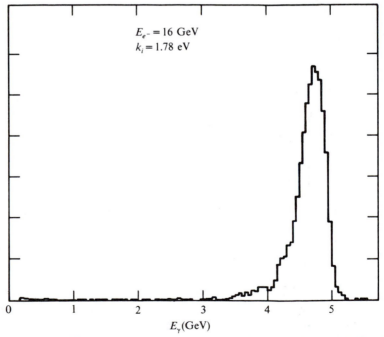

Figure 4-17 Photon energy spectrum produced by ruby laser radiation impinging on 16-GeV electrons. The ordinate is proportional to the number of photons produced. [From J. Ballam et al., *Phys. Rev.*, **D5**, 545 (1972)].

ISBN 0-8053-8601-7

4-10 BEAM-TRANSPORT APPARATUS

For experimental purposes it is frequently necessary to generate beams of particles originating from a target. The art of producing and controlling such beams is rather similar to the art of optics. The optical components, lenses and prisms, are replaced by magnetic lenses and deflecting magnets.

One of the most useful magnetic lenses is a *quadrupole* (Fig. 4-18). In the vicinity of the axis of the quadrupole **B** has the form

$$B_x = by \tag{4-10.1}$$

$$B_y = bx \tag{4-10.2}$$

$$B_z = 0 \tag{4-10.3}$$

satisfying Maxwell's equations and deriving from a scalar potential $V = -bxy$.

An ion moving in a region near the z axis and at a small inclination to this axis is subject to the forces

$$F_x = -(e/c)vbx \tag{4-10.4}$$

$$F_y = (e/c)vby \tag{4-10.5}$$

$$F_z = 0 \tag{4-10.6}$$

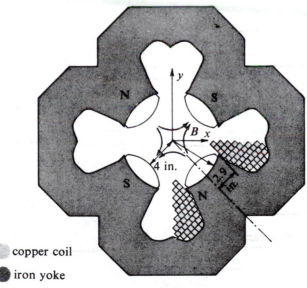

copper coil

iron yoke

Figure 4-18 Quadrupole magnet, showing the cross section, to scale, of one of the most satisfactory quadrupole varieties. [O. Chamberlain, *Ann. Rev. Nucl. Sci.*, **10**, 161 (1960).]

In our approximation

$$\frac{d}{dt} = v\,\frac{d}{dz} \qquad \text{and} \qquad \frac{d^2}{dt^2} = v^2\,\frac{d^2}{dz^2} \tag{4-10.7}$$

from which we have

$$\frac{d^2x}{dt^2} = -\frac{ev}{mc}\,bx = v^2\,\frac{d^2x}{dz^2} \tag{4-10.8}$$

or

$$\frac{d^2x}{dz^2} = -k^2x \tag{4-10.9}$$

where $k^2 = (e/cp)(\partial B_y/\partial x)$. Similarly, one obtains

$$\frac{d^2y}{dz^2} = +k^2y$$

The difference in sign in the x and y equations shows that a lens focusing in the x–z plane defocuses in the y–z plane and vice versa.

If at the entrance of the quadrupole ($z = 0$), $dx/dz = 0$, $x = x_0$, $dy/dz = y_0 = 0$, and the quadrupole has length l, the ion crosses the z axis at $z = l + (1/k)\cot kl$, as can be verified by integrating Eq. (4-10.9). Figure 4-19 shows that the action of the quadrupole in the x–z plane is the same as that of a lens of focal length

$$f = \frac{1}{k\,\sin kl} \tag{4-10.10}$$

and having its principal plane at a distance $-(1 - \cos kl)/(k\sin kl)$ from the image end of the quadrupole.

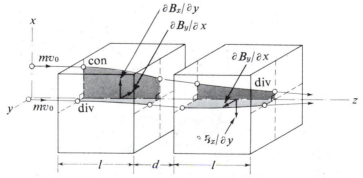

Figure 4-19 Double focusing by a pair of matched magnetic lenses in which the direction of the magnetic gradients is reversed. [From M. S. Livingston.]

In the *yz* plane the quadrupole acts as a diverging lens because Eqs. (4-10.4) and (4-10.5) have different signs. It is possible to combine two or more quadrupoles in such a way as to give a real focus in both the vertical and the horizontal direction (Fig. 4-19). Generally such foci do not coincide, and the system is astigmatic. This, however, is not always undesirable. Stigmatic systems may also be built, but they often have different magnifications in the *x* and *y* directions.

The combined action of several quadrupoles may be calculated by the following method. Consider first the *x* direction. A ray is then characterized by $x(z_0)$, $x'(z_0)$, where $dx/dz = x'$. The quantities $x(z)$, $x'(z)$ are linear functions of $x(z_0) = x_0$, $x'(z_0) = x'_0$, the coefficients a, b, c, d being functions of z.

$$x(z) = ax_0 + bx'_0 \tag{4-10.11}$$

$$x'(z) = cx_0 + dx'_0 \tag{4-10.12}$$

We can usefully represent this relation in the matrix form

$$\begin{pmatrix} x \\ x' \end{pmatrix} = \begin{pmatrix} a & b \\ c & d \end{pmatrix} \begin{pmatrix} x_0 \\ x'_0 \end{pmatrix} = X = MX_0 \tag{4-10.13}$$

The matrix M depends on the apparatus. For instance, for a free beam moving a distance b,

$$M = \begin{pmatrix} 1 & b \\ 0 & 1 \end{pmatrix} \tag{4-10.14}$$

Quadrupole lenses in the focusing and defocusing directions are represented between entrance and exit end by

$$\begin{bmatrix} \cos kl & \frac{1}{k}\sin kl \\ -k\sin kl & \cos kl \end{bmatrix} \quad \text{and} \quad \begin{bmatrix} \cosh kl & \frac{1}{k}\sinh kl \\ k\sinh kl & \cosh kl \end{bmatrix} \tag{4-10.15}$$

respectively.

A system is represented by the matrix product of the matrices corresponding to the components in the same order in which they are traversed by the beam. For example, if the beam traverses components M_1, M_2, M_3, etc., the matrix describing the apparatus is $M_3 \times M_2 \times M_1$ etc. $= M$.

The matrix technique is also useful in calculating the effect of a series of magnets in an alternating-gradient machine.

Rays traversing a complex system may be traced experimentally by using a wire held under a tension T through which a current i passes. Such a wire in a magnetic field assumes the stable configuration of the orbit of an ion of charge e and momentum p provided that

$$i/T = e/pc \tag{4-10.16}$$

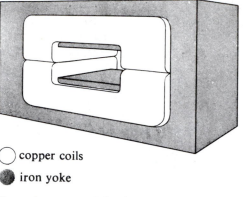

◯ copper coils

● iron yoke

Figure 4-20 General-purpose deflecting magnet. This arrangement, with so-called "window-frame" construction, has proved very useful because it provides a very uniform field, even close to the coils, and may be used at high-magnetic-flux densities. [O. Chamberlain, *Ann. Rev. Nucl. Sci.*, **10**, 161 (1960).]

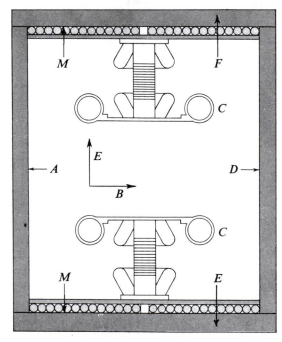

● iron

◯ copper

● stainless steel

Figure 4-21 Mass selector for a high-energy accelerator. The beam moves perpendicularly to the plane of the figure. The electric field is obtained by establishing a potential difference between the plates of condenser C. The magnet field is generated by the coils M. A and D act as pole faces, E and F as return paths.

ISBN 0-8053-8601-7

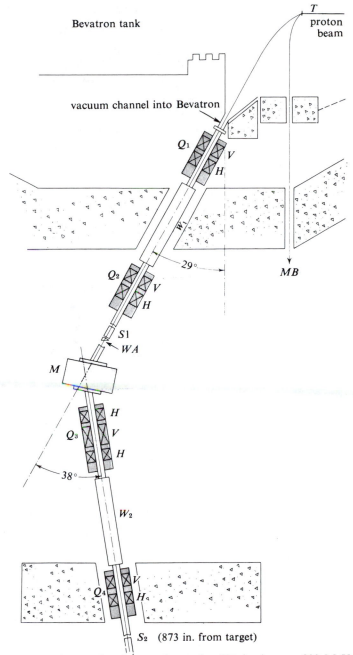

Figure 4-22 A complete beam selector for K^- having $p = 800$ MeV/c. T, target; $Q1$, $Q2$, $Q3$, $Q4$, 8-in. quadrupoles: V, vertical focusing; H, horizontal focusing; W_1, W_2, Wien filters; M, bending magnet; $S1$, $S2$, mass resolving slits; WA, wedge absorber. [Courtesy Lawrence Radiation Laboratory.]

ISBN 0-8053-8601-7

The magnetic systems thus far considered focus all particles having the same charge and momentum. In many experiments it is important to select particles of a given momentum and mass. This is accomplished by a combination of magnetic and electric deflection. In the Lorentz force

$$\mathbf{F} = e\left(\mathbf{E} + \frac{1}{c}\,\mathbf{v} \times \mathbf{B}\right) \tag{4-10.17}$$

the magnetic part depends on velocity, and the electric part does not. This permits distinguishing particles of the same momentum but of different velocity. A common method is to use electric and magnetic fields perpendicular to each other and to the direction of motion of the particle. When the electric and magnetic fields are in the ratio $E/B = v/c$, the particle traverses the velocity selector without deviation. Such an apparatus is called a *Wien filter*. Figure 4-20 shows a practical realization of a deflecting magnet, often used as a momentum selector. Figure 4-21 is a section of a Wien filter and Fig. 4-22 shows a complete setup for selecting a beam of particles having a given rest mass and velocity. In most laboratories having large accelerators computer programs are available which permit calculation of trajectories in a succession of magnetic fields and thus serve to plan beams for practical purposes.

BIBLIOGRAPHY

Baldinger, E., "Kaskadengeneratoren," in (Fl E), Vol. 44.

Blewett, M. H., "Characteristics of Typical Accelerators," *Ann. Rev. Nucl. Sci.*, **17**, 427 (1967).

Bruck, H., *Accélérateurs circulaires de particules*, Presses Universitaires de France, Paris, 1967.

Chamberlain, O., "Optics of High-Energy Beams," *Ann. Rev. Nucl. Sci.*, **10**, 161 (1960).

Cohen, B. L., "Cyclotrons and Synchrocyclotrons," in (Fl E), Vol. 44.

Courant, E. D., "Accelerators for High Intensities and High Energies," *Ann. Rev. Nucl. Sci.*, **18**, 435 (1968).

Green, G. K., and E. D. Courant, "The Proton Synchrotron," in (Fl E), Vol. 44.

Haeberli, W., "Sources of Polarized Ions," *Ann. Rev. Nucl. Sci.*, **17**, 373 (1967).

Herb, R. G., "Van de Graaff Generators," in (Fl E), Vol. 44.

Kerst, D. W., "The Betatron," in (Fl E), Vol. 44.

Lapostolle, P. M., and A. L. Septier, (eds.), *Linear Accelerators*, North-Holland, Amsterdam, 1970.

Livingston, M. S., and J. P. Blewett, *Particle Accelerators*, McGraw-Hill, New York, 1962.

McMillan, E. M., "Particle Accelerators," in (Se 59), Vol. III.

Panofsky, W. K. H., "High Energy Physics Horizons," *Physics Today*, **26**, June (1973).

Pellegrini, C., "Colliding-Beams Accelerators," *Ann. Rev. Nucl. Sci.*, **22**, 1 (1972).

Persico, E., E. Ferrari, and S. E. Segre, *Principles of Particle Accelerators*, Benjamin, New York, 1968.

Sanford, J. R., The FNAL Accelerator, to appear in *Ann. Rev. Nucl. Sci.*, (1976).

Septier, A. (ed.), *Focusing of Charged Particles*, Academic Press, New York, 1967.

Smith, L., "Linear Accelerators," in (Fl E), Vol. 44.

Steffen, K. G., *High Energy Beam Optics*, Wiley-Interscience, New York, 1964.

Wilson, R. R., "Electron Synchrotrons," in (Fl E), Vol. 44.

Major accelerators have users' manuals which are kept up to date and are indispensable in experiment planning.

ISBN 0-8053-8601-7

PROBLEMS

4-1   The loss of energy by an electron to radiation, given by Eq. (4-3.23), is associated with a radiation reaction force on the particle, given by

$$\mathbf{F}_L = - \frac{L}{2\pi r} \frac{\mathbf{v}}{v}$$

The components of the force that are transverse to the azimuthal direction produce damping. Calculate the e-folding time for this damping for electrons that have an energy of 10^9 eV and move in a magnetic field of 12 kG. Also calculate the power radiated per electron.

4-2   Linearized equations of motion for betatron oscillations, such as Eqs. (4-3.19) and (4-3.20), may be generalized by taking into account the slow changes in such parameters as mass, magnetic-field strength, orbit radius, and angular frequency which may occur during acceleration in various types of machines. Such changes are called *adiabatic* and lead in general to *adiabatic damping* of the oscillation amplitudes. It can be shown that the *action integral* J_i, given by

$$J_i = \oint p_i \, dq_i$$

(integrated over one cycle of an oscillation and neglecting slow changes) remains approximately constant when such changes occur. Assuming this, show that the amplitude of betatron oscillations in a betatron or electron synchrotron varies as the inverse square root of the magnetic-field strength B_0, for both nonrelativistic and relativistic particles.

4-3   Show that the c.m. energy of colliding particles of mass $m_{1,2}$, energy $E_{1,2}$, and momentum $p_{1,2}$ is

$$W = \left[2(E_1 E_2 + p_1 p_2 c^2) + (m_1^2 + m_2^2)c^4 \right]^{1/2}$$

Compare W for a 400-GeV proton colliding with a proton at rest with W for the same proton colliding with one having 10 GeV of energy and moving in the opposite direction.

4-4   Calculate the luminosity of two bunched colliding beams with an equal number of bunches per revolution, energy E, momentum p, and N particles per bunch. The beams circulate at frequency f, have area A, and have k bunches per beam. $L = fk(N^2/A)(c^2p/vE)$.

4-5   Calculate the luminosity for two beams, of width w, height h, and containing λ particles per unit length, that intersect at an angle 2α. The beams move in opposite directions with velocity v. Consider the case in which $v = c$.

$$\left(\text{Answer:} \qquad L = \frac{\lambda^2 v}{h \sin \alpha} (1 - \beta^2 \sin^2 \alpha)^{1/2} \right)$$

Express the result for an accumulation ring of radius R in which the particles move with frequency f.

4-6   Plan a magnetic-quadrupole lens having a 6-in. aperture. What is a reasonably attainable focal length for protons having $p = 400$ MeV/c?

4-7   Plan a separator for protons and K^+ mesons. The momentum of the particles is 500 MeV/c. The angular separation required is 0.01 rad and the maximum field obtainable is 5×10^4 V cm^{-1}.

4-8   Make a general plan for a 10-MeV proton cyclotron. Estimate the diameter of the pole pieces of the magnet, the frequency and power of the oscillator, and the thickness of the shield. From these data make an approximate cost estimate. Add the cost of the building. Also estimate what crew will be needed, the power bill, and the yearly costs of operating the machine.

ISBN 0-8053-8601-7

Radioactive Decay

Here we shall treat the laws of spontaneous radioactive decay, independent of the emission accompanying the transformation. This can be done because it happens that the law of decay is independent of the mechanism of the transformation. The results thus acquired apply to a great variety of cases and provide a phenomenological explanation of a vast category of experimental facts.

First, in Sec. 5-1 we shall neglect the fact that every substance contains an integral number of atoms and shall treat this number as a continuous variable. This procedure is legitimate if we deal with processes involving a great number of atoms. The continuum theory of radioactive decay is precise in the sense that if we treat many systems of initially identical radioactive atoms, the average number of atoms contained in the various systems at any subsequent time is given exactly by the theory. On the other hand, each system may depart from the average, and for the study of these departures it is necessary to take into account the discontinuous, atomic nature of matter.

The decay law of radioactive substances was first clearly formulated and applied by Rutherford and Soddy as a result of their studies on the radioactivity of various substances (notably thorium, thorium X, and the emanations), although in more or less explicit form, it had also been known to earlier investigators.

Emilio Segrè, Nuclei and Particles: An Introduction to Nuclear and Subnuclear Physics, Second Edition

ISBN 0-8053-8601-1

5-1 CONTINUUM THEORY—ONE SUBSTANCE

The fundamental law of radioactive decay can be formulated as follows: Given an atom, the probability that it will decay during the interval dt is λdt. The constant λ is called the *decay constant*. Dimensionally it is a reciprocal time, and it is characteristic of the given substance and of the mode of the decay. The constant is independent of the age of the atom considered and, as we shall see later, being a nuclear property, is not affected by any of the usual physical agents. This type of law is characteristic of random events and applies to all types of radioactive decay—alpha, beta, gamma, orbital electron capture, spontaneous fission; it is also applicable in the atomic process of light emission by excited atoms.

The simplest application of this law involves a single radioactive substance that has initially $N(0)$ atoms. $N(0)$ is a large number, by hypothesis, so that we may consider $N(t)$, the number of atoms at time t, to be a continuously variable quantity. Then, according to our fundamental law, $-dN$, the decrease during time dt in the number of atoms, is given by

$$-dN = \lambda N \, dt \qquad (5\text{-}1.1)$$

which, integrated with the condition that initially we have $N(0)$ atoms, gives

$$N(t) = N(0)e^{-\lambda t} \qquad (5\text{-}1.2)$$

Equation (5-1.2) is another formulation of the fundamental law of radioactive decay.

In practice one uses, in addition to the decay constant, its reciprocal $\tau = 1/\lambda$, called the *mean life*, and the time T in which the number of atoms initially present is reduced by a factor of 2. T is often call the *period*, or *half-life*, of the substance. The period is related to the decay constant and to the mean life by

$$e^{-\lambda T} = e^{-T/\tau} = \frac{1}{2} \quad \text{or} \quad \lambda T = T/\tau = \log 2 = 0.6931472 \qquad (5\text{-}1.3)$$

The term mean life is applied to τ because it is the average lifetime of the atoms. In fact, if we have initially $N(0)$ atoms, we shall have $N(t) = N(0)e^{-\lambda t}$ at time t, according to Eq. (5-1.2). Of these $N(t)\lambda \, dt$ will decay between times t and $t + dt$. The mean life is obtained by multiplying the last number by t, integrating with respect to dt between 0 and ∞, and dividing by the initial number of atoms present, $N(0)$:

$$\tau = \frac{1}{N(0)} \int_0^\infty \lambda t N(t) \, dt = \frac{1}{N(0)} \int_0^\infty N(0)e^{-\lambda t}\lambda t \, dt = \frac{1}{\lambda} \qquad (5\text{-}1.4)$$

It is also easily seen from Eq. (5-1.4) that if a radioactive substance continues to decay at its initial rate $N(0)\lambda$, it will all disappear in a time τ (Fig. 5-1).

ISBN 0-8053-8061-7

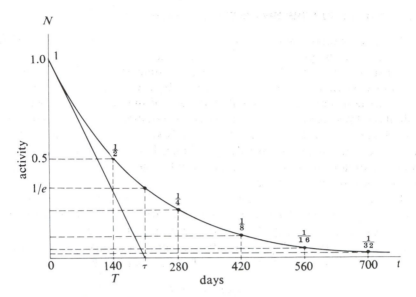

Figure   5-1   Decay of a radioactive substance.

If we plot $N(t)$ versus t graphically, we obtain an exponential and the tangent to the curve at $t = 0$ intercepts the t axis at a time τ. If we plot log $N(t)$ versus t, we obtain a straight line:

$$\log N(t) = \log N(0) - \lambda t$$

The slope of the straight line gives $-\lambda$. This second type of plot made directly on semilogarithmic paper is the most convenient and the most commonly used.

5-2   CONTINUUM THEORY—MORE THAN ONE SUBSTANCE

Very often one radioactive substance decays into another that is also radioactive. The two substances are then said to be genetically related; the first is called the *parent,* or *mother,* substance, the second, the *daughter* substance. The relation is not limited to parent and daughter but extends sometimes over many "generations." Sometimes one substance can decay by two processes, e.g., by alpha and beta emission, giving rise to two different daughter substances: this occurrence is called *dual decay,* or *branching.*

Examples of long chains of radioactive decays are offered by the natural radioactive families (see Fig. 5-2) and by the chains of successive beta decays typical of fission fragments.

Figure 5-2 shows that four alphas are expected from an atom of RdTh (^{228}Th) before reaching ThB (^{212}Pb), and their emission in succession is beautifully illustrated in Fig. 5-3. An atom of RdTh (^{228}Th) is embedded in a

ISBN 0-8053-8061-7

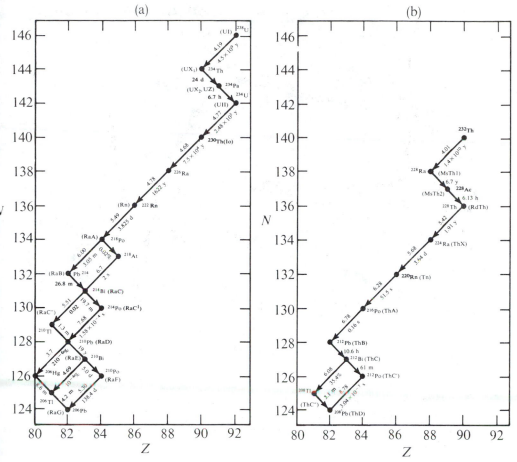

Figure   5-2   The natural radioactive families of (a) ^{238}U and (b) ^{232}Th in a Z–N diagram, with the associated energies and half-lives.

photographic emulsion, and each alpha particle emitted by it leaves its track in the emulsion.

We shall not consider in this section the case of branching, but we shall solve the following problem: Given that at time $t = 0$ we have $N_1(0)$, $N_2(0)$ atoms of the radioactive substances 1, 2, . . . , which are genetically related, find the number of atoms $N_1(t)$, $N_2(t)$, . . . present at any subsequent time.

Substance 1 decays according to the law expressed in Eq. (5-1.1): $dN_1 = -N_1\lambda_1 \, dt$. For every atom of substance 1 that disintegrates, an atom of substance 2 is formed. Hence the number of atoms of substance 2 varies for two reasons: it decreases because substance 2 decays, but it increases because the decay of substance 1 continuously furnishes atoms of substance 2. The

ISBN 0-8053-8061-7

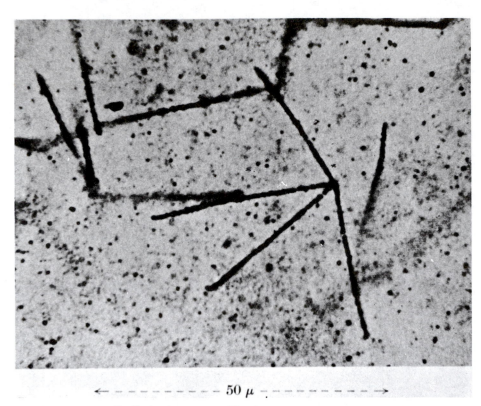

$$\leftarrow - - - - - - - - 50 \ \mu \ - - - - - - - - - \rightarrow$$

Figure 5-3 Radiothorium stars: The four tracks correspond to the alpha particles emitted by ^{228}Th, ^{224}Ra, ^{220}Rn, ^{216}Po. [C. F. Powell and G. Occhialini, *Nuclear Physics in Photographs*, Oxford University Press, London, 1947.]

net change is given by

$$\frac{dN_2}{dt} = \lambda_1 N_1 - \lambda_2 N_2 \tag{5-2.1}$$

where λ_1 and λ_2 are the decay constants of the first and second substances, respectively. For a third substance we have, in a similar way,

$$\frac{dN_3}{dt} = \lambda_2 N_2 - \lambda_3 N_3, \quad \text{etc.} \tag{5-2.2}$$

This system of differential equations can be solved by putting

$$N_1 = A_{11} e^{-\lambda_1 t}$$
$$N_2 = A_{21} e^{-\lambda_1 t} + A_{22} e^{-\lambda_2 t}$$
$$N_3 = A_{31} e^{-\lambda_1 t} + A_{32} e^{-\lambda_2 t} + A_{33} e^{-\lambda_3 t}$$
$$\cdots$$
$$N_k = A_{k1} e^{-\lambda_1 t} + A_{k2} e^{-\lambda_2 t} + \cdots + A_{kk} e^{-\lambda_k t} \tag{5-2.3}$$

ISBN 0-8053-8061-7

The constants A_{ki} are to be determined in such a way that the expressions of Eq. (5-2.3) satisfy the differential equations and the $N_k(0)$ have the prescribed initial values. Substituting the expressions from Eq. (5-2.3) into the differential equations (5-1.1), (5-2.1), (5-2.2), etc., we have

$$A_{ki} = A_{k-1,i} \frac{\lambda_{k-1}}{\lambda_k - \lambda_i} \tag{5-2.4}$$

This recursion formula is sufficient to determine all the A_{ki} with the exception of those with equal indexes.[1] These are determined by the initial conditions

$$N_k(0) = A_{k1} + A_{k2} + \cdots + A_{kk} \tag{5-2.5}$$

It will be noticed that in each of the expressions of Eq. (5-2.3) we have a sum of exponentials containing the decay constants of all the substances in the family preceding the one considered.

Some special cases of initial conditions are in practice very important, notably that in which only substance 1 is initially present and has $N_1(0)$ atoms initially. We have then, by a direct application of Eqs. (5-2.3) through (5-2.5),

$$N_1(t) = N_1(0)e^{-\lambda_1 t}$$

$$N_2(t) = N_1(0) \frac{\lambda_1}{\lambda_2 - \lambda_1} (e^{-\lambda_1 t} - e^{-\lambda_2 t})$$

$$N_3(t) = N_1(0)\lambda_1\lambda_2 \left[\frac{e^{-\lambda_1 t}}{(\lambda_2 - \lambda_1)(\lambda_3 - \lambda_1)} + \frac{e^{-\lambda_2 t}}{(\lambda_3 - \lambda_2)(\lambda_1 - \lambda_2)} \right.$$

$$\left. + \frac{e^{-\lambda_3 t}}{(\lambda_1 - \lambda_3)(\lambda_2 - \lambda_3)} \right] \tag{5-2.6}$$

This case occurs quite often, for example, in the active deposit of radium, where N_1, N_2, N_3 represent, respectively, the number of atoms of RaA, RaB, and RaC.

It is also important to see what happens to a mixture of radioactive substances left undisturbed for a long time. In Eq. (5-2.3) there will be terms containing the exponential with the smallest decay constant λ_s of the mixture. In a relatively short time all atoms of species 1, 2, . . . , $s - 1$ become atoms of species s, and N_s itself is given by $N_s(t) = \mathfrak{N}e^{-\lambda_s t}$, where $\mathfrak{N} = N_1(0) + \cdots + N_s(0)$. If t is large enough ($t \gg 1/\lambda_s$), and λ_{s+1}, λ_{s+2}, etc. are larger than λ_s, we have

$$N_1(t) = N_2(t) = N_{s-1}(t) = 0$$

$$N_s(t) = \mathfrak{N}e^{-\lambda_s t}$$

$$N_{s+1}(t) = \mathfrak{N} \frac{\lambda_s}{\lambda_{s+1} - \lambda_s} e^{-\lambda_s t} \tag{5-2.7}$$

---

[1] If two or more of the decay constants involved should accidentally be equal, Eq. (5-2.4) cannot be applied and Eq. (5-2.3) has to be modified by replacing the exponentials with equal λ with terms of the form $te^{-\lambda t}$, $t^2 e^{-\lambda t}$, etc.

ISBN 0-8053-8061-7

The ratio of the amount present of each of the substances following substance s to the amount of substance s itself is independent of time and is given by

$$\frac{N_t}{N_s} = \frac{\lambda_s \lambda_{s+1} \cdots \lambda_{t-1}}{(\lambda_{s+1} - \lambda_s)(\lambda_{s+2} - \lambda_s) \cdots (\lambda_t - \lambda_s)} \tag{5-2.8}$$

All substances decay according to the same time law, $e^{-\lambda_s t}$. When this situation obtains, we speak of *transient equilibrium*.

It sometimes happens that λ_s is very small compared with λ_t of all the substances following s in the radioactive family and that, during the interval under consideration, $e^{-\lambda_s t} \cong 1$. Then we write

$$N_t / N_s = \lambda_s / \lambda_t \tag{5-2.9}$$

and we speak of *secular equilibrium*. The interpretation of Eq. (5-2.9) is immediate: the number of atoms of substance s, $s + 1$, etc., disintegrating per unit time is the same, and hence the number of atoms present is inversely proportional to the decay constant. This situation occurs in ores containing uranium. The decay constant of ^{238}U is extremely small compared with those of the products following it in the radioactive families, and the ores have been undisturbed for a long enough time to attain radioactive equilibrium. For all practical purposes the number of atoms in such an ore is independent of time as long as we limit ourselves to periods of no more than a few thousand years, and Eq. (5-2.9) holds for the radioactive families. However, it must be pointed out that the inactive lead isotopes that terminate the radioactive families are not necessarily present in an amount corresponding to the age of the mineral, because geochemical processes may have removed either the lead or the uranium in a different way.

A brilliant experiment on radioactive equilibrium is the following: suppose that we have a solution containing ^{238}U in equilibrium with its daughter product UX_1 (^{234}Th), which has a period of 24.1 days. By a chemical procedure we separate, almost instantaneously, UX_1 from U. The UX_1 fraction then decays with the period 24.1 days. On the other hand, since this fraction and the mother solution, when considered together, must remain in equilibrium, there must grow in the mother solution an amount of UX_1 exactly equal to the amount disappearing from the separated fraction. This example was studied quantitatively by Rutherford and Soddy, and it helped to elucidate the theory of radioactive decay. The decay and growth curves (Fig. 5-4) were incorporated in Lord Rutherford's escutcheon.

Another important case occurring in practice is that of a radioactive substance (initially absent) formed at a constant rate. The differential equation for

ISBN 0-8053-8061-7

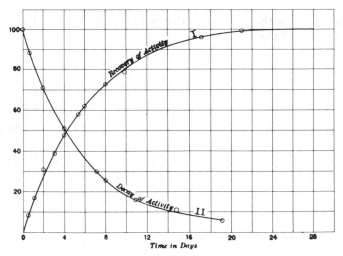

Figure 5-4 The original figure of decay of separated UX(^{234}Th) and of the recovery of the activity in the mother solution. [E. Rutherford.]

this process is

$$\frac{dN}{dt} = Q - \lambda N \tag{5-2.10}$$

where Q is the number of atoms formed per unit time. Its solution when $N(0) = 0$ is

$$N(t) = (Q/\lambda)(1 - e^{-\lambda t}) \tag{5-2.11}$$

where Q/λ represents the number of atoms that one would obtain for $t \to \infty$. It is sometimes called the *saturation number* of atoms and is indicated by N_∞.

5-3 BRANCHING

Several radioactive substances decay by more than one mechanism, for example, β^- and β^+ emission, alpha and beta emission. Let us consider alpha and beta branching and call the probability of alpha emission by one atom, in time dt, $\lambda_\alpha \, dt$ and the probability of beta emission by the same atom $\lambda_\beta \, dt$. Then the total probability of decay of the atom in time dt by either alpha or beta disintegration is $(\lambda_\alpha + \lambda_\beta) \, dt$. Hence

$$\frac{dN}{dt} = -(\lambda_\alpha + \lambda_\beta)N \tag{5-3.1}$$

ISBN 0-8053-8061-7

and the mean life of the substance, defined as the time in which the amount initially present is reduced from 1 to $1/e$, is

$$\tau = \frac{1}{\lambda_\alpha + \lambda_\beta} \tag{5-3.2}$$

The ratio of the number of alpha to beta particles emitted during a certain time is called *branching ratio* and is equal to $\lambda_\alpha/\lambda_\beta$. Sometimes the quantities $\tau_\alpha = 1/\lambda_\alpha$ and $\tau_\beta = 1/\lambda_\beta$ are used; these are called, not too accurately, the mean life for alpha and beta decay, respectively. From Eq. (5-3.2) it follows immediately that

$$\frac{1}{\tau} = \frac{1}{\tau_\alpha} + \frac{1}{\tau_\beta} \tag{5-3.3}$$

A similar terminology is used for other types of branching. Figure 5-2 gives some examples of alpha and beta branching.

5-4   SOME UNITS USED IN RADIOACTIVITY; DOSIMETRY

The number of atoms disintegrating per second, λN, in a given sample is often called the activity of the sample. Activities are generally measured in curies (abbreviated Ci). A sample undergoing 3.7×10^{10} disintegrations per second is said to have the activity of 1 Ci. The millicurie (mCi) and the microcurie (μCi) are 10^{-3} and 10^{-6} Ci, respectively. The origin of the number 3.7×10^{10} is as follows: In the early days of research on radioactivity the unit of activity was the amount of substance in equilibrium with 1 g of Ra and was called the curie in honor of P. and M. Curie. The unit was practical and convenient as long as all radioactive substances belonged to the Ra family. Later an attempt was made to determine the number of disintegrations per second in 1 Ci of a substance and, in particular, in 1 g of Ra. The results of many measurements gave approximately 3.7×10^{10} disintegrations per second. The definition of the curie was then changed to the number of disintegrations per second so as to extend its applicability to substances that do not belong to the radium family.

In many cases what is important is not the activity of a source but the ionization it produces under specified conditions of filtration of the radiation, etc. In this sense one may say that two sources are "equivalent" if they produce the same ionization in the specified conditions. From this point of view it is possible to measure the ionizing action of a radioactive source emitting gamma rays by specifying that it produces a certain number of *roentgens* per hour at a distance of 1 m (Rhm).

The roentgen (R) is defined as "that quantity of X or gamma radiation such that the associated corpuscular emission per 0.001293 g of air produces in air ions carrying 1 esu of quantity of electricity of either sign." The mass of

ISBN 0-8053-8061-7

air referred to is the mass of 1 cm^3 of dry air at 0°C and 1 atm pressure. If we assume that an average of 32.5 eV is expended to produce a pair of ions in air, we find that 1 R corresponds to the absorption of

$$\frac{32.5 \times 1.60 \times 10^{-12}}{4.80 \times 10^{-10}} = 0.108 \text{ erg cm}^{-3} \text{ of air} \tag{5-4.1}$$

or

$$\frac{0.108}{1.293 \times 10^{-3}} = 83.8 \text{ erg g}^{-1} \text{ of air} \tag{5-4.2}$$

or 6.77×10^4 MeV cm^{-3} of air.

The ionization produced by a certain number of X ray photons crossing a given volume of air depends on the rays' energy. In Fig. 5-5 we show the

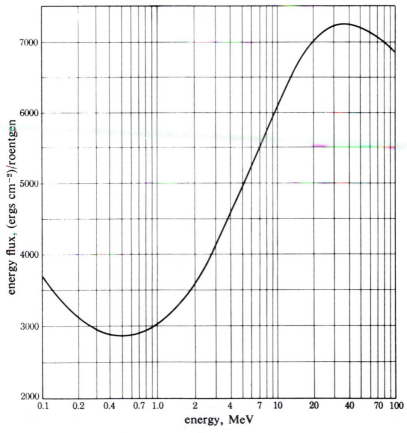

Figure 5-5   Energy flux per roentgen as a function of the energy of gamma radiation. [(Se 59).]

energy flux corresponding to a given ionization as a function of the energy of the gamma rays.

The roentgen is used in radiobiology and health physics as a unit of *exposure dose of X and γ rays.*

The effects of radiation, the biological effects in particular, are related to the amount of energy *deposited* by the radiation in the material under study (the tissue, in the case of biology).

The quantity of energy absorbed is referred to as the *radiation absorbed dose.* The universally used unit of absorbed dose is now the *rad.* An absorbed dose of 1 rad corresponds to the absorption of 100 ergs g^{-1} in the material under study. A radiation exposure of 1 R gives 0.83 rad in air according to Eq. (5-4.2). The same exposure for a substance with a different absorption coefficient would give a dose in rads proportional to the mass absorption coefficient. Soft tissue, air and water have comparable mass absorption coefficients for X rays.

As a rule of thumb, fluxes of 2×10^9 cm^{-2} 1 MeV photons or fluxes of 3×10^7 cm^{-2} minimum ionizing particles deposit 1 rad in crossing 1g cm^{-2} of carbon.

The energy absorbed within a given mass of material can rarely be measured directly. In special cases calorimetric methods have been developed, but in general, for biological materials and relatively small exposure doses, these are impracticable. However, the absorbed dose in a material is proportional to the ionization produced in the same mass and a convenient method of determining energy absorption in a material is the indirect one of measuring the ionization produced in a small gas-filled cavity in the material. The ionization produced in a cavity inside the material is related to the energy dissipated into the material by the Bragg–Gray relation, which can be written in the form

$$J = E/\rho w \qquad (5\text{-}4.3)$$

Here J is the number of ion pairs produced per unit volume of the gas, ρ is the ratio of the stopping power of the walls to the stopping power of the gas for the equilibrium radiation, E is the energy absorbed per unit volume of the walls, and w is the energy required to produce one ion pair. This relation is valid if the linear dimensions of the cavity are small compared with the range (in the gas filling the cavity) of the secondary corpuscular radiation (electrons) produced in the wall of the cavity. For a chamber where the gas and the walls have the same composition, ρ is simply the ratio of the densities of the wall and gas.

The biological effects of radiation are a function not only of the energy absorbed in the tissue but also of the type of radiation delivering that dose, the rate, and the organ receiving it. Whereas the measurement of the energy deposited by the radiation is a relatively simple problem in physics, the evaluation of its biological effects is a much more difficult one and is not even very well defined. For biological purposes, one introduces the concept of *relative biological effectiveness* (RBE). The RBE of a given radiation is the ratio of the absorbed dose of this radiation to the absorbed dose of a standard radiation (generally X rays of 200 kV peak) that produces the same biological effect on the organ under study.

$$(RBE)_{n,\,a} = D_x / D_n \qquad (5\text{-}4.4)$$

where $(RBE)_{n,\,a}$ is the relative biological effectiveness of a radiation n for the biological effect a and D_n and D_x are the absorbed doses of radiation n and of the standard radiation that produce the same amount of biological effect a. It is clear that there will be a different RBE for every type of radiation and every type of biological effect considered. The product of the absorbed dose multiplied by its RBE is called the *dose equivalent* (DE) and is measured in rems (rad equivalent man).

The RBE varies not only with the type of radiation but also with its energy; e.g., the RBE for thermal neutrons is different from the RBE of 1-MeV neutrons for the same biological effect. RBE values ranging from 1 to 20 and more have been measured. The RBEs are related to the way the radiation deposits energy at a microscopic level.

Another quantity used in dosimetry is the *linear energy transfer* (LET) of an ionizing radiation: it is the energy deposited per unit path length and it is measured in kiloelectron volts per micron. If one does not limit the deposition of energy to a small volume along the track, the energy absorbed along the track coincides with the energy lost and the LET coincides with the *specific energy loss*. Further biological studies have shown that the rate at which the dose is delivered, the geometry of the radiation, and other factors can influence the effects of a given radiation.

For radiation protection purposes the International Commission on Radiological Protection (ICRP) has recommended the use of the dose equivalent for quantifying the *risk* from radiation. The dose equivalent is calculated in practice as the product of the absorbed dose times the *quality factor* (QF) of the radiation. The QF is a dimensionless quantity that has been univocally related to the average LET of the radiation. In Table 5-1 are listed some recommends QF values. The ICRP also recommends, for protection purposes, maximum permissible doses (MPD); in Table 5-2 we show the present MPD values for external radiations. Their significance is that, given the present state of our knowledge of radiation effects, a person exposed to radiation within these limits has a negligible probability of suffering somatic or genetic injuries.

The rule of thumb number to remember for the maximum permissible occupational dose over the whole body is 5 rems yr^{-1} (100 mrems week^{-1}). The natural background at sea level and a latitude of about 50° is about 125 mrems yr^{-1} divided into 25 mrems yr^{-1} due to cosmic rays (charged particles

ISBN 0-8053-8061-7

TABLE 5-1 RECOMMENDED VALUES OF QF FOR DIFFERENT TYPES OF RADIATION[a]

LET $keV\mu^{-1}$ of water	Type of radiation	QF
3.5 or less	X, γ, and β rays of $E_{max} > 30$ keV	1
3.5–7	Very high-energy p and n	1–2
7–23	Neutron from thermal up to $\sim$ 20 keV	2–5
23–53	Fast neutrons and protons up to $\sim$ 10 MeV	5–10
53–175	Alpha particles and heavy recoil nuclei	10–20

[a] From ICRP Publication 9 (1966). In practice, rad × QF is often used in lieu of rems.

TABLE 5-2   SUMMARY OF DOSE LIMITS FOR INDIVIDUALS[a]

Organ or tissue	Maximum permissible doses for adults exposed in the course of their work rems in 1 yr	Dose limits for members of the public, rems in 1 yr
Gonads, red bone marrow[b]	5[c]	0.5
Skin, bone, thyroid	30	3
Hands and forearms; feet and ankles	75	7.5
Other single organs	15	1.5

[a] Values not to be exceeded if personal safety is to be assured.
[b] In the case of uniform irradiation, this also represents the whole body.
[c] Subject to the limitation that no more than 3 rems may be accumulated in any period of a quarter of a year. In special cases it will be justifiable to permit the quarterly quota to be repeated in each quarter of the year provided that the total dose accumulated at any age over 18 years does not exceed $5(N - 18)$ rems where N is the age in years.

and neutrons), an equal amount to gamma rays of cosmic origin, and 75 mrems yr^{-1} of radioactive origin from air and rocks.

The cosmic charged particles give about 1.7 ion pairs cm^{-3} sec^{-1} and in a counter about 1 count min^{-1} cm^{-2} sr^{-1}.

5-5   FLUCTUATIONS IN RADIOACTIVE DECAY—GENERAL THEORY

At the beginning of this chapter we spoke of radioactive decay from the point of view of a continuous change in the number of atoms. Since this

ISBN 0-805-38601-7

number is obviously an integer, it is clear that the theory previously given is only approximate. Although it holds exactly for the average values in the limit of a very large number of atoms, we may expect departures from it in cases in which the actual integral numbers of atoms or events are considered. It is the purpose of this section to treat the fluctuations or differences between the actual number of decaying atoms n and the average number $\langle n \rangle$.

The first problem is the following: We have a substance with an extremely long life (so long that we can neglect its average decay in the time considered) and emitting on the average $\langle n \rangle$ particles per second. What is the probability $P(n)$ that in a given second it emits n particles? This classical problem in the theory of probability is solved by *Poisson's formula*:

$$P(n) = (\langle n \rangle^n / n!)e^{-\langle n \rangle} \tag{5-5.1}$$

To prove this equation, we divide the time interval of 1 sec into K equal parts. K is an arbitrarily large number.

The probability of a disintegration occurring in any one of the K subintervals is then $\langle n \rangle / K$, and the probability that no disintegration occurs in any specified interval is $1 - (\langle n \rangle / K)$. The probability of a disintegration occurring in n, and only n, *specified* subintervals is

$$\left(\frac{\langle n \rangle}{K} \right)^n \left(1 - \frac{\langle n \rangle}{K} \right)^{K-n} \tag{5-5.2}$$

The first factor in Eq. (5-5.2) results from the requirement that n specified subintervals contain a disintegration and the second from the requirement that the remaining intervals $(K - n)$ do not contain a disintegration. If we now abandon the condition that the disintegrations must occur in *specified* subintervals and retain only the requirement that they must occur, we must sum Eq. (5-5.2) for all possible choices of the specified subintervals. These are equal in number to the combinations of K objects taken n at a time:

$$\binom{K}{n} = \frac{K(K-1), \cdots, (K-n+1)}{n!} = \frac{K!}{n!(K-n)!} \tag{5-5.3}$$

We thus have

$$P_K(n) = \binom{K}{n} \left(\frac{\langle n \rangle}{K} \right)^n \left(1 - \frac{\langle n \rangle}{K} \right)^{K-n} \tag{5-5.4}$$

This formula (Bernoulli distribution) can be reinterpreted in a more general way. We repeat many times a certain experiment that must have one of two results, E or F, which are mutually exclusive. The probability of E occurring in each trial is p; the probability of F occurring is $1 - p = q$. What is the probability, in K trials, that E occurs n times and hence that F occurs $K - n$

ISBN 0-8053-8061-7

times? The answer, as seen above, is

$$P_K(n) = \binom{K}{n} p^n q^{K-n} \tag{5-5.5}$$

Note that $P_K(n)$ is the term containing $p^n q^{K-n}$ in the binomial development of $(p + q)^K$, and hence

$$\sum_{n=0}^{K} P_K(n) = (p + q)^K = 1 \tag{5-5.6}$$

Equation (5-5.6) shows that the probabilities $P_K(n)$ are normalized correctly.

Before considering certain limiting cases of formula (5-5.3), we shall calculate the average $\langle n \rangle$ of the number of times that event E occurs in a series of K trials. Intuitively we expect it to be $\langle n \rangle = Kp$, and this is borne out by direct calculation. By definition,

$$\langle n \rangle = \sum_{n=0}^{K} nP_K(n) = \sum_{n=0}^{K} n \binom{K}{n} p^n q^{K-n} \tag{5-5.7}$$

On the other hand,

$$\frac{\partial}{\partial p}(p + q)^K = \frac{\partial}{\partial p} \sum_{n=0}^{K} \binom{K}{n} p^n q^{K-n} = \sum_{n=0}^{K} n \binom{K}{n} p^{n-1} q^{K-n} \tag{5-5.8}$$

and hence, upon using Eq. (5-5.7),

$$\langle n \rangle = p \frac{\partial}{\partial p}(p + q)^K = pK(p + q)^{K-1} = Kp \tag{5-5.9}$$

because $p + q = 1$.

We now calculate the average value of $(n - \langle n \rangle)^2$, i.e., $\langle (n - Kp)^2 \rangle$, which is called the *variance* by statisticians. Its square root is the *standard deviation*, indicated by σ. We have first

$$\sigma^2 = \langle (n - pK)^2 \rangle = \langle n^2 \rangle - 2pK\langle n \rangle + p^2 K^2$$

$$= \langle n^2 \rangle - p^2 K^2 = \langle n^2 \rangle - \langle n \rangle^2 \tag{5-5.10}$$

and $\langle n^2 \rangle$ is evaluated, in a way similar to that used above for $\langle n \rangle$, as follows:

$$\langle n^2 \rangle = \sum_{n=0}^{K} n^2 P_K(n) = \left(p^2 \frac{\partial^2}{\partial p^2} + p \frac{\partial}{\partial p} \right) \sum_{n=0}^{K} \binom{K}{n} p^n q^{K-n}$$

$$= \left(p^2 \frac{\partial^2}{\partial p^2} + p \frac{\partial}{\partial p} \right) (p + q)^K = K(K - 1)p^2 + Kp \tag{5-5.11}$$

ISBN 0-8053-8061-7

from which, remembering that $q = 1 - p$, we get

$$\sigma^2 = \langle n^2 \rangle - p^2 K^2 = - Kp^2 + Kp = Kpq \qquad (5\text{-}5.12)$$

or

$$\sigma = (Kpq)^{1/2} \qquad (5\text{-}5.13)$$

The last expression for the standard deviation is of considerable importance.

We now make use of the fact that K is an arbitrarily large number, and we pass to the limit for K tending to infinity. We apply the well-known formulas

$$\lim_{K \to \infty} \binom{K}{n} = \frac{K^n}{n!} \qquad \text{and} \qquad \lim_{K \to \infty} \left(1 + \frac{1}{K}\right)^K = e$$

or

$$\lim_{K \to \infty} \left(1 - \frac{\langle n \rangle}{K}\right)^{K - \langle n \rangle} = e^{\langle -n \rangle} \qquad (5\text{-}5.14)$$

Substituting these values in Eq. (5-5.4), we have

$$P(n) = \frac{\langle n \rangle^n}{n!} e^{-\langle n \rangle} \qquad (5\text{-}5.15)$$

where for $K \to \infty$ we have dropped the index K. This is the famous Poisson formula; it is illustrated in Fig. 5-6. From Eq. (5-5.15), remembering the development in power series of e^n, we also see immediately that

$$\sum_{n=0}^{\infty} P(n) = e^{-\langle n \rangle} \sum_{n=0}^{\infty} \frac{\langle n \rangle^n}{n!} = 1 \qquad (5\text{-}5.16)$$

which verifies that the sum of the probabilities for all possible numbers of disintegrations in a given time interval is 1.

If $\langle n \rangle$ is a large number, $P(n)$ has a sharp maximum in the vicinity of $n = \langle n \rangle$ and we may develop $\log P(n)$ in a power series of $n - \langle n \rangle$. Using *Stirling's asymptotic formula*, $\log(x!) = x \log x - x + \frac{1}{2} \log 2\pi x + \cdots$, we obtain for Eq. (5-5.15)

$$\log P(n) = n \log \langle n \rangle - \langle n \rangle - n \log n + n - \frac{1}{2} \log 2\pi n \qquad (5\text{-}5.17)$$

and, taking the first and second derivatives with respect to n,

$$\frac{d \log P}{dn} = \log \langle n \rangle - \log n - \frac{1}{2n} \qquad (5\text{-}5.18)$$

$$\frac{d^2 \log P}{dn^2} = -\frac{1}{n} + \frac{1}{2n^2} \qquad (5\text{-}5.19)$$

ISBN 0-8053-8061-7

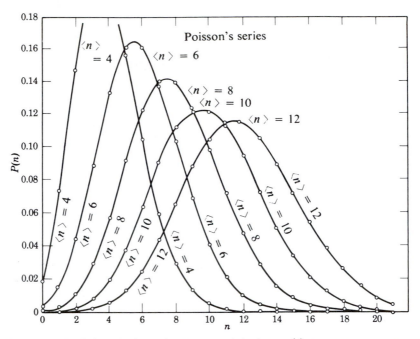

Figure 5-6 Poisson's formula: $P(n) = (\langle n \rangle^n / n!) e^{-\langle n \rangle}$ for different values of $\langle n \rangle$.

The zero of the first derivative confirms the existence of the maximum[1] near $n = \langle n \rangle$ (neglecting terms in $1/n$ with respect to $\log n$ and terms in $1/n^2$ with respect to terms in $1/n$); the second derivative can be used to write the first terms of the power series for $\log P(n)$,

$$\log P(n) = \log P(\langle n \rangle) + \frac{(n - \langle n \rangle)^2}{2!} \left[\frac{d^2}{dn^2} \log P(n) \right]_{n = \langle n \rangle}$$

$$= -\frac{1}{2} \log 2\pi \langle n \rangle - \frac{(n - \langle n \rangle)^2}{2 \langle n \rangle} \tag{5-5.20}$$

from which, passing from logarithms to numbers, we immediately get the famous *Gauss formula*,

$$P(n) = \frac{\exp\left[-(n - \langle n \rangle)^2 / 2 \langle n \rangle \right]}{(2\pi \langle n \rangle)^{1/2}} \tag{5-5.21}$$

(Examples of Poisson and Gauss distributions are shown in Fig. 5-7.) Recall-

---

[1] In a better approximation the maximum is near $n = \langle n \rangle - \frac{1}{2}$.

ISBN 0-805-38001-7

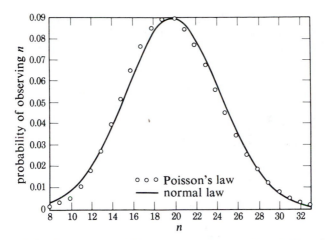

Figure 5-7 Comparison of Poisson's law for $n = 20$ with Gauss's law.

ing that

$$\int_{-\infty}^{\infty} e^{-x^2} dx = \pi^{1/2} \tag{5-5.22}$$

we readily see that

$$\int_{0}^{\infty} P(n) dn = 1 \tag{5-5.23}$$

which again expresses the normalization of the probability. Here we remember that $\langle n \rangle \gg 1$, and that the contributions to the integral for $n < 0$ are therefore negligible.

From Eq. (5-5.13), (5-5.15), or (5-5.21) it is possible to compute the root-mean-square (rms) deviation or standard deviation defined by Eq. (5-5.10). From Eq. (5-5.13), if K tends to ∞, p tends to 0 and q to 1; then Eqs. (5-5.13) and (5-5.9) give directly

$$\sigma^2 = Kp = \langle n \rangle \tag{5-5.24}$$

The same result is obtained if we insert for $P(n)$ the Poisson formula and use the identity

$$\sum_{n=0}^{\infty} n^2 \frac{\langle n \rangle^n}{n!} = \langle n \rangle (\langle n \rangle + 1) e^{\langle n \rangle} \tag{5-5.25}$$

which can be easily proved by comparing coefficients of equal powers of $\langle n \rangle$. Finally, the same result comes from the definition of σ^2 contained in Eqs. (5-5.10) and (5-5.11) and the Gaussian distribution [Eq. (5-5.21)] if we remember that

$$\int_{-\infty}^{\infty} x^2 e^{-x^2} dx = \tfrac{1}{2} \pi^{1/2} \tag{5-5.26}$$

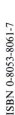ISBN 0-8053-8061-7

These expressions say that the average value of the square of the deviation from the average number of disintegrations to be expected in a certain time interval is equal to this same number, and hence

$$\sigma = \langle n \rangle^{1/2} \tag{5-5.27}$$

is known as the standard deviation of the distribution in n. Thus, if we have on the average 100 disintegrations per minute, we may expect deviations from this average, in every 1-min interval, such that the mean square of the deviations is again 100. Very often it is of great interest to know the relative standard deviation that results from Eq. (5-5.27),

$$\sigma / \langle n \rangle = 1 / \langle n \rangle^{1/2} \tag{5-5.28}$$

The relative standard deviation is inversely proportional to the square root of the number of counts used and is a measure of the precision of the determination.

One can also arrive at an estimate of σ by starting from its definition, Eq. (5-5.10), in which, however, we have not distinguished $\langle n \rangle$ from the "true" value of $\langle n \rangle$, which would be obtained from, for instance, an infinite number of observations. Let us indicate the true value of $\langle n \rangle$ by $\langle n_t \rangle$. In a sample giving n_i counts in each of N equally long time intervals, we should actually form

$$\sigma^2(n) = \frac{1}{N} \sum_{i=1}^{N} (n_i - \langle n_t \rangle)^2 \tag{5-5.29}$$

Since the value $\langle n_t \rangle$ is actually unobtainable, we replace it with

$$\langle n \rangle = \frac{1}{N} \sum_{i=1}^{N} n_i \tag{5-5.30}$$

and we assume, for the square of the standard deviation of the single measurement,

$$\sigma^2(n) = \frac{1}{N-1} \sum_{i=1}^{N} (n_i - \langle n \rangle)^2 \tag{5-5.31}$$

and, for the square of the standard deviation of the average,

$$\sigma^2(\langle n \rangle) = \frac{1}{N(N-1)} \sum_{i=1}^{N} (n_i - \langle n \rangle)^2 \tag{5-5.32}$$

The replacement of N in Eq. (5-5.29) by $N-1$ in Eq. (5-5.31) compensates for the fact that $\langle n_t \rangle$, the true value, has been replaced by $\langle n \rangle$. For large values of N, Eqs. (5-5.31) and (5-5.29) tend to coincide, and it is not actually

ISBN 0-8053-8061-7

important to distinguish between them, because as a matter of fact, the difference between them is of the order of magnitude of the error in σ^2. Justification of Eq. (5-5.31) and of the last statement will be found in standard books on probability and statistics.

Of course, the values of σ^2 thus obtained are themselves only approximate; we may have examples in which σ^2 as calculated by Eq. (5-5.31) comes out anomalously small. In fact, σ cannot be less, for our case, than $\langle n \rangle^{1/2}$; if our series of measures has given a lower value than this, we still must assume that the standard deviation is $\langle n \rangle^{1/2}$ and that we have obtained the lower value by accident. On the other hand, if we have found a standard deviation much larger than $\langle n \rangle^{1/2}$ we may suspect the existence of some cause of accidental errors.

We shall now illustrate these concepts by an actual example. A uranium sample was counted, and the number of alpha particles emitted in 1 min was recorded in 10 intervals, each of 1-min duration. The first column in Table 5-3 shows the actual number of counts observed; the second column, the difference between the average number of counts and the specific 1-min count; and the third column, the square of the numbers written in the second column. The average count for a 1-min period is 35 946.4, with the standard deviation of any single measurement given by $(311740/9)^{1/2} = 186$. If, according to Eq. (5-5.27), $\sigma = \langle n \rangle^{1/2}$, we find $\sigma = 189$, a satisfactory agreement. The average counting rate is 35 946.4 counts $\min^{-1}$, and the standard deviation of the average is given by $189/(10)^{1/2} = 60$, where 10 is the number of observations and the standard deviation of one observation is 189. In conclusion, the average counting rate of our sample is $35\ 946 \pm 60$ counts $\min^{-1}$.

TABLE 5-3   NUMERICAL EXAMPLE OF COUNTING OF A SAMPLE

n	$n - \langle n \rangle$	$(n - \langle n \rangle)^2$
36 076	130	16 900
35 753	− 193	37 249
35 907	− 39	1 521
36 116	170	28 900
35 884	− 62	3 844
36 136	190	36 100
35 741	− 205	42 025
35 640	− 306	93 636
36 124	178	31 684
36 087	141	19 881
$\langle n \rangle = 35\ 946.4$	$\Sigma 4$	$\Sigma 311\ 740$
	$\sigma(n) = (311\ 740/9)^{1/2} = 186$	

The question may be asked: What is the probability that, in the case of a counting rate A with a standard deviation σ, the deviation from the "exact" result $A*$ will be larger than $\epsilon\sigma$? The exact result is defined, as above, as the one obtained by repeating the measurement of A a very great number of times. The answer to this problem in the case of a Gaussian distribution is given in tables contained, for instance, in the *Handbook of Chemistry and Physics*. We quote here a few pertinent numbers.

The probability P of observing a counting rate differing from $A*$ by more than $\epsilon\sigma$ is given in Table 5-4. It will be noted that the probability of observing a counting rate differing from $A*$ by more than 0.6745σ is 0.5. This quantity is called the *probable error*.

TABLE 5-4   PROBABILITY OF A GIVEN MULTIPLE OF THE STANDARD DEVIATION

ϵ	0	0.6745	1	1.5	2.0	2.5	3	3.5	4
P	1	0.5000	0.3173	0.1336	0.0455	0.0124	0.0027	0.00046	0.000063

If we want to know the standard deviation of a function f of two or more independently observed quantities x_1 and x_2 affected by the standard deviation σ_1 and σ_2, we use the fundamental formula of the *propagation of errors*,

$$\sigma_f = \left[\left(\frac{\partial f}{\partial x_1} \right)^2 \sigma_1^2 + \left(\frac{\partial f}{\partial x_2} \right)^2 \sigma_2^2 \right]^{1/2} \tag{5-5.33}$$

Of frequent application are the formulas for the standard deviations of the sum, difference, product, and quotient of two quantities. If the observed quantities are numbers of counts, Eqs. (5-5.27) and (5-5.33) give

$$\sigma(n_1 \pm n_2) = (n_1 + n_2)^{1/2} \tag{5-5.34}$$

$$\sigma(n_1 n_2) = \left[n_1 n_2 (n_1 + n_2) \right]^{1/2} \tag{5-5.35}$$

$$\sigma\left(\frac{n_1}{n_2} \right) = \left(\frac{n_1}{n_2^2} + \frac{n_1^2}{n_2^3} \right)^{1/2} \tag{5-5.36}$$

and

$$\frac{\sigma(n_1 n_2)}{n_1 n_2} = \frac{\sigma(n_1/n_2)}{n_1/n_2} = \left(\frac{1}{n_1} + \frac{1}{n_2} \right)^{1/2} \tag{5-5.37}$$

where n_1 and n_2 are the observed numbers of counts. Expressions of the type of Eq. (5-5.36) are most rapidly obtained by use of the relation

$$\sigma(\log f) = \sigma(f)/f \tag{5-5.38}$$

which is an immediate consequence of Eq. (5-5.33).

5-6   FLUCTUATIONS IN RADIOACTIVE DECAY—APPLICATIONS

We shall now make some applications of the principles stated above, either in order to illustrate them or because the results are of practical importance:

1. Suppose that we have counted a sample that gives n_1 counts in a certain counter in time t_1; the background of the counter gave n_2 counts in time t_2. We want to know the counting rate of the sample and the background counting rate of the counter. The times t_1 and t_2 are exactly known. For the background we have

$$\nu_2 = \frac{n_2}{t_2} \pm \frac{n_2^{1/2}}{t_2} \tag{5-6.1}$$

(The expressions following the $\pm$ are always standard deviations.)
We have, for the activity plus the background,

$$\nu_1 = \frac{n_1}{t_1} \pm \frac{n_1^{1/2}}{t_1} \tag{5-6.2}$$

for the net activity,

$$\nu = \nu_1 - \nu_2 \tag{5-6.3}$$

and, for its standard deviation according to Eq. (5-5.33),

$$\sigma(\nu) = \left[\sigma^2(\nu_1) + \sigma^2(\nu_2)\right]^{1/2} = \left(\frac{n_1}{t_1^2} + \frac{n_2}{t_2^2}\right)^{1/2} = \left(\frac{\nu_1}{t_1} + \frac{\nu_2}{t_2}\right)^{1/2} \tag{5-6.4}$$

We may ask: What is the best way to apportion a fixed counting time $T = t_1 + t_2$ between the counting of the sample and the counting of the background? From Eq. (5-6.4) we obtain

$$\sigma^2(\nu) = \frac{\nu_1}{t_1} + \frac{\nu_2}{T - t_1} \tag{5-6.5}$$

and, by minimizing with respect to t_1,

$$\frac{\nu_1}{\nu_2} = \frac{t_1^2}{t_2^2} \tag{5-6.6}$$

2. We want to check the existence of an effect that slightly changes the counting rate of a device. When can we reasonably conclude that the effect is real?

Counts in conditions a, n_1; counts in conditions b, n_2; with counting time the same for both conditions and equal to t.

$$\nu_1 - \nu_2 = (n_1 - n_2)/t \tag{5-6.7}$$

$$\sigma(\nu_1 - \nu_2) = (n_1 + n_2)^{1/2}/t \tag{5-6.8}$$

The effect is probably real if

$$3\sigma(\nu_1 - \nu_2) \leqslant (\nu_1 - \nu_2)$$

or

$$n_1 - n_2 \geqslant 3(n_1 + n_2)^{1/2} \tag{5-6.9}$$

More precisely the probability P that $|\nu_1 - \nu_2| > \epsilon\sigma(\nu_1 - \nu_2)$ is given by Table 5-4. For $\epsilon = 3$ we have $P = 0.0027$; hence, the probability that the effect is not due to a statistical fluctuation is $1 - P = 0.9973$. This last number is often called the *confidence level* of the result.

3. A certain sample gives ν counts sec^{-1} on the average. What is the probability of finding an interval of t sec without counts? Equation (5-5.15) gives the answer if we make $n = 0$ and $\langle n \rangle = \nu t$,

$$P(0) = e^{-\nu t} \tag{5-6.10}$$

4. *Counting-loss problem.* Assume that a counting device is unable to register a pulse for a period τ after having registered a previous pulse (dead time). If we try to measure a sample giving rise to an average true counting rate ν_0, the device will show a counting rate $\nu < \nu_0$, because some of the pulses will occur during the dead time τ following each registered pulse and hence will not be registered. We want the relation between ν and ν_0. The total dead time per second is $\nu\tau$; the counts that should occur during this time but that cannot be registered are $\nu_0\nu\tau$, and from this

$$\nu = \nu_0 - \nu_0\nu\tau \tag{5-6.11}$$

or, upon assuming $\nu_0 - \nu \ll \nu$,

$$\nu_0 = \nu(1 + \nu\tau) \tag{5-6.12}$$

The practical procedure for correcting counting losses is to count two or more samples separately and then count them together. The count obtained from the sum of the samples is smaller than the sum of the counts obtained from the single samples because of the counting losses, and if Eq. (5-6.12) applies, the counting loss $\nu_0 - \nu$ is small compared with ν and the proportionality constant is determined empirically as stated above. It is not safe to correct for

ISBN 0-8053-8061-7

counting losses if these amount to more than about 20% of the counting rate; it is better to subdivide the samples, reduce the solid angle of the detecting device, or otherwise arrange to use lower counting rates. If we go to the other limiting case of a pulse rate $\nu_0 \gg 1/\tau$, the apparatus is blocked and the counting rate decreases with increasing ν_0 or it counts at a counting rate $1/\tau$ according to its detailed construction. Under these conditions the apparatus is obviously unsuitable for making measurements.

   5. *Fluctuations in ionization current.* Let us consider an ionization chamber containing a sample that emits alpha particles each of which gives a potential v to the collecting electrode. On the average ν particles per second are emitted, and we take a reading of the potential of the electrode every T sec. We compensate for the average charging effect of the source by using a leak resistor or by balancing it with another ionization chamber in opposition and an identical alpha source. The average potential reading $\langle p \rangle$ is then zero, but there are residual fluctuations of potential. We can calculate $\langle p^2 \rangle$ from Eq. (5-5.24). Calling n the actual number of particles emitted in time T and νT its average value, we have from Eq. (5-5.10)

$$\langle (p - \langle p \rangle)^2 \rangle = \langle (n - \nu T)^2 v^2 \rangle = v^2 \nu T \tag{5-6.13}$$

or, since $\langle p \rangle = 0$, owing to the compensating device,

$$\langle p^2 \rangle = v^2 \nu T \tag{5-6.14}$$

if the main effect is compensated with a resistor; or

$$\langle p^2 \rangle = 2v^2 \nu T \tag{5-6.15}$$

if we have two balanced chambers. In this equation $\langle p^2 \rangle$ and T are directly measurable, and an experiment of this type can be used to determine the quantity $v\nu$.

   Equations (5-6.14) and (5-6.15) are mainly qualitative because they do not consider the differences between the amounts of ionization produced in the various ionization acts and the electrical characteristics of the measuring instrument.

   The fluctuations of p have been the object of several studies in which the influence of the electrical characteristics of the detecting apparatus has also been considered. Although most of these studies were performed with long-period electrometers and experimental devices that are now obsolete, the method of evaluating the results has wide applicability. If a particle emitted at time 0 produces in the instrument a deflection at time t indicated by $f(t)$, where the specific function depends on the characteristics of the apparatus, and if the apparatus is linear, the deflection at time T produced by k particles

ISBN 0-8053-8061-7

emitted at times $t_1, t_2, t_3, \ldots$ is given by

$$\phi = \sum_k f(T - t_k) \tag{5-6.16}$$

From the deflection of the apparatus at time T in a series of many experiments, we can obtain an average deflection $\langle \phi \rangle$. To calculate this quantity and its standard deviation, call ν the average rate of emission of the particles. In a small time interval τ the probability of emission of a particle is $\nu\tau$.

The contribution $\delta\langle \phi \rangle$ to the average deflection due to particles emitted in time τ is

$$\delta\langle \phi \rangle = f(T - t)\nu\tau \tag{5-6.17}$$

and, upon passing to the limit,

$$\langle \phi \rangle = \sum \delta\langle \phi \rangle = \int_0^T \nu f(T - t)dt \tag{5-6.18}$$

The standard deviation of $\langle \phi \rangle$ is obtained by using the theorem of the calculus of probability contained in Eq. (5-5.13), which states that the square of the standard deviation of the number $\nu\tau$ is $\nu\tau$ multiplied by the complementary probability $1 - \nu\tau$. Moreover, since the number of particles emitted in the single time intervals τ are statistically independent, the contributions to the square of the standard deviation from the various time intervals are to be added. We have then for the standard deviation of the deflection

$$(\phi - \langle \phi \rangle)^2 = \sum \nu\tau(1 - \nu\tau) f^2(T - t) \tag{5-6.19}$$

or, upon passing to the limit for $\tau \to 0$,

$$\langle (\phi - \langle \phi \rangle)^2 \rangle = \int_0^T \nu f^2(T - t) \, dt \tag{5-6.20}$$

Equation (5-6.20) can be applied to specific cases; e.g., if $f(t) = 0$ for $t < 0$ and $f(t) = v e^{-\gamma t}$ for $t > 0$, we have, for $T \gg 1/\gamma$,

$$\langle (\phi - \langle \phi \rangle)^2 \rangle = v^2 \nu / 2\gamma \tag{5-6.21}$$

which is similar to Eq. (5-6.14).

Equation (5-6.21) is often applied in the determination of the conditions under which a heavily ionizing radiation (e.g., fission fragments) can be measured over a background of an intense radiation the single particles of

ISBN 0-8053-8061-7

which ionize only slightly (beta rays). If the resolving time of the instrument, including the effect of the electronic components, is τ and the number of ions obtainable from the heavily ionized particle is F, whereas each of the weakly ionizing particles, occurring with frequency ν, produces f ions, we must have

$$F \gg (\nu\tau)^{1/2} f \qquad (5\text{-}6.22)$$

This relation, however, can be used only for qualitative estimates, because a detailed study of the experimental setup is required for more precise evaluations.

5-7   METHOD OF MAXIMUM LIKELIHOOD

Often, in order to measure a half-life, we measure the time at which the nuclei in a sample disintegrate, and then analyze the data by plotting a histogram of the number of decays in successive equal time intervals. This practical procedure does not extract all the information available in the experiment, and one wants a method that puts to best use all the information collected. This purpose is achieved by the method of maximum likelihood, which we shall illustrate by an example. We must emphasize, however, that in order to use it we must make an a priori hypothesis that drastically affects the results. For instance, in the following example it is essential to assume that we have only one radioactive substance present and that the decay law is a simple exponential.

Under these hypotheses, having observed M disintegrations at times $t_1, t_2, \ldots, t_M \leqslant T = $ total time of observation, we want to know the most probable value of the decay constant λ of the substance. We proceed as follows: For each disintegrating nucleus of the substance the likelihood of observing a decay between time t and $t + dt$ is

$$G(\lambda, t)dt = \lambda e^{-\lambda t}dt(1 - e^{-\lambda T})^{-1} \qquad (5\text{-}7.1)$$

where the last factor takes into account the finite time of observation T.

The likelihood of observing decays at times $t_1, t_2, \ldots, t_M$ is then proportional to

$$G(\lambda, t_1, t_2, \ldots, t_M) = \lambda^M \prod_{i=1}^{M} e^{-\lambda t_i}(1 - e^{-\lambda T})^{-1} \qquad (5\text{-}7.2)$$

This quantity is a function of the parameter λ, and we ask for which value of λ $G(\lambda)$ has a maximum; the corresponding value of λ is the most likely one.

Explicity we may look for the maximum of log G. We then have for the maximum

$$0 = \frac{\partial \log G(\lambda, t_1, t_2, \ldots, t_M)}{\partial \lambda}$$

$$= -\frac{\partial}{\partial \lambda} \left[\sum_{i=1}^{M} \lambda t_i + M \log(1 - e^{-\lambda T}) - M \log \lambda \right] \tag{5-7.3}$$

$$0 = \sum_{i=1}^{M} t_i + MT(e^{\lambda T} - 1)^{-1} - \frac{M}{\lambda} \tag{5-7.4}$$

This equation may be solved for λ by iteration. In the limiting case where $\lambda T \gg 1$ it gives the obvious result

$$1/\lambda = \tau = \sum_{i=1}^{M} t_i / M \tag{5-7.5}$$

As a figure of merit for the value of the parameter λ obtained by solving Eq. (5-7.5), it is reasonable to assume

$$\mathcal{E}(\lambda) = \left[-\frac{\partial^2}{\partial \lambda^2} \log G(\lambda, t_1, \ldots, t_M) \right]^{-1/2} \tag{5-7.6}$$

We verify this result in the case in which G considered as a function of λ has a Gaussian distribution with standard deviation σ,

$$G = K \exp\left[-(\lambda - \lambda_0)^2 / 2\sigma^2 \right] \tag{5-7.7}$$

Then

$$\log G = \log K - \left[(\lambda - \lambda_0)^2 / 2\sigma^2 \right] \tag{5-7.8}$$

and

$$-\frac{\partial^2 \log G}{\partial \lambda^2} = \frac{1}{\sigma^2} \tag{5-7.9}$$

In this case $\mathcal{E}$ of Eq. (5-7.6) coincides with the standard deviation of λ.

The maximum likelihood method is commonly used in many problems. The question of the goodness of fit is thus important. The following example illustrates the χ^2 test, which is usually employed.

ISBN 0-8053-8061-7

Suppose we have a theoretical expression, containing unknown parameters, that we want to fit to experimental data. *We know that the theoretical expression is valid.* For instance, we have measured a differential cross section and we develop it in Legendre polynomials up to the fourth:

$$\frac{d\sigma}{d\theta} = a_1 P_1(\cos\theta) + a_2 P_2(\cos\theta) + a_3 P_3(\cos\theta) + a_4 P_4(\cos\theta). \quad (5\text{-}7.10)$$

We have measured $d\sigma/d\theta$ at N values of θ. We call these N quantities $x_1 \cdots x_N$ and their standard errors $\sigma_1 \cdots \sigma_N$. With four values of $a_1 \cdots a_4$ the likelihood function is

$$L(a_1 \cdots a_4) = \left((2\pi)^{N/2}\sigma_1 \cdots \sigma_N\right)^{-1} \exp\left[-\sum_1^N \frac{(x_i - \xi_i)^2}{2\sigma_i^2}\right] \quad (5\text{-}7.11)$$

where ξ_i is the theoretical value of $d\sigma/d\theta$ at the value of θ at which we obtained the measurement x_i, calculated using the values of the four parameters $a_1 \cdots a_4$. This likelihood function has a maximum for certain values of $a_1^* \cdots a_4^*$, which are then chosen as the result of the experiment. It is important to remember that the likelihood depends on the theoretical assumptions being valid. The maximum likelihood is reached by minimizing

$$\sum_1^N \frac{(x_i - \xi_i)^2}{2\sigma_i^2} = \frac{1}{2}\chi^2(a_1^* \cdots a_4^*) \quad (5\text{-}7.12)$$

The $\chi^2(a_1^* \cdots a_m^*)$ is thus defined.

If we have m adjustable parameters and N observed points, $n_D = N - m$ is called the number of degrees of freedom. Calculating $\chi^2(a_1^* \cdots a_m^*)$ with Eq. (5-7.12), we find in Fig. 5-8 the "confidence level" as a function of n_D and of χ^2. For instance, if we have measured $d\sigma/d\theta$ at 10 points and we have fitted the experimental results with four parameters $n_D = 6$, assuming a χ^2 of 5, the confidence level is 0.55. For more details see the current RPP or R.A. Fisher's *Statistical Methods for Research Workers* (Oliver & Boyd, Edinburgh, 1958).

5-8 METHODS OF MEASURING DECAY CONSTANTS

The simplest way to measure a decay constant is by direct application of Eq. (5-1.1) or (5-1.2). This is very simple if τ is a convenient interval of anywhere from about 1 min to a few years. Outside this range practical difficulties arise. Even if τ is convenient, however, it may happen that the radiation emitted in the decay is difficult to detect (e.g., in RaD), in which case it is sometimes possible to measure, not the decay of the substance under investigation, but the production of a daughter product, from which one can calculate τ of the mother substance.

ISBN 0-8053-8061-7

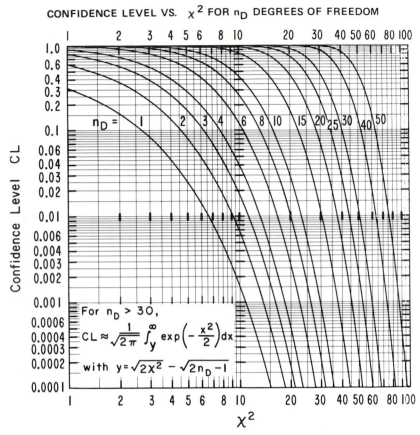

Figure 5-8   Confidence level versus χ^2 for n_D degrees of freedom. [From RPP 74.]

If τ is very short, it is still possible to employ mechanical methods in making the measurements (such as the container propelled by compressed air, used in piles), down to intervals of approximately 1 sec. For intervals of this duration, flow methods are effective if the substance is gaseous. The substance is produced in a certain place at a constant rate and is blown through a tube in a stream of gas with velocity v, measurable from the area of the tube and the rate of flow of the gas. The activity of the gas in the steady stream in the tube is measured at two points a distance d from each other. These activities are in the ratio $e^{-d/v\tau}$, from which τ can be obtained. Sometimes in experiments of this type the formation of an active deposit can be observed, and τ of the gas can be obtained from the distribution of the active deposit on the walls of the tube. Classical experiments by this method were performed by Rutherford. There have been many modifications of the

ISBN 0-8053-8061-7

method; noteworthy is the artifice employed by Jacobsen in which the velocity of recoiling atoms imparted by the emission of a particle is used.

A method in the same spirit, but using modern techniques by which one can measure the Doppler shift due to the velocity of a recoiling atom, makes it possible to reach 10^{-12} sec (Fig. 5-9). The atom formed at $t = 0$ moves with recoil velocity v from the thin target in which it was formed. If it emits a gamma ray in flight, its energy will be

$$E_S = E_0 \left[1 + (v/c) \cos \theta \right]$$

Nuclei surviving at a distance D from the source are $I = I_0 e^{-D/v\tau}$ where τ is the mean life of the nucleus and I_0 are the nuclei produced at the source. If one stops the nuclei in flight with a metal plunger, the stopped nuclei emit quanta of energy E_0. If the plunger is at a distance D from the source, one will observe $I_u = I_0 e^{-D/v\tau}$ unshifted gamma rays of energy E_0 and $I_S = I_0(1 - e^{-D/v\tau})$ gamma rays of energy E_S. Hence, $I_u/(I_S + I_u) = e^{-D/v\tau}$. D is measured directly, v may be calculated from the reaction giving rise to the nuclei or measured from E_S. One thus finds τ. For variations of this procedure and similar methods see the literature and Ch. 11 on the blocking method.

Electronic methods have been used for the direct measurement of very short lifetimes, down to $\tau \cong 10^{-9}$ sec. The principle of the methods requires that the decay to be studied be preceded by another event, which is used to actuate an electronic gate and establish the origin of time. The gate stays open for a time t. The probability of a count in the gate is then

$$k \int_0^t \lambda e^{-\lambda \xi} \, d\xi = k(1 - e^{-\lambda t}) \tag{5-8.1}$$

in which we have included in k geometric and efficiency factors, which, however, are constant. The apparatus is arranged to count the number of gates formed, as well as the number of gates during which a particle is detected. By measuring the ratio between the number of times a particle is detected while the gate is open and the total number of gates formed as a function of t, one obtains λ, from Eq. (5-8.1).

One may also keep the duration of the gate t constant but small compared with τ and vary the time T after the first triggering pulse at which the gate opens. This procedure is called *measurement of delayed coincidences*. The number of delayed coincidences per disintegration is proportional to $e^{-\lambda T}(1 - e^{-\lambda t})$ or, for $t \ll 1/\lambda$, to $\lambda t e^{-\lambda T}$ (Fig. 5-10).

Sometimes it is possible to feed the pulses of the detector into a cathode-ray oscilloscope with a calibrated time sweep. Pulses appear characteristically in pairs. By measuring the number of pairs corresponding to a time interval between t and $t + \Delta t$ as a function of t, one obtains an exponential curve

ISBN 0-8053-8061-7

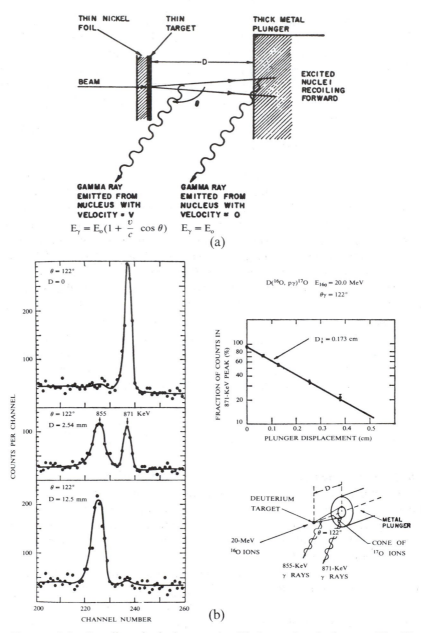

Figure 5-9   Recoil method of measuring lifetimes of excited states. The life-time measurement of the 0.871-MeV level of ^{17}O: The γ-ray spectra at the left in part (b) show the relative intensity of the 0.871- and 0.855-MeV peaks as the plunger displacement is increased. The logarithm of the relative intensity of the 0.871-MeV peak is plotted as a function of the plunger displacement in the graph at the right.

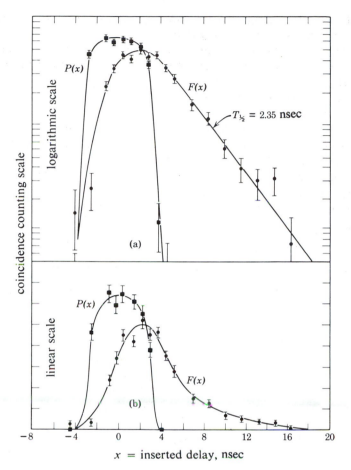

coincidence counting scale

logarithmic scale

$P(x)$

$F(x)$

$T_{\frac{1}{2}} = 2.35$ nsec

(a)

linear scale

$P(x)$

$F(x)$

(b)

x = inserted delay, nsec

Figure 5-10   Delayed coincidence resolution curve $F(x)$ for the 158–keV gamma ray of ^{199}Hg with a prompt curve $P(x)$ for comparison: (a) logarithmic scale; (b) linear scale. The half-life of the gamma ray can be measured by the slope of the right-hand part of log $F(x)$ in (a) or by the shift of the centroid of $F(x)$ to the right in (b). The result is $T_{1/2} = (2.35 \pm 0.20)$ nsec. The standard deviations of the points are indicated by vertical bars. [R. L. Graham and R. E. Bell, *Can. J. Phys.,*31, 377 (1953).]

$e^{-\lambda t}$, or, more directly, if one can neglect the background, the average time interval between the pulses gives τ.

For values of τ up to a few years, direct measurements are still practicable. For larger τ, direct measurement of the number of disintegrations undergone by a weighed amount of radioactive substance is generally used. Aside from the necessity of starting with a chemically and radioactively pure substance, the method is easily applied to alpha emitters where the absolute

ISBN 0-8053-8061-7

counting of the alpha particles does not offer great difficulties and where one alpha particle is emitted per disintegration.

Finally the condition of secular equilibrium [Eq. (5-2.9)] between two genetically related substances can also be used for the determination of the ratio of their decay constants. A classical example is that of radium and uranium: in an old, unaltered ore the ratio is 3.66×10^{-7} by weight (Boltwood's number).

The precision that has been attained up to now in measurements of λ is of the order, in the most favorable cases, of a few parts in 100 000, but the great majority of radioactive decay constants are not known to within 1%.

Extensive collections of data on radioactive decay are available in tabular and graphical form. For instance, an excellent table of isotopes is given by Lederer, Hollander, and Perlman (LHP 67); General Electric has also published a useful isotope chart (see the Bibliography).

5-9 CHRONOLOGICAL AND GEOLOGICAL APPLICATIONS

Radioactive decay has found very interesting applications to the problem of establishing dates of events remote in time.

On the archaeological time scale this has been established by ^{14}C dating. Cosmic rays in the atmosphere continuously form ^{14}C by nuclear reactions. ^{14}C itself is chemically combined with oxygen and finally finds its way into all organic material. Carbon in a living plant or animal has the same isotopic composition as atmospheric carbon. When the animal or plant dies, its exchange of carbon stops and in due course ^{14}C disappears with a half-life of 5730 years. Thus coal or petroleum has lost all its ^{14}C, whereas the carbon in a young tree has the same activity as atmospheric carbon (13.5 dis g^{-1} sec^{-1}). It is thus clear that one can date an object containing organic carbon by measuring the specific activity of the carbon. The sensitivity of the method sets a limit of about 30 000 years to the determinable ages. Older samples are too inactive to be accurately dated. A calibration curve is shown in Fig. 5-11. (There are some indications that carbon in wood produced today has a little lower specific activity than carbon derived from wood of a century ago. This has been attributed to the dilution of atmospheric carbon by inactive carbon originating from fossil fuels.)

On the geological time scale, radioactivity has given the most reliable information on absolute time. The substances commonly employed are uranium and its isotopes, thorium, potassium, and rubidium.

We shall briefly describe some of the methods used. First, however, let us define what we mean by the age of a rock. We do not know the early history of our planet, but at a certain time rocks reached their present composition, crystallizing and segregating their different components. A slow process of metamorphosis continues even today. Radioactive dating refers essentially to the time of formation of rocks. If at formation a rock contained a certain

ISBN 0-8053-8061-7

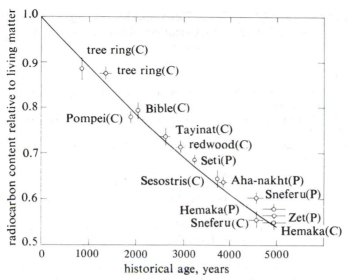

Figure 5-11 Calibration curve of ^{14}C method. Curve calculated from $T_{1/2} = 5568$ yr. Key: Samples of known age; (C), Chicago dates; (P), Pennsylvania dates (Ralph). [W. F. Libby, in *Les Prix Nobel en 1960*, Stockholm, 1961.] The presently accepted half-life of ^{14}C is 5730 ± 30 yr.

amount of uranium, it accumulated in it the helium produced by uranium decay. For each atom of ^{238}U that becomes ^{206}Pb, eight alpha particles are formed, and if these do not escape from the rock, the ratio between the uranium present and the helium accumulated gives the age of the rock. A similar procedure is applicable to thorium. The most important consideration is always to make certain that no helium or uranium has escaped the sample during the life of the rock. The uranium/lead ratio may be used in a similar way and is less subject than helium methods to errors caused by differences in diffusion or escape from the rocks of lead and uranium. Other methods use the $^{206}Pb/^{207}Pb$ isotopic ratio of radiogenic lead or the ratio of radiogenic lead to ^{204}Pb. K/Ar ratios and $^{87}Rb/^{87}Sr$ ratios are also used and are becoming increasingly important.

Finally, in some minerals containing traces of U one finds spontaneous fission tracks detectable by etching (see Sec. 3-7). One can then count the number of spontaneous fissions that have occurred in a particular specimen. The specimen is then exposed to slow neutrons, which produce many more fissions. From the neutron flux and the number of new fissions one obtains the U content of the specimen, and from this and the spontaneous fissions, its age. Occasionally it is possible to date the same rock by several methods; the results show reasonable agreement.

The oldest surface rocks have an age of about 3.7×10^9 yr. The same methods have been applied to the determination of the age of meteorites and the oldest of these are 4.5×10^9 yr. This should be the age of formation of the earth, as distinguished from the segregation of the surface rocks.

BIBLIOGRAPHY

Aldrich, L. T., and G. W. Wetherill, "Geochronology by Radioactive Decay," *Ann. Rev. Nucl. Sci.*, **8**, 257 (1958).

Annis, M., W. Cheston, and H. Primakoff, "On Statistical Estimation in Physics," *Rev. Mod. Phys.*, **2**, 818 (1953).

Dalrymple, G. B., and M. A. Lanphere, *Potassium–Argon Dating*, W. H. Freeman, San Francisco, 1969.

Eadie, W. T., D. Drijard, F. E. James, M. Roos, and P. Sadoulet, *Statistical Methods in Experimental Physics*, North-Holland, Amsterdam, 1971.

Feller, W., *An Introduction to Probability Theory and Its Application*, Wiley, New York, 1957.

Fisher, R. A. *Statistical Methods for Research Workers*, Oliver & Boyd, Edinburgh, 1958.

Gloyna, E. F., and J. O. Ledbetter, *Principles of Radiological Health*, Dekker, New York, 1969.

General Electric Co., *Nuclear Chart*, 11th ed., Schenectady N. Y., December 1972.

Hintenberger, H., "Mass Spectroscopic Researches in Nuclear Physics and Isotope Cosmology" in K. Ogata and T. Hayakawa (eds.), *Recent Developments in Mass Spectroscopy*, University Park Press, Baltimore, 1970.

Journal of Radiocarbon Dating.

Lal, D., and H. E. Suess, "The Radio-activity of the Atmosphere and Hydrosphere", *Ann. Rev. Nucl. Sci.*, **18**, 407 (1968).

Lederer, C. M., J. M. Hollander, and I. Perlman (LHP 67).

Libby, W. F., *Radiocarbon Dating*, 2nd ed., University of Chicago Press, Chicago, 1955.

Moyer, B. J., "Practical Control of Radiation Hazards in Physics Research," *Ann. Rev. Nucl. Sci.*, **8**, 327 (1958).

Particle Data Group (RPP 74).

Perlman, I., F. Asaro, and H. V. Michel, "Nuclear Applications in Art and Archaeology," *Ann. Rev. Nucl. Sci.*, **22**, 383 (1972).

Remsberg, L. P., "Determination of Absolute Disintegration Rates by Coincidence Methods," *Ann. Rev. Nucl. Sci.*, **17**, 347 (1967).

Reynolds, J. H., "Isotopic Abundance Anomalies in the Solar System," *Ann. Rev. Nucl. Sci.*, **17**, 253 (1967).

Schwarzschild, A. Z., and E. K. Warburton, "The Measurements of Short Nuclear Lifetimes," *Ann. Rev. Nucl. Sci.*, **18**, 265 (1968).

Solmitz, F. T., "Analysis of Experiments in Particle Physics," *Ann. Rev. Nucl. Sci.*, **14**, 375 (1964).

York, D., and R. M. Farquhar, *The Earth's Age and Geochronology*, Pergamon Press, London, 1972.

PROBLEMS

5-1   Calculate the volume of 1 Ci of radon at 0°C and 1 atm pressure.

5-2   Calculate the weight of 1 Ci of ^{239}Pu.

5-3   A piece of gold 1 mm thick is bombarded for 15 h by a slow neutron beam having 10^6 neutrons sec^{-1}. How many disintegrations per second of ^{198}Au occur in the sample 24 h after the end of bombardment?

5-4 ^{140}Ba decays into ^{140}La with a half-life of 12.5 days. ^{140}La has a half-life of 40 h. A sample initially contains pure ^{140}Ba. Call this amount 100 in arbitrary units. Using semilogarithmic paper, plot the amounts of Ba and La formed (as a function of time) over a period of 15 days.

5-5   An atomic explosion forms isotopes of ^{238}U by successive neutron capture. The capture cross sections are $\sigma(238)$, $\sigma(239)$, etc. The flux of neutrons through the material is φ, and it lasts a short time t. How much ^{250}U is formed per atom of ^{238}U?

5-6   As light a shield as possible is required for a very large source (1000 Ci) of a beta emitter having beta rays of 2 MeV, in order that a person can work near it safely. Give your suggestions.

5-7   Prove that for a Poisson distribution $\langle n^2 \rangle - \langle n \rangle^2 = \langle n \rangle$.

5-8   One milligram of iron does not emit any fission fragment for a month. The decay constant of iron for spontaneous fission has an even chance of being smaller than λ. Find λ, assuming every λ in a $d\lambda$ interval a priori equally probable.

5-9   A certain sample that gives a counting rate of 107 counts min^{-1} on a Geiger–Müller counter has been observed for 23 min. The background of the counter is 42 counts min^{-1}, based on an observation lasting 40 min. Find the probable net counting rate of the sample and its probable error.

5-10   In a given time T you want to measure the absorption coefficent of a substance for neutrons. The approximate value of this absorption coefficient is μ_0. The beam gives approximately N_0 counts sec^{-1} in the detector, with no absorber. Determine the thickness of absorption you will use and the apportionment of the time between measurements on a purely statistical basis. Consider also the case of a background of approximate value n_0 counts sec^{-1}.

5-11   A Geiger counter, after counting one particle, becomes dead for a time τ. A counting rate ν is observed for a certain sample; the true counting rate ν_0 is often calculated by the formula $\nu_0 = \nu(1 + \beta\nu)$. Justify the formula and find the relation between β and τ. The formula is an approximate one, valid if ν is close to ν_0.

5-12   Two ionization chambers are mounted in opposition on a compensating circuit. A certain sample is put in one chamber and gives a deflection of 180 arbitrary units min^{-1}. Two samples identical to it are put in the two chambers and the deflections obtained in 1-min periods are observed. The results are 0, $+40$, -12, -19, $+48$, -31, -25. How many ionization events occur in the chamber per minute, assuming that all ionization events give equal deflection?

5-13   In polarization measurements one scatters protons on a target. The protons may go to the left or right of the target. One is interested in the "asymmetry," that is, $p_L - p_R = \epsilon$, where p_L and p_R are the probabilities of scattering to the left and right. One measures the actual numbers scattered to the left, N_L, or to the right, N_R. Calling

$$p'_L = \frac{N_L}{N_L + N_R} \qquad p'_R = \frac{N_R}{N_L + N_R}$$

and

$$\epsilon' = p'_L - p'_R$$

one has approximately

$$p'_L - p'_R = p_L - p_R$$

In setting up an experiment it is desired that $|\epsilon' - \epsilon| < 0.005$ with 95% probability, or, as one says, a confidence level 0.95. How many counts should one collect?

5-14   Set up the equation for finding the value of α in an angular correlation distribution. Gamma rays are emitted in a cascade of 2, and we know that their angular correlation is given by

$$W(\varphi) = 1 + \alpha \cos^2 \varphi$$

M pairs have been observed making the angles $\varphi_1, \ldots, \varphi_M$. Try a numerical example and determine α and the confidence level.

5-15   A radiologist irradiates a hand with hard X rays from above. He positions two roentgen meters, one above and one under the hand. They both show 100 R. What is the dose received by the hand? Carefully specify the units.

5-16   A rad meter shows 1.1 rad day^{-1} in a radiation field of 3-MeV neutrons. How long would you work in that place?

5-17   Calculate the dose you would get by swimming in a swimming pool containing 0.1 mCi liter^{-1} of a substance emitting 0.7 of a gamma ray of 0.5 MeV energy and 0.9 of a gamma ray of 2 MeV energy per disintegration.

5-18   In a radiation field of 1 R of 20-KeV X rays, soft tissue is irradiated in the amount of about 1 rad. What is the dose for bone? Repeat the calculation for 1-MeV X rays.

ISBN 0-8053-8061-1

THE NUCLEUS

CHAPTER

6

Elements of the Nuclear Structure and Systematics

In order to obtain a picture of the nucleus in the ground state, we first measure its "global" properties: charge, mass, radius, spin, magnetic moment, and so on. Some of these, such as charge and mass, require refined techniques but present no conceptual difficulties. We cannot take too naive a view of the radius, thinking simply in terms of the radius of a macroscopic object, but we must be careful about its definition. Spin is quantized and, apart from being a quantum-mechanical vector, it offers no special conceptual difficulties. Magnetic and electric multipole moments in a quantum-mechanical system must also be defined; here the main idea is to take their expectation value for the case of maximum orientation in an external field. In this chapter we shall describe briefly the techniques used for these measurements.

The empirical findings must then be interpreted, although both historically and conceptually the two processes of accumulating empirical data and interpretating them go hand in hand. Having established the fundamental fact that nuclei consist of protons and neutrons, one might hope to repeat the feat accomplished by atomic physics in the interpretation of the atom, where the application of quantum mechanics to the Rutherford–Bohr models solved virtually all problems. We could set out to find the force between

Emilio Segrè, Nuclei and Particles: An Introduction to Nuclear and Subnuclear Physics, Second Edition

nucleons and from that derive all nuclear properties. This program is at present impossible for two reasons: first, the force between nucleons is not as simple as the Coulomb force and is not known precisely. Second, the mathematical problem of solving the Schrödinger equation for a nucleus is beyond our present powers, nor are there approximations that offer hopes of quantitative results of a precision comparable to that attainable in atomic problems.

The usual remedy in a situation of this kind is to introduce models that simulate, more or less accurately, nuclear behavior, and when necessary to introduce empirically parameters that help in the description. From the systematics of the results one then tries to improve the model. This procedure is very commonly employed in all experimental sciences.

As a consequence of this approach, we may find various models useful, depending on the property under consideration and on the mass number of the nucleus studied. The models, however, are not independent of each other. Ultimately they are all facets or approximations of a theory that does not yet exist. For this reason the study of limiting cases and of the relations between the models is especially important.

Some of the models, for example the liquid-drop static model, are very simple and require relatively few experimental data; however, they do not give detailed information, although they are valuable for overall systematic studies. Other models provide detailed information but only in special mass intervals or on special properties.

In this chapter we discuss the models as soon as the relevant empirical material has been treated. This sequence parallels to a certain extent the historical development. The interrelations between the models are set forth in Fig. 6-1, which also includes models, such as the optical model, to be discussed in later chapters. Although the knowledge of nuclear reactions is necessary for an understanding and appreciation of some of these models, they are included in Fig. 6-1 for completeness.

6-1   CHARGE

The Rutherford planetary model of the atom made possible, very early, an interpretation of X-ray spectra of many elements obtained by Moseley in 1913. It was apparent from the spectral data (Fig. 6-2), that the nucleus has a positive charge Ze, where Z is the atomic number and $|e| = 4.8032 \times 10^{-10}$ esu is the magnitude of the charge on the electron. (For greatest precision and errors on constants, consult current RPP.) The atomic number Z is exactly an integer; subelectronic charges have never been observed. The fact that single atoms are exactly neutral has been checked with great precision by sending a molecular beam of cesium or potassium atoms through an electric field. The total charge of the atom that resulted from these experiments is less than 10^{-18} electron charge. Furthermore, the electron–proton charge difference

ISBN 0-8053-8061-7

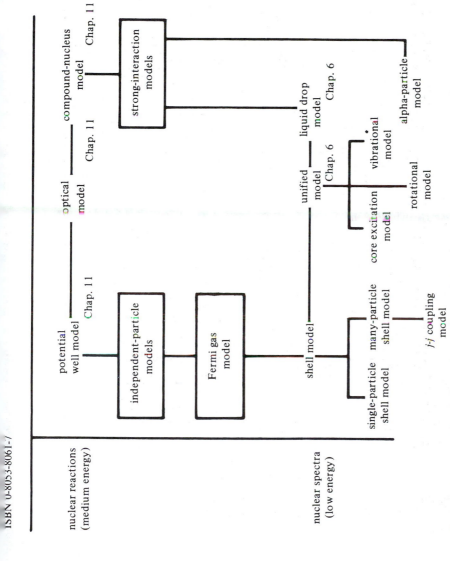

Figure 6-1 Models of the nucleus. The models are treated in different chapters as indicated in the figure. [From S. A. Moszkowski, in (F1 E).]

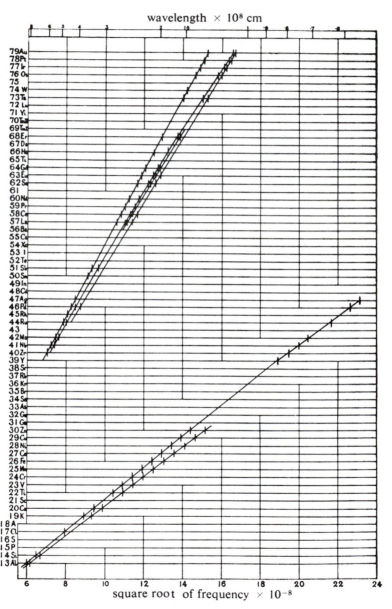

Figure   6-2   Original figure from a paper by H. G. J. Moseley [*Phil. Mag.*, **27**, 703 (1914)] showing the relation between Z and $\nu^{1/2}$ of X rays (K and L lines).

has been determined to be less than 5×10^{-19} electron charge and the charge of the neutron less that 5×10^{-19}, in the same units. Experiments on the average charge of atoms or molecules set limits about 10 times smaller. The identity of all electrons is further proved by the applicability to them of Pauli's exclusion principle: It has been verified that electrons emitted in beta decay cannot enter orbits already occupied by atomic electrons. This is an absolutely stringent test and rules out, for example, the possibility of a difference between orbital electrons and electrons emitted in beta decay. The identity of all protons has also be verified (see Sec. 6-5). The proton and the electron are assumed to be stable. Direct measurements on single protons give a mean life greater than 10^{21} yr and measurements on matter in bulk give a lower limit of 10^{30} yr. For the electron, direct measurements give a mean life greater than 10^{21} yr.

6-2 MASS

The study of atomic masses by the mass spectrograph shows that each nuclear species has a mass nearly equal to an integral multiple of the proton mass. The integer in question is called the mass number A. In 1961 the International Union of Pure and Applied Chemistry adopted as the unit of nuclear mass the twelfth part of the mass of the atom ^{12}C, and we shall use this unit consistently. Its absolute value is

$$1.660566 \times 10^{-24} \text{g} = 931.5016 \pm 0.0026 \text{ MeV}/c^2$$

The unit of mass ordinarily employed by physicists until 1960 was one giving ^{16}O = 16. Chemists used a unit of mass in which the average mass of the natural mixture of oxygen isotopes was 16. These two units and the new unit ^{12}C = 12 are in the ratio

$$1 \text{ mu}(^{12}\text{C} = 12) = 1.000317917 \text{ mu}(^{16}\text{O} = 16) = 1.000043 \text{ mu}(\text{O} = 16)$$

In Table 6.1 we give a few important masses in units of ^{12}C/12 and MeV.

TABLE 6-1   SOME IMPORTANT MASSES[a]

	^{12}C/12	MeV
^{12}C/12	1	931.5016
1 MeV	1.073535×10^{-3}	1
Electron	5.485580×10^{-4}	0.511003
Neutron	1.008665	939.5731
Proton	1.007276	938.2796
Deuterium atom	2.0141014	1876.14
Helium atom	4.002600	3728.44

[a]For errors and discussion see current RPP.

The quantity $A - M$, where M is the exact mass of the atom, is usually called the "mass defect," the quantity $M - A = \Delta$ is usually called the "mass excess," or "mass decrement" and the quantity

$$\frac{M - A}{A} = f \tag{6-2.1}$$

is called the "packing fraction" (Fig. 6-3).

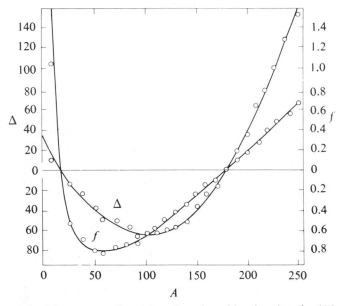

Figure 6-3 Mass excess ($\Delta = M - A$) and packing-fraction $f = (M - A)/A$ curve, based on mass-spectrographic and nuclear data for beta-stable nuclei. Circles represent experimental data points. The smooth curves are based upon an empirical mass formula. Ordinates are in millimass units. [A. S. Green, *Rev. Mod. Phys.*, **30**, 569 (1958).]

If, by a nuclear reaction, atoms 1 and 2 of masses M_1 and M_2 combine to form a third atom of mass M_3, then $(M_1 + M_2 - M_3)c^2$ is the binding energy that has escaped from the system, perhaps as gamma radiation, during the synthesis. Conversely it might be necessary to supply energy to atoms 1 and 2 in order to combine them.

These energy relations are expressed by the equation

$$M_1 + M_2 = M_3 + Q \tag{6-2.2}$$

a positive Q meaning that we obtain work by combining M_1 and M_2 into M_3,

ISBN 0-8053-8061-7

and a negative Q meaning that it takes work to combine M_1 and M_2 into M_3. For instance,

$$n + p = d + Q \qquad Q = 2.225 \text{ MeV} \qquad (6\text{-}2.3)$$

A neutron and a proton at rest combine to form a deuteron, the excess energy of 2.225 MeV escaping as a gamma ray and as recoil energy of the deuteron formed.

By convention the tables of masses give atomic masses. To obtain nuclear masses, one must subtract from them Z electron masses (m_e) and, to be exact, must also add the binding energy of the electrons. For instance,

$$M \text{ (hydrogen atom)} = M \text{ (proton)} + m_e - 13.60 \text{ eV}$$

In most equations the use of atomic or nuclear masses gives the same result except for the small and often negligible differences of electronic binding energy; if positrons are involved, however, the orbital electrons must be taken into account.

Thus, if an atom of mass $M(A, Z)$ undergoes a beta decay, we have

$$M(A, Z) = M(A, Z + 1) + Q$$

In this case $M(A, Z + 1)$ contains the rest mass of $Z + 1$ electrons, and Q is the maximum kinetic energy of the escaping electron (the neutrino then does not carry away any energy). If we have a positron decay,

$$M(A, Z) = M(A, Z - 1) + 2m_e + Q \qquad (6\text{-}2.4)$$

where Q is again the maximum kinetic energy of the positron. Clearly unless

$$M(A, Z) > M(A, Z - 1) + 2m_e \qquad (6\text{-}2.5)$$

positron emission is impossible (see also Chap. 9).

There are several methods for determining the mass of a nucleus. The most important are by means of the mass spectrometer and by energy measurements in nuclear reactions.

The mass spectrometer connected with earlier experiments by J. J. Thomson (1912), was developed by F. W. Aston (1920) and has been brought to a high degree of perfection. In order to determine the mass m of an ion of known charge ze, it is enough to determine ze/m. As is well known, a magnetic field produces on a charge ze moving in it a force,

$$\mathbf{F}_B = \frac{ze}{c} \mathbf{v} \times \mathbf{B} \qquad (6\text{-}2.6)$$

where $\mathbf{B}$ is the magnetic induction and $\mathbf{v}$ is the velocity of the particle. An

ISBN 0-8053-8601-7

electric field **E** gives a force

$$\mathbf{F}_E = ze\mathbf{E} \tag{6-2.7}$$

The orbit of a particle is affected by both fields, and a combination of these fields can be used to determine any two of the following three quantities: momentum p of the particle, velocity v, and its kinetic energy T. For instance, if we have a magnetic and an electric field perpendicular to each other and to the velocity of a particle and of magnitudes such that

$$\frac{v}{c} = \frac{E}{B} \tag{6-2.8}$$

then the particle moves in a straight line, since the electric and magnetic forces compensate each other. Adding two slits to the system, we have a *Wien filter*. Particles that have passed through such a Wien filter must all have the same velocity v. If we now pass the beam through a uniform magnetic field B and measure the radius r of the circular orbit described by the particles, we obtain the momentum of the particles. We then have

$$r = cp/zeB \tag{6-2.9}$$

The momentum and velocity yield the rest mass, $m = p/v$, or relativistically,

$$m = p/\beta\gamma c \tag{6-2.10}$$

There are many combinations of electric and magnetic fields suitable for mass spectrometers. Of particular importance are the focusing properties of the arrangement: it is desirable that particles emerging from one point into a small angle converge on one point or one line. It is even possible to obtain double focusing, whereby particles emerging within a small angle from a slit and having slightly different velocities converge in an image. Figure 6-4 shows a modern type of double-focusing mass spectrometer and Fig. 6-5 the lines obtainable. An interesting variation of these arrangements is the time-of-flight method, in which the velocity of an ion is determined by its period of revolution in a magnetic field.

Mass spectroscopy has many applications. Of greatest immediate interest to nuclear physics are precise mass determinations and the identification and abundance measurements of various isotopes.

The isotopic composition of an element as found in nature is usually fixed, because the different isotopes behave identically from the physico-chemical point of view. This is not always precisely true, although departures from the normal composition are generally minute, except when radioactive decay phenomena are involved, as in the case of radiogenic lead. For all

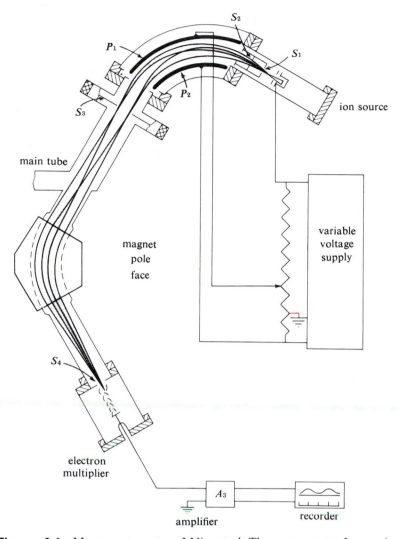

Figure 6-4 Mass spectrometer of Nier et al. The spectrometer focuses ions of a given mass; the focusing action is independent, within limits, of kinetic energy and of the entrance angle. Ions are produced in the ion source and accelerated by a voltage of about 40 kV. They are electrically deflected through an angle of 90 deg by a condenser P_1P_2 (radius 20 in.), and refocused on an electron multiplier by a magnetic field. The trajectories in the figure represent two beams of ions, of the same mass but different velocity and diverging from the source. The double-focusing action is demonstrated by the convergence of such beams at S_4. Two ions having a mass ratio m/m' and starting from rest follow exactly the same trajectory if all the voltages applied in the two cases are in the ratio $V'/V = m/m'$. Thus, measuring electrically the ratio V'/V for which two ions (e.g., CH_4^+, O^+) arrive at the same point gives their mass ratio. [Courtesy Prof. A. O. Nier.]

ISBN 0-8053-8601-7

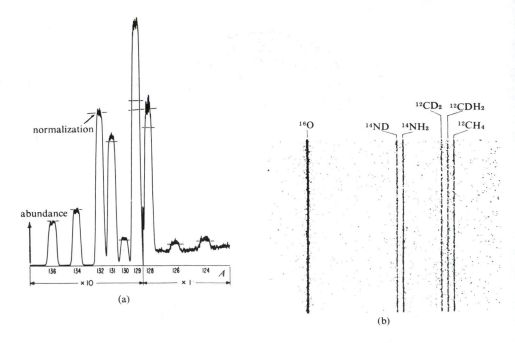

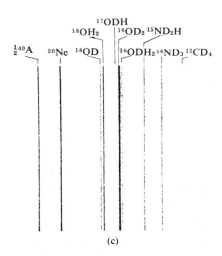

Figure 6-5 (a) Mass spectrum showing anomalous isotopic composition of Xe from a meteorite. The horizontal lines give the normal abundance of Xe isotopes in gas from the earth's atmosphere. The excess ^{129}Xe demonstrates the presence in the meteorite of a radioactive ^{129}I at the time of the meteorite's formation. [J. H. Reynolds, *Phys. Rev. Letters*, **4**, 8 (1960).] (b) and (c) Mass-spectra multiplet at $A = 16$ and $A = 20$ showing several molecules and ^{16}O. Resolution about 1/80,000. [From R. Bieri, E. Everling, and J. Mattauch, *Z. Naturfors.*, **10a**, 659 (1955).]

ISBN 0-8053-8061-7

practical purposes the chemical inseparability of isotopes is absolute and forms the basis of the important tracer method.

Mass determinations are now precise to the order of 1 part in 10^9. The quantity directly measured is the mass difference between two ions of approximately the same ze/m; one has, for example:

$$D_2 - He = 25600.232(8) \text{ micromass units} \qquad (6\text{-}2.11a)$$

$$D_2 - {}^1H^3H = 4329.257(3) \text{ micromass units} \qquad (6\text{-}2.11b)$$

$${}^{12}C_2{}^1H_4 - {}^{12}C^{16}O = 36386.01(24) \text{ micromass units} \qquad (6\text{-}2.12a)$$

$${}^1H^3H - He = 21271.073(10) \text{ micromass units} \qquad (6\text{-}2.12b)$$

(In the notation ${}^AX_n$, X stands for the chemical symbol, A for the mass number, n for the number of atoms in the molecule, and the whole symbol for the molecular mass.) Note that 1 micromass unit is about 1000 eV! The study of such "mass doublets," including the normalization ${}^{12}C = 12$, may be extended to many other nuclei and thus a table of atomic masses is built. Moreover, it is possible to integrate the mass spectrometric data with data derived from nuclear reactions. Thus, for instance, the mass of the neutron may be found by the measurement of the energy of the gamma rays emitted in the reaction

$$ {}^1n + {}^1H = {}^2H + \gamma \qquad (6\text{-}2.13)$$

if the mass of the deuteron and proton are known. In this case, if the neutron is captured at rest, we have

$$ {}^1n + {}^1H - {}^2H = Q = \hbar\omega + (\hbar^2\omega^2/2m_dc^2) \qquad (6\text{-}2.14)$$

where the last term is the recoil kinetic energy of the deuteron. As another example, consider an atom of mass M_1 and kinetic energy T_1 colliding with an atom of mass M_0 at rest. The reaction products are an atom of mass M_2, which escapes with a kinetic energy T_2 at an angle θ to the direction of the impinging particle, and an atom of mass M_3, which escapes with a kinetic energy T_3 at angle φ. The reaction is expressed by the equation

$$ M_1 + M_0 = M_2 + M_3 + Q \qquad (6\text{-}2.15)$$

The quantity Q may be measured if one knows M_1, M_2, M_3, the kinetic

ISBN 0-8053-8061-7

energies T_1, T_2, and θ. Conservation of energy and momentum gives

$$Q = (1/M_3)\left[(M_2 + M_3)T_2 - (M_3 - M_1)T_1 \right.$$

$$- 2(M_1 M_2 T_1 T_2)^{1/2} \cos\theta \left(1 + \frac{T_1}{2M_1 c^2}\right)^{1/2.}$$

$$\left. \times \left(1 + \frac{T_2}{2M_2 c^2}\right)^{1/2} + \frac{T_1^2 + T_2^2 - T_3^2}{2c^2} \right] \qquad (6\text{-}2.16)$$

where the two factors after $\cos\theta$ and the last term are relativistic corrections and may sometimes be omitted. A typical reaction that can be treated by Eq. (6-2.16) is

$$^1\text{H} + {}^6\text{Li} = {}^3\text{He} + {}^4\text{He} + Q \qquad Q = 4.023 \text{ MeV} \qquad (6\text{-}2.17)$$

The masses of many stable and unstable isotopes have been measured with high precision by a combination of mass-spectrometric measurements and measurements of energy in nuclear reactions.

Other branches of physics have also contributed to precise mass spectrometry. Microwave spectrometry in particular often permits the precision measurement of mass ratios of isotopes.

Another important application of mass spectrometry is the determination of the relative abundances of isotopes in a mixture. This measurement of the natural mixture of isotopes, combined with the masses of the single isotopes, allows ordinary chemical atomic weights to be found with a precision generally superior to that of chemical determinations. Even extremely rare isotopes with an abundance of only a few parts in 10^4 have been detected, and for many undetected ones low upper limits of the abundance have been set. In this field other branches of physics have contributed in a notable way: the important isotopes ^{2}H, ^{13}C, ^{15}N, and ^{18}O were discovered spectroscopically, and ^{3}He in the atmosphere was discovered by using the resonant properties of the cyclotron.

A. H. Wapstra and N. B. Gove have compiled an extensive table of masses, taking into account all available experimental data and adjusting them by a least-squares procedure. All of these data are of importance for the development of nuclear systematics. Some of the more striking results are as follows:

1. Nuclei with odd Z have only one or two stable isotopes.
2. Nuclei with odd Z and even A are unstable (the only exceptions are ^{2}H, ^{6}Li, ^{10}B, ^{14}N, and ^{180}Ta).

ISBN 0-8053-8601-7

3. In any group of isobars with A and Z odd, there are only one or two nuclei stable against beta decay.

Table 6-2, which refers to stable nuclei, shows the preference for the Z-even–A-even composition. A similar conclusion is reached by a study of nuclear abundances.

TABLE 6-2   NUMBER OF STABLE NUCLEI

	A even	A odd
Z even	156	48
Z odd	5	50

6-3   NUCLEAR RADII

The definition of a nuclear radius is somewhat arbitrary, because the result of a measurement depends on the phenomenon used to define the nuclear radius; however, all the results agree qualitatively and to a certain extent quantitatively also. This is especially remarkable considering the variety of methods used. Here we consider first the scattering of alpha particles, neutrons, and protons of widely different energies. The dominant force in these cases is of nuclear origin. Next we mention electron scattering, the isotope shift of spectral lines, mu-mesic atoms, and the electrostatic term in the nuclear masses. These involve only electric interaction. The concordance of the results shows among other things that at least approximately the bulk of nuclear matter has a constant specific charge, although there may be variations on the surface of the nucleus.

Historically the first measurement of nuclear size was made by Rutherford and Chadwick, who determined the angle of scattering at which the scattering cross section of alpha particles from a nucleus starts to show departures from Rutherford's law. There is a minimum distance of approach in a collision that corresponds to this angle, and the fact that the scattering law based on the pure electric Coulomb interaction breaks down is interpreted as evidence that at this minimum distance specific nuclear forces become effective. The alpha particle and the target nucleus have "touched" each other. In this way it was found by Rutherford that the nuclear radius of light elements could be represented by the formula

$$R = r_0 A^{1/3} \quad \text{with} \quad r_0 = 1.2 \times 10^{-13} \text{ cm} \qquad (6\text{-}3.1)$$

Although the method is a crude one, the result has been confirmed by later studies.

Using neutrons instead of charged particles as projectiles, we find the range of the nuclear forces directly. There are many such measurements. However, at low energies, where the de Broglie wavelength of the neutron is

ISBN 0-8053-8601-7

comparable to or larger than the nuclear radius, these measurements are not usable; at very high energies, nuclei are transparent to neutrons. Hence, intermediate energies (10–20 MeV) must be used. Figure 6-6 shows a graph of the nuclear cross section σ obtained from neutron measurements. Here we use the relation

$$\sigma = 2\pi R^2 \tag{6-3.2}$$

where σ is the total nuclear cross section. The justification of the factor 2 in Eq. (6-3.2) comes from the so-called diffraction scattering (see Chap. 11).

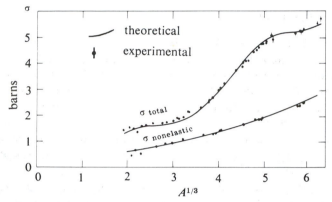

Figure   6-6   Experimental and theoretical total and nonelastic cross sections of 14-MeV neutrons as a function of $A^{1/3}$. [S. Fernbach, *Rev. Mod. Phys.*, **30**, 414 (1958).]

Diffraction scattering itself gives another method of determining the nuclear radius. If we assume for the sake of simplicity that the nucleus is a black disk of radius R, diffracting the de Broglie wave of the incident particle according to the laws of optics, we should obtain a differential scattering cross section (Ja 75)

$$\frac{d\sigma}{d\omega} = \frac{4\pi^2 R^2}{\sin^2\theta} J_1^2\left(\frac{R\sin\theta}{\lambdabar}\right) \tag{6-3.3}$$

where $J_1(x)$ is the Bessel function of order unity and $\lambdabar = \hbar/mv$ is the de Broglie wavelength of the incident particle. The function $J_1(x)$ has the first zero at

$$x = 0.610 \times 2\pi \tag{6-3.4}$$

ISBN 0-8053-8061-7

and the asymptotic forms

$$J_1(x) \rightarrow \left(\frac{2}{\pi x} \right)^{1/2} \sin\left(x - \frac{\pi}{4} \right) \quad \text{and} \quad J_1(x) = \frac{x}{2} \left(1 - \frac{x^2}{8} + \cdots \right) \quad (6\text{-}3.5)$$

for large and small x, respectively. The first minimum of scattering thus occurs for an angle such that

$$\sin \theta = 0.610 \times 2\pi (\lambdabar / R) \quad (6\text{-}3.6)$$

and from its measurement we have another method of obtaining R (Fig. 6-7).

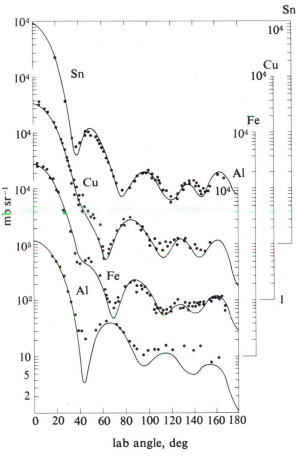

Figure 6-7 Experimental and theoretical differential cross sections for 14-MeV neutrons scattered from Sn, Cu, Fe, and Al. The experimental data presented are not completely corrected for multiple scattering, nor have angular and energy resolutions been taken into account. [S. Fernbach, *Rev. Mod. Phys.*, **30**, 414 (1958).]

ISBN 0-8053-8601-7

In addition to the specific nuclear methods mentioned above, it is possible to deduce the nuclear radius from the scattering of high-energy electrons on the assumption that the electron is subject to electromagnetic interactions only. This gives the distribution of the electric charge of the nucleus, and the result is again approximately represented by Eq. (6-3.1), with $r_0 = 1.2$ F.

It will be noted that Eq. (6-3.1) is an approximation, in that the nucleus is not a homogeneous sphere but has a density distribution decreasing gradually near the surface. An electric-charge density distribution (Saxon) that is widely used and is in agreement with experiment is

$$\rho(r) = \frac{\rho_1}{\exp[(r - C)/Z_1] + 1} \tag{6-3.7}$$

with

$$C = 1.07A^{1/3} \times 10^{-13} \text{ cm} \quad \text{and} \quad Z_1 = 0.545 \times 10^{-13} \text{ cm} \tag{6-3.8}$$

ρ_1 is determined by the normalization condition

$$Ze = \int_0^\infty \rho(r) 4\pi r^2 \, dr \tag{6-3.9}$$

(Fig. 6-8). This formula is valid, with small variations of C, for any nucleus having $A > 30$. A measure of the surface thickness is provided by the difference of the radii for which $\rho(r)$ is 0.1 and 0.9 of its maximum value. For the Saxon distribution this is $4.4Z_1 \sim 2.4 \times 10^{-13}$ cm.

Electron scattering measurements give the most detailed information at present obtainable on the nuclear-charge distribution. Using high-energy electrons, one has de Broglie wavelengths small enough to permit exploration of the nuclear structure; at 1 GeV, ƛ is already 1.95×10^{-14} cm.

The experimental data obtained are $d\sigma/d\omega$ as a function of the scattering angle θ; we shall consider here only elastic scattering. The calculations may be carried out to various degrees of approximation. As a first orientation consider the scattering of an electron of velocity v and momentum p, from a fixed point nucleus in the Born approximation. For the potential $U(r) = -Ze^2/r$, one obtains Rutherford's scattering formula

$$\frac{d\sigma_1}{d\omega} = \left(\frac{Ze^2}{2pv} \right)^2 \frac{1}{\sin^4(\theta/2)} = |f_1(\theta)|^2 \tag{6-3.10}$$

Now describe the nucleus as a uniformly charged sphere of radius R. For $r > R$ the potential is $-Ze^2/r$. For $r < R$, by Gauss's theorem we see immediately that the force must be proportional to r and the potential must have the form $a + br^2$. The constants are determined by the continuity of force and potential at $r = R$ obtaining for $r < R$, $U = -(Ze^2/2R)[3 - (r/R)^2]$.

ISBN 0-8053-8061-7

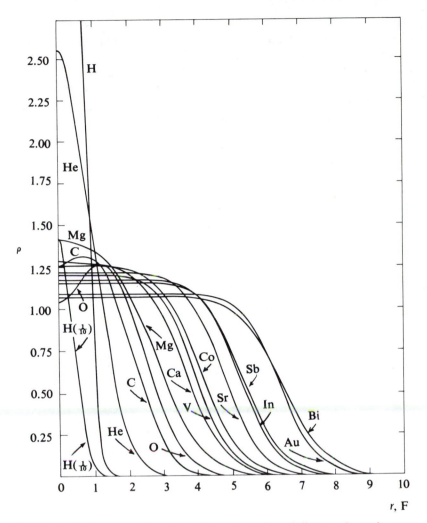

Figure   6-8   Nuclear charge density as a function of distance from the center of the nucleus found by electron scattering methods. Ordinate unit: 10^{19} C cm^{-3}. [R. Hofstadter, *Ann. Rev. Nucl. Sci.*, **7**, 231 (1957).]

By calculating (in the Born approximation) the scattering for the new potential, we find

$$\frac{d\sigma_2}{d\omega} = |f_2(\theta)|^2 = \frac{d\sigma_1}{d\omega}\left[(\sin qR - qR\cos qR)\frac{3}{q^3R^3}\right]^2$$

$$= \frac{d\sigma_1}{d\omega}|F(q)|^2 \tag{6-3.11}$$

with $\hbar q = 2p\sin(\theta/2)$; note that q is a reciprocal length. The ratio between

ISBN 0-8053-8601-7

$f_1(\theta)$ and $f_2(\theta)$ is $F(q)$; that between $d\sigma_2/d\omega$ and $d\sigma_1/d\omega$ is $|F(q)|^2$. $F(q)$ is called a *form factor*; it gives the ratio between the amplitude of scattering by an extended charge distribution and a point scatterer.

Measurements of $d\sigma_2/d\omega$ compared with Eq. (6-3.10) gives values of the nuclear radius, which enters as a parameter in $F(q)$. We generalize this calculation to an arbitrary spherical charge distribution.

The matrix element entering Born's formula

$$M_{if} = \int \exp(i\mathbf{q} \cdot \mathbf{r}) U(\mathbf{r}) \, d\mathbf{r} \qquad (6\text{-}3.12)$$

may be evaluated for the more general case of a spherical charge distribution $\rho(\mathbf{r})$. We normalize $\rho(\mathbf{r})$ by requiring that $\int_0^\infty \rho(\mathbf{r}) \, d\mathbf{r} = 1$, and write

$$U(\mathbf{r}) = Ze^2 \int \frac{\rho(\mathbf{R})}{|\mathbf{r} - \mathbf{R}|} \, d\mathbf{R} \qquad (6\text{-}3.13)$$

Introducing this expression in Eq. (6-3.12), one has

$$M_{if} = Ze^2 \int \int \exp(i\mathbf{q} \cdot \mathbf{r}) \frac{\rho(\mathbf{R})}{|\mathbf{r} - \mathbf{R}|} \, d\mathbf{r} \, d\mathbf{R} \qquad (6\text{-}3.14)$$

We now call $\mathbf{r} - \mathbf{R} = \mathbf{s}$ and μ the cosine of the angle between $\mathbf{q}$ and $\mathbf{s}$. The integration with respect to $d\mathbf{r}$ at $\mathbf{q}$ and $\mathbf{R}$ fixed gives $d\mathbf{r} = 2\pi \, d\mu \, s^2 \, ds$ and

$$M_{if} = Ze^2 \int \frac{\exp(iq\mu s)}{s} 2\pi s^2 \, ds \, d\mu \int \exp(i\mathbf{q} \cdot \mathbf{R}) \rho(\mathbf{R}) \, d\mathbf{R} \qquad (6\text{-}3.15)$$

The first integral was discussed in Sec. 2-2 and is $4\pi Ze^2/q^2$; the second is by definition $F(q)$. Born's formula then gives

$$\frac{d\sigma}{d\omega} = \frac{1}{4\pi^2 \hbar^4} \frac{p^2}{v^2} |M_{if}|^2 \qquad (6\text{-}3.16)$$

$$\frac{d\sigma}{d\omega} = \frac{d\sigma_1}{d\omega} |F(q)|^2 \qquad (6\text{-}3.17)$$

The cross section is thus factored into the point-scattering cross section multiplied by the form factor

$$F(q) = \int \rho(\mathbf{R}) \exp(i\mathbf{q} \cdot \mathbf{R}) \, d\mathbf{R} \qquad (6\text{-}3.18)$$

Note that the normalization of ρ gives $F(0) = 1$. In Eq. (6-3.17) the form factor appears as the three-dimensional Fourier transform of the charge density.

ISBN 0-8053-8601-7

Expanding the exponential under the integral in Eq. (6-3.14) we have

$$F(q) = \int \rho(\mathbf{R}) \left[1 + i\mathbf{q} \cdot \mathbf{R} - \frac{(\mathbf{q} \cdot \mathbf{R})^2}{2!} + \cdots \right] d\mathbf{R}$$

$$= 1 - \frac{q^2}{6} \int R^2 \rho(\mathbf{R}) \, d\mathbf{R} + \cdots \tag{6-3.19}$$

The last integral is $\langle R^2 \rangle$, the mean-square charge radius. A measurement of $F(q)$ for small q (large de Broglie wavelength) gives in successive approximations the momenta $\int R^n \rho(\mathbf{R}) \, d\mathbf{R}$ of the charge distribution. One can also say that $F(q)$ crudely represents the charge effective in the collision, or the amount of charge contained in a sphere of radius $1/q$.

The nuclear charge densities shown in Fig. 6-8 derive from the analysis of electron scattering experiments on the nuclei in question.

For a relativistic Dirac electron, including spin and recoil effects on a charged spinless point nucleus of mass M, the scattering formula is

$$\left(\frac{d\sigma}{d\omega} \right)_{\text{point}} = \left(\frac{Ze^2}{2E_0} \right)^2 \frac{1}{\sin^4(\theta/2)} \frac{\cos^2(\theta/2)}{1 + (2E_0/Mc^2)\sin^2(\theta/2)} \tag{6-3.20}$$

where θ is the laboratory scattering angle, E_0 the total energy of the electron in the laboratory, and M the mass of the nucleus. Here the last factor originates from the kinematics of nuclear recoil. The extended nuclear charge distribution will again give rise to a form factor, and we write

$$\frac{d\sigma}{d\omega} = \frac{d\sigma}{d\omega}_{\text{point}} |F_1(q)|^2 \tag{6-3.21}$$

The index 1 on F reminds us that it refers only to electric charge, neglecting possible magnetic effects. F_1 is given again by Eq. (6-3.19).

A hypothetical charge distribution can be tested against experiment by comparing the observed $d\sigma/d\omega$ with the one calculated through Eqs. (6-3.18) and (6-3.21) and the assumed charge distribution.

The proton and the neutron are of special interest. Even for them it has been possible to measure form factors. But in order to obtain meaningful results, it is also necessary to take into account the magnetic moments of the particles. Scattering of high-energy (up to many GeV) electrons on protons reveals the "meson clouds" surrounding the nucleons and gives very important information on the structure of the nucleons. This subject is treated in Chap. 18. Here we say only that as a first approximation, the electric-charge density of the proton may be represented by an exponential,

$$\rho = \rho(0)e^{-r/a} \tag{6-3.22}$$

ISBN 0-8053-8601-7

with $a = 0.23 \times 10^{-13}$ cm. The root of the mean-square radius for this distribution is $\sqrt{12}\, a = 0.8 \times 10^{-13}$ cm. The value of this density distribution formula is chiefly qualitative.

Another line of evidence concerning the nuclear radius derives from measurements of the electrostatic energy of the nucleus due to the accumulation of the positive charge in the nuclear volume. If all the charge were uniformly distributed over the surface of the nucleus, this energy would be $Z^2 e^2/2R$. Assuming that the charge is uniformly distributed in the nuclear volume, we obtain

$$\frac{3}{5} Z^2 \frac{e^2}{R} = \text{electrostatic energy} \tag{6-3.23}$$

There are pairs of nuclei having the same A such that either nucleus can be obtained from the other by transforming all the neutrons of one into protons and all the protons into neutrons. Examples of these *mirror nuclei*, as they are called, are ^{3}H and ^{3}He, ^{11}B and ^{11}C, ^{13}C and ^{13}N, and ^{21}S and ^{21}P. There is reason to believe that the specific nuclear forces acting in such nuclei are practically identical and that the energy difference, apart from the different rest mass of neutron and proton, is entirely of electrostatic origin. When the nuclei differ by one unit in Z, Eq. (6-3.23) gives, for the difference in their electrostatic energy,

$$\frac{3}{5} \frac{(2Z + 1)e^2}{R} = \Delta E \tag{6-3.24}$$

and if they also have $Z = (A \pm 1)/2$, respectively, as, for example, ^{11}C and ^{11}B, we rewrite Eq. (6-3.24) as

$$\Delta E = \frac{3e^2}{5} \frac{A}{R} = \frac{3}{5} \frac{e^2}{r_0} A^{2/3} \tag{6-3.25}$$

if, as we have repeatedly assumed,

$$R = r_0 A^{1/3} \tag{6-3.26}$$

From this relation it is clear that a plot of ΔE versus $A^{2/3}$ should give a straight line. From its slope we obtain $r_0 = 1.30 \times 10^{-13}$ cm (Fig. 6-9). The departures from the straight line are due to finer effects (shell closure), which will be discussed later on. All these methods give results in agreement with Eq. (6-3.1), with constants r_0 varying between 1.2 and 1.5×10^{-13} cm for all but the lightest nuclei.

Nuclear radii are also obtained from the study of the departure from the simple Bohr hydrogen-atom formula that is observed in the mu-mesic atoms. (See Sec. 6-4.)

ISBN 0-8053-8601-7

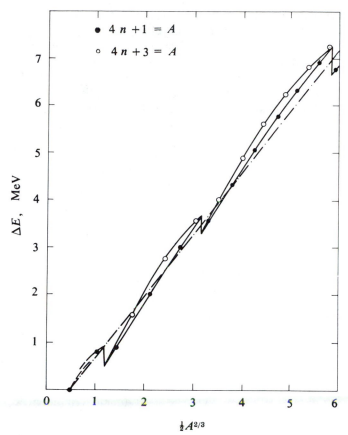

Figure 6-9 Experimental Coulomb energy differences for mirror nuclei as a function of $\frac{1}{2}A^{2/3}$. The dashed-and-dotted line shows the simple classical approximation for $r_0 = 1.30$ F. The curved lines are based on more refined models. [O. Kofoed-Hansen, *Rev. Mod. Phys.*, **30**, 449 (1958).]

Equation (6-3.1) and its refinements cannot be properly applied to the proton, neutron, deuteron, or lightest nuclei. We have already mentioned the case of the proton. The deuteron will be discussed in detail in Chap. 10; it has an especially loose structure, and its "radius," defined in Chap. 10, is 4.32×10^{-13} cm.

6-4 MESIC AND EXOTIC ATOMS

This section, inserted here for convenience, is relevant to several subjects treated in different parts of this book.

When a negatively charged particle (muon, pion, kaon, antiproton, Σ^-, etc.) stops in a material of atomic number Z, the particle is captured by the nucleus and forms a hydrogen-like atom. Because the Bohr radius $n^2\hbar^2/Ze^2m$

ISBN 0-8053-8601-7

$= a$ is inversely proportional to the mass of the particle, and the particles considered are much heavier than the electron, the particle orbit is much smaller than the electronic orbits and the electron shielding is mostly negligible. In jumping from one orbit to the other, the system emits electromagnetic radiation, mostly in the X-ray frequency range.

As long as the Bohr radius is large compared with the nuclear radius the energy levels for muons are given by Dirac's formula:

$$E = mc^2 \left[1 + \left(\frac{\alpha Z}{n - k + (k^2 - \alpha^2 Z^2)^{1/2}} \right)^2 \right]^{-1/2} - mc^2 \qquad (6\text{-}4.1)$$

Here $k = j + \frac{1}{2}$ for particles of spin $\frac{1}{2}$ (and $k = l$ for particles of spin 0), α is $e^2/\hbar c = (1/\text{``137''})$, m is the reduced mass of the particle and nucleus, and n is the total quantum number. For $\alpha^2 Z^2 < 1$, this formula can be developed in powers of αZ obtaining to terms in $\alpha^4 Z^4$,

$$E = - \frac{mc^2 Z^2}{2n^2} \alpha^2 \left[1 + \frac{\alpha^2 Z^2}{n} \left(\frac{1}{j + \frac{1}{2}} - \frac{3}{4n} \right) \right] \qquad (6\text{-}4.2)$$

This equation gives, for the separation of a doublet having a given n and $j = l \pm \frac{1}{2}$,

$$\Delta E = \frac{\alpha^4 Z^4}{2n^3} mc^2 \frac{1}{l(l+1)} \qquad (6\text{-}4.3)$$

For a particle having an anomalous magnetic moment $\mu = (1 + g_1)(e\hbar/2m'c)$ where m' is the mass of the particle and μ its magnetic moment, ΔE is multiplied by $(1 + 2g_1)$. The measurement of ΔE has shown the equality of the $\bar{p}, p$ magnetic moments to an accuracy of about 1% and may become an important method for measuring magnetic moments of other particles.

The radiative transition probabilities are obtainable by scaling those of hydrogen. In addition, the particle may lose energy through the Auger effect (i.e., by transferring it to one of the electrons in the atom). Since the heavy-particle orbit is very small compared to electron orbits, for these it is as if the nucleus underwent an electric dipole transition, which has a definite conversion coefficient (see Chap. 8). The result of these deexcitation processes is qualitatively to push the heavy particles into orbits with the maximum possible angular momentum ($n = l + 1$).

As the particle passes to lower and lower orbits, the overlap of its wave function with the nucleus becomes important. We must then distinguish the different particles. Muons interact only electromagnetically and the levels are perturbed by the finite nuclear size. In fact, if the orbit is smaller than the nuclear radius, the potential and the levels are those of a harmonic oscillator (see p. 222). The study of the perturbation of the hydrogenic levels by the finite nuclear size gives information on the latter. The observations confirm the nuclear radius $r_0 A^{1/3}$ and give a value for r_0 of approximately 1.2 F.

ISBN 0-8053-8601-7

For strongly interacting particles the situation is more complicated. As soon as the wave function overlaps the nucleus appreciably, the orbital particle reacts with the nucleus. If the probability of this reaction is large compared with the radiation probability, the level broadens and the X radiation disappears. The series are thus truncated at a certain n. The shifts and broadening of the levels give information on the particle–nucleus interaction.

The position of the unperturbed levels (generally those with high n) gives the reduced mass of the particle because all hydrogenous spectra are similar and the frequencies scale with the mass of the particle. We infer the mass from the observed scaling factor. This method of mass determination in some cases has given the most precise mass values.

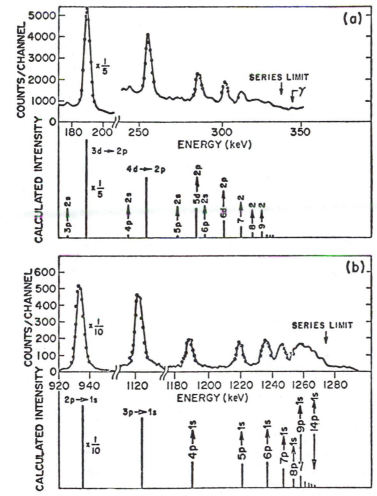

Figure 6-10 (a) Balmer series $(nd \rightarrow 2p)$ and (b) Lyman series $(np \rightarrow 1s)$ of muonic transitions in Ti. (D. Kessler et al. *Phys. Rev. Lett.* **18**, 1179 (1967).)

ISBN 0-8053-8601-7

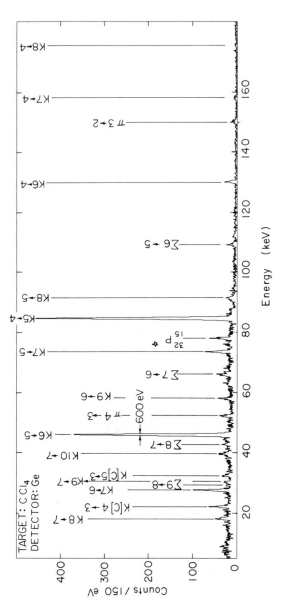

Figure 6-11 Spectra obtained by K^- mesons stopping in CCl_4. The lines marked $K10 \rightarrow 7$ are kaonic lines corresponding to a transition $n' = 10 \rightarrow n'' = 7$ in Cl. Other lines are of pionic or Σ^- origin. The π^- and the Σ^- are secondary and are obtained by the reaction $K^- + N \Rightarrow \Sigma + \pi$ in nuclei. One line is due to the excited ^{32}P nucleus. [Courtesy C. E. Wiegand.]

ISBN 0-8053-8601-7

ISBN 0-8053-8601-7

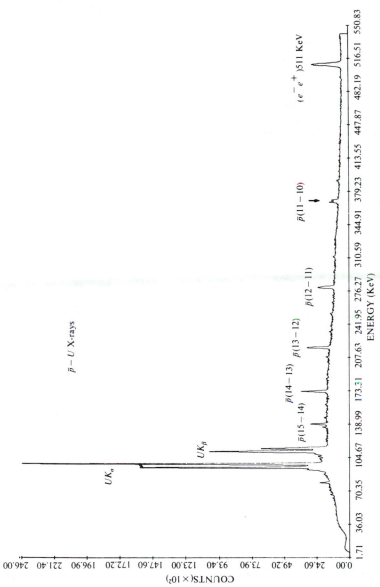

Figure 6-11a Spectrum of an antiprotonic uranium atom. Note the doublet in the 11–10 line from which the magnetic moment of $\bar{p}$ is obtained. [Courtesy C. S. Wu.]

The main contributions to the departure of the energy of mesic atoms from the simple hydrogen formula stem from nuclear interactions, vacuum polarization, electron screening, and finite nuclear size. One can always find levels for which these corrections are small and calculable, and thus use the spectroscopic data to find the reduced mass and the mass of the orbital particle. Experimentally the art has been refined by using solid-state detectors. Figure 6-10 shows a muonic spectrum, Fig. 6-11 a spectrum containing pionic, kaonic, and Σ-mesic lines, and Fig. 6-11a a spectrum containing antiprotonic lines.

6-5   NUCLEAR STATISTICS

Systems containing identical particles show, according to quantum mechanics, a peculiar behavior that affects many of their properties. We shall treat here the simplest cases of direct interest to nuclear physics.

Given two identical particles (e.g., two alphas, two electrons, two protons, etc.), we write the Schrödinger equation for the system:

$$(\hbar^2/2m)(\nabla_1^2 + \nabla_2^2)u(r_1, r_2) + \left[E - V(r_1, r_2) \right]u(r_1, r_2) = 0 \quad (6\text{-}5.1)$$

or, symbolically,

$$H(1, 2)u(r_1, r_2) = Eu(r_1, r_2) \quad (6\text{-}5.2)$$

indicating by r_1 or 1, r_2 or 2, all coordinates, including the spin of particles 1 and 2.

We now find that the solutions are degenerate. That is, if $u(r_1, r_2)$ is a solution of the problem for a certain eigenvalue, so necessarily is $u(r_2, r_1)$ for the same eigenvalue. This follows from the fact that

$$H(1, 2) = H(2, 1) \quad (6\text{-}5.3)$$

because of the indistinguishability of identical particles. Hence a linear combination

$$U(r_1, r_2) = \alpha u(r_1, r_2) + \beta u(r_2, r_1) \quad (6\text{-}5.4)$$

is also a solution for the same eigenvalue. The choice of constants $\alpha = \pm\beta$, real, has the important property that if $\alpha = \beta$,

$$U(r_1, r_2) = U(r_2, r_1) \quad (6\text{-}5.5)$$

If $\alpha = -\beta$,

$$U(r_1, r_2) = -U(r_2, r_1) \quad (6\text{-}5.6)$$

ISBN 0-8053-8601-7

In the first case, the eigenfunction is symmetric with respect to the exchange of the particles; in the second, it is antisymmetric. If the system is initially described by a symmetric or by an antisymmetric eigenfunction, it will be described by the same type of eigenfunction forever, because the change of the eigenfunction is determined by a symmetric hamiltonian, and this ensures the preservation of the symmetry or antisymmetry. In order to see this, observe that according to Schrödinger's equation

$$-\frac{\hbar}{i}\frac{\partial \psi}{\partial t} = H\psi \tag{6-5.7}$$

and hence the change in time of ψ, $\partial\psi/\partial t$ is a symmetric or antisymmetric function together with $H\psi$. But since H is symmetric, $H\psi$ has the same symmetry as ψ and consequently the time variation of ψ also has the same symmetry as ψ. Observation demonstrates that natural particles have eigenfunctions that are either symmetric or antisymmetric with respect to the exchange of two identical particles. For instance, electrons, protons, neutrons, and, in general, particles with spin $\frac{1}{2}$ all have eigenfunctions that are antisymmetric with respect to the exchange of the coordinates, whereas particles with spin 0, 1, or, in general, integral spin, like alpha particles or deuterons, have eigenfunctions that are symmetric with respect to the exchange of the coordinates. These facts are among the most important in quantum mechanics and have far-reaching consequences. For instance, consider two noninteracting identical particles. The hamiltonian then separates into the sum of the two terms

$$H(1) + H(2) = H(1, 2) = H(2, 1) \tag{6-5.8}$$

If

$$u_1(1), u_2(1), u_3(1), \ldots$$
$$u_1(2), u_2(2), u_3(2), \ldots$$

are the eigenfunctions of $H(1)$, $H(2)$, etc., corresponding to eigenvalues E_1, E_2, E_3, . . . and normalized in such a way that

$$\int |u_1|^2 \, d\tau = \int |u_2|^2 \, d\tau = \cdots = 1 \tag{6-5.9}$$

then the eigenfunction of $H(1, 2)$ belonging to the eigenvalue $E_n + E_m$ is degenerate and has the form

$$u_{nm}(1, 2) = 2^{-1/2}\left[u_n(1)u_m(2) \pm u_m(1)u_n(2) \right] \tag{6-5.10}$$

where we must take the plus sign if the eigenfunctions are symmetric or the minus sign if they are antisymmetric.

ISBN 0-8053-8601-7

If only antisymmetric solutions are admissible, for m equal to n, $u(1, 2)$ vanishes. Hence it is not possible for two identical antisymmetric particles to be in the same state. This is the original formulation of Pauli's exclusion principle.

In statistical mechanics, particles having antisymmetric eigenfunctions are governed by Fermi–Dirac statistics, and particles having symmetric eigenfunctions by Bose–Einstein statistics. For this reason, particles having antisymmetric wave functions are called fermions, and particles having symmetric wave functions are called bosons.

Whether a particle is a boson or a fermion depends on its spin. The empirical evidence accumulated over a long period of years shows that bosons have integral spin and fermions have half-integral spin. This striking connection between spin and statistics has very deep theoretical reasons which were finally elucidated by Pauli (1940). Before Pauli's argument it was important to establish empirically whether a given nucleus was a boson or a fermion. This can be done in several ways. One of the historically most important and most ingenious methods is based on the analysis of alternating intensities in the rotational spectra of homonuclear diatomic molecules.

● We give here a somewhat oversimplified version of the molecular physics involved. Consider the eigenfunction of a diatomic molecule having identical nuclei. It can be factored, approximately, into four eigenfunctions: one containing the electronic coordinates, a second the relative distance of the two nuclei and corresponding to the vibration of the nuclei, a third the rotational coordinates of the nuclei, and a fourth the nuclear spin coordinates:

$$u = u_{el}u_{vib}u_{rot}u_{spin} \tag{6-5.11}$$

where u_{el} depends on the coordinates of the electrons. Its symmetry character on the exchange of the nuclei can be analyzed by a detailed study of the molecule. For the sake of definiteness, let us assume that it is symmetric. The u_{vib} factor is symmetric because it depends only on the magnitude of the distance between the nuclei. The rotational part of the eigenfunction is the eigenfunction of a free rotator in space,

$$u_{rot} = P_l^m (\cos \theta)e^{im\varphi} \tag{6-5.12}$$

where θ and φ are the colatitude and the longitude of the axis joining the two nuclei. The function u_{rot} is symmetric for even l and antisymmetric for odd l. This follows from the fact that $P_l^m(\cos \theta)e^{im\varphi}$ is a spherical harmonic of order l that is r^{-l} times a polynomial in x, y, z, of degree l, homogeneous and harmonic. The interchange of the nuclei corresponds to a change in the variables of the function: θ goes to $\pi - \theta$ and φ to $\pi + \varphi$; or, in cartesian coordinates, x, y, z change sign. This leaves the function unchanged if l is even or changes its sign if l is odd.

In order to analyze the fourth factor we need spin eigenfunctions. For spin $\frac{1}{2}$ the functions have only two values (spin up or spin down), indicated as $\alpha \equiv$ spin up or $\beta \equiv$ spin down. The particle considered is indicated as the argument of the function. Thus, $\alpha(2)$ means spin up for the second particle. For a complete explanation we refer the reader to texts on quantum mechanics. In the case of two spin $\frac{1}{2}$ particles, the last factor concerning the nuclear spins can then be written as follows: If the two spins are parallel, the eigenfunction u_{spin} has one of the three following forms corresponding to the three triplet states:

$$\alpha(1)\alpha(2); \qquad 2^{-1/2}[\alpha(1)\beta(2) + \alpha(2)\beta(1)]; \qquad \beta(1)\beta(2) \qquad (6\text{-}5.13)$$

If the spins are antiparallel, the eigenfunction has the form

$$2^{-1/2}[\alpha(1)\beta(2) - \alpha(2)\beta(1)] \qquad (6\text{-}5.14)$$

The triplet states are therefore symmetric, and the singlet state is antisymmetric with respect to the exchange of the spins only. If the whole eigenfunction has to be antisymmetric with respect to the exchange of the nuclei, it is clear that eigenfunctions with l odd must be associated with triplet states. Conversely, states of even l are all associated with singlet states. It follows that the statistical weights of states with l odd are three times as large as those of states with l even.

In the emission or absorption of electromagnetic radiation the symmetry character of the level must be preserved, as shown at the beginning of this section. Moreover, electromagnetic radiation, in practice, produces transitions that change only the factor of the eigenfunction containing ordinary coordinates ($E1$ transitions). It follows that spectral lines are divided between those connecting states that are symmetric with respect to the exchange of ordinary (not spin) coordinates of the nuclei (symmetric lines) and those connecting states antisymmetric with respect to the exchange of ordinary coordinates (antisymmetric lines). The intensity of the spectral lines, other things being equal, is proportional to the statistical weight of the states. From this it follows that, in a band in which the rotational states do not change, we find a characteristic alternation of intensities. In the example considered above lines corresponding to l odd would be three times as strong as those corresponding to l even. It is clear that in this argument it is essential to assume that the two nuclei are fermions; if they were bosons, the lines with even l would have been stronger. Similar alternations of intensities are shown in the Raman effect of homonuclear diatomic molecules. An example ($^{14}N_2$) is shown in Fig. 6-12.

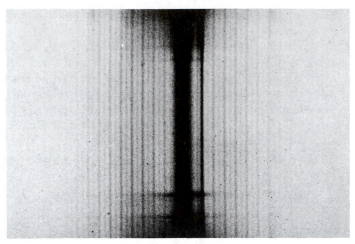

Figure 6-12 Alternating intensities in the Raman spectrum of the $^{14}N_2$ molecule (as observed by F. Rasetti).

The reasoning may be generalized to include the treatment of an arbitrary nuclear spin. The statistical weights of neighboring rotational states are in the ratio

$$(I+1)/I = \rho \qquad (6\text{-}5.15)$$

For fermions the greater weight belongs to the states that are antisymmetric with respect to the exchange of the spatial coordinates (spin thus excluded) of the identical nuclei. For bosons the smaller weight belongs to the same type of state. The ratio of the intensities giving ρ of Eq. (6-5.15) may be used to determine I. Note that for $I = 0$, half the levels are missing. ●

The main results of this and other methods are that all nuclei with A even have an integral spin and are bosons; all nuclei with odd A have a half-integral spin and are fermions. These rules have been of great importance in establishing the model of the nucleus formed by neutrons and protons.

6-6   THE NUCLEUS AS A FERMI GAS

Until about 1931 nuclear models treated the proton and the electron as fundamental constituents of the nucleus. These particles were the simplest known to exist in the free state; and electrons, since they are ejected by nuclei in beta decay, could reasonably be expected to be present inside nuclei. A nucleus of mass number A and charge Z was supposed to contain A protons and $A - Z$ electrons. This hypothesis accounted for the charge and mass of

ISBN 0-8053-8601-7

the nucleus, because electrons are very light and the Z electrons did not greatly affect the nuclear mass. It was the most natural hypothesis in many respects, but serious objections were discovered about 1930. In 1929 Rasetti showed by a study of the Raman spectrum of the $^{14}N_2$ molecule (Fig. 6-12) that ^{14}N is a boson (see Sec. 6-5) and hence must contain an even number of fermions. It is clear that according to the electron–proton hypothesis ^{14}N contains 14 protons and 7 electrons that is, 21 fermions, in contradiction to experiment, which requires it to contain an even number of fermions. A second difficulty is that a nucleus supposed to contain an odd number of fermions, namely ^{14}N, was found experimentally to have an integral spin instead of the expected half-integral spin.

In addition, the confining of the electron to a region in space of nuclear dimensions (10^{-13} cm) presents serious difficulties because, through the uncertainty principle of quantum mechanics, it implies very high kinetic energy for the electron, and an implausibly high potential to bind the electron.

All these difficulties were eliminated by Chadwick's discovery, in 1932, of the free neutron, of mass nearly equal to that of the proton. Immediately thereafter Ivanenko and Heisenberg independently proposed a nuclear model containing neutrons and protons as fundamental constituents. This model assumes that the neutron is a fermion of spin $\frac{1}{2}$. Nuclei contain Z protons and $A - Z = N$ neutrons and thus, in the case of ^{14}N, 7 neutrons and 7 protons (i.e., an even number of fermions), which agrees with the boson character of ^{14}N and with its integral spin. Moreover, the relatively large mass of the neutron and proton removes the difficulties created by the uncertainty principle. The neutron–proton model of the nucleus has been universally accepted and forms the basis of all studies of nuclear structure.

The simplest of all models neglects the nucleon–nucleon interaction, or better, represents it by a potential well that determines the size of the nucleus. The nuclear system, when A is not too small, is complicated enough to warrant the use of statistical mechanics, provided one keeps in mind that in ordinary circumstances the ensembles of fermions considered are in a state of extreme degeneracy and thus need to be treated by Fermi statistics. Consider a nucleus as a collection of free neutrons and protons enclosed in a sphere of radius R or volume Ω. The neutrons and protons are fermions and thus two identical particles must be in different quantum states according to the exclusion principle. The number of states corresponding to a momentum smaller than P_F for protons or for neutrons is

$$n = \frac{2}{(2\pi\hbar)^3} \ \frac{4}{3} \ \pi\Omega P_F^3 \qquad (6\text{-}6.1)$$

which is obtained by dividing the phase space into cells of volume $(2\pi\hbar)^3$ and assigning two particles to a cell, in order to take into account the statistical

ISBN 0-8053-8601-7

weight of states of particles of spin $\frac{1}{2}$. At complete degeneracy (i.e., for the ground state), we have

$$P_F^{\text{proton}} = (3\pi^2)^{1/3}\hbar\left(\frac{Z}{\Omega}\right)^{1/3} = \hbar k_F^{\text{proton}} \tag{6-6.2}$$

$$P_F^{\text{neutron}} = (3\pi^2)^{1/3}\hbar\left(\frac{A-Z}{\Omega}\right)^{1/3} = \hbar k_F^{\text{neutron}} \tag{6-6.3}$$

In crude approximation $Z = N = A/2$; furthermore if one puts $\Omega = (4/3)\pi R^3$, with $R = r_0 A^{1/3}$, one has $\Omega = 4.18 r_0^3 A$, and expressing r_0 in fermis one finds

$$P_F = \frac{297}{r_0} \text{ MeV}/c \tag{6-6.4}$$

or $k_F = 1.50/r_0$ F^{-1} independent of A. The corresponding kinetic energy $(P_F^2/2M = E_F)$ is called the *Fermi energy*. For $r_0 = 1.3$ F, E_F is 28 MeV. This is the maximum kinetic energy of a neutron bound in the nucleus. If the binding energy of the last nucleons is 8 MeV, the potential energy must then be 36 MeV (Fig. 6-13).

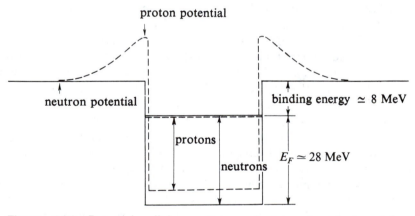

Figure 6-13 Potential well for neutrons and protons in a heavy nucleus, showing the Fermi level. Note the difference between neutron and proton wells.

For protons the situation is similar except for the potential barrier. However, a nucleus contains fewer protons than neutrons and hence the maximum kinetic energy of the protons is less than that of the neutrons. The total energies for neutron or proton are the same; otherwise the nucleus would transform a neutron into a proton, or vice versa, by a beta process. We must thus conclude that the potential energy for protons is smaller than that for neutrons, an effect in part due to Coulomb repulsion.

ISBN 0-805-3801-1

The expression for the kinetic energy of all the protons in a nucleus, considered as a degenerate gas at absolute zero, is

$$E_0^{\text{protons}} = \frac{2\Omega}{(2\pi\hbar)^3} \int_0^{P_F} \frac{p^2}{2M} 4\pi p^2 \, dp = \frac{\pi^{4/3} 3^{5/3}}{10} \left(\frac{Z}{\Omega}\right)^{2/3} Z \frac{\hbar^2}{M} = \frac{3}{5} E_F Z$$

(6-6.5)

with a similar formula for neutrons.

The Fermi-type energy distribution of nucleons in a heavy nucleus is directly shown by (e, e') scattering. (See Fig. 6-14.)

The Fermi gas model also gives useful, if crude, information on nuclear excited states. (See Chap. 11.) A refinement of the free particle model takes into account, semiphenomenologically, some of the interactions neglected in the Fermi gas: surface effects, Coulomb repulsion, and the tendency to have

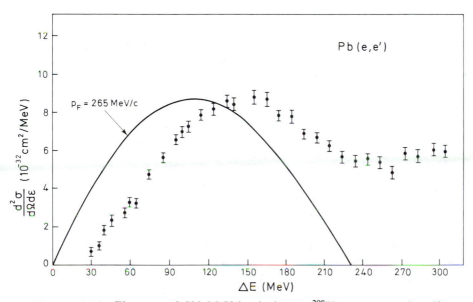

Figure 6-14 Electrons of 500 MeV impinging on ^{208}Pb are scattered at 60 deg. The experimental points show the differential cross section as a function of energy loss. The solid curve shows the theoretical result for a Fermi distribution having $P_F = 265$ MeV/c corresponding to $r_0 = 1.15$ F. P_F determines the width of the distribution. The position of the maximum is shifted from the theoretical one because the effective mass of the nucleon in nuclear matter is different from that of a free nucleon. (See Sec. 6-13) [From A. Bohr and B. R. Mottelson, *Ann. Rev. Nucl. Sci.*, **23**, 363 (1973).] The large cross sections observed for energy losses $\Delta E \gtrsim 240$ MeV cannot be accounted for in terms of quasi-elastic scattering from a Fermi gas. In this energy region, important contributions arise from meson production as well as from the presence of nucleons with high momenta, resulting from violent short-range interactions.

ISBN 0-8053-8601-7

an equal number of protons and neutrons. This is done, for instance, in the liquid-drop model and may be justified, to an extent, from first principles. Here, however, we treat it on an empirical basis.

6-7   THE LIQUID DROP MODEL

This very simple model can reproduce many important features of nuclei with few empirical parameters. If we plot the number of neutrons contained in the nucleus as a function of Z for all stable nuclei, we obtain the diagram in Fig. 6-15, sometimes called the *Segrè chart*. Postponing the consideration of fine points, we see that there is a relation between Z and N, because representative points are restricted to a small region, one would say almost a curve, in the Z–N plane. An important requirement for a nuclear model is thus to account for this regularity.

Figure 6-3 shows the packing fraction and mass decrement as a function of A. The packing fraction and mass decrement are smooth functions of A. If the packing fraction were zero, the binding energy per nucleon would be about 7.6 MeV, which is the average mass excess of neutron and proton. The actual values found indicate that the average binding energy per nucleon varies between about 6 and 8 MeV, as shown in Fig. 6-16.

The fact that both the binding energy per nucleon and the density of nuclear matter are almost independent of A shows the resemblance of nuclei to liquid droplets, where the heat of vaporization and the density of the liquid are independent of the size of the droplet. Pursuing this analogy rather literally, we shall try to express the mass of the stable nuclei as a function of A and Z, in agreement with the experimental facts.

The largest term in the mass will clearly be

$$ZM_p + NM_n = (A - Z)M_n + ZM_p \tag{6-7.1}$$

However, since nuclei are bound, the nuclear mass must be smaller than this quantity. Since the binding energy per nucleon is almost constant, for a first approximation we should subtract from $(A - Z)M_n + ZM_p$ a positive quantity proportional to A in order to represent the total binding energy. Thus, we add to Eq. (6-7.1) the quantity

$$- a_{vol}A = M_1 \qquad a_{vol} > 0 \tag{6-7.2}$$

The droplet analogy suggests the possibility of the existence of surface effects. Actually a nucleon near the nuclear surface is not expected to be bound as strongly as a nucleon in the interior, because it has nucleons on only one side instead of all around. In subtracting $a_{vol}A$ we have thus over-corrected, and we must add a term proportional to the number of nucleons on the nuclear surface. We know that the nuclear radius is proportional to $A^{1/3}$; hence, the nuclear surface is proportional to $A^{2/3}$, and this surface

ISBN 0-8053-8061-7

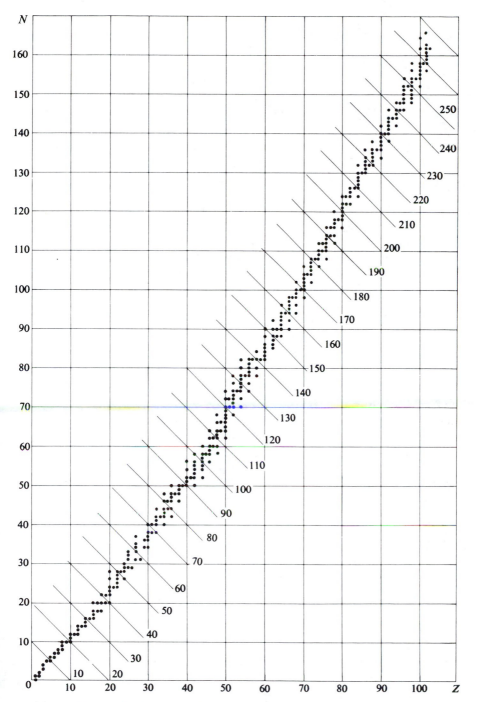

Figure 6-15 Chart of all beta-stable nuclei in a Z–N plane.

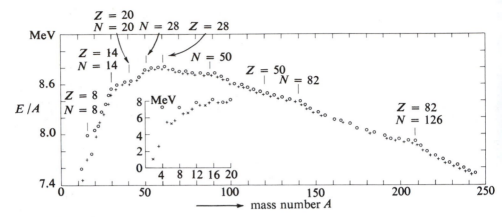

Figure 6-16 Binding energy per nucleon of the most beta-stable isobars as a function of A. Points refer to even-even nuclides, crosses to nuclides of odd mass. The positions of magic numbers are indicated. Insert: Detailed picture for the lowest masses, with the four known beta-stable odd-odd nuclides indicated by oblique crosses. [From A. H. Wapstra.]

effect will give a contribution to the mass, $+a_{surf}A^{2/3}$, of opposite sign from $-a_{vol}A$. We thus add a term

$$a_{surf}A^{2/3} \qquad a_{surf} > 0 \tag{6-7.3}$$

If we examine Fig. 6-15 we see that there is a tendency for Z to equal N, at least for light nuclei. For heavy nuclei Z becomes smaller than N, because the Coulomb repulsion requires energy to accumulate charge in the nuclear volume. We want to express the fact that there is a tendency to favor $N = Z$. This can be done by adding to the mass a term proportional to A and depending on N/Z, with a minimum for $N = Z$. We want this term to be proportional to A by analogy with a liquid in which, given the composition, the energy is proportional to the amount of liquid. A simple expression satisfying these conditions is

$$M_3 = \tfrac{1}{4} a_{sym} A \frac{\left[1 - (Z/N)\right]^2}{\left[1 + (Z/N)\right]^2}$$

$$= \tfrac{1}{4} a_{sym} \frac{(N - Z)^2}{A} = a_{sym} \frac{\left[(A/2) - Z\right]^2}{A} \qquad a_{sym} > 0 \tag{6-7.4}$$

Another term takes into account the Coulomb repulsion between the protons in the nucleus. It will be positive and of the form

$$M_4 = \xi \frac{Z^2 e^2}{r_0 A^{1/3}} \tag{6-7.5}$$

where ξ is a number depending on the radial dependence of the charge distribution; e.g., for a uniform density $\xi = \frac{3}{5}$.
We prefer to write

$$M_4 = \frac{3}{5} \frac{Z^2 e^2}{R_c} \tag{6-7.6}$$

with $R_c = 1.24 A^{1/3}$ F, one has $M_4 = 0.6965 Z^2 A^{-1/3}$ (in MeV).
Finally, there is a term $\delta(A - Z, Z)$ that takes into account the increased stability when N and Z are even, as opposed to cases where either Z or N (or both) is odd (see Sec. 6-2). The final formula for the *atomic mass*, neglecting the binding energy of the electrons, is thus

$$M(A, Z) = (A - Z)M_n + ZM_p + Zm_e - a_{\text{vol}}A + a_{\text{surf}}A^{2/3}$$

$$+ a_{\text{sym}} \frac{\left[(A/2) - Z \right]^2}{A} + \frac{3}{5} \frac{e^2 Z^2}{R_c} + \delta \tag{6-7.7}$$

Equation (6-7.7) is called the *Weizsäcker mass formula*. We know that $M_n = 939.573$, $M_p = 938.280$, and $m_e = 0.511$ in MeV (Table 6-1).
The determination of the constants a_{vol}, a_{surf}, a_{sym}, and R_c must take into account as many facts as possible. For instance, we could take all the measured masses and fit them by a least-squares adjustment of the constants. However, we must use not only the masses, but also other data, such as the energy necessary for producing fission (see Chap. 11), because some coefficients are insensitive to mass but are affected strongly by a different type of information. Considerable effort has been spent on this determination. A set of constants often used is given in Table 6-3, where $\delta(e, o)$ means δ for N even, Z odd, etc.

TABLE 6-3 CONSTANTS FOR THE WEIZSÄCKER MASS FORMULA IN MeV

a_{vol}	a_{surf}	a_{sym}	$\delta(e, o)$	$\delta(o, e)$	$\delta(e, e)$	$\delta(o, o)$
15.67	17.23	93.15	0	0	$-12A^{-1/2}$	$+12A^{-1/2}$

To study the δ term further, consider isobaric nuclei, which can be transformed into one another by beta decay (A constant, $\Delta Z = +1$) or by orbital electron capture or positron emission (A constant, $\Delta Z = -1$). If Eq. (6-7.7) contained all but the last term, as in the case for A odd, where the δ term is zero, the masses of a series of isobars, considered as a function of Z, would have a minimum for a certain value of $Z = Z_0$ and would lie on a parabola on a Z–M plane (Fig. 6-17). The only stable nucleus of mass A would then be the one with $Z = Z_0$. Rarely, it may happen that two neighboring nuclei with $Z = Z_0$ and $Z_0 + 1$ or $Z = Z_0$ and $Z_0 - 1$ have very nearly the same energy; thus, whereas one is stable, the other has such a long life as to be semistable.

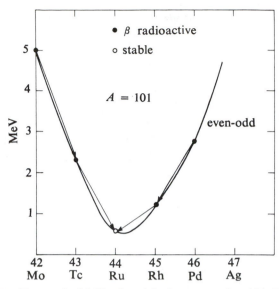

Figure 6-17 Energy (in MeV) of nuclei of mass number 101. The zero point on the energy scale is arbitrary.

Strictly speaking, however, only one nucleus of a given odd A is expected to be stable. On examining a table of isotopes we find that this is true.

The situation is different for nuclei with even A. Here we find that there may be several stable isobars, which however have values of Z differing by at least two units. The last term of Eq. (6-7.7) describes this situation. If N and Z are both odd, the masses are increased by δ above the dotted line (Fig. 6-18), which represents Eq. (6-7.7) without the δ term. If N and Z are both even, the mass is decreased by δ and lies on the lower line. It is clear from the figure that in these circumstances there may be more than one beta-stable isotope of mass A, and that such isotopes must differ by at least two units in Z. These stability rules account for, among other things, the fact that there are no stable isotopes of elements 43 (Tc) and 61(Pm). Figure 6-19 shows the empirical values of δ and our approximate formula.

The average accuracy of Eq. (6-7.7) is about 2 MeV except at places where there are strong shell effects (see Sec. 6-13).

An interesting application of the mass formula is the determination of the nuclear radius by using it as a parameter in a_{surf} and a_{vol} and fitting the constants to the experimental masses. Green obtained a value determined solely by mass measurements of $r_0 = 1.237 \times 10^{-13}$ cm, which is in excellent agreement with the other methods previously mentioned.

Equation (6-7.7) is very useful any time that a panoramic view of nuclear properties is needed. For instance, it gives the relation' between A and Z for stable nuclei: by setting $(\partial M / \partial Z)_A = 0$, we obtain the Z for which a series of

ISBN 0-8053-8601-7

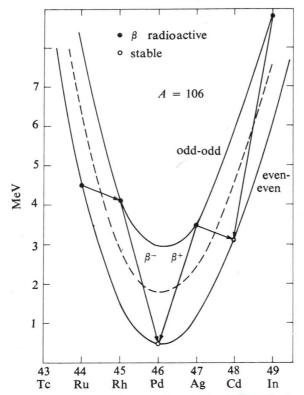

Figure 6-18   Energy (in MeV) of nuclei of mass number 106. The zero point on the energy scale is arbitrary.

isobars has the minimum mass. Using Eq. (6-7.7) we have

$$ -0.7825 - \frac{2a_{sym}}{A}\left(\frac{A}{2} - Z\right) + \frac{3}{5}\frac{2Ze^2}{R_c} = 0 \qquad (6\text{-}7.8) $$

which, recalling that $A = Z + N$, we can interpret as the equation of the "curve" of Fig. 6-15. Other interesting applications of Eq. (6-7.7) involve the calculations of the energy released in the fission of heavy nuclei and the calculations of the limits (Fig. 7-7) of alpha stability (see Chap. 7); however, Eq. (6-7.7) has to be further refined for many practical purposes, such as fission studies.

Swiatecki, Myers, and others have carefully analyzed the mass formula, trying to establish it on a more rigorous basis as an expansion of the nuclear hamiltonian in powers of $A^{1/3}$. They have also taken into account some additional effects, such as the compressibility of nuclear matter, change of surface

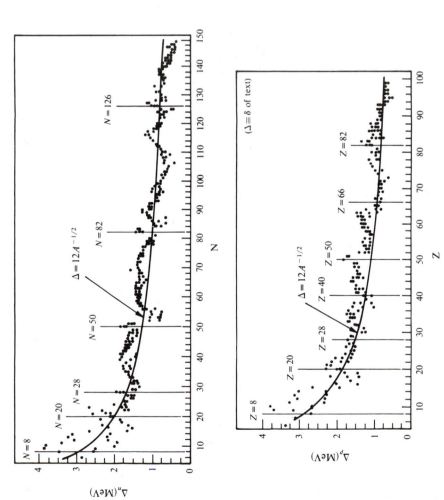

Figure 6-19 The odd-even mass differences for neutrons and protons ($\Delta \equiv \delta$ of text) are based on the analysis of N. Zeldes, A. Grill, and A. Simievic, *Mat. Fys. Skr. Dan. Vid. Selsk.*, **3**, No. 5 (1967).

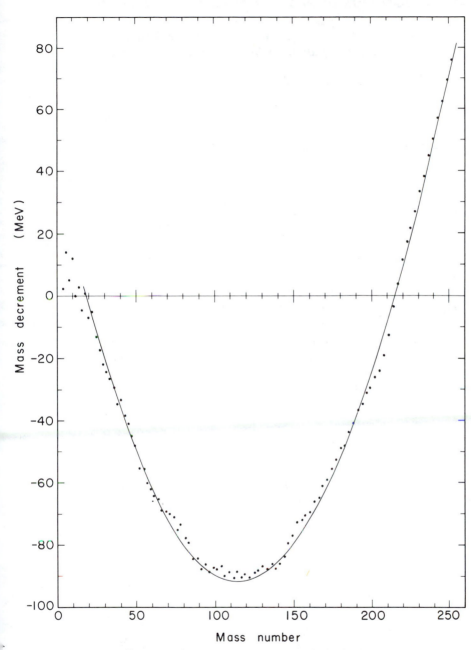

Figure 6-20 The mass decrements of 97 beta-stable nuclei are compared with the smooth curve coresponding to the liquid-drop part of the mass formula [Eq. (6-7.9)]. Note that the overall trend of the decrements is reproduced throughout the periodic table, including the light nuclei. The scatter of the points is due to shell effects. Myers W. D. and W. J. Swiatecki, *Nucl. Phys.*, **81**, 1 (1966).

tension with A, and most important, shell and deformation effects. For a spherical nucleus their equation (in MeV) gives for the atomic mass

$$E = \left[M - Z(M_p + m_e) - N M_n \right] c^2$$

$$= -15.68A + 18.66A^{2/3} + 28.1(A - 2Z)^2 A^{-1}(1 - 1.18A^{-1/3})$$

$$+ 0.717Z^2 A^{-1/3}(1 - 1.69A^{-2/3}) + \text{pairing term} + \text{shell effects} \qquad (6\text{-}7.9)$$

The term in $(N - Z)^2$ is the symmetry energy and $1.18A^{-1/3}$ is a correction to it for the surface energy. The term $0.717Z^2A^{-1/3}$ is the Coulomb energy and $1.69A^{-2/3}$ a correction to it due to exchange forces. The pairing term and the shell effects have to be added separately. Figure 6-20 shows E, as calculated from Eq. (6-7.9), without pairing and shell effects, compared with the experimental data. We refer the reader to the literature for the shell, deformation, and pairing terms.

A different approach to the calculation of nuclear masses, valid in limited regions of the Z–N plane, uses charge independence and mass relations between nuclei forming i-spin multiplets (see Chap. 11). This basic idea may be practically exploited by forming linear combinations of nuclear masses which must have a null sum. Figure 6-21 shows two particularly useful cases. The masses located at the sites marked $+$ or $-$ are added or subtracted and the sum is zero. In symbols, Fig. 6-21a gives the relation

$$M(N + 2, Z - 2) - M(N, Z) + M(N, Z - 1)$$

$$- M(N + 1, Z - 2) + M(N + 1, Z) - M(N + 2, Z - 1) = 0$$

and a similar one is obtained from Fig. 6.21b. These relations are valid for $A > 16$ and $N \geqslant Z$ and N even for $N = Z$. For $N \leqslant Z$ the corresponding figures have N, Z in the upper left-hand corner. These and similar more

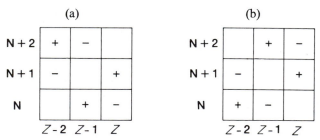

Figure 6-21 A portion of the Z–N plane. The masses of nuclei placed at spots marked $+$ or $-$, when summed algebraically with the signs indicated, add to zero; hence, if one mass is unknown, it can be determined from the other masses. The figures are valid for $N \geqslant Z$ and $A > 16$. [Adapted from G. T. Garvey et al., *Rev. Mod. Phys.*, **41**, S1 (1969)]

complicated relations predict masses of unknown nuclei with an accuracy of about 200 keV. The previously mentioned table of Wapstra and Gove (1971) should be consulted for numerical values of nuclear masses.

6-8 ELECTRIC MOMENTS OF NUCLEI

Nuclei contain moving electric charges, and their energy depends on the electric and magnetic fields in which they may be located. This energy dependence reveals some simple parameters of the whole nucleus. The most important are the charge, electric and magnetic dipoles, and electric quadrupole.

The fields may be due to externally applied fields or to the electrons of the atom in which the nucleus is located. Magnetic fields due to atomic electrons are directed along $\mathbf{J}$, the atomic total angular momentum. On the other hand, the nuclear angular momentum and nuclear magnetic moment are directed along a vector $\mathbf{I}$. Both external and atomic fields are important. External fields may be measured and controlled with great precision, but are practically limited to 10^5 G; internally generated fields are known with less precision, but are stronger.

The electric energy of the nuclear charges is given by

$$W = \sum_{i=1}^{Z} e\varphi(x_i, y_i, z_i) \tag{6-8.1}$$

where φ is the potential at the coordinates x_i, y_i, z_i of the ith proton. Assuming the center of the nucleus to be at the origin and denoting $(\partial\varphi/\partial x)_0$ by φ_x, etc., we can write the energy as

$$W = \sum_{1}^{Z} {}_i e\left[\varphi(0) + (\varphi_x x_i + \varphi_y y_i + \varphi_z z_i) + \tfrac{1}{2}\left(\varphi_{xx} x_i^2 + 2\varphi_{xy} x_i y_i + \cdots \right) \right] \tag{6-8.2}$$

The coefficient of $\varphi(0)$ is $Ze = \int\rho(x, y, z)\, d\tau$ where with ρ we indicate the electric charge density. Without loss of generality we may assume $\varphi_x = \varphi_y = 0$, that is, that the field is directed in the z direction. The quantity $d_z = \sum_i e z_i = \int\rho(x, y, z)z\, d\tau$ is the z component of the electric dipole moment.

To apply the concepts of classical electrostatics to nuclei, we give the quantum-mechanical expression for the charge density. For a system of nucleons

$$\rho(x, y, z) = \sum_{1}^{A} {}_i e_i P_i(x, y, z) \tag{6-8.3}$$

where e_i are the charge of the ith nucleon and P_i is the probability of finding it at x, y, z.

ISBN 0-8053-8601-7

The nucleus is described by $\psi(\mathbf{r}_1, \ldots, \mathbf{r}_A)$ where $\mathbf{r}_i$ are the coordinates of the ith nucleon. The Schrödinger ψ is in a configuration space of $3A$ dimensions. Now

$$P_i(x, y, z) = \int |\psi(\mathbf{r}_1, \ldots, \mathbf{r}_A)|^2 \, d\mathbf{r}_1 \ldots d\mathbf{r}_{i-1} \, d\mathbf{r}_{i+1} \ldots d\mathbf{r}_A \qquad (6\text{-}8.4)$$

where the integral extends over all space for all coordinates except $\mathbf{r}_i$ and $\mathbf{r}_i$ has components x, y, z.

Substituting Eq. (6-8.4) in (6-8.3), we have

$$d_z = \sum_1^A e_i \int z_i |\psi(\mathbf{r}_1, \ldots, \mathbf{r}_A)|^2 \, d\mathbf{r}_1 \ldots d\mathbf{r}_A \qquad (6\text{-}8.5)$$

Every term of the sum is zero if $\psi(\mathbf{r}_1, \ldots, \mathbf{r}_A)$ has a definite parity; that is, if

$$\psi(\mathbf{r}_1, \ldots, \mathbf{r}_A) = \pm \psi(-\mathbf{r}_1, \ldots, -\mathbf{r}_A) \qquad (6\text{-}8.6)$$

(where $+$ denotes an even function and $-$ an odd one), because in either case $|\psi(\mathbf{r}_1, \ldots, \mathbf{r}_A)|^2$ is an even function and $z_i|\psi(\mathbf{r}_1, \ldots, \mathbf{r}_A)|^2$, as the product of an odd function times an even one, is odd and has a vanishing integral over all space.

Nuclear wave functions have a definite parity except for a possible minute influence of the weak interaction, which would introduce parity-nonconserving terms into the hamiltonian. However, a definite parity is a sufficient but not necessary condition for the vanishing of the electric dipole (and all odd-order) moments. Current theories also predict a vanishing electric dipole moment for all nuclei if all motions in the nucleus are "time reversible" (see Chap. 9).

Direct experiments on the electric dipole moment of the neutron have given (Ramsey, 1969) a moment $\mu/e < 5 \times 10^{-23}$ cm where μ is the dipole moment and e the charge of the electron. The moment is likely to be exactly zero and the measurement is essentially an upper limit.

The third term in Eq. (6-8.2) gives the electric quadrupole moment. In order to evaluate it, assume a potential having cylindrical symmetry around the z axis. Such would be the potential produced by atomic electrons with the z axis coincident with J. Then the condition

$$\nabla^2 \varphi = 0 \qquad (6\text{-}8.7)$$

gives

$$\varphi_{zz} = -2\varphi_{xx} = -2\varphi_{yy} \qquad \varphi_{xy} = \varphi_{xz} = \varphi_{yz} = 0 \qquad (6\text{-}8.8)$$

ISBN 0-8053-8601-7

and the expression for the energy becomes

$$W_Q = \tfrac{1}{2}\varphi_{zz}\int\rho(\mathbf{r})\left[z^2 - \tfrac{1}{2}(x^2 + y^2)\right]d\tau \tag{6-8.9}$$

but $x^2 + y^2 + z^2 = r^2$ and hence

$$W_Q = \tfrac{1}{2}\varphi_{zz}\int\rho(\mathbf{r})\left[\tfrac{3}{2}z^2 - (r^2/2)\right]d\tau \tag{6-8.10}$$

● In this formula it is desirable to express the integral as a nuclear characteristic. To this end, we introduce a nuclear coordinate system ξ, η, ζ with the ζ axis in the direction of the axis of symmetry of the nucleus. We shall call the electric density expressed as a function of nuclear coordinates ρ_n and

$$eQ = \int\rho_n(3\zeta^2 - r^2)\,d\tau = 3Q_{\zeta\zeta} - Q_{rr} \tag{6-8.11}$$

the nuclear quadrupole moment, and we use it to express the integral of Eq. (6-8.10). Observe first that, the nuclear-charge distribution being symmetric around ζ, we have

$$\int\rho_n\xi^2\,d\tau = eQ_{\xi\xi} = eQ_{\eta\eta} \tag{6-8.12}$$

and

$$\int\rho_n\xi\eta\,d\tau = Q_{\xi\eta} = Q_{\eta\zeta} = Q_{\xi\zeta} = 0 \tag{6-8.13}$$

The relations between the x, y, z and ξ, η, ζ give

$$\xi^2 + \eta^2 + \zeta^2 = x^2 + y^2 + z^2 = r^2 \tag{6-8.14}$$

whence

$$Q_{rr} = Q_{xx} + Q_{yy} + Q_{zz} = Q_{\xi\xi} + Q_{\eta\eta} + Q_{\zeta\zeta} \tag{6-8.15}$$

Moreover,

$$z = \xi\cos\xi z + \eta\cos\eta z + \zeta\cos\zeta z \tag{6-8.16}$$

with

$$\cos^2\xi z + \cos^2\eta z + \cos^2\zeta z = 1 \tag{6-8.17}$$

Calculating z^2 and Q_{zz} by using Eqs. (6-8.12), (6-8.13), and (6-8.16) we have

$$Q_{zz} = Q_{\xi\xi}(1 - \gamma^2) + Q_{\zeta\zeta}\gamma^2 \tag{6-8.18}$$

ISBN 0-8053-8601-7

where

$$\gamma = \cos \zeta z = \cos IJ = \cos \theta \tag{6-8.19}$$

From Eq. (6-8.10) we obtain, by Eqs. (6-8.12), (6-8.14), and 6-8.11),

$$3Q_{zz} - Q_{rr} = Q\left(\tfrac{3}{2}\gamma^2 - \tfrac{1}{2}\right) = QP_2(\cos \theta) \qquad \bullet \tag{6-8.20}$$

Equation (6-8.20) inserted in Eq. (6-8.10) finally yields the expression for W_Q,

$$W_Q = (eQ/4)\varphi_{zz}P_2(\cos \theta) \tag{6-8.21}$$

The nuclear quadrupole moment Q obviously vanishes for a spherically symmetric nucleus. Quantum-mechanically we must define Q in a more precise way as the expectation value

$$\langle 3z^2 - r^2\rangle$$

for a given wave function. The wave function to be chosen is the one for which $M_I = I$, that is, the one corresponding to the maximum alignment of the total angular momentum along the z axis. We see then that according to quantum mechanics Q vanishes also for $I = \tfrac{1}{2}$, as can be found by calculating its average value, mentioned above. In order to do this, note that, using laboratory coordinates,

$$z = \frac{\mathbf{r} \cdot \mathbf{I}}{I} = \frac{xI_x + yI_y + zI_z}{I} \tag{6-8.22}$$

Equations (6-9.1) and (6-9.2) for $I = \tfrac{1}{2}$ give

$$I_x^2 + I_y^2 + I_z^2 = I(I+1) = \tfrac{3}{4}$$

$$I_x^2 = I_y^2 = I_z^2 = \tfrac{1}{4} \tag{6-8.23}$$

$$I_xI_y + I_yI_x = 0, \text{ etc.}$$

Using these relations in Eq. (6-8.22) we find

$$z^2 = \frac{1}{4}(x^2 + y^2 + z^2)\frac{4}{3} = \frac{r^2}{3} \tag{6-8.24}$$

that is, $3z^2 - r^2 = 0$, or using Eq. (6-8.20),

$$Q = 0 \tag{6-8.25}$$

ISBN 0-8053-8601-7

A positive Q indicates a cigar-shaped nucleus and a negative Q a lens-shaped nucleus.

Equation (6-8.21) is semiclassical, and the spherical harmonic $P_2(\cos\theta)$ can be written

$$\frac{3}{2}\left(\frac{\mathbf{I}\cdot\mathbf{J}}{IJ}\right)^2 - \frac{1}{2} = \frac{\frac{3}{2}C^2 - 2I^2J^2}{4I^2J^2}$$

To transform it into a quantum-mechanical expression (Casimir, 1936), the spherical harmonic must be replaced by

$$P_2(\cos\theta)\to\frac{\frac{3}{2}C(C+1) - 2I(I+1)J(J+1)}{I(2I-1)J(2J-1)} \tag{6-8.26}$$

where C is given in Eq. (6-9.9) as $F(F+1) - I(I+1) - J(J+1) = 2\mathbf{I}\cdot\mathbf{J}$. We have, in conclusion,

$$W_Q = \frac{B}{4}\frac{\frac{3}{2}C(C+1) - 2I(I+1)J(J+1)}{I(2I-1)J(2J-1)} \tag{6-8.27}$$

with

$$B = eQ\varphi_{zz}(0) \tag{6-8.28}$$

Equation (6-8.27) seems to give an indeterminate result for $J=\frac{1}{2}$ or $I=\frac{1}{2}$, but in these cases $W_Q = 0$ because either φ_{zz} or Q vanishes.

In order to obtain Q from the energy, we need to know $\varphi_{zz}(0)$. This is an atomic problem for which, to date, only approximate solutions have been found.

The spectral terms of an atom having a nucleus with $Q\neq 0$ are shifted by the amount W_Q and thus do not follow the interval rule [Eq. (6-9.12)]. A detailed spectroscopic study, in either the optical or the microwave region, can give values of B and, if $\varphi_{zz}(0)$ can be calculated, of Q. Other methods using matter in bulk, similar to the Bloch–Purcell methods for μ_I, can also be applied to find B (Dehmelt and Kruger).

Molecular-beam methods have also been extensively applied, and through them Q of the deuteron has been measured with great precision. We have limited our discussion to methods for directly measuring magnetic moments and electric quadrupole moments of nuclei, making no assumptions from the nuclear model. There are other methods, which we shall not describe but which allow us to relate certain experimental data (such as the cross section for Coulomb excitation; see Chap. 8) to nuclear moments, provided that a given nuclear model is valid. These methods can be used either to check the model, if one measures all the quantities involved, or to measure the moment, if one accepts the model.

ISBN 0-8053-8601-7

6-9   SPIN AND MAGNETIC MOMENTS I

Definitions and Measurements

Many nuclei show an intrinsic angular momentum, or "spin." This is always a multiple of $\hbar$ for nuclei of even mass number and always an odd multiple of $\hbar/2$ for nuclei of odd mass number. The spin in $\hbar$ units is indicated by the vector **I**. We must remember here that **I** has the properties of a quantum-mechanical angular momentum vector. In particular the measurement of $\mathbf{I}^2$ always gives the result $I(I + 1)$, and the measurement of one component of **I**, say I_z, can give as a result any of the numbers I, $I - 1, \ldots, -I$. A component of **I**, such as I_z and I^2, can be measured simultaneously, but I_x and I_z or two other different components of **I** are not compatible observables. The main quantum-mechanical properties of angular momentum are embodied in the commutation relations:

$$I_x I_y - I_y I_x = iI_z \tag{6-9.1}$$

$$I_z I^2 - I^2 I_z = 0 \tag{6-9.2}$$

eigenvalue of $I^2 = I(I + 1)$

eigenvalues of $I_z = I, I - 1, \ldots, -I$

Associated with the spin is a magnetic moment. The magnitude of the magnetic moment μ_I is not quantized and can take any value. The natural unit for measuring magnetic moments is the nuclear magneton

$$\frac{|e|\hbar}{2M_p c} = \mu_N = 0.5050823 \times 10^{-23} \text{ erg G}^{-1}$$

$$= \frac{\mu_B}{1836.151} = 3.15245 \times 10^{-12} \text{eV G}^{-1}$$

where M_p is the mass of the proton. In addition to the magnetic moment we shall often consider the nuclear g_I number, defined by

$$\frac{\mu_I}{\mu_N} = \mathbf{I} g_I \tag{6-9.3}$$

and the gyromagnetic ratio γ_I, defined by

$$\mu_I = \gamma_I \hbar I \tag{6-9.4}$$

Note that by convention $g_I > 0$ means nuclear spin and magnetic moment parallel to each other. By a similar convention we call the magnetic moment due to electrons $\mu_J/\mu_B = \mathbf{J} g_J$. The g_J are mostly negative. For instance, an

ISBN 0-8053-8601-7

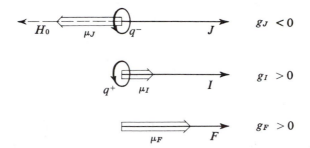

Figure 6-22 Sign conventions for **I**, **J**, μ, **H**, and g numbers. $\mathbf{F} = \mathbf{J} + \mathbf{I}$.

electron in an s state has $g_J = -2$ (Fig. 6-22). The sign conventions used in the literature vary.

Historically, nuclear spins and magnetic moments were first postulated by Pauli in 1924, although he thought of them as due to orbital motions. The electron spin was postulated by Uhlenbeck and Goudsmit in 1925 as an intrinsic angular momentum $\frac{1}{2}\hbar$ associated with a magnetic moment $e\hbar/2mc$. The spin $\frac{1}{2}\hbar$ of the proton was recognized by Dennison in 1927. Nuclear magnetic 2^3 and 2^5 poles have been observed. Dirac in 1931 hypothesized a magnetic monopole of magnitude $(\hbar c/e^2)(g/2)e$ with g integral number. Extensive search has given no convincing evidence for it. Magnetic poles of order 2^p with $p > 0$, even, are precluded by a parity argument.

The magnetic nuclear moment gives rise, classically, to an energy

$$W_{\text{mag}} = -\boldsymbol{\mu}_I \cdot \mathbf{H}(0) = -\mu_I(\mathbf{I}/I) \cdot \mathbf{H}(0) \tag{6-9.5}$$

where the second equality reminds us that μ is directed along **I**. If **H** is due to the motion of the atomic electrons, it will be directed parallel to **J** and Eq. (6-9.5) can be rewritten as

$$W_{\text{mag}} = -\mu_I H(0) \frac{\mathbf{I} \cdot \mathbf{J}}{IJ} = -\mu_I H(0) \cos IJ \tag{6-9.6}$$

where the cosine is to be understood quantum-mechanically.

In order to evaluate this formula for the atomic case, consider in addition to **I** and **J** the total angular momentum **F** (Fig. 6-23) of the atom, including the nucleus. We then have

$$\mathbf{F} = \mathbf{I} + \mathbf{J} \tag{6-9.7}$$

or, squaring and isolating $\mathbf{J} \cdot \mathbf{I}$,

$$\mathbf{I} \cdot \mathbf{J} = IJ \cos IJ = \tfrac{1}{2}(F^2 - I^2 - J^2) \tag{6-9.8}$$

ISBN 0-8053-8601-7

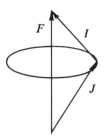

Figure 6-23 Vector model for an atom having nuclear spin. **J** is the total angular momentum of the electrons, **I** the nuclear spin, and **F** the total angular momentum of the atom.

The quantum-mechanical vectors **F**, **I**, and **J** give as eigenvalues for F^2 the numbers $F(F+1)$, etc. Substituting into Eq. (6-9.8) we find for the value of $IJ \cos \mathbf{IJ}$,

$$\tfrac{1}{2}\left[F(F+1) - I(I+1) - J(J+1)\right] = C/2 \qquad (6\text{-}9.9)$$

Equation (6-9.9) inserted in Eq. (6-9.6) gives

$$\Delta W_{I,J} = \frac{-\mu_I}{2}\frac{H(0)}{IJ}\left[F(F+1) - I(I+1) - J(J+1)\right] = \frac{A}{2}C \qquad (6\text{-}9.10\text{a})$$

with

$$A = -\mu_I\frac{H(0)}{IJ} \qquad (6\text{-}9.10\text{b})$$

The magnetic dipole moment manifests itself in many phenomena, most directly in the hyperfine structure of spectral lines. Atomic spectral lines, when examined by means of high-resolution apparatus, often show splitting of the order of magnitude of a fraction of a wave number (Fig. 6-24). Different isotopes may radiate light at slightly different spectral frequencies (isotope shift). However, we are concerned here with structures shown in the

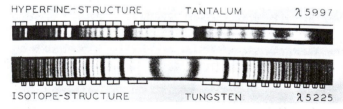

HYPERFINE−STRUCTURE   TANTALUM   λ5997

ISOTOPE-STRUCTURE   TUNGSTEN   λ5225

Figure 6-24 Hyperfine structure in the visible spectrum of tantalum. Isotope structure in the visible spectrum of tungsten. [Original from Harvey E. White.]

ISBN 0-8053-8601-7

spectra of monoisotopic substances. The energy levels giving rise to them show what is called a "hyperfine" structure (hfs), to distinguish it from the fine structure due to the electron spin–orbit coupling. The hfs is attributed to the magnetic energy of the magnetic moment of the nucleus immersed in the magnetic field due to the atomic electrons, given by Eq. (6-9.10 a).

The vector **F** is fixed in space and takes the values

$$F = I + J, I + J - 1, \ldots, |I - J| \tag{6-9.11}$$

These are $2J + 1$ or $2I + 1$ values, whichever is smaller. Terms with different **F** vectors have slightly different energies and form an hfs multiplet. There is a close analogy between an ordinary multiplet in Russell–Saunders (spin–orbit) coupling and an hfs multiplet. The vectors correspond to each other as in Table 6-4. The energy levels of the hfs multiplet are given by Eq. (6-9.10a), which is exactly analogous to the equation that gives the levels for an ordinary spin–orbit coupling of atomic multiplets. The sign of A is such that when $H(0)$ and **J** are antiparallel, as usually happens (because of the negative electron charge), then for a positive nuclear moment the energy is lowest when μ_I is parallel to $H(0)$ and hence antiparallel to **J**.

**TABLE 6-4   CORRESPONDENCE
BETWEEN FINE STRUCTURE AND
HYPERFINE   STRUCTURE**

Ordinary multiplet	Hfs multiplet
L	J
S	I
J	F

If we introduce in Eq. (6-9.10a) the values of F given by Eq. (6-9.11), we obtain the levels of the hfs multiplet. These obey the interval rule, which states that the intervals in an hfs multiplet are in the ratios

$$I + J : I + J - 1 \cdots |I - J| + 1 \tag{6-9.12}$$

It is clear that the analysis of an hfs multiplet will give the value of I simply by counting the number of components if $J \geqslant I$; if $J < I$, application of the interval rule or measurement of the intensity of spectral lines gives I if J is known.

The first indication of nuclear electric quadrupole moments came when it was found that the interval rule did not apply to the hfs multiplet (Schmidt and Schüler, 1935).

The calculation of $H(0)$ is a problem of atomic physics; the solution is simple in the case of a single s electron outside a closed shell, as, for instance,

ISBN 0-8053-8601-7

in the ground state of an alkali atom. This case is of practical importance. One obtains

$$H(0) = -\tfrac{8}{3}\pi\mu_B|\psi(0)|^2 \tag{6-9.13}$$

where μ_B is the Bohr magneton. (The minus sign means that the direction of the spin and the direction of J are antiparallel.) Equation (6-9.13) can be made plausible by a semiclassical argument as follows: We assume that the Schrödinger ψ is associated not only with an electric-charge density but also with a magnetization intensity; an element of volume $d\tau$ has associated with it a magnetic moment $-\mu_B|\psi(\mathbf{r})|^2\,d\tau\,\mathbf{k}$, where $\mathbf{k}$ is a unit vector in the spin direction. We must now calculate the interaction energy between the nuclear magnetic moment and the electronic spin.

● Assume that the nucleus is a small, charged, rotating sphere. It then has in its interior a uniform field $\mathbf{B}$ directed parallel to the rotational axis. Outside the nucleus the field is that of a dipole of magnetic moment μ_I oriented in the direction of the rotational axis. Considering a line of force along the rotational axis and applying to it the relation

$$\oint \mathbf{B}\,dl = 4\pi i \tag{6-9.14}$$

one finds that

$$BV = \frac{8\pi}{3}\,\mu_I \tag{6-9.15}$$

where V is the volume of the sphere.

The contribution to the interaction energy of the electron spin with the nucleus coming from regions outside the nucleus is zero, as can be seen by matching contributions from volumes at $\mathbf{r}$ and at $-\mathbf{r}$. On the other hand, the interaction inside the nuclear sphere is

$$W_{\text{mag}} = -\mathbf{B}\cdot\mathbf{k}V|\psi(0)|^2\mu_B \tag{6-9.16}$$

where we have assumed that inside the nucleus

$$\psi(r) \cong \psi(0) \tag{6-9.17}$$

Replacing BV in Eq. (6-9.16) by its expression, Eq. (6-9.15), we have

$$W_{\text{mag}} = -\frac{8\pi}{3}\,\mu_I\cdot\mathbf{k}|\psi(0)|^2\mu_B \tag{6-9.18}$$

and comparing with Eq. (6-9.5), we obtain Eq. (6-9.13). This argument obtains if we consider the nuclear magnetic moment as originating from a current loop. A permanent magnetic dipole would give a different answer. A

ISBN 0-8053-8601-7

rigorous proof of Eq. (6-9.18) is based on Dirac's theory. It is useful to express $\psi(0)$ in a simple way that is a fair approximation. Such an expression [see, e.g., (Ko 58)] is

$$|\psi(0)|^2 = \frac{1}{\pi a^3} \frac{Z(1+z)^2}{n_{\text{eff}}^2} \qquad (6\text{-}9.19)$$

where a is Bohr's hydrogen-atom radius, Z is the atomic number of the nucleus, z is the degree of ionization of the atom, and n_{eff} is the effective quantum number. The accuracy of this formula for $|\psi(0)|^2$ is of the order of 10%. ●

As an example we give in Table 6-5 some values of $H(0)$ for the ground state of the alkali atoms where the field is due to one s electron and for the ^2P states where the field is due to a p electron.

TABLE 6-5   MAGNETIC FIELD AT THE NUCLEUS PRODUCED BY THE ATOMIC ELECTRONS

Atom	$^2S_{1/2}$, n	$H(0)$, G	n	$H(0)$, G $^2P_{1/2}$	$H(0)$, G $^2P_{3/2}$
H	1	1.74×10^5			
Li	2	1.3×10^5			
Na	3	4.4×10^5	3	4.2×10^4	2.5×10^4
K	4	6.3×10^5	4	7.9×10^4	4.6×10^4
Rb	5	1.3×10^6	5	1.6×10^5	8.6×10^4
Cs	6	2.1×10^6	6	2.8×10^5	1.3×10^5

Thus far we have considered the nucleus as interacting only with the atomic electrons, in the absence of external fields. It is important also to consider how the hfs multiplet is affected by the perturbation caused by a constant external magnetic field H_e. We must distinguish two limiting cases in which the external field is such that

$$\frac{|\mu_B g_J \mathbf{J} \cdot \mathbf{H}_e|}{|\mu_I \cdot \mathbf{H}(0)|} \gg 1 \text{ or } \ll 1 \qquad (6\text{-}9.20)$$

The first case corresponds to the atomic Zeeman splitting of a level large compared with the hfs splitting; the second corresponds to the Zeeman splitting small compared with hfs splitting. For the first limiting case (strong field) the vector diagram is that shown in Fig. 6-25a, and the magnetic energy is

$$\Delta W = (-m_I g_I \mu_N - m_J g_J \mu_B) H_e \qquad (6\text{-}9.21)$$

where the first term is negligible compared with the second. To the magnetic

ISBN 0-8053-8601-7

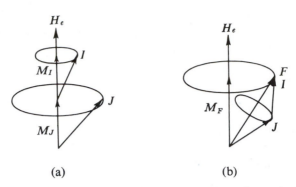

Figure 6-25 (a) Vector diagram of an atom with nuclear spin in a strong magnetic field. The interactions $\mathbf{I} \cdot \mathbf{H}_e$ and $\mathbf{J} \cdot \mathbf{H}_e$ are large compared to the $\mathbf{I} \cdot \mathbf{J}$ interaction. I and J precess independently around H_e. (b) Vector diagram of an atom with nuclear spin in a weak magnetic field. The interaction between $\mathbf{J}$ and $\mathbf{H}_e$ is small compared to the $\mathbf{J} \cdot \mathbf{I}$ interaction. I and J precess around F, which in turn precesses around H_e.

energy must be added the energy of the interaction between I and J. The approximate result is

$$\Delta W = - m_J g_J \mu_B H_e + A m_I m_J \qquad (6\text{-}9.22)$$

In the second case (weak field) the vector diagram of the atom is given by Fig. 6-25b. The hfs multiplet undergoes a Zeeman effect with a magnetic energy given by

$$\Delta W = - H_e F \cos F H_e \left[(J/F) \mu_B g_J \cos JF + (I/F) \mu_N g_I \cos IF \right]$$

$$\cong - \mu_B g_F m_F H_e \qquad (6\text{-}9.23)$$

where m_F indicates, as usual, $F \cos F H_e$. The second term in parentheses is about 2000 times smaller than the first and can be neglected. Using the quantum-mechanical expression for the cosines, we can then write

$$g_F = g_J \frac{F(F+1) + J(J+1) - I(I+1)}{2F(F+1)} \qquad (6\text{-}9.24)$$

Figure 6-26 gives the exact value of the magnetic energy, calculated for all values of the magnetic field.

The atomic magnetic moment can be defined by

$$\mu_{\text{eff}} = \frac{\partial W}{\partial H_e} \qquad (6\text{-}9.25)$$

ISBN 0-8053-8601-7

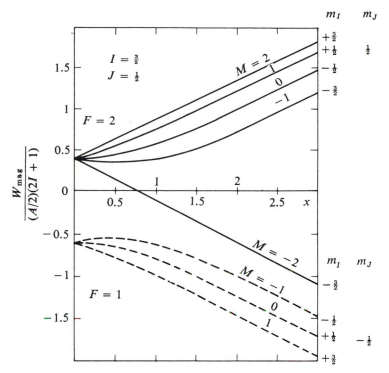

Figure 6-26 Variation of energy levels with magnetic field H of an atom in a $^2S_{1/2}$ state for a nuclear spin of $\frac{3}{2}$. The solid and dashed lines correspond to the levels arising from the states $F = 2$ and $F = 1$, respectively. The parameter x is $(-\mu_J J + \mu_I I)H/\Delta W \sim \mu_B H/\Delta W$, where ΔW is the energy difference between states $F = 2$ and $F = 1$ at zero magnetic field: $\Delta W = (A/2)(2I + 1)$; $M = m_I + m_J = m_F$.

This is given in Fig. 6-27 for the cesium atom. For a given hfs multiplet the atomic magnetic moment is a function of F, m_F, and H, and for certain values of the field it may vanish.

Radiative transitions between different hfs levels are subject to selection rules similar to those governing ordinary multiplets (refer to Fig. 6-28). In the case of electric dipole radiation (see Chap. 8), in the absence of an external field, we have

$$\Delta F = \pm 1, 0 \qquad (6\text{-}9.26)$$

no $0 \rightarrow 0$ transition

In the presence of an external magnetic field, if the field is weak, we have

$$\Delta m_F = 0 \qquad (6\text{-}9.27)$$

ISBN 0-8053-8601-7

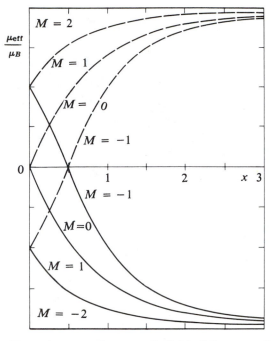

Figure   6-27   Dependence on the magnetic field of the magnetic moment of an atom in a $^2S_{1/2}$ state with a nuclear spin of $\frac{3}{2}$. The solid lines refer to M levels of the $F = 2$ state and the dashed lines to $F = 1$ levels. x has the same meaning as in Fig. 6-26.

for components polarized parallel to the field (π components) and

$$\Delta m_F = \pm 1 \tag{6-9.28}$$

for components polarized perpendicular to the field (σ components). In the case of a strong field we have

$$\Delta m_I = 0, \ \pm 1 \tag{6-9.29}$$

$$\Delta m_J = 0 \text{ for } \pi \text{ components} \tag{6-9.30}$$

$$\Delta m_J = \pm 1 \text{ for } \sigma \text{ components} \tag{6-9.31}$$

In every case the parity must change. All this is shown in Fig. 6-28. For magnetic dipole radiation we have the same selection rules, except that the parity must not change. Optical transitions occur between different hfs multiplets. Transitions within an hfs multiplet cannot be obtained by electric dipole because of the parity selection rule. They are magnetic dipole and their frequency is in the microwave region.

ISBN 0-8053-8601-7

ISBN 0-8053-8601-7

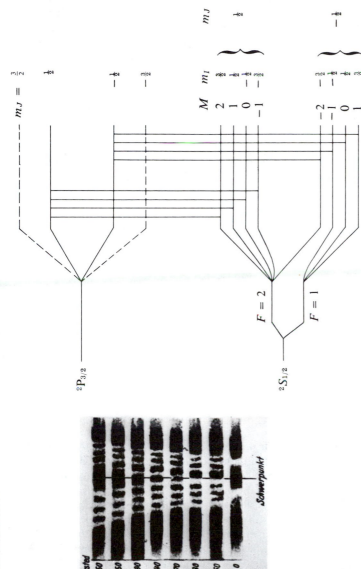

Figure 6-28 The π components ($\Delta m_J = 0$) of the $^2S_{1/2} - {}^2P_{3/2}$ (5890) line of Na as observed in absorption. The nuclear spin $I = \frac{3}{2}$ splits the terms as indicated in the right side of the figure. Note that in the $^2P_{3/2}$ term the hfs is so small that the strong external field case always applies. The observations have been made for various values of an external field H as indicated in the left part of the figure. [D. A. Jackson and H. Kuhn, *Proc. Roy. Soc. (London)*, **167**, 210 (1938).]

In the preceding pages we briefly outlined the spectroscopic method for measuring spin, magnetic moment, and electric quadrupole moments. This method was the first to be used and is still important. Over the years, however, different techniques have been invented. In particular, O. Stern, I. I. Rabi, and others have developed methods based on the application of molecular beams. These are beams of molecules moving in a vacuum (Fig. 6-29 a); as the figure indicates, the molecules leave the oven O, are collimated by the slit S, and are finally detected at the detector D. The detector may have a variety of forms; for instance, it may be a hot wire that ionizes impinging molecules, or a Pirani manometer in which the molecules are trapped and which shows an increase in pressure, or it may be a detector of radioactivity if

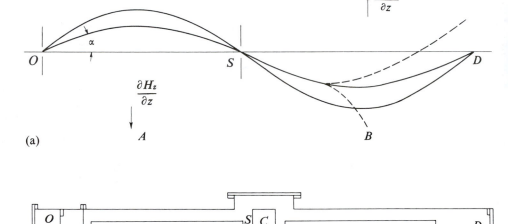

(a)

(b)

Figure 6-29 Schematic diagram of a molecular-beam apparatus. The two solid curves in part (a) indicate the paths of two molecules having different moments and/or velocities and whose moments are not changed during passage through the apparatus. For them the deviations in A are exactly compensated by those in B. The two dashed curves in the region of the B magnet indicate the possible changes in path for one of these molecules if its component of magnetic moment has been either increased or decreased in the region of the C field. It is then lost to the beam and the intensity dips. The motion in the z direction has, in each of the curves, been greatly exaggerated. [J. B. M. Kellogg and S. Millman, *Rev. Mod. Phys.*, **18**, 323 (1946).]

the substance used is radioactive. The molecule in the free beam can be subjected to forces and hence can be deflected. In the original experiment of Stern and Gerlach silver atoms were sent into an inhomogeneous magnetic field and were deflected in the z direction by a force

$$F = \mu_{eff} \frac{\partial H_z}{\partial z} \tag{6-9.32}$$

where $\partial H_z / \partial z$ is called the inhomogeneity of the field. This force produces a deflection that can be calculated from elementary mechanics; we have immediately

$$\ddot{z} = \mu_{eff} \frac{\partial H_z}{\partial z} \frac{1}{M} \tag{6-9.33}$$

$$z = \tfrac{1}{2} \mu_{eff} \frac{\partial H_z}{\partial z} \frac{1}{M} \left(\frac{x}{v} \right)^2 \tag{6-9.34}$$

where x is the length of the field, M the mass of the molecule, and v its velocity.

Now, if molecules are deflected from the beam, the beam intensity decreases and this decrement is experimentally detectable. Subjecting the molecules to a magnetic field H_z having an inhomogeneity $\partial H_z / \partial z$ thus results, in general, in a decrease of intensity in the undeflected beam. However, if H_z is such that $\mu_{eff}(H_z) = 0$, the intensity does not decrease and, as a function of H, it has a maximum value because the beam is undeflected. The field value at which this occurs can be seen in Fig. 6-27 for a special case. This experiment thus gives $\mu_B H_z / \Delta W$ or A. In order to find μ_I, we still need information about the atomic field of the nucleus, according to Eq. (6-9.10).

6-10   SPIN AND MAGNETIC MOMENTS (MEASUREMENTS) II

In order to obtain μ_I by the methods given in Sect. 6-9 we need to know the atomic magnetic field at the nucleus. Much higher precision is obtainable by other methods that dispense with this knowledge. These involve measuring the interactions of the nucleus with an externally applied, directly measurable field. The method is illustrated in Fig. 6-29. A beam is deflected in the inhomogeneous magnetic field A and refocused on the detector D by the inhomogeneous field B. Between the A and B fields is a C field that has a constant component H_z and a component perpendicular to it, variable in time, proportionally to $\sin \omega t$. This component induces transitions between various levels of the hfs multiplet, according to the selection rules given in Sec. 6-9, when the frequency ω is such that resonance occurs—in other words, such that the energy difference between the levels is equal to the energy of the quanta of the field, $\hbar\omega$. Equation (6-9.21) gives, for instance, in the case of a strong

ISBN 0-8053-8601-7

field, the resonance condition

$$\hbar\omega = g_I \mu_N H_z \, \Delta m_I \qquad (6\text{-}10.1)$$

When such transitions occur, the B magnet may not refocus, because the molecule has changed magnetic moment and the beam arriving at D is weakened. As can be seen from Eq. (6-10.1) the g_I factor can be determined solely from the external field, H_z and ω.

This method can give extremely precise results. In particular, the diatomic molecules of the hydrogen isotopes have been analyzed in great detail by this method (Kellogg, Rabi, Ramsey, and Zacharias, 1934). The same principle has also been applied by Alvarez and Bloch (1939) to the free neutron. They polarized and analyzed a beam of neutrons by passing it through magnetized iron (see Chap. 12). The polarized neutrons were made to precess by a known magnetic field, and from the rotation of the plane of polarization these workers calculated the neutron magnetic moment. Molecular-beam methods have been used extensively for radioactive nuclei, using for detection the radioactivity deposited on a catcher target.

● Assuming that the reader has some familiarity with Pauli's spin theory, we treat as an example a typical problem corresponding to the flipping over of a spin $\frac{1}{2}$ in a molecular-beam experiment. The spin is described by an eigenfunction $\begin{vmatrix} s_1 \\ s_2 \end{vmatrix}$ where s_1 and s_2 are complex numbers. If the spin points in the direction defined by polar angles θ, φ, we can take, for example,

$$s_1 = \cos(\theta/2), \qquad s_2 = -ie^{i\varphi}\sin(\theta/2) \qquad (6\text{-}10.2)$$

The Schrödinger equation for the motion of the spin is $i\hbar\dot{\psi} = H\psi$ where H is the spin hamiltonian.

The spin is associated with a magnetic moment

$$(ge\hbar/2mc)\mathbf{s} = (e\hbar/2mc)\boldsymbol{\sigma} = \mu_B \boldsymbol{\sigma} \qquad (6\text{-}10.3)$$

where $\boldsymbol{\sigma}$ is the Pauli vector matrix and we have taken $g = -2$. The hamiltonian is $-\mu_B \boldsymbol{\sigma} \cdot \mathbf{B}$ where $\mathbf{B}$ is the magnetic field of components B_x, B_y, B_z. Substituting in the Schrödinger equation, we have

$$i\hbar \frac{\partial}{\partial t} \begin{vmatrix} s_1 \\ s_2 \end{vmatrix} = -\mu_B \begin{vmatrix} B_z & B_x - iB_y \\ B_x + iB_y & -B_z \end{vmatrix} \begin{vmatrix} s_1 \\ s_2 \end{vmatrix} \qquad (6\text{-}10.4)$$

We consider now the case of a field with a rotating component such that $B_z = \text{constant} = B_0$,

$$B_x + iB_y = B_1 \exp(i\omega t) \qquad (6\text{-}10.5)$$

ISBN 0-8053-8601-7

Introducing the frequencies $\omega_0 = (\mu_B/\hbar)B_0$, $\omega_1 = (\mu_B/\hbar)B_1$, and substituting in Eq. (6–10.4), we have

$$\dot{s}_1 = i\left[s_1\omega_0 + s_2\omega_1 \exp(-i\omega t)\right] \tag{6-10.6a}$$

$$\dot{s}_2 = i\left[s_1\omega_1 \exp(i\omega t) - s_2\omega_0\right] \tag{6-10.6b}$$

These equations can be solved by putting

$$s_1 = L \exp(ip_1 t) + M \exp(ip_2 t) \tag{6-10.7}$$

$$s_2 = \left[\frac{p_1 - \omega_0}{\omega_1} L \exp(ip_1 t) + \frac{p_2 - \omega_0}{\omega_1} M \exp(ip_2 t)\right] \exp(i\omega t) \tag{6-10.8}$$

where L, M are determined from the initial conditions and

$$p_1 = \left[\left(\frac{\omega}{2} + \omega_0\right)^2 + \omega_1^2\right]^{1/2} - \frac{\omega}{2} \tag{6-10.9}$$

$$p_2 = -\left[\left(\frac{\omega}{2} + \omega_0\right)^2 + \omega_1^2\right]^{1/2} - \frac{\omega}{2} \tag{6-10.10}$$

Assuming the initial condition $s_1(0) = 1$, $s_2(0) = 0$ and remembering the normalization condition $|s_1|^2 + |s_2|^2 = 1$, we obtain exact formulas for $s_1(t)$, $s_2(t)$. The probability of finding the spin pointing in the $-z$ direction is $|s_2(t)|^2$.

Calculation gives

$$|s_2(t)|^2 = \frac{4\omega_1^2}{(\omega + 2\omega_0)^2 + 4\omega_1^2} \sin^2\left\{\left[(\omega + 2\omega_0)^2 + 4\omega_1^2\right]^{1/2} \frac{t}{2}\right\} \tag{6-10.11}$$

For $B_z \gg (B_x^2 + B_y^2)^{1/2}$ and ω very close to the precession frequency due to B_0, $\omega = -2\omega_0$ one has:

$$|s_2(t)|^2 = \sin^2 \omega_1 t \tag{6-10.12}$$

This is a situation close to the experimental one in molecular-beam or nuclear induction experiments.

One method for measuring magnetic moments, developed by Bloch and Hansen (1946) and independently by Purcell (1946), employs bulk amounts of matter. A simplified schematic version of the form developed by Bloch is shown in Fig. 6-30. Suppose that we have some ordinary water and that we

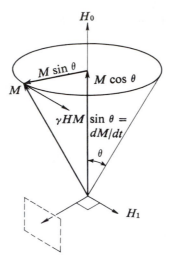

Figure 6-30 Schematic diagram of spin precession in a magnetic field and the Bloch method of measuring magnetic moments.

consider the hydrogen nuclei in a constant magnetic field H_0 in the z direction. Because of the Boltzmann factor there are more protons oriented with their magnetic moment parallel than antiparallel to the field. This gives rise to a very small average magnetization of matter in bulk, given by Langevin's theory,

$$P_0 = N\mu_I^2 \frac{H}{3kT} \frac{I+1}{I} = \chi_0 H_0 \qquad (6\text{-}10.13)$$

where N is the number of nuclei per cubic centimeter and P_0, the magnetization, is the magnetic moment per unit volume. This magnetization would be imperceptibly small if we tried to measure it by a static-field method. If, however, we apply a field that rotates in a plane perpendicular to H_0, we can render it observable. Consider classically the mechanical angular momentum per unit volume $\mathbf{M}$ associated with P_0; according to Eq. (6-9.4) it is

$$\gamma \mathbf{M} = \mathbf{P}_0 \qquad (6\text{-}10.14)$$

with γ equal to the gyromagnetic ratio.

The equation of motion for $\mathbf{M}$ (per unit volume) is

$$\frac{d\mathbf{M}}{dt} = \text{torque} = \gamma \mathbf{M} \times \mathbf{H} \qquad (6\text{-}10.15)$$

Bloch showed that the quantum-mechanical treatment in Eqs. (6-10.2) to (6-10.12) gives a result equivalent to Eqs. (6-10.14) and (6-10.15) if we inter-

ISBN 0-8053-8601-7

pret $\mathbf{M}$ as the average value of the angular momentum and $\gamma\mathbf{M}$ as the average value of the magnetic moment.

For constant H_0 we find, taking the dot product of both sides by $\mathbf{M}$, that $d(M^2)/dt = 0$ or $M^2 = $ constant; similarly, taking the dot product by $\mathbf{H}$ we have $\mathbf{M} \cdot \mathbf{H} = $ constant. The conclusion is that $\mathbf{M}$ precesses around $\mathbf{H}_0$ with a period T such that [Fig. 6-30 and Eq. (6-10.15)]

$$T\gamma MH_0 \sin\theta = 2\pi M \sin\theta \qquad (6\text{-}10.16)$$

that is, with a circular frequency

$$\gamma H_0 = \omega_L \qquad (6\text{-}10.17)$$

If we use a reference system with the z axis on H_0 and rotating around H_0 with a frequency ω_L, the motion is the same as that which would occur without $\mathbf{H}_0$ in a nonrotating system (Larmor's theorem). Now add a small field $\mathbf{H}_1$ normal to $\mathbf{H}_0$ and rotating with a frequency ω. If $\omega \neq \omega_L$, the field $\mathbf{H}_1$ will sometimes tend to increase θ and will sometimes tend to decrease it, having on the average no effect. If $\omega = \omega_L$, this field will follow the vector $\mathbf{M}$ and will act on it as a constant field, tending to vary the angle θ. Thus for $\omega = \omega_L$ the angle θ after a while will become large, and $\mathbf{M}$ will precess on cones of increasing aperture, executing complete rotations about the z axis until it becomes antiparallel to z. From there it again starts to precess until it once more becomes parallel to the z axis, and so on (Fig. 6-31). In practice the small field $\mathbf{H}_1$ is not a rotating field but an alternating field in the y direction, varying as $\mathbf{H}_1 \cos\omega t = \frac{1}{2}\mathbf{H}_1(e^{i\omega t} + e^{-i\omega t})$. This field is equivalent to two fields of amplitude $\frac{1}{2}\mathbf{H}_1$ rotating in opposite directions in the x–y plane. Of these only one is effective if $\omega = \omega_L$; the other can be neglected because it is out of resonance. When $\mathbf{M}$ forms a large angle with $\mathbf{H}_0$, it can induce an emf of frequency ω_L in a fixed coil lying in a plane parallel to $\mathbf{H}_0$ and perpendicular to $\mathbf{H}_1$. The induced emf is detected, and from the value of $\mathbf{H}_0$ and of the frequency ω_L one finds, using Eq. (6-10.17), the value of γ.

In the Purcell method the resonance is detected by measuring the radio frequency-energy absorbed in the material placed in a magnetic field. This absorption has a sharp maximum when the radio-frequency coincides with the ω_L.

In this oversimplified description we have not yet considered the very important role of the relaxation times of the substance. The study of relaxation times and other peculiarities of these phenomena has extensive ramifications in chemistry, in studies of molecular structure, and in solid-state physics. The method has undergone many developments and is now extensively applied even in the measurement of magnetic fields (using the known magnetic moment of hydrogen). Figure 6-32 is a diagram of a typical apparatus for measuring magnetic moments by the Purcell method.

ISBN 0-8053-8601-7

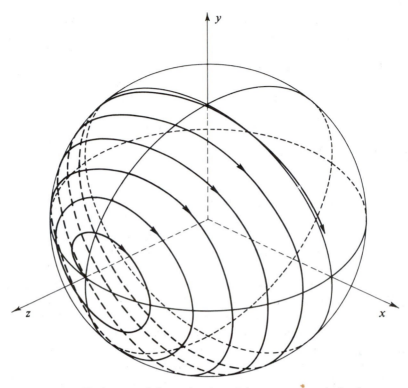

Figure 6-31 Trajectory of the end point of the magnetic polarization vector of hydrogen nuclei subject to an alternating field at the resonance frequency. The strong field is in the z direction, the alternating field in the x or y direction. The relaxation time is infinite. [From (Fl E).]

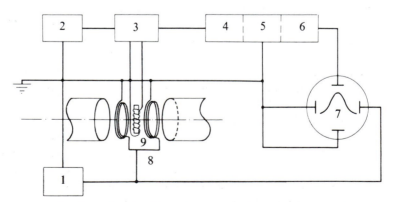

Figure 6-32 Principle of a Purcell apparatus for observing nuclear precession motions: 1, low-frequency generator; 2, high-frequency generator; 3, bridge; 4, high-frequency amplifier; 5, high-frequency rectifier; 6, low-frequency amplifier; 7, oscilloscope; 8, Helmholtz coils; 9, solenoid on sample.

ISBN 0-8053-8601-7

The methods described above give the spin and magnetic moment of stable and radioactive nuclei. Another group of methods applies only to radioactive nuclei, and especially to short-lived states (see also Chap. 8). We shall explain them by a simplified example that will show the essential principles of the method. Suppose that the nucleus has three levels connected by two gamma rays in cascade and that the intermediate state has a mean life τ. With two counters in coincidence we find the probability that the two gamma rays of the cascade will form an angle between θ and $\theta + d\theta$. This probability has the form

$$P(\theta)\,d\theta = A + B\cos^2\theta + \cdots \qquad (6\text{-}10.18)$$

a series with even powers of $\cos\theta$. For the sake of simplicity we shall stop at the term in $\cos^2\theta$. The existence of an angular correlation is explained by the fact that in the first gamma emission the nucleus is left with its spin in a direction correlated with the direction of the outgoing gamma. For instance, in a macroscopic dipole radio antenna the quanta are emitted preferentially in a direction perpendicular to the antenna. Thus the emission of the first quantum gives some information on the nuclear orientation. The second quantum is then emitted anisotropically with respect to the first, because an oriented nucleus emits, in general, gamma rays in a pattern which is oriented with respect to the spin.

Now introduce a magnetic field **H** perpendicular to the plane in which we measure the correlation. This field will force the spins of the intermediate state to precess with an angular velocity

$$\omega_I = g_I \mu_N H / \hbar \qquad (6\text{-}10.19)$$

and during the mean life of the intermediate state τ the spin will rotate through an angle $\omega_I \tau$. The angular correlation will thus change, the angle θ being replaced by $\theta - \omega_I \tau$, and at the same time a damping of the correlation occurs, because different nuclei emit the gamma rays at different times. $P(\theta)$ is replaced by

$$P(H, \theta, T) = \int_0^T P(\theta - \omega_I t) e^{-t/\tau}\, \frac{dt}{\tau} \Big/ \int_0^T e^{-t/\tau}\, \frac{dt}{\tau} \qquad (6\text{-}10.20)$$

where T is the time during which the apparatus is sensitive after the emission of the first quantum. By varying T and H, it is possible to determine, in favorable cases, both τ and g_I. The method has many variations. It is useful in dealing with problems in solid-state physics because in the field H one must also consider crystalline fields. It has been applied to ^{111}Cd and several other radioactive nuclides (Fig. 6-33).

Finally it has become possible, in some cases, to observe the nuclear Zeeman effect directly by using nuclear "recoilless" emission (see Chap. 8).

ISBN 0-8053-8601-7

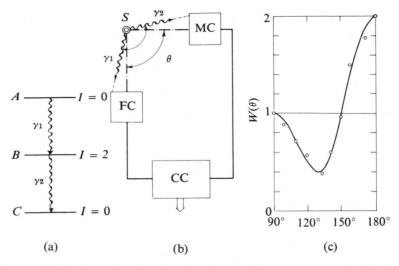

Figure 6-33 Example of a directional correlation measurement: (a) level diagram; (b) apparatus: S, source; FC, fixed counter; MC, movable counter; CC, coincidence circuit; (c) result. [H. Frauenfelder in (Si 55).]

6-11   NUCLEAR POLARIZATION

Nuclei with $I \neq 0$ may be oriented in such a way that the substates with different m have different populations. For instance, in the case $I = \frac{1}{2}$ we may have $a_{1/2}$ nuclei with spin pointing up and $a_{-1/2}$ nuclei with spin pointing down relative to an external magnetic field. We call the expression

$$P = \frac{a_{1/2} - a_{-1/2}}{a_{1/2} + a_{-1/2}} \tag{6-11.1}$$

the "polarization" of our system of nuclei.

For arbitrary values of I we consider the quantities

$$f_1 = \frac{1}{I} \sum_{m_I} m_I a_{m_I}$$

$$f_2 = \frac{1}{I^2} \left[\sum_{m_I} m_I^2 a_{m_I} - \frac{1}{3} I (I + 1) \right] \tag{6-11.2}$$

$$f_3 = \frac{1}{I^3} \left[\sum_{m_I} m_I^3 a_{m_I} - \frac{1}{5} (3I^2 + 3I - 1) \sum_{m_I} m_I a_{m_I} \right] \text{ etc.} \tag{6-11.3}$$

where the a_{m_I} are normalized so as to make

$$\sum_{m_I} a_{m_I} = 1, \qquad -I \leqslant m_I \leqslant I \tag{6-11.3}$$

ISBN 0-8053-8061-7

The number f_k is called the *degree of orientation of order k*. If all a_{m_I} are equal, all $f_k = 0$. For spin I, only the f_k with $k \leqslant 2I$ may differ from zero. In particular, for $I = \frac{1}{2}$, f_1 is the polarization. For larger spins one may have

$$a_{m_I} = a_{-m_I} \qquad (6\text{-}11.4)$$

for every m_I, but $a_{m_I} \neq a_{m'_I}$ ($|m_I| \neq |m'_I|$). When this occurs, all f_k with k odd vanish, but some of the f_k with k even are different from zero, and the system is "aligned." For instance, a system of deuterons all with $m_I = 0$ is aligned but not polarized.

The energy differences associated with nuclear orientation are generally of the order of magnitude of $\mu_N H$, where H is an external magnetic field. The thermal agitation determines energy fluctuations of order kT, and when kT is comparable with $\mu_N H$, the nuclear orientation will be destroyed by the thermal agitation. For $H = 20 \times 10^4$ G, $\mu_N H/kT \cong 10^{-3}/T$, which indicates the necessity of using temperatures in the range of 10^{-2} to 10^{-3}°K. These may be achieved by methods of adiabatic demagnetization. The simplest method, often called the *brute-force method*, is to cool the material in a magnetic field (Gorter and Simon, 1934).

For spin $\frac{1}{2}$ nuclei, the ratio of the number of nuclei oriented parallel or antiparallel to an external field H at a temperature T is given by Boltzmann's law,

$$\frac{a_{1/2}}{a_{-1/2}} = \exp\left(g_I \mu_N \frac{H}{kT} \right) \qquad (6\text{-}11.5)$$

and the polarization obtainable is approximately

$$f_1 = g_I \mu_N \frac{H}{2kT} \qquad (6\text{-}11.6)$$

The nuclear relaxation time, often inconveniently long, and the low temperature required make this method technically difficult.

Indirect methods (Gorter, 1948; Rose, 1949) proceed by polarizing the atomic magnetic moments of paramagnetic salts. These are about 1000 times larger and easier to polarize than the nuclear moments. They produce, in turn, oriented magnetic fields of the order of 10^6 G at the nucleus, and these fields produce the nuclear orientation.

It is possible to obtain alignment but not polarization by using the anisotropic properties of crystals; at sufficiently low temperatures, the nuclear moments set themselves parallel to a crystal axis, but without producing any net polarization. For instance, the nuclear electric quadrupole moment will align with an ionic local electric field due to the lattice of the crystal. Methods of this type have been tested successfully in several cryogenic laboratories (Pound and Bleaney, 1951).

ISBN 0-8053-8061-7

There are also dynamic processes of orientation that make it possible, when the appropriate electromagnetic radiation is applied, to increase the population of one of the substates of an hfs multiplet and thus achieve a nuclear polarization. This field is being rapidly developed, and we must refer to current literature for more details.

If the oriented nuclei are radioactive, the alignment or the polarization may be checked by observing the anisotropy of the radiations emitted. The technique is one of great and increasing importance.

6-12   VALUES OF SPIN, MAGNETIC MOMENTS, AND QUADRUPOLE MOMENTS

In Table 6-6 we summarize some of the results of the measurement of spins, magnetic moments, and electric quadrupole moments. Note the striking rule that all even-even nuclei in the ground state have zero spin and hence no electric or magnetic moments. For this reason they have been omitted from Table 6-6.

Turning to the nuclei with A odd, beginning with the proton, we should expect the proton to have, according to Dirac's theory, the magnetic moment of exactly $1\mu_N$ (nuclear magneton), but instead we find an experimental value of 2.793, and for the neutron, which should have no magnetic moment whatsoever, the value is -1.913. It is interesting to note that the difference between the magnitudes of the magnetic moment of the neutron and the proton is approximately $1\mu_N$. This might be explained, qualitatively at least, by the hypothesis that the magnetic moment of the nucleon is composed of two parts: one a Dirac moment, of $1\mu_N$ for the proton and 0 for the neutron; the other an additional moment resulting from mesonic effects. The part attributable to the mesonic effects is then $2.793 - 1 = 1.793$ for the proton and -1.913 for neutron. This may be qualitatively explained as originating from the virtual dissociation of the nucleon into a "bare Dirac nucleon" and a meson. The mesonic orbital motion gives rise to a magnetic moment of $+\mu_\pi$ if the meson is positive and of $-\mu_\pi$ if the meson is negative. We now write the dissociations of proton and neutron as

$$p \rightarrow n + \pi^+$$

$$n \rightarrow p + \pi^-$$

and the fraction of the time during which the nucleon is dissociated as τ. We then have, indicating by μ_p, μ_n the magnetic moments of proton and neutron,

$$\mu_p = +\mu_\pi \tau + (1 - \tau)1 = (\mu_\pi - 1)\tau + 1$$

$$\mu_n = (1 - \mu_\pi)\tau$$

ISBN 0-8053-8061-7

This relation would give

$$\mu_p + \mu_n = 1$$

whereas experimentally one finds

$$\mu_p + \mu_n = 0.88$$

The agreement is surprisingly good for such a simple argument. One of the problems of meson theory is to account precisely for the magnetic moment of the proton and neutron. If we assume $\mu_\pi = e\hbar/2m_\pi c$, τ comes to about 0.3.

The deuteron has a magnetic moment which is almost exactly the algebraic sum of the magnetic moment of the proton and neutron. Remembering that the deuteron has spin 1 and that it is composed of a neutron and a proton with parallel spins and no relative angular momentum (s state), one should not be suprised at the result for the magnetic moment. The next problem is to study why the magnetic moment is not exactly equal to the sum of the two moments. The discrepancy is reasonably accounted for by the admixture of d states into the ground state of the deuteron, as will be seen in Chap. 10. Turning to heavier nuclei, we find in several cases values of the magnetic moment that are fairly easily explained. For example, in the case of ^{3}He and ^{3}H we should expect to find the magnetic moment of a neutron in the first and of a proton in the second, because the identical nucleons have opposite spin and the only effective nuclide is the unpaired one.

A general regularity has been observed by Schmidt (1937) and can be explained by assuming that odd-A nuclei are composed of a core forming a closed shell plus one nucleon. The closed shell must contain an even number of neutrons and an even number of protons and has no angular momentum or magnetic moment. To this shell we now add one nucleon with angular momentum l and spin $\frac{1}{2}$. Clearly the total I of the nucleus is then equal to $l \pm \frac{1}{2}$. According to the vector model, $\mathbf{l}$ and $\mathbf{s}$ combine to form $\mathbf{I}$. Each one precesses around $\mathbf{I}$ independently, and the magnetic moment in nuclear magnetons along $\mathbf{I}$ is

$$\mu_I = g_l \frac{\mathbf{l} \cdot \mathbf{I}}{I} + g_s \frac{\mathbf{s} \cdot \mathbf{I}}{I} = I g_I \tag{6-12.1}$$

The scalar product of $\mathbf{l} \cdot \mathbf{I}$ is given by

$$s^2 = l^2 + I^2 - 2\mathbf{l} \cdot \mathbf{I} \tag{6-12.2}$$

or quantum-mechanically by

$$\frac{\mathbf{l} \cdot \mathbf{I}}{I^2} = \frac{l(l+1) + I(I+1) - s(s+1)}{2I(I+1)} \tag{6-12.3}$$

ISBN 0-8053-8061-7

TABLE 6-6   SELECTED EXAMPLES OF VALUES OF NUCLEAR SPINS, MAGNETIC DIPOLE MOMENT, AND ELECTRIC QUADRUPOLE MOMENT[a]

Z	Element	A	Half-life	Spin	Magnetic dipole moment μ	Electric quadrupole moment Q
0	n	1	11.7 min	$\frac{1}{2}$	− 1.913165	
1	H	1		$\frac{1}{2}$	2.792782	
		2		1	0.857420	0.002875
		3	12.262 yr	$\frac{1}{2}$	2.978897	
2	He	3		$\frac{1}{2}$	− 2.127577	
3	Li	6		1	0.822034	− 0.000644
		7		$\frac{3}{2}$	3.256372	− 0.0366
4	Be	9		$\frac{3}{2}$	− 1.17746	0.053
5	B	10		3	1.800598	0.08472
9	F	19		$\frac{1}{2}$	2.628391	
11	Na	22	2.602 yr	3	1.746	
		23		$\frac{3}{2}$	2.217558	0.101
13	Al	27		$\frac{5}{2}$	3.641339	0.140
15	Cl	35		$\frac{3}{2}$	0.821821	− 0.08249
		36	3.00×10^5 yr	2	1.28539	− 0.0180
19	K	39		$\frac{3}{2}$	0.391434	0.049
		40	1.267×10^9 yr	4	− 1.2981	− 0.061
		41		$\frac{3}{2}$	0.214874	0.060
		42	12.361 h	2	− 1.1424	
		43	22.2 h	$\frac{3}{2}$	0.163	
35	Br	76	16.1 h	1	± 0.5480[b]	∓ 0.27[b]
		77	58.0 h	$\frac{3}{2}$		
		79		$\frac{3}{2}$	2.105534	0.293
						$\Omega = 0.116$
		80	17.57 min	1		
		80^m	4.42 h	5		
		81		$\frac{3}{2}$	2.269628	0.27
						$\Omega = 0.129$
		82	35.344 h	5	1.6264	0.76
43	Tc	99	2.14×10^5 yr	$\frac{9}{2}$	5.6807	0.34

and a similar expression is obtained for $(\mathbf{s} \cdot \mathbf{I})/I^2$. Inserting them in Eq. (6-12.1) we find

$$g_I = g_s \frac{I(I+1) + s(s+1) - l(l+1)}{2I(I+1)}$$

$$+ g_l \frac{I(I+1) + l(l+1) - s(s+1)}{2I(I+1)} \tag{6-12.4}$$

ISBN 0-8053-8061-7

TABLE 6-6 (*continued*)

Z	Element	A	Half-life	Spin	Magnetic dipole moment μ	Electric quadrupole moment Q
49	In	109	4.2 h	$\frac{9}{2}$	5.53	0.89
		110^m	4.9 h	7		
		111	2.81 days	$\frac{9}{2}$	5.53	0.87
		113		$\frac{9}{2}$	5.5229	0.846
		113^m	99.47 min	$\frac{1}{2}$	-0.21050	
		114^m	49.51 d	5	4.7	
		115	6×10^{14} yr	$\frac{9}{2}$	5.5348	0.861
		115^m	4.50 h	$\frac{1}{2}$		
		116^m	54.12 min	5	4.21	
55	Cs	127	6.25 h	$\frac{1}{2}$	1.46	
57	La	138	1.12×10^{11} yr	5	3.7073	0.51
71	Lu	176	3.27×10^{10} yr	7	3.184	8.0
72	Hf	177		$\frac{7}{2}$	0.7902	4.5
		178		0		
		178^m	4.3 sec	8		
		179		$\frac{9}{2}$	0.6382	5.1
		180^m	5.5 h	8		
73	Ta	181		$\frac{7}{2}$	2.361	3.9
80	Hg	197	64.14 h	$\frac{1}{2}$	0.524061	
		197^m	23.8 h	$\frac{13}{2}$	-1.021228	1.61
		199		$\frac{1}{2}$	0.502707	
92	U	233	1.553×10^5 yr	$\frac{5}{2}$	0.55	3.5
		235	7.13×10^8 yr	$\frac{7}{2}$	0.34	4.1
94	Pu	239	24.390 yr	$\frac{1}{2}$	0.200	
95	Am	241	432.7 yr	$\frac{5}{2}$	1.58	4.9

[a]The unit of magnetic moment is the nuclear magneton. The electric quadrupole moments are measured in units of 10^{-24} cm^2. A recent, complete table of the same quantities, containing references to the original literature, is given by V. S. Shirley and C. M. Lederer, "Hyperfine Interactions studied in Nucl. Reactions and Decay" (E. Karlson and R. Wäppling Ed.) Almquist Int., Stockholm, 1975. The symbol Ω represents magnetic octupole moment. [b]Magnetic moments and electric quadrupole moment have opposite sign.

From Eqs. 6-12.1 and 4 we obtain, for $I = l + \frac{1}{2}$,

$$\mu_I = \tfrac{1}{2} g_s + l g_l \tag{6-12.5}$$

and for $I = l - \frac{1}{2}$,

$$\mu_I = -\tfrac{1}{2} g_s \frac{2l-1}{2l+1} + g_l \frac{(l+1)(2l-1)}{2l+1} \tag{6-12.6}$$

ISBN 0-8053-8061-7

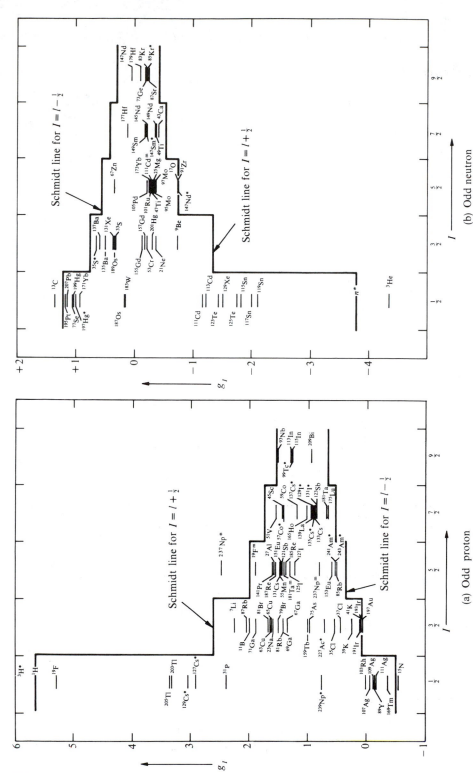

Figure 6-34 Schmidt lines and experimental nuclear magnetic g-factors (*, radioactive nuclei; *m*, metastable states).

The two formulas can be combined to give

$$\mu_I = I\left[g_l \pm \frac{1}{2l+1}\left(g_s - g_l \right)\right] \tag{6-12.7}$$

where the upper sign is valid for $I = l + \frac{1}{2}$ and the lower for $I = l - \frac{1}{2}$. The g_l and g_s values are given in Table 6-7.

TABLE 6-7 g NUMBERS FOR NEUTRON AND PROTON

	Neutron	Proton
g_l	0	1
g_s	–3.826	5.586

The experimental values of nuclear moments are plotted in Fig. 6-34, together with the values predicted by Eq. (6-12.7).

The observed values do not fall exactly on the Schmidt lines but lie between them. It is not surprising, however, that a theory based on such a crude model fails to give quantitative results; in fact it is rather encouraging to find that the agreement is as good as it is.

6-13   SHELL MODEL

In the liquid-drop model we have emphasized the properties of "nuclear matter" and have said nothing about single nucleons. This is a great departure from the atomic model, where the emphasis is on the motion of the electrons in the field provided by the nucleus.

The atomic model, however, has been so successful that one is tempted to find a way to extend at least some of its features to the nucleus. Indeed, there are reasons for suspecting that nucleons have well-defined individual orbits. Among them is the fact that the Fermi gas, which neglects the interaction between nucleons, is at least partially successful. This means that nucleons are relatively free in the potential well and the Schmidt lines (Sec. 6-12) point to the existence of orbits.

To describe the potential well a little more fundamentally, consider the hamiltonian of the nucleus

$$H = \sum_1^A T_i + \frac{1}{2}\sum_{i \neq k} v_{ik}(\mathbf{r}_{ik}) \tag{6-13.1}$$

where the first term is the kinetic energy of the nucleons and the second the

ISBN 0-8053-8061-7

interaction between them, neglecting three or more nucleon forces; the second term is then written as

$$\sum_i V_0(\mathbf{r}_i) + \left[\frac{1}{2} \sum_{i \neq k} v_{ik}(\mathbf{r}_{ik}) - \sum_i V_0(\mathbf{r}_i) \right] \qquad (6\text{-}13.2)$$

and the $V_0(\mathbf{r}_i)$ is chosen so as to make the last bracket (residual interaction) possibly small and the vector $\mathbf{r}_i$ gives the coordinates of the ith nucleon with respect to the nuclear center. A reasonable $V_0(\mathbf{r}_i)$ because of the short range of the nuclear force may have the same $\mathbf{r}$ dependence as the density distribution of the nucleus.

One may then require that the density obtained give rise to the potential with which one started. This self-consistency requirement was introduced by Hartree in atomic physics and is called the *Hartree self-consistent field*. A further refinement derives from the antisymmetrization of the self-consistent wave function (Hartree–Fock).

The nucleon–nucleon forces are to an extent momentum dependent and as a consequence so is $V_0(\mathbf{r}_i)$. This function can then be developed into a power series of momentum

$$V_0(\mathbf{r}_i) = V_{00}(\mathbf{r}_i) + \alpha p_i^2 + \cdots \qquad (6\text{-}13.3)$$

Only even powers are present because when the sign of p_i is changed, $V_0(\mathbf{r}_i)$ does not change sign.

The term of V_0 containing p^2 may be incorporated into the kinetic energy term of the hamiltonian by writing

$$T_i = \frac{p_i^2}{2m^*} = \frac{p_i^2}{2m} + \alpha p_i^2 \quad \text{or} \quad \frac{1}{2m^*} = \frac{1}{2m} + \alpha \qquad (6\text{-}13.4)$$

The quantity m^* is called the effective mass, and in first approximation we simply account for the velocity dependence of $V_0(\mathbf{r})$ by replacing m in the kinetic energy by m^* (cf. Fig. 6-14).

Experiment and theory suggest that the hamiltonian also contains a spin–orbit coupling term

$$\frac{\lambda^2}{r} \frac{\partial V_0}{\partial r} \mathbf{l} \cdot \mathbf{s} = V_s(r) \mathbf{l} \cdot \mathbf{s} \qquad (6\text{-}13.5)$$

where $\mathbf{l}$ and $\mathbf{s}$ are the orbital and spin angular momenta, respectively, and λ is a length to be fixed by experiment. Thus the approximate single-particle hamiltonian is

$$H = \sum_i \left(\frac{p_i^2}{2m^*} + V_{00}(\mathbf{r}_i) + V_s(\mathbf{r}_i) \mathbf{l}_i \cdot \mathbf{s}_i \right) \qquad (6\text{-}13.6)$$

ISBN 0-8053-8061-7

The difference between Eq. (6-13.6) and Eq. (6-13.1) is the residual interaction. We will discuss its effects later.

According to the shell model, nucleons move in definite orbits in this potential well. It would be desirable to derive a suitable form of $V_0(r)$ from the nucleon–nucleon interaction, and many physicists have worked in this direction with a fair degree of success. Even without going into this complicated subject we must point out a serious difficulty in the shell model. How can a nucleon move in an orbit in nuclear matter?

Using free particle cross sections, we would expect a nucleon mean free path in nuclear matter to be short compared with the distance required before one can speak of an orbit. Pauli's principle gives a partial answer to this difficulty by inhibiting collisions within a nucleus when the final states that the colliding nucleons should reach are already occupied. The argument also shows qualitatively that nucleons near the surface of the Fermi momentum sphere are more susceptible to collisions than those near the center of the sphere.

We close with a remark on the potential V_{00} of Eq. (6-13.6). The nucleus, with its strong central force, gives rigidity to the atomic structure. In a nucleus there is no fixed center of force. The surface tension (compare Sec. 6-7) tends to make the nucleus spherical. However, this is a small effect. The nucleus as a whole, then, is fairly easily deformed and nonspherical nuclei are well known.

Numerous experimental facts speak for a shell model. For instance, for certain numbers of neutrons or protons, called *magic numbers*, nuclei exhibit special characteristics of stability reminiscent of the properties shown by noble gases among the atoms. In the latter case the closure of a shell ($K, L, M, \ldots$ shells for 2, 10, 18, $\ldots$ electrons) is responsible for the peculiarities of the noble gases, such as their chemical inertness, diamagnetism, and high ionization potential. M. Mayer, and Haxel, Jensen, and Suess (1949), surmised that nuclear shells might close at some magic numbers that are revealed in nuclear phenomena.

We shall mention here some of the evidence supporting the "magic" character of the numbers 2, 8, 20, 28, 50, 82, and 126. The fundamental fact is that the 3rd, 9th, 21st, 29th, 51st, 83rd, and 127th nucleons are especially loosely bound. The decrease from the average binding energy of these nucleons is approximately 1 or 2 MeV (see Figs. 6-35, 6-36). We can note this effect by observing the jumps in the differences between the experimental values of the masses and the values predicted by the semiempirical mass formula. The differences are particularly large for values of A that correspond to magic numbers of neutron and proton. An exaggerated example is ^{4}He, in which the second neutron or the second proton is bound with an energy of about 20 MeV and a third neutron or proton is not bound at all. Other shells at low magic numbers present less-marked anomalies, but the doubly magic ^{16}O is well known for its stability. The evidence for the magic character of the

ISBN 0-8053-8061-7

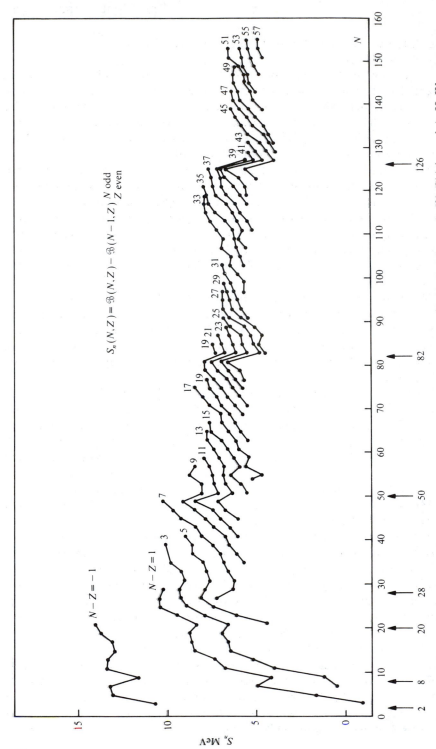

Figure 6-35 Neutron separation energies S_n taken from the compilation by J. H. E. Mattauch, W. Thiele, and A. H. Wapstra, *Nuclear Phys.*, **67**, 1 (1965).

$$S_n(N,Z) = \mathcal{B}(N,Z) - \mathcal{B}(N-1,Z) \quad \begin{array}{l} N \text{ odd} \\ Z \text{ even} \end{array}$$

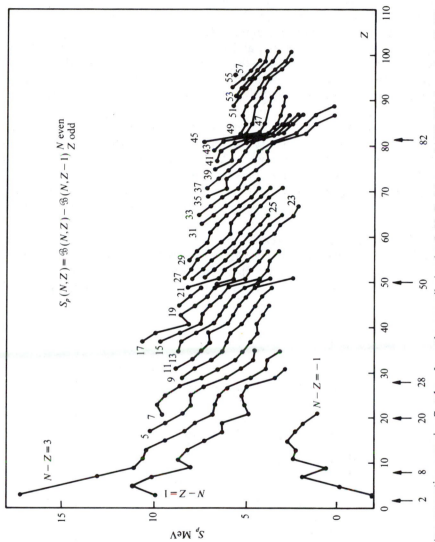

Figure 6-36 Proton separation energies S_p taken from the compilation by J. H. E. Mattauch, W. Thiele, and A. H. Wapstra, *Nuclear Phys.*, **67**, 1 (1965).

ISBN 0-8053-8061-7

higher numbers is less direct. To cite some of the relevant facts, nuclei with 50 protons or 50 neutrons are especially abundant, and isotones with 28, 50, or 82 neutrons extend over a larger range of Z than the neighboring isotonic series. Similarly, tin ($Z = 50$) has the greatest number of stable isotopes, 10.

The magic character of the number 126 for neutrons is strikingly shown in alpha decay (Fig. 7-10). When alpha emission separates the 126th neutron from a nucleus—necessarily together with the 125th neutron—the energy of alpha particles is markedly lower than when the 128th and 127th neutrons are ejected. This is explained by the sudden decrease in binding energy for neutrons after the 126th.

Similar phenomena occur in beta decay. Consider the energy released in beta decay, during which process the difference between the number of neutrons and the number of protons in the nucleus decreases by two units. We find rather regular curves if we plot this energy as a function of A for a given value of $N - Z$ in the original nucleus. However, obvious jumps appear in the curves when a nucleus with a magic number of neutrons or protons is involved.

Neutron-capture cross sections are particularly low for nuclei containing 50, 82, or 126 neutrons, because the neutron to be captured will be only slightly bound and this indirectly decreases probability of capture (Fig. 6-37).

The shell model also accounts for another prominent feature of nuclear

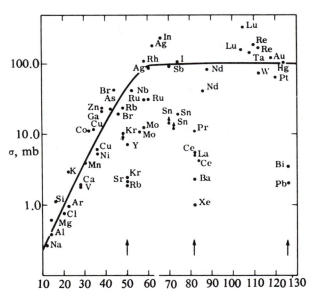

Figure 6-37 Neutron capture cross sections as a function of neutron number of target nucleus. Ordinate, logarithmic scale.

ISBN 0-8053-8061-7

spectra, the separation between the $I^P = 0^+$ ground state of even-even nuclei and the first excited 2^+ state. This state represents the most easily excited state, and if the shell is not closed it is obtained by rearranging the nucleons without changing their shell. On the other hand, this is not possible for a filled shell; it is then necessary to send one nucleon to a higher shell and thus the jump in energy is larger. For instance, the first excited states for $^{112}_{48}$Cd and $^{116}_{52}$Te are 0.61 MeV above the ground state, but the first excited state for $^{114}_{50}$Sn is 1.30 MeV above the ground state. All three nuclei have 64 neutrons, but Sn has 50 protons, which form a closed shell, whereas 48 or 52 do not.

The phenomena mentioned above are consequences of the jumps in binding energy shown in Figs. 6-35 and 6-36. The relation of the nuclear spin and the magnetic moment to the shell structure of nuclei has already been mentioned. There are many more facts in nuclear physics that affirm the shell structure or find a plausible explanation on the basis of this model.

Assuming that the nucleons move in a square potential well, we can establish a series of levels, but with such a potential or simple modifications thereof we find that the shells do not close at the magic numbers. To account for the experimental results it is necessary to introduce the additional hypothesis of a strong spin–orbit coupling. [Eq. (6-13.5)].

The single-particle coordinates are usually referred to a fixed center. In fact, they should be referred to the center of mass of the system, which approach may produce, especially in light nuclei, appreciable changes in the wave function.

The order of levels shown in Fig. 6-38, with spin–orbit coupling taken into account, is fundamental to the interpretation of nuclear phenomena according to the shell model. Actually there are minor differences between the case of an unpaired neutron and that of an unpaired proton. However, one may neglect them in a first approximation and use Fig. 6-38 for neutrons and protons alike. Then the l and j of the extra nucleon outside the closed even-even shell are predictable from the shell model (Fig. 6-38), and thus the systematics of nuclear spins and magnetic moments is accounted for by the shell model, because the spins are simply the spin of the unpaired nucleon and the magnetic moments are obtained from Eq. (6-12.7).

For example, ^{17}O has an extra neutron, the ninth, that should be in a state $d_{5/2}$; and the spin of ^{17}O is $5/2$, as expected. Its magnetic moment, lying very close to the Schmidt line, also confirms this interpretation.

The shell model can also be used to give information on the spin of odd-odd nuclei. We require here the "Nordheim rules." Define the Nordheim number as $N = j_p - l_p + j_n - l_n$, where p and n refer to the odd proton and neutron, respectively.

The rules say that if $N = 0$, then $I = |j_n - j_p|$ and that if $N = \pm 1$, I is either $|j_n - j_p|$ or $j_n + j_p$. There are exceptions to these rules, however, especially among light nuclei.

ISBN 0-8053-8061-7

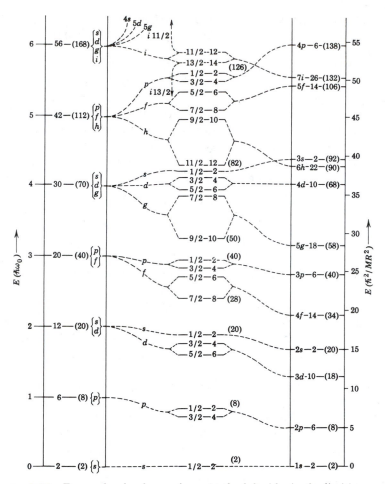

Figure 6-38 Energy levels of a nucleus. At the left side, in the limiting case of a parabolic potential, we have the harmonic oscillator levels; at the right side, the limiting case of a square well; in the center, the nuclear case. The integral number on the level gives the number of sublevels contained in it (statistical weight). The numbers in parentheses are the total number of sub-levels below the energy indicated by the level, in other words, the number of nucleons in a nucleus necessary to fill all states below, and inclusive to, the level considered. The half-integral numbers give the angular momentum $j = l \pm \frac{1}{2}$ of the nucleons in the level. The parity of the levels is even or odd, as $\Sigma_i \, l_i$. A magic number corresponding to a closed shell appears whenever the jump between adjacent levels is particularly large. The actual filling of orbits in nuclei differs somewhat from the scheme of this figure. The neutron and proton levels are slightly different, because the Coulomb repulsion is present only for protons.

ISBN 0-8053-8601-1

6-14 THE PAIRING INTERACTION

●Up to now we have discussed mainly the effects of the average nuclear potential, including spin–orbit coupling. We turn now to the residual interaction, giving only some of the main results. The theory has close analogy to the Bardeen–Cooper–Schrieffer theory of superconductivity; for a more complete treatment, see the literature (e.g., Brown, 1971.)

The most important component of the residual interaction is a very short-range attractive force between identical nucleons, and it is thus especially important for nucleons with overlapping wave functions. Such nucleons are in an s state with respect to each other and form a pair whose total angular momentum is zero. Even-even nuclei have all their nucleons paired in this fashion and the total resultant angular momentum for the nucleus as a whole is zero.

It is a good approximation to consider the residual interaction only for a pair of nucleons having the same quantum numbers except for the magnetic quantum number m, which has opposite signs for the two members of a pair. We denote all four quantum numbers of an orbit by k; $-k$ means the same quantum numbers except that m has been changed into $-m$. Only a pair of nucleons close to the Fermi surface in momentum space are likely to be affected by the residual interaction, because there are unoccupied levels close to them able to accommodate scattered particles. The residual interaction thus sends pairs with quantum numbers k, $-k$ to states k', $-k'$, all states being in the vicinity of the Fermi surface. We first consider a system with many levels, almost forming a continuum. The case of few orbits will be mentioned later.

We call the number of nucleons in a given state k the *occupation number*; because of the exclusion principle, this number is either 1 or 0. The average occupation number of state k is indicated by v_k^2 and it varies between 0 and 1.

Neglecting the residual interaction, the occupation number for each orbit is given by the dotted line in Fig. 6-39. Every orbit below a certain energy is occupied; above this energy, all orbits are empty. The energy dividing the occupied from the empty levels is called the Fermi energy and denoted by the letter λ. When the residual interaction is taken into account, the occupation numbers do not go sharply to zero at the Fermi energy; on the contrary, there is an interval 2Δ in which the occupation numbers go from one to zero, as shown by the solid line in Fig. 6-39. In other words the ground state of the nucleus does not correspond to a single spectroscopic configuration, but to a linear combination of many.

One can show that the average occupation numbers as a function of energy are given by

$$v_k^2 = \frac{1}{2}\left\{1 - \frac{\varepsilon_k - \lambda}{\left[(\varepsilon_k - \lambda)^2 + \Delta^2\right]^{1/2}}\right\} \tag{6-14.1}$$

ISBN 0-8053-8061-7

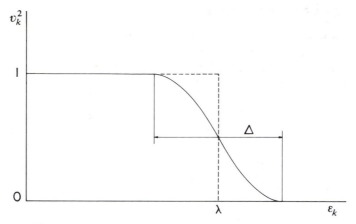

Figure   6-39   Occupation number versus energy in a Fermi gas. Dotted line, without residual interaction; solid curve, with residual interaction. The Fermi energy is at λ.

The quantity λ is practically the Fermi energy; Δ depends on the interaction. For $\Delta = 0$ the solid line becomes the dotted line, as we can easily verify. For $\Delta \neq 0$, $v_k^2 = 1$ for $\varepsilon_k - \lambda \ll \Delta$ and $v_k^2 = 0$ for $\varepsilon_k - \lambda \gg \Delta$. For $\varepsilon_k = \lambda$, $v_k^2 = \frac{1}{2}$. Along with v_k^2 we consider the quantity $1 - v_k^2 = u_k^2$, which gives the probability that a given level is empty. The condition $\sum_k v_k^2 = n$, where n is the number of nucleons of a certain kind, is used to determine λ and gives the result that λ depends on Δ and is the Fermi energy.

The residual interaction has matrix elements connecting states k, $-k$ with k', $-k'$ and we indicate the average value of such matrix elements by

$$G = \overline{\langle k', -k'|V|k, -k \rangle} \tag{6-14.2}$$

The quantity Δ is connected to G by the relation

$$\Delta = G \sum_k u_k v_k \tag{6-14.3}$$

Inserting in it the values of v_k and u_k [Eq. (6-14.1)], we obtain

$$1 = \tfrac{1}{2} G \sum_k \left[(\varepsilon_k - \lambda)^2 + \Delta^2 \right]^{-1/2} \tag{6-14.4}$$

To solve this equation for Δ, we transform the sum into an integral, introducing ρ, the number of levels per energy interval near the Fermi energy; we obtain

$$1 = \tfrac{1}{2} G \int_{\lambda - s}^{\lambda + s} \frac{\rho \, d\varepsilon}{\left[(\varepsilon - \lambda)^2 + \Delta^2 \right]^{1/2}} = G\rho \sinh^{-1}\!\left(\frac{s}{\Delta} \right) \tag{6-14.5}$$

ISBN 0-8053-8601-7

The integral is extended to an energy interval around the Fermi energy in which the pairing energy is effective. We thus obtain

$$\Delta = \frac{s}{\sinh{(1/\rho G)}} \qquad (6\text{-}14.6)$$

or

$$\Delta = 2s\,\exp(-1/\rho G) \qquad (6\text{-}14.7)$$

for $\rho G \ll 1$. Approximate values of ρ may be $\approx A/30$ in MeV^{-1} and of s 1 MeV. Δ may be obtained experimentally by observing the lowest excited state of an even-even nucleus or the energy difference δ of Table 6-3.

The lowest excited state of an even-even nucleus (apart from collective states, such as rotational or vibrational ones) is obtained by breaking a pair and moving one of the nucleons to an excited orbit. We thus create a "hole" and an unpaired particle. These are called *quasiparticles*. Their energy above the ground state of the nucleus is given by

$$E_k = \left[(\varepsilon_k - \lambda)^2 + \Delta^2\right]^{1/2} \qquad (6\text{-}14.8)$$

and for $\varepsilon_k - \lambda \gg \Delta$ it is essentially the same as $\varepsilon_k - \lambda$, but for $|\varepsilon_k - \lambda| \ll \Delta$ it is Δ and cannot go to zero. In an even-even nucleus the lowest intrinsic state above the ground state is a two-quasiparticle state and its energy is

$$E_{k1} + E_{k2} = \left[(\varepsilon_{k1} - \lambda)^2 + \Delta^2\right]^{1/2} + \left[(\varepsilon_{k2} - \lambda)^2 + \Delta^2\right]^{1/2} > 2\Delta \quad (6\text{-}14.9)$$

For the lowest state $\varepsilon_{k1} - \lambda$ and $\varepsilon_{k2} - \lambda$ are small compared with Δ; nevertheless the minimum excitation is $> 2\Delta$. This is called the *pairing gap*.

In the ground state of odd-A nuclei there is an unpaired nucleon; in odd-odd nuclei there are two unpaired nucleons. Therefore, an even-odd nucleus will have a mass exceeding that of an even-even nucleus by the mass of the added quasiparticle, and odd-odd nuclei will exceed the mass of even-even nuclei by the mass of two quasiparticles. This energy is, in good approximation, $12A^{-1/2}$ MeV, which thus is identified with δ of Eq. (6-7.7).

The pairing energy is also related to the superfluidity of nuclear matter, which reduces the moment of inertia of deformed nuclei as observed in rotational bands, and to other phenomena.

Configuration mixing may also be treated by considering the occupation numbers of the levels in a spherical well. The occupation number v_j^2 for a degenerate orbit is here defined as the number of particles in a certain orbital, p, divided by the maximum possible number. The orbital is specified by giving n, l, j but not m, and the maximum possible number of particles in that orbital is $2j + 1$, corresponding to the possible values of m. The occupation

ISBN 0-8053-8601-7

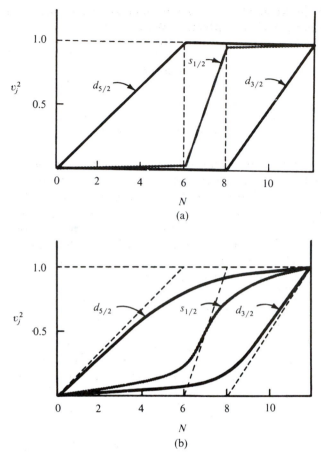

Figure 6-40 (a) The occupation numbers v_j^2 versus N, the number of nucleons in the $n = 3$ shell, in the approximation that there are no residual interactions. (b) v_j^2 versus N corrected for the existence of residual interactions. The dashed lines are a reproduction of part (a), and the curves showing the effects of residual interactions are essentially a rounded-off version of the lines in part (a). From B. L. Cohen, Concepts of Nuclear Physics, McGraw-Hill Co. New York 1971.

number is $p/(2j + 1) = v_j^2$. The number v_j^2 obviously varies between 0 and 1. Without residual interaction, adding nucleon pairs to a closed shell would produce a linear increase of v_j^2. In reality this does not occur; Fig. 6-40a,b compares the ideal and real situations. The nuclear ground state in the real case corresponds to a linear combination of several configurations and the occupation numbers may be approximated by Eq. (6-14.1) where λ is determined by the subsidiary condition that the number of nucleons outside a

ISBN 0-8053-8011-7

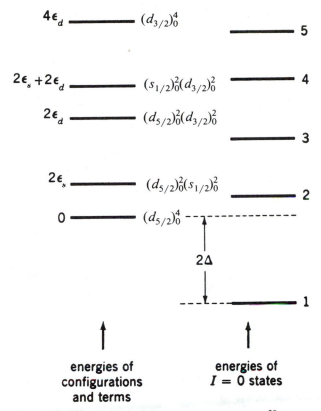

$$4\epsilon_d \quad \underline{\hspace{2cm}} \quad (d_{3/2})_0^4 \qquad \underline{\hspace{1.5cm}} \quad 5$$

$$2\epsilon_s + 2\epsilon_d \quad \underline{\hspace{2cm}} \quad (s_{1/2})_0^2 (d_{3/2})_0^2 \qquad \underline{\hspace{1cm}} \quad 4$$

$$2\epsilon_d \quad \underline{\hspace{2cm}} \quad (d_{5/2})_0^2 (d_{3/2})_0^2$$

$$\underline{\hspace{1.5cm}} \quad 3$$

$$2\epsilon_s \quad \underline{\hspace{2cm}} \quad (d_{5/2})_0^2 (s_{1/2})_0^2 \qquad \underline{\hspace{1cm}} \quad 2$$

$$0 \quad \underline{\hspace{2cm}} \quad (d_{5/2})_0^4$$

$$2\Delta$$

$$\underline{\hspace{1.5cm}} \quad 1$$

energies of   **energies of**
configurations $I = 0$ **states**
and terms

Figure   6-41   Energies of some $I = 0$ terms and states in ^{20}O (four neutrons in the $n = 3$ shell). These are the terms in which all neutrons are coupled in pairs of zero angular momentum. The term energies shown at the left are just the sum of the energies of occupied orbits. (Note that the assumed unperturbed level order $d_{5/2}$, $s_{1/2}$, $d_{3/2}$ is different from that in Fig. 6-38.) The energy gap 2Δ is also shown.

closed shell is given by

$$\sum v_j^2 (2j + 1) = N \tag{6-14.10}$$

The parameter Δ^2 is the same as before. The energy of an orbit is also changed according to Eq. (6-14.8), where the energy of the previous closed shell is assumed to be zero. As an example, consider the 12 neutrons of ^{20}O; eight are in the $1s_{1/2}$ and $2p_{1/2, 3/2}$ orbits. The remaining four are expected to go into $3d_{5/2}$ orbits and form a configuration $(3d_{5/2})^4$ $I = 0$. (We use here spectroscopic notation in which the exponent is the number of nucleons in orbits characterized by the quantum numbers in parentheses.) They do this with a vengeance because the residual interaction produces a pairing gap by

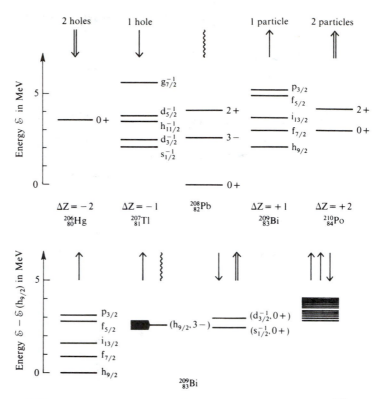

Figure 6-42 Elementary excitations based on the ground state of $^{208}_{82}$Pb. The spectra of the nuclei around $^{208}_{82}$Pb can be described in terms of elementary excitations based on the ground state of $^{208}_{82}$Pb, which has closed shells of neutrons and protons and acts as the "vacuum state" for the excitations.

Top: Fermion-like excitations involving the addition or removal of a single proton ($\Delta Z = +1$ or $\Delta Z = -1$), and boson-like excitations involving correlated pairs of protons ($\Delta Z = \pm 2$) as well as collective excitations in $^{208}_{82}$Pb itself. The latter type of quanta can be expressed in terms of coherent particle–hole excitations with a resulting density oscillation approximately corresponding to that of a surface vibration. The energy scale employed in the figure involves a linear term in ΔZ, so chosen that the lowest one-particle and one-hole excitations ($h_{9/2}$ and $s_{1/2}^{-1}$) have the same ordinate. Additional elementary excitations not shown in the figure involve changes in neutron number ($\Delta N = \pm 1, \pm 2$).

Bottom: Low-energy spectrum of $^{209}_{83}$Bi. In addition to the one-particle excitations shown at the left, the spectrum exhibits excitations involving a single particle or a single hole combined with a collective excitation. The configuration ($h_{9/2}, 3^-$) gives rise to a multiplet of states with total angular momentum $3/2, 5/2, \ldots, 15/2$ that have all been identified within an energy region of a few hundred keV. The configurations involving a hole and a pair quantum with $I^P = 0^+$ each give rise to only a single state. At an excitation energy of about 3 MeV, a rather dense spectrum of two-particle one-hole states sets in, as indicated at the right in the figure. [From A. Bohr and B. R. Mottelson, *Ann. Rev. Nucl. Sci.*, **23**, 363 (1973).]

ISBN 0-8053-8061-1

lowering the energy of this level. Figure 6-41 illustrates the situation. Without residual interaction, the energies would be as at the left in the figure. Terms in different configurations with the same I and parity form linear combinations, and the energy is altered. The result is shown at the right in Fig. 6-41. Without residual interaction the occupation numbers would vary as in Fig. 6-40a. The residual interaction smoothes the curves, giving the occupation numbers in Fig. 6-40b.

All this is valid for even-even nuclei. If there are unpaired particles, as in even-odd nuclei or in excited states of even-even nuclei, one has quasiparticles, and the situation is more complicated. Nuclear spectroscopy is by now a vast subject; these samples are presented only to give an idea of the problems it considers. Figure 6-42, taken from Bohr and Mottelson (BM 69), is an eloquent illustration of the subject.●

6-15 COLLECTIVE NUCLEAR MODEL

The shell model has been most successful in explaining a number of nuclear features, some of which have been mentioned and some of which will be considered later. However, it does not provide a complete description of the nucleus. It is particularly successful in the case of nuclei composed of a closed shell plus one or a few additional nucleons. In the closed-shell configuration the nucleus is spherical. The addition of one or more nucleons produces only small deformations. However, midway between closed shells the situation is different. The nuclei depart appreciably from the spherical form, and collective motions involving many nucleons become important.

This is clearly seen, for instance, in nuclear quadrupole moments (Fig. 6-43). Near a closed shell the values obtainable by ascribing the quadrupole to a single particle outside a shell agree reasonably well with experiment; but in the middle of the shell the quadrupole moments are many times larger than can be accounted for by a single particle. Rainwater (1950) suggested that the single particle deforms the whole nucleus, and that the observed quadrupole results from the collective deformation of many orbits. Very crudely, the effect would be similar to that produced by a small heavy ball rotating inside a rubber balloon. The pressure produced by the centrifugal force of the small ball would determine the deformation of the rubber envelope. The nucleons then move in a potential that is not spherically symmetrical. We thus have two types of motion: motion of the entire nucleus as though all the nucleons occupied an ellipsoidal box that might rotate or even deform itself by vibrations; and motions of the nucleons inside the box. The two types of motion are coupled more or less strongly to each other. The mathematical development of these ideas (A. Bohr, B. Mottelson, and others from 1952 on) is the basis for the "collective model," which has been particularly successful for $A \sim 24$, $150 < A < 190$, and $A > 230$. From the point of view of the simplest shell model, the excitation of suitable orbits gives rise to states that are similar

ISBN 0-8053-8001-7

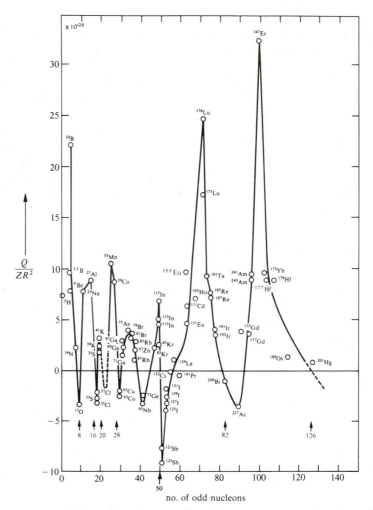

Figure 6-43 Reduced nuclear quadrupole moments as a function of the number of odd nucleons. The quantity Q/ZR^2 gives a measure of the nuclear deformation independent of the size of the nucleus.

to vibrational states, just as a harmonic motion may be classically considered the superposition of two coplanar circular motions in opposite directions. Such vibrational states appear also in nuclei, when there are several nucleons outside a closed shell. The simplest vibration for a spherical nucleus is one that changes it periodically to a lens form, to a sphere, to a cigar, back to a sphere, etc. For a cigar-shaped nucleus there are two lowest types of vibration: β vibrations, in which the major axis changes length periodically, the cigar becoming thinner or fatter; and γ vibrations which keep the length of the cigar constant but squeeze it alternately in a plane containing the major axis (z) of the cigar so as to make it elliptical in cross section. The major axis of this ellipse points alternately in the x and y directions.

An example of analysis of a complex motion by decomposing it into simpler components has long been known in the case of diatomic molecules, the motions of which are naturally classified as rotational, vibrational, and electronic, in order of increasing energy. As typical magnitudes in a molecule, we may take 0.005 eV for a rotational level spacing, 0.1 eV for a vibrational level spacing, and 2 eV for an electronic level spacing. The motions are clearly separable and, correspondingly, the wave function can be factored into rotational, and electronic terms. Mathematically this means that the hamiltonian is composed of three additive parts containing: (1) electronic coordinates (2) vibrational coordinates, and (3) rotational coordinates. The wave function is then the product of three wave functions (compare Sec. 6-4), each containing the respective coordinates.

We may attempt a similar approximate separation of motions for the nucleus. We consider first a central "core" containing the nucleons of a closed shell. Outside this core there are n nucleons forming a "cloud." If $n = \pm 1$ (we consider one nucleon missing from a closed shell as $n = -1$), we have the case to which the shell model applied best. For such a nucleus the excited states are given by the excited levels of the single nucleon (quasiparticle) of the cloud (Fig. 6-38).

For larger even values of n the nucleons in the cloud are subject to short-range residual interactions responsible for orienting the spins of pairs of identical nucleons. Under their influence nuclei tend to a spherical form and zero spin. There are also long-range components of the residual interaction which tend to produce an overlap of the eigenfunctions of *all* n particles. They push the nucleons in certain special directions, thereby deforming the cloud.

In this case the potential in which the nucleons move ceases to be spherically symmetric, as Fig. 6-44 shows.

In the simplest case of one particle moving in a nonspherical potential, the hamiltonian is

$$H = \frac{p^2}{2m} + V_0(r) + V_2(r)P_2(\cos\theta) + V_{so}\mathbf{l}\cdot\mathbf{s} \qquad (6\text{-}15.1)$$

where θ is the polar angle of the particle with respect to the nuclear symmetry axis. $V_2(\mathbf{r})$ is the first term of an expansion in spherical harmonics. If the nucleus takes an ellipsoidal shape defined by $\Delta R/R = \delta$, then $V_2(\mathbf{r})$ for $\delta \ll 1$ is

$$-\frac{2}{3}\frac{dV_0(r)}{dr}r\delta$$

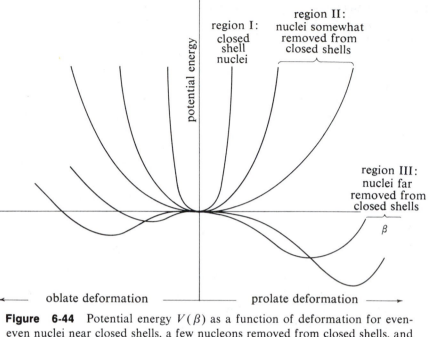

Figure 6-44 Potential energy $V(\beta)$ as a function of deformation for even-even nuclei near closed shells, a few nucleons removed from closed shells, and far removed from closed shells where the nucleus is stabilized in a nonspherical shape. The curves are schematic only, and all of them refer to the potential energy of the ground state.

Corresponding to this change of potential, the energy levels of the shell model lose their degeneracy and shift as shown in the examples of Fig. 6-45; the corresponding wave functions have been extensively studied by Nilsson (1955).

The shift of the energy levels as a function of deformation gives rise to many crossings of levels and the situation becomes very complex (see Fig. 6-46). A remarkable possible consequence of the deformation is the occurrence of magic numbers different from those valid for a spherical nucleus. This effect may have important consequences for the stability of superheavy elements, fission phenomena, etc. Figure 6-47 shows a simple theoretical illustration of the single-particle energy levels for an anharmonic oscillator of potential energy

$$V(x, y, z) = \frac{M}{2}\left[\omega_z^2 z^2 + \omega_\perp^2 \left(x^2 + y^2\right)\right] \tag{6-15.2}$$

The single-particle states are specified by the integral numbers n_3, $n_\perp$, giving the number of quanta in vibrational states. For $\omega_z = \omega_\perp$ there is a high

ISBN 0-8053-8601-1

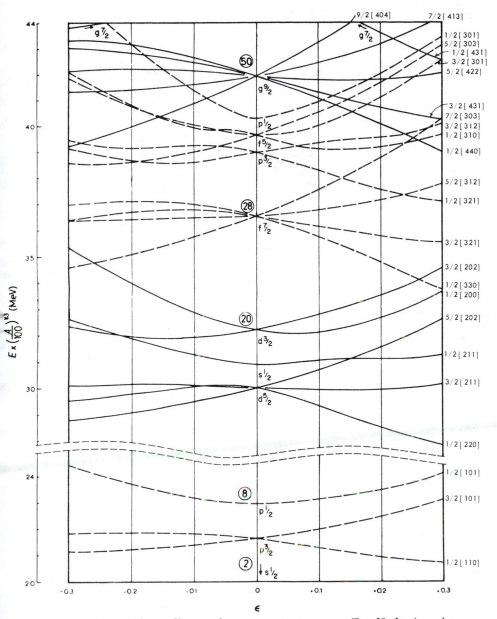

Figure 6-45 Nilsson diagram for protons or neutrons, $Z \leqslant 50$. It gives the calculated energies of nuclear levels as a function of ellipsoidal deformation, defined by the eccentricity coordinate ϵ, approximately equal to $\Delta R / R$. Positive values of ϵ correspond to prolate (cigar) shape. The energy scale for $Z \leqslant 50$ is given in units of MeV $(A/100)^{1/3}$; for $Z \geqslant 50$ (Fig. 6-46) the energy is given in units of $\hbar\omega_0$, the oscillator quantum for the deformed potential. The positions of levels at zero deformation have been adjusted empirically. Even parity levels are given as solid lines, odd parity levels as dashed lines. [From (LHP 67).]

297

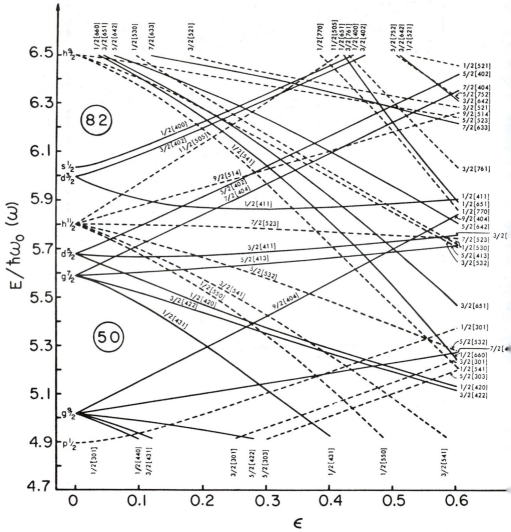

Figure 6-46 Nilsson diagram for odd protons, $50 \leqslant Z \leqslant 82$. [From (LHP 67).]

degree of degeneracy and large energy gaps appear after having used 2, 8, 20, etc. spin-$\frac{1}{2}$ particles to fill the levels. Thus, 2, 8, 20, etc. are magic numbers, as indicated at the left in the figure. As $\epsilon = (\omega_\perp - \omega_z)/\bar{\omega}$ increases, the degeneracy is removed, but it reappears for simple rational values of ϵ. The magic numbers for $\epsilon = 0.6$ (i.e., 2, 4, 10, 16, 28, etc.) are indicated in the figure.

The levels considered thus far are relatively low (several MeV at most). At much higher energies the core itself may be excited. Such levels are found up

ISBN 0-8053-8601-7

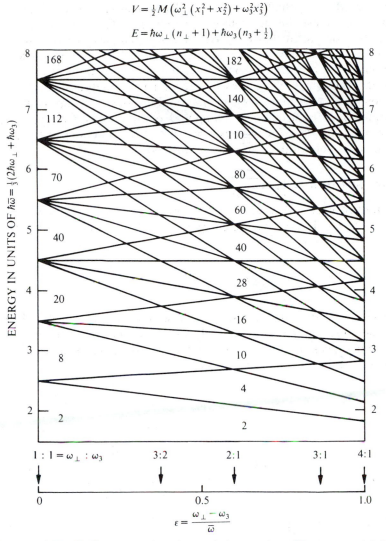

Figure 6-47 Shell structure in anisotropic harmonic oscillator potential. The single-particle energy levels, as a function of deformation, in a prolate axially symmetric oscillator potential are shown. The single-particle states can be specified by the number of quanta n_3 and $n_\perp$, and each energy level has a degeneracy $2(n_\perp + 1)$ due to the spin and the degeneracy in the motion perpendicular to the axis. [From J. R. Nix, *Ann. Rev. Nucl. Sci.*, **22**, 65 (1972).]

to about 20 MeV, at which level gamma absorption may cause vibration of all the protons against the neutrons (see Chap. 11). There are other ways of exciting the core, for instance, by electron or alpha-particle bombardment.

The distinction between the different types of motion in nuclei is not always clear-cut, because the coupling of the motions is often strong. In other words, if, for example, the energy difference between vibrational levels is not large compared with that between rotational levels, it is not possible to consider rotation as a structure superimposed on vibrations. Consequently, the expression of the wave function as a product of the functions of the different coordinates is not always a good approximation.

6-16　ROTATIONAL LEVELS

Deformed nuclei corresponding to large *even* values of n show prominent rotational levels. These have the energies $E_I = (\hbar^2/2\mathcal{I})I(I+1)$ and the spacings

$$\Delta E_{I,\,I-2} = \frac{\hbar^2}{\mathcal{I}}(2I-1) \qquad (6\text{-}16.1)$$

characteristic of rotational levels. I is an even integral number $\geqslant 0$. However, for $\mathcal{I}$ one must consider not the total moment of inertial of the nucleus but, very roughly speaking, that part associated with the nuclear deformation. For a spherical nucleus $\mathcal{I}$ is zero (Fig. 6-48).

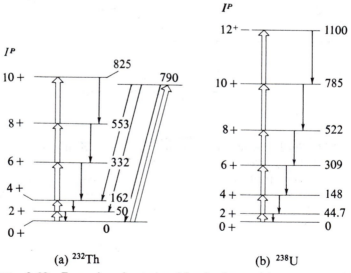

(a) ^{232}Th　　　　　　(b) ^{238}U

Figure 6-48 Examples of rotational levels observed in even-A nuclei of (a) ^{232}Th and (b) ^{238}U by the Coulomb excitation process. (Energy in KeV)

ISBN 0-805-8061-7

For *odd n* the situation is more complicated. Single particles in the nucleus having an axis of rotational symmetry z', possess angular momenta that precess rapidly around z', preserving, however, a constant component Ω_p in this direction. The values of Ω_p are half-integral numbers, positive or negative. States with Ω_p differing only in sign are degenerate because they correspond to rotations differing only in sense. The sum of all Ω_p values is called Ω. In the ground state of even-even nuclei the nucleons are paired, in couples, with Ω_p numbers of opposite sign so that $\Omega = 0$.

The rotations of a symmetric nucleus are described by quantum vectors **I**, **K**, and **M** (see Fig. 6-49). The first is the angular momentum. Vector **K** is the component of **I** along the symmetry axis z' and M is the component of **I** in a fixed arbitrary direction. The component of **I** perpendicular to **K** is called **R**, and is due to the rotation of the nucleus as a whole. In the rotational ground state of nuclei **K** = **Ω**.

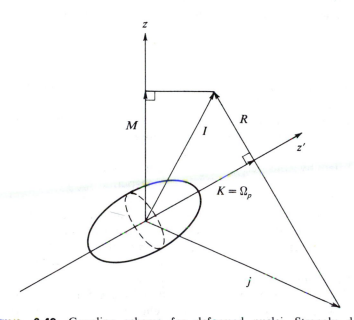

Figure 6-49 Coupling scheme for deformed nuclei. Strongly deformed nuclei still have an axis of symmetry (indicated by z'). The angular momentum properties are characterized by three vectors, constants of the motion: **I**, **M**, and **K**. **I** and **M** are the total angular momentum and its component along a fixed axis z. **K** is the component of the angular momentum of a single nucleon **j** in the direction of z'. It is thus the same as Ω_p and is also the component of **I** in the direction of the nuclear axis of symmetry z'. The collective rotational angular momentum **R** is perpendicular to z'. Since there are no collective rotations around a symmetry axis, K is a constant in a rotational band and represents an intrinsic angular momentum. For even-even nuclei, $K = 0$ in the ground state.

ISBN 0-8053-8061-7

The energy levels of the system are those of a symmetrical top, and the moment of inertia $\mathcal{I}$ that appears in the formula is left as a parameter to be determined experimentally.

With reference to Fig. 6-49 we see that the rotational energy levels are given by

$$E_{\text{rot}} = \frac{\hbar^2}{2\mathcal{I}} \langle \mathbf{R}^2 \rangle \tag{6-16.2}$$

but $\mathbf{I} = \mathbf{j} + \mathbf{R}$, and therefore, by the cosine rule and insertion of the quantum-mechanical expectation values, we have

$$\langle R^2 \rangle = \left[I(I+1) + j(j+1) - 2I_{z'}j_{z'} - 2I_y j_y - 2I_x j_x \right] \tag{6-16.3}$$

The quantity $I_{z'}j_{z'}$ is simply K^2, as is apparent from Fig. 6-49. The last two terms give zero except in the important case where $K = \frac{1}{2}$. If $K \neq \frac{1}{2}$, we obtain for the rotational part of the energy the well-known formula

$$E_{\text{rot}} = \frac{\hbar^2}{2\mathcal{I}} I(I+1) \tag{6-16.4}$$

In this formulas, we repeat, $\mathcal{I}$ represents an empirically determined moment of inertia associated with the nuclear deformation.

In the ground state of an even-even nucleus $\Omega = 0$, as we have seen above, K is also zero, and hence I vanishes, as has always been found. Furthermore, by exciting the lowest rotational states, we find only even values of I. This can be justified by considering the Sommerfeld quantization rule $\oint p \, dq = 2\pi n\hbar$, where the integral is extended over one period of the motion. In the case considered here, the angular momentum and the angle of rotation are the conjugate variables and the period of the motion is a rotation by π, not by 2π, because such a rotation about I is sufficient to restore the original configuration. This symmetry thus has the consequence that I can take only even values; $I = 0, 2, 4, \ldots$ with even parity. We show in Fig. 6-48 the energy levels corresponding to nuclear rotation for some even-even nuclei. These levels are given with excellent approximation by the formula (6-16.4), and thus for $I = 0, 2, 4, 6$, etc., the rotational energies are in the ratios $E_2 : E_4 : E_6 : E_8 = 6:20:42:72$ (see Fig. 6-48).

In odd-A nuclei with $K \neq \frac{1}{2}$, or in excited states of even-even nuclei with $\Omega \neq 0$, the ground state has

$$I_0 = K = \Omega \tag{6-16.5a}$$

and the rotational spectrum has

$$I = I_0, \, I_0 + 1, \, I_0 + 2, \text{ etc.} \tag{6-16.5b}$$

The $K = \frac{1}{2}$ case is more complicated, and we omit it here.

ISBN 0-8053-8601-7

We have said repeatedly that the effective moment of inertia appearing in the rotational levels is due to nuclear deformation and thus may vary on increasing I, producing departures from Eq. (6-16.4) ascribable to a change in $\mathcal{I}$. We have thus a variable moment of inertia.

Classically the rotational energy $E_I = I^2/2\mathcal{I}$, the angular momentum $I = \mathcal{I}\omega$, and the angular velocity $\omega = dE_I/dI$ are continuous variables. The rotational energy may be represented in higher approximation by

$$E_I = \{I^2/2\mathcal{I}(\Delta)\} + V(\Delta) \tag{6-16.6}$$

where Δ is a parameter, depending on I, describing the nucleus (e.g., its deformation, viscosity, etc.). For each I, E_I is minimum on variation of $\mathcal{I}$, keeping I constant. This gives

$$\frac{dE_I}{d\mathcal{I}} = -\frac{I^2}{2\mathcal{I}^2} + \frac{dV}{d\mathcal{I}} = 0 \quad \text{or} \quad \frac{dV}{d\mathcal{I}} = \frac{1}{2}\omega^2 \tag{6-16.7}$$

Considerable success is obtained with the special formula (variable moment of inertia, or VMI model)

$$E_I(\mathcal{I}) = \frac{\hbar^2}{2\mathcal{I}} I(I+1) + \frac{1}{2}C(\mathcal{I} - \mathcal{I}_0)^2 \tag{6-16.8}$$

corresponding to $V(\Delta) = \frac{1}{2}C(\mathcal{I}_1 - \mathcal{I}_0)^2$. In this formula C and $\mathcal{I}_0$ are two parameters to be empirically determined. The moment of inertia $\mathcal{I}_I$ is then chosen so as to minimize E_I with the result that $C(\mathcal{I} - \mathcal{I}_0) = (\hbar^2/2\mathcal{I}^2)I(I+1)$. This determines $\mathcal{I}_I$ for each value of I as a function of $\mathcal{I}_0$ and C. Instead of C one can use the parameter σ (softness) $= d\log\mathcal{I}/dI$, which for $I = 0$ is $\hbar^2/2C\mathcal{I}_0^3$. The values of $\mathcal{I}_0$ and σ for several nuclei are given in Fig. 6-50, together with experimental and calculated energy levels. When σ is large and $\mathcal{I}_0$ small, the spectrum approaches a vibrational one; when $\mathcal{I}_0$ is large and σ small, it approaches a rotational one. Figure 6-51 shows $\mathcal{I}_0$ as a function of Z and N.

It is instructive to plot $\mathcal{I}$ versus ω^2, the square of the angular velocity. First, however, we must define ω more accurately. Classically $\omega = dE/dI$ and we may take for it $\Delta E_{I,(I-2)}/2\hbar$. For a definition of $\mathcal{I}$ we may use the relation $\mathcal{I} = \hbar^2(2I - 1)/\Delta E_{I,(I-2)}$, which follows immediately from Eq. (6-16.1). Thus both ω and $\mathcal{I}$ are defined as functions of the directly observable $\Delta E_{I,I-2}$. Note that ω corresponds to half the frequency of the radiation emitted in the jump from I to $I - 2$, as is to be expected on the basis of symmetry arguments given previously. Equations (6-16.6, 7, 8) combined would give a straight line for the plot of $\mathcal{I}$ versus ω^2 and this obtains up to about $I = 14$ (Fig. 6-52). The striking departures occurring at higher I indicate drastic changes in the nuclear matter and remind us of the change from a superfluid to an ordinary liquid. Such a transition would greatly increase the moment of inertia.

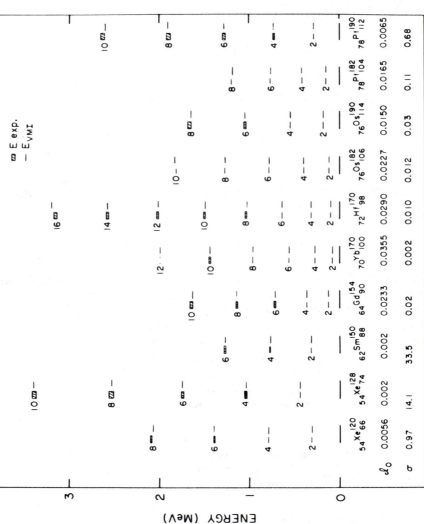

Figure 6-50 Experimental and calculated ground-state bands of some even-even nuclei. For each nucleus, the experimental energies are shown on the left and the calculated energies on the right. The values of the parameters ℓ_0 and σ corresponding to each nucleus are listed at the bottom of the figure. [From M. A. J. Mariscotti, G. Scharff-Goldhaber, and B. Buck. *Phys. Rev.* **178**, 1864

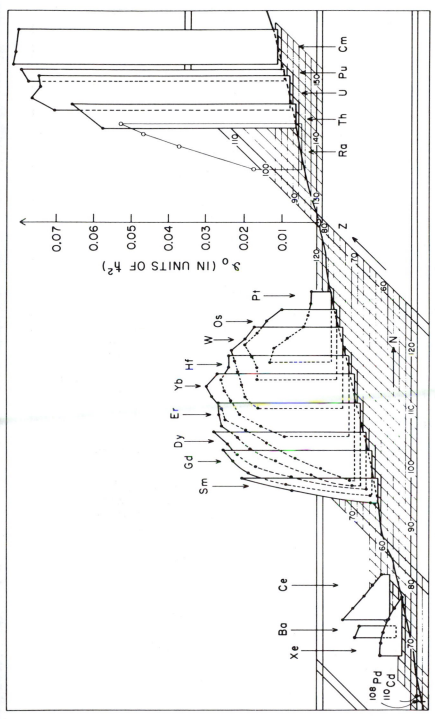

Figure 6-51 Calculated ground state moment of inertia $\mathcal{J}_0$ of even-even nuclei as a function of Z and N. Only those nuclei with at least three known levels (2^+, 4^+, 6^+) of the band are included, with the exception of the Ra isotopes, for which only the 2^+ and 4^+ states are known. The latter are included to show the transition to the well-deformed heavy elements. [From M. A. J. Mariscotti, G. Scharff-Goldhaber, and B. Buck, *Phys. Rev.*, **178**, 1864 (1969).]

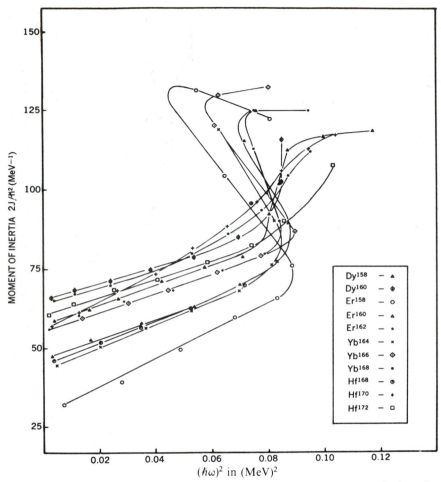

Figure 6-52 Variation of moment of inertia with the angular velocity of rotation squared, ω^2. Ordinate is $2\mathcal{J}/\hbar^2 = (4I-2)/\Delta E_{I,\,I-2}$; abscissa is $\hbar^2\omega^2$ $=(\Delta E_{I,\,I-2}/2)^2$. These results were obtained from inbeam-spectroscopy studies of the ground-state rotational bands. The first point on the left is the $2\rightarrow 0$ transition, the second represents the $4\rightarrow 2$, and so on for the other points. [From J. H. Hamilton and I. A. Sellin, *Phys. Today*, **26**, No. 4, 42 (1973).]

It is at once clear that a deformed nucleus will have an electric quadrupole moment, and thus a connection between $\mathcal{J}$ and Q is apparent and affords a test of the model. This test is especially interesting considering that the large values of certain electric quadrupole moments were the starting point of the idea of collective motions. The first step in evaluating $\mathcal{J}$ is to express the nuclear deformation by a deformation parameter β, defined

ISBN 0-8053-8061-1

through the expansion of the nuclear surface in spherical harmonics. Calling $R(\theta)$ the nuclear radius in a direction at angle θ to the symmetry axis, we write

$$R(\theta) = R_0(1 + \beta Y_{20}(\cos\theta) + \cdots) \qquad (6\text{-}16.9)$$

or, stopping at Y_{20},

$$R(\theta) = R_0\left[1 + \beta\left(\frac{5}{4\pi}\right)^{1/2}\left(\tfrac{3}{2}\cos^2\theta - \tfrac{1}{2}\right)\right] \qquad (6\text{-}16.10)$$

For instance, in ellipsoidal nuclei the deformation parameter is approximately related to the difference between the semiaxes ΔR by

$$\frac{\Delta R}{R} = \frac{3}{2}\left(\frac{5}{4\pi}\right)^{1/2}\beta = \delta \qquad (6\text{-}16.11)$$

Early theories attempted to relate the moment of inertia $\mathcal{I}$ to the deformation parameter β by the model of an irrotational fluid, giving $\mathcal{I}$ proportional to β^2. As the data on actual deformation (obtained from electric quadrupole moments and transition probabilities) accumulated, it appeared that the moment of inertia to be used in Eq. (6-16.1) lay between the values predicted by the rigid-body and the irrotational-fluid models. Moments of inertia close to experimental ones have now been calculated by using wave functions of single nucleons, with correlations between different nucleons (see Fig. 6-53). To connect the nuclear deformation with the electric quadrupole moment, remember that we must find the expectation value

$$\langle\rho_n(3z^2 - r^2)\rangle = \int\psi^*(3z^2 - r^2)\psi\,d\tau \qquad (6\text{-}16.12)$$

To compute it we must state which ψ are to be used. In Sec. 6-9 we specified that we must have $M_I = I$, that is, maximum alignment of the nucleus as a whole. This gives the quadrupole moment Q.

The expectation value for a nonspherical nucleus is best computed in two steps. We define an intrinsic quadrupole moment Q_0 with respect to the rotational-symmetry axis by using as the eigenfunction not the complete eigenfunction, but only the factors that do not involve the rotation of the nucleus as a whole. In the classical limit Q_0 is simply

$$Q_0 = \int\rho_n(3\xi^2 - r^2)\,d\tau \qquad (6\text{-}16.13)$$

For example, for a rotation ellipsoid it has the value

$$Q_0 = \frac{4}{5}ZR\,\Delta R \qquad (6\text{-}16.14)$$

ISBN 0-8053-8061-7

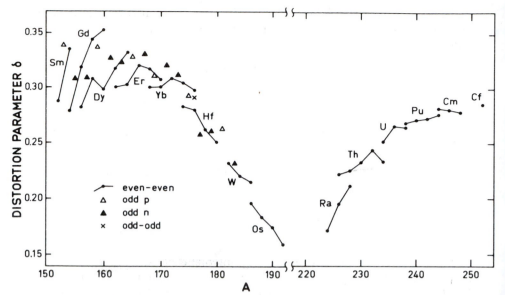

Figure 6-53 Comparison of equilibrium ground-state nuclear deformation calculated from Nilsson wave functions with experimental values. The solid line represents calculated values of the equilibrium deformation for the odd-A nuclei along the valley of beta stability. The experimental data correspond to δ values obtained by means of the equation

$$Q_0 = \tfrac{4}{5}\delta Z R_0^2 \left(1 + \tfrac{1}{2}\delta + \cdots\right)$$

using measured Q_0 values. The latter are based on observed $E2$ transition probabilities, from A. Bohr and B. Mottelson, *Nuclear Structure*, Vol. 2 (1975).

or, introducing β,

$$Q_0 = \frac{4}{5} Z R^2 \frac{3}{2} \left(\frac{5}{4\pi}\right)^{1/2} \beta \qquad (6\text{-}16.15)$$

Next we pass from Q_0 to Q, multiplying Q_0 by a "projection factor" that is a function of I and K. This projection factor depends on the model chosen.

By using the coupling scheme of Fig. 6-49, we obtain (by a rather complicated calculation)

$$Q = Q_0 \frac{3K^2 - I(I+1)}{(I+1)(2I+3)} \qquad (6\text{-}16.16)$$

which in the ground state, when $K = I$, reduces to

$$Q = Q_0 \frac{I(2I-1)}{(I+1)(2I+3)} \qquad (6\text{-}16.17)$$

ISBN 0-8053-8601-7

Measurement of Q thus provides a direct means of finding the deformation parameter β of Eq. (6-16.11). We can then use it for calculating, under given assumptions, $\mathfrak{I}$ and compare the result with the experimental values of $\mathfrak{I}$.

Equation (6-16.17) again shows that for $I = \frac{1}{2}$, Q is zero even for a finite Q_0. Physically this is due to the fact that if Q_0 cannot be oriented with respect to the total angular momentum I, it averages out to zero owing to the nuclear motions. This result is shown in another form by the argument given in Eqs. (6-8.22) to (6-8.25). The interrelation between the phenomena in which Q or Q_0 play an important role is illustrated in Fig. 6-54.

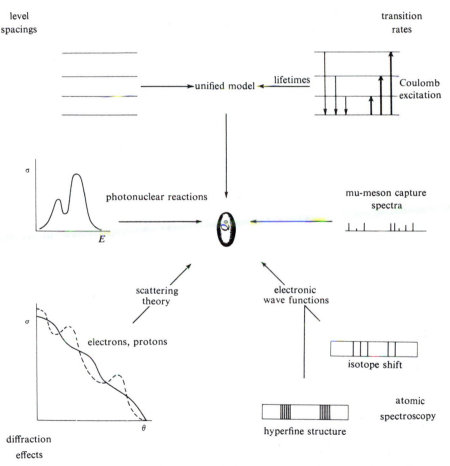

Figure 6-54 Schematic summary of various experimental conditions leading to the intrinsic nuclear quadrupole moment Q_0, and the theoretical concepts involved to obtain Q_0 from its experimental manifestations. [G. Temmer, *Rev. Mod. Phys.*, **30**, 498 (1958).]

ISBN 0-8053-8601-7

The intrinsic quadrupole moment manifests itself not only through Q but also in several other phenomena, such as the isotope shift of spectral lines, the nuclear scattering of electrons or protons, and most prominently, the probability of exciting low rotational levels of a nucleus by the electric field of a heavy particle passing near it (Coulomb excitation; see Chap. 8). This probability is proportional to the square of the matrix element of the electric quadrupole connecting the two states having $I = 0, 2$, indicated by $Q_{0,2}$. Plotting this matrix element versus $\frac{1}{2}(\mathcal{I}_0 + \mathcal{I}_2) = \mathcal{I}_{0,2}$, we obtain points that lie on a curve $Q_{0,2} = k\mathcal{I}_{0,2}^{1/2}$ with $k = 39.4 \pm 2.6 \times 10^{-24}$ cm^2 keV$^{1/2}$. (See Fig. 6-55.)

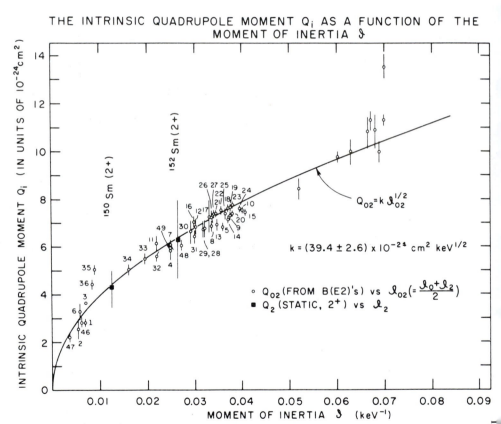

THE INTRINSIC QUADRUPOLE MOMENT Q_i AS A FUNCTION OF THE MOMENT OF INERTIA $\mathcal{I}$

Figure 6-55 Correlation between intrinsic quadrupole moments (absolute values) and moments of inertia given by the VMI model. The points (open circles) refer to the transition quadrupole moments Q_{02}. The ordinates are obtained from the $B(E2, 2 \rightarrow 0)$ values and the corresponding abscissas are $\mathcal{I}_{02} = \frac{1}{2}(\mathcal{I}_0 + \mathcal{I}_2)$. [From M. A. J. Mariscotti, G. Scharff-Goldhaber, and B. Buck, *Phys. Rev.*, **178**, 1864 (1969).]

ISBN 0-8053-8601-1

6-17 CONCLUDING REMARKS

We have given a very incomplete sample of current ideas on nuclear models. The subject has developed tremendously in the last decade and the theory has progressed so far as to often permit detailed quantitative calculations, similar in many respects to those of atomic and molecular spectroscopy. We show in Fig. 6-56 regions of the $N-Z$ plane to which various models are applicable. The relations of the models to each other have also been clarified and derived from first principles. The underlying physical ideas are common to othermany-fermion systems, such as metals, molecules, and atoms. Table 6-8 shows some of these relations

TABLE 6-8 INTRINSIC AND COLLECTIVE MOTIONS IN METALS, MOLECULES, AND NUCLEI[a]

Property	Metals	Molecules	Nuclei
Single-particle state occupied by	Electrons	Electrons	Nucleons
Slowly varying parameters that affect motion of single particles	Nuclear lattice constants	Nuclear positions	Form of the nucleus as a whole and its deformations
Intrinsic motions	The electrons move in the periodic field of the lattice	The electrons move in the field of the nuclei	The nucleons move independently in the field formed by the nucleons themselves
Energy levels	Brillouin zones	Electronic levels	Shell levels
Collective motions	Lattice vibrations	Vibrations and rotations	Energies of nucleons in deformed potential
Energy levels	Specific heat	Energy localized in nuclei	Rotational and vibrational energy of the collective motion
Coupling of intrinsic to collective motion	Weak	Strong	Moderate to strong
$\omega_{coll}/\omega_{part}$	>1	$\ll 1$	$\leqslant 1$

[a] From Moszkowski, with some changes.

Even stereochemistry has inspired an interesting and simple model, although one of limited application. This is the alpha-particle model applicable to light nuclei having the same number of neutrons and protons, this number being a multiple of 4. Evidently one can think of these nuclei ^8Be, ^{12}C, ^{16}O, etc.) as being composed of ^4He nuclei. The interesting feature here is that such a simple idea can be used to predict successfully several properties of these nuclei (see Fig. 6-57). We cannot further enlarge upon these rapidly

ISBN 0-8053-8601-7

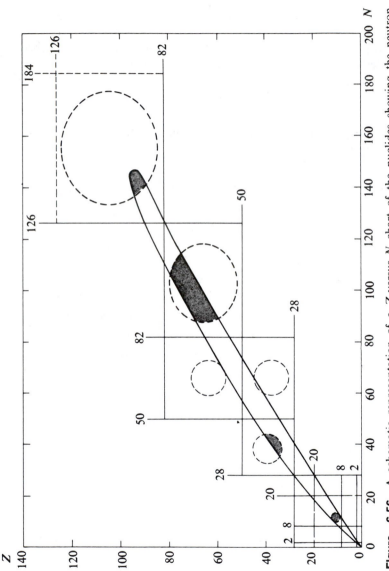

Figure 6-56 A schematic representation of a Z-versus-N chart of the nuclides showing the neutron (vertical lines) and proton (horizontal lines) closed-shell lines. The outer, irregular boundary encloses those nuclei having a half-life longer than 1 min. The line of beta stability runs approximately down the center of this area. The groups of nuclides where rotational spectra of the type predicted for deformed nuclei have been observed are indicated by shading. Additional regions where it may be possible to find such nuclei are also indicated by dashed lines. [From E. Marshalek, L. W. Person, and R. K. Sheline, *Rev. Mod. Phys.*, **35**, 108 (1963).]

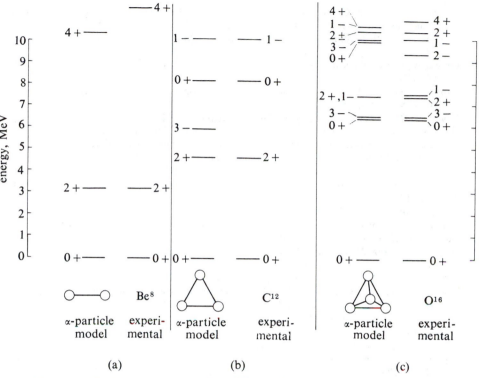

Figure 6-57 Level schemes of ^{8}Be, ^{12}C, and ^{16}O and a comparison with the alpha-particle model. [From (Fl E).]

developing subjects. However, it is noteworthy that they demonstrate once again the formal relations existing between apparently remote branches of physics.

BIBLIOGRAPHY

Abragam, A., *The Principles of Nuclear Magnetism*, Clarendon Press, Oxford, 1961.

Backenstoss, G., "Pionic Atoms," *Ann. Rev. Nucl. Sci.*, **20**, 467 (1970).

Bainbridge, K. T., "Charged Particle Dynamics and Optics, Relative Isotopic Abundances of the Elements, Atomic Masses," in (Se 59), Vol. I.

Bohr, A., and B. R. Mottelson (BM 69).

Brown, G. E., *Unified Theory of Nuclear Models and Forces*, North-Holland, Amsterdam, 1971.

Cohen, B. L., *Concepts of Nuclear Physics*, McGraw-Hill, New York, 1971.

DeShalit, A., and I. Talmi, *Nuclear Shell Theory*, Academic Press, New York, 1963.

Donnelly T. W. and Walecka J. D. "Electron scattering and nuclear structure," *Ann. Rev. Nuc. Sci.* **25**, 329 (1975).

Duckworth, H. E., *Mass Spectroscopy*, Cambridge University Press, New York, 1958.

Fernbach, S., "Nuclear Radii as Determined by Scattering of Neutrons," *Rev. Mod. Phys.*, **30**, 414 (1958).

ISBN 0-8053-8601-7

Garvey, G. T., W. J. Gerace, R. L. Jaffe, I. Talmi, and I. Kelson, "Set of Nuclear Mass Relations and a Resultant Mass Table," *Rev. Mod. Phys.*, **41**, S1 (1969).

Geschwind, S., "Determination of Atomic Masses by Microwave Methods," in (Fl E), Vol. 38.1.

Green, A. E. S., "Nuclear Sizes and the Weizsäcker Mass Formula," *Rev. Mod. Phys.*, **30**, 569 (1958).

Hintenberger, H., "High-Sensitivity Mass Spectroscopy in Nuclear Studies," *Ann. Rev. Nucl. Sci.*, **12**, 435 (1962).

Hodgson, P. E., *Nuclear Reactions and Nuclear Structure*, Clarendon Press, Oxford, 1971.

Hofstadter, R., *Nuclear and Nucleon Structure*, Benjamin, New York, 1963.

Jeffries, C. D. *Dynamic Nuclear Orientation*, Wiley-Interscience, New York, 1963.

Kelly, F. M., "Determination of Nuclear Spins and Magnetic Moments by Spectroscopic Methods," in (Fl E), Vol. 38.1.

Kofoed-Hansen, O., "Mirror Nuclei Determinations of Nuclear Size," *Rev. Mod. Phys.*, **30**, 449 (1958).

Kopfermann, H. (Ko 58).

Laukien, G., "Nuclear Magnetic High Frequency Spectroscopy," in (Fl E), Vol. 38.1.

Marshalek, E., L. W. Person, and R. K. Sheline, "Systematics of Deformation of Atomic Nuclei," *Rev. Mod. Phys.*, **35**, 108 (1963).

Mayer, M. G., and J. H. D. Jensen (MJ 55).

Moszkowski, S. A., "Models of Nuclear Structure," in (Fl E), Vol. 39.

Myers, W. D., and W. J. Swiatecki, "The Macroscopic Approach to Nuclear Masses and Deformations," *Ann. Rev. Nucl. Part. Sci.*, **32**, 309 (1982).

Nemirowsky, P. E., *Nuclear Models*, Spon Ltd., London, 1963.

Nierenberg, W. A. "Measurement of the Nuclear Spins and Static Moments of Radioactive Isotopes," *Ann. Rev. Nucl. Sci.*, **7**, 349 (1957).

Preston, M. A., and R. K. Bhaduri, *Structure of the Nucleus*, Addison Wesley, Reading Mass, 1975.

Ramsey, N. F. (Ra 56).

Seaborg, G. T., "Elements beyond 100," *Ann. Rev. Nucl. Sci.*, **18**, 53 (1968).

Shirley, D. A., "Thermal Equilibrium Nuclear Orientation," *Ann. Rev. Nucl. Sci.*, **16**, 89 (1966).

Siegbahn, K. (Si 65).

Sorensen, R. A. "Nuclear Moment of Inertia at High Spin", *Rev. Mod. Phys.* **45**, 353 (1973).

Wapstra, A. H., and N. B. Gove, "The 1971 Atomic Mass Evaluation," *Nucl. Data Tables*, **9**, 267 (1971).

Wiegand, C. E., and R. Seki, "Kaonic and Other Exotic Atoms," *Ann. Rev. Nucl. Sci.*, **25**, 241 (1975).

Wu, C. S., and L. Wilets, "Muonic Atoms and Nuclear Structure," *Ann. Rev. Nucl. Sci.*, **19**, 527 (1969).

PROBLEMS

6-1 Show that ions of the same charge starting from rest in static electric fields follow the same trajectory, irrespective of their mass.

6-2 Show that ions of the same charge starting from rest in any combination of static electric and magnetic fields follow the same trajectories if their mass is multiplied by a fixed number K while all electric fields are multiplied by $1/K$ and the magnetic fields are left unchanged.

6-3 When neutrons are captured by ^{1}H to form ^{2}D, gamma rays of 2.230 ± 0.005 MeV are observed. Find the mass of the neutron.

ISBN 0-8053-8601-7

6-4 The following two nuclear reactions have been observed:

$$^{14}N + {}^4He = {}^{17}O + {}^1H - 1.26 \text{ MeV}$$

$$^{16}O + {}^2H = {}^{14}N + {}^4He + 3.13 \text{ MeV}$$

When ^{16}O is bombarded with ^2H, $^{17}O + {}^1H$ is also formed. Calculate the energy of the protons ejected at 90 deg and at 0 deg with respect to the direction of the deuterons, in the laboratory system, as a function of the deuteron energy.

6-5 Find the threshold for the (γ, n) reaction on ^{14}N from the following data:

^{13}N is a positron emitter with an upper energy limit of 1.2 MeV.

Mass of ^{14}N = 14.00307   Mass of ^{13}C = 13.00335

6-6 Find the threshold for the (p, n) reaction of ^{63}Cu given that

$$^{63}Cu + {}^1H = {}^{63}Zn + n$$

$$^{63}Zn = {}^{63}Cu + e^+ + \nu$$

and the upper limit for the positron energy is 2.3 MeV.

6-7 At what value of Z does the lowest Bohr orbit of a μ meson fall just inside the nuclear radius?

6-8 Given that the proton has a root-mean-square radius of 0.8 F, and assuming a uniform charge distribution (*a*) calculate the radius R for the proton, (*b*) what value of momentum transfer (express as $\hbar q$ in MeV/c) is required to observe a decrease from the point-charge cross section by a factor of 10? What is the minimum electron energy required to produce this effect?

6-9 The form factors in electron scattering can be written as

$$F(q) = \int \rho e^{-i\mathbf{q}\cdot\mathbf{r}} \, d\tau$$

Show that for a spherically symmetric $\rho(r)$ this reduces to

$$F(q) = \frac{4\pi}{q} \int_0^\infty \rho r \sin(qr) \, dr$$

Further, show that

$$F(q) = 1 - \frac{q^2\langle r^2\rangle}{6} \cdots$$

where

$$\langle r^2 \rangle = 4\pi \int_0^\infty \rho r^4 \, dr$$

6-10 Calculate $F(q)$ for the distribution

$$\rho = \rho_0 \quad 0 < r < R$$
$$\rho = 0 \quad r > R$$

6-11 Calculate $F(q)$ for

$$\rho = \rho_0 e^{-\alpha r}, \qquad \rho = \rho_0 \exp(-\alpha^2 r^2),$$

and

$$\rho = \rho_0 \exp(-\alpha r)/r.$$

6-12   Show that two identical nuclei of spin I give rise to $(2I + 1)^2$ linearly independent spin eigenfunctions. Of these $I(2I + 1)$ are antisymmetric and $(I + 1)(2I + 1)$ are symmetric with respect to nuclear exchange. Apply to the case $I = 1$ and construct the eigenfunctions.

6-13   Show that the electrostatic energy of a uniformly charged sphere of radius R is $\frac{3}{5}(Q^2/R)$, where Q is the total charge of the sphere.

6-14   Apply the mass formula to the isobaric pairs ^{113}(Cd–In), ^{187}(Os–Re), ^{123}(Sb–Te) and discuss what types of activities you would expect.

6-15   With the mass formula calculate the energy to be expected in a uranium fission. Compare the instantaneous energy release to that due to beta and gamma activity of the fission products. Assume that U splits into equal fragments.

6-16   Calculate the magnetic field at the nucleus of a hydrogen atom and of an atom of Fr in their ground states.

6-17   Discuss methods for producing elements with $Z > 100$, taking into account α, β, and spontaneous fission decay. Use a table of masses for the new isotopes.

6-18   Calculate the surface energy of a nucleus, assuming that the binding energy of a nucleon B is the surface energy corresponding to the small bulge produced in the nuclear surface by a nucleon before it escapes the nucleus.

6-19   Suppose we consider a proton as a uniformly dense sphere of radius $R = 1 \times 10^{-13}$ cm. (a) What angular velocity is needed to give it an angular momentum of $(3/4)^{1/2}\hbar$? (b) What rotational kinetic energy does this correspond to? (c) How many amperes are going around the axis of rotation of the proton?

6-20   (a) Consider two particles of masses m_1, m_2 revolving in a circular orbit about their common center of mass. If their charges are e_1, e_2, respectively, calculate the gyromagnetic ratio of the system; neglect spins. (b) Apply this result to the calculation of the orbital g factor, g_l, for the following systems: (i) The neutron and proton in a deuterium nucleus. (ii) A μ^- meson (charge e^-, mass $= 207m_e$) bound to a proton.

6-21   Back and Wulff [Z. *Physik*, **66**, 31 (1930)] measured the hyperfine structure of the line 3775, $7s^2S_{1/2} - 6p^2P_{1/2}$ of ^{205}Tl and found three components, as shown in the accompanying figure.

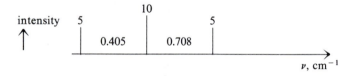

Deduce all that you can about the spin and magnetic moment of the nucleus. The $^2P_{1/2}$ has a larger term split than $^2S_{1/2}$. Calculate the Zeeman effect of this hyperfine structure.

6-22   Show that for an atom having $J = \frac{1}{2}$, the hfs energy levels in an external field H are given by

$$W_M = -\frac{\Delta W}{2(2I + 1)} - \frac{\mu_I}{I}Hm \pm \frac{\Delta W}{2}\left(1 + \frac{4m}{2I + 1}x + x^2\right)^{1/2}$$

where $\Delta W = (A/2)(2I + 1) = W(F = I + \frac{1}{2}) - W(F = I - \frac{1}{2}) = 2\pi\hbar\,\Delta\nu$, $x = [(-\mu_J/J) + (\mu_I/I)]H/\Delta W$, and the upper sign is used for $F = I + \frac{1}{2}$, the lower for $F = I - \frac{1}{2}$ (Breit–Rabi formula). Show that this exact relation has as limiting cases Eqs. (6-9.22) and (6-9.23). From the Breit–Rabi formula calculate μ_{eff}.

6-23   Figure out how the apparatus of Fig. 6-32 works.

ISBN 0-8053-8601-7

6-24 An atomic-beam experiment was performed to measure the spin of 3.2-h ^{112}Ag. A resonance of the flop-in type, that is, a transition from a state $F' = F'' = I + \frac{1}{2}$, $M'_F = -I + \frac{1}{2}$, to $M''_F = -I - \frac{1}{2}$, was observed for a frequency of 5.825 MHz. In the same beam ^{85}Rb gave the same type of resonance for a frequency of 4.685 MHz. I of ^{85}Rb is $\frac{5}{2}$. Find I of ^{112}Ag.

6-25 With the same substances (^{112}Ag and ^{85}Rb) but at a higher magnetic field than in the previous problem, the same type of transition was observed at frequencies of 91.739 MHz for ^{85}Rb and 204.920 MHz for ^{112}Ag. Calculate A and $\Delta\nu$ for ^{112}Ag. $g_J = -2.00238$ for Rb and -2.00233 for Ag, $\Delta\nu$ for ^{85}Rb is 3035.7 MHz.

6-26 For ^{107}Ag, $\Delta\nu = -1712.56$ MHz, $I = \frac{1}{2}$, $g_I = -0.2261$ nuclear magnetons. Using the previous results, find g_I and μ_I for ^{112}Ag.

6-27 Show that for a homogeneous ellipsoid of semiaxes a, a, b the quadrupole moment is given by $Q = \frac{2}{5}(b^2 - a^2)$.

6-28 The nuclear electric quadrupole moment due to one proton is defined as the expectation value of $\langle 3z^2 - r^2 \rangle = Q$ for the state in which $M_I = I$ (which is the state of maximum alignment).

Giving an explicit form to the eigenfunction of the proton,

$$R(r)Y_l^l(\theta, \varphi)\alpha$$

where R is the radial function, Y_l^l the spherical harmonic, and α the spin function, show that for $I = l + \frac{1}{2}$,

$$Q = -\frac{2l}{2l+3}\langle r^2\rangle = -\frac{2I-1}{2I+2}\langle r^2\rangle$$

The same expression of Q as a function of I is valid for $I = l - \frac{1}{2}$.

6-29 Show that the values of

$$Q(M) = \langle 3z^2 - r^2 \rangle$$

for $M \neq I$ are given by

$$Q(M) = Q(I)\frac{3M^2 - I(I+1)}{I(2I-1)}$$

6-30 Show that for a rotational ellipsoid of small eccentricity and uniform charge density, $Q_0 = \frac{4}{5}ZR\,\Delta R$ and that the parameters β, δ and the eccentricity of the ellipsoid are related by

$$\pm e = \frac{|a^2 - b^2|^{1/2}}{a} = \left(\frac{2\Delta R}{R}\right)^{1/2} \qquad \frac{\Delta R}{R} = \delta = \frac{3}{4}\left(\frac{5}{\pi}\right)^{1/2}\beta$$

where $2a$ is the length of the rotational axis and $2b$ the length of the other two axes. Signs are such that for a cigar shape $e > 0$ and for a lens shape $e < 0$; $\Delta R = a - b$.

6-31 Calculate the alignment coefficients f_1, f_2 for a system of deuterons at $T = 10^{-2}$ °K in a field of 10^5 G.

6-32 A dynamic scheme of proton polarization envisages hydrogen atoms in a magnetic field giving levels as illustrated below, where the magnetic moment of the electron is opposite to H in (1) and (2) and the magnetic moment of the proton is parallel to H in (1) and opposite in (2). Find the orientations of these moments in levels (3) and (4). Now consider a tuned radio frequency (calculate its frequency) that, given sufficient intensity, equalizes the populations of states (2) and (3). Calculate the populations of states (1) and

(4), taking them to be in thermal equilibrium with (3) and (2), respectively. From this show that the nuclear polarization is $-\tanh(\delta/2kT) \approx -\delta/2kT$.

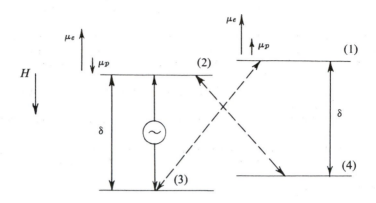

6-33   Consider two particles with magnetic moments μ_1 and μ_2, each having the same spin **s**. Calculate their total magnetic moment in a state of zero orbital angular momentum and total spin **I**. (Express the result in terms of μ_1, μ_2, s, and I.) What is the magnetic moment when $I = 0$?

6-34   (a) Suppose that we try to interpret the $\frac{3}{2}$He nucleus (spin $\frac{1}{2}$) as a two-body system consisting of a deuteron bound to a proton, in a state of zero orbital angular momentum (s state). From the known deuterium magnetic moment (0.857 μ_N) and proton moment (2.79 μ_N), calculate the $\frac{3}{2}$He magnetic moment expected on this model. (b) What is the Schmidt model prediction for $\frac{3}{2}$He, assuming the odd nucleon to be in an $^2S_{1/2}$ state? (c) Which agrees better with the measured value of -2.13 μ_N?

6-35   For the following nuclei the spins and parities of the ground state are given. Justify the observed values by shell-model considerations:

$$^3\text{He}\left(\tfrac{1}{2}+\right) \qquad ^{21}\text{Ne}\left(\tfrac{3}{2}+\right) \qquad ^{27}\text{Al}\left(\tfrac{5}{2}+\right) \qquad ^{38}\text{K}(3+) \qquad ^{66}\text{Ga}(1+) \qquad ^{69}\text{Ga}\left(\tfrac{3}{2}-\right)$$

$$^{209}\text{Bi}\left(\tfrac{9}{2}-\right) \qquad ^{210}\text{Bi}(1-)$$

Remember that s, d, g nucleons have $+$ parity and p, f, etc., have $-$ parity. [From (P 62).]

6-36   Show that Nordheim rules forbid the existence of odd-odd nuclei with ground states $0+$ or $1-$. Find some exceptions. [From (P 62).]

6-37   Describe the physical conditions under which $I = K \rightarrow \infty$ or $I \rightarrow \infty$, with K constant and small, and explain why in the first case $Q = Q_0$ and in the second $Q = -Q_0/2$. Note in the second case the opposite sign of Q and Q_0 and explain it.

6-38   Show that a distribution of electric charges of density ρ, symmetric around the z axis and limited to a region $r < R$, gives outside of this region a potential

$$V(r, \theta) = \frac{1}{r} \sum_n \frac{a_n}{r^n} P_n(\cos\theta)$$

and that $a_0 = \int \rho \, d\tau$; $a_1 = \int \rho z \, d\tau$; $a_2 = \int \rho \frac{1}{2}(3z^2 - r^2) \, d\tau$; where the integral is extended to the region occupied by the charges.

6-39   Adapt the treatment of spin–orbit interaction in (Fe 61, lectures 27 and 28) to a J–I interaction and calculate the hyperfine structure of ^{43}K in a magnetic field varying from 0 to 10 kG.

ISBN 0-805-38601-7

7

Alpha Emission

Nuclei subjected to nuclear bombardment are known to emit many types of heavy particles, such as neutrons, protons, alpha particles, and deuterons. We shall not consider bombardment in this chapter but shall concentrate instead on spontaneous processes, that is, processes arising from metastable states that have a relatively long life. These involve only the emission of alpha particles and spontaneous fission, as far as heavy particles are concerned. The emission of delayed neutrons is really an instantaneous process; its apparent slowness is caused by the fact that it must be preceded by the emission of a beta ray. The emission of delayed neutrons will be treated in Chap. 11.

The study of alpha-particle emission is a rather large and important chapter of nuclear physics. The accumulation of experimental material, which continues to increase at a rapid pace, has helped considerably in the formulation and refinement of nuclear models. The theory still leaves many interesting problems unsolved, mainly those of trying to understand the hindrance factors (see Sec. 7-2) and to correlate them to specific models.

7-1 INTRODUCTION: BARRIER PENETRATION

Why are alpha particles emitted by nuclei? The answer to this question is found in the release of nuclear energy accompanying the reactions. The alpha

Emilio Segrè, Nuclei and Particles: An Introduction to Nuclear and Subnuclear Physics, Second Edition

particle has such a large mass defect that emission of an alpha particle is generally energetically advantageous. This feature is clearly shown by the semiempirical mass formula given in Sec. 6-7. It is also apparent from the formula that alpha radioactivity is expected only for heavy nuclei; however, the sharpness of the boundary of the zone of alpha radioactivity at $Z = 83$ is an effect of the shell structure of the nucleus (see Sec. 6-11). A few alpha emitters—^{190}Pt, ^{174}Hf, ^{152}Gd, ^{147}Sm, and ^{144}Nd—occur at lower Z (Fig. 7-1).

A few cases of delayed proton emission are known. They are mostly due to a previous positron emission or K capture followed by an instantaneous proton emission completely analogous to the delayed neutron emission. There are, however, cases of genuine delay due to the potential barrier effect. A case occurs in an excited state of ^{53}Co, emitting protons of 1.57 MeV, with a half-life of 0.24 sec.

If we want to treat alpha emission by classical arguments, we face a serious difficulty: by scattering alpha particles on heavy nuclei and observing whether or not they follow Rutherford's law, it is possible to map the electrostatic potential to which the alpha particle is subject as a function of its

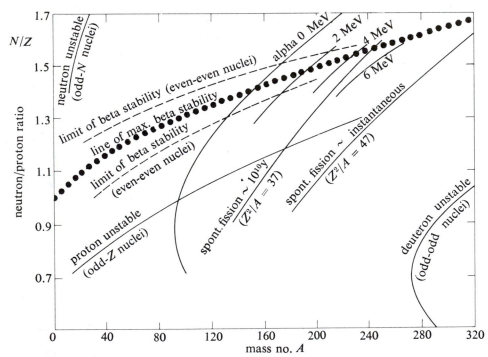

Figure 7-1 Nuclear stability limits predicted by the liquid-drop model. The curves are calculated from the semiempirical mass formula of Chap. 6. [After G. C. Hanna, from (Se 59).]

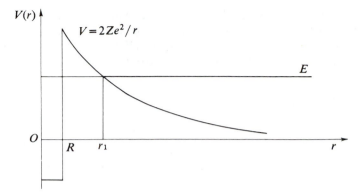

Figure 7-2 Electrostatic potential energy of an alpha particle as a function of the distance from the center of the nucleus.

distance from the center of the nucleus (Fig. 7-2). We thus find a potential energy

$$V(r) = \frac{zZe^2}{r} \qquad \text{for} \qquad r > R \qquad (7\text{-}1.1)$$

where $z = 2$ for alpha particles.

For $r < R$, nuclear forces become effective, and the potential is not simply the electrostatic one. However, we might try to represent the attractive force that retains the alpha particle in the nucleus by a potential well of suitable depth. The total energy E of the alpha particle is known, because we can measure its kinetic energy directly once it leaves the nucleus. If we try to reconcile these facts with classical mechanics, we confront a paradox. We know that alpha particles remain in the nucleus a long time—a very long time compared with the period necessary for them to cross the nucleus if they move with a velocity corresponding to their energy outside the nucleus. One would conclude that if the potential barrier preventing their escape is low enough, they should be emitted in a time of the order of magnitude of the nuclear radius divided by their velocity. Conversely, if the potential barrier is insurmountable with the energy available, they should never be able to leave the nucleus.

This paradox was resolved by Gamow and by Condon and Gurney (1929), who, treating the problem quantum-mechanically, showed that there is a finite probability of escape even in the case for which classical mechanics would predict an absolutely impenetrable barrier and hence nuclear stability. In order to show the essential points of the argument without becoming involved in mathematical difficulties, we shall proceed in several steps. First, we shall use a semiclassical argument developed by von Laue, which will give an estimate of the decay constant of an alpha emitter. Second, we shall study

ISBN 0-8053-8601-7

some important characteristic features of the quantum-mechanical problem.

It is known that if we try to find a stationary solution to the one-dimensional problem of a potential energy barrier (Fig. 7-3), we find that even a barrier higher than the total energy of the particles has a finite transparency.

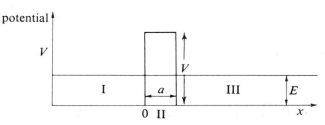

Figure 7-3 A barrier of thickness a, corresponding to the potential energy V. The solutions of Eq. (7-1.2) for the zones I, II, and III are given in the text.

Assume a barrier of the form shown in Fig. 7-3 with width a. We shall consider only the case of interest to us, $V > E$. A set of solutions to Schrödinger's equation

$$\frac{d^2u}{dx^2} + \frac{2m}{\hbar^2}(E - V)u = 0 \qquad (7\text{-}1.2)$$

in zones I, II, and III is

$$u_{\mathrm{I}} = A_{\mathrm{I}}e^{ikx} + B_{\mathrm{I}}e^{-ikx} \qquad -\infty < x < 0 \qquad (7\text{-}1.3)$$

$$u_{\mathrm{II}} = A_{\mathrm{II}}e^{+Kx} + B_{\mathrm{II}}e^{-Kx} \qquad 0 < x < a \qquad (7\text{-}1.4)$$

$$u_{\mathrm{III}} = A_{\mathrm{III}}e^{ikx} + B_{\mathrm{III}}e^{-ikx} \qquad a < x < \infty \qquad (7\text{-}1.5)$$

with

$$k = \frac{(2mE)^{1/2}}{\hbar} = \frac{p}{\hbar} = \frac{1}{\lambda} \qquad (7\text{-}1.6)$$

$$K = \frac{(2m|E - V|)^{1/2}}{\hbar} \qquad (7\text{-}1.7)$$

where p is the momentum and λ the de Broglie wavelength. We must determine the A's and B's in such a way that

$$u \quad \text{and} \quad u' = \frac{du}{dx} \qquad (7\text{-}1.8)$$

ISBN 0-8053-8601-7

are continuous, as known from elementary quantum mechanics. Moreover, in region III we want to have only particles moving to the right, which means

$$B_{III} = 0 \qquad (7\text{-}1.9)$$

Calculation gives at $x = a$

$$A_{III}e^{ika} = A_{II}e^{Ka} + B_{II}e^{-Ka} \qquad (7\text{-}1.10)$$

$$ikA_{III}e^{ika} = K\left(A_{II}e^{Ka} - B_{II}e^{-Ka}\right) \qquad (7\text{-}1.11)$$

Solving for A_{II} and B_{II} one obtains

$$A_{II} = \frac{A_{III}}{2}\left(1 + \frac{ik}{K}\right)e^{(ik-K)a} \qquad (7\text{-}1.12)$$

$$B_{II} = \frac{A_{III}}{2}\left(1 - \frac{ik}{K}\right)e^{(ik+K)a} \qquad (7\text{-}1.13)$$

At the other side of the barrier for $x = 0$ one has

$$A_I + B_I = A_{II} + B_{II} \qquad (7\text{-}1.14)$$

$$A_I - B_I = (A_{II} - B_{II})(K/ik) \qquad (7\text{-}1.15)$$

We should eliminate A_{II} and B_{II} from Eqs. (7-1.12), (7-1.13), (7-1.14), and (7-1.15), and find the relation among A_I, amplitude of an incoming wave; B_I, amplitude of a reflected wave; and A_{III}, amplitude of a transmitted wave. In practical cases $Ka \gg 1$, and this condition, equivalent to small transparency of the barrier, permits us to simplify the formulas considerably. In fact, if $Ka \gg 1$, we can neglect A_{II} compared with B_{II} and obtain

$$2A_I = B_{II}\left(1 - \frac{K}{ik}\right) = \frac{A_{III}}{2}\left(1 - \frac{ik}{K}\right)\left(1 - \frac{K}{ik}\right)e^{(ik+K)a} \qquad (7\text{-}1.16)$$

The ratio

$$\frac{|A_{III}|^2}{|A_I|^2} = \frac{16k^2K^2\exp(-2Ka)}{(k^2 + K^2)^2} \sim e^{-2Ka} = T \qquad (7\text{-}1.17)$$

gives the "transmissivity" of the barrier, that is, the flux transmitted for unity incident flux. To calculate the reflectivity

$$R = \frac{|B_I|^2}{|A_I|^2} \qquad (7\text{-}1.18)$$

one must use the exact solution of the problem. As is to be expected, one finds

$$R + T = 1 \qquad (7\text{-}1.19)$$

In the case of a variable potential the transparency is approximately expressed by a simple generalization of Eq. (7-1.17),

$$T = \exp\left[-\frac{2}{\hbar} \int_R^{r_1} (2m|E - V|)^{1/2} \, dx \right] = e^{-2G} \qquad (7\text{-}1.20)$$

The integral is to be extended through the region in which $E - V < 0$, that is, to the forbidden region of classical mechanics.

Passing to the tridimensional case without angular momentum and for the Coulomb potential,

$$V = zZe^2/r \qquad (7\text{-}1.21)$$

we have

$$E = zZe^2/r_1 \qquad (7\text{-}1.22)$$

where E is the kinetic energy of the alpha particle for large separation. We also define the barrier height as

$$B = Zze^2/R \qquad (7\text{-}1.23)$$

We can then calculate G explicitly, finding

$$
\begin{aligned}
G &= \frac{(2m)^{1/2}}{\hbar} \int_R^{r_1} \left(\frac{Zze^2}{r} - E \right)^{1/2} dr \\
&= \frac{(2mZze^2 r_1)^{1/2}}{\hbar} \left[\arccos\left(\frac{R}{r_1} \right)^{1/2} - \left(\frac{R}{r_1} - \frac{R^2}{r_1^2} \right)^{1/2} \right] \qquad (7\text{-}1.24)
\end{aligned}
$$

This formula may be rewritten using $x = R/r_1 = E/B$ and $v = (2E/m)^{1/2}$ as

$$G = \left(\frac{1}{\hbar} \right)\left(\frac{2m}{E} \right)^{1/2} zZe^2 \left[\arccos(x^{1/2}) - x^{1/2}(1 - x)^{1/2} \right] \qquad (7\text{-}1.25)$$

or

$$G = \frac{2zZe^2}{\hbar v} \gamma(x) \qquad (7\text{-}1.26)$$

ISBN 0-8053-8601-7

with

$$\gamma(x) = \arccos(x^{1/2}) - x^{1/2}(1 - x)^{1/2}$$

The function $\gamma(x)$ has been tabulated and is found, for example, in Bethe (1937), and in Perlman and Rasmussen (Fl E). For $x \ll 1$ it is approximately $\gamma(x) \to (1/2)\pi - 2x^{1/2} + \cdots$. For $x = E/B \ll 1$, G can be crudely approximated by

$$G \approx \frac{\pi z Z e^2}{\hbar v} - \frac{2e}{\hbar}(2zZmR)^{1/2} \qquad (7\text{-}1.27)$$

From the transparency we pass to the decay constant of the nucleus by assuming that an alpha particle moves inside the potential well with a certain velocity v_0 and hence hits the wall $v_0/2R$ or, in a tridimensional case, v_0/R times per second. At each "hit" it has the probability T of leaking out; hence the decay constant

$$\lambda = \frac{v_0}{R} e^{-2G} \qquad (7\text{-}1.28)$$

Here v_0 and R can be inferred crudely from the velocity of the alpha particle outside the nucleus and from any estimate of the nuclear radius, but the really important factor is e^{-2G}.

The velocity v_0 is of the order of 10^9 cm sec^{-1}, and R is about 10^{-12} cm. Hence, the alpha particle may make 10^{21} attempts per second to escape; e^{-2G} must range from 10^{-13} to 10^{-39} to encompass the decay constants of short-lived substances such as ThC$'$(^{212}Po) with a period of 3.0×10^{-7} sec, and long-lived substances such as ^{238}U with a period of 4.5×10^9 yr. The factor v_0/R might vary a little from nucleus to nucleus, but it is clear that a variation of 26 powers of 10 can come only from the transparency, where $2G$ in the exponential is a function of R and E.

In the one-dimensional model the angular momentum is automatically zero; however, in the real case the centrifugal potential energy $\hbar^2 l(l + 1)/mr^2$ ($l\hbar$ is the angular momentum) must be added to the Coulomb potential. Its effect is generally very small in the case of alpha decay, and it has been neglected in the previous analysis.

Using the approximate expression of G [Eqs. (7-1.24) and (7-1.27)], we can write, for $E/B \ll 1$,

$$\log \lambda = \frac{-(2mB)^{1/2}}{\hbar} R \left[\pi \left(\frac{B}{E} \right)^{1/2} - 4 \right] + \log \frac{v_0}{R} \qquad (7\text{-}1.29)$$

where we have neglected terms containing positive powers of E/B. Remembering the dependence of B on R [Eq. (7-1.23)] and the fact that R is proportional to $A^{1/3}$, we recognize that for a series of isotopes we can expect that log

ISBN 0-8053-8601-7

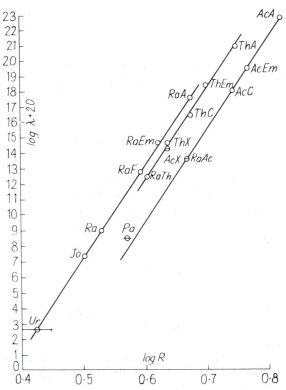

Figure 7-4 Original figure of Geiger and Nuttall showing the connection between range and decay constant in alpha-particle decay. [From E. Rutherford, J. Chadwick, and C. D. Ellis, *Radiations from Radioactive Substances*, Cambridge University Press, New York, 1930.]

λ will be approximately a linear function of $E^{-1/2}$. A numerical formula for the half-life T (in years) given by Taagepera and Nurmia is

$$\log_{10} T = 1.61(ZE^{-1/2} - Z^{2/3}) - 28.9 \qquad (7\text{-}1.30)$$

where E is in MeV and Z refers to the daughter substance. It is derived from Eq. (7-1.29) using suitable semiempirical formulas for the dependence of R on Z. One can use Eqs. (7-1.25) and (7-1.28) to calculate R from the experimentally known decay constants and energies of the alpha particles. This has been done extensively and one finds that the R value calculated from ground transitions of even-even nuclei show great uniformity, corresponding to values of about $1.5 \times 10^{-13} A^{1/3}$ cm. Nuclei below the neutron closed shell of 126 show rates slower by an order of magnitude, hence they give effective R values somewhat smaller than other nuclei do. Decay rates of odd nuclei tend to be smaller than would be predicted by systematics of even-even rates.

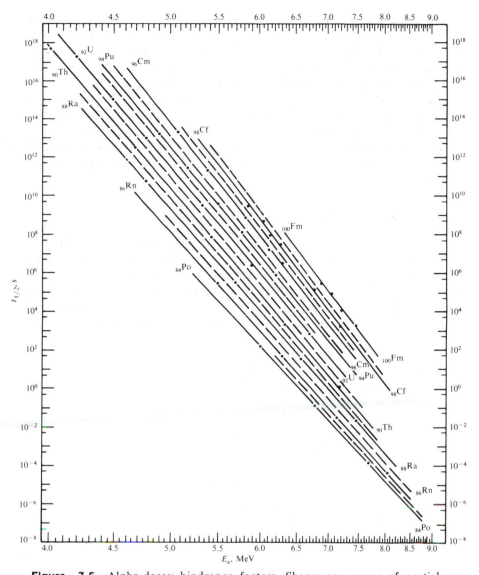

Figure 7-5 Alpha-decay hindrance factors. Shown are curves of partial half-life versus alpha energy and Z for unhindered transitions. The solid lines for even-proton alpha emitters are drawn through experimental points for the ground-state transitions of even-even nuclei. Dashed lines for odd-proton alpha emitters are calculated with use of nuclear radii averaged from neighboring even-even nuclei. To calculate an approximate hindrance factor, divide the experimental partial half-life of the alpha group by the value derived from the appropriate curve.

ISBN 0-8053-8601-7

This method of determining the nuclear radius was one of the earliest used. It gives a value higher than others by about 20%. It should be remembered, however, that the definition of nuclear radius used here applies to a special model. One could argue that the radius measured is near the sum of the alpha-particle radius plus the nuclear radius itself.

Geiger and Nuttall as early as 1911 had plotted the logarithm of the half-life versus the range of the alpha particles, observing a remarkable regularity (Fig. 7-4). The range is in fair approximation proportional to $E^{1.5}$, and hence the old Geiger–Nuttall plot is in principle similar to the plot of Fig. 7-5. It shows in a striking way the importance of energy in determining the decay constant.

7-2   FINE STRUCTURE OF ALPHA SPECTRA

Alpha decay usually involves a nucleus in the ground state. However, many alpha emitters show a line spectrum of alpha particles (Rosenblum, 1929), because there are several final levels for the alpha transition. This explanation is confirmed by the finding of gamma rays of energy corresponding to the energy difference between the alpha lines. An example of this situation is given in Fig. 7-6. It will be noted that the lower levels of ^{234}U have

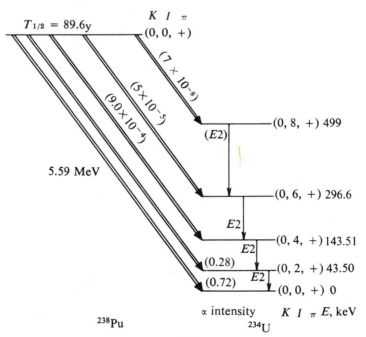

Figure   7-6   Partial decay scheme of ^{238}Pu showing alpha transitions starting from one level and ending in different levels. [Data from LHP 67.]

ISBN 0-805-38601-1

angular momenta 0, 2, 4 and energy separations in the ratio expected for rotational levels of a rotator, as was mentioned in Sec. 6-15.

The different alpha lines have intensities determined primarily by their energy, which affects the transparency of the barrier, and to a lesser extent by the influence on the barrier of the angular momentum. In specific cases, other factors, not all clearly understood, alter the decay constant expected in the simple penetration picture.

In the case of ^{212}Po (ThC') and ^{214}Po (RaC') we have an instance of a structure in the upper level. The alpha decay is so rapid that it can compete with the gamma transitions between the upper levels. We thus have "exceptionally long-range alpha particles" (Figs. 7-7 and 7-8). They are very rare, and they afford a method of measuring gamma-decay rates, if one knows the alpha-decay rate. One has for the upper state

$$\frac{\lambda_\gamma}{\lambda_\alpha} = \frac{N_\gamma}{N_{lr}} \qquad (7\text{-}2.1)$$

The ratio N_γ/N_{lr} is the same as the ratio N_α/N_{lr} between the number of normal and long-range alpha particles, because each gamma ray is immediately followed by a "normal" alpha particle. The ratio N_α/N_{lr} may be measured directly. The probability per unit time of alpha emission by the excited state, λ_α, may be estimated from the energy of the transition and other factors, and λ_γ is then obtained from Eq. (7-2.1). The λ_γ obtained in the case of ThC' are of the order of 10^{12} sec^{-1}. The λ_α are about 0.21×10^7 sec^{-1} for the ground state and 0.9×10^8 sec^{-1} and 10^{10} sec^{-1} for the states at 0.73 and

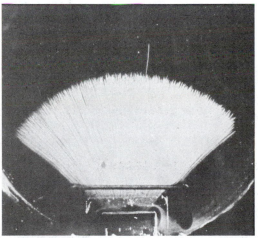

Figure 7-7 Excited states of alpha emitters: long range from ^{214}Po(RaC') and alpha rays from ^{214}Bi(RaC). [K. Philipp, *Naturwiss.*, **14**, 1203 (1926).]

ISBN 0-8053-8601-7

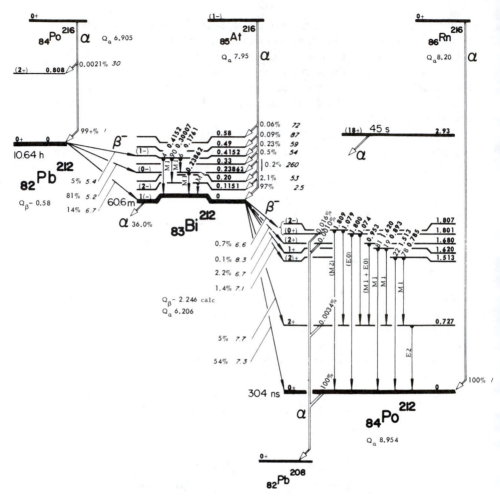

Figure 7-8 Level scheme and decay scheme of ^{212}Po and neighboring nuclei. [From (LHP 67).]

1.8 MeV. The method is not quantitative, but it is interesting and historically important.

7-3 SYSTEMATICS OF ALPHA DECAY

The great accumulation of experimental material on alpha decay in the last decade has made possible the development of an elaborate systematics of alpha decay.

The ground states of even-even nuclei show the largest decay constants, corresponding to simple barrier penetration with no change in angular momentum (Fig. 7-5). Other transitions are "hindered" in varying degrees. The hindrance factor is the factor by which the observed half-life is greater than one calculated according to the behavior of even-even nuclei.

The introduction of hindrance is thus equivalent to considering a "reduced half-life" in which the energy effect is taken into account. It is very similar to the ft values and reduced gamma widths used in beta and gamma decays (see Chaps. 8 and 9).

One factor in hindrance is the change of angular momentum l, which adds to the Gamow factor G a term

$$\frac{l(l+1)\hbar}{R(2mB)^{1/2}} \tag{7-3.2}$$

This term for $l = 4$ increases the mean life for the case $R = 9.3 \times 10^{-13}$ cm, $Z = 88$, $A = 226$ by a factor of 6.7. It is thus relatively unimportant.

The study of hindrances is useful in connection with nuclear models. Even-even nuclei, between ground states, where both states have $I = 0$ and are even, have, by definition, a hindrance of 1. Even-even nuclei often have "rotational levels" 2^+, 4^+, 6^+, etc. The hindrances for transitions to these levels increase rapidly and are only in part accounted for by the effect of angular momentum.

Even-odd nuclei, when they depart from sphericity, show more than one system of rotational bands. Among the levels there are then some that correspond to the same configuration in the upper and lower states, except for the additional alpha particle in the upper state. In this last case the transition to the base level is practically unhindered, and it is believed that it does not involve the unpaired nucleon at all. A striking example is given by ^{241}Am (Fig. 7-9).

A factor important in determining the hindrance is the probability of finding a preformed alpha particle in the nucleus. Indeed the striking probability of Eq. (7-1.28) might occasionally be considerably overestimated, because the nucleons might not be in the proper configuration corresponding to an alpha particle. The influence of this effect is shown conspicuously in the branching ratios of the fine structure of the alpha decay of ^{211}Po (Mang, 1957), where the probability of finding an alpha-particle configuration can be calculated to a good approximation by using the shell model. Departures of the nucleus from the spherical form may also produce hindrance. Furthermore, in a cigar-shaped nucleus the alpha particles tend to come out from the regions of highest curvature, as directly shown by experiments with oriented nuclei.

Finally, huge discontinuities in the energy–mass number diagram occur in connection with closed shells for all nuclei in which $Z < 84$, $N < 128$, or both (Fig. 7-10). These discontinuities, however, are explained by energy changes due to the shell model and do not involve any anomaly in the alpha decay.

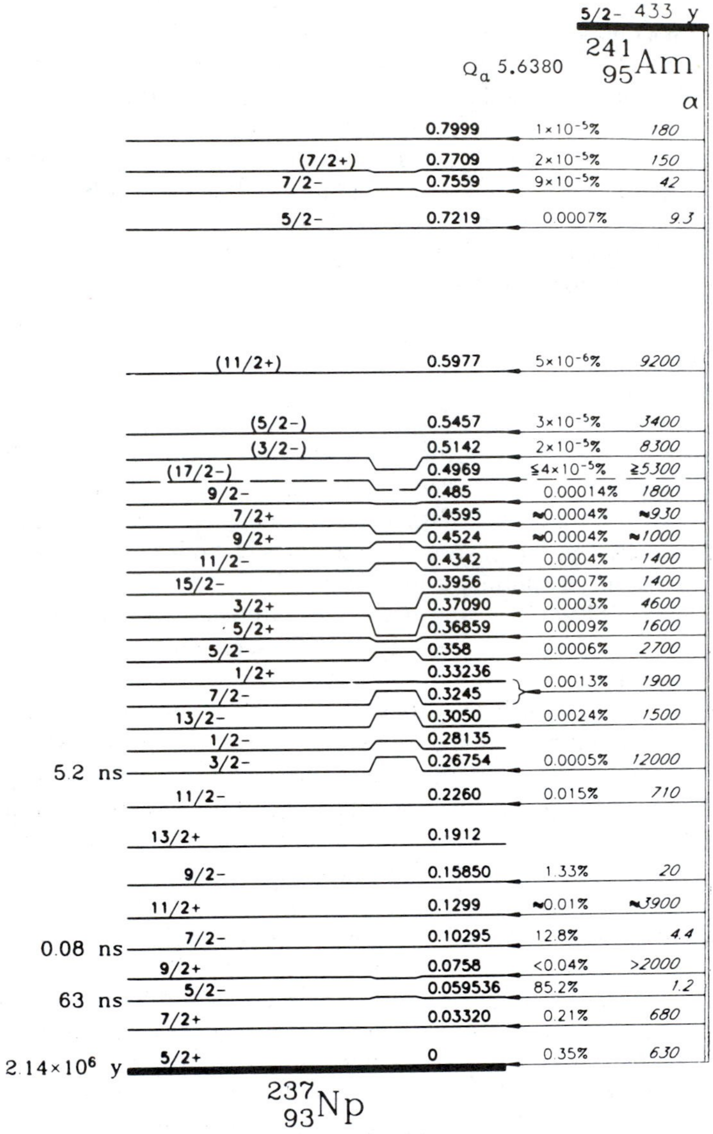

Figure 7-9 Partial decay scheme of ^{241}Am. The total energy available for alpha decay to ^{237}Np is 5.638 MeV. The alpha structure, from the lower level, shows transitions to $5/2^-$, $7/2^-$, $9/2^-$, etc. with small hindrance and transitions to $5/2^+$, $7/2^+$, etc. with large hindrance. (The figure in italics at the right is the hindrance.) The first leave the unpaired proton undisturbed, the others require a rearrangement.

ISBN 0-8053-8601-7

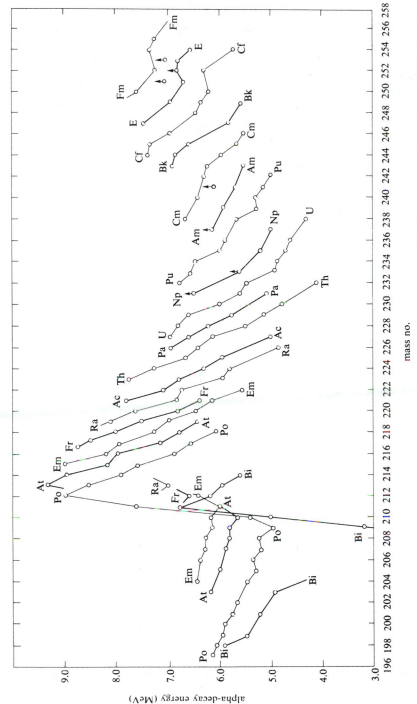

Figure 7-10 Alpha-decay energy versus mass number for the heavy elements. The lines connect isotopic nuclei. The breaks in the lines of Bi, Po, At, Em, and Fr are due to shell effects associated with the magic number 126 (neutrons).

7-4   VIRTUAL BINDING

We shall now give a supplementary treatment of the problem of a particle in a well that will show qualitatively features sometimes called *virtual binding*. Although the example we shall treat is oversimplified, we shall try to preserve the essential physical features of the phenomenon, which is very important in several phases of nuclear physics (see also Flügge and Marschall, 1971).

We shall again treat a one-dimensional problem, with a potential as shown in Fig. 7-11. At $x = 0$ and at $x = l$ we have infinitely high potential walls. At $x = a$ we have an extremely thin but very high wall, such that

$$\frac{2m}{\hbar^2} \int_{a-\epsilon}^{a+\epsilon} V(x - a) \, dx = \frac{1}{g} \tag{7-4.1}$$

($V(x)$ is a Dirac delta function).

The Schrödinger equation

$$u'' + \frac{2m}{\hbar^2} (E - V)u = 0 \tag{7-4.2}$$

gives, on integration over a small interval containing $x = a$,

$$u'_{II} - u'_{I} + \frac{2m}{\hbar^2} \int_{a-\epsilon}^{a+\epsilon} [E - V(x - a)]u(x) \, dx = 0 \tag{7-4.3}$$

$$u'_{II} - u'_{I} = \frac{u(a)}{g} \tag{7-4.4}$$

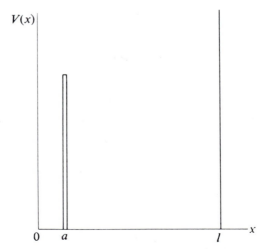

Figure 7-11 A box of length l divided at a by a barrier of the form of a delta function to give a simple example of virtual binding.

where u_I', u_{II}' are the derivatives of $u(x)$ left and right of a. We thus have a discontinuity in $u'(x)$ at $x = a$. We also have the boundary conditions for $u(x)$,

$$u_I(0) = u_{II}(l) = 0 \tag{7-4.5}$$

and

$$u_I(a) = u_{II}(a) \tag{7-4.6}$$

In general,

$$u(x) = A \sin kx + B \cos kx \tag{7-4.7}$$

with

$$\hbar k = (2mE)^{1/2} \tag{7-4.8}$$

We can satisfy the boundary conditions for $u(x)$ by putting

$$u_I(x) = A \sin kx \qquad x \leqslant a \tag{7-4.9}$$

$$u_{II}(x) = B \sin k(l - x) \qquad x \geqslant a \tag{7-4.10}$$

with

$$B = A \, \frac{\sin ka}{\sin k(l - a)} \tag{7-4.11}$$

The condition of Eq. (7-4.4) gives the eigenvalues of k. We find from Eqs. (7-4.4) and (7-4.6)

$$\cot ka + \cot k(l - a) = -1/kg \tag{7-4.12a}$$

or

$$k \sin kl = -(1/g) \sin ka \sin k(l - a) \tag{7-4.12b}$$

These relations give the eigenvalues of k and the relative amplitudes of A and B. Instead of discussing them in general, we shall treat the case $kg \ll 1$, which corresponds to a small transparency of the barrier. If

$$kg = 0 \text{ exactly} \tag{7-4.13}$$

we have two separate compartments. The boundary conditions then require that $u(x)$ be zero at 0, a, l, and we have for the eigenvalues of k

$$\sin k_n a = 0 \tag{7-4.14}$$

$$k_n = n\pi / a \tag{7-4.15}$$

$$\sin k_m (l - a) = 0 \tag{7-4.16}$$

$$k_m = m\pi / (l - a) \tag{7-4.17}$$

with n, m integral numbers (to avoid complications we assume $k_n \neq k_m$). The normalized eigenfunctions corresponding to Eq. (7-4.14) are

$$u_{\mathrm{I}} = \left(\frac{2}{a} \right)^{1/2} \sin \frac{n\pi x}{a} \qquad u_{\mathrm{II}} = 0 \tag{7-4.18}$$

whereas those corresponding to Eq. (7-4.16) are

$$u_{\mathrm{II}} = \left(\frac{2}{l - a} \right)^{1/2} \sin \frac{m\pi (l - x)}{l - a} \qquad u_{\mathrm{I}} = 0 \tag{7-4.19}$$

Equation (7-4.18) asserts that the particle is somewhere in the left compartment, and certainly not in the right compartment. Equation (7-4.19) asserts that the particle is not in the left compartment, but somewhere in the right one. The two compartments are completely separate, and the energy levels of the whole system are the ensemble of the energy levels of the two component parts.

We shall now consider the case of $kg \neq 0$, but small. We shall also make the assumption that $l \gg a$, which implies that the levels corresponding to Eq. (7-4.18) are widely separated, whereas those corresponding to Eq. (7-4.19) almost form a continuum. The model thus has the essential features of a nucleus containing an alpha particle (zone I) confined by a barrier of small transparency. The nucleus itself is contained in a large box, $l \gg a$, the purpose of which is only to avert mathematical complications arising from the continuous spectrum of eigenvalues that occurs for an infinite l.

We are mainly interested in what happens to Eq. (7-4.18), which corresponds to the physical conditions of an alpha particle in a nucleus. The eigenvalue k of Eq. (7-4.14) is now changed to

$$k_n^{(1)} = k_n + \epsilon_n^{(1)} = (n\pi / a) + \epsilon_n^{(1)} \tag{7-4.20}$$

and we determine ϵ_n by satisfying the condition expressed in Eq. (7-4.12) up to terms in ϵ_n. Now Eq. (7-4.12) for $k_n^{(1)} = (n\pi / a) + \epsilon_n^{(1)}$ gives

$$\cot\left(k_n + \epsilon_n^{(1)}\right)(l - a) + \frac{\cos k_n a - \epsilon_n^{(1)} a \sin k_n a}{\sin k_n a + \epsilon_n^{(1)} a \cos k_n a} = -1/k_n g$$

ISBN 0-8053-8601-7

where we have developed $\cos(k_n + \epsilon_n^{(1)})a$ to first order in $\epsilon_n^{(1)}$. The first term on the left is finite and negligible, the second gives, for $k_n = n\pi/a$, $1/\epsilon_n^{(1)}a$ and tends to infinity, as does the right-hand side of the equation. One thus has

$$-\epsilon_n^{(1)} = \frac{k_n g}{a} \qquad \epsilon_n \tag{7-4.21}$$

The corresponding eigenfunction in the left compartment is

$$u_I(x) = A \sin k_n^{(1)}x = A(\sin k_n x + \epsilon_n^{(1)}x \cos k_n x) \tag{7-4.22}$$

It joins smoothly with the eigenfunction in the right compartment:

$$u_{II}(x) = A(-1)^n \frac{\epsilon_n^{(1)}a}{\sin k_n(l-a)} \sin k_n(l-x)$$

$$= -A(-1)^n k_n g \frac{\sin k_n(l-x)}{\sin k_n(l-a)} \tag{7-4.23}$$

These eigenfunctions represent a stationary state in which there are currents from left to right and from right to left in both compartments. The eigenfunction is large in the left compartment and small but finite in the right one.

The amplitude in the left compartment is approximately A; in the right compartment

$$\frac{A k_n g}{\sin k_n(l-a)} \tag{7-4.24}$$

The amplitude in the left compartment is hence much greater than the amplitude in the right compartment. However, this situation represents an exceptional case. As pointed out before, the eigenvalues of k for which this situation obtains are widely separated, the distance between them being approximately π/a, as seen from Eq. (7-4.20). There are also the eigenvalues of k that correspond to Eq. (7-4.19), which are quite frequent, being spaced approximately π/l apart. For them the amplitude is large in the right compartment and small in the left compartment.

For energies

$$\frac{\hbar^2 k_n^2}{2m} = \frac{\hbar^2}{2m}\left(\frac{n\pi}{a}\right)^2 \tag{7-4.25}$$

or energies in their immediate vicinity, the ratio of the probabilities per unit length of finding the particle in the left rather than in the right compartment is of the order of $1/k^2 g^2$. For other energies this ratio is $k^2 g^2$. The states

ISBN 0-8053-8601-7

corresponding to the exceptional energies $E = (\hbar^2/2m)(n\pi/a)^2$ are called *virtual states*.

Suppose now that in a virtual state we could suppress the wave traveling from right to left in the right, large compartment, leaving the other waves unaffected. The state would not be stationary any more, and the amplitude in the small compartment must decrease, because it is not replenished by the incoming wave. The decrease in amplitude in the left compartment is such as to maintain the current from left to right in the large compartment.

The initial situation would be described by

$$u_{\mathrm{I}} = A \sin k_n x \tag{7-4.26}$$

and

$$u_{\mathrm{II}} = -A(-1)^n \frac{g k_n}{\sin k_n(l-a)} \frac{e^{-ik_n(l-x)}}{2i} \tag{7-4.27}$$

where Eq. (7-4.27) is obtained from Eq. (7-4.23) by writing the $\sin k_n(l-x)$ as the sum of two exponentials and suppressing the one corresponding to the incoming wave.

The outgoing flux corresponding to $u_{\mathrm{II}}(x)$ is

$$\varphi = \tfrac{1}{2} A^2 g^2 k_n^2 \left(\frac{\hbar k_n}{m} \right) \tag{7-4.28}$$

where we have averaged over $\sin^2 k_n(l-a)$, replacing it by $\tfrac{1}{2}$, and used the relation

$$v_n = \hbar k_n/m \tag{7-4.29}$$

This means that we consider a small interval of k in the neighborhood of k_n. The level k_n is indeed broad, as will be seen later, and this justifies our procedure.

We now consider A slowly variable with time, corresponding to the leaking of the particle from the small compartment to the large one. The principle of conservation of matter requires that the outgoing flux φ be equal to the rate of decrease of the probability of finding the particle in the left compartment:

$$-\frac{d}{dt} A^2 \int_0^a \sin^2 k_n^{(1)} x \, dx = \varphi \tag{7-4.30}$$

and

$$-\frac{1}{2} \frac{dA^2}{dt} = \frac{1}{2} A^2 g^2 k_n^2 \frac{v_n}{a} \tag{7-4.31}$$

ISBN 0-8053-8601-7

or

$$-\frac{dA^2}{A^2} = g^2 k_n^2 \frac{v_n}{a}\, dt$$

This last formula can be easily interpreted semiclassically. $dA^2/A^2\, dt$ is the decay constant of the particle in the small box, and this is equal to the number of collisions on the wall per unit time v/a multiplied by the transparency of the barrier, here $g^2 k^2$. We return thus to the point of view expressed in Eq. (7-1.28), but with considerably deeper insight into the situation.

The fact that the amplitude decreases exponentially in time can be expressed mathematically by writing the complete Schrödinger function as

$$\psi(x, t) = u(x)e^{-(iE/\hbar)t}e^{-(\lambda/2)t} \tag{7-4.32}$$

or by formally considering a complex eigenvalue of the energy $E - (i\lambda\hbar/2)$, where λ is the decay constant.

A more elaborate, complete, and rigorous treatment, which brings us to the same result, is to construct, out of stationary solutions by superposition, a $\psi(x, 0)$ that is zero outside the small compartment and then to follow its evolution in time by using the time-dependent Schrödinger equation. It is noteworthy that, in order to construct the initial $\psi(x, 0)$, it is necessary to superimpose states of different energies, lying in an interval of the order of ΔE around the virtual state, with ΔE related to the mean life τ by

$$\tau\, \Delta E = \hbar \tag{7-4.33}$$

The relation of Eq. (7-4.33) to the uncertainty principle is clear. We have at our disposal for measuring the energy of the virtual state a time τ and hence $\tau\, \Delta E = \hbar$.

In practice, for many natural alpha emitters, τ is of the order of seconds or more. Hence ΔE is of the order of 10^{-27} erg, which means practically an infinitely sharp energy interval. But in nuclear reactions one meets situations where τ is extremely short and thus produces large ΔE. For example, $\Delta E = 1$ eV corresponds to $\tau = 6 \times 10^{-16}$ sec, and times as short as 10^{-22} sec occur.

BIBLIOGRAPHY

Ajzenberg-Selove, F. (AS 60).
Bethe, H. A., "Nuclear Physics," *Rev. Mod. Phys.*, **9**, 69 (1937).
Flügge, S., and H. Marschall, *Practical Quantum Mechanics*, Springer, New York, 1971.
Mang, H. J. "Alpha Decay" *Ann. Rev. Nucl. Sci.*, **14**, 1 (1964).
Perlman, I., and F. Asaro, "Alpha Radioactivity," *Ann. Rev. Nucl. Sci.*, **4**, 157 (1954).
Perlman, I., and J. O. Rasmussen, "Alpha Radioactivity," in (Fl E), Vol. 42.
Rasetti, F., *Elements of Nuclear Physics*, Prentice-Hall, Englewood Cliffs, N. J., 1936.
Rasmussen, J. O., "Alpha Radioactivity," in (Si 65).
Stephens, F. S., "The Study of Nuclear States Observed in Alpha Decay," in (AS 60).

ISBN 0-8053-8601-7

PROBLEMS

7-1 Consider a particle of mass M and energy E in the potential

$$V = V_0 > 0 \quad 0 < x < a$$
$$V = 0 \qquad x < 0 \qquad \text{or} \quad x > a$$

Show that transmission through the potential barrier is approximated by

$$T = \frac{16E}{V_0} \exp\left[\frac{-2a}{\hbar} (2MV_0)^{1/2} \right]$$

7-2 Show that for constant V, Eq. (7-1.20) gives approximately the same result as Eq. (7-1.17).

7-3 In beta decay, the emission of low-energy positrons is inhibited by the Coulomb barrier. Show that for very low-energy positrons ($E \sim 100$ keV), the inhibition factor is approximately $e^{-2\pi Ze^2/\hbar v}$, where v is the positron speed and Z is the atomic number of the daughter nucleus.

7-4 The considerations used in alpha-decay theory also apply to nuclear reactions in which charged particles coming from the outside must penetrate the Coulomb barrier to interact with the nucleus. Compute the penetration probability (i.e., the transmission) of a 1-MeV proton through the Coulomb barrier surrounding a $^{238}_{92}$U nucleus.

7-5 The centrifugal force of a spinning nucleus makes it more stable with respect to alpha decay. Show by a drawing of the potential barrier why this is the case.

7-6 $^{212}_{84}$Po decays by emission of an 8.8-MeV alpha particle with a half-life of 0.3×10^{-6} sec. Estimate the probability that an 8.8-MeV alpha particle, in a head-on collision with a $^{208}_{82}$Po nucleus, will penetrate its Coulomb barrier.

7-7 The nucleus $^{216}_{84}$Po has a half-life of 0.16 sec for the emission of an alpha particle of energy 6.77 MeV. From the theory of alpha decay and these measured quantities, estimate the radius of this nucleus.

7-8 Compute the hindrance factors for the rotational band in ^{238}Pu and compare with the experimental numbers of Fig. 7-6.

7-9 Justify Eq. (7-1.30) including its numerical coefficients.

7-10 Prove Eq. (7-4.20).

ISBN 0-8053-8601-7

CHAPTER

8

Gamma Emission

Gamma rays were discovered very early among the radiations emitted by nuclei, and their electromagnetic nature was established at the same time as that of X rays (von Laue, 1912). The study of gamma rays has always played an important role in nuclear physics. They yield information on the energy and quantum numbers of nuclear states; and for this reason they are, like other nuclear radiations, a powerful tool in analyzing nuclear phenomena. In particles physics we find that some particles, the neutral pi meson, for instance, can convert themselves into pairs of gamma rays. This fact is sufficient to give considerable information about the original particle, as we shall see in Chap. 13. We must here distinguish two aspects of this study: one is essentially electromagnetic theory, the other its application to nuclear problems. In this book we are mainly concerned with the second aspect; however, we shall also treat some areas that are, in the main, the subject of electromagnetic theory. For further details and a reference book on electromagnetism consult, for example (Ja 75).

8-1 INTRODUCTION

We shall start with a semiclassical description of the radiation process. We imagine that the nucleus consists of a charge–current distribution confined to

Emilio Segrè, Nuclei and Particles: An Introduction to Nuclear and Subnuclear Physics, Second Edition

ISBN 0-8053-8601-7

a region about the nuclear origin and undergoing periodic motion, whose frequency ω is related to the energies involved in a nuclear transition between two levels by $\omega = (E_1 - E_2)/\hbar$. We shall, as far as is possible, apply to the radiating system concepts taken from classical electromagnetism, which we shall translate into their quantum-mechanical equivalents. A fundamental relation in electromagnetism gives the interaction between a current and the vector potential:

$$\int j_\mu A_\mu \, dx_1 \, dx_2 \, dx_3 \, dx_4$$

where j is the four vector of components $\mathbf{j}$, $i\rho/c$ with $\mathbf{j}$ and ρ current and charge density, and A is the four vector $\mathbf{A}$, $i\varphi$ (vector and scalar potentials) and $x_4 = ict$. This expression can be simplified and leads to the usual electric dipole formula Eq. (8-1.5), which is an approximation valid for $v \ll c$. A rigorous but less intuitive theory is based on the quantum theory of radiation.

As we shall see shortly, if the wavelength of the electromagnetic radiation considered is large compared with nuclear dimensions, the treatment is much simplified by using the "long-wavelength" approximation. This approximation is valid for energies up to several MeV, that is, for most nuclear gamma rays. However, precise quantitative calculations are seldom possible because of insufficient knowledge of nuclear wave functions. An exception is the photodisintegration of the deuteron, which will be treated in Chap. 10; nuclear photoreactions will be considered in Chap. 11. Here we shall deal primarily with emission and absorption of gamma rays occurring in transitions between nuclear levels of low or moderate excitation.

Remember that for a system of periodically moving charges, located at the origin of the coordinates, we can distinguish an induction zone defined by $r \ll \lambdabar = c/\omega$, in which the electric and magnetic fields can be calculated from the instantaneous velocity and position of the charges, and a "radiation" zone for $r \gg \lambdabar$, where retardation effects must be considered. Here r is the distance from the radiator, $\lambda = 2\pi \lambdabar$ is the wavelength of the emitted radiation, ω is the angular frequency of motion, and c is the velocity of light.

In the radiation zone, the electric and magnetic fields $\mathcal{E}$ and $\mathcal{H}$ in Gaussian units are related by the following fundamental equations:

$$|\mathcal{E}| = |\mathcal{H}| \quad \mathcal{H} \cdot \mathcal{E} = 0 \quad \mathcal{E} \cdot \mathbf{r} = \mathcal{H} \cdot \mathbf{r} = 0 \tag{8-1.1}$$

Moreover, they decrease as $1/r$ and give rise to a Poynting vector,

$$\mathbf{S} = \frac{\mathcal{E} \times \mathcal{H}}{4\pi} c \tag{8-1.2}$$

decreasing as $1/r^2$, which ensures a constant flow of energy toward infinity. A detailed study of the electric dipole radiation gives the following expressions for $\mathcal{E}$ in the radiation zone: $\mathcal{E}$ is directed along the meridian, and its

ISBN 0-8053-8601-7

magnitude is

$$\mathcal{E}_\theta = \frac{\sin\theta}{rc^2}\,\ddot{p}\left(t - \frac{r}{c}\right) \tag{8-1.3}$$

where $p(t - r/c)$ is the retarded value of the electric dipole moment directed in the z direction and located at the origin. The dipole varies in time according to $p = p_0 \cos \omega t$ and θ is the angle between r and the direction of the dipole. $\mathcal{H}$ is equal in magnitude to $\mathcal{E}$ and is directed along the parallels (Figs. 8-1 and 8-2). The average power radiated according to Eq. (8-1.2), with $d\Omega$ the element of solid angle, is

$$\langle W \rangle = \frac{c}{4\pi}\int \frac{\mathbf{r}}{r}\cdot(\mathcal{E}\times\mathcal{H})r^2\,d\Omega$$

$$= \int \frac{\sin^2\theta}{c^3 4\pi}\langle(\ddot{p})^2\rangle\,d\Omega = \frac{2}{3}\frac{\langle(\ddot{p})^2\rangle}{c^3} = \frac{\omega^4}{3c^3}p_0^2 \tag{8-1.4}$$

This is a particular case of Larmor's important formula,

$$\langle W \rangle = \frac{2}{3}\frac{e^2\langle \mathbf{a}^2\rangle}{c^3} \tag{8-1.5}$$

where $\mathbf{a}$ is the acceleration of the charge e. This formula is valid for charges moving with velocity small compared to c.

Consider now two identical dipoles oriented in the z direction, having the same frequency ω but opposite phase, and shifted with respect to each other

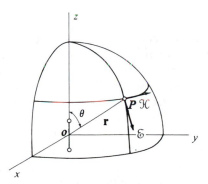

Figure 8-1 Schematic diagram of the electric dipole field in the radiation zone for a given $(\mathbf{r}, \theta, t)$. The field has a cylindrical symmetry with respect to the z axis. The direction of the electric field is the tangent to a meridian, while the direction of the magnetic field is the tangent to a parallel for the sphere of radius r.

ISBN 0-8053-8601-7

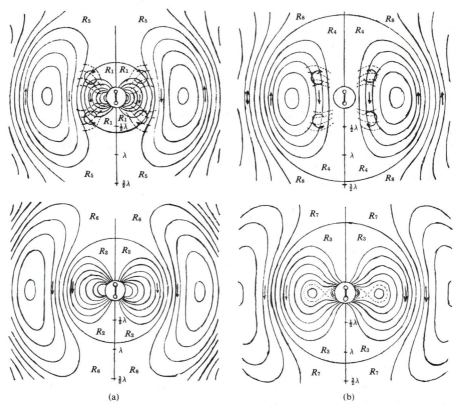

(a)   (b)

Figure   8-2 Hertz's original figures. Electric field lines produced by an oscillating dipole at various times as a function of r and θ. These pictures show mainly the induction zone $(r \ll \lambda)$ and the transition to the radiation zone $(r \gg \lambda)$.

by a length $\Delta x \ll \lambdabar$. The system is called a quadrupole (Fig. 8-3). The dipole moment of the system is zero at all times, but the system nevertheless radiates energy. The field $\mathcal{E}_Q$ generated by the quadrupole at each point of space is equal to $-(\partial \mathcal{E}/\partial x)\Delta x = \mathcal{E}_Q$, where $\mathcal{E}$ is the field produced by one of the dipoles alone. It is easily seen that on the x axis the quadrupole field is $\mathcal{E}(\Delta x/\lambdabar)$ and that the same relation obtains as an order of magnitude in all directions. The ratio of the energy radiated by our quadrupole to that radiated by one of the constituent dipoles alone is of the order of $(\Delta x/\lambdabar)^2$. However, the angular distribution of the intensity per unit solid angle is radically different from that of the dipole. For a nuclear system $\Delta x/\lambdabar$ is a number of the order of magnitude of the nuclear dimensions divided by $\lambdabar$. Using the relations

$$\lambdabar = (197/E\,(\mathrm{MeV})) \times 10^{-13} \ \mathrm{cm} \tag{8-1.6}$$

ISBN 0-8053-8601-7

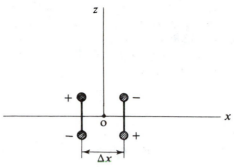

Figure 8-3 A configuration of electric charges having a net electric quadrupole but no monopole or dipole.

and

$$R = 1.2A^{1/3} \times 10^{-13} \text{ cm}$$

we find

$$R/\lambdabar = 6.1 \times 10^{-3}A^{1/3}E \text{ (MeV)} \tag{8-1.7}$$

For low levels with $E = 0.1 - 1$ MeV and for $A^{1/3}$ of several units, the ratio $R/\lambdabar$ is of the order of 10^{-2}. The quadrupole radiation is thus about 10^4 times weaker than the dipole radiation and is important only when the dipole radiation is forbidden (i.e., has intensity zero).

The important condition $\Delta x/\lambdabar \ll 1$ can be transformed to illustrate another aspect of its physical meaning. For a nucleon moving in a nucleus of radius R we have the order-of-magnitude relation

$$\omega \approx v/R \tag{8-1.8}$$

where v is the velocity of the nucleon and hence

$$\Delta x/\lambdabar \approx v/c \tag{8-1.9}$$

Electric dipole radiation is indicated as $E1$, electric quadrupole radiation as $E2$, and so on; in general, El stands for a 2^l pole.

In addition to the electric radiation, we must also consider magnetic multipole radiation. We may think of the magnetic dipole as a small loop of alternating current or a variable magnetic dipole. From the magnetic dipole, giving $M1$ radiation we pass by shifting two loops to magnetic quadrupole ($M2$) radiation, and so on.

In atoms the comparison between the amplitude of electric and magnetic dipole radiation gives $E1/M1 = ea/\mu_B = $ "137"$/2$ where a is the Bohr radius

$\hbar^2/me^2$ and μ_B is the Bohr magneton $e\hbar/2mc$. For nuclei the amplitude of the electric dipole field produced by a charge e confined to a volume $\sim R^3$ is proportional to $eR/r\lambdabar^2$ according to (8-1.3). The uncertainty relation applied to the coordinate of the charge gives

$$Rmv \approx \hbar \tag{8-1.10}$$

or

$$R = \hbar/mv \tag{8-1.11}$$

where m and v are the mass and velocity of the charge. Hence, the electric field due to the electric dipole is of the order of

$$\mathscr{E}^{(E)} \approx \frac{e\hbar}{mv} \frac{1}{r\lambdabar^2} \tag{8-1.12}$$

The nuclear magnetic moment originating from the same system of charges is of the order of $e\hbar/mc$. The associated electric field is therefore proportional to

$$\mathscr{E}^{(M)} \approx \frac{e\hbar}{mc} \frac{1}{r\lambdabar^2} \tag{8-1.13}$$

$$\frac{\mathscr{E}^{(E)}}{\mathscr{E}^{(M)}} = \frac{c}{v} \tag{8-1.14}$$

which means by comparison with Eq. (8-1.9) that the electric field due to the magnetic dipole is comparable to that due to the electric quadrupole. This estimate is also true for higher-order multipoles and is not altered by taking into account the intrinsic magnetic moment, which is of the same order of magnitude as the orbital one.

Returning to electric dipole radiation, we see that its emission is determined by the electric dipole moment of the radiating system, which is a vector of components

$$\Sigma ex, \quad \Sigma ey, \quad \text{and} \quad \Sigma ez \tag{8-1.15}$$

where the sum is extended to all the nuclear charges. To calculate the transition probabilities quantum-mechanically, the quantities x of the classical formulas must be replaced by matrix elements,

$$x \to x_{if} = \int \psi_f^* x \psi_i \, d\tau \tag{8-1.16}$$

where ψ_f and ψ_i are the wave functions of the final and initial states, respectively.

ISBN 0-8053-8061-7

Thus, for example, Eq. (8-1.4) gives the average power radiated by a linear dipole as

$$\langle W \rangle = \frac{4e^2\omega^4}{3c^3} |x_{if}|^2 \tag{8-1.17}$$

Note that p_0^2 has been replaced by $4e^2|x_{if}|^2$.[1] Dividing by the energy content of one quantum $\hbar\omega$, we obtain the radiative decay constant

$$\lambda_\gamma = \frac{4}{3\hbar} \frac{e^2}{c^3} \omega^3 |x_{if}|^2 \tag{8-1.18}$$

It is often possible to jump from one quantum-mechanical stationary state to another by different types of electromagnetic radiation, but it turns out that if $\lambda \gg R$, only one or two types of radiation are important. The particular type depends on the angular momentum and "parity" of the stationary states considered. We thus have selection rules, which we shall discuss shortly.

We shall now develop the classical theory by looking first at the electromagnetic field and afterward at its source.

The main purpose of the analysis of the electromagnetic fields in electric and magnetic multipoles is to separate the different parts with definite parity and angular momentum so as to be able to establish selection rules with the help of the conservation theorems. Using this approach, we derive from Maxwell's equations in vacuum the vector equations for the electric and magnetic fields. Let $\mathcal{E}$ and $\mathcal{H}$ be the electric and magnetic time-dependent fields,

$$\mathcal{E}(x, y, z, t) = \mathbf{E}(x, y, z)e^{-i\omega t} \tag{8-1.19}$$

and

$$\mathcal{H}(x, y, z, t) = \mathbf{H}(x, y, z)e^{-i\omega t} \tag{8-1.20}$$

---

[1]This substition may be justified by the quantum theory of radiation. A heuristic argument considers large quantum numbers for which classical and quantum theory must give the same results. Classically the transition probability is

$$\lambda_\gamma = \frac{\langle W \rangle}{\hbar\omega} = \frac{\omega^3}{3\hbar c^3} p_0^2 = \frac{2}{3} \frac{e^2\omega E}{\hbar c^3 m}$$

In the last passage we have used the relation

$$E = m\omega^2 p_0^2/2e^2$$

between energy and amplitude of the dipole for an harmonic oscillator.
According to quantum theory

$$|X_{nn-1}|^2 = \hbar n/2m\omega \simeq E/2m\omega^2$$

and we obtain the correct result for λ_γ by replacing p_0^2 with $4e^2|x_{if}|^2$ in the classical expression for λ_γ.

ISBN 0-8053-8061-7

Maxwell's equations as shown in electricity give the relations

$$\nabla^2 \mathbf{E} + k^2 \mathbf{E} = 0 \quad \text{and} \quad \nabla^2 \mathbf{H} + k^2 \mathbf{H} = 0 \tag{8-1.21}$$

where $k = \omega/c$, with the subsidiary condition $\nabla \cdot \mathbf{E} = \nabla \cdot \mathbf{H} = 0$. We try to solve these equations in polar coordinates. The procedure is to obtain a complete set of solutions of the vector equations by solving first the simpler scalar equation

$$\nabla^2 \Phi_l^m + k^2 \Phi_l^m = 0 \tag{8-1.22}$$

This is accomplished by setting

$$\Phi_l^m(r, \theta, \varphi) = f_l(kr) Y_l^m(\theta, \varphi) \tag{8-1.23}$$

where $f_l(kr)$ is a Hankel function defined in terms of the ordinary Bessel function of half-odd-integer order as

$$f_l(kr) = \left(\frac{\pi}{2kr} \right)^{1/2} \left[J_{l+1/2}(kr) + i N_{l+1/2}(kr) \right] \tag{8-1.24}$$

and $Y_l^m(\theta, \varphi)$ is the spherical harmonic. Note that the spherical harmonics form an orthonormal set obeying the relation

$$\int Y_{l'}^{*m'}(\theta, \varphi) Y_l^m(\theta, \varphi) \, d\Omega = \delta_{l'l} \delta_{m'm} \tag{8-1.25}$$

Application of the operators

$$\mathbf{L} = -i(\mathbf{r} \times \nabla) \quad \text{and} \quad \frac{-i}{k}(\nabla \times \mathbf{L}) \tag{8-1.26}$$

to Φ_l^m gives the vectors

$$\mathbf{F}_{lm}^{(0)} = \mathbf{L}\Phi_l^m \quad \text{and} \quad \mathbf{F}_{lm}^{(1)} = \frac{-i}{k}(\nabla \times \mathbf{L})\Phi_l^m \tag{8-1.27}$$

which are solutions of Eq. (8-1.21) if we put

$$\mathbf{H}_{lm} = -\mathbf{F}_{lm}^{(0)} \qquad \mathbf{E}_{lm} = \mathbf{F}_{lm}^{(1)} \tag{8-1.28}$$

for an electric multipole El or

$$\mathbf{H}_{lm} = \mathbf{F}_{lm}^{(1)} \qquad \mathbf{E}_{lm} = \mathbf{F}_{lm}^{(0)} \tag{8-1.29}$$

for a magnetic multipole Ml.

The calculations necessary to show these results are found in (BW 52) and (Ja 75). Note that the operator **L** is essentially the angular momentum operator of quantum mechanics.

The relation between the fields of an electric multipole and those of a magnetic multipole is simple. We pass from one to the other by interchanging electric and magnetic fields and changing the sign of the electric field. This transformation is called a "dual" transformation and is expressed by

$$\mathscr{E}' = -\mathscr{H} \qquad \mathscr{H}' = \mathscr{E} \tag{8-1.30}$$

where the primed field is the dual of the unprimed one.

We can now determine what kind of source located at the origin of the coordinates would give a field described by Eqs. (8-1.28) and (8-1.29). We would find in the case of Eq. (8-1.28) that an electric 2^l pole is necessary, whereas the field of Eq. (8-1.29) would be generated by a 2^l magnetic pole.

El and Ml fields have different symmetry properties, and this fact is important in establishing selection rules. For example, let us consider an $E1$ and an $M1$ field. These are generated by charges or currents, and we shall consider the fields produced at a time t and at a point $\mathbf{r}(x, y, z)$ by a distribution of moving charges having the coordinates $\mathbf{s}_i$.

The fields are a function of $\mathbf{r}$ and $\mathbf{s}$, where the $\mathbf{s}$ are all the coordinates of the charges. We can thus write $\mathscr{E}(\mathbf{r}, \mathbf{s})$, $\mathscr{H}(\mathbf{r}, \mathbf{s})$. Now suppose that we change the coordinates of the charges from $\mathbf{s}$ to $-\mathbf{s}$, which means that we reflect the position of the charges with respect to the origin. For an electric dipole field we have (cf. Fig. 8-4a and b)

$$\mathscr{E}(\mathbf{r}, \mathbf{s}) = -\mathscr{E}(\mathbf{r}, -\mathbf{s})$$

$$\mathscr{H}(\mathbf{r}, \mathbf{s}) = -\mathscr{H}(\mathbf{r}, -\mathbf{s}) \tag{8-1.31}$$

If, on the other hand, we look at the same electric dipole field at points $\mathbf{r}, -\mathbf{r}$ without changing the source, we have (Fig. 8-4a and b)

$$\mathscr{E}(\mathbf{r}, \mathbf{s}) = +\mathscr{E}(-\mathbf{r}, \mathbf{s})$$

$$\mathscr{H}(\mathbf{r}, \mathbf{s}) = -\mathscr{H}(-\mathbf{r}, \mathbf{s}) \tag{8-1.32}$$

Fields obeying Eqs. (8-1.31) and (8-1.32) are called *odd*.

On the other hand, for a magnetic dipole field we have (cf. Fig. 8-4c and d)

$$\mathscr{E}(\mathbf{r}, \mathbf{s}) = \mathscr{E}(\mathbf{r}, -\mathbf{s})$$

$$\mathscr{H}(\mathbf{r}, \mathbf{s}) = \mathscr{H}(\mathbf{r}, -\mathbf{s}) \tag{8-1.33}$$

because reflection through the origin does not change the sense of rotation of

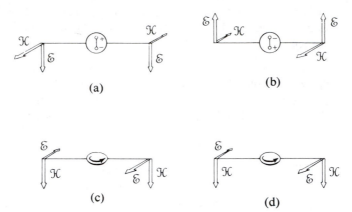

Figure 8-4 Dual fields of the electric and magnetic dipoles. (a) and (b) show the electric dipole field and illustrate its symmetry properties on reflection of the source, or of the observation point through the origin. (c) and (d) show the magnetic dipole field in like manner. Comparison of (a) with (c) shows how the dual transformation changes the $E1$ field into the $M1$ field and vice versa.

the charges. Changing $\mathbf{r}$ to $-\mathbf{r}$ (Fig. 8-4c and d) gives

$$\mathcal{E}(\mathbf{r}, \mathbf{s}) = -\mathcal{E}(-\mathbf{r}, \mathbf{s})$$

$$\mathcal{H}(\mathbf{r}, \mathbf{s}) = +\mathcal{H}(-\mathbf{r}, \mathbf{s}) \qquad (8\text{-}1.34)$$

Fields, such as the magnetic dipole field, that obey Eqs. (8-1.33) and (8-1.34) are called *even* fields. From Eqs. (8-1.31) to (8-1.34) it follows that for all radiation fields, even or odd,

$$\mathcal{E}(\mathbf{r}, \mathbf{s}) = -\mathcal{E}(-\mathbf{r}, -\mathbf{s})$$

$$\mathcal{H}(\mathbf{r}, \mathbf{s}) = \mathcal{H}(-\mathbf{r}, -\mathbf{s}) \qquad (8\text{-}1.35)$$

Equation (8-1.35) is a consequence of the fact that the combined operation $\mathbf{r} \rightarrow -\mathbf{r}$, $\mathbf{s} \rightarrow -\mathbf{s}$ is an inversion of all coordinates. $\mathcal{E}$, a polar vector, then changes sign, whereas $\mathcal{H}$, an axial vector, or pseudovector, does not. (See also Chap. 9).

From the definition of Eqs. (8-1.28) to (8-1.30) it follows that dual fields have opposite parity.

The classical electric and magnetic multipoles, located at the source, which generate the fields of Eqs. (8-1.28) and (8-1.29) have expressions of the form

$$Q_{lm} = \int \rho(r) r' Y_l^m (\theta, \varphi) \, d\tau \quad \text{for} \quad El \qquad (8\text{-}1.36)$$

ISBN 0-8053-8061-7

and

$$M_{lm} = - \frac{1}{c(l+1)} \int r^l Y_l^m (\theta, \varphi) \nabla \cdot (\mathbf{r} \times \mathbf{j}) \, d\tau \quad \text{for} \quad Ml \quad (8\text{-}1.37)$$

where the integral number m, $(-l \leqslant m \leqslant l)$, characterizes the $2l + 1\, El$ or Ml independent multipole fields of order l, $\rho(r)e^{-i\omega t}$ is the electric charge density, $\mathbf{j}$ is the electric-current density at r related to ρ by

$$i\omega\rho = \nabla \cdot \mathbf{j}(\mathbf{r}) \quad (8\text{-}1.38)$$

and the integral is extended to the whole region containing the charges. The continuity equation (8-1.38) connects a time variation of ρ with a current density. Hence, electric and magnetic radiation are present together. However, they cannot be of the same order, as we shall see presently.

Let us first consider electric multipole transitions. We shall assume that we have only one charged particle of coordinates $\mathbf{r}$. The results obtainable in this very special case are true in general. To obtain the transition probabilities in quantum mechanics, we replace $\rho(\mathbf{r})$ by

$$e\psi_f^* (\mathbf{r})\psi_i(\mathbf{r}) \quad (8\text{-}1.39)$$

for the single-particle case. If A particles are involved, a procedure similar to that discussed in Sec. 6-8 is used.

In the special case of $E1$, it is easily recognized that Eq. (8-1.36) gives, apart from constants, the matrix elements of Eqs. (8-1.12) and (8-1.13). To see this it suffices to replace the spherical harmonics by their expressions in cartesian coordinates. This equivalence is true for any l.

8-2 SELECTION RULES

If, for a certain order of multipole, the matrix elements that determine the transition probability vanish exactly, the transition is forbidden and that multipole component of the electromagnetic field is absent. We thus have *selection rules*.

A simple selection rule derives immediately from the parity properties of the function ψ. For instance, for any central system we must have

$$\psi(\mathbf{r}) = \pm \psi(-\mathbf{r}) \quad (8\text{-}2.1)$$

This relation results from the fact that, for a system in which the potential depends on $|\mathbf{r}|$ only (central system), the hamiltonian does not change if we change $\mathbf{r}$ into $-\mathbf{r}$. If a solution of the corresponding Schrödinger equation is

ISBN 0-8053-8061-7

not degenerate, we then have

$$H(\mathbf{r})\psi(\mathbf{r}) = E\psi(\mathbf{r})$$

$$H(-\mathbf{r})\psi(\mathbf{r}) = E\psi(\mathbf{r})$$

$$H(\mathbf{r})\psi(-\mathbf{r}) = E\psi(-\mathbf{r}) \qquad (8\text{-}2.2)$$

which implies that $\psi(\mathbf{r}) = k\psi(-\mathbf{r})$, with k constant. If we now perform the operation of inversion (change from $\mathbf{r}$ to $-\mathbf{r}$) twice, which obviously gives the identity, we have

$$\psi(\mathbf{r}) \rightarrow \underset{\text{1st inv.}}{k\psi(-\mathbf{r})} \rightarrow \underset{\text{2nd inv.}}{k^2\psi(\mathbf{r})} = \psi(\mathbf{r}) \quad \text{or} \quad k = \pm 1 \qquad (8\text{-}2.3)$$

If $k = 1$, the eigenfunction is called *even*, or of parity $+1$; if $k = -1$, it is called *odd*, or of parity -1. If the eigenfunction is degenerate, one can always form linear combinations $\psi(\mathbf{r}) \pm \psi(-\mathbf{r})$ that have the property indicated in Eq. (8-2.3).

The product of two even functions or of two odd functions is even; the product of an even function multiplied by an odd one is odd. The integral over all space of an odd function is zero, because the contributions to the integral of elements of volume at $\mathbf{r}$ and $-\mathbf{r}$ cancel each other out.

The function $r^l Y_l^m(\theta, \varphi)$ is itself even or odd, according to whether l is even or odd, irrespective of m (see Sec. 6-4).

Assume now that $\psi_i(\mathbf{r})$ and $\psi_f(\mathbf{r})$ have the same parity. Then the integrand in Eq. (8-1.36) will have the parity $(-1)^l$, and the integral will vanish for l odd. Similarly, if ψ_i and ψ_f have opposite parity, the integral will vanish for l even.

We thus have the selection rule that El radiation for l odd (even) is accompanied by a (no) change of parity in the eigenfunction. We shall designate by $\Delta\pi = 1$ (or yes), $\Delta\pi = 0$ (or no) a transition with or without a change of parity.

In addition to the selection rules connected with the change of parity, there are very important selection rules associated with the change of angular momentum of the nucleus. We cannot give here a general proof for these rules, but we shall give examples to show how they arise, and then state the general case.

If the nucleon intrinsic spin is neglected, the initial and final states of the nucleus have angular momenta l' and l'', respectively, with l integer, and z components of the angular momenta m' and m'', respectively. The initial and final states will thus have eigenfunctions of the form

$$u_i = f_{n'l'}(r) Y_{l'}^{m'}(\theta, \varphi) \qquad (8\text{-}2.4)$$

and

$$u_f = f_{n''l''}(r) Y_{l''}^{m''}(\theta, \varphi) \qquad (8\text{-}2.5)$$

If we consider electric dipole radiation ($E1$), the matrix elements will be given by Eq. (8-1.36) and will contain integrals of the type

$$\int Y_{l''}^{m''*}(\theta, \varphi) Y_1^{1, 0, -1}(\theta, \varphi) Y_{l'}^{m'}(\theta, \varphi)\, d\Omega \qquad (8\text{-}2.6)$$

It suffices then to remember that $Y_1^{0, \pm 1}$ are proportional to $\cos\theta$, $\sin\theta e^{\pm i\varphi}$, and the general form of Y_l^m, to see immediately that the integrand will contain a factor $e^{i(m' - m'')\varphi}$ or $e^{i[(m' - m'') \pm 1]\varphi}$ and thus that the integral will vanish unless

$$m' = m'' \qquad (8\text{-}2.7)$$

or

$$m' = m'' \pm 1 \qquad (8\text{-}2.8)$$

Similarly one finds that the integral vanishes also unless

$$l' = l'' \pm 1 \qquad (8\text{-}2.9)$$

by using properties of $Y_l^m(\theta, \varphi)$ mentioned in Problem 8-8. Now, if we recall the meaning of l and m, we see that in the emission of electric dipole radiation the total angular momentum of the nucleus changes by one unit and that its z component changes by zero or one unit. This argument may be extended to noninteger spin and arbitrary radiation, with the result that for El radiation we must have

$$|J' - J''| \leqslant l \leqslant J' + J'' \qquad (8\text{-}2.10)$$

In this chapter J', J'' are the total angular momentum in the initial and final nuclear states. [For the proof refer to (Ma 57) or (BW 52).] Equation (8-2.10) has the geometrical interpretation that it must be possible to construct a triangle of sides J', J'', and l. The principle of conservation of angular momentum suggests that the outgoing wave of El light carries an angular momentum of magnitude l with respect to the center of mass of the nucleus. This is borne out by direct calculation of the density of angular momentum for the quantized electromagnetic field.

Thus far we have been concerned only with electric multipoles. Magnetic multipoles arise from the electric currents caused by the motions of the charges in the nucleus and from the intrinsic magnetic moments connected

ISBN 0-8053-8061-7

with the spins. Remember in this connection that the neutron, although neutral, has a magnetic moment. In the simplest case of the dipole due to the motion of a spinless charged particle, the z component of the matrix element for magnetic radiation is proportional to

$$\int \psi_f^* (xp_y - yp_x)\psi_i \, d\tau \tag{8-2.11}$$

because the operator $xp_y - yp_x = L_z$ is proportional to the component of the magnetic moment produced by the electric current associated with the transition. This is to be compared to the electric dipole matrix element given by Eq. (8-1.16). We shall shortly see the effect of the intrinsic magnetic moment.

In the specific case mentioned above, the $M1$ matrix element is not zero only if ψ_f^*, ψ_i have the same parity, because the operator L_z does not change the parity of ψ_i. Moreover, as with $E1$ transitions, selection rules for $M1$ are $\Delta l = \pm 1$ and $\Delta m = \pm 1, 0$, as can be easily verified.

The selection rules for magnetic radiation can be generalized and can be interpreted as representing conservation of angular momentum if one associates with $M1$, $M2$, $M3$ radiation 1, 2, 3 units of angular momentum, just as in the case of $E1$, $E2$, and $E3$. In fact, from the relation between the fields of El and Ml expressed by Eqs. (8-1.28) and (8-1.29) and the expression for the density of momentum $\mathbf{p}$ in an electromagnetic field,

$$\mathbf{p} = \frac{\mathcal{E} \times \mathcal{H}}{4\pi c} \tag{8-2.12}$$

we see that the magnitude of the density of momentum of El and Ml is the same, and hence the magnitude of the density of angular momentum is also identical in the two cases.

In Table 8-1 we sum up the results for the lowest-order radiation possible between two states of angular momentum J', J'' and parity π', π''. It must be remembered that the triangular relation [Eq. (8-2.10)] must always be obeyed. Hence, the table gives conditions that are necessary but not sufficient for radiative transitions. Thus, for example, transitions from $J' = 0$ to $J'' = 0$ are always forbidden, and transitions $J' = \frac{1}{2}$ to $J'' = \frac{1}{2}$, $\Delta\pi = 0$ cannot occur as

TABLE 8-1   SELECTION RULES FOR ELECTROMAGNETIC MULTIPOLE RADIATION

	$E1$	$E2$	$E3$	$E4$	$\cdots$
$\Delta\pi$	1	0	1	0	
$|\Delta J| \leqslant$	1	2	3	4	$\cdots$

	$M1$	$M2$	$M3$	$M4$	$\cdots$
$\Delta\pi$	0	1	0	1	
$|\Delta J| \leqslant$	1	2	3	4	$\cdots$

ISBN 0-8053-8601-7

$E2$, because the triangular relation is violated. Transitions from $J' = 0$ to $J'' = 0$ can occur only by mechanisms different from electromagnetic radiation, namely, by the emission of conversion electrons or by the formation of electron–positron pairs.

In practice the types of radiation that have been observed up to the present are $E1$ to $E6$ inclusive and $M1$ to $M5$ inclusive. In almost all cases, except the pairs $E2$–$M1$ and $E1$–$M2$, only a single type of radiation occurs in a given transition.

8-3   TRANSITION PROBABILITIES

If we generalize the results obtained in Sec. 8-2, the structure of the complete formulas for the transition probabilities will appear plausible. The detailed calculation (BW 52) yields

$$\lambda^{(E)}(l, m) = \frac{8\pi(l+1)}{\hbar l\left[(2l+1)!!\right]^2}\left(\frac{\omega}{c}\right)^{2l+1}|\mathfrak{Q}_{lm} + \mathfrak{Q}'_{lm}|^2 \qquad (8\text{-}3.1)$$

$$\lambda^{(M)}(l, m) = \frac{8\pi(l+1)}{\hbar l\left[(2l+1)!!\right]^2}\left(\frac{\omega}{c}\right)^{2l+1}|\mathfrak{M}_{lm} + \mathfrak{M}'_{lm}|^2 \qquad (8\text{-}3.2)$$

Here l is the order of the transition, and $n!!$ means $1 \cdot 3 \cdot 5 \cdots n$. The first formula is valid for El and the second for Ml radiation. $\mathfrak{Q}_{lm}$ and $\mathfrak{M}_{lm}$ are the parts of the matrix element containing the ordinary coordinates, and $\mathfrak{Q}'_{lm}$ and $\mathfrak{M}'_{lm}$ are the parts of the matrix elements containing the intrinsic magnetic moment.

We can write them formally as

$$\mathfrak{Q}_{lm} = e\sum_{1}^{Z} {}_k\int r_k^l Y_l^{m*}(\theta_k, \varphi_k)\psi_f^*\psi_i \, d\tau \qquad (8\text{-}3.3)$$

$$\mathfrak{M}_{lm} = -\frac{1}{l+1}\frac{e\hbar}{Mc}\sum_{1}^{Z} {}_k\int r_k^l Y_l^{m*}(\theta_k, \varphi_k)\nabla\cdot(\psi_f^*\mathbf{L}_k\psi_i)\, d\tau \qquad (8\text{-}3.4)$$

$$\mathfrak{Q}'_{lm} = -\frac{i(\omega/c)}{l+1}\frac{e\hbar}{2Mc}\sum_{1}^{A} {}_k\int \mu_k r_k^l Y_l^{m*}(\theta_k, \varphi_k)\nabla\cdot(\psi_f^*\mathbf{r}_k \times \sigma_k\psi_i)\, d\tau \qquad (8\text{-}3.5)$$

$$\mathfrak{M}'_{lm} = -\frac{e\hbar}{2Mc}\sum_{1}^{A} {}_k\int \mu_k r_k^l Y_l^{m*}(\theta_k, \varphi_k)\nabla\cdot(\psi_f^*\sigma_k\psi_i)\, d\tau \qquad (8\text{-}3.6)$$

Here the symbols have the following meanings: i and f indicate initial and final states; k is the number of the nucleon; 1 to Z for protons, $Z + 1$ to A for

neutrons; ψ refers to the eigenfunction of the whole nucleus; the vector oper-
ator $\mathbf{L}_k$ is $-i\mathbf{r}_k \times \nabla_k$; the μ_k are the magnetic moments of the nucleons in
units of $e\hbar/2Mc$; the $\boldsymbol{\sigma}$ are Pauli matrix operators. The spherical harmonics
are evaluated for the position of each nucleon in turn. The $\lambda(l, m)$ averaged
over the initial m' states and summed over the final m'' states correspond to
the average transition probability from an unpolarized source irrespective of
the polarization of the emitted radiation. We call

$$B(l, J_i, J_f) \tag{8-3.7}$$

the expression

$$\frac{1}{2J_i + 1} \sum_{m'} \sum_{m''} |\mathfrak{L}_{lm} + \mathfrak{L}'_{lm}|^2 \tag{8-3.8}$$

or

$$\frac{1}{2J_i + 1} \sum_{m'} \sum_{m''} |\mathfrak{M}_{lm} + \mathfrak{M}'_{lm}|^2 \tag{8-3.9}$$

averaged and summed up as indicated. We introduce correspondingly the
expression

$$\lambda(l) = \frac{1}{2J_i + 1} \sum_{m'} \sum_{m''} \lambda(l, m) \tag{8-3.10}$$

and we have

$$\lambda(l) = \frac{8\pi(l + 1)}{l[(2l + 1)!!]^2} \left(\frac{\omega}{c} \right)^{2l+1} \frac{1}{\hbar} B(l, J_i \to J_f) \tag{8-3.11}$$

If we need to distinguish electric and magnetic transitions, we shall use $\lambda(El)$,
$B(El)$, etc.

The evaluation of formulas such as Eqs. (8-3.3) to (8-3.11) requires a
detailed knowledge of the nucleus. Detailed calculations are possible only for
low-lying states and for some simple models; a notable case is that of only
one nucleon radiating. In particular we shall consider a nucleus of odd A
according to the shell model. Its angular momentum is then due to the odd
nucleon alone, and we assume j–j coupling. The radiation is emitted only
because the single nucleon changes orbit. We write the interesting part of the
eigenfunction as

$$\psi_i = R_{nl'}(r)\phi_{j, l', m'}(\theta, \varphi) \tag{8-3.12}$$

and a similar one for ψ_f.

By inserting Eq. (8-3.12) into the multipole formulas (8-3.3) to (8-3.6), the parts containing the spherical harmonics can be integrated and factored out. They give a result that we indicate by

$$S(J_i, J_f, l) \tag{8-3.13}$$

The numerical value of this expression has been explicitly calculated (Moszkowski, 1951) and tabulated. In general, it is of the order of magnitude of unity. The matrix element $\mathfrak{Q}_{lm}$, $\mathfrak{Q}'_{lm}$ and $\mathfrak{M}_{lm}$, $\mathfrak{M}'_{lm}$ have the dimensions of er^l and $(e\hbar/Mc)r^{l-1}$, respectively. The integrals expressing them are extended over the nuclear volume; thus the variable r appearing in them has the order of magnitude of the nuclear radius R. This suggests introducing dimensionless quantities

$$\mathfrak{Q}(El, J_i, J_f) = (\mathfrak{Q}_{lm} + \mathfrak{Q}'_{lm})/eR^l$$

and

$$\mathfrak{M}(Ml, J_i, J_f) = (\mathfrak{M}_{lm} + \mathfrak{M}'_{lm})/\left[(e\hbar/Mc)R^{l-1}\right]$$

summed over m'' and averaged over m'. One then obtains for the transition probabilities

$$\lambda(El) = \left(\frac{e^2}{\hbar c}\right)\frac{l+1}{l}\frac{\omega}{[(2l+1)!!]^2}\left(\frac{\omega R}{c}\right)^{2l}(2J_f + 1)S|\mathfrak{Q}(El)|^2 \tag{8-3.14}$$

and

$$\lambda(Ml) = \left(\frac{e^2}{\hbar c}\right)\frac{l+1}{l}\frac{\omega}{[(2l+1)!!]^2}\left(\frac{\omega R}{c}\right)^{2l}(2J_f + 1)S\left(\frac{\hbar}{McR}\right)^2|\mathfrak{M}(Ml)|^2 \tag{8-3.15}$$

In the case in which a single proton changes state in the transition,

$$\mathfrak{Q}(El) = \int_0^\infty R_i(r)\left(\frac{r}{R}\right)^l R_f^*(r)r^2\,dr \tag{8-3.16}$$

and

$$\mathfrak{M}(Ml) = \left(\mu_p l - \frac{l}{l+1}\right)\int_0^\infty R_i(r)\left(\frac{r}{R}\right)^{l-1} R_f^*(r)r^2\,dr \tag{8-3.17}$$

where $\mu_p = 2.79$, the magnetic moment of the proton in nuclear magnetons,

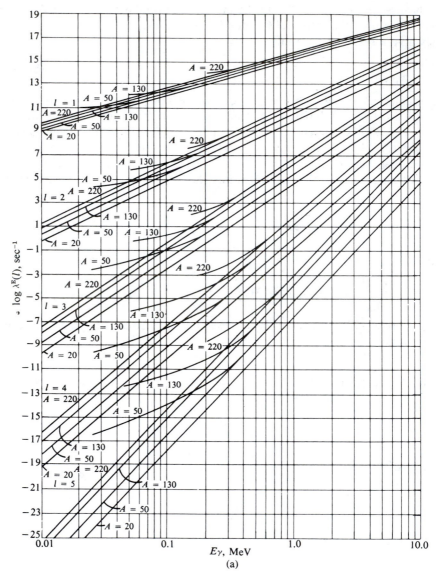

Figure 8-5 The transition probability for gamma-ray emission (as a function of the transition energy E_γ in MeV) based on the single-particle model. Part (a) plots the transition probability for El radiation ($l = 1, \ldots, 5$) for nuclei of mass 20, 50, 130, and 220 according to the formula

$$\lambda(El) = \frac{4(l+1)}{l[(2l+1)!!]^2}\left(\frac{3}{3+l}\right)^2\left(\frac{E_\gamma}{140}\right)^{2l+1} A^{2l/3}\frac{mc^2}{\hbar}$$

ISBN 0-805-38601-7

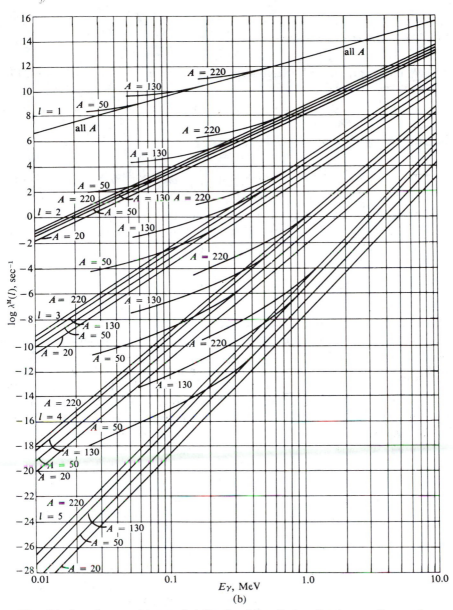

Part (b) plots the transition probability for *Ml* radiation ($l = 1, \ldots, 5$) according to the formula

$$\lambda(Ml) = \frac{0.088(l+1)}{l[(2l+1)!!]^2} \left(\frac{3}{2+l} \right)^2 \left(\frac{E_\gamma}{140} \right)^{2l+1} A^{(2l-2)/3} \left(\frac{\mu_p l}{2} - \frac{l}{l+1} \right)^2 \frac{mc^2}{\hbar}$$

The formulas neglect the factor $S(J_i, J_f, l)$ of Eq. (8-3.13) and assume

$$R = 1.40 A^{1/3} F$$

The curves represent the additional contributions to the total transition probability by the internal conversion process. [From E. U. Condon and H. Odishaw, *Handbook of Physics*, McGraw-Hill, New York 1967.]

and R_i and R_f are the radial eigenfunctions of the initial and final states. For the single-neutron case

$$\mathfrak{L}(El) = 0 \tag{8-3.18}$$

because the neutron has no charge, and

$$\mathfrak{M}(Ml) = \mu_n l \int R_i(r)\left(\frac{r}{R}\right)^{l-1} R_f^*(r)r^2\, dr \tag{8-3.19}$$

The radial integrals may be approximated by assuming $R_{nl}(r) = $ constant $= (3/R^3)^{1/2}$ from $r = 0$ to $r = R$, as required by normalization, and zero for $r \gg R$. We then have immediately, for example,

$$\mathfrak{L}(El) = \int R_f^*(r)\frac{r^l}{R^l}R_i(r)r^2\, dr = \frac{3}{3+l} \tag{8-3.20}$$

and similar expressions for the other matrix elements.

The transition probabilities have been evaluated on the usual assumption that $R = r_0 A^{1/3}$. They are shown in Fig. 8-5. The transition probabilities depend strongly on the energy of the transition through the factor ω^{2l+1} and on A, which enters through the radius R at the power $2l$. The approximations involved are crude, and one cannot expect good numerical agreement. Important effects such as the recoil of the rest of the nucleus except the nucleon considered have been entirely neglected, and an oversimplified model has been used. We gain an impression of the measure of agreement between the experimental results and our schematization, comparing the values of the experimental mean life for gamma transitions with the values predicted by Eqs. (8-3.14) and (8-3.15). To facilitate the comparison, we use the "reduced mean life," or "comparative mean life," that is, the mean life corrected by the factors due to transition energy and nuclear size (Fig. 8-6).

Experimentally the mean life of a gamma emission can be measured directly down to approximately 10^{-10} sec. Indirect measurements involve the observation of Coulomb excitation (see Sec. 8-7) and the observation of level width either through resonance fluorescence or otherwise. Here one can reach values of about 10^{-12} sec. The direct observation of the level width is a valid method for extremely short times. In particle physics, levels of several MeV in width are common and 1 MeV corresponds to 6.58×10^{-22} sec.

To show the influence of the type of motion on the gamma transition probabilities, we shall mention a case almost opposite to the one-particle model: the liquid-drop model. According to this model, neutrons and protons in a nucleus are bound in such a way that the local composition of nuclear matter is practically constant. The electric center of charge then has the same coordinates as the center of mass and cannot move under the action of internal forces, because of the principle of the conservation of momentum. It

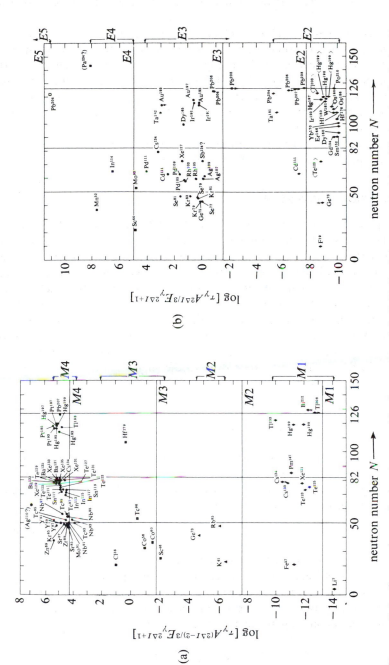

Figure 8-6 Comparison of experiment with the "reduced half-life" of gamma transition. The "reduction" should eliminate the influence of energy and ΔI. [M. Goldhaber and J. Weneser, *Ann. Rev. Nucl. Sci.*, **5**, 1 (1955).] The various transitions are expected to fall in the regions labeled $E2$, $E3$, etc. at the right.

follows that the electric dipole moment is zero, and dipole radiation is strictly forbidden. This situation exists in many low-energy transitions in which the dipole radiation is much weaker than would be expected on the basis of Eq. (8-3.14).

At high energy (~20 MeV) there seems to be a different type of motion, in which all protons together oscillate relative to all neutrons. This motion gives rise to a large electric dipole moment that enables the nuclei to absorb electromagnetic energy strongly in that frequency range and hence gives rise to large cross sections for the (γ, n), (γ, p) reactions. The absorption has the character of a broad resonance, sometimes called the *giant* resonance, and will be treated in Chap. 11.

Finally, special types of surface motion in the nucleus occasionally favor electric quadrupole radiation (see Secs. 6-12 and 8-7).

8-4   INTERNAL CONVERSION

In the preceding section we have considered the transition of a nucleus from one level to another by emission of electromagnetic radiation. This and, given sufficient energy, pair production would be the only ways to execute the transition for an isolated nucleus deprived of all its atomic electrons. The presence of the electrons makes possible a different process: the nucleus can lose its excitation and transfer it directly to one of the atomic electrons which is ejected with a kinetic energy equal to the energy of the gamma transition minus the binding energy of the electron. Electrons ejected by this mechanism are called *conversion electrons*, and one speaks of conversion in the K shell, in the L shell, etc., according to the shell vacated by the conversion electron. The conversion coefficient

$$\frac{N_e}{N_\gamma} = \alpha \tag{8-4.1}$$

is the ratio between the average number of electrons and the average number of gamma rays emitted in connection with a given transition. It is possible also to distinguish *partial conversion coefficients* according to the shell from which the electron is taken. Thus one has

$$\alpha_K + \alpha_L \cdots = \alpha \tag{8-4.2}$$

where α_K, α_L, etc., are the partial conversion coefficients (Fig. 8-7).

The possibility of decay of an excited state by internal conversion adds to its decay constant; thus the total decay constant λ is equal to the sum of the partial decay constants λ_γ and λ_e for gamma emission and conversion electron emission. As a consequence of Eq. (8-4.1) one has

$$\lambda = \lambda_\gamma + \lambda_e = \lambda_\gamma(1 + \alpha) \tag{8-4.3}$$

ISBN 0-805-3-8b01-7

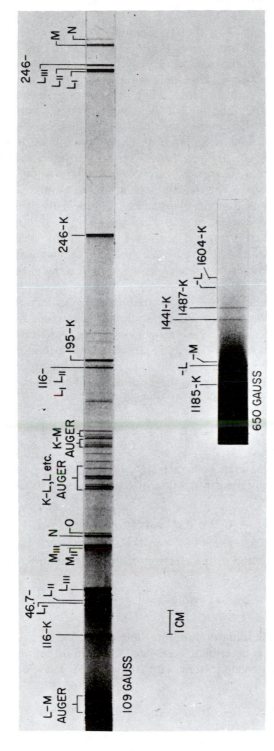

Figure 8-7 Internal conversion spectrum of ^{209}At, ^{210}At, ^{211}At, showing K, L, M, and N conversion. Numbers are energies in KeV. [J. W. Mihelich, A. W. Schardt, and E. Segrè, *Phys. Rev.*, **95**, 1508 (1954).]

A direct experimental proof of the last relation is given by the decay of an excited state in ^{99}Tc, which proceeds with a slightly longer mean life if some of the surrounding electrons are removed by chemical bonding.

If there are two radioactive emissions in rapid succession, e.g., β (negative electrons) and gamma emission, the energy of the conversion electrons tells which emission is the first. Starting with a nucleus of atomic number Z, if the beta emission precedes the gamma emission, the binding energies of the conversion electrons correspond to an atom with atomic number $Z + 1$ and not Z (L. Meitner and H. J. von Baeyer, 1919). Similar considerations obtain for alpha emission or orbital electron capture associated with gamma emission.

The internal conversion coefficients depend on the atomic number of the nucleus, on the energy, and on the character of multipolarity of the transition, but not on the specific nuclear model. Thus their study is a powerful aid to the classification of nuclear levels. Their values as a function of the energy, of the type of radiation, and of A have been extensively tabulated. We shall now calculate one of the simplest possible cases, which will bring out the essentials of the phenomenon.

Suppose that the nucleus is in an excited state from which it can pass to the ground state by the emission of $E1$ radiation. The nucleus can then be compared with an electric dipole of frequency ω. The presence of this dipole may induce transitions from the ground state of the atom to an excited state; specifically, the K electrons, which are in a $1s$ state, can be brought by dipole radiation to a p state, possibly in the continuum. The probability of this transition is calculable with golden rule No. 2,

$$w = \frac{2\pi}{\hbar} |M_{if}|^2 \frac{d\,\mathfrak{N}}{dE} \tag{8-4.4}$$

We have to evaluate the matrix element M_{if} and the density of the accessible final states. Call the initial eigenfunction of the electron in the $1s$ state

$$\psi_i(\mathbf{r}, t) = u_i(r) \exp\left(-\frac{iE_i}{\hbar} t\right) \tag{8-4.5}$$

and the final eigenfunction of the electron in the continuum

$$\psi_f(\mathbf{r}, t) = u_f(\mathbf{r}) \exp\left(-\frac{iE_f}{\hbar} t\right) \tag{8-4.6}$$

The transition from the initial to the final state is induced by the electric field of the nucleus, which is described as an electric dipole of moment $\mathbf{P}$ directed along the z axis and varying in time with frequency ω. The electric potential of this dipole is

$$V(\mathbf{r}, t) = p_0 \frac{\cos\theta}{r^2} \cos\omega t = p_0 \frac{\cos\theta}{r^2} \frac{1}{2}\left(e^{i\omega t} + e^{-i\omega t}\right) \tag{8-4.7}$$

ISBN 0-8053-8601-7

where θ is the angle between $\mathbf{r}$ and the z axis. The matrix element of the induced transitions is

$$M_{if} = e \int \psi_f^* (\mathbf{r}, t) V \psi_i (\mathbf{r}, t) \, d\tau \qquad (8\text{-}4.8)$$

M_{if} is of appreciable magnitude only if

$$|E_i - E_f| = \hbar\omega \qquad (8\text{-}4.9)$$

Moreover, since u_i corresponding to an s state does not contain θ, while V contains the factor $\cos\theta = P_1(\cos\theta)$, ψ_f^*, when expanded in spherical harmonics, will contribute to the integral only through the term that also contains $P_1(\cos\theta)$, all the other terms being orthogonal to $u_i V$. This shows that the transition will occur only to p levels. The density of the final states in Eq. (8-4.4) must thus be limited to the density of the p states.

At this point we can already see the important qualitative conclusion mentioned above. The probability of internal conversion and the radiation probability are both proportional to p_0^2; hence this quantity will disappear from the internal conversion coefficient. The internal conversion coefficient is a function of the energy of the transition and of the atomic number of the atom, because they appear in the initial and final wave functions, but not of the complicated nuclear wave functions. For other multipole fields the same qualitative conclusion obtains: the internal conversion coefficients depend on the character of the radiation (El or Ml), the atomic shell in which it occurs, the atomic number, and the energy.

Note that this result is not absolutely exact. There are some additional effects depending on nuclear size and internal motions, which in special cases affect the conversion coefficient appreciably. In the main, however, it is correct to assume that the internal conversion coefficient is a purely atomic property.

In order to take the calculation a little further in our simple example, we take as the wave function of the final state that of a free electron and expand it in spherical harmonics (see Appendix A). We are, however, interested only in the p-wave component, because all other components give zero matrix elements with the initial s state. We thus write

$$u_f = N \frac{\cos\theta}{(kr)^{1/2}} J_{3/2}(kr) \qquad (8\text{-}4.10)$$

or, asymptotically for large kr,

$$u_f = -N \cos\theta \left(\frac{2}{\pi k^2 r^2} \right)^{1/2} \cos kr \qquad (8\text{-}4.11)$$

ISBN 0-8053-8601-7

We have indicated by N a normalization factor, which we evaluate by enclosing the system in a very large sphere of radius R and using the asymptotic expression for the eigenfunction. We find

$$N = k\left(\frac{3}{4R}\right)^{1/2}$$

(8-4.12)

For the initial state we take the hydrogen-type s-wave function

$$u_i = \frac{1}{\pi^{1/2}}\left(\frac{Z}{a_0}\right)^{3/2}\exp\left(-\frac{Zr}{a_0}\right) \quad \text{with} \quad a_0 = \frac{\hbar^2}{me^2}$$

(8-4.13)

The matrix element is then

$$M_{if} = p_0(\cos \omega t)ek\left(\frac{3}{4R}\right)^{1/2}\frac{1}{\pi^{1/2}}\left(\frac{Z}{a_0}\right)^{3/2}$$

$$\times \int_0^\infty \exp\left(-\frac{Zr}{a_0}\right)\frac{\cos\theta}{r^2}\frac{J_{3/2}(kr)}{(kr)^{1/2}}\cos\theta\, d\tau$$

(8-4.14)

$$= p_0(\cos \omega t)\left(\frac{4\pi}{3R}\right)^{1/2}ek\left(\frac{Z}{a_0}\right)^{3/2}I$$

(8-4.15)

with

$$I = \int_0^\infty \exp\left(-\frac{Zr}{a_0}\right)\frac{J_{3/2}(kr)}{(kr)^{1/2}}\, dr$$

(8-4.16)

The density of the final states must be limited to p states only. From the asymptotic expression [Eq. (8-4.11)] and the condition $u_f(R) = 0$ we find the quantization condition

$$kR = (n + \tfrac{1}{2})\pi$$

(8-4.17)

with n an integral number. Thus in the k interval Δk there are

$$R\frac{\Delta k}{\pi} = \Delta\mathfrak{N}$$

(8-4.18)

states. From this equation we get

$$\rho = \frac{d\mathfrak{N}}{dE} = \frac{R}{\hbar\pi v}$$

(8-4.19)

ISBN 0-8053-8001-7

Taking into account the statistical weights of the p states and the averaging over M_{if} required in Eq. (8-4.4), and combining Eqs. (8-4.15) and (8-4.19), we obtain for two K electrons

$$\lambda_e = \frac{14\pi}{\hbar} \, p_0^2 \, \frac{e^2 k^2}{3} \left(\frac{Z}{a_0} \right)^3 \frac{I^2}{\hbar v} \tag{8-4.20}$$

On the other hand λ_γ is given by

$$\lambda_\gamma = \frac{1}{3} \frac{p_0^2 \omega^3}{\hbar c^3} \tag{8-4.21}$$

according to Eq. (8-1.4) and hence the internal conversion coefficient is

$$\alpha = \frac{4\pi}{\hbar} \frac{k^2 e^2}{v} \left(\frac{Z}{a_0} \right)^3 \frac{c^3}{\omega^3} I^2 \tag{8-4.22}$$

A closed-form evaluation can be obtained in the special case of $a_0/Z \gg 1/k$, which means that the energy of the transition is very large compared with the electron binding energy. We shall assume, moreover, that the ejected electron is not relativistic. To be consistent, then, we assume for the electron that $mv^2/2 \cong (\hbar k)^2/2m \cong \hbar\omega$.

The integral I can be calculated by elementary means on the assumption that $e^{-Zr/a_0} = 1$, and we have

$$I = \int_0^\infty J_{3/2}(kr) \, \frac{dr}{(kr)^{1/2}} = \left(\frac{2}{\pi k^2} \right)^{1/2} \tag{8-4.23}$$

Replacing in Eq. (8-4.22) with the approximations mentioned above, we have

$$\alpha_K = \frac{8}{\hbar} \frac{e^2 m^{1/2}}{(2\hbar\omega)^{1/2}} \left(\frac{Z}{a_0} \right)^3 \frac{c^3}{\omega^3}$$

$$= \frac{1}{2} Z^3 \left(\frac{e^2}{\hbar c} \right)^4 \left(\frac{2mc^2}{\hbar\omega} \right)^{7/2} \tag{8-4.24}$$

This formula, valid under the hypothesis mentioned for dipole radiation, may be extended to El radiation, giving

$$\alpha_K^l = Z^3 \left(\frac{e^2}{\hbar c} \right)^4 \frac{l}{l+1} \left(\frac{2mc^2}{\hbar\omega} \right)^{l+(5/2)} \tag{8-4.25}$$

ISBN 0-8053-8601-7

The approximations used here are too crude to give valuable numerical results. However, it is possible to obtain good accuracy by employing relativistic wave functions and other necessary refinements. The extension to higher electric and magnetic multipoles becomes increasingly cumbersome. Typical numerical results are shown in Figs. 8-8, 8-9, and 8-10. Extensive tables of internal conversion coefficients are reported in the literature, for instance in (Se 59) (LHP 67), and (Si 65).

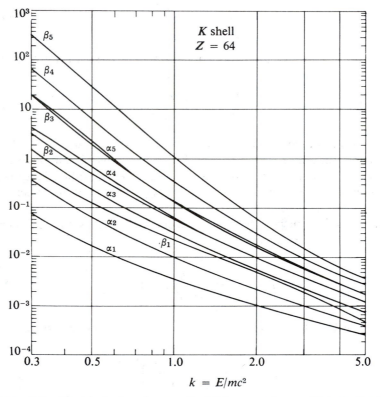

Figure 8-8   Electric (α_L) and magnetic (β_L) conversion coefficients for the K shell and for $Z = 64$. Energy scale gives $E/mc^2 = k$. [From (Si 65).]

The measurement of internal conversion coefficients is performed by counting the number of gamma rays, for instance, with a scintillation counter, and the number of conversion electrons with a Geiger–Müller counter or photographically, often with the help of beta spectrographs (Fig. 8-10). The measurement of the ratios of the conversion coefficients for the different X-ray levels $\alpha_K : \alpha_{L_I} : \alpha_{L_{II}}$, etc., can be made with a beta spectrometer without

ISBN 0-8053-8601-7

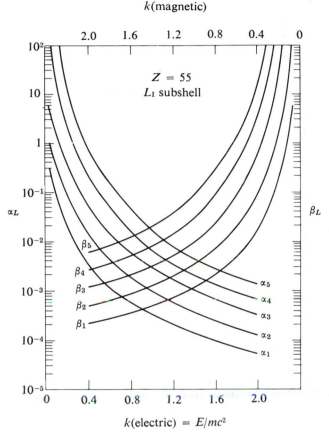

Figure 8-9 Electric and magnetic conversion coefficients for the L_I subshell and for $Z = 55$. For clarity the electric and magnetic curves have been separated by using the separate energy scales (in units of E/mc^2) given at the bottom and top of the figure. [From (Si 65).]

measuring gamma rays. The ratios by themselves give valuable information for classifying the type of radiation.

In some cases ($J = 0 \rightarrow J = 0$ transitions), gamma emission is forbidden in all orders and the emission of atomic electrons or electron–positron pairs is the deexcitation mechanism.

8-5 NUCLEAR ISOMERISM

The selection rules described in Sec. 8-2 can slow down electromagnetic transitions to such a point that the excited state has a very long mean life, "very long" meaning from 0.1 sec to years. In this case, the excited state is

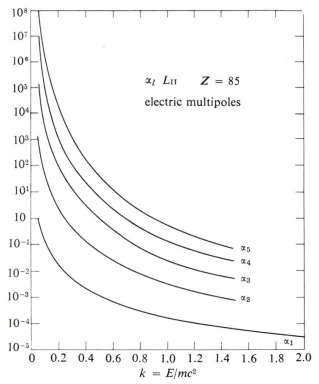

Figure 8-10   Electric conversion coefficients for the L_{II} subshell, $Z = 85$. Energy scale in E/mc^2. [From (Si 65).]

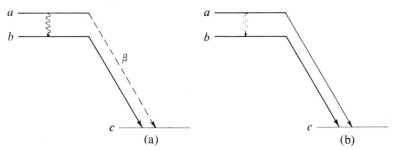

Figure 8-11   Types of isomerism. In (a) the upper level decays prevalently by gamma emission to the ground state, which decays by beta emission. In (b) upper and lower states decay independently by beta emission.

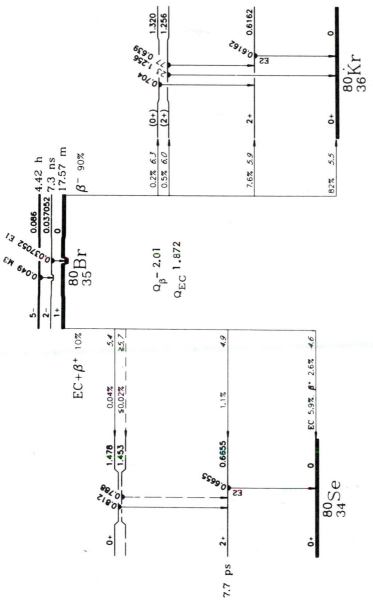

Figure 8-12 Energy-level diagram for ^{80}Br, illustrating isomeric transition. All energies in MeV.

called a *metastable* or an *isomeric* state in analogy with the chemical isomers. It is clear from the definition that the limit of 0.1 sec is completely arbitrary. There are gamma transitions of mean lives ranging from 10^{-16} to 10^{8} sec; thus the point at which one starts to call a state metastable is arbitrary.

The phenomenon of nuclear isomerism was discovered in ^{234}Pa by O. Hahn (1921). The explanation in terms of forbidden gamma transitions is due to von Weizsäcker (1936).

Often nuclear isomerism accompanies beta transitions, as indicated schematically in the typical level diagrams of Fig. 8-11. In Fig. 8-11a the isomeric transition between levels *a* and *b* is very probable compared with the beta transition between *a* and *c*. If $\lambda_{ia} \ll \lambda_{\beta b}$, and $\lambda_{ia} \gg \lambda_{\beta a}$, the substance

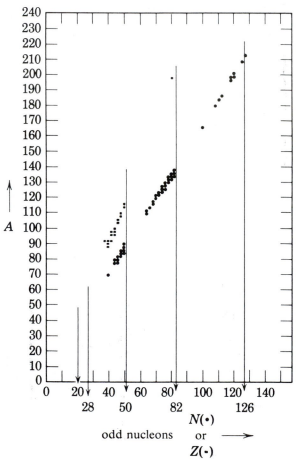

Figure 8-13 Distribution of long-lived isomers of odd mass number *A* plotted against the number of odd nucleons (*N* or *Z*). (Isomer islands.) [M. Goldhaber and J. Weneser, *Ann. Rev. Nucl. Sci.*, **5**, 1 (1955).]

ISBN 0-8053-8601-1

exhibits the beta spectrum typical of level b, with the decay constant λ_{ia}. ^{80}Br is an example, and its level diagram is illustrated in Fig. 8-12. In other cases (Fig. 8-11b) levels a and b decay as independent substances.

Isomeric transitions, being highly forbidden, must correspond to large ΔJ and small energy. Both circumstances favor high internal conversion, and this is another characteristic of isomeric transitions. In fact, the large internal conversion may be used to separate nuclear isomers chemically. In the case of ^{80}Br, for instance, the 4.4-h excited level emits a 49-keV gamma ray, which is highly converted and leaves in an inner shell a vacancy subsequently filled by an outer electron. This process continues until one of the valence electrons is used. If the ^{80}Br atom is bound in an organic compound, loss of the valence electron sets it free as a Br$^-$ ion, which can be chemically separated by precipitation of AgBr. It is thus possible to separate the nuclei that have undergone isomeric transition from the others.

To be metastable an excited level must differ from lower energy levels by three or more units of J; the radiation emitted is thus $E3$, $M3$, or of higher multipolarity. The condition mentioned is satisfied only for $A \geqslant 39$, and there are no isomers of the light elements. Even at higher A, nuclear isomeric states are not spread uniformly among all nuclei but are preferentially concentrated in "islands" of nuclei with Z or N just below the magic numbers 50, 82, and 126, and even A. Isomers with both N and Z even are very rare (Fig. 8-13).

These facts are accounted for satisfactorily by the shell model. First, in an even-even nucleus, the excitation of a nucleon involves the pairing energy, which is too large to allow the formation of isomers. Second, the islands of isomerism are explained by a study of Fig. 6-36. For low A up to 40 nucleons the orbits involved have $j \leqslant 5/2$, and there are no possibilities of large spin differences between energetically close orbits. Shortly before the numbers 50, 82, or 126, there are energetically neighboring orbits with $j = 1/2$, $9/2$; $1/2$, $11/2$; $1/2$, $13/2$; and these give rise to the "islands of isomers."

8-6 ANGULAR CORRELATIONS IN GAMMA EMISSION

Often, when two gamma rays are emitted in rapid succession by the same nucleus, one finds that the directions of emission of the two rays are correlated. This means that upon assuming the direction of the first gamma as the z axis, the probability of the second falling into the element of solid angle $d\omega$ at an angle θ with the z axis is not constant but depends on the angle θ. This type of correlation, already mentioned in Chap. 6, is not restricted only to γ–γ emission. The principles that we shall repeat here may be generalized to other cases.

Suppose that we have nuclei emitting light quanta through electric dipole transitions. We place the nuclei in a magnetic field, which orients them in such a way that the electric dipole is along the z axis. We know that no quanta will be emitted in the direction of the z axis and that the maximum probability of emission will be in the x–y plane. Conversely, in the absence of an orienting field, the fact that a gamma quantum is emitted in a certain direction tells us that the direction is *not* the direction of the electric dipole and makes it a priori more probable that the electric dipole is perpendicular to the direction of the first quantum. A second quantum is thus less likely to be emitted in a direction perpendicular to the first quantum than in any other. This argument can be made quantitative. The probability of finding the angle θ between successively emitted gamma rays is

$$P(\theta)\,d\omega = A(1 + \cos^2\theta)\,d\omega \qquad (8\text{-}6.1)$$

where $d\omega$ is the element of solid angle and

$$A = \frac{3}{16\pi} \qquad (8\text{-}6.2)$$

is a normalization constant.

The quantum-mechanical treatment of the correlation for more complicated cases shows that $P(\theta)$ is a polynomial of even degree in $\cos\theta$. This is apparent from the fact that $P(\theta)$ and $P(\pi - \theta)$ must be equal if parity is conserved, that is, if the correlation is not altered on reflection through the origin, as is the case for electromagnetic interactions. The degree of the polynomial in $\cos^2\theta$ and its coefficients depend on the spins of the three states involved and on the character of the two radiations connecting them.

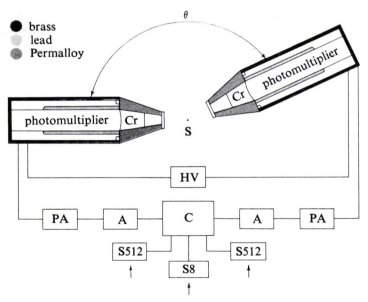

Figure 8-14 Apparatus for measuring angular correlations of gamma rays. PA, preamplifiers; A, amplifiers; C, coincidence circuit, 0.1-μsec resolution; S512, scale of 512; S8, scale of 8; Cr, crystals; S, source.

Angular correlations are not limited to γ–γ cascades. They are observed in beta emission followed by gamma emission, in gamma emission followed by emission of Auger electrons, and in other cases. The subject has become vast, but the formulas giving the correlations are similar in all cases. For a detailed treatment see (Si 65).

In Chap. 6 we discussed the effect of a magnetic field on angular correlations. A typical apparatus is illustrated in Fig. 8-14. Despite the apparent simplicity of the equipment, considerable care must be exercised in order to avoid the numerous sources of error, such as scattering of radiation from one counter into the other, finite solid angles, and the efficiency of counters. (See also Fig. 6-32).

8-7 COULOMB EXCITATION

Nuclear levels can be excited by bombarding nuclei with charged heavy particles such as protons or alpha particles. This method of study is experimentally preferable to the use of electrons where there are background difficulties from the bremsstrahlung. Keeping the energy of the heavy charged particles below the nuclear Coulomb barrier, we avoid specific nuclear reactions, and the backgrounds due to unwanted effects are small. Projectiles used have some MeV of energy; the levels excited, some tenths of MeV of energy. The electric field of the projectile induces transitions in the target, and the effect can be calculated semiclassically by considering the field produced by the projectile moving in its hyperbolic trajectory as a perturbation on the nucleus. The matrix elements involved are the multipole electric moments of the nucleus connecting the initial and final states. The calculation is rather involved, but the underlying physical concepts are simple. For the method of calculation to be valid, the orbit concept must be applicable; hence the de Broglie wavelength $\lambdabar$ of the projectile must be small compared with the distance of closest approach to the target. Now calling a half this distance, one has

$$a = \frac{Z_1 Z_2 e^2}{mv^2} \tag{8-7.1}$$

where Z_1, Z_2 are the atomic numbers of projectile and target and m, v are mass and velocity of the projectile. The condition $a/\lambdabar \gg 1$ is then

$$\eta \equiv \frac{a}{\lambdabar} = \frac{Z_1 Z_2 e^2}{\hbar v} \gg 1 \tag{8-7.2}$$

Furthermore, the collision must be nonadiabatic; otherwise no transition occurs. This means that the collision time a/v must be short compared with the

ISBN 0-8053-8601-7

nuclear periods τ to be excited, where

$$\tau = \frac{1}{\omega} = \frac{\hbar}{\Delta E} \tag{8-7.3}$$

and ΔE is the excitation energy of the target. If we indicate by E the kinetic energy of the projectile, the condition of nonadiabatic collision gives

$$\xi = \frac{a}{v} \frac{\Delta E}{\hbar} = \frac{Z_1 Z_2 e^2}{\hbar v} \frac{\Delta E}{2E} \ll 1 \tag{8-7.4}$$

When both conditions [Eqs. (8-7.2] and (8-7.4)] are satisfied, calculation gives for the total cross section for Coulomb excitation the approximate result for $\Delta E / E \ll 1$:

$$\sigma_{El} = \left(\frac{Z_1 e}{\hbar v} \right)^2 a^{-2l+2} B\left(El, J_i \rightarrow J_f \right) f_{El}\left(\xi \right) \tag{8-7.5}$$

and

$$\sigma_{Ml} = \left(\frac{Z_1 e}{\hbar c} \right)^2 a^{-2l+2} B\left(Ml, J_i \rightarrow J_f \right) f_{Ml}\left(\xi \right)$$

Note that the transition is characterized by its multipolarity and that the matrix element in B is the same as the one entering into the spontaneous emission formula (8-3.11). The function $f(\xi)$ takes into account details of the orbit and is tabulated. For

$$\xi \ll 1 \qquad f_{E2} = 1 \tag{8-7.6}$$

One of the most important cases of Coulomb excitation is that of $E2$ excitation of rotational levels 0^+, 2^+, 4^+, etc., in even-even nuclei with

$$\Delta J = 2 \tag{8-7.7}$$

The matrix elements derived from experiment are sometimes many times larger than can be accounted for by single-particle models and are due to collective motions. They are connected with the intrinsic quadrupole moment Q_0 of such nuclei; e.g., for a $0^+ \rightarrow 2^+$ transition,

$$B(E2) = \frac{5}{16\pi} Q_0^2 \tag{8-7.8}$$

We see here a connection through the intrinsic quadrupole moment between the moment of inertia, the transition probability, and the observable electric quadrupole moment (cf. Sec. 6-12).

The intensities of $E2$ $0^+ \rightarrow 2^+$ rotational transitions can be calculated from the static nuclear quadrupole moments. In some heavy nuclei, notably ^{238}U and ^{230}Th, one finds that the intensity of the $E4$ $0^+ \rightarrow 4^+$ transition is detectable. This indicates a 16-pole permanent moment of the ground state. This moment is of the order of $1 \cdot eb^4$. It shows up also in alpha decay where it accounts for the low hindrance of some $\Delta J = 4$ transitions and in α inelastic scattering at about 40 MeV.

8-8 NUCLEAR FLUORESCENCE

In atomic physics, resonance and fluorescence radiation are important and easily observed phenomena. We would also expect to observe resonance radiation from nuclei; however, many early attempts to detect it failed. The reason for this is that the nuclear absorption lines are very narrow and thus absorb very little radiation from a continuum. On the other hand, if one tries to excite resonance radiation by the corresponding emission light as in the optical case, the recoil of the emitting nucleus shifts the light out of resonance with the absorber. Let us first consider a strictly monochromatic line. If the initial excited state has an energy E above the ground states in the laboratory system, we have, by the principles of conservation of energy and momentum,

$$\hbar\omega + \frac{p^2}{2m} = E \qquad (8\text{-}8.1)$$

with

$$p = \hbar\omega/c \qquad (8\text{-}8.2)$$

where ω is the frequency of the light emitted and p and m are the recoil momentum and the mass of the nucleus supposed initially at rest. These relations give approximately

$$\hbar\omega = E\left(1 - \frac{E}{2mc^2}\right) \qquad (8\text{-}8.3)$$

On the other hand, for absorption we need

$$\hbar\omega = E\left(1 + \frac{E}{2mc^2}\right) \qquad (8\text{-}8.4)$$

to conserve energy and momentum. We see that, to have resonance, we must multiply the frequency of emission by

$$\sim \left(1 + \frac{E}{mc^2}\right) \qquad (8\text{-}8.5)$$

ISBN 0-8053-8601-7

or

$$1 + \frac{\hbar\omega}{mc^2} \tag{8-8.6}$$

to find the necessary absorption frequency.

However, spectral lines have a natural width $\delta\omega$ associated with the mean life τ of the excited state by

$$\delta\omega = 1/\tau \tag{8-8.7}$$

If this width is large compared to $\hbar\omega^2/mc^2$, the emission line will overlap the absorption line sufficiently to produce resonance radiation. This is the usual case in visible light. If, however, the natural width is insufficient to produce the desired overlap, as is the case in nuclear gamma rays, one can still modify the frequency, as seen by the absorber, with the help of the Doppler effect. If the source moves toward the absorber with velocity

$$v = \hbar\omega/mc \tag{8-8.8}$$

the Doppler shift compensates the recoil effects and resonance is obtained. For instance, in the case of ^{198}Hg, $\hbar\omega = 0.41$ MeV, and v must be equal to 0.67×10^5 cm sec^{-1}, which is attainable by mechanical means or by thermal agitation in a hot vapor. Actually nuclear resonance radiation has been observed (Moon, 1951) by using a source of radioactive gold 198 in rapid motion with respect to a mercury resonator. Recoil from a previous nuclear decay or reaction may also impart the necessary velocity to the nucleus. The study of the intensity of the resonance radiation as a function of velocity of the source can be made to yield information on τ, which, in the case of ^{198}Hg, is of the order of 10^{-11} sec.

Resonance absorption has been demonstrated in ^{191}Ir by Mössbauer (1958) by a different system. He used a ^{191}Os source and ^{191}Ir as the absorber, both cooled to a low temperature. The source decays by beta emission to an excited state at 129 keV of the stable ^{191}Ir. The half-life of this state is 1.3×10^{-10} sec and its natural width $(\delta\omega/\omega)_{nat} = 4 \times 10^{-11}$. The recoil energy of the free nucleus would be 0.047 eV, corresponding to a Doppler shift $(\Delta\omega/\omega)_{Doppler} = 3.6 \times 10^{-7}$. However, when the nucleus is bound in a crystal its motion may be crudely compared to that of an oscillator. If the recoil energy is insufficient to raise the oscillator from its ground state to the first excited state, no energy can be transferred to the crystal lattice, and we have recoilless emission as if in Eqs. (8-8.3 and 4) the mass were infinite and emission and absorption frequency coincided.

To understand some of the features of the Mössbauer effect, consider an oversimplified model of a nucleus of mass M bound in a harmonic oscillator

ISBN 0-8053-8601-7

potential of frequency $\nu = \omega/2\pi$. The hamiltonian of the system is

$$H = H_{nucl} + \frac{p^2}{2M} + \frac{M\omega^2}{2} x^2 \tag{8-8.9}$$

The eigenvalues are then

$$E_{a,n} = \epsilon_a + \hbar\omega(n + \tfrac{1}{2}) \tag{8-8.10}$$

where ϵ_a are the nuclear energy levels and n is an integral number. The eigenfunctions are $\psi_a(\mathbf{r})u_n(x)$ where the first factor is nuclear and the second refers to the harmonic oscillator. The energy of the gamma emitted in the transition $a \to b$ is

$$E_\gamma = \epsilon_a - \epsilon_b + \hbar\omega(n' - n'') \tag{8-8.11}$$

On the emission of the gamma ray the oscillator momentum changes abruptly, because of recoil, by the amount

$$q = E_\gamma/c \cong (\epsilon_a - \epsilon_b)/c \tag{8-8.12}$$

Assume initially the oscillator is in its ground state, eigenfunction $u_0(x)$, and write the transform of this eigenfunction in momentum space:

$$v_0(p) = (2\pi\hbar)^{-1/2} \int u_0(x) \exp(ipx/\hbar)\, dx \tag{8-8.13}$$

The recoil at the gamma emission changes all momenta in the oscillator by q and the eigenfunctions after the emission is, in momentum space, $v_0(p - q)$; from this function in momentum space we obtain the function in coordinate space immediately after the emission:

$$u'(x) = (2\pi\hbar)^{-1/2} \int v_0(p - q) \exp(-ipx/\hbar)\, dp = u_0(x)\exp(-iqx/\hbar) \tag{8-8.14}$$

and the probability of finding the oscillator in the ground state is, according to the principles of quantum mechanics (Fe 61, pp. 40–43),

$$P_{00} = \left| \int u_0(x)u_0^*(x) \exp(iqx/\hbar)\, dx \right|^2 \tag{8-8.15}$$

where the subscripts on P denote n' and n''.

ISBN 0-8053-8601-7

Upon inserting the harmonic oscillator wave function in Eq. (8-8.15), one obtains

$$P_{00} = \exp(-q^2/2M\hbar\omega) \tag{8-8.16}$$

One can calculate P_{nn} or P_{nm} similarly.

Equation (8-8.16) shows that recoilless emission is likely only if the recoil energy $q^2/2M$ is small compared with $\hbar\omega$, the energy corresponding to the excitation jumps in the oscillator.

One may describe a crystal as an ensemble of oscillators of the same frequency ω_E, as Einstein did in his theory of specific heat.

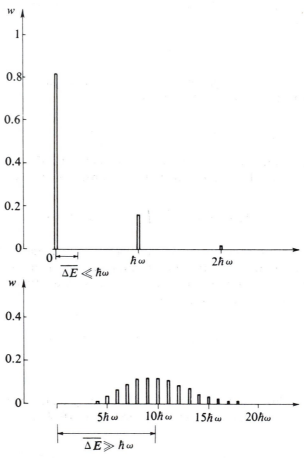

Figure 8-15 Relative probability for a gamma-ray transition simultaneous with the excitation of 1, 2, ..., n oscillators in the crystal lattice for the Einstein model. The figures refer to two values of the ratio between the average recoil energy $\Delta E = E_0^2/2Mc^2$ and $\hbar\omega_E$, $T = 0°K$. [After R. L. Mössbauer, *Les Prix Nobel en 1961*, Stockholm, 1962.]

ISBN 0-8053-8601-7

For $T \neq 0°$K the probability of finding an oscillator in the nth energy state is given by the Boltzmann factor $\exp(-\beta E_n)/\Sigma_n \exp(-\beta E_n)$ with $\beta = 1/kT$ and the probability of recoilless emission is

$$f = \Sigma_n \exp(-\beta E_n)P_{nn}/\Sigma_n \exp(-\beta E_n) \qquad (8\text{-}8.17)$$

Figure 8-15 shows the probability of finding n oscillators excited after the transition for $T = 0°$K.

The actual spectrum of the lattice vibrations is much more complicated than in the Einstein schematization and as a better approximation one uses a Debye spectrum, replacing $\hbar\omega_E$ by $k\Theta$, where Θ is the Debye temperature. The development of these ideas in a quantitative form gives a formula for the probability f of recoilless emission:

$$f = \exp\left\{ -\frac{3}{2} \frac{\hbar^2\omega^2/2mc^2}{k\Theta} \left[1 + 4\left(\frac{T}{\Theta}\right)^2 \int_0^{\Theta/T} \frac{x\,dx}{e^x - 1} \right] \right\}$$

$$\cong \exp\left\{ -\frac{3}{2} \frac{\hbar^2\omega^2/2mc^2}{k\Theta} \left[1 + \frac{2}{3}\left(\frac{\pi T}{\Theta}\right)^2 \right] \right\} \qquad (T \ll \Theta) \quad (8\text{-}8.18)$$

The first term is independent of the temperature and shows that, even at absolute zero, the fraction of recoilless decays is large only if the recoil energy of the free nucleus is small compared with $k\Theta$. The probability of recoilless decay decreases with increasing temperature and it is negligible for temperatures large compared with Θ (Fig. 8-16). Similar effects occur in the absorption process.

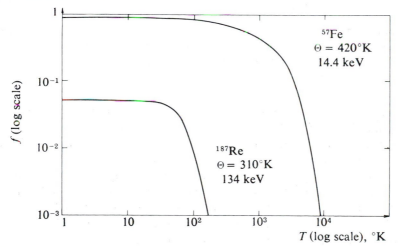

Figure 8-16 Fraction of recoilless transitions in iron or rhenium as a function of the temperature. [R. L. Mössbauer, *Ann. Rev. Nucl. Sci.*, **12**, 123 (1962).]

The experiments usually involve a source and an absorber of the same substance and the measurement of the amount of radiation absorbed. If one moves the absorber with respect to the source, the frequency of the absorption line changes by an amount

$$\Delta\omega/\omega = v/c \qquad (8\text{-}8.19)$$

and less radiation is absorbed, because the overlap of the emission and absorption curves is less complete (Figs. 8-17 and 8-18). It is thus possible to analyze the structure, or form, of a line by using "Doppler spectrometry."

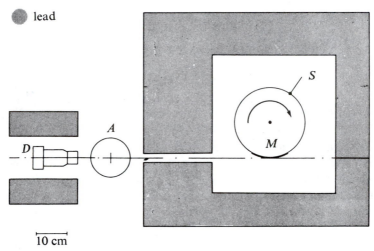

lead

10 cm

Figure 8-17 Experimental arrangement: A, cryostat of absorber; S, rotating cryostat with source; D, scintillation detector; M, region in which the source is seen from D. [R. L. Mössbauer, *Naturwiss.*, **45**, 538 (1958).]

A quantitative illustration of the form of an absorption line as a function of the temperature of the lattice is given in Fig. 8-19. For an emission line one would have the same figure, but it would be reflected on the spike of the line, corresponding to the frequency of the recoilless transition.

The absorption of the radiation is naturally accompanied by the emission of fluorescent radiation from the absorbing nucleus and by the related emission of conversion electrons, etc. One of the examples of recoilless radiation most often studied is a line of 14.4-keV energy emitted by ^{57}Fe in an $M1$ transition. It has a mean life of 10^{-7} sec and hence

$$\left(\frac{\Delta\omega}{\omega}\right)_{\text{nat}} \sim 3 \times 10^{-13} \qquad (8\text{-}8.20)$$

ISBN 0-8053-8601-7

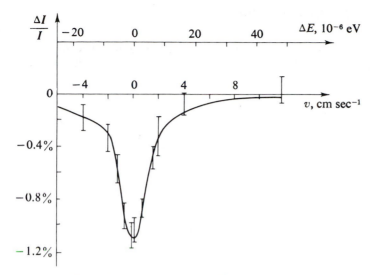

Figure 8-18 Fluorescent absorption in ^{191}Ir as a function of the relative velocity between source and absorber. The upper scale on the abscissa shows the Doppler energy that corresponds to the velocity on the lower scale. T = 88°K. [R. L. Mössbauer, *Naturwiss.*, **45**, 538 (1958).]

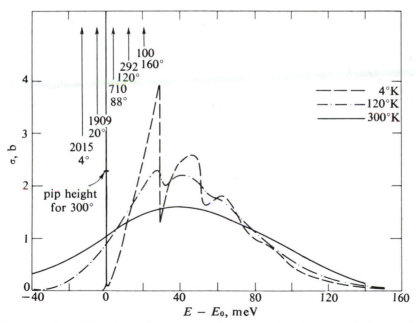

Figure 8-19 The absorption cross section per nucleus in a crystal of natural iridium, for a monochromatic gamma-ray beam and phonons having a Debye spectrum. The arrows give the cross section at zero relative velocity for the temperatures indicated. [W. M. Visscher, *Ann. of Phys.* **9**, 194 (1960).]

ISBN 0-8053-8601-7

The Doppler effect corresponding to this width is achieved with a velocity of about 0.01 cm sec^{-1} (Fig. 8-20); still sharper is a 13.3 keV line in ^{73}Ge.

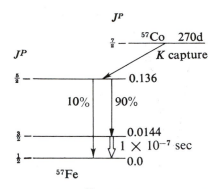

Figure 8-20   Decay scheme of ^{57}Fe. The data to the left of the levels are their spins and parities, those to the right their energies. The times shown are the half-lives.

Such an unprecedented degree of monochromaticity has already permitted the observation of the nuclear Zeeman effect (Fig. 8-21) and it is an important tool in several fields of physics (e.g., gravitational red shift and solid-state investigations; Fig. 8-22).

An elegant application of Mössbauer spectroscopy is found in the so-called isomeric effect. Nuclei in an excited state have a slightly different radius than in the ground state. The electronic cloud surrounding the nucleus produces an energy shift, with respect to the energy for a point nucleus, given by

$$\Delta W = (2/5)\pi Ze^2 r_n^2 |\psi(0)|^2 \tag{8-8.21}$$

where r_n is the nuclear radius. If the Mössbauer emitter and absorber are immersed in different materials, giving $\psi_e(0)$ for the emitter and $\psi_a(0)$ for the absorber, and if the nuclear radii for the excited and ground states are r_n^* and r_n, respectively, one finds a difference between the energy of the line absorbed and emitted, and can be measured by the Doppler shift. (See Fig. 8-23.) The shift is

$$\Delta E = (2/5)\pi Ze^2 \left(r_n^{*2} - r_n^2\right)\left(|\psi_a(0)|^2 - |\psi_e(0)|^2\right) \tag{8-8.22}$$

ISBN 0-8053-8061-7

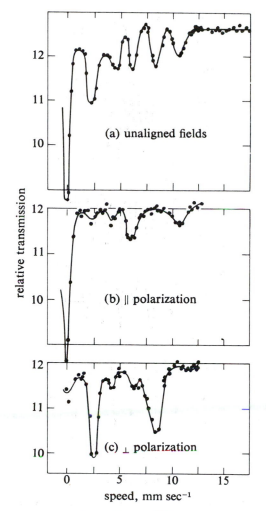

Figure 8-21 Fluorescent absorption in ^{57}Fe showing hyperfine structure. Plotted is the observed relative transmission as a function of the velocity between source and absorber. (a) Randomly oriented magnetic field in source and absorber. (b) Source and absorber fields aligned parallel to one another. (c) Source and absorber fields aligned perpendicular to one another. [S. S. Hanna, J. Heberle, C. Littlejohn, G. J. Perlow, R. S. Preston, and D. H. Vincent, *Phys. Rev. Letters*, **4**, 177 (1960).]

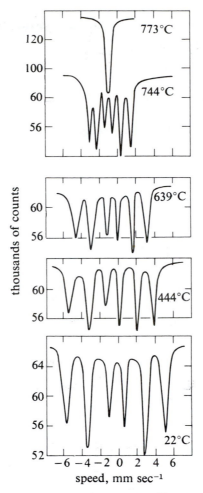

Figure 8-22 Absorption spectra for metallic ^{57}Fe between room temperature and the Curie temperature. The single-line source (^{57}Co in Cu) was at the temperature of liquid nitrogen. [R. S. Preston, S. S. Hanna, and J. Heberle, *Phys. Rev.*, **128**, 2207 (1962).]

ISBN 0-8053-8601-7

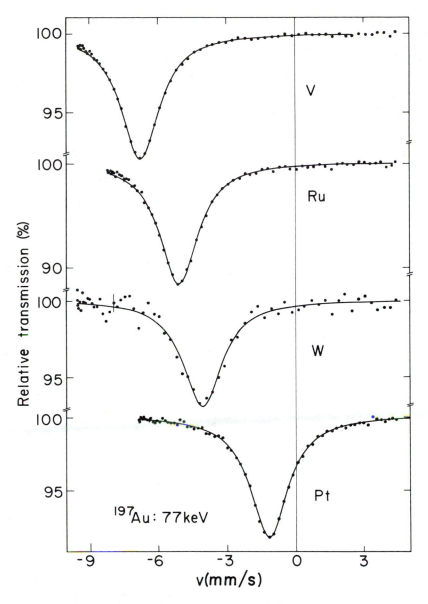

Figure 8-23 Mössbauer absorption spectra of the 77-keV gamma rays of
^{197}Au, measured at 4.2°K with sources of dilute impurities of ^{197}Pt in V, Ru,
W, and Pt, using a single-line gold metal absorber. [G. Kaindl and D. Salomon
LBL 1666 see also *Phys. Rev. B*, **8**, 1912 (1973).]

BIBLIOGRAPHY

Ajzenberg-Selove, F. (AS 60).

Alburger, D. E., "Nuclear Isomerism," in (Fl E), Vol. 42.

Alder, K., A. Bohr, T. Huus, B. R. Mottelson, and A. Winther, "Study of Nuclear Structure by Electromagnetic Excitation with Accelerated Ions," *Rev. Mod. Phys.*, **28**, 432 (1956).

B. Arad and G. Ben David, "Nuclear Studies with Neutron Capture γ-rays", *Rev. Mod. Phys.* **45**, 230 (1973).

Blatt, J. M., and V. Weisskopf (BW 52).

Burhop, E. H., *The Auger Effect and Other Radiationless Transitions*, Cambridge University Press, New York, 1952.

Church, E. L., and J. Weneser, "Nuclear Structure Effects in Internal Conversion," *Ann. Rev. Nucl. Sci.*, **10**, 193 (1960).

Deutsch, M., and O. Kofoed-Hansen, "Gamma-Rays," in (Se 59), Vol. III.

Devons, S., and J. B. Goldfarb, "Angular Correlations," in (Fl E), Vol. 42.

Frauenfelder, H., "Angular Correlation of Nuclear Radiation," *Ann. Rev. Nucl. Sci.*, **2**, 129 (1952).

Frauenfelder, H., *The Mössbauer Effect*, Benjamin, New York, 1962.

Goldhaber, M., and J. Weneser, "Electromagnetic Transitions in Nuclei," *Ann. Rev. Nucl. Sci.*, **5**, 1 (1955).

Greenwood, N. N. and T. C. Gibb, *Mössbauer Spectroscopy*, Harper & Row, New York, 1971.

Jackson, J. D. (Ja 75).

Mayer, M., and H. D. Jensen (MJ 55).

Mössbauer, R. L., "Recoilless Nuclear Resonance Absorption," *Ann. Rev. Nucl. Sci.*, **12**, 123 (1962).

Preston, M. A. (P 62).

Raghavan, R. S., and L. Pfeiffer, "Observation of the High Resolution Mössbauer Resonance in ^{73}Ge," *Phys. Rev. Letters*, **32**, 512 (1974).

Rose, H. J., and D. M. Brink, "Angular Distribution of Gamma Rays in Terms of Phase-Defined Reduced Matrix Elements," *Rev. Mod. Phys.*, **39**, 306 (1967).

Rose, M. E., *Internal Conversion Coefficients*, North-Holland, Amsterdam, 1959.

Rose, M. E., *Multipole Fields*, Wiley, New York, 1955.

Siegbahn, K. (Si 65).

Stelson, P. H., and F. K. McGowan, "Coulomb Excitation," *Ann. Rev. Nucl. Sci.*, **13**, 163 (1963).

Yoshida, S., and L. Zamick, "Electromagnetic Transitions and Moments in Nuclei," *Ann. Rev. Nucl. Sci.*, **22**, 121 (1972).

PROBLEMS

8-1 Calculate from Eq. (8-1.5) the mean life of a nuclear excited state that may decay by electric dipole radiation. Develop a numerical example and compare it with atomic radiation.

8-2 Draw the figures corresponding to Fig. 8-4 for the electric quadrupole indicated in Fig. 8-3.

8-3 An almost spherical surface defined by $R(\theta) = R_0[1 + \beta P_2(\cos \theta)]$ has inside it a total charge Q uniformly distributed. The small parameter β varies harmonically in time with frequency ω. Keeping only lowest-order terms in β and making the long-wavelength approximation, calculate the nonvanishing multipole moments, the angular distribution of radiation, and the power radiated [From (Ja 75); compare Sec. 6-12 for collective nuclear motions.]

ISBN 0-805-8601-1

8-4 Suppose that part of the electromagnetic radiation emitted by a nucleus arises from an intrinsic magnetization:

$$\mathbf{M}(\mathbf{r}', t) = \mathbf{M}_\omega(\mathbf{r}')e^{-i\omega t} + \text{complex conjugate}$$

(a) Obtain the general formula for the total radiation rate from such a system
(b) Show that the electric dipole contribution from this magnetization is zero.

8-5 Let us assume that we can crudely represent a nucleus as two oscillating currents in opposite directions, as shown in the figure below.

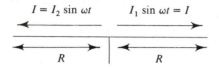

(a) If $I_1 = 2(e\omega/c)$, $I_2 = e\omega/c$ determine the multipole nature of the radiation, and obtain expressions for the angular distribution and the total decay rate. (b) Do the same for $I_1 = I_2 = e\omega/c$. (c) In part (b) let $R = 5 \times 10^{-13}$ cm, $\hbar\omega = 1$ MeV, and determine the mean life of the excited state of the nucleus.

8-6 $^{60}_{28}$Ni has an excited state that decays to a lower excited state and then to the ground state in two successive electric quadrupole transitions. On the basis of this information, what are the possible spin and parity assignments for these two excited states of the Ni nucleus?

8-7 The lithium nucleus ^{7}Li emits a 0.48-MeV gamma ray in a transition that goes from an excited state with angular momentum $\frac{1}{2}$, odd parity, to the ground state with angular momentum $\frac{3}{2}$, odd parity. (a) What are the possible choices for the multipolarity and nature of the emitted radiation? (b) Of these possibilities which one is likely to be the principal contributor to the transition? (c) Estimate the lifetime of the excited state.

8-8 Using the formulas

$$\left(\frac{8\pi}{3}\right)^{1/2} Y_1^1 Y_l^{m-1} = \left[\frac{(l+m)(l+m+1)}{(2l+1)(2l+3)}\right]^{1/2} Y_{l+1}^m$$

$$- \left[\frac{(l-m)(l-m+1)}{(2l+1)(2l-1)}\right]^{1/2} Y_{l-1}^m$$

$$\left(\frac{4\pi}{3}\right)^{1/2} Y_1^0 Y_l^m = \left[\frac{(l+1)^2 - m^2}{(2l+1)(2l+3)}\right]^{1/2} Y_{l+1}^m$$

$$+ \left[\frac{l^2 - m^2}{(2l+1)(2l-1)}\right]^{1/2} Y_{l-1}^m$$

$$\left(\frac{8\pi}{3}\right)^{1/2} Y_1^{-1} Y_l^{m+1} = \left[\frac{(l-m)(l-m+1)}{(2l+1)(2l+3)}\right]^{1/2} Y_{l+1}^m$$

$$- \left[\frac{(l+m)(l+m+1)}{(2l+1)(2l-1)}\right]^{1/2} Y_{l-1}^m$$

calculate explicitly the integral (8-2.6) entering $E1$ radiation. Prove the formulas given above from the definition of $Y_l^m(\theta, \varphi)$ and properties of $P_l(x)$ as given in (Sc 68).

8-9   Discuss the conditions of energy and quantum numbers under which a nucleus can decay by electron–positron emission.

8-10 ^{89}Y has an excited level 0.915 MeV above the ground state; it decays to the ground state with a half-life of about 16 sec. The initial state has spin $\frac{9}{2}$, the final state spin $\frac{1}{2}$, and there is a parity change in the transition. (a) What is the lowest multipole order which can contribute? Calculate the expected rate and compare it with the experimental result. (b) Suppose that we wanted to check whether, in the ground state, there was any mixture of other angular momenta than $\frac{1}{2}$ (i.e., check angular momentum conservation). Set a rough upper limit to the amplitude of an angular momentum $= \frac{3}{2}$ component in the final state, using as experimental data only the measured lifetime and transition energy. Assume strict parity conservation.

8-11   Plan an experiment to measure the internal conversion coefficient for the lines of ^{80}Br.

8-12   In a mu-mesic atom the transition from the $2p$ to the $1s$ mesic orbit may occur by radiation or by an Auger process, in which an electron of the atom is emitted. Estimate the ratio of the probability of the second process with respect to the first one. Note the analogy to internal conversion.

8-13   Draw a diagram illustrating why the correlation between two γ emissions can depend only on even powers of $\cos \theta$.

8-14   Three states having spins J_1, J_2, J_3 are connected by two gamma rays γ_1, γ_2 having an electric or magnetic multipolarity l_1, l_2. Show (or make a plausible argument) that the angular correlation between them does not depend on the energy of γ_1 and γ_2 or on their electric or magnetic character. Moreover, the cascade $J_1 \xrightarrow{l_1} J_2 \xrightarrow{l_2} J_3$ gives the same correlation as the cascade $J_3 \xrightarrow{l_2} J_2 \xrightarrow{l_1} J_1$. If $J_2 = 0$, there is no correlation.

8-15   Using protons or alpha particles, one produces Coulomb excitation in a thin target. Adjust the velocities of the projectiles so that they have the same ξ in both cases. What is the ratio of the cross sections and how does it depend on the electric multipole order?

8-16   Consider the resonance absorption of the 0.014-MeV gamma ray from ^{57}Fe. (*a*) If there were no Mössbauer effect, and resonance absorption were to be observed by moving the source with respect to the absorber, how large would the relative speed have to be? How well controlled would it have to be (that is, what speed variations would destroy the effect?) (*b*) Considering the Mössbauer effect, suppose that the excited state has a magnetic moment of 0.5 nuclear magneton, the ground state has none, and the nucleus is in a field of 10^5 G. For what values of the relative velocity of emitter and absorber will absorption peaks be observed? (*c*) Would you expect the 0.136-MeV gamma ray from the higher state of ^{57}Fe to give rise to recoil-free emission? If not, why not?

8-17   Prove Eqs. (8-8.21) and (8-8.22).

8-18   Prove Eq. (8-8.16).

ISBN 0-8053-8061-7

CHAPTER

Beta Decay

Beta decay is a *nuclear* transformation accompanied by the emission of an electron. The same name has been applied to other types of transformations, notably the emission of positrons or the capture by the nucleus of orbital electrons. During nuclear transformations, electrons may be emitted by an *atom* for different reasons, for instance, as conversion electrons in a gamma transition. Beta decay does not refer to such secondary emission of orbital electrons.

9-1 INTRODUCTION

In the early studies of radioactivity there was considerable confusion as to the origin of the electrons observed in nuclear transformations. A clear distinction between nuclear electrons, or beta rays, and conversion electrons was made in 1919, when Chadwick showed that, besides the monoenergetic line spectrum of conversion electrons, there was also a continuous spectrum of disintegration electrons. This continuous spectrum at once presented a serious difficulty: the decay is a transition between two definite states, yet the kinetic energy of the electron is not always the same. To uphold the principle of energy conservation, it is necessary to account for the energy which does not

Emilio Segrè, Nuclei and Particles: An Introduction to Nuclear and Subnuclear Physics, Second Edition

ISBN 0-8053-8601-7

appear as kinetic energy of the electrons. Many hypotheses were formulated
to account for this missing energy; for instance, it was suggested that an
associated gamma emission carried off the energy missing in the beta
spectrum. Careful investigations, with calorimeters that should have observed
any known radiation, failed to reveal any energy besides that visible in the
continuous beta spectrum.

It was also found that the beta spectrum has a definite upper energy limit
corresponding to the total energy available in the disintegration. For instance,
if we consider the transformation of ^{3}H into ^{3}He, the mass spectrographic
data for the nuclear masses of the two substances shows us that the reaction

$$^3H = {}^3He + e^- + Q \tag{9-1.1}$$

balances exactly if we make Q the upper limit of the kinetic energy of the
electrons of the continuum. Similarly, there are several cases of substances
that decay by alternative branches, such as an alpha emission followed by a
beta or a beta emission followed by an alpha. The two branches start from
the same nucleus and arrive at the same nucleus; hence, the energy liberated
in either branch must be the same. The sum of the energy of the alpha
particles plus the upper limit of the beta energy, not the energy of the single
beta rays, is equal in the two branches of the disintegration diagram (see Fig.
9-1).

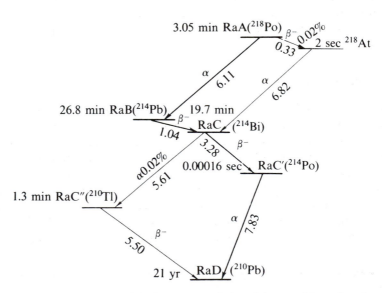

Figure 9-1 Cycles in which the energy of the alpha particles, plus that of
the upper limit of the beta spectrum, balance, along two different paths. Decay
scheme of the radium active deposit. Details of the complex radiations are not
shown. The total disintegration energy of each mode of decay is given in MeV.

Beta decay presents other grave difficulties as well, because it apparently violates not only the principle of conservation of energy but also the principle of conservation of angular momentum and the rule for the statistics of a composite system. For example, if the beta decay were

$$^3H \rightarrow {}^3He + e^-$$

the angular momentum of ^3H is $\frac{1}{2}$ and the angular momentum on the right hand side is 0 or 1 from the spins, plus an integer from a possible orbital angular momentum; but in any case the total is an integer. As far as the statistics are concerned, ^3H is a fermion and the system $^3He + e^-$ is a boson.

All these difficulties are eliminated by assuming the existence of an additional undetected particle which accompanies the electron in beta decay and carries off an amount of energy equal to the difference between the observed energy for a given electron and the maximum energy of the beta spectrum. This new particle, postulated by Pauli as early as 1930, is the neutrino.

To preserve not only the principle of energy conservation but also the principle of conservation of electric charge and of angular momentum and the rules governing statistics, we must ascribe certain properties to the neutrino. First of all, conservation of charge requires that the neutrino be electrically neutral; also, if the particle were not neutral it would leave a detectable ionization. This last consideration ensures that the particle's magnetic moment, if present, must also be extremely small (actually $< 10^{-6} \mu_N$). The balance of energy in the case of the $^3H \rightarrow {}^3He$ transformation shows that the neutrino rest mass is less than 500 eV. A limit of 60 eV is derived from the shape of the beta spectra. In fact, the neutrino's rest mass is, in all likelihood, exactly 0, as will be shown by arguments to be discussed later. The balance of angular momenta in beta decay requires a spin $\frac{1}{2}$ for the neutrino; and in order to satisfy the statistical requirements, the neutrino must be a fermion. This can be seen from the fact that ^3H and ^3He are both fermions and transform into each other by the emission of an electron, which is a fermion, and a neutrino, which therefore must also be a fermion. In beta decay it is possible to measure neutrino momentum by observing the nuclear recoil from the combined neutrino and electron emission, and the result is consistent with $m_\nu = 0$.

All these properties of the neutrino were ascribed to it in order to preserve the validity of some of the fundamental conservation laws of physics. In addition, a particle with these properties can be made to account quantitatively for the phenomena of beta decay (Fermi, 1934).

At this point we shall anticipate some results, the importance of which will be shown later. A neutrino is a fermion of charge 0 and as such has an antiparticle that may or may not be identical to it. The particle–antiparticle relation assigns to the antineutrino the same mass, spin, charge, and magnitude of magnetic moment as those of the neutrino. Moreover, in all processes where a neutrino is emitted, an antineutrino can be absorbed with a physically possible result; for example, $n \rightarrow p + e^- + \bar{\nu}$ (beta decay) becomes $n + \nu \rightarrow p + e^-$, a possible result in which a neutrino of suitable energy colliding with a neutron generates a proton and an electron. By definition the antineutrino is the particle emitted in beta decay, for instance, when a neutron becomes a proton; the neutrino is the particle emitted in positron decay or in orbital electron capture.

At first sight it might appear that the neutrino and the antineutrino are strictly indistinguishable. The question whether neutrino and antineutrino are the same particle or not can be solved only by experiment. Experiments to be described later answer this question in a startling way. The spin of the neutrino is always antiparallel to its momentum; the spin of the antineutrino is parallel to its momentum. This property gives to the neutrino a "handedness" (in the sense that a screw has), introducing a fundamental asymmetry in its behavior. It is clear, then, that neutrino and antineutrino are different.

The handedness of the neutrino is best described by using the concept of helicity. Helicity is the scalar product of the spin and the momentum divided by the product of the modulus of these quantities:

$$\mathcal{H} = \text{helicity} = \frac{\mathbf{p} \cdot \boldsymbol{\sigma}}{|\mathbf{p}||\boldsymbol{\sigma}|} \tag{9-1.2}$$

For the neutrino the helicity has the value -1, which means that its spin vector is antiparallel to its momentum direction. To compare it to a screw, it has the sense of a left-handed screw (ordinary screws are right-handed). The antineutrino has helicity $+1$ and may be compared to an ordinary (right-handed) screw.

If we assign a fixed helicity to the neutrino and to the antineutrino, we must also assume that the rest mass of both is exactly 0; thus they always move with velocity c, and it is impossible to overtake them by transformation to a faster-moving frame. If this were not so, a neutrino would have opposite helicity in systems moving with a velocity greater or smaller than that of the neutrino itself, contrary to the hypothesis of fixed intrinsic helicity.

If we assume the hypotheses of (1) mass 0 for the neutrino and antineutrino and (2) neutrino and antineutrino helicities of -1 and $+1$, respectively, it is possible to develop a two-component theory of the neutrino (Weyl, 1929; Lee and Yang; Landau; A. Salam, 1957) which is based on a degeneracy of Dirac's theory for a spin $\frac{1}{2}$ particle occurring when the rest mass of the particle is 0. In the ordinary Dirac theory there are four components, corresponding to two possible spin orientations of positive or negative energy (or particle and antiparticle). According to the two-component theory, half the states are suppressed, and the particle is associated with one helicity ($\mathcal{H} = -1$), the antiparticle with the other ($\mathcal{H} = +1$). This theory describes a particle having the two properties listed, and in many respects it is the simplest of the theories that account for the known phenomena of beta decay. It requires nonconservation of parity, as will be explained, and was rejected until this fact was experimentally demonstrated.

9-2 EXPERIMENTS ON THE NEUTRINO

The neutrino is the most elusive of the nuclear particles, and most of the evidence concerning it is indirect. We have already mentioned how its energy appears in beta decay as a deficit in the electron kinetic energy. Associated with this energy is a momentum that can be determined by measuring the vector momentum of the decay electron and the vector momentum of the recoil nucleus. Their sum must be equal and opposite to the vector momentum of the undetectable neutrino. On the assumption of 0 rest mass for the neutrino, as indicated above, the momentum of the neutrino is E_ν/c and is calculable from the energy of the other two particles. The main difficulty in experiments of this type is that the momentum of the recoil nucleus in beta decay is of the order of 1 MeV/c, and hence the recoil energy, even in favorable cases, is only about 100 eV. This makes accurate measurements very difficult.

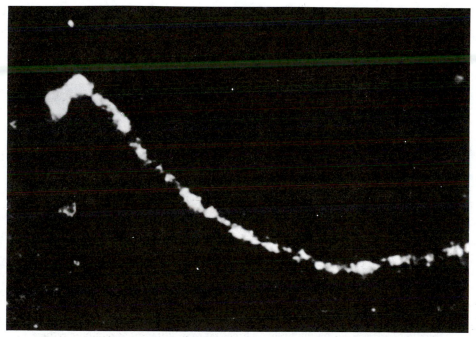

Figure 9-2 Recoil of ^{6}He in beta decay. [J. Csikay and A. Szalay, *Nuovo Cimento, Suppl.*, Padova Conference, 1957.]

Recoil experiments have been performed with the neutron, ^{6}He, ^{7}Be, ^{19}Ne, ^{23}Ne, ^{35}A, and ^{152m}Eu. In some cases attempts have been made to observe the nuclear recoil in a low-pressure cloud chamber. A picture obtained with ^{6}He is shown in Fig. 9-2. Most of the measurements require very difficult and ingenious techniques, involving the further acceleration of the recoil nucleus in an external field. Figure 9-3 shows an experimental arrangement used and Fig. 9-4 the results obtained.

In the case of K capture the recoil situation is simpler than in beta decay. Only a monoenergetic neutrino is emitted, and hence the recoil nuclei all have the same momentum. However, even in a very favorable case, the transformation of ^{7}Be to ^{7}Li, the recoil energy is only 57.3 eV.

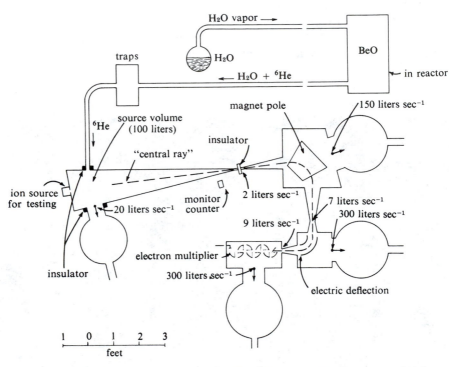

Figure 9-3 Experimental apparatus. The ^{6}He produced in a reactor by the ^{9}Be$(n, \alpha)^6$He reaction is carried by a continuous stream of water vapor to the laboratory where the vapor is removed and the ^{6}He is left to decay in the conical source volume. A proportional counter monitors the source activity. Recoil ^{6}Li ions undergo magnetic and electrostatic analysis and are detected by a secondary electron multiplier. Three stages of differential pumping reduce the background of atoms which decay near the detector. [From C. H. Johnson, F. Pleasonton, and T. A. Carlson, *Phys. Rev.*, **132**, 1149 (1963).]

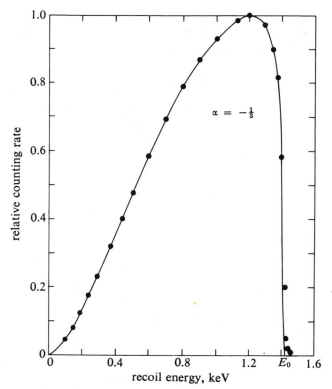

Figure 9-4 Spectrum of singly charged ⁶Li ions from the beta decay of ⁶He as a function of the average recoil energy of the ions transmitted by the analyzers. Ions were accelerated to about twice their recoil energy before analysis. Uncertainties from counting statistics are less than the point sizes.

The form of this spectrum depends on the angular correlation between the momenta of the electron p and of the neutrino q. In the case of ⁶He the correlation is approximately $P(\mathbf{p}, \mathbf{q}) \, dE \, d\omega = \text{const } pEq^2[1 + \alpha(cp/E)\cos\theta] \, dE \, d\omega$, where E is the total energy of the electron, θ is the angle between p and q, and $d\omega$ is the element of solid angle. The constant α is important because its value indicates the type of interaction effective in the decay (see Sec. 9-8). The theoretical curve is plotted for $\alpha = -1/3$ with the normalization constant and the end point E_0 chosen to give a good fit of theory to experiment. [C. H. Johnson, F. Pleasonton, and T. A. Carlson, *Phys Rev.*, **132**, 1149 (1963).]

The properties ascribed to the neutrino to satisfy the conservation laws of momentum and energy are verified by the recoil experiments. This verification is of fundamental importance, but it does not add to our knowledge of the neutrino. After all, the particle was initially postulated in order to satisfy the conservation laws.

Another type of experiment, in which the direct interaction of the free neutrino is observed, was performed by Cowan and Reines in 1953. It demon-

ISBN 0-805-86011-7

strates properties of the neutrino which go beyond those required by the conservation laws and which are implied in Fermi's theory. In this sense it gives independent proof of the existence of the neutrino. The reaction

$$n \rightarrow p + e^- + \bar{\nu} \tag{9-2.1}$$

may be inverted by bombarding protons with antineutrinos, causing the protons to emit a positron and become neutrons.

$$\bar{\nu} + p \rightarrow n + e^+ \tag{9-2.2}$$

This reaction was observed with a nuclear reactor as the source of antineutrinos. A large vat containing a hydrogenous substance with Cd added was bombarded by antineutrinos and the following sequence of events was recorded: The positron is emitted, and its annihilation radiation, which follows rapidly ($\sim 10^{-9}$ sec), is observed. The neutron itself is captured much later (10^{-5} sec) by the Cd and capture gamma rays are then seen (Figs. 9-5 and 9-6). This sequence is sufficiently characteristic to be unmistakable. The cross section for the process indicated by Eq. (9-2.2) can be estimated by detailed balance (see Chap. 11) from the observed mean life of the neutron (17.3 min) and is expected to be of the order of 10^{-43} cm^2 for antineutrinos originating in fission. This number agrees with the experimental observations of Cowan and Reines. Note that the cross section is energy dependent; at energies of several GeV it reaches values of the order of 10^{-38} cm^2. This increase follows from the increased density of final states, which varies as p_ν^2.

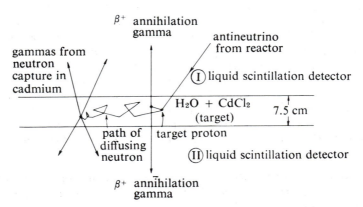

Figure 9-5 Schematic diagram of neutrino detection. An antineutrino from a reactor produces a neutron and a positron by the reaction $\bar{\nu} + p \rightarrow n + e^+$. The positron is detected by its annihilation with an electron. The neutron is detected by the gamma rays emitted upon its capture. [After C. L. Cowan and F. Reines.]

ISBN 0-8053-8601-7

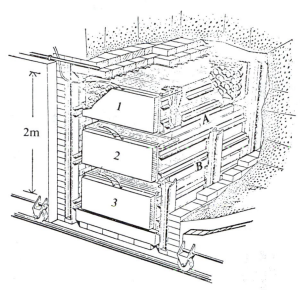

Figure 9-6 Sketch of detectors inside their lead shield. The detector tanks marked 1, 2, and 3 contained 1400 liters of liquid scintillator solution. The scintillations were viewed in each tank by 110 5-in. photomultiplier tubes. The white tanks *A* and *B*, about 28 cm deep, contained 200 liters of water–cadmium chloride target.

Finally the question of the identity or nonidentity of neutrino and antineutrino has been tested experimentally in two other ways, aside from the determination of the helicity. The first is double beta decay under the assumption of a difference between neutrino and antineutrino. Consider three nuclei, such as ^{130}Te, ^{130}I, and ^{130}Xe. Their masses are such that ^{130}Te (Fig. 9-7) could conceivably transform into ^{130}Xe with the simultaneous emission of two electrons and two antineutrinos having 2.54 MeV of energy. The whole process, a second-order process with respect to the beta decay, is highly improbable. The mean life would be of the order of $10^{22 \pm 2}$ yr, and the two electrons would not balance the momentum between them.

Now consider the antineutrino as identical to the neutrino. ^{130}Te may then transform directly into ^{130}Xe, with the emission of only the two electrons, which balance the momentum between them. We may think that, in the second-order process giving rise to this transformation, the antineutrino was emitted and reabsorbed, the reabsorption counting as the second emission of the previous case. Under this hypothesis the half-life turns out to be $10^{16 \pm 2}$ yr, much shorter than in the previous case.

The only certain cases of double beta decay were demonstrated in mass spectrographic experiments by Inghram, Reynolds, and others. They showed that minerals containing Te had an anomalous content of Xe, and from the

ISBN 0-8053-8601-7

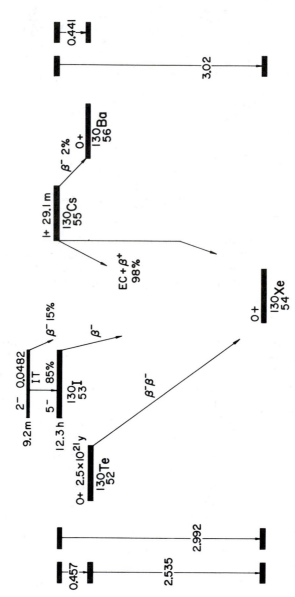

Figure 9-7 Level diagram for double beta decay of ^{130}Te. Energy scales in MeV.

ISBN 0-8053-8601-7

measurements found a half-life for ^{130}Te decaying into ^{130}Xe of $10^{21.34 \pm 0.12}$ yr. Similarly, for ^{82}Se decaying into ^{82}Kr they found a half-life of 10^{20} yr. These observations and negative results on ^{76}Ge, ^{96}Zr, ^{150}Nd, ^{124}Sn, and ^{48}Ca favor the nonidentity of the neutrino and antineutrino. Furthermore, if and only if the antineutrino is identical to the neutrino, the reaction

$$\bar{\nu} + {}^{37}\text{Cl} \rightarrow {}^{37}\text{Ar} + e^- \qquad (9\text{-}2.3)$$

is the inverse of

$$^{37}\text{Cl} \rightarrow {}^{37}\text{Ar} + e^- + \bar{\nu} \qquad (9\text{-}2.4)$$

and therefore the bombardment of ^{37}Cl with antineutrinos should be effective in producing ^{37}Ar. Again the experimental results are against the identity of the two particles (Davis, 1955). Thus the ν and $\bar{\nu}$ appear to be two different particles.

However, known solar nuclear reactions generate a neutrino (not $\bar{\nu}$) flux on the earth, which should give a detectable rate for reaction (9-2.3) with incident neutrinos. Careful study by Davis et al. has failed to detect such an Ar formation; the reason for this failure is still unclear, although it is probably of astrophysical origin.

Neutrinos appear, not only in beta decay, but also in the decay of pi mesons, mu mesons, and other particles (see Chap. 13). The question then arises whether all neutrinos are identical or whether there are different species. The example of the electron and the mu meson points to the possibility of particles that have the same interactions and the same spin and charge but which differ in some other way, in the electron–muon case, in mass. In other words, there seems to be some other, still unknown property (see Chaps. 12 and 13), which is needed to characterize a particle completely. Experiments (Lederman, Schwartz, Steinberger, et al., Brookhaven, 1962) have indeed shown that the neutrino of beta decay is different from the neutrino arising in the principal mode of pi-meson decay. The latter type of neutrino was used to bombard aluminum in a spark chamber. It was found that the pi-meson neutrino can produce mu mesons, but not electrons. If there were only one species of neutrino, electrons and muons would be produced in about the same abundance. Hence, the result of this experiment is taken to indicate the occurrence of at least two different kinds of neutrinos. (In what follows, "neutrinos" always means a neutrino arising in beta decay, unless otherwise indicated.)

Cross sections for interactions of neutrinos with electrons or nucleons increase with the neutrino momentum, and high-energy neutrinos are much easier to detect than beta-decay neutrinos. (For $E_\nu = 1$ GeV, $\sigma \cong 10^{-38}$ cm^2.) Figure 9-8 shows neutrino collisions obtained by using as projectiles ν coming from the decay in flight of high-energy muons or pions.

Figure   9-8   Event produced by interaction of an electron-neutrino, $\nu_e : \nu_e + n \to p + e^-$. The incident beam consists mostly of muon-neutrinos ν_μ with a very small admixture of $\nu_e (\sim \frac{1}{2} \%)$ from the 3-body decays in flight, $K^+ \to \pi^0 + e^+ + \nu_e$. The high energy electron secondary is recognized by the characteristic shower it produces by the processes of bremsstrahlung and pair-production. The chamber diameter is 1.1 m, and the radiation length in CF_3Br is 0.11 m.

The relative numbers of events giving electron and muon secondaries is consistent with the calculated fluxes of ν_e and ν_μ in the beam, and thus confirms conservation of muon number. (Courtesy, CERN Information Service.)

9-3   ENERGETICS OF BETA DECAY

Before proceeding to develop the theory of beta decay, we shall point out the energetics of the reactions involved. We shall consider the emission of electrons, the emission of positrons, and the capture of orbital electrons separately. We shall use atomic masses M_Z and nuclear masses N_Z. The relation between the atomic mass and the nuclear mass is

$$M_Z = N_Z + Zm - B_Z \qquad (9\text{-}3.1)$$

where B_Z is the binding energy of the totality of the atomic electrons and m is the mass of the electron. This binding energy and, especially, the difference between binding energies that will appear in the final equations are generally small compared with the beta-ray energies.

The energy balance for beta decay of a bare nucleus is

$$N_Z = N_{Z+1} + m + Q \qquad (9\text{-}3.2)$$

where m is the electron (or positron) mass and Q the kinetic energy of the electron, neutrino, and recoil nucleus, the last being almost always negligible. The equivalent form for atomic masses is

$$M_Z = M_{Z+1} + (B_{Z+1} - B_Z) + Q \qquad (9\text{-}3.3)$$

The energy balance for positron emission is

$$N_Z = N_{Z-1} + m + Q \qquad (9\text{-}3.4)$$

or, with atomic masses,

$$M_Z = M_{Z-1} + 2m + (B_{Z-1} - B_Z) + Q \qquad (9\text{-}3.5)$$

Finally the energy equation for orbital electron capture is

$$N_Z + m - B_Z^K = N_{Z-1} + Q \qquad (9\text{-}3.6)$$

or

$$M_Z = M_{Z-1} + (B_{Z-1} - B_Z) + B_Z^K + Q \qquad (9\text{-}3.7)$$

Here B_Z^K means the binding energy of the K electron if the electron captured by the nucleus is in a K shell. It is apparent that if

$$B_Z^L < (M_Z + B_Z) - (M_{Z-1} + B_{Z-1}) - Q < B_Z^K \qquad (9\text{-}3.8)$$

the case might present itself in which K capture is impossible but L capture might occur. In orbital electron capture Q is the energy of the neutrino plus the very small recoil energy of the nucleus. Figure 9-9 shows the neutrino

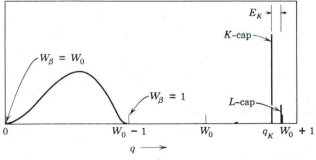

Figure   9-9   Neutrino spectrum from combined β^+ decay and K and L capture. Abscissa, momentum or energy of neutrino. W means total energy. W_0 is the disintegration energy in mc^2 units. $W_\beta = W_0$ corresponds to the upper limit of the positron spectrum; $W_\beta = 1$ corresponds to positrons at rest. [From (Se 59).]

ISBN 0-8053-8601-7

energy spectrum in a case when all three—positron emission, K capture, and L capture—occur.

9-4 CLASSIFICATION OF INTERACTIONS. PARITY

In physics four types of forces act:

1.   The electromagnetic interactions—familiar from macroscopic and atomic physics and responsible for phenomena such as Coulomb scattering and gamma emission.

2.   The specific strong nuclear interactions—responsible for phenomena such as neutron–proton scattering.

3.   The weak interactions—responsible for beta decay and allied phenomena.

4.   The gravitational interactions, which are negligible and will not be considered here.

The electromagnetic interactions are treated by Maxwell's equations and the quantum-mechanical generalizations of these equations. The charge of the electron is the universal constant that appears whenever electromagnetic interactions are considered. Instead of the elementary charge, one often considers the dimensionless quantity $e^2/\hbar c \cong 1/\text{``137''}$ as representative of the coupling strength of the electromagnetic interaction. The ordinary photons are the quanta of the electromagnetic field; they are bosons of zero rest mass.

The most typical example of a strong interaction is the Yukawa interaction:

$$N \to N + \pi \tag{9-4.1}$$

in which a nucleon transforms into a nucleon and a pion. Whenever this type of interaction occurs, another universal constant appears which performs a role similar to that of the electric charge in the case of the electromagnetic field. This constant, expressed in a dimensionless fashion as $f^2/\hbar c$, has a value close to one. The strong interactions have a short range; that is, their force acts within a characteristic distance of the order of 10^{-13} cm. This fact, as Yukawa pointed out, forces the quanta of the field to have a finite rest mass; the quanta are identified with the various mesons (pions, kaons, etc.) which are bosons of integral spin.

The weak interaction is a new force in nature, as Faraday would have said. It was introduced by Fermi in 1934 in a form suitable for beta decay. His main ideas have, however, survived, suitably generalized. The full scope of the weak interaction is now very considerable and will be restudied in Chaps. 14 and 19 in the sections on particle physics. Historically, beta decay was the first manifestation of the weak interaction to be understood; it is also the most important in nuclear physics. After studying beta decay in some detail, in the subsequent discussions of particle physics we shall again take up weak interactions, generalizing the results of this chapter. Some of the particles participating in beta decay are leptons (e, ν) and some are hadrons (n, p); for this reason beta decay is called semileptonic. There are also leptonic weak interactions (e.g., $\mu \rightarrow e + \nu + \bar{\nu}$), in which all the participants are leptons, and hadronic weak interactions (e.g., $\Lambda \rightarrow p + \pi^-$), in which all the participants are hadrons.

Beta decay is characterized by the participation of four fermions, free or bound in a nucleus. Examples are

$$n \rightarrow p + e^- + \bar{\nu} \qquad \text{(beta emission)} \qquad (9\text{-}4.2)$$

$$^{11}\text{C} \rightarrow {}^{11}\text{B} + e^+ + \nu \qquad \text{(positron emission)} \qquad (9\text{-}4.3)$$

$$^{7}\text{Be} + e^- \rightarrow {}^{7}\text{Li} + \nu \qquad (K \text{ capture}) \qquad (9\text{-}4.4)$$

Weak interactions have a coupling constant g or, in more complete theories, several coupling constants that give effects much smaller than those of strong and electromagnetic interactions. In c.g.s. units the value of g is 1.43 10^{-49} erg $\cdot$ cm^3. Because these dimensions are different from those of the charge of the electron, erg$^{1/2}$ cm$^{1/2}$, it is not possible to compare g with e directly. We may consider the dimensionless numbers $e^2/\hbar c = 1/\text{"137"}$ and the dimensionless combination $G = (g/mc^2)(\hbar/mc)^{-3} = gm^2 c\hbar^{-3}$ whose numerical value obviously depends on the choice of m. If we chose the *protonic mass* $G = 1.03 \cdot 10^{-5}$. If we use the pionic mass that gives $\hbar/mc$ of nuclear dimensions, we find for $g(\hbar/mc)^{-3}$ about 31 eV. Note that two point electrons at the same distance have an electrostatic energy of about 1 MeV.

In 1956 the study of certain mesons (then called θ and τ) presented a great puzzle. Experimentally they were found to have the same mass, half-life, and spin, suggesting that they were the same particle. On the other hand, they decayed into pion systems of opposite parity; hence, if parity was conserved they had to be different. The acuteness of the dilemma led Lee and Yang to question the conservation of parity in weak interactions. It was then found that all weak interactions seem to have in common the nonconservation of parity (Lee and Yang, 1956). This is a major difference between weak interactions on the one hand and strong and electromagnetic interactions on the other.

Conservation of parity can be described at different levels of sophistication. In the simplest terms it can be defined as follows: If the mirror image of a physical phenomenon represents another possible physical situation, the phenomenon is said to conserve parity. All phenomena involving strong and electromagnetic interactions alone *do* conserve parity. An example, shown in Fig. 9-10, will make the meaning of these statements clear. The upper half of

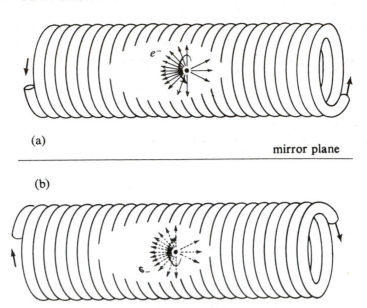

(a)

_____ mirror plane

(b)

Figure 9-10 (a) Beta decay of polarized cobalt. (b) Mirror image of experiment (a). This is not what happens in fact, because if the current in the solenoid is reversed [which is equivalent to going from (a) to (b)], the beta rays go preferentially to the right. On the other hand, if one observes gamma rays, the probability of emission at an angle θ or $\pi - \theta$ with the axis of the solenoid is the same, and thus parity is conserved for gamma emission.

the figure shows the intensity of beta emission from cobalt placed in a magnetic field generated by the coil. The magnetic field orients the nuclear spin of cobalt in such a way that the current of the solenoid and the current in the nucleus generating its magnetic moment are parallel. The lower part of the figure gives the mirror image of the upper part, which *is not* what is observed. In fact the lower part of the figure can be realized experimentally as far as solenoid and nucleus are concerned by simply inverting the direction of the current in the solenoid. What happens then is that the maximum intensity goes from left to right and not from right to left, as indicated by the dashed arrows in the figure (mirror image of the upper half). Thus parity is not conserved. Conservation of parity would require equal probability of emission of beta rays to the right or to the left of the figure. This is what happens for gamma-ray emission.

The invariance on mirroring was assumed, incorrectly, to have a logical necessity a priori, such as is usually accepted for the invariance of the result of an experiment repeated in different places, in different orientations in space, or at different times. The experiment shows that this invariance on mirroring is not verified in beta decay.

ISBN 0-8053-8061-1

We now present these remarks in a different form by noting that the vectors used in physics are of two different kinds. The distinction becomes clear when we consider the same physical quantity in two coordinate systems derived from each other by reflection through the origin,

$$x \to -x' \qquad y \to -y' \qquad z \to -z' \qquad (9\text{-}4.5)$$

If a vector such as displacement, velocity, momentum, acceleration, electric field, etc., has components

$$X \qquad Y \qquad Z$$

in the first (unprimed) system, then in the second system the reference of the same vector has components

$$X' = -X \qquad Y' = -Y \qquad Z' = -Z \qquad (9\text{-}4.6)$$

Vectors of this type are called polar vectors. Consider now the angular momentum

$$\mathbf{L} = \mathbf{r} \times \mathbf{p} \qquad (9\text{-}4.7)$$

and remember the definition of vector product. In components one has $L_x = yp_z - zp_y$ and the similar formulas for L_y and L_z. In the primed system both $\mathbf{r}$ and $\mathbf{p}$ change sign. Hence the components of $\mathbf{L}$ in the primed and unprimed systems are the same. This result derives also from the geometrical definition of vector product, which involves the convention of the right-hand rule or its equivalent.

A vector such as $\mathbf{L}$ is called a pseudovector or axial vector. Torque, angular momentum, spin, magnetic field, and magnetic moments are examples of pseudovectors. Vectors and pseudovectors behave in the same way for translations or rotations of the coordinate system; it is only on inversion that they behave differently.

The scalar product of two polar vectors or of two axial vectors is a number, invariant upon reflection of the coordinate system, and is called a true scalar, or simply a scalar. The scalar product of a polar vector and an axial vector is a number that changes sign upon inversion of the coordinate system. Such numbers are called pseudoscalars.

The result of a physical measurement is always a number, which, however, may behave as a scalar or as a pseudoscalar. The mirror reflection mentioned above leaves true scalars unchanged, but it changes the sign of pseudoscalars. Hence if the image and the object must be indistinguishable, all pseudoscalars must vanish. The observation of pseudoscalars different from 0 thus implies the breakdown of parity conservation. On the other hand, until a pseudoscalar has been observed, no information on the conservation of parity is available.

It was pointed out by Lee and Yang in 1956 that all experiments on weak interactions performed up to that time were not designed to observe pseudo-scalar quantities. Actually Cox in 1928 had observed a longitudinal polarization of the electrons in beta decay; that is, the quantity

$$\mathbf{v} \cdot \mathbf{I} \tag{9-4.8}$$

had not been 0. This observation was disbelieved at the time and later forgotten.

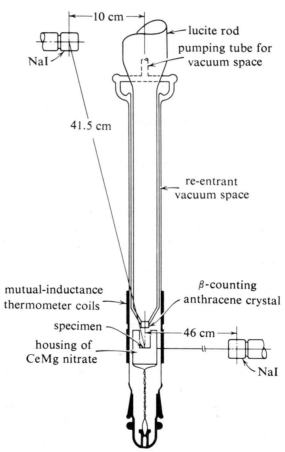

Figure   9-11 Schematic diagram of lower portion of the cryostat used for polarization of ^{60}Co nuclei, the specimen used. The beta particles are detected by a thin anthracene crystal $\sim$2 cm above the ^{60}Co source. The scintillations from the crystal are transmitted through the lucite rod to a photomultiplier tube above. The two NaI counters (one in the equatorial plane and one near the polar position) were used to measure the gamma-ray anisotropy and thus the amount of polarization of the ^{60}Co. [From C. S. Wu, E. Ambler, R. W. Hayward, D. D. Hoppes, and R. P. Hudson, *Phys. Rev.*, **105**, 1413 (1957).]

ISBN 0-8053-8061-7

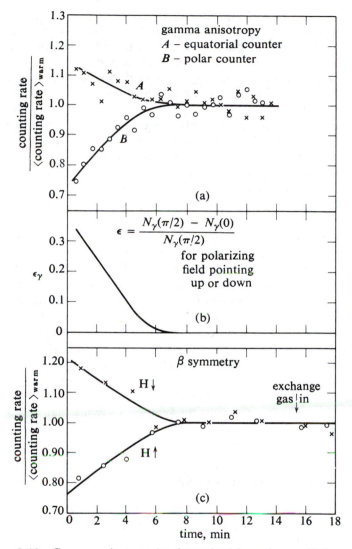

Figure 9-12 Gamma anisotropy (as determined from the two NaI counters) and beta asymmetry for the polarizing field pointing up and down as a function of time. The times for disappearance of the beta and gamma asymmetry coincide; this is the warm-up time. The warm-up time for the sample is approximately 6 min and the counting rates for the warm unpolarized sample are independent of the field direction. [From C. S. Wu, E. Ambler, R. W. Hayward, D. D. Hoppes, and R. P. Hudson, *Phys. Rev.*, **105**, 1413 (1957).]

ISBN 0-8053-8061-7

In 1957 an experiment specifically designed to detect a pseudoscalar in beta decay was set up by Wu, Ambler, Hayward, Hoppes, and Hudson. Using polarized cobalt-60 nuclei, they found that the direction of emission of electrons in the transformation to nickel-60 is preferentially opposite to the spin direction. Thus the expectation value of the pseudoscalar $\langle \mathbf{v} \cdot \mathbf{I} \rangle$, where $\mathbf{I}$ is the nuclear spin and $\mathbf{v}$ the electron velocity, was measured and found to be different from 0.

The discovery of this startling phenomenon was made by the experiment illustrated in Fig. 9-11. The cobalt nuclei were oriented by cooling in a magnetic field (Chaps. 6 and 10). The extremely low temperature ($0.01°K$) required was obtained by adiabatic demagnetization, and the degree of orientation ascertained by observing the anisotropy in the emission of the gamma rays. The beta rays were then counted, with the results shown in Fig. 9-12. If parity were conserved, the intensity pattern would be symmetric with respect to a horizontal plane passing through the source, as the gamma-ray pattern is.

Mathematically, the principle of conservation of parity rules out certain terms in the expression of the hamiltonian, which is the operator that governs the time dependence of the wave function, because these terms would not be invariant on a reflection through the origin or, more explicitly, if x, y, and z were changed into $-x, -y$, and $-z$, respectively; thus the parity of the wave function would be changed as time went on. For example, in the expression of the energy of the electromagnetic field, terms like $\mathbf{H}^2$ and $\mathbf{E}^2$ (scalars) are permissible, but a term such as $\mathbf{E} \cdot \mathbf{H}$ (pseudoscalar) would change sign on inversion of the coordinates and is not permissible if parity is to be conserved. In strong and electromagnetic interactions, terms that do not conserve parity are absent from the hamiltonian. By analogy all terms that do not conserve parity were omitted in writing the hamiltonian for weak interactions until experiment showed that they must be present.

9-5 FERMI'S THEORY OF BETA DECAY

In constructing his theory of beta decay, Fermi was guided by the analogy to electromagnetic gamma-ray emission. In the latter, a nucleus passes from an excited state to the ground state, creating a photon. In beta decay, a neutron is transformed into a proton, and an electron and an antineutrino are created. The object of a theory of beta decay is to give a dynamical description of this process.

Fermi assumed the conservation of parity, but his calculations involved only scalar quantities, and it turns out that the results of his theory still stand in large measure, in spite of the fundamental changes produced by the discovery of the nonconservation of parity.

The main results of Fermi's theory are the explanation, by the introduction of only one new parameter (the g coupling constant) of:

ISBN 0-8053-8601-1

1. The form of beta spectra (i.e., number of electrons per energy interval).
2. The relation between the maximum energy of the decay and the mean life.
3. The classification of beta transitions and the establishment of selection rules.

The fundamental ideas of Fermi's theory are general enough to be easily adapted to changes of the interaction (e.g., to parity nonconservation), still preserving the validity of many of his results. Furthermore, the theory can be extended to encompass all forms of weak interactions.

The starting point is golden rule No. 2, which gives the transition probability

$$w = \frac{2\pi}{\hbar} |H_{if}|^2 \rho \qquad (9-5.1)$$

To apply this formula we need to know the expression for the matrix element between the initial and final states and to calculate the density of the final states:

$$\rho = \frac{dN}{dW} \qquad (9-5.2)$$

The last assignment will be performed shortly.

It is found that the density of final states completely determines the form of the beta spectrum, the matrix element in allowed transitions being a constant independent of the energy of the electrons. The matrix element must be guessed at and this, of course, represents the most difficult part of the theory. We postulate that the wave functions of the four particles intervening in the reaction appear linearly in H_{if}. This postulate, which excludes the appearance of higher powers or derivatives of the wave functions, is justified by experience and suggested by analogy with the electromagnetic field. Relativistic invariance is assumed. The matrix element, by definition, has the form

$$H_{if} = \int \psi_{\text{fin}}^* H \psi_{\text{in}} \, d\tau \qquad (9-5.3)$$

Here H is the operator representing the unknown interaction. This operator is more general than the ordinary operators of nonrelativistic quantum mechanics, because the initial and final states do not refer to the same particles. The operator must be able to "destroy" the initial particles and "create" the final ones. The initial wave function $\psi_{\text{in}} = u_i$ describes the nucleus before decay; the final eigenfunction ψ_{fin} is the product of u_f (describing the nucleus in the final state) times ψ_e times $\psi_{\bar{\nu}}$, the wave functions of the electron and antineutrino appearing in the final state. The unstarred states are

ISBN 0-8053-8601-7

destroyed in the transition, and the starred states are created. Moreover, the emission of a neutrino and the absorption of an antineutrino of opposite momentum are equivalent, and we may replace $\psi_{\bar{\nu}}^*$ by ψ_ν to make the equations more symmetrical. To a good approximation (for exceptions see below) ψ_e and $\psi_{\bar{\nu}}$ are constant throughout the nuclear volume.

The matrix element for an interaction with a fixed potential field is

$$M_{if} = \int \psi_f^*(\mathbf{r}) V(\mathbf{r}) \psi_i(\mathbf{r}) \, d\mathbf{r}$$

For two particles interacting it is

$$M_{if} = \int \psi_f^*(\mathbf{r}_a) \psi_f^*(\mathbf{r}_b) V(\mathbf{r}_a - \mathbf{r}_b) \psi_i(\mathbf{r}_a) \psi_i(\mathbf{r}_b) \, d\mathbf{r}_a \, d\mathbf{r}_b$$

If the potential $V(\mathbf{r}_a - \mathbf{r}_b)$ has an extremely short range, we may replace it with a delta function $g\delta(\mathbf{r}_a - \mathbf{r}_b)$, and the expression for the matrix element is then written as

$$H_{if} = g \int u_f^* u_i \psi_e^* \psi_\nu \, d\tau \tag{9-5.4}$$

Here the integral is extended only over the nuclear coordinates and the nuclear wave functions take into account the transformation of a neutron into a proton in the nucleus. Thus it would be simply $\psi_e^* \psi_\nu$ if a neutron became a proton, without any other appreciable change in the nuclear wave function, as, for example, in the ${}^3\text{H} \rightarrow {}^3\text{He}$ transformation. A more formal description of the nuclear integral is obtainable with the help of the isotopic spin operators (see Chap. 10), but we shall omit this development here.

The new universal constant g, characteristic of beta decay, has the dimensions of energy $\times$ volume. Its numerical value can be determined by experiment, as we shall see shortly, and is of the order of 10^{-49} erg cm³.

In the first study we shall assume that the wave functions of the electron and the neutrino (or antineutrino) have the form (neglecting spin)

$$\psi_e(\mathbf{r}) = \Omega^{-1/2} e^{i\mathbf{k}_e \cdot \mathbf{r}} \qquad \psi_\nu(\mathbf{r}) = \Omega^{-1/2} e^{i\mathbf{k}_\nu \cdot \mathbf{r}} \qquad \mathbf{k} = \mathbf{p}/\hbar \tag{9-5.5}$$

where Ω is the volume of a big box in which we enclose the system for normalization purposes. The use of nonrelativistic expressions and the neglecting of spins are certainly objectionable features. Nonetheless the results obtained are useful for orientation, and they represent a first approximation to the relativistic theory. By assuming the plane wave form for the wave function of the electron and neutrino, we have neglected their possible interactions with the nucleus; this is an exceedingly good approximation for the neutrino, and a fair approximation for the electrons, which are actually subject to electromagnetic forces from the nuclear charge. The correction for the latter effect

ISBN 0-8053-8601-7

is mathematically involved but does not offer difficulties in principle; it will later be taken into account numerically.

We now observe that for the momenta occurring ordinarily in beta decay,

$$k^{-1} = \hbar/p \sim 2 \times 10^{-11} \text{ cm} \tag{9-5.6}$$

is large compared with the nuclear dimensions over which the integrals appearing in the matrix element extend; we can thus write

$$e^{i\mathbf{k}\cdot\mathbf{r}} = 1 + i\mathbf{k}\cdot\mathbf{r} - \cdots \tag{9-5.7}$$

and the second term will be, in the nuclear region, between 0.1 and 0.01. We then stop with the first term of the expansion unless it should happen that the corresponding matrix element becomes exactly 0, in which case we must consider higher terms. This classification generates transitions of various orders (allowed, first forbidden, etc.) closely analogous to the multipole expansion of electromagnetic radiation. We shall consider only allowed transitions, for which we then write, according to Eqs. (9-5.4) and (9-5.5),

$$H_{if} = g \frac{M_{if}}{\Omega} \tag{9-5.8}$$

We have represented by M_{if} the integral $\int u_f^* u_i \, d\tau$, about which we shall make no hypothesis at present, and have introduced a volume Ω in which the system is enclosed in order to simplify the normalization. M_{if} is dimensionally a pure number.

What we have said up to now concerning H_{if} is still vastly oversimplified. We mention here the necessary improvements: (1) The u_i, u_f, ψ_e, ψ_ν must be treated relativistically with Dirac spinors. This brings about the possibility of several different interactions (cf. Sec. 9-8). (2) H_{if} must cover not only electron emission but also positron emission and orbital electron capture. This is achieved by adding to H_{if} its hermitian conjugate. (3) The interaction as written is parity conserving; in order to describe transitions that do not conserve parity, further terms have to be added to it.

Leaving these points for the present, we must now consider the density of final states, that is, the number of states per energy interval of the total energy W. Three bodies participate in the disintegration: the final nucleus, the electron, and the neutrino. Energy and momentum must be conserved. The momenta of the three particles are generally of comparable magnitudes and balance to 0, but the energy taken up by the recoil nucleus is very small compared with the energy of the electron and the neutrino, and we can neglect it.

We can thus say that the sum of the energies of electron and neutrino, E_e and E_ν, is equal to the total disintegration energy W, without worrying about

ISBN 0-8053-8601-7

the energy imparted to the nucleus by conservation of momentum,

$$W = cp_\nu + \left(m^2 c^4 + c^2 p_e^2 \right)^{1/2} = E_\nu + E_e \tag{9-5.9}$$

We now want the number of states of the electron having a momentum in the interval between p_e and $p_e + dp_e$ irrespective of the momentum of the neutrino. The number of states in which the neutrino has a momentum in the interval dp_ν and the electron has a momentum in the interval dp_e is

$$dN_e \, dN_\nu = \frac{16\pi^2 \Omega^2}{(2\pi\hbar)^6} \, p_\nu^2 \, dp_\nu p_e^2 \, dp_e \tag{9-5.10}$$

We eliminate dp_ν and p_ν from Eq. (9-5.10) by noting that for constant p_e Eq. (9-5.9) gives

$$W - E_e = E_\nu = cp_\nu \tag{9-5.11}$$

and

$$dW = dE_\nu = c \, dp_\nu \tag{9-5.12}$$

which, replaced in Eq. (9-5.10), yields

$$dN_e \frac{dN_\nu}{dW} = \frac{16\pi^2 \Omega^2}{(2\pi\hbar)^6 c^3} \, (W - E_e)^2 p_e^2 \, dp_e \tag{9-5.13}$$

This is the density of final states, for which the electron has a momentum between p_e and $p_e + dp_e$ irrespective of the neutrino momentum, while the total energy is between W and $W + dW$. It is the appropriate density to be used in Eq. (9-5.1), which then gives for the probability of a disintegration per unit momentum yielding an electron of momentum between p_e and $p_e + dp_e$

$$w(p_e)dp_e = \frac{2\pi}{\hbar c} \, g^2 |M_{if}|^2 \frac{(4\pi)^2}{(2\pi\hbar)^6} \frac{(W - E_e)^2}{c^2} \, p_e^2 \, dp_e \tag{9-5.14}$$

The volume Ω disappears, as it must, by cancellation in Eq. (9-5.14).

We may ask also what the total decay rate λ is, irrespective of the electron momentum. This is obtained from Eq. (9-5.14), by integrating with respect to dp_e from 0 to the maximum possible momentum of the electron,

$$p_{e\,\text{max}} = (1/c)(W^2 - m^2 c^4)^{1/2} \tag{9-5.15}$$

giving

$$\lambda = \frac{g^2 |M_{if}|^2}{2\pi^3 c^3 \hbar^7} \int_0^{p_{e\,\text{max}}} \left[W - \left(m^2 c^4 + c^2 p_e^2 \right)^{1/2} \right]^2 p_e^2 \, dp_e \tag{9-5.16}$$

ISBN 0-8053-8601-7

It is convenient to measure energies in units of mc^2 and momenta in mc, putting

$$E_e = \epsilon mc^2 \qquad (9\text{-}5.17)$$

$$p_e = \eta mc \qquad (9\text{-}5.18)$$

$$W = \epsilon_0 mc^2 \qquad (9\text{-}5.19)$$

$$p_{e\,max} = \eta_0 mc \qquad (9\text{-}5.20)$$

Equation (9-5.14) takes the form

$$w(\eta)d\eta = \frac{g^2 m^5 c^4}{2\pi^3 \hbar^7} |M_{if}|^2 (\epsilon_0 - \epsilon)^2 \eta^2 \, d\eta \qquad (9\text{-}5.21)$$

or

$$w(\eta)d\eta = \frac{g^2 m^5 c^4}{2\pi^3 \hbar^7} |M_{if}|^2 \left[(1 + \eta_0^2)^{1/2} - (1 + \eta^2)^{1/2} \right]^2 \eta^2 \, d\eta \qquad (9\text{-}5.22)$$

Actually we have oversimplified our discussion on several counts besides the matrix element. For instance, we neglected the Coulomb interaction between nucleons and electrons when we assumed a plane wave for the electron [Eq. (9-5.5)]. This can be corrected by introducing the $\psi_e^2(0)$ calculated numerically with the eigenfunctions of positive energy (continuous spectrum). The factor thus obtained is a function of the product of the charge of the particle emitted, $\pm e$, multiplied by the nuclear charge Ze of the final nucleus and E. We shall call it $F(Z, \epsilon)$ for electrons. For positron emitters Z must be taken as negative.

An approximate expression of $F(Z, \epsilon)$ is

$$F(Z, \epsilon) = 2\pi n \left[1 - \exp(-2n\pi) \right]^{-1} \qquad (9\text{-}5.23)$$

where n is $Ze^2/\hbar v_e$, and v_e is the velocity of the electron far from the nucleus. For n small compared with 1, $F(Z, \epsilon)$ is unity.

Equation (9-5.21) is thus corrected to

$$w(\eta) \, d\eta = \frac{g^2 m^5 c^4 |M_{if}|^2}{2\pi^3 \hbar^7} F(Z, \epsilon)(\epsilon_0 - \epsilon)^2 \eta^2 \, d\eta \qquad (9\text{-}5.24)$$

or its equivalent

$$w(\epsilon) \, d\epsilon = \frac{g^2 m^5 c^4}{2\pi^3 \hbar^7} |M_{if}|^2 F(Z, \epsilon)(\epsilon_0 - \epsilon)^2 \epsilon\eta \, d\epsilon \qquad (9\text{-}5.25)$$

ISBN 0-8053-8601-7

Since M_{if} is independent of p_e and p_ν (at least for allowed transitions), we have from Eq. (9-5.24) the "form" of the beta spectrum; that is, the probability that an electron is emitted in a momentum interval $d\eta$ around η. From (Eq. 9-5.25) we have the similar expression for the probability of emission in an energy interval $d\epsilon$. This form is shown in Fig. 9-13. Often one plots

$$\left[\frac{w(\eta)}{\eta^2 F(Z, \epsilon)} \right]^{1/2} \text{ versus } (\epsilon_0 - \epsilon) \tag{9-5.26}$$

From Eq. (9-5.24) it is clear that this plot should give a straight line (Kurie plot, Fig. 9-14). Departures from the straight line are attributed to a dependence of M_{if} on p_e such as occurs in forbidden transitions according to Eq. (9-5.7) (Fig. 9-15).

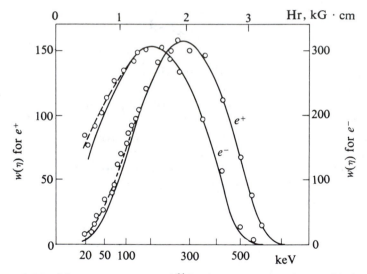

Figure 9-13 Momentum spectra of ^{64}Cu electrons and positrons. Circles are experimental points, solid curves are theoretical. Curves not normalized. [C. S. Wu and R. D. Albert, *Phys. Rev.*, **75**, 315 (1949).]

In this discussion we have assumed that the neutrino has a zero rest mass. If this were not the case, Eqs. (9-5.9) to (9-5.13) would be slightly different, and the form of the beta spectrum, especially near the upper energy limit, would be characteristically affected (Fig. 9-16). The experimental data from this type of information helps to put an upper limit on the neutrino mass of 60 eV.

ISBN 0-8053-8061-7

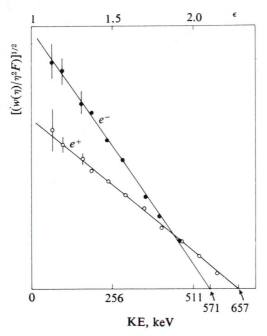

Figure 9-14 The Kurie plots of the ^{64}Cu beta spectra. Both positron and electron spectra are shown. End points are at 571 keV for e^- and at 657 keV for e^+. Points are not normalized. [G. E. Owen and C. S. Cook, *Phys. Rev.*, **76**, 1726 (1949).]

The decay constant λ is obtained by integration of Eq. (9-5.24) from 0 to η_0 [or integration of Eq. (9-5.22) for $F(Z, \epsilon) = 1$]. The result for $F(Z, \epsilon) = 1$ is

$$\lambda = \frac{g^2 m^5 c^4}{2\pi^3 \hbar^7} |M_{if}|^2 f(\eta_0) \tag{9-5.27}$$

where

$$f(\eta_0) = \int_0^{\eta_0} \left[(1 + \eta_0^2)^{1/2} - (1 + \eta^2)^{1/2} \right]^2 \eta^2 \, d\eta$$

$$= -\frac{1}{4} \eta_0 - \frac{1}{12} \eta_0^3 + \frac{1}{30} \eta_0^5$$

$$+ \frac{1}{4} (1 + \eta_0^2)^{1/2} \log \left[\eta_0 + (1 + \eta_0^2)^{1/2} \right] \tag{9-5.28}$$

which has the limiting forms

$$f(\eta_0) \cong \frac{1}{30} \eta_0^5 \qquad \eta_0 \gg 5 \tag{9-5.29}$$

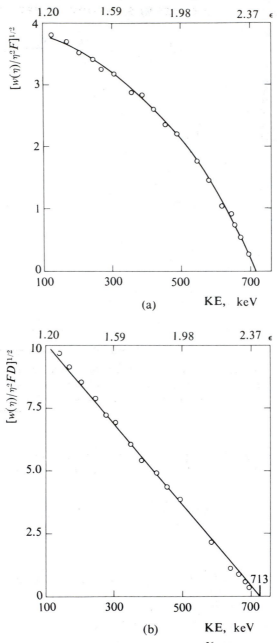

Figure 9-15 (a) The conventional Kurie plot for ^{36}Cl beta spectrum is given. (b) Forbidden Kurie plot for ^{36}Cl beta spectrum. The correction factor used in this case was

$$D \propto (\epsilon_0 - \epsilon)^4 + (10/3)(\epsilon^2 - 1)(\epsilon_0 - \epsilon)^2 + (\epsilon^2 - 1)^2$$

(curves not normalized). [C. S. Wu and L. Feldman, *Phys. Rev.*, **76**, 693 (1949).]

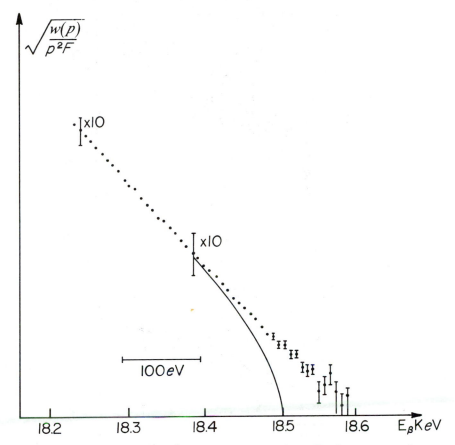

Figure 9-16 Kurie plot of ^{3}H. The position of the end point corresponds to an end-point energy of 18.6 keV. The solid line corresponds to a neutrino mass of 100 eV. [From E. K. Bergkvist, *Nucl. Phys.*, **39B**, 317 (1972).]

and

$$f(\eta_0) \cong \frac{2}{105} \eta_0^7 \qquad \eta_0 \ll 0.5 \tag{9-5.30}$$

For quantitative calculations and Z not too small the Coulomb correction of the electron wave function must be considered. From Eq. (9-5.25) we have, on integration,

$$\lambda = \frac{g^2}{2\pi^3} \frac{m^5 c^4}{\hbar^7} |M_{if}|^2 f(Z, \epsilon_0) \tag{9-5.31}$$

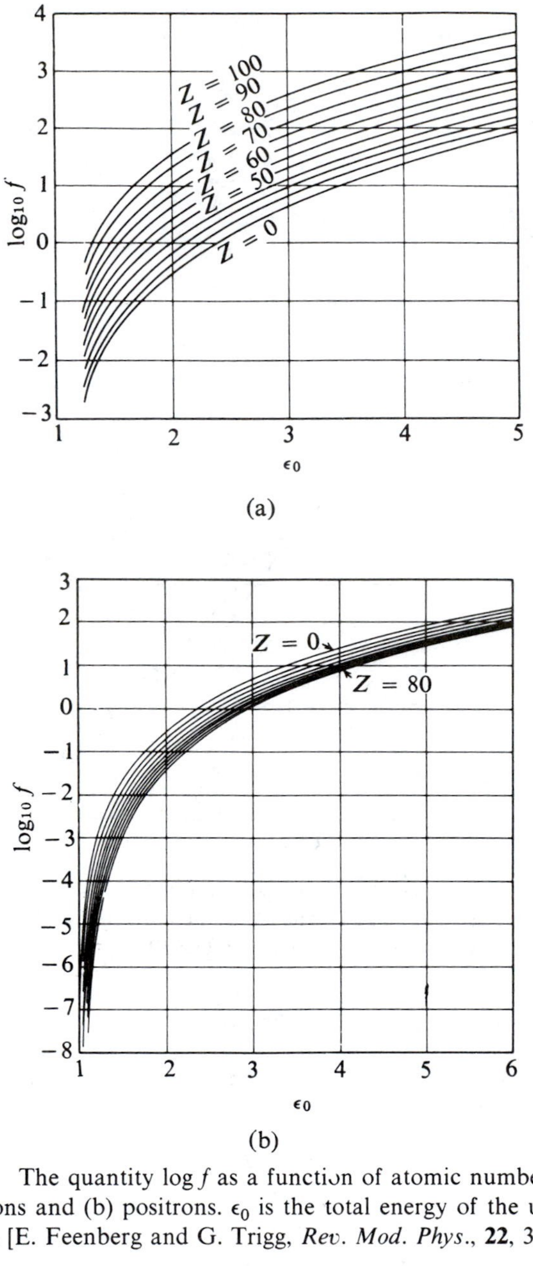

(a)

(b)

Figure 9-17 The quantity $\log f$ as a function of atomic number and energy for (a) electrons and (b) positrons. ϵ_0 is the total energy of the upper limit of the spectrum. [E. Feenberg and G. Trigg, *Rev. Mod. Phys.*, **22**, 399 (1950).]

or introducing the dimensionless constant $G = (g/mc^2)(mc/\hbar)^3$

$$\lambda = \frac{G^2}{2\pi^3} \frac{mc^2}{\hbar} |M_{if}|^2 f(Z, \epsilon_0) \qquad (9\text{-}5.32)$$

where $f(Z, \epsilon_0)$ is now a complicated function which includes the Coulomb correction. Figure 9-17 shows a typical example of $f(Z, \epsilon_0)$.

Equation (9-5.31) could be used to determine g or G from the energy and decay constant of a beta emitter if $|M_{if}|^2$ were known. To anticipate a result to be discussed later, in some simple cases the expression $|M_{if}|^2$ may be evaluated and from this we obtain, by using the experimental values of f/λ, the value of $g = 1.435 \times 10^{-49}$ erg cm^3 or the value $1.026 \ 10^{-5}$ for G, using for m the proton mass.

9-6 MATRIX ELEMENT

To obtain from mean-life experiments a more quantitative insight into the beta interaction we must reduce the data in such a way as to separate the influence of the energy of the transition, the atomic number involved, etc., from the relevant nuclear data.

Equation (9-5.31) can be rewritten as

$$\frac{\text{const}}{|M_{if}|^2} = f(Z, \epsilon_0)t \qquad (9\text{-}6.1)$$

where t is the half-life of beta decay. On the right we have a directly measurable quantity, t and f, a function of Z and ϵ_0; on the left we find $1/|M_{if}|^2$, which depends on the nuclear structure. Equation (9-6.1) allows us to extract from the experimental data, which is affected by the energy of the decay and Z, the information about $|M_{if}|^2$ that is relevant to the nuclear problem. Some examples of values of ft for typical decay are given in Table 9-1. The quantity ft is sometimes called the *comparative half-life*. Nomograms for the rapid evaluation of ft are given in Fig. 9-18.

From the table it is apparent that ft varies over a great range of values and that small values of ft, that is, large matrix elements, occur only if

$$\Delta I = \pm 1, 0 \qquad (9\text{-}6.2)$$

without change of parity. This is strongly reminiscent of the selection rules of electromagnetic radiation. A qualitative nonrelativistic argument gives some insight into the situation. In the interaction [Eqs. (9-5.4) and (9-5.8)] $\psi_\nu(0)$ and $\psi_e(0)$ appear: these are different from 0 only for states having total angular momentum $j = \frac{1}{2}$. Nonrelativistically we could say that only s states have $\psi(0)$ different from zero.

TABLE 9-1   SOME TYPICAL BETA DECAYS[a]

De-caying nucleus	Initial spin and parity	Product nucleus	Final spin and parity	$T_{1/2}$	E max, MeV	Log(ft)	$\|M_{GT}\|^2$	$\|M_F\|^2$
n	$\frac{1}{2}^+$	H	$\frac{1}{2}^+$	10.61 ± 0.16 min	0.782	3.0473	3	1
^{3}H	$\frac{1}{2}^+$	^{3}He	$\frac{1}{2}^+$	12.26 yr	0.01861	3.0535	3	1
^{6}He	0^+	^{6}Li	1^+	0.797 sec	3.508	2.77	2	0
^{14}O	0^+	^{14}N*	0^+	71.36 sec	1.813	3.49513	0	2
^{26}Al*	0^+	^{26}Mg	0^+	6.374 sec	3.208	3.48940	0	2
^{34}Cl	0^+	^{34}S	0^+	1.565 sec	4.46	3.49693	0	2
^{14}C	0^+	^{14}N	1^+	5730 yr	0.156			
^{39}Ar	$\frac{7}{2}^-$	^{39}K	$\frac{3}{2}^+$	269 yr	0.565	9.03		
^{38}Cl	2^-	^{38}Ar	0^+	37.3 min	4.91	8.15		
^{22}Na	3^+	^{22}Ne	0^+	2.62 yr	1.82	11.9		
^{10}Be	0^+	^{10}B	3^+	1.6×10^6 yr	0.56	12.08		
^{40}K	4^-	^{40}Ca	0^+	1.26×10^9 yr	0.63	15.60		
^{115}In	$\frac{9}{2}^+$	^{115}Sn	$\frac{1}{2}^+$	6×10^{14} yr	0.50	23.0		

[a] The first few nuclei, from the neutron to ^{34}Cl, have allowed transitions. Moreover, the matrix elements may be computed on the basis of simple and reliable models.
^{14}C shows an allowed transition with an exceptionally small matrix element.
^{39}Ar and ^{38}Cl with $\Delta I = 2$ and parity change are first forbidden transitions. Their spectrum departs from the simple Fermi form, and their ft value is high.
^{22}Na and ^{10}Be with $\Delta I = 3$ and no change in parity are second forbidden transitions.
^{40}K has $\Delta I = 4$ and change in parity and is a third forbidden transition.
^{115}In has the largest known ft value.

Relativistically l is no longer a good quantum number; however, j is, and it must have the value $\frac{1}{2}$ for both antineutrino and electron in order that $\psi(0) \neq 0$.

Since electron and antineutrino are emitted without orbital angular momentum coming from practically a point source ($R/\lambda \ll 1$, with R the nuclear radius and λ the de Broglie wavelength of the lepton), they carry away no total angular momentum if they are emitted with spins antiparallel (singlet states) or a total angular momentum of 1 if they are emitted with parallel spins (triplet states). In the first case, the spin of the nucleus cannot change at all in beta decay. We thus have the selection rule due to Fermi,

$$\Delta I = 0 \quad \text{(F selection rule)} \tag{9-6.3}$$

In the second case, the *vector* difference between initial and final angular momentum must be 1; thus

$$\Delta I = \pm 1 \text{ or } 0, \text{ but no } 0 \rightarrow 0 \quad \text{(GT selection rule)} \tag{9-6.4}$$

This selection rule is due to Gamow and Teller. In both the F and GT cases, the nuclear eigenfunction must not change parity.

The interactions that give rise to F and GT selection rules are different. However, experiment shows that there are allowed transitions of the type

$$\Delta I = 1 \tag{9-6.5}$$

that obey GT selection rules but are forbidden by F selection rules. Such is the decay

$$^6\mathrm{He} \rightarrow {}^6\mathrm{Li} + e^- + \bar{\nu} \tag{9-6.6}$$

There are also allowed transitions of the $0 \rightarrow 0$ type that are allowed by F selection rules but forbidden by GT selection rules. For example,

$$^{14}\mathrm{O} \rightarrow {}^{14}\mathrm{N}^* + e^+ + \nu \quad \text{and} \quad {}^{26}\mathrm{Al} \rightarrow {}^{26}\mathrm{Mg}^* + e^+ + \nu \tag{9-6.7}$$

and

$$^{34}\mathrm{Cl} \rightarrow {}^{34}\mathrm{S}^* + e^+ + \nu \tag{9-6.8}$$

Hence we must conclude that both types of interaction are operative. Of course, many transitions (e.g., $n \rightarrow p + e^- + \bar{\nu}$) are allowed by both selection rules, since the rules are not mutually exclusive.

Forbidden transitions occur because the finite extent of the nucleus requires us to consider not simply $\psi(0)$ but also the ψ of the electron and neutrino throughout the entire volume occupied by the nucleus. Thus p, d, and higher waves become evident. In other words, the higher terms in the plane wave development,

$$e^{i\mathbf{k}\cdot\mathbf{r}} = 1 + i\mathbf{k}\cdot\mathbf{r} - \frac{(\mathbf{k}\cdot\mathbf{r})^2}{2} + \cdots \tag{9-6.9}$$

are to be considered even though they are of the order of R/λ or smaller. Another reason for the occurrence of forbidden transitions is the relativistic effects in the nuclear wave functions, which are of the order of $v/c \sim 0.1$. When the higher terms of the plane wave are considered, in forbidden transitions, they introduce powers of p_e and p_ν into the matrix element M_{if}. As a consequence the form of the beta spectrum changes, and the ordinary Kurie plot departs from a straight line. The plot can be reduced to a straight line by introducing the proper factor, which takes into account the dependence of M_{if} on p_e (Fig. 9-15). If the ordinary Kurie plot is not straight, the transition is not allowed; for possible exceptions consult (WM 66). The converse, however, is not always true, and special types of forbidden transitions may give rise to straight Kurie plots.

Selection rules for forbidden transitions depend on the term of the interaction producing the transition. For instance, if a transition is caused by the

ISBN 0-8053-8601-7

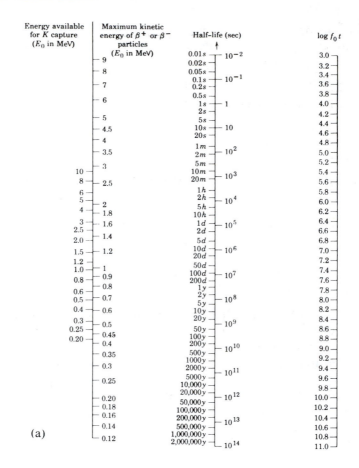

Energy available for K capture (E_0 in MeV)	Maximum kinetic energy of β^+ or β^- particles (E_0 in MeV)	Half-life (sec)	$\log f_0 t$

(a)

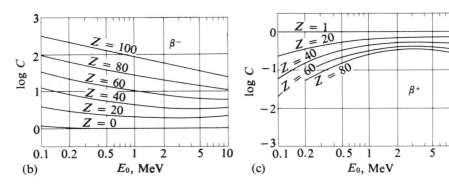

(b)

(c)

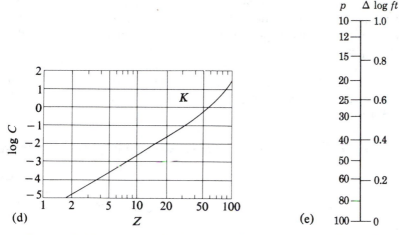

ISBN 0-8053-8061-7

Figure 9-18 Rapid method for calculating $\log_{10}(ft)$ values. [From S. A. Moszkowski, *Phys. Rev.*, **82**, 35 (1951).]

The following figures permit the rapid calculation of $\log(ft)$ for a given type of decay, given energy, half-life, etc. Notation: E_0 for $\beta^{\pm}$ emission is the maximum kinetic energy of the particles in MeV; E_0 for K electron capture is the disintegration energy in MeV. When a β^+ emission and K electron capture go from and to the same level, E_0 for K capture = E_0 for β^+ emission $+ 1.02$ MeV. Z is the atomic number of the initial nuclei, t is the total half-life, and p is the percentage of decay occurring in the mode under consideration. When no branching occurs, $p = 100$.

Procedures for obtaining $\log(ft)$:

1. First obtain $\log(f_0 t)$, using part (a). E_0 is read off the left-hand side of the E_0 column for K electron capture, and off the right-hand side for $\beta^{\pm}$ emission. Put a straight edge over the given values of E_0 and t and note where it crosses the column of $\log(f_0 t)$ values.

2. Then read off $\log(C)$ from parts (b), (c), and (d) for β^-, β^+, and K electron capture, respectively.

3. Get $\Delta \log(ft)$ from part (e) if $p < 100$. When $p = 100$, $\Delta \log(ft) = 0$.

4. $\log(ft) = \log(f_0 t) + \log(C) + \Delta \log(ft)$.

For details concerning the construction, significance, and range of usefulness of these graphs, see original paper.

second term of Eq. (9-6.9), the leptons are emitted in a wave having $l = 1$ and changing parity. The Fermi selection rules then give as second approximation $\Delta I = \pm 1, 0$ (except $0 \to 0$); change of parity, yes. The Gamow–Teller rules, on the other hand, give $\Delta I = \pm 2, \pm 1, 0$ (except $0 \to 0$; $\frac{1}{2} \to \frac{1}{2}$; $0 \to 1$); change of parity, yes. It turns out that the inclusion of terms of the first order in v/c gives the same selection rules. Transitions forbidden by the selection rules (9-6.3) and (9-6.4) but allowed by the rules given above are first forbidden. [For more details on forbidden transitions refer to (WM 66).]

Some typical numerical examples are given in Table 9-1. Note that for forbidden transitions the evaluation of ft must take into account the proper form of the spectrum, which, as pointed out before, may be different from the normal Fermi spectrum.

Up to now we have accounted for two main results: the form of the beta spectrum and the relation between energy and lifetime (ft values). We shall now consider other, more refined experiments, most of which have been performed since 1956.

9-7   FURTHER EXPERIMENTS ON THE WEAK INTERACTION

Further experiments relevant to the analysis of the weak interaction are those on:

1. The asymmetry of beta emission by polarized nuclei.
2. The measurement of the helicity of the neutrino.
3. The observation of $\boldsymbol{\sigma}_e \cdot \mathbf{p}_e$ (helicity of the electron).
4. The observation of the electron–neutrino angular correlation.
5. The simultaneous measurement of the angular correlation of neutron spin direction, electron momentum, and antineutrino momentum in neutron decay.

We shall limit ourselves to qualitative arguments and to the main points, without following the historical order. (1) The first experiment demonstrated the nonconservation of parity and has already been discussed.

(2) We shall discuss an experiment that has established the important fact that the neutrino has helicity -1, or is left-handed (M. Goldhaber, Grodzins, and Sunyar, 1958). The experiment (Fig. 9-19) is as follows: ^{152m}Eu undergoes K capture to become ^{152}Sm. The K capture is followed by emission of a gamma ray of 961 keV. ^{152m}Eu and the ground state of ^{152}Sm have spin 0. The angular momentum carried off by the neutrino and by the gamma ray must therefore balance the angular momentum brought to the nucleus by the K capture, whose magnitude is $\frac{1}{2}$. Let the axis of quantization be the direction of motion of the gamma quantum, and m_γ, m_ν, and m_K represent the components (in this direction) of the angular momentum of the gamma ray, the neutrino, and the captured electron, respectively. Clearly, then,

$$m_\gamma = \pm 1 \qquad m_\nu = \pm \tfrac{1}{2} \qquad m_K = \pm \tfrac{1}{2} \tag{9-7.1}$$

We also have

$$m_\gamma + m_\nu + m_K = 0 \tag{9-7.2}$$

and this leads to two possibilities:

$$m_\gamma = 1 \qquad m_\nu = -\tfrac{1}{2} \qquad m_K = -\tfrac{1}{2} \tag{9-7.3}$$

ISBN 0-8053-8601-7

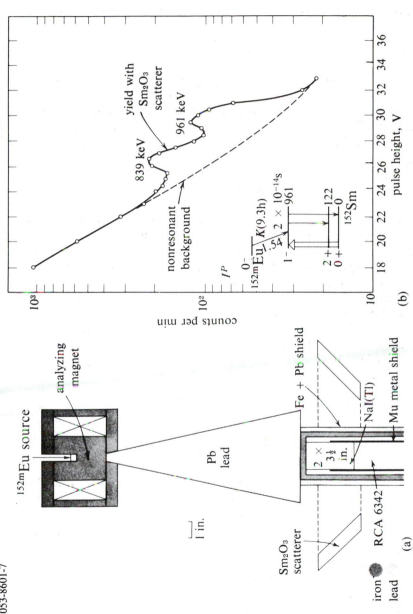

Figure 9-19   The Goldhaber–Grodzins–Sunyar experiment to determine the helicity of the neutrino. (a) The experimental arrangement. The analyzing magnets select a circular polarization of gamma rays. The Sm_2O_3 preferentially scatters nuclear fluorescence radiation into the sodium iodide detector. (b) Level schemes of ^{152}Sm. The radiation following K capture excites the 961-keV state of ^{152}Sm, as indicated by the double arrow. This state decays, emitting the 961- and 839-keV lines that are observed.

or

$$m_\gamma = -1 \qquad m_\nu = \tfrac{1}{2} \qquad m_K = \tfrac{1}{2} \qquad\qquad (9\text{-}7.4)$$

Observation of the circular polarization of the 961-keV gamma ray indicates that

$$m_\gamma = -1 \qquad\qquad (9\text{-}7.5)$$

$$m_\nu = +\tfrac{1}{2} \qquad \text{or} \qquad 2\mathbf{m}_\nu \cdot \hat{\mathbf{p}}_\gamma = 1 \qquad\qquad (9\text{-}7.6)$$

To determine the helicity of the neutrino, we still have to find its momentum. It will be shown shortly that the neutrino's momentum is opposite to that of the gamma ray. From this we conclude that the helicity of the neutrino is negative:

$$\mathcal{H} = 2\hat{\mathbf{p}}_\nu \cdot \mathbf{m}_\nu = -1 \qquad\qquad (9\text{-}7.7)$$

To show that the momentum of the neutrino is opposite to that of the gamma ray, we note that nuclear resonance scattering of gamma rays occurs only if the source moves toward the scatterer (cf. Chap. 8) at the time of the gamma emission. This happens only if the preceding ejection of the neutrino has occurred in the opposite direction, because then the nuclear recoil moves the nucleus in the direction of the gamma ray. The use of gamma rays that are resonance scattered by an Sm target ensures the condition mentioned above. A -1 helicity for the neutrino is thus directly proved. A similar, but less clear-cut, result is obtained by using ^{203}Hg (Palathingal, 1970); it shows that the antineutrino is right-handed. This is consistent with an intrinsic difference between ν and $\bar\nu$, as discussed in Sec. 9-2.

For point 3, helicity of the electrons in beta decay, several measurements have been made on ^{32}P, ^{60}Co, ^{22}Na, ^{66}Ga, ^{68}Ga, ^{198}Au, and others. The result is a value for the helicity of $+v/c$ for positrons and of $-v/c$ for electrons in all cases examined. In the case of ^{60}Co (Lazarus and Greenberg, 1970) the helicity is $-v/c$ (1.01 ± 0.02). The principle of the experiments is to scatter beta rays of chosen momentum on a thin foil of magnetized iron. Two electrons of each iron atom are oriented parallel to the magnetizing field, which is parallel or antiparallel to the momentum of the impinging beta rays. The electron–electron scattering depends on the mutual spin orientations being greater for antiparallel than for parallel spin. The same is true for electron–positron collisions in the relativistic case (Fig. 9-20). (See Si 65.)

It is therefore possible by changing the direction of magnetization in the iron to determine the degree of longitudinal polarization of the beta rays. We thus have the important results summarized in Table 9-2; these helicities are the maximum possible for ν and $\bar\nu$, and they are consistent with the two-component theory of the neutrino (see Sec. 9-8). Hence we can define the

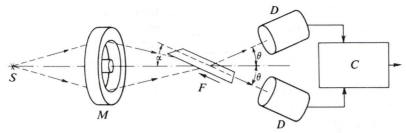

Figure 9-20 Basic arrangement for the measurement of electron and positron helicity by Møller and Bhabha scattering. S, source; M, monochromator; F, thin iron foil magnetized in the direction of arrow; D, detectors; C, coincidence. Electrons scattered by polarized electrons of the foil are detected, using the coincidences between the two electrons. The cross section depends on the mutual orientation of the spins. A change of the direction of magnetization in the foil thus affects the scattering cross section if the incident electrons are longitudinally polarized. [From J. D. Ullman, H. Frauenfelder, H. J. Lipkin, and A. Rossi, *Phys. Rev.*, **122**, 536 (1961).]

TABLE 9-2 HELICITIES

Particle	ν	$\bar{\nu}$	e^-	e^+
Helicity	-1	1	$-v/c$	v/c

neutrino as a left-handed particle and the antineutrino as a right-handed particle.

The v/c electron helicity is related to the analysis of that particle's average velocity as compared with its instantaneous velocity. In Dirac's theory the operator $\dot{x} = (i/\hbar)(Hx - xH)$ has the eigenvalues $\pm c$; hence, an observation of $\dot{x}$ can give only the results $\pm c$. This is related to the uncertainty relation $\Delta E \, \Delta t = \hbar$ because in order to measure the instantaneous velocity $\dot{x}$ with great precision, we must do it in a very short time; hence, ΔE goes to infinity and one finds the velocity c. The average velocity, usually observed, is defined through the momentum by $c^2 p / W = \langle \dot{x} \rangle$ and results from a back and forth motion with $\dot{x} = \pm c$ (Zitterbewegung). In order to find the average velocity $\langle \dot{x} \rangle$, call a^2 and $1 - a^2$ the probabilities of finding $\dot{x} = +c$ and $\dot{x} = -c$, respectively. One must then have

$$a^2 c - (1 - a^2)c = \langle \dot{x} \rangle \tag{9-7.8}$$

We may assume that for $\dot{x} = c$ the polarization in the direction of motion reaches its maximum value, 1. The observed value is the result of the spin flipping from $+1$ to -1 as the velocity changes from $+c$ to $-c$. The average longitudinal component of the spin is then

$$a^2 - (1 - a^2) = \langle \dot{x} \rangle / c \tag{9-7.9}$$

Experiments of type 4 show that in beta decay electrons are generated together with antineutrinos and positrons are generated together with neutrinos. The relation of this fact to the angular correlation between positron and neutrino momentum may be seen in the case of a beta transition occurring without change of angular momentum. The neutrino and the positron, having helicities of opposite sign, must tend to move parallel in order not to carry away angular momentum; hence they will show an angular correlation. Similar qualitative arguments are valid in other cases.

Experiment 5 is one of the most informative on beta decay; because of its importance it is discussed in Sec. 9-9.

There are also several types of correlation experiments relevant to the form of the beta interaction, such as the correlation between the direction of beta emission, that of a subsequent gamma emission, and the helicity of the particles involved. We will briefly describe one example only.

Because of the nonconservation of parity, beta emission from polarized nuclei depends on the angle between the nuclear spin and electron momentum. Conversely, a nucleus that emits an electron is left in a certain state of polarization, that is, in a state with a definite quantum number m, assuming the electron momentum as the axis of quantization. Gamma emission from this state is circularly polarized (think, e.g., of a state $J = 1$, $m = 1$ going to $J' = 0$, $m' = 0$). We thus expect that the probability of gamma emission depends on the angle between the beta and gamma rays as well as on the helicity of the gamma. Theory gives for the emission probability

$$W_\pm (\theta) = 1 \pm A(v/c) \cos \theta$$

where θ is the angle between the beta and gamma rays, v the electron velocity, and the $+$ and $-$ refer to the gamma helicity $+1$ or -1 respectively. A depends on the nature of the beta interaction. The helicity of the gammas is measured from Compton scattering on magnetized iron.

Weak interactions in nuclei manifest themselves not only in beta decay but also in some phenomena that, though small in magnitude, are of considerable interest. Because weak interactions do not conserve parity, they mix, by perturbation, states of different parities; hence, the wave function of a state has the form $\psi = \psi^+ + \mathcal{F}\psi^-$ where by $+$ or $-$ we indicate the parity of the states. These wave functions of indefinite parity permit transitions that would be strictly forbidden by parity conservation. Examples include a weak alpha transition from a $J^P = 2^-$ of ^{16}O to the ground state of ^{12}C and gamma transitions in several nuclei (^{175}Lu, ^{180}Hf, ^{181}Ta, etc.) in which one observes circular polarization in the gamma rays emitted by unpolarized nuclei or angular asymmetry in the gamma emission from polarized nuclei. From these observations one derives the magnitude of $\mathcal{F}$, which is of the order of 10^{-6} to 10^{-7}.

9-8 THEORY OF THE BETA INTERACTION

We now come to an important point of the theory of beta decay, which, of necessity, can be explained only by the use of fairly advanced relativistic quantum mechanics. Rather than leave this argument untouched, we shall try to give some idea of the subject, referring for more complete treatment and proofs to (WM 66).

We have already pointed out that the matrix element of beta decay (basically the transformation of a neutron into a proton with emission of an electron and antineutrino) contains four wave functions *linearly*. The particles involved are all fermions of spin $\frac{1}{2}$, and their eigenfunctions should be four-component spinors in relativistic (Dirac) theory. This is certainly the case for the nucleons and the electron; the fact that the neutrino has a mass 0 and that the beta interaction is not parity conserving makes possible a simpler description of that particle (with two components only). We shall touch on this description later. For the time being, if each particle is described by a spinor and we want the interaction hamiltonian to be a scalar, we must combine the four spinors two at a time in forms that may be a scalar, vector, tensor, axial vector, or pseudoscalar; then we combine two of these pairs of scalars, vectors, etc., in such a way as to form a scalar.

In his original work Fermi explicitly chose a form in which the beta radiation is generated proportionally to the leptonic current. This choice was suggested by the analogy with electromagnetic radiation, which is also generated in proportion to the electric current. However, the electromagnetic interaction is a long-range force (potential $1/r$), whereas the weak interaction occurs only on contact, that is, when the wave functions of the interacting particles overlap. From the point of view of Yukawa (range $= \hbar/mc$), the light quanta have rest mass 0, whereas those of the weak interaction are very massive.

● The four-current producing the transition in Dirac's theory is written as

$$J_\mu = \bar{\psi}_p \gamma_\mu \psi_n \qquad (9\text{-}8.1)$$

where the γ_μ are the 4×4 Dirac matrices defined by

$$\gamma_5^2 = \gamma_\mu^2 = 1 \qquad \gamma_\mu \gamma_\nu + \gamma_\nu \gamma_\mu = 2\delta_{\mu\nu} \qquad \mu, \nu = 1 \cdots 4$$

$$\gamma_5 \gamma_\mu + \gamma_\mu \gamma_5 = 0 \qquad (9\text{-}8.2)$$

or explicitly, for instance:

$$\gamma_k = \begin{vmatrix} 0 & -i\sigma_k \\ i\sigma_k & 0 \end{vmatrix} \quad k = 1, 2, 3 \qquad \gamma_4 = \begin{vmatrix} 1 & 0 \\ 0 & -1 \end{vmatrix} \qquad \gamma_5 = \begin{vmatrix} 0 & -1 \\ -1 & 0 \end{vmatrix}$$

$$(9\text{-}8.3)$$

where the σ_k are the 2×2 Pauli matrices and 1 is the 2×2 unit matrix. The ψ_p, ψ_n are the proton and neutron four-component Dirac wave functions, and $\bar{\psi}$ is $\psi^\dagger \gamma_4$ where $\psi^\dagger$ means the hermitian conjugate (h.c.) of ψ.

Fermi then assumed as the hamiltonian for the beta-decay interaction

$$H = g\left(\bar{\psi}_p \gamma_\mu \psi_n\right)\left(\bar{\psi}_e \gamma_\mu \psi_\nu\right) + \text{h.c.} \equiv gV \qquad (9\text{-}8.4)$$

The factors in this interaction are four-vectors, hence the name vector interaction. This form of interaction produces the F selection rules. Since there are allowed transitions obeying GT selection rules, we must have other terms in the interaction to produce them. One of the other possible choices is

$$H = g\left(\bar{\psi}_p \gamma_\mu \gamma_5 \psi_n\right)\left(\bar{\psi}_e \gamma_\mu \gamma_5 \psi_\nu\right) + \text{h.c.} \equiv gA \qquad (9\text{-}8.5)$$

The factors in this interaction are axial vectors, hence the name axial vector interaction. A linear combination of these two is also admissible.

There are other possible choices, for example, scalar, giving rise to F selection rules, and tensor, giving rise to GT selection rules (see Fig. 9-21).

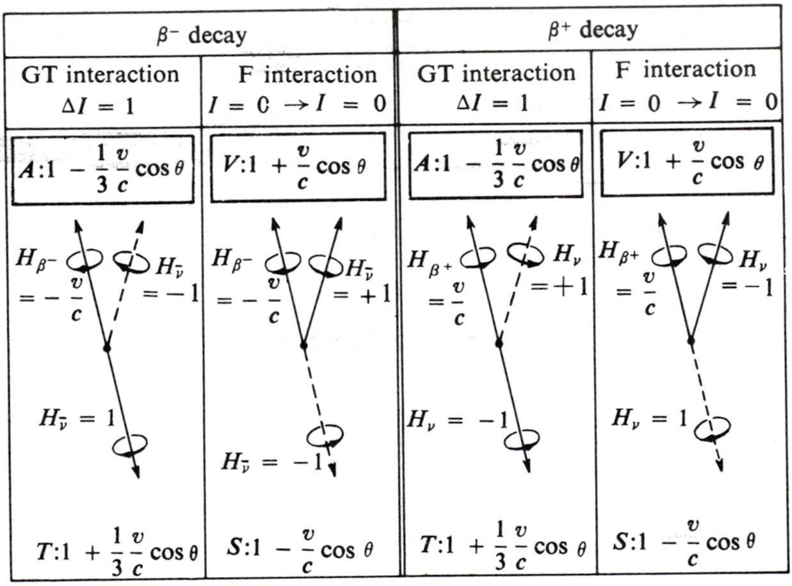

Figure 9-21 Correlations of neutrino helicities and beta interactions. The interactions and helicities realized in nature are framed in squares and represented by solid lines; other alternative interactions not realized in nature are unframed and represented by dashed lines. Angle θ is the angle between p_ν and p_e. The factor $\frac{1}{3}$, in the correlation when $\Delta I = 1$ is due to the three orientations of the initial and final I. For instance, if $I_i = 1$ and $I_f = 0$, then $M_i = \pm 1, 0$ and $M_f = 0$; the effect is to dilute the angular correlation.

But these choices predict a wrong angular correlation between the direction of the neutrino and the direction of the electron. Therefore we discard them. To see this point, at least qualitatively, consider the fact that in an F transition a right-handed antineutrino (helicity $+1$) must be accompanied by an electron of opposite spin, because together they carry off zero angular momentum. Electrons, according to Table 9-2, are generated with negative helicity; thus the momenta of electron and antineutrino must tend to be parallel, as predicted by calculations based on vector interaction. Calculations based on scalar interaction favor antiparallelism between the momenta and make the interaction unacceptable. ●

In GT transitions electrons and antineutrinos have parallel spin, because together they carry away one unit of angular momentum. The antineutrino has helicity $+1$, the electron is generated with negative helicity, and hence they must tend to have opposite momenta, as calculations based on axial-vector interaction predict. Calculations based on tensor interaction favor parallelism between the momenta and make the interaction unacceptable.

The situation is illustrated schematically in Fig. 9-21 and Table 9-3. Experimental verification is secured by measuring the electron momentum and the nuclear recoil. The neutrino momentum is then calculated on the principle of momentum conservation. When the electron and antineutrino tend to move parallel to each other, large values of the nuclear-recoil momentum are favored. When the electron and antineutrino tend to move in opposite directions, small nuclear recoils are favored.

TABLE 9-3   CORRELATION BETWEEN MOMENTA AND HELICITIES

Momenta of $e^{\pm}$ and ν or $\bar{\nu}$	Parallel	Opposite
GT	Tensor	Axial vector
Fermi	Vector	Scalar

Observing the recoil nucleus energy or momentum distribution in the pure GT transition occurring, for example, in ^{6}He, and comparing it with the calculated one, we conclude that GT transitions are associated with the axial-vector interaction (see Fig. 9-4). From similar arguments and experiments for ^{35}Ar, for example, we conclude that F transitions are due to the vector interactions. A linear combination

$$g(C_A A + C_V V) \tag{9-8.6}$$

is also admissible.

Expression (9-8.6) for the hamiltonian is not, however, broad enough to describe beta decay, because the expression is parity conserving. Lee and

ISBN 0-8053-8061-7

Yang have generalized Eq. (9-8.6), obtaining parity violation by writing

$$H = g \sum_{i=V,A} \left(\bar{\psi}_p O_i \psi_n\right)\left[\bar{\psi}_e O_i (C_i + C_i' \gamma_5)\psi_\nu\right] + \text{h.c.} \qquad (9\text{-}8.7)$$

where $O_V = \gamma_\mu$ is the V operator of Eq. (9-8.4) and $O_A = \gamma_\mu \gamma_5$ is the A opera-tor of Eq. (9-8.5). This hamiltonian brings about parity nonconservation by the simultaneous presence of the two terms in the square brackets. If either C_i or C_i' vanishes, the hamiltonian conserves parity.

Consider now the behavior of H when we perform the operation of space inversion (P), when we replace a particle by its antiparticle (charge conjuga-tion C), or when we invert the direction of time (T).

Investigation of Eq. (9-8.7) shows that these operations leave its form unchanged except for the coefficients (complex numbers) C and C'. These are changed by the transformation indicated in Table 9-4.

TABLE 9-4   CHANGES OF COEFFICIENTS C_i AND C_i' UNDER DIFFERENT TRANSFORMATIONS

Transformation	C	C'	Condition for invariance of Eq. (9-8.7)
Space inversion (P)	C	$-C'$	$C' = 0$ or $C = 0$
Charge conjugation (C)	C^*	$-C'^*$	C real and C' imag. or C imag. and C' real
Time reversal (T)	C^*	C'^*	C and C' real
TCP	C	C'	Automatically satisfied

There is an important theorem, due to Schwinger, Lüders, and Pauli, which states that under very broad conditions, and certainly for theories developed up to now, the hamiltonian is invariant under the operation CPT (the order of C, T, P is unimportant). This theorem is trivial for electromagnetic and strong interactions, where the hamiltonian is invariant for each operation, but becomes very important in weak interactions, where invariance for P is cer-tainly violated. We shall postulate that in beta decay invariance under time reversal is preserved, and hence that invariance for the product CP is also preserved. Its meaning is easily illustrated with reference to Fig. 9-10. Part (b) of this figure *becomes correct* as drawn, provided cobalt is replaced by its charge conjugate, that is, by a nucleus derived from ordinary cobalt by re-placing each proton by an antiproton and each neutron by an antineutron. Although this experiment is at present practically unfeasible with nuclei, simi-lar experiments with positive and negative mesons support CP invariance.

The decay of mesons is similar to beta decay and here two pairs of charge-conjugate particles are known: the positive and negative pion and the positive and negative muon (see Chaps. 14 and 15). For each pair the decay

ISBN 0-8053-8061-1

process of particle and antiparticle are mirror images of each other. Thus in these cases the *CP* transformation gives a result in agreement with experiment. This, together with the *CPT* theorem, proves the *T* invariance in these cases.

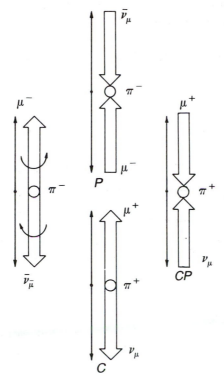

Figure 9-22 Schematic illustration of the effect of *C* and *P* operations on the most common pion decay, $\pi^- \to \mu^- + \bar{\nu}_\mu$. The initial π^- decay is transformed by the *C* or *P* operations into physically unrealized situations. The *CP* operation describes the physically realized π^+ decay. Arrows indicate momenta, double arrows spins. The helicity of $\bar{\nu}_\mu$ is +1, that of ν_μ is −1. The pion spin is zero.

Because *T* invariance is very likely to be correct in beta decay and because it simplifies further discussion appreciably, we shall make use of it. In one case however, *CP* invariance has been proved to be violated. This is the decay of a neutral meson called K_L. We believe that this decay is accompanied by a *T* violation that salvages *CTP* invariance. Up to now this case remains unique. We shall discuss it in Sec. 19-5.

Time-reversal invariance requires, as stated above, that coefficients C_A, C_A' and C_V, C_V' be real. Another important relation among the $C_{A,V}$ coefficients

ISBN 0-8053-8061-7

derives from the helicities given in Table 9-2. These values produce the relations

$$C_A = C_A' \qquad C_V = C_V' \tag{9-8.8}$$

which is equivalent to accepting a two-component theory of the neutrino. The hamiltonian can thus be written as

$$H = g \sum_{i=A,\,V} C_i\left(\bar{\psi}_p O_i \psi_n\right)\left[\bar{\psi}_e O_i (1 + \gamma_5)\psi_\nu\right] + \text{h.c.} \tag{9-8.9}$$

To complete the analysis we must next find the ratio C_A/C_V and the universal constant g. A study of the beta decay of the neutron and of some other especially suitable nuclei answers these questions.

9-9 QUANTITATIVE STUDY OF SOME MATRIX ELEMENTS

To obtain numerical results, we first introduce a considerable simplification by treating the nuclear part of the matrix element in a nonrelativistic approximation. This is permissible because v_n/c is relatively small. We need to consider, then, only the two large components of the four-component nuclear eigenfunctions. The matrix elements corresponding to vector interaction in this approximation are simply

$$\left|\int u_f^* u_i \, d\tau\right|^2 = |M_F|^2 \tag{9-9.1}$$

whereas those corresponding to axial-vector interaction are

$$\left|\int u_f^* \boldsymbol{\sigma} u_i \, d\tau\right|^2 = |M_{GT}|^2 \tag{9-9.2}$$

where $\boldsymbol{\sigma}$ is an axial-vector operator whose components are the Pauli spin matrices.

Transitions of the first type, having nuclear matrix element M_F, are Fermi transitions; those of the second type are Gamow–Teller transitions. It is clear from the orthogonality of the nuclear wave functions that M_F vanishes if initial and final states are different (e.g., have different I values), except for the transformation of a neutron into a proton, and that M_{GT} also requires a spin flip in order not to vanish. Furthermore, the initial and final state (in each case) must have the same parity.

Here we see how the parity-nonconserving beta decay imposes a condition on the relative parity of the initial and final nuclear state.

We may go one step further and try to evaluate M_F and M_{GT} in specific cases. We note also that for practical purposes one needs an $\overline{M}_{if}$, where the

ISBN 0-8053-8061-7

bar indicates average over initial and sum over final z components of angular momentum.

The simplest cases are those in which a single neutron becomes a proton, or vice versa, without any other change in the eigenfunction. Examples are the decay of the free neutron without spin flip and the decay of ^{14}O to an excited state of ^{14}N, both having $I = 0$. For the first case $|M_F|^2 = 1$; for the second, $|M_F|^2 = 2$. For the decay of the free neutron with spin flip

$$\overline{|M_{GT}|^2} = 3 \tag{9-9.3}$$

The evaluation of these matrix elements follows from Eqs. (9-9.1) and (9-9.2), but we shall not develop the relevant calculations. We have assumed that the interactions operative in beta decay are of the A and V type only. The first gives rise to GT selection rules, the second to F selection rules.

Under this assumption we can calculate $(gC_V)^2$ from a transition allowed only by F selection rules and for which $\overline{|M_F|^2}$ is calculable. Such is the decay of ^{14}O. One then has (compare Eq. 9-5.27)

$$g^2 C_V^2 = \frac{2\pi^3}{\overline{|M_F|^2}} \frac{\hbar^7}{m^5 c^4} \frac{\log 2}{ft} \tag{9-9.4}$$

which, with $\overline{|M_F|^2} = 2$ and with the observed ft of 3127 sec (see Table 9-1, which includes some small radiative corrections), gives

$$gC_V = 1.4029 \pm 0.0022 \times 10^{-49} \text{ erg cm}^3 \tag{9-9.5}$$

For the neutron we can write similarly

$$ft = \frac{1}{g^2} \frac{2\pi^3 \hbar^7}{m^5 c^4} \frac{\log 2}{C_V^2 \overline{|M_F|^2} + C_A^2 \overline{|M_{GT}|^2}} \tag{9-9.6}$$

with $\overline{|M_F|^2} = 1$, $\overline{|M_{GT}|^2} = 3$, and with an observed ft of 1115 sec (see Table 9-1). The ratio between ft(neutron) and $ft(^{14}$O) then gives

$$\frac{2C_V^2}{C_V^2 + 3C_A^2} = 0.3566 \tag{9-9.7}$$

or

$$\frac{C_A^2}{C_V^2} = 1.53 \tag{9-9.8}$$

The study of neutron beta decay also gives information as to the relative sign of C_V and C_A.

ISBN 0-8053-8061-7

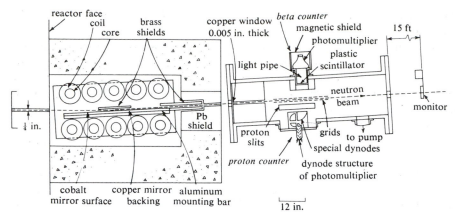

Figure 9-23 Horizontal section through the center of the apparatus used for studying the decay of polarized neutrons. [M. T. Burgy, V. E. Krohn, T. B. Novey, G. R. Ringo, and V. L. Telegdi, *Phys. Rev.*, **120**, 1829 (1960).] For a more recent version of this experiment see B. G. Erozolimsky, *Soviet J. Nucl. Phys.*, **11**, 583 (1970), and A. I. Steinberg et al., *Bull. Am. Phys. Soc.*, **19**, 514 (1974).

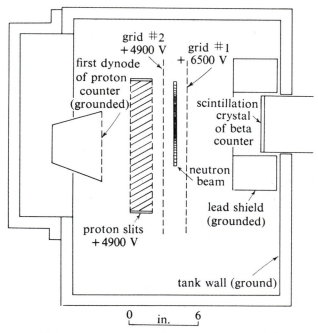

Figure 9-24 Detector arrangement for measurement of the correlation between the directions of neutron spin and antineutrino momentum. This is a cross section of the detector part of Fig. 9-23.

ISBN 0-8053-8061-7

It has been possible to measure experimentally the $\mathbf{p}_e$–$\mathbf{I}$ and $\mathbf{p}_\nu$–$\mathbf{I}$ angular correlation in neutron and ^{19}Ne decay. A beam of polarized neutrons (Figs. 9-23 and 9-24) (see Chaps. 10 and 12) passes into an evacuated tank, and both the electrons and the recoil protons coming from the decay in flight are observed. It is found that there is a very small negative angular correlation between $\mathbf{I}$ and $\mathbf{p}_e$ and a strong positive correlation between $\mathbf{I}$ and $\mathbf{p}_\nu$.

More quantitatively, indicate by $\mathbf{I}$ the unit vector in the direction of the neutron spin, by $\mathbf{v}$ the electron velocity divided by c, and by $\hat{\mathbf{q}}$ the unit vector in the direction of the neutrino momentum. One has an emission probability proportional to

$$1 + a\mathbf{v} \cdot \hat{\mathbf{q}} + \mathbf{I} \cdot \{A\mathbf{v} + B\hat{\mathbf{q}} + D\mathbf{v} \times \hat{\mathbf{q}}\}$$

and similarly for ^{19}Ne decay. Table 9-5 gives the values found for A, B, and D and the $V - A$ predictions. This is evidence that the coefficients C_A and C_V have opposite sign, as indicated in Fig. 9-25. Their magnitude is not very

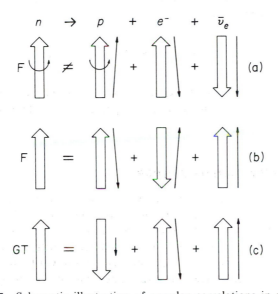

Figure 9-25 Schematic illustration of angular correlations in neutron beta decay. Double arrows: spin directions; single arrows: momenta directions. In (a) and (b) nucleon spin is unchanged. Since the helicities are $h(\bar{\nu}) = +1$; $h(e) = -(v/c) \cong -1$, the relative orientations of lepton spins and momenta must be as shown. Conservation of linear and angular momenta requires the proton to have the momenta indicated. Experiment shows $C_A \cong C_V$ (i.e., the proton is also left-handed). Thus alternative (a) actually does not occur. If C_A were equal to $-C_V$, alternative (b) would not occur. Total amplitude is obtained by superposing amplitudes corresponding to (b) and (c), which yields $A \cong 0$, $B \cong 1$, $a \cong 0$.

ISBN 0-8053-8061-7

TABLE 9-5   NEUTRON AND ^{19}Ne DECAY PARAMETERS

	$V - A$	Neutron	^{19}Ne
$T_{1/2}$		10.80 ± 0.16 min	17.36 ± 0.06 sec
E		0.78 MeV	2.206 MeV
A	0	0.115 ± 0.008	-0.039 ± 0.002
B	$1(-1)$	1.01 ± 0.04	-0.90 ± 0.13
D	0	< 0.0025	0.001 ± 0.003
a	0	-0.1 ± 0.1	

different: one has

$$\frac{|C_A|}{|C_V|} = 1.24 \pm 0.01 \tag{9-9.9}$$

The g constant derived from the ^{14}O, ^{26}Al, ^{34}Cl, and the other beta decays ft parameter has the value

$$g = 1.4149 \pm 0.0022 \times 10^{-49} \text{ erg cm}^3 \tag{9-9.10}$$

We summarize in Table 9-6 some of the most important experiments relevant to beta decay, their immediate interpretation, and their theoretical consequences.

9-10 CONSERVATION OF LEPTONS

We assume that the reader knows the nomenclature and some of the relevant facts about muons and the other particles discussed in Chaps. 13 and 14. In particular, recall that "leptons" are fermions, which are lighter than the nucleon (i.e., the neutrinos, electrons, and muons and their charge conjugates or antiparticles). For the time being, no distinction will be made between the two kinds of neutrinos (see Sec. 9-2). Each particle is assigned a leptonic number as follows:

$$l = 1 \text{ for } e^-, \mu^-, \nu \equiv \nu_L \text{ (helicity} = -1)$$

$$l = -1 \text{ for } e^+, \mu^+, \bar{\nu} \equiv \nu_R \text{ (helicity} = +1)$$

$$l = 0 \text{ for all other particles} \tag{9-10.1}$$

ISBN 0-8053-8601-1

TABLE 9-6 MAIN FEATURES OF BETA DECAY

Experiment	*Immediate consequence*	*Theory*
Form of beta spectrum	Three-body decay	Neutrino mass, statistics, spin; general form of theory
^{60}Co asymmetry	Nonconservation of parity and no charge conjugation invariance	
Inverse beta decay and double beta decay	Distinction between neutrino and antineutrino	
ft of ^{14}O* $(0^+ \rightarrow 0^+)$ transition	F selection rules	Interaction V or S, or both
ft of ^{6}He $(0^+ \rightarrow 1^+)$ transition	GT selection rules	Interaction T or A, or both
ft of neutron $(\frac{1}{2}^+ \rightarrow \frac{1}{2}^+)$	Allowed by F and GT selection rules	$|C_{GT}|^2/|C_F|^2 = 1.53$
Recoil spectrum in F transition $(^{35}$Ar)	Angular correlation $e\nu$ $(1 + a_F(v/c)\cos\theta_{e\nu})$ with $a_F = +1$	Interaction V
Recoil spectrum in GT transition $(^6$He and ^{23}Ne)	Angular correlation $e\nu$ $(1 + a_{GT}(v/c)\cos\theta_{e\nu})$ with $a_{GT} = -\frac{1}{3}$	Interaction A
Recoil in K capture ^{152m}Eu$\rightarrow^{152}$Sm and in beta decay ^{203}Hg$\rightarrow^{203}$Tl	Helicity of $\nu = -1$ Helicity of $\bar\nu = +1$	
Electron polarization measurements: ^{60}Co, ^{22}Na, ^{34}Cl, etc.	$H = +v/c$ for e^+ $H = -v/c$ for e	$C_S = -C_S'; C_V = C_V'$ $C_T = -C_T'; C_A = C_A'$
Decay of polarized neutrons and ^{19}Ne	Intensity $\sim 1 + a\hat{\mathbf{q}}\cdot\boldsymbol{\beta} +$ $\mathbf{I}\cdot(A\boldsymbol{\beta} + B\hat{\mathbf{q}} +$ $D\boldsymbol{\beta} \times \hat{\mathbf{q}})$	$D = 0$, time-reversal invariance
	Values of A, B give $C_A \cong C_V$; others $C_i = 0$; $a \cong 0$ gives $|C_A| \cong |C_V|$	
Spectra of ^{12}B and ^{12}N	Form of spectra	Conserved vector current

$\mathbf{I}$ = spin of decaying nucleus; $\hat{q}$ = unit vector in the direction of neutrino velocity; $\boldsymbol{\beta} = \mathbf{v}/c$ of electron.

ISBN 0-8053-8061-7

Writing the typical weak interaction reactions in the forms

$$
\begin{array}{ll}
n \rightarrow p + e^- + \bar{\nu} & p \rightarrow n + e^+ + \nu \\
\pi^+ \rightarrow \mu^+ + \nu & \pi^- \rightarrow \mu^- + \bar{\nu} \\
\mu^+ \rightarrow e^+ + \nu + \bar{\nu} & \mu^- \rightarrow e^- + \nu + \bar{\nu} \\
\mu^- + p \rightarrow n + \nu & \mu^+ + n \rightarrow p + \bar{\nu} \\
K^+ \rightarrow \mu^+ + \nu & K^- \rightarrow \mu^- + \bar{\nu} \\
\pi^+ \rightarrow e^+ + \nu & \pi^- \rightarrow e^- + \bar{\nu} \\
e^+ + e^- \rightarrow 2\gamma &
\end{array}
\tag{9-10.2}
$$

we see that the number of leptons on the right side of the equation appears to equal that on the left. This equality is called the law of conservation of leptons. It is valid regardless of the interaction operative in the reaction.

The leptonic number of one particle, the e^-, say, is arbitrary. But once it is decided that l for e^- is $+1$, the conservation of leptons assigns to $\bar{\nu}$—defined by the first of Eqs. (9-10.2)—the leptonic number -1, and to ν (defined by the second) the leptonic number $+1$, because e^+ according to the last equation (9-10.2) has $l = -1$. Similarly, other leptonic numbers are uniquely determined. Equations (9-10.2) have a physical content; for example, the helicities of neutrino and antineutrino differ and the consequent angular correlations are sometimes observable.

The origin of the law of conservation of leptons, as well as of the similar law of conservation of nucleons, is unknown. Although no exception to lepton conservation is known, one fails to observe certain reactions that one would expect if the assignments in Eqs. (9-10.2) were correct and *complete*. For instance, the reactions

$$
\mu^- \rightarrow e^- + \gamma \qquad \text{or} \qquad \mu^\pm \rightarrow e^\pm + e^\pm + e^\mp
\tag{9-10.3}
$$

would be obtained from

$$
\mu^\pm \rightarrow e^\pm + \bar{\nu} + \nu
\tag{9-10.4}
$$

by the suppression of the neutrino–antineutrino pairs. In fact, the branching ratios are less than 2×10^{-8} and 1.3×10^{-7}, respectively. This raised the suspicion that there might be more than one kind of neutrino. Direct experiment (see Sec. 9-2) has now shown that this is the case.

The discovery of two types of neutrinos, ν_e and ν_μ, sharpens the classification of leptons. These are of two types: electronic, comprising e^-, ν_e, and their antiparticles; and muonic, comprising μ^-, ν_μ, and their antiparticles. The two kinds of leptons, electronic and muonic, are conserved separately. Then reactions such as $\mu^- \rightarrow e^- + \gamma$ or $\mu^- \rightarrow e^+ + e^- + e^-$ are forbidden because the muonic number (number of muonic leptons − number of muonic

ISBN 0-8053-8061-7

antileptons) or the similar electronic number is not conserved. Strong direct evidence for two kinds of neutrinos is offered also by the observation at high energy of production reactions such as $\nu_\mu + n \rightarrow p + \mu^-$. This reaction is produced from neutrinos originating in $\pi^+ \rightarrow \mu^+ + \nu_\mu$ decay. Such neutrinos are unable to produce the reaction $\nu_\mu + p \rightarrow n + \mu^+$, which is forbidden if ν_μ and $\bar{\nu}_\mu$ are distinct. Furthermore, they do not produce the reaction $\nu_\mu + n \rightarrow p + e^-$, which would be permitted if ν_μ and ν_e were not different and separately conserved. (See Fig. 9-8.)

9-11 UNIVERSAL FERMI INTERACTION

It has been noted by many physicists that the phenomenon of the decay of the mu meson into an electron and a $\nu + \bar{\nu}$ pair is closely analogous to ordinary beta decay. For instance, the decay of the muon may be described in a manner very closely resembling the description of neutron decay if we replace

$$n \quad p \quad e^- \quad \bar{\nu}_e$$

by

$$\mu^- \quad \nu_\mu \quad e^- \quad \bar{\nu}_e \tag{9-11.1}$$

and use an interaction of the form

$$g(V - A) \tag{9-11.2}$$

The value of g in the mu-meson decay differs no more than 3% from the value derived from neutron decay, provided that we write the interaction for the neutron as

$$g\left(V - \frac{C_A}{C_V} A\right) \quad \text{with} \quad \frac{C_A}{C_V} = -1.239 \pm 0.011 \tag{9-11.3}$$

It is then very tempting to assume Eq. (9-8.9) as the fundamental law of beta decay. It is possible that there are enough corrections in the beta decay of the neutron, due to mesonic effects, to account for the different coefficients of V and A appearing in Eq. (9-11.3).

Feynman and Gell-Mann have developed some theoretical arguments to explain why the coefficient of V is unaffected in passing from the system $npe\bar{\nu}$ to the system $\mu e\nu\bar{\nu}$, whereas the coefficient of A is modified. The change is attributed to pionic effects that cancel out exactly for the V part and affect only the A part.

The Feynman and Gell-Mann hypothesis, called the *conserved vector current* theory, gives an especially elegant formulation of the beta interaction,

ISBN 0-8053-8601-7

based on analogies with electromagnetism, and yields automatically the $V - A$ form of the weak interaction. Important experimental support for this theory has come from experiments by Lee, Mo, and Wu (1963). They have shown that the beta-ray spectra in the decays of ^{12}B and ^{12}N to ^{12}C take a form that can be accounted for quantitatively by the conserved vector current theory.

A rare and particularly interesting mode of decay of the pion is

$$\pi^+ \to \pi^0 + e^+ + \nu_e$$

If we compare the particles involved to

$$n \to p + e^- + \bar{\nu}_e$$

the analogy with ordinary beta decay is apparent. The observed decay, has a branching ratio of the order of 10^{-8}, in agreement with the result of a calculation, based on the conserved current hypothesis. Here we conclude our discussion of the subject of beta decay. The weak interactions in general will be reconsidered in Chap. 19.

BIBLIOGRAPHY

Burgy, M. T., V. E. Krohn, T. B. Novey, G. R. Ringo, and V. L. Telegdi, "Measurements of Spatial Asymmetries in the Decay of Polarized Neutrons," *Phys. Rev.*, **120**, 1829 (1960).

Commins, E. (Com 73).

Cowan, C. L., and F. Reines, "Detection of the Free Neutrino," *Phys. Rev.*, **92**, 830 (1953).

Danby, G., J. M. Gaillard, K. Goulianos, L. M. Lederman, N. Mistry, M. Schwartz, and J. Steinberger, "Observation of High Energy Neutrino Reactions and the Existence of Two Kinds of Neutrinos," *Phys. Rev. Letters*, **9**, 36 (1962).

Deutsch, M., and O. Kofoed-Hansen, "Beta-Rays," in (Se 59), Vol. III.

Feinberg, G., and L. M. Lederman, "The Physics of Muons and Muon Neutrino," *Ann. Rev. Nucl. Sci.*, **13**, 431 (1963).

Fermi, E., "Tentativo d'una teoria dei raggi beta," *Nuovo Cimento*, **11**, 1 (1934).

Goldhaber, M., L. Grodzins, and A. W. Sunyar, "Helicity of Neutrinos," *Phys. Rev.*, **109**, 1015 (1958).

Henley, E. M. "Parity and Time Reversal Invariance in Nuclear Physics," *Ann. Rev. Nucl. Sci.*, **19**, 367 (1969).

Inghram, M. G., and J. H. Reynolds, "Double Beta-Decay of ^{130}Te," *Phys. Rev.*, **78**, 822 (1950).

Konopinski, E. J., *The Theory of Beta Radioactivity*, Oxford University Press, New York, 1966.

Lee, T. D., and C. S. Wu, "Weak Interactions," *Ann. Rev. Nucl. Sci.*, **15**, 381 (1965), **16**, 471 (1966).

Lee, T. D., and C. N. Yang, "Question of Parity Conservation in Weak Interactions," *Phys. Rev.*, **104**, 254 (1956).

Lipkin, H. J., *Beta Decay for Pedestrians*, Wiley-Interscience, New York, 1962.

Morita, M., *Beta Decay and Muon Capture*, Benjamin, New York, 1973.

Preston, M. A. (P 62).

Reines, F., "Neutrino Interactions," *Ann. Rev. Nucl. Sci.*, **10**, 1 (1960).

ISBN 0-8053-8601-7

Siegbahn, K. (Si 65).

Wu, C. S., E. Ambler, R. W. Hayward, D. D. Hoppes, and R. P. Hudson, "Experimental Test of Parity Conservation in β-Decay," *Phys. Rev.*, **105**, 1413 (1957).

Wu, C. S., and S. A. Moszkowski (WM 66).

PROBLEMS

9-1 Suppose that the cross section for the interaction of neutrinos with either protons or neutrons is 10^{-43} cm^2 per nucleon. Taking the mean density of the earth as 5 g cm^{-3} and the radius as 6×10^6 m, what is the probability that a neutrino will pass through the interior of the earth *without* interacting along the way?

9-2 The nucleus $^{62}_{30}$Zn can decay either by positron emission or by K capture. The maximum kinetic energy for the positron is 0.66 MeV. (*a*) Calculate the maximum neutrino energy in positron decay. (*b*) Calculate the neutrino energy emitted in K capture. Neglect recoil and electron binding-energy corrections.

9-3 (*a*) A certain number of nuclei can decay by electron emission, by positron emission, and by electron capture. Give arguments to show that only odd Z, even A nuclei can have this property. (*b*) $^{64}_{29}$Cu is such a nucleus. From the atomic masses given below, calculate (i) the maximum β^+ and β^- kinetic energies, (ii) the neutrino energy in electron capture, and (iii) the kinetic energy of the recoil nucleus in electron capture. Relevant atomic masses:

$$
\begin{array}{ll}
^{64}_{29}\text{Cu} & 63.92976 \text{ amu} \\
^{64}_{28}\text{Ni} & 63.92796 \text{ amu} \\
^{64}_{30}\text{Zn} & 63.92914 \text{ amu}
\end{array}
$$

9-4 In what energy interval is L capture possible, K capture impossible? See whether you can find any example of this phenomenon.

9-5 ^{187}Re decays by electron emission into ^{187}Os. The energy of the decay is 2 keV. As a consequence, the interactions between the decay electron and the atomic electrons (negligible in ordinary cases) significantly affect the decay constant. Estimate the effects.

9-6 Calculate the allowed *energy* distribution of beta-decay electrons, assuming that the neutrino has a very small rest mass m_ν. Show that for energies near the upper limit, the shape of the distribution changes markedly from that obtained in the case of $m_\nu = 0$.

9-7 Estimate the difference between the decay constants of neutral ^{7}Be and twice-ionized ^{7}Be, using even crudely approximate wave functions. Suppose that it were possible to produce ^{7}Be^{3+}. What would be its mean life?

9-8 Show that the muon and neutron decay give the same value of g.

9-9 Suppose we want to show that in beta decay only one neutrino is emitted (and not, say, two). Calculate the electron momentum distribution when two neutrinos are emitted. (*Note:* In this case, the momentum of the electron does not determine the momenta of the two neutrinos, and in the phase-space distribution both the electron momentum and one of the neutrino momenta enter as variables. To get the expected electron distribution, it is necessary to integrate over values of the neutrino momentum.)

9-10 Plot the form of a beta spectrum as a function of momentum or energy. How does one pass from the data obtained directly from a magnetic spectrometer to a Kurie plot?

9-11 Calculate the GT matrix element for the neutron and show that it is 3.

9-12 Calculate the F matrix element for the neutron and show that it is 1.

9-13 Show that from Fig. 9-13 one obtains the Kurie plot of Fig. 9-14.

9-14 In beta decay the electron momentum is often of the order of $p = m_e c$. Calculate the order of magnitude of the nuclear recoil energy and show that it is small compared to the electron energy.

ISBN 0-8053-8601-7

CHAPTER 10

The Two-Body Systems and Nuclear Forces

In this chapter we shall begin to deal with specific nuclear forces. The nature of these forces is still incompletely understood, and it is impossible to present them in a closed deductive form, as one can electromagnetic forces. Instead we shall use a semiempirical approach, starting from the simplest facts. The scattering of neutrons on protons and the binding of the nuclei show immediately that nonelectromagnetic forces must be involved. Scattering experiments show also that these forces have short range; that is, their "sphere of influence" has a radius of the order of 10^{-13} cm. Their intensity over such short distances is large compared with that of electric forces, otherwise nuclei would disintegrate under the Coulomb repulsion of the protons.

The facts, summarized in the semiempirical mass formula of Chap. 6, also allow us to draw several other conclusions in regard to nuclear forces, most notably that the binding energy of a nucleus is proportional to the number of nucleons and that the density of nuclear matter is approximately constant. This leads us to conclude that nuclear forces have a "saturation property" similar to the one exhibited by the forces that act between molecules in solids and liquids.

Emilio Segrè, Nuclei and Particles: An Introduction to Nuclear and Subnuclear Physics, Second Edition

ISBN 0-8053-8601-7

The Yukawa theory postulates that nuclear forces are due to a field, the quanta of which are the pi mesons. From this point of view the first step in studying nuclear forces should be the investigation of the pion–nucleon interaction. There are, however, many more particles that mediate the nucleon–nucleon interaction, and a derivation of the nuclear forces from field theory is at present impossible.

To surmount the various difficulties one at a time, we shall start with the phenomenologically simplest problem: the interaction between two nucleons. The study of the neutron–proton system, in which there are no complications raised by electromagnetic forces (the purely magnetic spin–spin interaction is negligible) or by Pauli's principle, will be our first step. We shall at first neglect spin, although later we shall be forced to introduce it in order to account for the experimental facts. Comparison of the results for the neutron–proton system with those for the proton–proton system will lead us to introduce the concept of isotopic spin. Finally we shall consider the extension of the results obtained to systems containing more than two nucleons. The treatment will be essentially phenomenological. The connection with meson theory will be sketched in Chap. 15.

10-1 THE DEUTERON

The empirical facts about the deuteron are the following: binding energy, 2.2246 MeV; spin 1; magnetic moment, 0.85742 μ_N; electric quadrupole moment, 2.875×10^{-27} cm^2. Moreover, we have extensive and precise data on the neutron–proton scattering cross section and on the photodisintegration of the deuteron over a large interval of energies.

The simplest possible model of a bound neutron-proton system (i.e., of a deuteron) is obtained by considering an attractive force between them, having a square-well potential of radius r_0 and depth V_0. The binding energy gives a relation between r_0 and V_0 (Fig. 10-1).

Because we shall find that the potential depends on the relative spin orientation, we shall indicate its depth by V_t for triplet states and by V_s for singlet states.

The value of the spin of the deuteron and the value of the magnetic moment, which is nearly the sum of the magnetic moments of the proton and neutron, show that the orbital motion in the ground state of the deuteron has angular momentum 0, and that the spins are parallel. Thus, we expect the deuteron ground state to be a $^3S_1$ state. However, this assumption cannot be strictly true, because it would rule out an electric quadrupole moment, the expectation value of

$$\langle {}^3S_1 | 3z^2 - r^2 | {}^3S_1 \rangle$$

being 0 for a spherically symmetric eigenfunction. We must conclude that our

ISBN 0-8053-8601-7

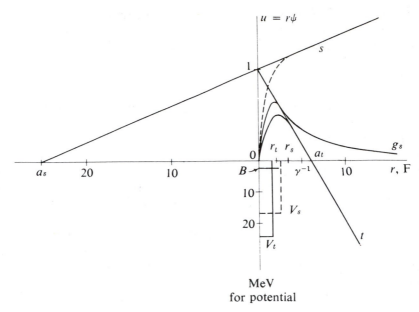

Figure 10-1 Potential well and eigenfunction for rectangular well: s, singlet-state eigenfunction, zero energy; t, triplet-state eigenfunction, zero energy normalized by requiring $u(0) = 1$; g_s, ground-state eigenfunction (not normalized). $B = 2.225$ MeV.

radius of well $r_t = 2.02$ F $V_t = -36$ MeV

radius of well $r_s = 2.60$ F $V_s = -14$ MeV

$\gamma^{-1} = 4.32$ F

assumption is only approximately correct. In fact, the eigenfunction of the ground state of the deuteron assigns a probability of 0.07 to finding a $^3D_1$ state and a probability of 0.93 to finding a $^3S_1$ state.

There might also be a finite probability of finding a $^3P_1$ state of opposite parity, because weak interactions do not conserve parity, but quantitatively this effect is negligible.

The Schrödinger equation for the system in a $^3S_1$ state is then

$$\frac{d^2u}{dr^2} + \frac{2\mu}{\hbar^2}(E - V_t)u = 0 \tag{10-1.1}$$

where μ is the reduced mass $M_p M_n / (M_p + M_n) \cong M/2$, $u = r\psi(r)$, and r is the distance between the nucleons. If E is negative, the state is bound and $-E$ is the binding energy. The solution of Eq. (10-1.1) for the square-well

ISBN 0-805-3-8601-7

potential is

$$u = A \sin k_t r \qquad r \leqslant r_0 \qquad k_t = \frac{1}{\hbar} \left[M (E - V_t) \right]^{1/2}$$

$$u = B e^{-\gamma (r - r_0)} \qquad r \geqslant r_0 \qquad \gamma = \frac{1}{\hbar} (M|E|)^{1/2} \qquad (10\text{-}1.2)$$

and the required continuity of u and u' at $r = r_0$ gives

$$\frac{u'(r_0)}{u(r_0)} = k_t \cot k_t r_0 = -\gamma$$

and

$$A = \frac{B \left(k_t^2 + \gamma^2 \right)^{1/2}}{k_t} \qquad (10\text{-}1.3)$$

The binding energy of the deuteron inserted in Eq. (10-1.2) gives

$$\gamma = 2.32 \times 10^{12} \text{cm}^{-1}$$

The continuity relation gives V_t if r_0 is known. Experiments on neutron–proton collisions at an energy of several MeV give an approximate value of r_0 of the order of 10^{-13} cm = 1 F. Electron scattering experiments have given (see Chap. 6) a radius of the same order of magnitude. For the sake of definiteness we shall assume that $r_0 = 2.8$ F, which corresponds to twice the Compton wavelength $\hbar/m_\pi c$ of the pion. This is a reasonable value from the point of view of meson theory (see Chap. 15).

Insertion of this value in Eq. (10-1.2) gives

$$V_t = -21 \text{ MeV} \qquad (10\text{-}1.4)$$

Eigenfunctions that take into account all low energy data are plotted in Fig. 10-1. The normalization requires

$$4\pi A^2 \int_0^{r_0} \sin^2 k_t r \, dr + 4\pi B^2 \int_{r_0}^{\infty} e^{-2\gamma (r - r_0)} \, dr = 1 \qquad (10\text{-}1.5)$$

With the values of r_0 and γ as given above, the second integral is about twice as large as the first. Hence, the nucleons in the deuteron spend only one third of the time within the range of the nuclear force; two thirds of the time they are at a distance larger than r_0. A crude but simple approximation to the normalized wave function is

$$u(r) = \left(\frac{\gamma}{2\pi} \right)^{1/2} e^{-\gamma r} \qquad (10\text{-}1.6)$$

ISBN 0-8053-8601-7

We can call $1/\gamma$ the "radius" of the deuteron (4.32 F). A good part of the results obtained for the rectangular well are independent of this special potential form. An elegant mathematical development without unnecessary assumptions about the form of the potential well will be found in Appendix B.

10-2   LOW-ENERGY NEUTRON–PROTON SCATTERING

We can use the value of the depth of the potential well obtained above to calculate neutron–proton scattering at low energy, that is, under conditions such that only s waves are important $(\lambdabar \gg r_0)$. Numerically $\lambdabar = 9.10/E^{1/2}$, where $\lambdabar$ is in fermis and E is the laboratory kinetic energy in MeV.

Measurements of the total n–p cross section have been made at many energies because they are practically and theoretically important. Figure 10-2 gives the results. The substances employed are always hydrogenous compounds or liquid hydrogen: it is impractical to use atomic hydrogen. As it turns out, when the energy of the neutrons is smaller than or comparable to the energy of the chemical bond of the hydrogen atom, the molecule acts as

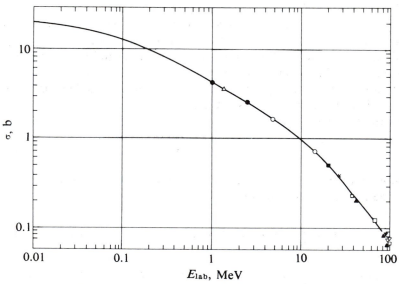

Figure   10-2   Total cross section of hydrogen atoms for neutron energies between 10^4 eV and 100 MeV. The total cross section in this energy interval is practically identical to the elastic cross section. At low energies the elastic cross section tends to 20.3 b and the capture cross section is 0.33 $[2200/v(\mathrm{m}\ \mathrm{sec}^{-1})]$ b, until one reaches energies comparable to the energy of the chemical bond of hydrogen (1 eV). Log–log plot. Energies and cross sections in the laboratory system.

ISBN 0-8053-8601-7

a whole and the scattering cross section is increased, by a factor of approximately 4, over the value that would obtain for a free atom. The reason for the fourfold increase will be discussed in Chap. 12. To avoid complications due to the chemical bond, we shall compare cross sections at energies of several electron volts. In this region we have a value of 20.3 b for $1 < E < 1000$ eV, constant within 1 or 2%. Here $\lambdabar$ is still very large compared with the range of nuclear forces; a condition which ensures that there is only s-wave ($l = 0$) scattering. The result of the comparison is that the experimental scattering cross section seems incompatible with the potential well given above for the ground state ($^3S_1$) of the deuteron.

The following will justify the statements made above. Using a result of scattering theory (see Appendix A), we have

$$\sigma = 4\pi \lambdabar^2 \sin^2 \delta_0 \tag{10-2.1}$$

where δ_0 is the s-wave phase shift. This can easily be calculated for a square well. If $E \ll V(r)$, we can approximate the wave function with two sine curves, one of which will correspond to the inside of the well and will have a wavelength much shorter than that of the exterior part of the wave function. In other words, inside the well ($r < r_0$) the wave function will be curved. Outside the well ($r > r_0$) it will become practically a straight line. The slope of the straight line at $r = r_0$ is obtained by writing the function u for $r < r_0$ where it obeys Eq. (10-1.1) for zero energy ($E \to 0$),

$$u'' - \frac{2\mu V_t}{\hbar^2} u = 0 \quad - V_t = 21 \text{ MeV} \quad \text{for} \quad r < r_0 \tag{10-2.2}$$

Its solution is

$$u = A \sin k_t r$$

with

$$k_t = \frac{(-2\mu V_t)^{1/2}}{\hbar} = \frac{(21)^{1/2}}{6.4 \times 10^{-13}} \text{ cm}^{-1} \cong 7.1 \times 10^{12} \text{ cm}^{-1} \tag{10-2.3}$$

The index t on k and V reminds us that we are referring to the triplet state (ground state of the deuteron).

For $r > r_0$ the wave function in the limit of zero energy is $C(r - a_t)$; it thus intercepts the r axis at abscissa $r = a_t$. The quantity a_t is called the (triplet) *scattering length*. (Some writers use a definition of scattering length differing in sign from this one. This is especially true of recent work in high-energy physics.) Joining the two functions at $r = r_0$, we have

$$\frac{u'(r_0)}{u(r_0)} = k_t \cot k_t r_0 = \frac{1}{r_0 - a_t} \tag{10-2.4}$$

ISBN 0-8053-8601-7

or

$$a_t = r_0 - \frac{1}{k_t} \tan k_t r_0 = r_0\left(1 - \frac{1}{k_t r_0} \tan k_t r_0\right) \qquad (10\text{-}2.5)$$

Consideration of Fig. 10-1 shows that the phase shift δ_0 is simply $-a_t/\lambdabar$. The straight line representing u is the limit for $\lambdabar \to \infty$ of a sine curve with wavelength $2\pi\,\lambdabar$; and this sine curve starts, not from the origin, but from a point of abscissa a_t. The sine function is thus

$$\sin\left(\frac{r}{\lambdabar} - \frac{a_t}{\lambdabar}\right) \qquad (10\text{-}2.6)$$

which means a phase shift

$$\delta_0 = \frac{-a_t}{\lambdabar} \qquad (10\text{-}2.7)$$

and hence a limiting value for $\lambdabar \to \infty$ of the cross section

$$\sigma_t = 4\pi\,\lambdabar^2 \sin^2 \delta_0 \to 4\pi a_t^2 \qquad (10\text{-}2.8)$$

Using the value of k_t corresponding to the potential derived from the binding energy of the deuteron ($V_t = -21$ MeV) and the usual $r_0 = 2.8$ F, we obtain $\sigma = 4.4$ b, instead of the experimental value of 20.36 ± 0.05 b. Reasonable variations cf r_0 do not modify this result appreciably. We have, therefore, an apparent contradiction.

The way out was indicated by Wigner (1935), who noted that the scattering occurs not only in the triplet state, but in the singlet state as well. The binding energy of the deuteron gives information only about the triplet state; if the forces are spin dependent, the singlet potential may be different from the triplet one. We use then the observed cross section to determine the singlet potential.

In collisions where the statistical weights determine the relative probability of singlet and triplet states (i.e., for unpolarized beams on unpolarized targets), the singlet and triplet states are represented in the ratio of 1 : 3, as shown in Sec. 6-4. The cross section will then be

$$\sigma = \tfrac{1}{4}\sigma_s + \tfrac{3}{4}\sigma_t \qquad (10\text{-}2.9)$$

where σ_s and σ_t indicate the cross section in singlet and triplet states. We can use the calculated value $\sigma_t = 4.4$ b and the experimental information on σ to calculate σ_s, and hence V_s. To fit the experimental data, we must have (in barns)

$$20.36 = \tfrac{3}{4}(4.4) + \tfrac{1}{4}\sigma_s \qquad (10\text{-}2.10)$$

ISBN 0-8053-8601-7

from which $\sigma_s = 4\pi a_s^2 = 68$ b and $|a_s| = 23.7$ F. This value of a_s can be inserted in Eq. (10-2.5) or better, in its equivalent for a_s, to find

$$k_s = \frac{(2\mu|V_s|)^{1/2}}{\hbar} = 5.28 \times 10^{12} \text{ cm}^{-1} \quad \text{and} \quad V_s = -11.5 \text{ MeV} \quad (10\text{-}2.11)$$

The sign of a_s is not determined by the simple measurement of the cross section, which gives only a_s^2. To obtain the sign, one uses interference effects appearing in scattering from ortho- and parahydrogen, and one finds that a_s is negative, whereas the sign of a_t is positive (see Chap. 12).

The scattering-length concept can be extended to energies other than 0 by defining the scattering length through the relation

$$-ka(k) = \tan \delta_0 \qquad (10\text{-}2.12)$$

which for $k \to 0$ agrees with Eq. (10-2.7). With this definition of $a(k)$, Eq. (10-2.8) gives

$$\sigma = \frac{4\pi}{k^2} \frac{1}{1 + \cot^2 \delta_0} = \frac{4\pi}{k^2 + a^{-2}(k)} \qquad (10\text{-}2.13)$$

Equation (10-2.13) is convenient and shows that the knowledge of $a(k)$ is sufficient to determine completely the s-wave scattering.

In general $a(k)$ can be approximated by a simple and important formula that is the generalization for $k \neq 0$ of Eq. (10-2.7).

$$\frac{1}{a(k)} = -k \cot \delta_0 = \frac{1}{a(0)} - \frac{1}{2} r_0 k^2 + \cdots \qquad (10\text{-}2.14)$$

where r_0 is called the *effective range* of the potential (see Appendix B). Equation (10-2.14) is only approximate, but its next term would be in $k^4 r_0^3$ with a small coefficient (~ 0.10). The physical meaning of the effective range corresponds to the mean distance of interaction between neutron and proton. The effective range is not to be confused with the width of a rectangular well [r_0 of Eq. (10-2.5)] or the length parameter in other forms of the potential, although for a given potential the two quantities are related and often of nearly the same magnitude. It is noteworthy that in this excellent approximation the whole neutron-proton scattering at low energy is described by two parameters: scattering length and effective range. Low-energy measurements cannot provide more than these two numbers.

The effective range approximation has been generalized to extend to waves with $l \neq 0$ and to systems other than the nucleon-nucleon. It is a powerful way of describing many scattering phenomena. It can also be formally extended to the bound state of the deuteron, for which γ [Eq. (10-1.2)] corresponds to $1/a_t(k)$, as Fig. 10-1 shows. For the bound state k is imaginary and equal to $i|\gamma|$, and Eq. (10-2.14) becomes

$$\gamma = a^{-1} + \tfrac{1}{2}r_0\gamma^2 + \cdots \qquad (10\text{-}2.15)$$

If the last term is negligible compared with the others, then the radius of the deuteron is equal to the scattering length. This refers, of course, to the bound triplet state.

Detailed comparisons with experiments with all refinements on the approximations give (Noyes, 1972):

$$a_t = 5.42 \text{ F} \qquad r_{0t} = 1.73 \text{ F}$$

$$a_s = -23.71 \text{ F} \qquad r_{0s} = 2.73 \text{ F}$$

(For errors and discussion see original paper.) These values summarize all the information obtainable from low-energy (< 10 MeV) neutron–proton scattering. The value of $\gamma = 0.232 \times 10^{13}$ cm^{-1} is obtainable from these data and Eq. (10-2.15). Figure 10-3 shows the separate contributions of singlet and triplet scattering computed from the foregoing constants and the experimental total cross section.

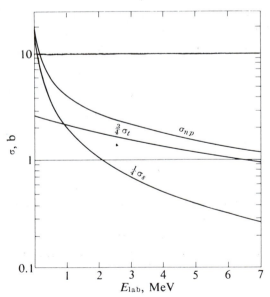

Figure 10-3 The neutron–proton scattering cross section σ_{np} has two components: the singlet scattering ($\tfrac{1}{4}\sigma_s$) and the triplet scattering ($\tfrac{3}{4}\sigma_t$). The two contributions, as well as their sum, are shown in the figure. Ordinate in log scale. [From (BW 52).]

ISBN 0-8053-8001-7

10-3 PROTON–PROTON SYSTEM AND SCATTERING

In passing from the neutron–proton to the proton–proton system we find two main differences: (1) to the specific nuclear forces we must add the Coulomb repulsion; (2) the identity of the protons entails some quantum-mechanical complication (Mott, 1930). We shall first consider point 2.

A first important consequence of Pauli's principle for the proton–proton system is that it can exist only in states of ^1S, ^3P, ^1D, etc. To prove this statement, consider the spherical-harmonic part of the eigenfunction. On exchange of the protons, or, which is the same, on changing θ into $\pi - \theta$, the spherical harmonic is multiplied by $(-1)^l$, where l is the orbital angular momentum. As a consequence the spin eigenfunction must be singlet for l even or triplet for l odd so that the complete eigenfunction can change its sign as required by Pauli's principle.

Furthermore, we have seen that the scattering cross section is related to $f(\theta)$ of Appendix A by

$$\frac{d\sigma}{d\omega} = |f(\theta)|^2 \tag{10-3.1}$$

Consider now two *distinguishable* particles of the same mass. In the center-of-mass system, the probability that either will be scattered through an angle θ is

$$|f(\theta)|^2 + |f(\pi - \theta)|^2 \tag{10-3.2}$$

(see Fig. 10-4). However, if the particles are *identical*, their waves interfere;

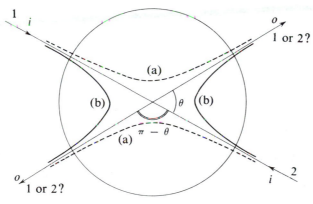

Figure 10-4 Scattering of identical particles. Incoming particles 1 and 2 move in directions marked i, outgoing in directions marked o. Classically one can follow the trajectories and distinguish case (a), corresponding to scattering by an angle θ, from case (b), corresponding to scattering by an angle $\pi - \theta$. Quantum-mechanically, trajectories are meaningless. One must consider wave functions symmetric or antisymmetric with respect to the exchange of the particles.

ISBN 0-8053-8061-7

and instead of summing the squares of the amplitudes, we must first sum the amplitudes and then square them. Moreover, the eigenfunction must be symmetric or antisymmetric with respect to the exchange of the particles, according to whether they are bosons or fermions (see Chap. 6). The portion of the eigenfunctions containing only spatial coordinates, without spin, will then be

$$f(\theta) \pm f(\pi - \theta) \tag{10-3.3}$$

Exchange of the two proton coordinates means exchanging θ with $\pi - \theta$, as shown in Fig. 10-4. If the plus sign is used in Eq. (10-3.3), the expression is symmetric with respect to the exchange of the coordinates; if the minus sign is used, the expression is antisymmetric with respect to the same exchange. We have already seen that the spins give a symmetric eigenfunction in the triplet state and an antisymmetric eigenfunction in the singlet states [Eqs. (6-4.13) and (6-4.14)]. For protons the total eigenfunction must be antisymmetric; hence the triplet states will be associated with

$$f(\theta) - f(\pi - \theta) \tag{10-3.4}$$

and the singlet states will be associated with

$$f(\theta) + f(\pi - \theta) \tag{10-3.5}$$

The scattering cross section in triplet states is thus

$$\frac{d\sigma_t}{d\omega} = |f(\theta) - f(\pi - \theta)|^2 = |f(\theta)|^2 + |f(\pi - \theta)|^2 - 2 \operatorname{Re}[f(\theta)f^*(\pi - \theta)]$$
$$\tag{10-3.6}$$

and in singlet states it is

$$\frac{d\sigma_s}{d\omega} = |f(\theta) + f(\pi - \theta)|^2 = |f(\theta)|^2 + |f(\pi - \theta)|^2 + 2 \operatorname{Re}[f(\theta)f^*(\pi - \theta)]$$
$$\tag{10-3.7}$$

At very low energy the Coulomb repulsion is the most important force acting in the p–p system because the nucleons do not approach sufficiently to feel the nuclear forces. The Coulomb scattering can then be calculated exactly.

We shall present only the result of the calculation of $f(\theta)$:

$$f(\theta) = \frac{e^2}{Mv^2 \sin^2(\theta/2)} \exp\left(-i\eta \log \sin^2 \frac{\theta}{2}\right) \tag{10-3.8}$$

with $\eta = e^2/\hbar v$. M is the proton mass and v the relative velocity of the

ISBN 0-8053-8001-7

protons. Replacing Eq. (10-3.8) in Eqs. (10-3.6) and (10-3.7), we obtain a σ_s and a σ_t. In the absence of polarization these must be added with the statistical weights $\frac{1}{4}$ and $\frac{3}{4}$, as in the n–p case, because all waves are operative in Coulomb scattering and the relative weight of the existing singlet and triplet states are in the ratio 1 to 3. The final result, in the center-of-mass system, is

$$\left(\frac{d\sigma}{d\omega}\right)_c = \left(\frac{e^2}{Mv^2}\right)^2\left[\frac{1}{\sin^4(\theta/2)} + \frac{1}{\cos^4(\theta/2)} - \frac{\cos\left[\eta\log\tan^2(\theta/2)\right]}{\cos^2(\theta/2)\sin^2(\theta/2)}\right]$$

$$(10\text{-}3.9)$$

The first two terms on the right represent the classical Rutherford scattering; the third term is a quantum-mechanical interference term. A similar term, but of opposite sign, occurs in the scattering of alpha particles in helium, as was demonstrated experimentally by Chadwick, Blackett, and Champion (1930). Note that the numerator of the third term is nearly 1 for protons of energy larger than 1 MeV and for angles not too close to 0 or to 90 deg. A striking example of the effect of the identity between target and projectile is shown in Fig. 10-5. For nonidentical particles the third term of Eq. (10-3.9) would be missing and the equation would give the classical Rutherford scattering.

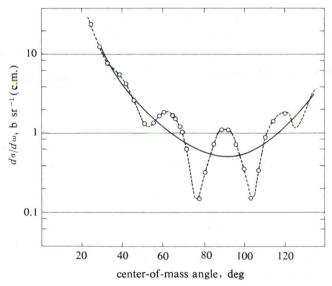

Figure 10-5 Carbon–carbon elastic scattering. Typical angular distributions for ^{12}C elastically scattered from carbon at a c.m. energy of 5 MeV. The Rutherford and Mott scattering predictions (solid and dashed curves) differ only in that the latter includes the quantum-mechanical interference term arising from the identity of the incident and target nuclei. [From A. D. Bromley, J. A. Kühner, and E. Almqvist, *Phys. Rev. Letters*, **4**, 365 (1960).]

ISBN 0-8053-8061-7

Thus far we have neglected specific nuclear forces. To include them, at low energy, we need consider only $^1S_0$ waves. $^3S_1$ waves are ruled out by Pauli's principle and higher angular momenta are inoperative. We then calculate $d\sigma_t(\theta)/d\omega$ using Eq. 10-3.6 and $f(\theta)$ from Eq. 10-3.8. For $d\sigma_s/d\omega$ we use Eq. 10-3.7 and $f(\theta)$ is modified by the nuclear forces into

$$f_{\delta_0}(\theta) = f(\theta) + i(e^{2i\,\delta_0} - 1)/2k$$

where the last term contains the phase shift δ_0 of the $^1S_0$ wave. The cross section is $\frac{1}{4}$

$$\frac{1}{4}\frac{d\sigma_s}{d\omega} + \frac{3}{4}\frac{d\sigma_t}{4\,d\omega}$$

and the calculation gives a term $(d\sigma/d\omega)_c$ as before, a term linear in $\sin 2\delta_0$, and a term containing $\sin^2\delta_0$:

$$\left(\frac{d\sigma}{d\omega}\right) = \left(\frac{d\sigma}{d\omega}\right)_c + \left(\frac{e^2}{Mv^2}\right)^2\left[-\frac{2\hbar v\sin 2\delta_0}{e^2\sin^2\theta} + \left(\frac{\hbar v}{e^2}\right)^2\sin^2\delta_0\right]$$

We have here assumed that the argument of the exponential is small ($E > 1$ MeV for $0.1 < \theta \ll \pi$). This can be compared with experiment to find δ_0.

The experimental data are analyzed further along lines similar to those used in neutron–proton scattering but are complicated considerably by the presence of the Coulomb force. A scattering length and an effective range for the $^1S_0$ state are derived, and then one asks the interesting question: What value would they have if it were possible to suppress the Coulomb interaction, leaving the specific nuclear force unchanged? The result would be about

$$
\begin{array}{ll}
\text{Scattering length:} & -17.2 \text{ F} \\
\text{Effective range:} & 2.65 \text{ F}
\end{array}
\tag{10-3.10}
$$

Direct experiments on neutron–neutron scattering are impractical but the scattering length and the effective range have been found by careful analysis of the $d(n, 2n)p$ and other reactions. The scattering length is -17.0 ± 1 F and the effective range is 2.84 ± 0.03 F. These values are close to the corresponding values for the n–p system. Indeed the similarities between n–p and p–p scattering were among the first indications (Breit, 1936) that the n–p and p–p nuclear forces might be identical.

It must be noted that the negative scattering length implies that, even apart from the electrostatic repulsion, the $^1S_0$ state of the p–p and n–n systems is unbound, as is the corresponding state of the n–p system.

10-4   CHARGE INDEPENDENCE OF NUCLEAR FORCES—ISOTOPIC SPIN

The virtual identity of the n–p and p–p forces, shown in the low-energy scattering experiments mentioned above, manifests itself also in the properties of mirror nuclei, such as ^3H and ^3He, ^7Li, ^7Be, etc., which are obtained one from the other by transforming all neutrons into protons, and vice versa. It is true that Coulomb forces are necessarily different in mirror nuclei, but apart from that, the nuclear levels show a remarkable similarity, as seen in Fig. 10-6. If we examine mirror nuclei of $N = Z \pm 1$, we find that, once correction is made for the electrostatic energy, the nuclei have the same mass. More precisely, we should expect, for the mass difference between two nuclei with $N = Z \pm 1$,

$$\Delta M = \frac{3e^2(2Z + 1)}{5R} = \frac{3e^2(2Z + 1)}{5r_0 A^{1/3}} \tag{10-4.1}$$

where R is the nuclear radius. That this is indeed true is shown in Fig. 10-6. Strictly speaking, the similarity of the levels of mirror nuclei of the type $N = Z \pm 1$ tells us only that there is equality between n–n and p–p forces. More convincing are the arguments derived from nuclei with A even, such as ^{6}He, ^{6}Li, and ^{6}Be; ^{8}Li, ^{8}Be, and ^{8}B: or ^{14}C, ^{14}N, and ^{14}O; some of their energy levels are shown in Figs. 10-6 to 10-8. The correspondence between levels is clear, and the differences in energy are accounted for mainly by Coulomb effects. (See also Sec. 6-3).

Facts of this type confirm Heisenberg's and Condon and Cassen's (1932) hypothesis that the specific nuclear forces should be considered "charge-independent." The forces between n–p, p–p, and n–n, apart from electromagnetic effects, are assumed to be the same. This assumption is stronger than the assumption of charge symmetry, which postulates the equality of n–n and p–p forces but says nothing concerning n–p forces. There are many other phenomena supporting the charge-independence hypothesis and extending it to pi-meson physics as well. On the other hand, we must remember that charge independence is only approximate, because it obviously does not take into account electromagnetic effects or the neutron–proton mass difference.

To exploit the hypothesis fully we treat it by the isotopic spin formalism, which is its appropriate mathematical formulation. A nucleon is endowed with another degree of freedom besides the ordinary ones of coordinates and spin, and the corresponding internal variable, called isotopic spin, isospin, or i spin, can take only two values (i.e., is "dichotomic"). The nucleon is a proton or a neutron, depending upon its value.

We already have a model of dichotomic variables, the ordinary spin for a particle of spin $\frac{1}{2}$ in the Pauli treatment, assumed known to the reader. The i-spin function can take two values, that we shall write as

$$\begin{vmatrix} 1 \\ 0 \end{vmatrix} \equiv \pi \quad \text{or} \quad \begin{vmatrix} 0 \\ 1 \end{vmatrix} \equiv \nu \tag{10-4.2}$$

and the i-spin operators are

$$\tau_1 = \begin{vmatrix} 0 & 1 \\ 1 & 0 \end{vmatrix} \quad \tau_2 = \begin{vmatrix} 0 & -i \\ +i & 0 \end{vmatrix} \quad \tau_3 = \begin{vmatrix} 1 & 0 \\ 0 & -1 \end{vmatrix} \quad 1 = \begin{vmatrix} 1 & 0 \\ 0 & 1 \end{vmatrix}$$

$$\tag{10-4.3}$$

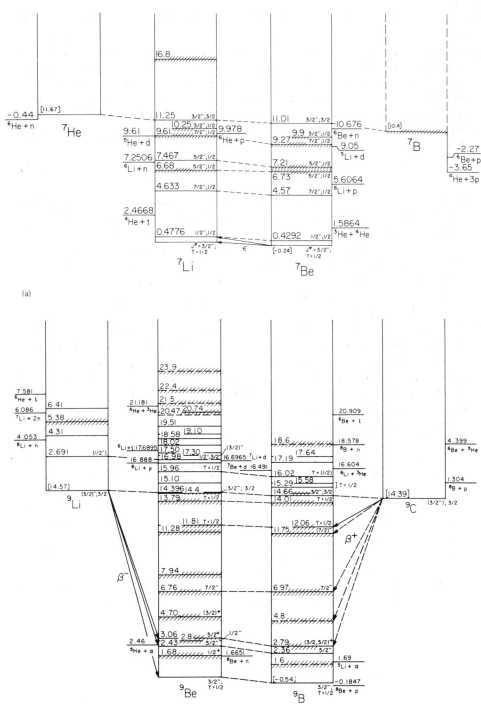

(a)

(b)

Figure 10-6 Mirror nuclei. Nuclear levels in mirror nuclei ($N = Z \pm 1$): (a) ^{7}Li, ^{7}Be; (b) ^{9}Be, ^{9}B.

460

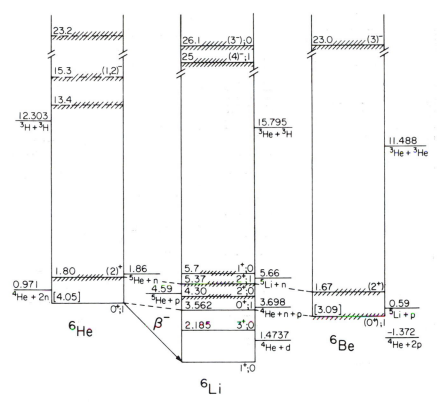

Figure 10-7 Nuclear levels in ^{6}He, ^{6}Li, and ^{6}Be.

They obey the same commutation relations and are a perfect analog to Pauli's spin matrices. The isotopic spin components are thus $\frac{1}{2}\tau_i = t_i$. The $\begin{vmatrix} 1 \\ 0 \end{vmatrix}$ state corresponds to a proton, the $\begin{vmatrix} 0 \\ 1 \end{vmatrix}$ to a neutron.

It is easily verified that

$$\frac{1}{2}(\tau_1 - i\tau_2)\begin{vmatrix} 1 \\ 0 \end{vmatrix} = \tau_- \begin{vmatrix} 1 \\ 0 \end{vmatrix} = \begin{vmatrix} 0 \\ 1 \end{vmatrix}; \qquad \tau_- \begin{vmatrix} 0 \\ 1 \end{vmatrix} = 0 \qquad (10\text{-}4.4)$$

$$\frac{1}{2}(\tau_1 + i\tau_2)\begin{vmatrix} 1 \\ 0 \end{vmatrix} = \tau_+ \begin{vmatrix} 1 \\ 0 \end{vmatrix} = 0; \qquad \tau_+ \begin{vmatrix} 0 \\ 1 \end{vmatrix} = \begin{vmatrix} 1 \\ 0 \end{vmatrix} \qquad (10\text{-}4.5)$$

$$\tau_3 \begin{vmatrix} 1 \\ 0 \end{vmatrix} = \begin{vmatrix} 1 \\ 0 \end{vmatrix}; \qquad \tau_3 \begin{vmatrix} 0 \\ 1 \end{vmatrix} = -\begin{vmatrix} 0 \\ 1 \end{vmatrix} \qquad (10\text{-}4.6)$$

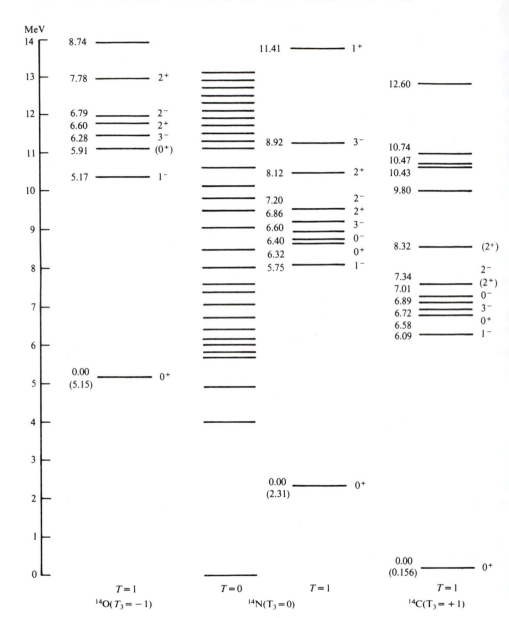

Figure 10-8   The level schemes for the nuclei with $A = 14$. The relative energies represent atomic masses. [From (BM 69).]

The operator τ_+ transforms a neutron into a proton, and the operator τ_- transforms a proton into a neutron. The third component of the i spin has the value $+\frac{1}{2}$ for the proton, $-\frac{1}{2}$ for the neutron.

If we take two nucleons, the total i spin of the system is composed of the i spins of the individual nucleons, as it is for ordinary spin. We have, therefore, a total i spin $\mathbf{T} = \mathbf{t}(1) + \mathbf{t}(2)$. The eigenvalues of $\mathbf{T}^2$ are $T(T+1)$, where $t(1) + t(2) \geqslant T \geqslant |t(1) - t(2)|$. In our case T is 1 or 0, and $T^2 = 2$ or 0. We shall refer to the two states as triplet, or $T = 1$, and singlet, or $T = 0$. Similarly for the third component of $\mathbf{T}$ we have $T_3 = t_3(1) + t_3(2)$, with eigenvalues 1, 0, -1.

The corresponding eigenfunctions of T^2 and T_3 are

$$\left. \begin{array}{c} \pi(1)\pi(2) \\[2em] 2^{-1/2}\left[\pi(1)\nu(2) + \pi(2)\nu(1)\right] \quad \text{for} \quad T^2 = 2 \\[2em] \nu(1)\nu(2) \end{array} \right\} \quad T_3 \left\{ \begin{array}{r} = \quad 1 \\[2em] = \quad 0 \\[2em] = \ -1 \end{array} \right. \qquad (10\text{-}4.7)$$

$$2^{-1/2}\left[\pi(1)\nu(2) - \pi(2)\nu(1)\right] \quad \text{for} \quad T^2 = 0 \qquad T_3 \ = 0 \qquad (10\text{-}4.8)$$

The complete eigenfunction for the two-nucleon system is now written $f(\mathbf{r}, \mathbf{s}_1, \mathbf{s}_2, \mathbf{t}_1, \mathbf{t}_2)$, with $\mathbf{r}$ the relative coordinates of the two nucleons, $\mathbf{s}_1, \mathbf{s}_2$ their spins, and $\mathbf{t}_1, \mathbf{t}_2$ their i spins. This expression can be factored as $f_r(\mathbf{r})f_\sigma(\mathbf{s}_1, \mathbf{s}_2) f_\tau(\mathbf{t}_1, \mathbf{t}_2)$ if we neglect, in the hamiltonian, interactions between i spin and spin, spin and coordinates, etc., and in general between degrees of freedom of a different kind. The functions f_σ and f_τ take the forms of Eqs. (6-5.13), (6-5.14), (10-4.7), and (10-4.8).

Pauli's principle may now be formulated by saying that the eigenfunctions must be antisymmetric with respect to the exchange of *all* coordinates, including i-spin coordinates.

For a p–p or an n–n system the i-spin factor of the eigenfunction is $\pi(1)\pi(2)$ or $\nu(1)\nu(2)$, symmetric with respect to the exchange of 1 with 2. Hence the remainder of the eigenfunction must be antisymmetric with respect to the exchange of spatial and spin coordinates and we thus have the usual requirement of Pauli's principle for identical particles.

As an application to the n–p system, note that, for the deuteron in the $^3S_1$ state, $f_r(r)f_\sigma(\mathbf{s}_1, \mathbf{s}_2)$ form a function symmetric with respect to interchange of nucleon 1 and 2. Hence f_τ has the form of Eq. (10-4.8) and is antisymmetric. It belongs to the singlet state of i spin. It is easily shown in somewhat more general terms that for a two-nucleon system

$$l + S + T = \text{odd integer} \qquad (10\text{-}4.9)$$

where l is the orbital angular momentum, S the spin, and T the i spin of the system.

The total charge of a system of nucleons is related to the eigenvalue of T_3, T_3' by

$$\frac{Q}{e} = T_3' + \frac{A}{2} \tag{10-4.10}$$

where A is the mass number or the number of nucleons.

The total isotopic spin has the remarkable property of being a constant of the motion for specific nuclear forces, in the same way that the total angular momentum is a constant of the motion for an isolated system. Although this conservation principle is only approximate (being violated, for instance, by electromagnetic interactions), it nevertheless proves to be most important, especially in particle physics. At present its foundation is empirical.

Coming back to nuclear forces, we see that the collision experiments demonstrate that these forces are the same in the $^1S_0$ state for proton–proton and neutron–proton systems, both of which correspond to $T = 1$, but each of which corresponds to different T_3. We postulate that the forces depend on T but not on T_3.

● It is known from quantum mechanics (Sc 68) that rotation around an axis in the **n** direction by an angle θ is represented by the unitary operator

$$R(\mathbf{n}, \theta) = e^{i\mathbf{n}\cdot\mathbf{J}\theta} \tag{10-4.11}$$

where **J** is the angular momentum and **n** a unit vector of components α, β, γ. For $J = \frac{1}{2}$ the operator takes the explicit form

$$R(\mathbf{n}, \theta) = \exp\left[i(\mathbf{n}\cdot\boldsymbol{\sigma})(\theta/2)\right] \tag{10-4.12}$$

where $\boldsymbol{\sigma}$ are the Pauli matrices. Developing the exponential, we have

$$R(\mathbf{n}, \theta) = 1 + \frac{1}{1!}\frac{i\theta}{2}(\alpha\sigma_x + \beta\sigma_y + \gamma\sigma_z)$$

$$+ \frac{1}{2!}\left(\frac{i\theta}{2}\right)^2(\alpha\sigma_x + \beta\sigma_y + \gamma\sigma_z)^2 + \cdots \tag{10-4.13}$$

and taking into account the commutation properties of the σ this gives

$$R(\mathbf{n}, \theta) = \cos\frac{\theta}{2}1 + i\mathbf{n}\cdot\boldsymbol{\sigma}\sin\frac{\theta}{2}$$

$$= \begin{vmatrix} \cos\dfrac{\theta}{2} + i\gamma\sin\dfrac{\theta}{2} & (i\alpha + \beta)\sin\dfrac{\theta}{2} \\ (i\alpha - \beta)\sin\dfrac{\theta}{2} & \cos\dfrac{\theta}{2} - i\gamma\sin\dfrac{\theta}{2} \end{vmatrix} \tag{10-4.14}$$

ISBN 0-8053-8061-7

The matrix σ_x except for a factor i thus represents the rotation by 180 deg around the x axis ($\alpha = 1, \theta = \pi$). Similar relations obtain for the other components of σ. What has been said for **J** is valid also for **T** and rotations in i-spin space. Charge independence is mathematically equivalent to invariance with respect to rotation in i-spin space, because rotation leaves **T** invariant and changes only T_3. One can also say that the hamiltonian commutes with T; $TH - HT = 0$; or that **T** is a constant of the motion. ●

This postulate is the mathematical formulation of the principle of charge independence of nuclear forces and can be generalized with useful results. In nuclear physics proper the postulate gives rise to approximate selection rules that forbid transitions between states of different i spin under the action of nuclear forces. A remarkable example is afforded by comparing the results of deuteron and proton bombardment in the light nuclei. In deuteron bombardment the isotopic spin cannot change, because the deuteron has an i spin of 0. In proton bombardment the i spin can change by $\frac{1}{2}$. If we bombard ^{14}N [in its ground state ($T = 0$)] with protons or deuterons, it is possible to excite the state at 3.95 MeV, which has i spin 0. On the other hand, the state at 2.35 MeV, which has i spin 1, can be excited by proton bombardment (initial i spin $\frac{1}{2}$, final i spin $1 \pm \frac{1}{2}$) but not by deuteron bombardment (initial i spin 0, final i spin 1).

Similarly it is impossible to form ^{10}B in a state at 1.74 MeV which has $T = 1$ by bombarding ^{12}C with deuterons, but the same state can be reached by bombarding ^{13}C with protons. In the first case the i spin of ^{12}C and deuteron is 0, in the second the i spin of proton and ^{13}C is 1.

A stringent test of i-spin conservation in nuclear reactions (Barshay and Temmer, 1964) is given by the angular distribution of a reaction of the type $A + B = C + C'$ where B has $T = 0$ and C and C' belong to the same i-spin multiplet. (See below). Zero i-spin for B means that the system is in a pure i-spin state with $T = T_A$. Particles C and C' in the i-spin formalism are the same particle except for different values of their third component of i-spin. Particles C and C' are both fermions or both bosons, according to the parity of their mass number. In a given T state, the scattered amplitude must then contain only waves with l even or l odd, but not l even and l odd at the same time. Only even powers of $\cos \theta$ appear in the intensity (the square of the amplitudes) and thus the intensity is symmetric with respect to a plane at 90 deg to the initial direction in the center-of-mass system. This is verified with good accuracy in Fig. 10-9, that refers to the reaction

$$^{4}\text{He} + d = {}^{3}\text{He} + {}^{3}\text{H} \tag{10-4.15}$$

Another very interesting type of application of the i-spin concept is that indicated in Figs. 10-6 and 10-7. It will be noted that a level with a certain value of T occurs in $2T + 1$ isobars, corresponding to the possible values of

ISBN 0-8053-8601-1

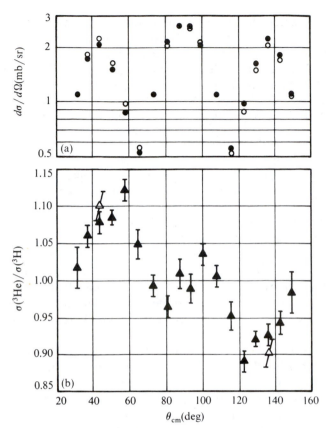

Figure   10-9   Top: measured differential cross sections for the process ^{4}He + ^{2}H → ^{3}He + ^{3}H using an 82-MeV α beam. Open circles, ^{3}He yields; closed circles, ^{3}H yields. Relative errors are smaller than the size of the data symbols and the absolute scale is uncertain to ±10%. Bottom: angular dependence of the ratio of ^{3}He to ^{3}H yields. The Barshay–Temmer theorem requires this ratio to be 1.0 at all angles. [From E. E. Gross, E. Newman, W. J. Roberts, R. W. Rutkowsky, and A. Zucker, *Phys. Rev. Letters*, **24**, 473 (1970).]

T_3. For example, the ground level of ^{6}Li has $T = 0$ and occurs only in this nucleus, but the excited level at 3.56 MeV has $T = 1$ and occurs in three nuclei corresponding to the ground levels of ^{6}He and ^{6}Be. All corresponding levels with the same T also have the same I and parity, irrespective of T_3, as is to be expected from the hypothesis of charge independence of nuclear forces. They form an i-spin multiplet (Wigner, 1940).

States belonging to the same i-spin multiplet have essentially the same wave function, or more precisely, would have the same wave function if it were possible to nullify the Coulomb interaction. This interaction may be expressed by a term in the hamiltonian of the form

$$H_c = \sum_{i<j} \left(\tfrac{1}{2} + t_3^{(i)} \right) \left(\tfrac{1}{2} + t_3^{(j)} \right) \frac{e^2}{r_{ij}} \tag{10-4.16}$$

where the sum is extended over all nucleons. The product of the two parentheses is 1 for two protons, 0 otherwise. This part of the hamiltonian may be rewritten (by the technique of irreducible tensors or by direct verification) as

$$H_c = \sum_{i<j} \left(\tfrac{1}{4} + \tfrac{1}{3} \mathbf{t}^{(i)} \cdot \mathbf{t}^{(j)} \right) \frac{e^2}{r_{ij}} + \sum_{i<j} \tfrac{1}{2} \left(t_3^{(i)} + t_3^{(j)} \right) \frac{e^2}{r_{ij}}$$

$$+ \sum_{i<j} \left(t_3^{(i)} t_3^{(j)} - \tfrac{1}{3} \mathbf{t}^{(i)} \cdot \mathbf{t}^{(j)} \right) \frac{e^2}{r_{ij}} \tag{10-4.17}$$

The first term is a scalar in i-spin space, the second is a vector, and the third transforms under rotation in i-spin space as $Y_2^m(\theta, \varphi)$ transforms under rotation in ordinary space, that is, as a tensor of the second order. The Coulomb energy is the diagonal term of H_c, given by

$$E_c = \langle \alpha, T, T_3 | H_c | \alpha, T, T_3 \rangle = E_c^{(0)} + E_c^{(1)} T_3 + E_c^{(2)} \left[3T_3^2 - T(T+1) \right] \tag{10-4.18}$$

where $E_c^{(k)}$ are functions of A, T, and the other quantum numbers α of the state but not of T_3. The nuclear mass then becomes

$$M(A, T, T_3) = a + bT_3 + cT_3^2 \tag{10-4.19}$$

For an i-spin multiplet with $T \geqslant 3/2$ the constants may be determined from three masses and one finds the masses of the other members of the multiplet. This has been done for many nuclei or nuclear levels having A up to about 40 and $T = 3/2$ or 2, thus checking Eq. (10-4.19). The accuracy of agreement is of the order of 10 keV for $7 \leqslant A \leqslant 37$.

Although electromagnetic interactions do not conserve i-spin, one can obtain i-spin selection rules for electromagnetic radiation (Radicati, 1952). As an example consider $E1$ transitions. The electric dipole moment comes from the operator

$$D = \sum_1^A e_k z_k = e \sum_1^A \left(\tfrac{1}{2} + t_3^{(k)} \right) z_k = \frac{e}{2} \sum_1^A z_k + e \sum_1^A t_3^{(k)} z_k \tag{10-4.20}$$

This is verified by noting that $\tfrac{1}{2} + t_3^{(k)}$ is 0 for neutrons and 1 for protons.

The formula contains a term depending only on the center of mass of the nucleus which cannot cause transitions between different states and a second term which is a vector in i-spin space and thus can cause only transitions for which $|T_i - T_f| \leqslant 1$, according to the triangle rule. This i-spin selection rule can be generalized to all electromagnetic transitions. In self-conjugate nuclei having $N = Z$, $T_3 = 0$ and all transitions with $\Delta T = 0$ are forbidden, as can be seen by an application of the Wigner–Eckart theorem (Sc 68). The intensities of the lines in the level diagram in Fig. 10-10 show the operation of these selection rules.

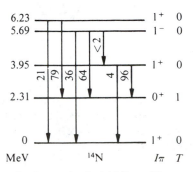

Figure   10-10   The selection rule forbidding $E1$ transitions between two $T = 0$ states can be tested by the decay of the 5.69-MeV, $T = 0$ level in ^{14}N; it is found that the transitions to the ground state and the 3.95-MeV, $T = 0$ level are weaker by an order of magnitude than the allowed transition to the 2.31-MeV, $T = 1$ level. (Note that, for the same nuclear matrix element, the transition rate for an $E1$ transition varies as the cube of the transition energy.) [From (BM 69).]

Isospin also enters in beta decay matrix elements. The operator $t_1^{(k)} \pm it_2^{(k)}$, transforming a neutron into a proton, is also a vector in i-spin space and thus $\Delta T = \pm 1$, 0. Furthermore, since beta decay changes the charge by one unit, we also have $\Delta T_3 = \pm 1$.

Even for large T, as found in heavy nuclei, the same situation obtains. In $^{117}_{51}$Sb the ground state has $-T_3 = \frac{1}{2}(N - Z) = 15/2$ and hence $T \geqslant 15/2$. However, there are other levels with $T = 17/2$. These levels show up as resonances in p–p and p–n bombardments of $^{116}_{50}$Sn ($T = 8$). They are analogs of the low levels of $^{117}_{50}$Sn, which also have $T = 17/2$, except that they are shifted by the Coulomb energy and by the neutron–proton mass difference. All this is shown in Fig. 10-11.

The applications of i-spin in nuclear physics are numerous and pervasive; for more details, consult (W 69). Isospin is of major importance in particle physics, as we shall see in Part III.

ISBN 0-805-38601-1

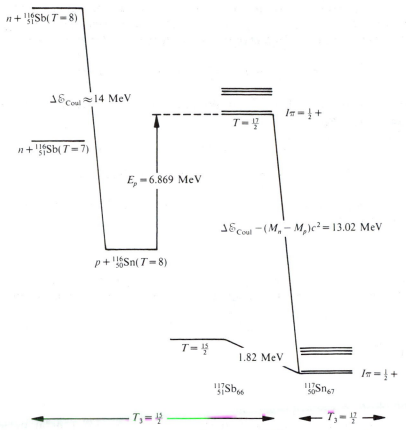

Figure 10-11 Level spectrum of $^{117}_{51}$Sb ($T_3 = 15/2$). The $T = 17/2$ states have been observed as resonances in the $^{116}_{50}$Sn(p, p) and $^{116}_{50}$Sn(p, n) processes. The observed resonances $T = 17/2$ are isobaric analogs of the low-lying levels in $^{117}_{50}$Sn ($T = 17/2$). The data in the figure show the striking similarity of the energy spacing and spin parity quantum numbers of the $T = 17/2$ levels in the two nuclei. The difference in binding energy between the $T_3 = 15/2$ and the $T_3 = 17/2$ levels of the $T = 17/2$ multiplet corresponds to the Coulomb energy difference. [From (BM 69).]

10-5 SPIN-DEPENDENT AND TENSOR FORCES

Thus far, forces between two nucleons have been ,assumed to admit a potential, function of the distance between the nucleons. This gives rise to central forces. We have seen that the forces are spin dependent. This is best expressed mathematically with the help of the Pauli matrices $\boldsymbol{\sigma} = 2\mathbf{s}$ and by using the operator $\boldsymbol{\sigma}_1 \cdot \boldsymbol{\sigma}_2$, which we shall meet on several occasions. It is

ISBN 0-8053-8601-7

interesting to note its effect on the singlet or triplet eigenfunctions. The eigen-functions of $(\sigma_1 + \sigma_2)^2$ and $\sigma_{1z} + \sigma_{2z}$ are also eigenfunctions of $\sigma_1 \cdot \sigma_2$, with eigenvalues -3 and 1. In fact, considering that $\sigma_x^2 = \sigma_y^2 = \sigma_z^2 = 1$, we have

$$(\sigma_1 + \sigma_2)^2 = \sigma_1^2 + \sigma_2^2 + 2\sigma_1 \cdot \sigma_2 = 6 + 2\sigma_1 \cdot \sigma_2$$

The eigenvalue of $(\sigma_1 + \sigma_2)^2$ for triplet states is 8. For singlet states the eigenvalue of $(\sigma_1 + \sigma_2)^2$ is 0. Hence

$$8 = 6 + 2\sigma_1 \cdot \sigma_2 \qquad \sigma_1 \cdot \sigma_2 = 1 \text{ for triplet states}$$

$$0 = 6 + 2\sigma_1 \cdot \sigma_2 \qquad \sigma_1 \cdot \sigma_2 = -3 \text{ for singlet states}$$

For an immediate application of this result we write the most general central force potential as

$$V_1(r) + V_2(r)\sigma_1 \cdot \sigma_2 \tag{10-5.1}$$

where $V_1(r)$ and $V_2(r)$ are functions of the distance between the nucleons.

It follows then that the potential for singlet and triplet states is given by

$$V_s = V_1 - 3V_2$$

and

$$V_t = V_1 + V_2$$

respectively.

In 1941 Wigner and Eisenbud showed that the most general nucleon–nucleon potential (under the restrictions that the potential depends only on the distance r between the nucleons, and on their spins, so that the total momentum, angular momentum, and parity are constants of the motion) has the form

$$V_1(r) + V_2(r)\sigma_1 \cdot \sigma_2 + V_3(r)S_{12} \tag{10-5.2}$$

with

$$S_{12} \equiv \left[\frac{3(\sigma_1 \cdot \mathbf{r})(\sigma_2 \cdot \mathbf{r})}{r^2} - \sigma_1 \cdot \sigma_2 \right] \tag{10-5.3}$$

The third term is identical in structure to the term that would represent the interaction between two magnetic dipoles at distance $\mathbf{r}$. Naturally it depends on the orientation of the two spins with respect to $\mathbf{r}$.

The actual presence of the third term [Eq. (10-5.3)] is demonstrated by the finite quadrupole moment of the deuteron. The argument runs as follows: If

ISBN 0-8053-8601-7

the eigenfunction of the deuteron were exactly a $^3S_1$ state i.e., spherically symmetric, the electric quadrupole moment would be 0. With a potential such as Eq. (10-5.1) for two nucleons, parity, $I^2 = (\mathbf{L} + \mathbf{S})^2$, I_z, L^2, L_z, S^2, and S_z are constants of the motion, as can be directly verified. Thus, a $^3S_1$ state would not mix with any other state, and the electric quadrupole moment of the deuteron would be 0. The term $V_3(r)S_{12}$ leaves only parity, I^2, I_z, and S^2, but not L^2 and L_z, as exact constants of the motion. [(For proofs of these statements see (El 66).] This has the consequence that the ground state of the deuteron will have as its eigenfunction a mixture of $^3S_1$ and $^3D_1$, the latter being the only state with the same parity and I as $^3S_1$. The average value of the electric quadrupole moment is not 0, because of the terms

$$\langle \psi_{^3S_1} | 3z^2 - r^2 | \psi_{^3D_1} \rangle \tag{10-5.4}$$

which appear in its computation with the complete eigenfunction.

The admixture of the $^3D_1$ state also influences the magnetic moment of the deuteron. An analysis of all the data leads to the conclusion that the eigenfunction must have approximately the form

$$0.96\psi_{^3S_1} + 0.26\psi_{^3D_1} \tag{10-5.5}$$

with considerable uncertainty about the value of the admixture. This means that the deuteron in the ground state spends about 7% of the time in a $^3D_1$ state.

There is no a priori reason why the forces acting between nucleons may not be velocity dependent. In fact, there is good experimental evidence for a term of the form

$$V_l(r)\mathbf{l} \cdot \mathbf{S} \quad \text{with} \quad \mathbf{l} = \mathbf{r} \times \mathbf{p}, \tag{10-5.6}$$

in the potential between two nucleons, where the velocity appears through the momentum p. There are no other possible terms, linear in p, which satisfy the usual conservation laws.

10-6 NUCLEON–NUCLEON FORCES; EXCHANGE FORCES

It is highly desirable to be able to account for the main properties of the nuclei, as embodied, for instance, in the mass formula, by deriving them from the nucleon–nucleon interaction. This program has been the object of very extensive and difficult investigations (Brueckner, 1955; Goldstone, Bethe, etc.), which up to now have been only partially successful. The difficulties are twofold: our imperfect knowledge of the nucleon–nucleon force, and the mathematical complexity of the multibody problem.

However, we can see at once that simple attractive potentials between nucleons are inadequate for the explanation of nuclear binding. Consider a

ISBN 0-8053-8601-7

nucleus of mass number A. The total energy of this nucleus would be $T + U$, where T is the kinetic energy of all nucleons and U their potential energy. Under our hypothesis of attractive potentials between each nucleon pair, the potential energy would be equal to the number of nucleons pairs, which is $A(A - 1)/2$, times a function of the average distance between nucleons. On the other hand, the kinetic energy is dominated by the exclusion principle. If we enclose a gas of fermions in a sphere of radius R, the particles in their lowest quantum state will occupy the lowest levels, one per level, according to the exclusion principle. For each level there is a kinetic energy, and the whole assembly will therefore have total kinetic energy, which is easily calculated and which we find to be proportional to $A^{5/3}R^{-2}$ (see Chap. 6).

In the ground state $T + U$ will be minimum, and we determine the only parameter available, the nuclear radius R, by this condition. For large A the potential energy is proportional to A^2, and the kinetic energy is proportional to $A^{5/3}$. If we choose R so as to minimize the potential energy, on varying R the corresponding increase of kinetic energy will be insufficient to counterbalance the increase of potential energy, and the binding will be maximum near the value of R that minimizes the potential energy (i.e., for R equal to the range of nuclear forces, irrespective of A). In other words, the terms proportional to A^2 prevail over those proportional to $A^{5/3}$, and it is enough to minimize the potential energy in order to obtain the ground state. All nuclei would have the same radius, approximately equal to the range of nuclear forces. This is contrary to experiment, because we know that the nuclear radii are proportional to $A^{1/3}$ and not constant for large A. Moreover, the total binding energy for large A is proportional to A, and not to A^2, as one would conclude from the arguments given above.

Nuclear forces must have a property (similar to that of the chemical valence forces) that brings about "saturation." Such a force would attract a small number of nucleons but become repulsive for a larger number. This type of interaction can be achieved in the simplest way by assuming a force which is strongly repulsive at a very short distance (repulsive core) and attractive at a suitable range of distances. There is good evidence for the existence of such forces, with the repulsive core having a radius of about 0.4 F. This core is in fact chiefly responsible for the constant density of nuclear matter.

However, in addition to the repulsive core, there are other mechanisms which produce saturation. Chemical forces show very clear examples of saturation: two hydrogen atoms combine to form a hydrogen molecule, but a third atom is not bound by such a molecule. The quantum-mechanical reason for this effect is well known. In the hydrogen molecule the two electrons overlap, or better, are in the same spatial orbit, although with opposite spin. A third electron cannot occupy this orbit, since it is prevented by Pauli's principle.

ISBN 0-8053-8061-7

This situation suggests some analogies to the case of nuclei. The strong binding (28.11 MeV) of the alpha particle gives important qualitative indications on this point. Consider, hypothetically, that each nucleon has bonds qualitatively similar to chemical valence bonds. In the case of D, ^{3}H, and ^{4}He, we have, respectively, two particles and one shared bond, three particles and three shared bonds, and four particles and six shared bonds. From the mass defects we find the energy associated with each bond to be 2.22, 2.83, and 4.72 MeV. A fifth particle is not bound at all, since both ^{5}He and ^{5}Li are unstable. That saturation is reached for a structure of four nucleons, two neutrons, and two protons can be interpreted as indicating a strong interaction when the nucleons have overlapping eigenfunctions as far as space coordinates are concerned, irrespective of spin.

The strong binding of alpha particles was also used by Wigner to show the short range of nuclear forces. His argument was that in an alpha particle there are (1) three bonds per nucleon, and the nucleons come closer together than in the other light nuclei and (2) being closer together and the nuclear forces having a short range, the nucleons interact more strongly.

"Exchange forces" can produce saturation. They are typically quantum mechanical in nature, and they occur as follows: The eigenfunction of a neutron and a proton can be written

$$\psi(r_1 s_1, r_2 s_2) = \psi(1, 2) \tag{10-6.1}$$

where r_1, s_1 are the coordinate and spin of the neutron and r_2, s_2 are the coordinate and spin of the proton. The expectation value of the potential is obtained by the usual rule,

$$\langle U \rangle = \int \psi(1, 2)^* U(1, 2)\psi(1, 2)\, d\tau_1\, d\tau_2 \tag{10-6.2}$$

where the possible spin dependence of the potential is taken into account because $U(1, 2) \equiv U(r_1 s_1, r_2 s_2)$. However, one may also replace the potential U by more general operators which exchange the two particles. The expectation value of the potential energy will then be, not

$$\langle U \rangle = \int \psi(1, 2)^* U(1, 2)\psi(1, 2)\, d\tau_1\, d\tau_2 \tag{10-6.3}$$

as for ordinary forces, but

$$\int \psi(2, 1)^* U(1, 2)\psi(1, 2)\, d\tau_1\, d\tau_2 \tag{10-6.4}$$

The ordinary function U is replaced by the operator UP_H, where P_H exchanges the particles, or, formally, the coordinate *and* spin variables of the first and second particles.

ISBN 0-8053-8061-7

One can also think of operators that exchange, not both spin and coordinate of the particles, but only the coordinate and not the spin or only the spin and not the coordinate. These operators give the following explicit results:

$$P_H \psi(r_1 s_1, r_2 s_2) = \psi(r_2 s_2, r_1 s_1) \quad \text{Heisenberg} \qquad (10\text{-}6.5)$$

$$P_M \psi(r_1 s_1, r_2 s_2) = \psi(r_2 s_1, r_1 s_2) \quad \text{Majorana} \qquad (10\text{-}6.6)$$

$$P_B \psi(r_1 s_1, r_2 s_2) = \psi(r_1 s_2, r_2 s_1) \quad \text{Bartlett} \qquad (10\text{-}6.7)$$

named after the physicists who first suggested them. Ordinary forces are generally called Wigner forces.

It is possible to show that some exchange forces can give rise to saturation. More specifically, of the three types of exchange forces mentioned above, the Heisenberg forces would give special stability to the deuteron, and the Majorana forces to the alpha particles; the Bartlett forces do not give saturation. We conclude that exchange forces, at least in part of the Majorana type, are present.

The existence of exchange forces can be shown directly by high-energy neutron–proton scattering, as was pointed out by Wick (1933). Experiments on n–p scattering at 90 MeV showed that neutrons impinging on protons give rise, preferentially in the forward direction, to protons (Figs. 10-12 and 10-13).

Calculating classically, with an ordinary force, we estimate the maximum momentum transfer in a collision between two particles of the same mass to be (average force) × (duration of collision). If the radius of action of the potential of depth U is r, the duration of the collision is r/v (where v is the relative velocity of the particles) and the average force is U/r. The momentum transfer is of the order of $\Delta p = U/v$. If the neutron stops and the proton escapes forward, $\Delta p \cong 2p$, where p is the momentum of the impinging neutron. However, $U/v \ll p$, if we assume that $U = 20$ MeV, $Mv^2 \sim 500$ MeV as in the experimental conditions. This contradiction is removed by the hypothesis that during the collision the particles exchange their charge and that the real momentum transfer for the protons escaping forward is very small.

● The same result is obtained in quantum mechanics by considering the collision in the Born approximation. Here the scattered wave has the form $f(\theta)(e^{ikr}/r)$, with

$$f(\theta) = \frac{m}{2\pi \hbar^2} \int \exp(-i\mathbf{k}_f \cdot \mathbf{r}) U(r) \exp(i\mathbf{k}_i \cdot \mathbf{r}) \, d\tau \qquad (10\text{-}6.8)$$

where m is the reduced mass, $\mathbf{r} = r_1 - r_2$, and θ, $\mathbf{k}_i$, $\mathbf{k}_f$ are in the center-of-mass system.

The integral for a short-range $U(r)$ is appreciable only if $\mathbf{k}_i - \mathbf{k}_f \cong 0$, because otherwise the function $\exp[i(\mathbf{k}_i - \mathbf{k}_f)\mathbf{r}]$ oscillates so rapidly over the region for which $U(r)$ is appreciable that the result averages to 0. Now $\mathbf{k}_i - \mathbf{k}_f$

ISBN 0-8053-8061-7

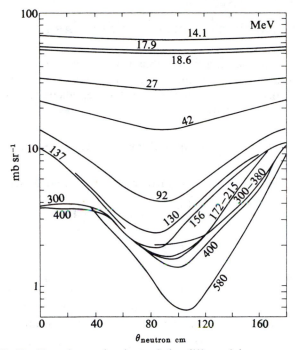

Figure 10-12 Experimental values of the differential neutron–proton cross section at various energies. Cross sections and angles refer to the center-of-mass system, energies to the laboratory system.

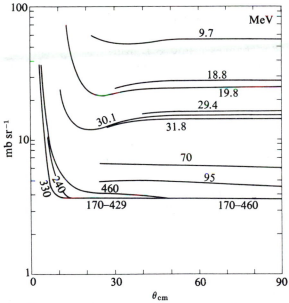

Figure 10-13 Experimental values of the differential proton–proton cross section at various energies up to 460 MeV. Cross sections and angles refer to the center-of-mass system, energies to the laboratory system.

ISBN 0-8053-8061-7

= 0 indicates forward scattering. We thus find again that the impinging neutrons are scattered forward, contrary to experiment. Upon introducing an exchange force, instead, the potential $U(r)$ is replaced by $U(r)P_M$, which gives, in place of Eq. (10-6.8),

$$f_e(\theta) = \frac{m}{2\pi\hbar^2} \int \exp(+i\mathbf{k}_f \cdot \mathbf{r}) U(r) \exp(i\mathbf{k}_i \cdot \mathbf{r}) \, d\tau \qquad (10\text{-}6.9)$$

because the P_M operator changes particles 1 and 2 and hence transforms $\mathbf{r}$ into $-\mathbf{r}$.

The $f_e(\theta)$ due to exchange forces is thus large when

$$\mathbf{k}_f + \mathbf{k}_i = 0 \qquad (10\text{-}6.10)$$

that is, when the neutron is scattered backward and the proton forward. The experimental result of Fig. 10-12 demonstrates clearly two maxima in the forward and backward directions at high energies. We conclude that both normal and exchange forces are present and that they are of comparable intensities.

Exchange forces have the interesting feature of changing sign according to the value of the angular momentum of the two nucleons. To see this, consider the eigenfunction of the two nucleons in the center-of-mass system (neglecting spin and noncentral forces),

$$\psi(r, \theta, \varphi) = f(r) Y_l^m (\theta, \varphi) \qquad (10\text{-}6.11)$$

The operator P_M changes $\mathbf{r}$ into $-\mathbf{r}$, which is the same as leaving r unchanged, but changing θ into $\pi - \theta$ and φ into $\varphi + \pi$. This has the result of multiplying $Y_l^m(\theta, \varphi)$ by $(-1)^l$; hence

$$P_M \psi = (-1)^l \psi \qquad (10\text{-}6.12)$$

and the result of any calculation involving $U(r)P_M$ is the same as that obtainable from $(-1)^l U(r)$, where l is the angular momentum of the operand on which $U(r)P_M$ operates.

Taking into account the spin, we must distinguish the operators P_M, P_H, and P_B. They give the following potentials:

$$UP_M = (-1)^l U \qquad (10\text{-}6.13)$$

$$UP_H = (-1)^{l+S+1} U \qquad (10\text{-}6.14)$$

$$UP_B = (-1)^{S+1} U \qquad (10\text{-}6.15)$$

where S is the resultant spin of the two nucleons. ●

ISBN 0-8053-8061-7

The operators $P_i (i = M, H, B)$ can also be written explicitly, using the operators σ and τ relative to spin and i spin. For instance,

$$P_B = \tfrac{1}{2}(1 + \boldsymbol{\sigma}_1 \cdot \boldsymbol{\sigma}_2) \quad P_H = -\tfrac{1}{2}(1 + \boldsymbol{\tau}_1 \cdot \boldsymbol{\tau}_2) = -P_\tau \quad (10\text{-}6.16)$$

where P_τ is the operator exchanging i spins. Assuming charge independence and remembering that parity, i spin, and S^2 are always constants of the motion, we can classify the velocity-independent forces. Consider separately the ordinary and the tensor forces. Each type has its own dependence on r, according to the even or odd parity of the state and to the total spin and i spin of the state. However, tensor forces do not exist for singlet spin states, because the operator S_{12} applied to a singlet eigenfunction gives identically zero. We thus have six functions of r to be determined. If we include $\mathbf{l} \cdot \mathbf{S}$ forces (operating only in triplet states) as well, we have eight functions.

These have been determined, at least approximately, by fitting a large number of empirical data, such as the properties of the deuteron and the results of measurements on nucleon–nucleon scattering, including polarization (see the next section). When possible, meson theory has also been used, mainly to determine the outer fringes of the potentials by the so-called one-pion exchange theory, or OPEP (Gammel, Thaler; Signell, Marshak; Breit and the Yale group). The results are illustrated by the example in Fig. 10-14.

Figure 10-14 (*see overleaf*) Yale nucleon–nucleon potentials. A potential (Yale) that reproduces the scattering experiments and phase shifts in nucleon–nucleon scattering. The form used is

$$V = V_c + V_T S_{12} + V_{LS} (\mathbf{L} \cdot \mathbf{S}) + V_q \left[(\mathbf{L} \cdot \mathbf{S})^2 + \mathbf{L} \cdot \mathbf{S} - \mathbf{L}^2 \right]$$

The last term is omitted in the figures. V_c is different for even and odd states and also depends on S. We thus have four curves for V_c. V_T and V_{LS} are different from zero only for triplet states and also give four curves, depending on the parity of the states. (a) The singlet even-parity potential $^1V_c{}^+$, as a function of $x = r m_\pi c / \hbar$. The short vertical line is at the hard-core radius corresponding to $x_c = 0.35$. The potential is $+\infty$ for $x < x_c$. Different scales are used for $x < 1.0$ and $x > 1.0$. (b) The singlet odd-parity potential $^1V_c{}^-$, as a function of x. Other conventions as in (a). (c) The triplet even-parity central potential $^3V_c{}^+$, as a function of x. Other conventions as in (a). (d) The triplet even-parity potential function of x, $^3V_T{}^+$, which multiplies the tensor operator S_{12}. Other conventions as in (a). (e) The triplet even-parity potential function of x, $^3V_{LS}{}^+$, which multiplies the spin–orbit operator $\mathbf{L} \cdot \mathbf{S}$. Other conventions as in (a). (f) The triplet odd-parity central potential $^3V_c{}^-$, as a function of x. Other conventions as in (a). (g) The triplet odd-parity potential function of x, $^3V_T{}^-$, which multiplies the tensor operator S_{12}. Other conventions as in (a). (h) The triplet odd-parity potential function of x, $^3V_{LS}{}^-$, which multiplies the spin–orbit operator $\mathbf{L} \cdot \mathbf{S}$. Other conventions as in (a). [Courtesy of G. Breit.]

ISBN 0-8053-8061-7

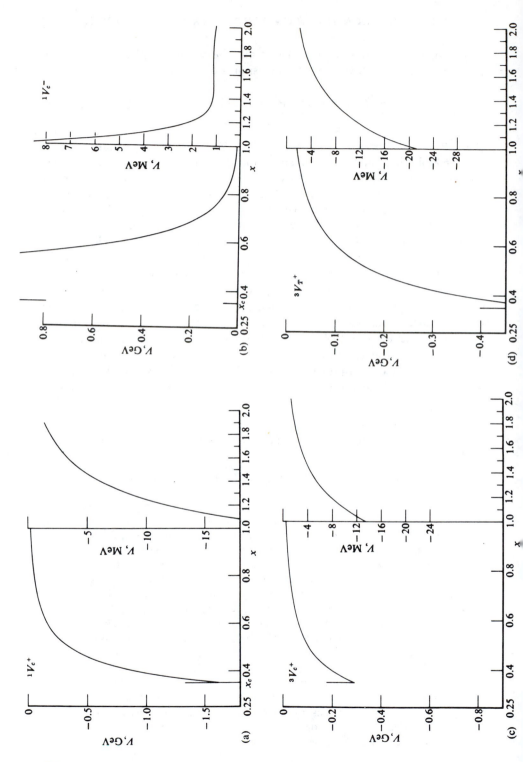

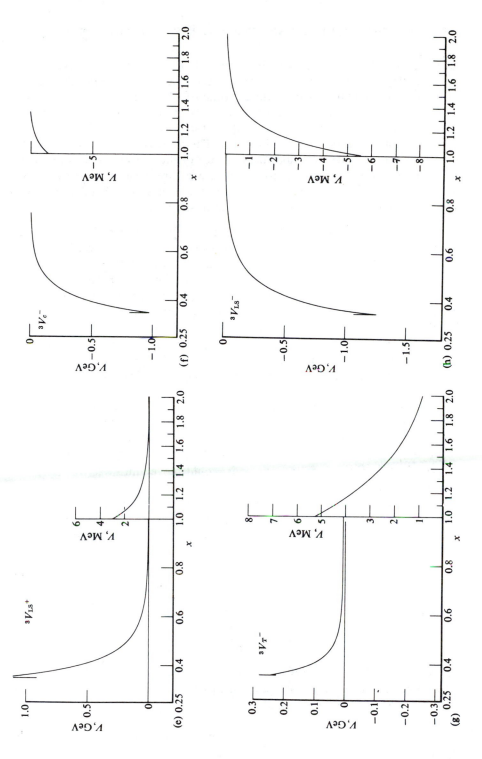

These potentials do not yet give results in perfect agreement with experiment nor are they based on an exact theory; nevertheless, they reproduce fairly well the experimental collision data up to 200 MeV. Their qualitative features are important for explaining nuclear models in terms of nucleon–nucleon forces. An important characteristic of these forces is the strong dependence of the forces on the parity of states involved. The force is stronger in even states than it is in odd states. The spin dependence of the central force is relatively weak. Both central and tensor forces have a central repulsive core of about 0.4 F in radius and an outside attractive region of 1 to 2 F in radius. The spin–orbit potential has a much shorter radius.

Brueckner and numerous other investigators, among them Goldstone and Bethe, have been able to derive from the forces described above the properties of a hypothetical nuclear matter. Assume this matter to consist of a very large assembly of neutrons and protons interacting with each other according to the nuclear forces described above; however, the electrical repulsion between protons is neglected. The assembly is so large that surface effects may be also neglected. The results of the calculation show that the average binding energy per particle would be approximately 15.7 MeV and the average distance between neighboring nucleons $d = 1.8$ F and hence the density $d^{-3} = 0.17 \times 10^{39}$ nucleons cm^{-3}, or 2.85×10^{14} g cm^{-3}. Note that $d = 1.61 r_0$ [r_0 of Eq. (6-3.1)].

The Brueckner–Goldstone–Bethe theory is complicated; one of the most lucid accounts is given by B. D. Day (1967) and a recent and complete one by Bethe (1971). The theory is beyond the scope of this book.

A simplified account of the arguments used, according to Weisskopf, follows. In this nuclear matter, nucleons would move almost as free particles (compare Sec. 6-12) because the Pauli principle inhibits low-energy collisions leading to final stages that are already occupied. Therefore, nuclear matter can be crudely approximated by a degenerate gas occupying the nuclear volume. The average kinetic energy per particle in a Fermi gas, given in Sec. 6-6 is

$$T = \frac{B}{d^2} \qquad \text{with} \quad B = \frac{3^{5/3}\pi^{4/3}\hbar^2}{2^{2/3}10m} \tag{10-6.17}$$

or, numerically,

$$T(\text{MeV}) = \frac{75}{d^2} \qquad (d \text{ in F})$$

However, in the nuclear case the hard core of the nuclear potential gives an effect reminiscent of the "covolume" in the van der Waals equation of state, and one gets in closer approximation

$$T = \frac{B}{(d-c)^2} \qquad \text{with} \quad c \cong 0.22\, d \tag{10-6.18}$$

ISBN 0-8053-8061-7

The potential energy per nucleon U, assuming that the nucleons form an almost perfect gas, would be given by

$$- U = \frac{4\pi}{3} \frac{Vb^3}{d^3} \tag{10-6.19}$$

where the V is the depth and b is the range of the even central forces, which are the most important ones and can be taken from Fig. 10-14. We justify this equation by considering V as a perturbation on the unperturbed wave uniformly occupying the volume d^3. A better expression for the potential energy is obtained by multiplying Eq. (10-6.19) by an $f(d)$ and writing $- U = Cf(d)/d^3$. Summing the two expressions we can find the energy per particle:

$$E = \frac{B}{(d - c)^2} - \frac{C}{d^3} f(d) \tag{10-6.20}$$

and hence the value d_0 for which the energy is a minimum. It has been pointed out by Weisskopf that, near $d = d_0$, $f(d)$ is approximately $\frac{1}{2}(d/d_0)$. From Eq. (10-6.20) we find, by setting to zero the derivative with respect to d, the equilibrium d_0 and the average binding energy per nucleon (15 MeV).

Equation (10-6.20) also allows us to calculate the compressibility $\frac{1}{2} d^2(\partial^2 E/\partial d^2)$ of nuclear matter and the "symmetry energy":

$$\frac{1}{2} \frac{\partial^2 E}{\partial y^2} \quad \text{with} \quad y = \frac{N - Z}{N + Z}$$

The compressibility is about 100 MeV and the symmetry energy 30 MeV. These calculations are important because they connect the nucleon–nucleon force to observable properties of nuclear matter; however, they are not precise and they depend on the potential chosen, which is not uniquely determined.

Pure neutron matter may be present in certain pulsating stars (pulsars). They seem to consist mostly of neutrons except near the surface, which contains ordinary nuclei and electrons. When the density exceeds about $3 \cdot 10^{11}$ g cm^{-3}, one has neutron matter. The density of neutron stars is of the order of 10^{14} g cm^{-3}.

10-7 POLARIZATION; HIGH-ENERGY NUCLEAR SCATTERING

High-energy nuclear scattering, for a laboratory kinetic energy above 50 MeV, involves, besides s waves, increasingly higher-order waves, with $l = 1, 2, 3$, etc. The angular distribution is no longer spherically symmetric, and analysis of the experimental data becomes increasingly complicated. It is possible

ISBN 0-8053-8061-7

to obtain a description of the scattering through phase shifts (see Appendix A). However, to obtain the phase shifts it is not enough to measure angular distributions of the scattering only: it is necessary to consider the polarization of the beams as well. A complete treatment of this subject is beyond the scope of this book, and we shall limit ourselves to the simplest ideas and facts.

We treat first the scattering of a nucleon on a spinless center, which is much simpler than nucleon–nucleon scattering and provides an experimental tool for many polarization experiments. Nucleons may be polarized, or polarized beams analyzed, by elastic scattering on a spinless center provided that the interaction producing the scattering is spin dependent, as when it contains a term $\mathbf{l} \cdot \mathbf{S}$. To polarize a beam of protons, we may accelerate polarized protons obtainable at low energies with defined molecular-beam techniques. We may also scatter unpolarized protons on a target (which for simplicity we shall assume to be spinless and very heavy compared to the proton) and select the beam scattered under an angle θ_1. By scattering the polarized beam on a target, under an angle θ_2, we find that the scattered intensity depends not only on θ_2 but also on angle φ, between the two planes of scattering, and that it has the form

$$I = A(\theta_2) + B(\theta_2) \cos \varphi \tag{10-7.1}$$

The quantity B vanishes, unless the beam incident on the second target is polarized. To define φ precisely, consider the vector momenta of the incident and scattered particle in each scattering and call

$$\mathbf{n}_1 = \frac{\mathbf{p}_i \times \mathbf{p}_s}{|\mathbf{p}_i \times \mathbf{p}_s|} \tag{10-7.2}$$

where the index 1, 2 means first, second scattering, and i and s incoming and scattered. We have then

$$\mathbf{n}_1 \cdot \mathbf{n}_2 = \cos \varphi \tag{10-7.3}$$

We define the beam polarization as the ratio

$$P = \frac{N^+ - N^-}{N^+ + N^-} \tag{10-7.4}$$

where N^+ means the number of particles with spin up and N^- the number of particles with spin down. The up–down direction is here assumed to be perpendicular to the direction of the plane of scattering. This definition is equivalent to saying that P is the expectation value of the component of $\boldsymbol{\sigma}$ in the direction $\mathbf{n}$. If we scatter an unpolarized beam on a spinless target, the beam scattered under an angle θ acquires a polarization P that is a function

ISBN 0-8053-8061-7

of θ, of the energy of the beam, and of the particles involved. This polarization is often called the polarizing power. In other words, P is the polarization, after scattering, of an initially unpolarized beam.

Let us indicate by $p(+L+)$ the probability that an incident particle with spin up will be scattered to the left with spin up and by $p(+L-)$ the probability that the same particle will be scattered to the left with spin down. We have eight similar probabilities altogether. These, however, are not independent, as we can see by observing that rotation by 180 deg around the incident-beam direction brings $p(+L+)$ into $p(-R-)$; hence the two quantities must be equal.

Also, in the specific case of a particle of spin $\frac{1}{2}$ scattering on a spinless center, the spin component normal to the scattering plane cannot flip, or $p(+L-)=0$.

As a consequence we can write for our case

$$p(+L+)=p(-R-)=\tfrac{1}{2}(1+P)$$

$$p(-L+)=p(+R-)=0$$

$$p(+L-)=p(-R+)=0$$

$$p(-L-)=p(+R+)=\tfrac{1}{2}(1-P)$$

To measure the polarization of a beam, consider first the scattering of an unpolarized beam on a spinless center (Fig. 10-15). The number of particles scattered left and right with spin up or down is given in Table 10-1.

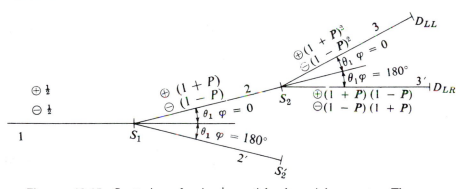

Figure 10-15 Scattering of spin $\frac{1}{2}$ particles by spinless center. The unpolarized beam 1 is scattered by scatterer S_1. The intensities scattered in directions 2 and 2' are identical, but the scattered beams are polarized. Beam 2 scattered at S_2 now has different intensities in beams 3 and 3'. The number of particles with spin up or down in the different beams is proportional to the quantities indicated by $\oplus$ and $\ominus$. It has been assumed for simplicity that $P_1 = P_2 = P$ and factors $\frac{1}{2} p_1$ and $\frac{1}{2} p_1 p_2$ have been omitted.

ISBN 0-8053-8061-7

TABLE 10-1   BEAM INTENSITIES IN SIMPLE AND DOUBLE SCATTERING

Beam	Spin[a]	Intensity[b]
Incident	+	$\frac{1}{2}$
	−	$\frac{1}{2}$
Scattered once left (L)	+	$\frac{1}{2} p_1(1 + P_1)$
	−	$\frac{1}{2} p_1(1 - P_1)$
Scattered once right (R)	+	$\frac{1}{2} p_1(1 - P_1)$
	−	$\frac{1}{2} p_1(1 + P_1)$
Scattered twice left (LL)	+	$\frac{1}{2} p_1 p_2(1 + P_1)(1 + P_2)$
	−	$\frac{1}{2} p_1 p_2(1 - P_1)(1 - P_2)$
Scattered twice—first left,	+	$\frac{1}{2} p_1 p_2(1 + P_1)(1 - P_2)$
then right (LR)	−	$\frac{1}{2} p_1 p_2(1 - P_1)(1 + P_2)$

[a] Plus sign means spin up; minus sign, spin down.
[b] p and P are functions of energy and θ (see Fig. 10-15).

The intensity left or right, after the first scattering, will be the same, but the polarization of the beam scattered to the left will be

$$\frac{(1 + P_1) - (1 - P_1)}{(1 + P_1) + (1 - P_1)} = P_1 \tag{10-7.5}$$

On a second scattering of the "left" beam, the intensity of scattering to the left and right will be different. If in first and second scattering, energy and θ are the same ($P_1 = P_2 = P$), we shall have an intensity to the left proportional to

$$L = (1 + P)^2 + (1 - P)^2 = 2(1 + P^2) \tag{10-7.6}$$

and on the right an intensity proportional to

$$R = 2(1 + P)(1 - P) = 2(1 - P^2) \tag{10-7.7}$$

Hence

$$\epsilon = \frac{L - R}{L + R} = \frac{2 + 2P^2 - 2 + 2P^2}{2 + 2P^2 + 2 - 2P^2} = P^2 \tag{10-7.8}$$

The quantity ϵ is called the asymmetry. This ratio $(L - R)/(L + R)$ is directly measurable, and therefore we may obtain $|P|$ from it, although not the sign of P. The sign may be determined by studying the interference of nuclear scattering with Coulomb scattering, or by slowing down the polarized beam and using some processes of low-energy nuclear physics in which the sign of the polarization is predictable (Marshall, 1955).

ISBN 0-8053-8601-7

The quantity P is by definition the expectation value of the component σ in the direction z perpendicular to the scattering plane. If parity is conserved in the scattering of an unpolarized beam on a spinless target, the direction of σ must be defined by the two momenta $\mathbf{p}_i$ and $\mathbf{p}_s$, which are polar vectors. However, σ is an axial vector, and the only way of forming an axial vector with two polar vectors is to set

$$\sigma \propto \mathbf{p}_i \times \mathbf{p}_s$$

which shows that σ must be perpendicular to the plane of scattering. We conclude that in the case under consideration

$$P = \langle \sigma_z \rangle = \langle \sigma \rangle$$

However, one may have a beam with $\langle \sigma \rangle$ in an arbitrary direction, for instance, by accelerating polarized protons. If we want a complete measurement of $\langle \sigma \rangle$, we must measure its three components. Taking the x axis in the direction of the beam and y and z perpendicular to it, we obtain, from scattering experiments in the x–y and x–z planes, $\langle \sigma_z \rangle$ and $\langle \sigma_y \rangle$. To obtain $\langle \sigma_x \rangle$ we deflect the beam magnetically or electrically so that the spin rotates (see Sec. 2-13). In the deflected beam the original $\langle \sigma_x \rangle$ has transverse components, which can be measured by a scattering experiment. In this way one determines the vector $\langle \sigma \rangle$.

To understand the left–right asymmetry produced by nuclear scattering of polarized protons, consider the density distribution ρ of nuclear matter and a nucleon traveling through it. Nuclear forces are short range, and the net force acting on a nucleon within the nucleus is 0. Near the surface, however, the nucleon is subject to a net force, because it has nuclear matter only on one side. In this region the energy (scalar) may depend only on σ, $\mathbf{p}$, and $\nabla \rho$ through the scalar that can be formed with the axial vector σ and the two polar vectors $\mathbf{p}$ and $\nabla \rho$. This scalar is proportional to

$$\sigma \cdot (\nabla \rho \times \mathbf{p}) \tag{10-7.9}$$

Here

$$\nabla \rho = \frac{\mathbf{r}}{r} \frac{d\rho}{dr} \tag{10-7.10}$$

where $\mathbf{r}$ is a vector from the center of the nucleus to the point considered. This is the expression for the spin–orbit coupling energy. It may be rewritten as

$$V = \text{const} \; \frac{1}{r} \frac{d\rho}{dr} \, \sigma \cdot \mathbf{l} \tag{10-7.11}$$

where $\mathbf{l} = \mathbf{r} \times \mathbf{p}$ is the angular momentum.

ISBN 0-8053-8601-7

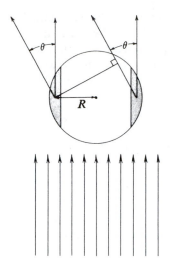

Figure 10-16 Beam of polarized protons (spin up) impinging on nucleus. The shaded area represents the region where the nuclear potential is distorted by the spin–orbit coupling. The figure is purely schematic.

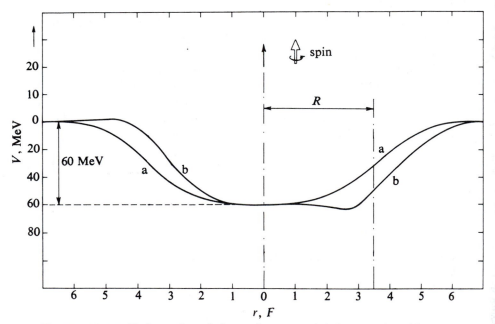

Figure 10-17 Deformation of the nuclear potential due to spin-orbit coupling for a spin-up particle directed perpendicularly into the drawing: (a) without spin–orbit coupling, (b) with spin–orbit coupling.

According to the shell model of the nucleus, the sign of the coefficient of the spin–orbit coupling should be negative (Figs. 10-16 and 10-17).

Now let us use an optical model (see Chap. 11), adding the spin–orbit coupling to the complex potential $(U_1 + iU_2)$. For a first approximation use Born's method, which makes the scattering amplitude from each point of the scatterer proportional to the local value of the potential. If the incident wave has the equation $e^{i\mathbf{k}\cdot\mathbf{r}}$ and contains only particles with spin up, the wave scattered through a certain small angle θ has different amplitudes when scattered from the right and left sides of the nucleus; for on one side the spin–orbit coupling V increases the scattering potential and on the other side decreases it.

The scattering from the two sides of the nucleus also has a phase difference of $\pm kR\theta$ from the scattering produced by the center of the nucleus. As a result, the wave scattered through a small angle θ has an amplitude

$$(U_1 + iU_2)\exp(ikr) - D\,\exp\big[ik(r - R\theta)\big] + D\,\exp\big[ik(r + R\theta)\big]$$

$$= (U_1 + iU_2 + 2iD\,\sin k\theta R)\exp(ikr) \quad (10\text{-}7.12)$$

where D is the volume of the potential depression and rise near the nuclear edge, relative to the total nuclear volume, multiplied by the potential prevailing at the nuclear edge. R is the nuclear radius. In this expression we have taken into account the phase differences and the amplitude of the wavelets originating in the scattering nucleus. The intensity of the scattered beam, obtained from the modulus square of Eq. (10-7.12), is proportional to

$$U_1^2 + U_2^2 + 4D^2 \sin^2 k\theta R + 4U_2 D \sin k\theta R \quad (10\text{-}7.13)$$

It is the last term that generates a left–right asymmetry in the scattering intensity. Note that in order to produce an effect, U_2 must be different from zero (i.e., there must be nuclear absorption), and that the sign of D determines whether particles are scattered more often to the right or left. At small angles, particles with spin up go to the left if the sign of the spin–orbit coupling is the one required by the shell model.

The considerations given here to show the mechanism by which the asymmetry arises are qualitative. A more refined theory is beyond the scope of this book. However, some details of the methods for treating polarized beams are presented in Appendix C, where it is shown that the scattering of a spin $\frac{1}{2}$ particle on a spinless center is described by three numbers, each a function of energy and scattering angle. One could take as three such numbers, for instance, the differential scattering cross section, a number $\langle \sigma \rangle$ measuring the degree of polarization of a scattered beam when the incident beam is unpolarized, and one of the Wolfenstein parameters shown in Fig. 10-18. Their measurement requires a triple scattering experiment or a polarized beam and a double scattering experiment. In Fig. 10-18 the first and last scattering are omitted because they are not necessary in principle.

ISBN 0-8053-8601-7

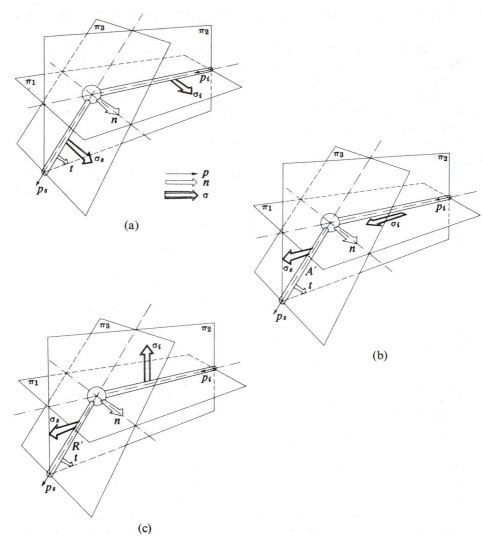

Figure 10-18 Definition of the Wolfenstein coefficients. In this figure π_2 is the plane of scattering. Planes π_1 and π_3 contain the incident and scattered beam and are perpendicular to π_2. $\mathbf{n}$, $\mathbf{p}_i$, $\mathbf{p}_s$ are unit vectors. Remember that $\mathbf{n} \propto \mathbf{p}_i \times \mathbf{p}_s$ and define unit vector $\mathbf{t} \propto \mathbf{n} \times \mathbf{p}_s$. Assume always $|\langle \sigma_i \rangle| = 1$. In (a) $\langle \sigma_i \rangle$ is in the $\mathbf{n}$ direction, and $\langle \sigma_s \rangle \cdot \mathbf{n} = (P + D)/(1 + P)$ defines D. Here P is the polarizing power of the target. Note that for a spinless target D is always 1.

In (b) $\langle \sigma_i \rangle$ is in the direction of $\mathbf{p}_i$. A is defined by $\langle \sigma_s \rangle \cdot \mathbf{t} = A$. One can also measure A' defined by $\langle \sigma_s \rangle \cdot \mathbf{p}_s = A'$.

In (c) $\langle \sigma_i \rangle$ is in the direction of $\mathbf{p}_i \times \mathbf{n}$. R is defined by $\langle \sigma_s \rangle \cdot \mathbf{t} = R$. One can also measure R' defined by $\langle \sigma_s \rangle \cdot \mathbf{p}_s = R'$.

To observe quantities such as $\langle \sigma_s \rangle \cdot \mathbf{t}$ one must perform a scattering in a plane containing $\mathbf{p}_s$ and perpendicular to $\mathbf{t}$. To observe components of $\langle \sigma_s \rangle$ parallel to $\mathbf{p}_s$ one must first turn $\langle \sigma_s \rangle$ (for instance, by magnetic deflection; see Sec. 2-13), so that they become transverse to $\mathbf{p}_s$.

ISBN 0-8053-8601-7

● Nucleon–nucleon scattering experiments, involving two particles of spin $\frac{1}{2}$, are much more complicated to analyze than the scattering of a nucleon on a spinless center. To completely describe the scattering, one requires in general eleven parameters that are functions of energy and angle, instead of the previous three. The number is reduced to nine if the nucleons are identical, or if one assumes charge independence. To obtain them one has to perform a number of experiments to determine the scattering cross section, polarization, Wolfenstein parameters, and, possibly, the correlation coefficients. Correlation coefficients may be obtained by starting with an unpolarized beam and unpolarized target and determining the expectation values of quantities such as $\langle \sigma_{1i}\sigma_{2k} \rangle = C_{ik}$, where 1 and 2 refer to the incident and scattered particles, i and k, to the direction in which the polarizations are observed (see Fig. 10-19). For instance, C_{pn} gives the expectation value that the projectile is polarized in the p direction and the recoil in the n direction. The quantities C_{ik} form a tensor that also determines the cross section when a beam polarized in direction i impinges on a target polarized in direction k. To completely determine the nine parameters mentioned above, it is possible to choose different sets of measurements: for an unpolarized beam on an unpolarized target a theoretically possible set is $d\sigma/d\omega$, P, D, R, A, A'; C_{nn}, C_{qp}, C_{pp}, C_{qq}. With a polarized beam and an unpolarized target, or an unpolarized beam and a polarized target, it suffices to measure fewer quantities, and still fewer with a polarized beam and a polarized target. (For a detailed discussion see Schumacher and Bethe.)

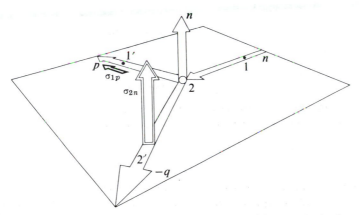

Figure 10-19 Unpolarized beam of nucleons (1) hits unpolarized target (2) and escapes with momentum **p**. The target recoils with momentum $-\mathbf{q}$. C_{pq} is defined as the value of $\langle \sigma_{1p}\sigma_{2n} \rangle$, where σ_{1p} is the component of the spin of the scattered nucleon (1) in the **p** direction and σ_{2n} is the component of the recoiling target nucleon (2) in the **n** direction. The figure is drawn for the laboratory system.

ISBN 0-8053-8601-7

To give an idea of the type of experimental data obtainable, Figs. 10-12 and 10-13 show the differential scattering cross section as a function of θ for neutron–proton and proton–proton. Figure 10-20 gives the total (elastic + inelastic) cross section as a function of energy for the same systems. Figure 10-21 shows typical polarization curves in proton–proton scattering extended to higher energies.

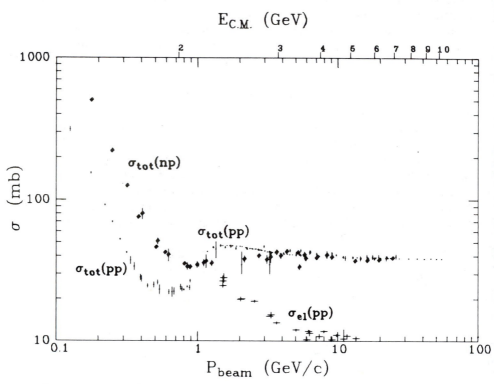

Figure 10-20 The neutron–proton total and the proton–proton total and elastic cross sections as a function of energy. [From (RPP 74).]

In analyzing nucleon–nucleon scattering, all available data are used to compute a set of phase shifts as functions of energy, for the different component waves. This is usually done by choosing a random set of phase shifts and calculating back (using high-speed computers) the observable quantities, and then changing some phase shifts and systematically improving the fit of the initial set in each trial. Often, such studies ultimately produce several sets of of acceptable phase shifts, and it may be difficult to obtain unambiguous results.

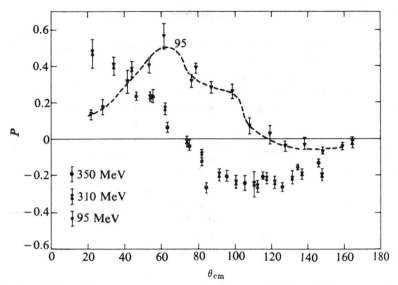

Figure 10-21 Neutron–proton measurements of $P(\theta)$ at different energies. [From M. H. McGregor, M. J. Moravcsik, and H. P. Stapp, compilation in *Ann. Rev. Nucl. Sci.*, **10**, 291 (1960).]

The connection between theory and experiment usually occurs at the computation of phase shifts. For instance, meson theories of nuclear forces or potentials are tested to see whether they reproduce the selected phase shifts and hence the experimental data. There are reasonably satisfactory sets of phase shifts up to about 300 MeV, where waves up to $L = 5$ and $J = 6$ have been calculated. Another approach, different from the calculation of the phase shifts, entails the calculation of a "scattering matrix" (see Chap. 11) having 4×4 complex elements, which, however, are not independent and reduce, as we said, to nine numbers. Each of them is a function of energy and angle. The two approaches have different advantages and supplement each other. ●

10-8 CAPTURE OF SLOW NEUTRONS BY HYDROGEN

Among the two-body problems that can be treated with simple approximations are the capture of slow neutrons by hydrogen and the reverse reaction, the photodisintegration of the deuteron, at least at low energies. They are interesting not only for their significance as two-body problems, but also as examples of electromagnetic nuclear phenomena. The reactions treated are

$$n + p \rightleftharpoons d + \hbar\omega \tag{10-8.1}$$

Read from right to left one has the photodisintegration; left to right, neutron capture.

In the slow neutron capture reaction we must reach the only bound state of the deuteron, which is almost entirely $^3S_1$, starting from an s state of the continuum. The $^1S_0$ state of the continuum can pass to a $^3S_1$ state by magnetic dipole radiation according to the selection rules of Chap. 8. The $^3S_1$ states of the continuum do not give rise to radiative capture, as we shall see shortly.

The following is a classical model of the physical mechanism of the capture: The magnetic moment of the deuteron is not pointed in the same direction as the sum of the spins, because the two spins carry different magnetic moments (see Fig. 10-22). Hence, in an n–p system in the $^3S_1$ state, there is a resulting magnetic moment which precesses around the fixed total angular momentum and can, therefore, radiate. The radiation by a magnetic dipole (cf. Chap. 8) is completely analogous to that of an electric dipole moment, and the transition probability per unit time is

$$\lambda = \frac{4}{3}\frac{\omega^3}{\hbar c^3}|\mu_{fi}|^2 \tag{10-8.2}$$

where μ_{fi} is the matrix element of the magnetic moment between the initial and final states.

The final state has three eigenfunctions corresponding to the three possible values of the z component of the spin,

$$\psi_f = \frac{u(r)}{r}\begin{cases} \alpha(n)\alpha(p) & S_z = 1 \\ 2^{-1/2}[\alpha(n)\beta(p) + \alpha(p)\beta(n)] & S_z = 0 \\ \beta(n)\beta(p) & S_z = -1 \end{cases} \tag{10-8.3}$$

where $\alpha(n)$ means neutron spin up, $\beta(n)$ means neutron spin down, etc., and, approximately [see Eq. 10-1.6],

$$u(r) = \left(\frac{\gamma}{2\pi}\right)^{1/2}e^{-\gamma r} \tag{10-8.4}$$

with

$$\gamma = (1/\hbar)(M|E|)^{1/2}$$

(where $-E$ is the binding energy of the deuteron and M is the mass of the nucleon), as discussed in Sec. 10-1.

ISBN 0-8053-8061-7

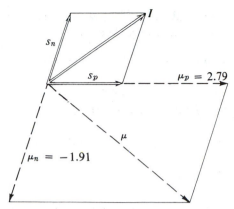

Figure 10-22 Angular momenta and magnetic moments in the neutron–proton capture. The whole figure rotates around the I axis.

The eigenfunction of the initial state is asymptotically, for large r,

$$\psi_i \underset{r \to \infty}{\to} \frac{j(r)}{r} \frac{1}{\sqrt{2}} \big[\alpha(n) \beta(p) - \alpha(p) \beta(n) \big] \qquad (10\text{-}8.5)$$

$$j(r) \to \frac{1}{k} \sin(kr + \delta_0) \qquad (10\text{-}8.6)$$

where δ_0 is the phase shift for the s wave, produced by the singlet potential (see Appendix A).

This expression is valid for $r \gg$ radius of potential well. The normalization of ψ_i is such that in the incident wave there is one particle per unit volume ($\rho = 1$). The matrix element is then

$$\mu_{if} = \int \psi_f^* \boldsymbol{\mu} \psi_i \, d\tau \qquad (10\text{-}8.7)$$

where the magnetic-moment operator $\boldsymbol{\mu}$ is

$$\boldsymbol{\mu} = \mu_p \boldsymbol{\sigma}_p + \mu_n \boldsymbol{\sigma}_n = \tfrac{1}{2}(\mu_n + \mu_p)(\boldsymbol{\sigma}_n + \boldsymbol{\sigma}_p) + \tfrac{1}{2}(\mu_n - \mu_p)(\boldsymbol{\sigma}_n - \boldsymbol{\sigma}_p) \qquad (10\text{-}8.8)$$

with μ_p and μ_n the magnetic moments of proton and neutron and $\boldsymbol{\sigma}$ the Pauli spin operator. Clearly $\boldsymbol{\mu}$ operates only on the spin-dependent part of the eigenfunction. Note that the term containing the factor $\mu_n + \mu_p$ vanishes in a singlet state because there

$$\boldsymbol{\sigma}_n + \boldsymbol{\sigma}_p = 0 \qquad (10\text{-}8.9)$$

The second term also would vanish if μ_n were equal to μ_p. As an example we now calculate the z component of μ. We have

$$\sigma_n^{(z)}\alpha(n) = \alpha(n) \qquad \sigma_n^{(z)}\beta(n) = -\beta(n) \qquad \text{etc.} \qquad (10\text{-}8.10)$$

Hence

$$\left(\sigma_n^{(z)} - \sigma_p^{(z)}\right)2^{-1/2}\left[\alpha(n)\beta(p) - \alpha(p)\beta(n)\right]$$

$$= 2^{-1/2}\left[\alpha(n)\beta(p) + \alpha(p)\beta(n) + \alpha(n)\beta(p) + \alpha(p)\beta(n)\right]$$

$$= 2^{+1/2}\left[\alpha(n)\beta(p) + \alpha(p)\beta(n)\right] \qquad (10\text{-}8.11)$$

In words, $\sigma_n^{(z)} - \sigma_p^{(z)}$ applied to the spin part of the singlet eigenfunction gives twice the eigenfunction for $^3S_1$, $S_z = 0$ and is orthogonal to those with $S_z = \pm 1$.

The matrix element is, therefore,

$$\mu_{if}^{(z)} = (\mu_n - \mu_p)\int_0^\infty \frac{u^*(r)j(r)}{r^2}\, 4\pi r^2\, dr \qquad (10\text{-}8.12)$$

Note that $u^*(r)$ is calculated with the triplet potential prevailing in the ground state and $j(r)$ with the singlet potential for the continuum. Thus they are not orthogonal. If the state in the continuum were a triplet state, $j(r)$ would be orthogonal to $u^*(r)$ and the matrix element would vanish, as we stated at the beginning of this section. The calculation of μ_x and μ_y in the case of unpolarized neutrons and unpolarized target gives the same result as that of μ_z.

Using the relation between λ and the cross section ($\sigma\rho v = \lambda$) and the normalization $\rho = 1$, we obtain from Eqs. (10-8.2) and (10-8.12)

$$\sigma_c = \frac{16\pi^2}{v}\frac{\omega^3}{\hbar c^3}(\mu_p - \mu_n)^2 I^2 \qquad (10\text{-}8.13)$$

where I is given by Eq. (10-8.14).

Here we have introduced the factor $\frac{1}{4}$ because only one fourth the initial n–p systems are in the singlet state, three fourths being in the triplet state. Another factor 3 derives from the fact that

$$\langle(\mu_p - \mu_n)_x\rangle^2 = \langle(\mu_p - \mu_n)_y\rangle^2 = \langle(\mu_p - \mu_n)_z\rangle^2$$

and that the transition probabilities add to each other, without interference terms.

We must still evaluate the integral

$$I = \int_0^\infty u^*(r)j(r)\, dr \qquad (10\text{-}8.14)$$

Replacing the expressions for $u(r)$, $j(r)$, we have

$$I = \frac{1}{k} \left(\frac{\gamma}{2\pi} \right)^{1/2} \int_0^\infty e^{-\gamma r} \sin(kr + \delta_0) \, dr$$

$$= \frac{1}{k} \left(\frac{\gamma}{2\pi} \right)^{1/2} \frac{k \cos \delta_0 + \gamma \sin \delta_0}{k^2 + \gamma^2} \tag{10-8.15}$$

To obtain a formula directly comparable to experiment, it is convenient to introduce the singlet scattering length a_s, which is related to the phase shift δ_0 by

$$k \cot \delta_0 = -1/a_s \tag{10-8.16}$$

[see Eq. (10-2.14)]. Expressing $\sin \delta_0$ and $\cos \delta_0$ through $\cot \delta_0$, we obtain

$$I = \left(\frac{\gamma}{2\pi} \right)^{1/2} \frac{1 - \gamma a_s}{(\gamma^2 + k^2)(1 + k^2 a_s^2)^{1/2}} \tag{10-8.17}$$

For orientation remember the numerical values

$$k = Mv/\hbar = 2.20 \times 10^9 (E_{\mathrm{eV}})^{1/2} \ \mathrm{cm}^{-1}$$

$$\gamma = 0.232 \times 10^{13} \ \mathrm{cm}^{-1} \tag{10-8.18}$$

$$a_s = -23.7 \ \mathrm{F}$$

Up to a few keV, $k \ll \gamma$, $ka_s \ll 1$; therefore, approximately

$$I = \left(\frac{\gamma}{2\pi} \right)^{1/2} \frac{1 - \gamma a_s}{\gamma^2} \tag{10-8.19}$$

Substituting this expression in Eq. (10-8.13) and measuring the magnetic moments in units of $e\hbar/2Mc$, we obtain the approximate formula

$$\dot{\sigma}_c = \frac{(4\pi)^2}{v} \frac{\omega^3}{c^3 \hbar} \left(\frac{e\hbar}{2Mc} \right)^2 (\mu_p - \mu_n)^2 \frac{\gamma}{2\pi} \frac{(1 - \gamma a_s)^2}{\gamma^4} \tag{10-8.20}$$

Note that $\hbar\omega$, the center-of-mass energy of the photon, is related to the laboratory velocity of the neutron captured by a proton at rest by

$$\hbar\omega \left(1 + \frac{\hbar\omega}{4Mc^2} \right) = \frac{M}{4} v^2 + |E| \tag{10-8.21}$$

where $|E| = 2.224$ MeV is the binding energy of the deuteron. It is clear that, for low-energy neutrons,

$$\hbar\omega \cong |E| \tag{10-8.22}$$

Remembering also that $\gamma = (M|E|)^{1/2}/\hbar$, we obtain from Eq. (10-8.20) the expression

$$\sigma_c = \frac{6.6 \times 10^4 \text{ b}}{v(\text{cm sec}^{-1})} \tag{10-8.23}$$

which agrees very well with experiments at low energy.

10-9   PHOTODISINTEGRATION OF THE DEUTERON

In the process

$$\hbar\omega + d \rightarrow n + p \tag{10-9.1}$$

at low energy, we have transitions from the bound $^3S_1$ state to the $^1S_0$ state of the continuum. These are inverse transitions from those considered in Sec. 10-8. They are especially important at very low energies (up to a few tenths of an MeV above threshold) because the p states of the continuum require too high a relative velocity to be attained near the threshold. As the energy of the gamma ray increases, the electric dipole transitions from $^3S_1$ to $^3P_{0, 1, 2}$ become predominant. The simple considerations to be developed here are valid only for energies up to about 10 MeV. At higher energies there are further effects difficult to evaluate.

The photodisintegration due to magnetic dipole radiation is obtained from Eq. (10-8.23) by applying the rule (see Chap. 11) relating inverse reactions

$$\frac{\sigma(1 \rightarrow 2)}{g_2 p_2^2} = \frac{\sigma(2 \rightarrow 1)}{g_1 p_1^2} \tag{10-9.2}$$

where g and p are the statistical weight and momentum of the final state for the cross section in the numerator of Eq. (10-9.2).

In our case, for capture, the final state has photons of momentum $\hbar\omega/c$; for photodisintegration the final state has nucleons of momentum $\hbar k$. Thus we have

$$\frac{c^2 \sigma_c}{g_{ph+d}(\hbar\omega)^2} = \frac{\sigma_{\text{dis}}}{g_{n+p}\hbar^2 k^2} \tag{10-9.3}$$

ISBN 0-8053-8601-1

Now we express σ_c through Eqs. (10-8.13) and (10-8.15), taking the exact value of I given in Eq. (10-8.17), and we obtain

$$\sigma_{dis} = \frac{2\pi}{3} \frac{e^2}{\hbar c} \left(\frac{\hbar}{Mc}\right)^2 (\mu_n - \mu_p)^2 \frac{k\gamma(1 - \gamma a_s)^2}{(k^2 + \gamma^2)(1 + k^2 a_s^2)} \tag{10-9.4}$$

and

$$\sigma_{dis} = \frac{2\pi}{3} \frac{e^2 \hbar}{M^2 c^3} (\mu_p - \mu_n)^2 \frac{(\omega_0)^{1/2}(\omega - \omega_0)^{1/2}\left[(\hbar\omega_0)^{1/2} + W_0^{1/2}\right]^2}{\omega[\hbar(\omega - \omega_0) + W_0]} \tag{10-9.5}$$

where W_0 is defined by

$$|a_s| = \frac{\hbar}{(MW_0)^{1/2}}$$

just as $1/\gamma = \hbar/(M|E|)^{1/2}$ and $\hbar\omega_0 = |E|$. The electric dipole part requires the consideration of $S \to P$ transitions. We omit the calculation, limiting ourselves to the final result,

$$\sigma(E1) = \frac{8\pi}{3} \frac{e^2}{\hbar c} \frac{k^3 \gamma}{(k^2 + \gamma^2)^3} \tag{10-9.6}$$

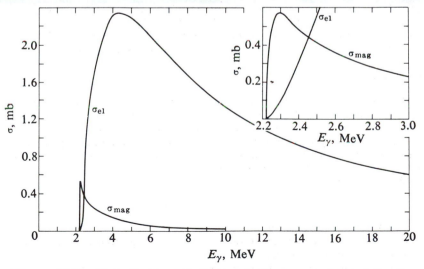

Figure 10-23 Photoelectric and photomagnetic cross sections of the deuteron. The inset shows the region near threshold energy much magnified.

ISBN 0-8053-8061-7

or

$$\sigma(E1) = \frac{8\pi}{3} \frac{e^2}{Mc\omega_0} \left[\frac{\omega_0(\omega - \omega_0)}{\omega^2} \right]^{3/2} \tag{10-9.7}$$

These cross sections are illustrated in Fig. 10-23.

BIBLIOGRAPHY

Bethe, H. A., "Theory of Nuclear Matter," *Ann. Rev. Nucl. Sci.*, **21**, 93(1971).

Breit, G., E. U. Condon, and R. D. Present, "Theory of Scattering of Protons by Protons," *Phys. Rev.* **50**, 825 (1936).

Day, B. D., "Elements of the Brückner–Goldstone Theory of Nuclear Matter," *Rev. Mod. Phys.* **39**, 719 (1967).

Faissner, H., "Polarisierte Nucleonen," *Ergeb. Exakt. Naturw.*, **32** 180 (1959).

Gomes, L. C., J. D. Walecka, and V. F. Weisskopf, "Properties of Nuclear Matter," *Ann. Phys.*, **3**, 241 (1958).

Heisenberg, W., "Über den Bau der Atomkerne. I," *Z. Physik*, **77**, 1 (1932).

Hulthen, L., "The Two-Nucleon Problem," in (Fl E), Vol. 39.

Majorana, E., "Über die Kerntheorie," *Z. Physik*, **82**, 137 (1932).

McMaster, W. H., "Matrix Representation of Polarization," *Rev. Mod. Phys.*, **33**, 8 (1961).

Moravcsik, M. J., *The Two-Nucleon Interaction*, Oxford University Press, New York, 1964.

Nigam, B. P., "The Two-Nucleon Interaction," *Rev. Mod. Phys.*, **35**, 117 (1963).

Preston, M. A. and R. K. Badhuri, *Structure of the Nucleus,* Addison Wesley, Reading Mass., 1975.

Schwinger, J., and E. Teller, "The Scattering of Neutrons by Ortho- and Parahydrogen," *Phys. Rev.*, **52**, 286 (1937).

Wigner, E., "On the Consequences of the Symmetry of the Nuclear Hamiltonian on the Spectroscopy of Nuclei," *Phys. Rev.*, **51**, 106 (1937).

Wigner, E., "Isotopic Spin-A Quantum Number for Nuclei," *Proc. Robert A. Welch Found. Conf. Chem. Res., 1st, Houston* 1957 (publ. 1958), p. 67.

Wilson, R., *The Nucleon–Nucleon Interaction*, Wiley-Interscience, New York, 1963.

Wolfenstein, L., "Polarization of Fast Nucleons," *Ann. Rev. Nucl. Sci.*, **6**, 43 (1956).

PROBLEMS

10-1   Show that if deuterons are scattered by protons, the maximum scattering angles in the center-of-mass and laboratory systems are 120 deg. and 30 deg., respectively, but that if protons are scattered by deuterons, the maximum angle in both systems is 180 deg.

10-2   Plot $u = r\psi$ for the neutron–proton system at the positive energy of 1 eV in the triplet state. Indicate in the same figure the eigenfunction of the ground state of the deuteron and the corresponding rectangular potential.

10-3   Verify that for a potential in the neutron–proton system

$$V_1(r) + V_2(r)\boldsymbol{\sigma}_1 \cdot \boldsymbol{\sigma}_2$$

$\mathbf{I}^2$, parity, $\mathbf{L}^2$, and $\mathbf{S}^2$ are constants of the motion.

10-4   Show that the expectation value of the electric quadrupole moment for a neutron–proton system in the $^3S_1$ state is zero.

10-5   If neutron and proton were bound in the deuteron in the $^1P_1$ or $^3P_1$ state, what would be the magnetic moment of the deuteron?

10-6   Calculate the effective range r_0 for a rectangular well of depth V_0 and width a.

ISBN 0-8053-8601-7

10-7   Starting from the equation $k_t \cot k_t r_0 = -\gamma$, and assuming $E \ll V_0$, show that

$$\cot k_t r_0 \approx -(|E|/V_t)^{1/2}$$

and

$$V_t r_0^2 \approx \frac{\pi^2 \hbar^2}{4M}$$

10-8   Assuming $n-p$ and $n-n$ forces to be identical, what would be the $n-n$ scattering cross section at low energy (0 to 1000 eV)?

10-9   Show that for $\lambda \gg r_0$ (range of forces)

$$\sigma = \frac{4\pi}{\gamma^2 + k^2} \qquad k = \frac{1}{\lambda}$$

10-10   Show that for $E = 10$ MeV this value agrees with experiment, but that for $E = 1$ eV there is no agreement. Explain why.

10-11   By what factor would the neutron capture cross section of hydrogen change if the singlet scattering length were positive (i.e., the singlet state were bound)?

10-12   Calculate the phase shift δ_0 for an impenetrable sphere of radius R. Compare its cross section to its geometrical area.

10-13   Neutrons of 100-eV energy, polarized with spin in the z direction, move along the x axis hitting protons having spin in a direction $\mathbf{n}$. Find the total collision cross section.

10-14   For a proton–proton system there exist only the states ^1S, ^3P, ^1D, etc. Prove this statement and give the i spin of each state. Do the same thing for a neutron–proton system.

10-15   List the states with $J = 4$ for the proton–proton and neutron–proton system and classify them according to spin and parity. Indicate states that can mix with each other.

10-16   What states may exist for an alpha–alpha and a deuteron–deuteron system?

10-17   Write the scattering amplitude for two deuterons in a state of spin 2 and in one of spin 1.

10-18   Show that the energy of two magnetic dipoles of moments μ_1, μ_2 is

$$V = \left[\frac{3(\mu_1 \cdot \mathbf{r})(\mu_2 \cdot \mathbf{r})}{r^2} - \mu_1 \cdot \mu_2 \right] \frac{1}{r^3}$$

where $\mathbf{r}$ is the distance between the two dipoles. Calculate V for the $n-p$ system and show by an order-of-magnitude estimate that it cannot account for the experimentally measured mixing of s and d states.

10-19   Show that if S_{12} is the tensor operator, then

$$S_{12} f(r) \alpha(1) \alpha(2) = f(r) \{ a Y_2^0 \alpha(1) \alpha(2)$$

$$+ b Y_2^1 [\alpha(1) \beta(2) + \beta(1)\alpha(2)] + c Y_2^2 \beta(1) \beta(2) \}$$

where $Y_l^m(\theta, \varphi)$ are spherical harmonics, and determine the coefficients a, b, c.

10-20   Show that

$$P_B = \tfrac{1}{2}(1 + \sigma_1 \cdot \sigma_2)$$

and that

$$P_H = -P_\tau = -\tfrac{1}{2}(1 + \tau_1 \cdot \tau_2)$$

where P_τ is an operator that exchanges only i spin between the nuclei [Eq. (10-6.16)].

ISBN 0-8053-8061-7

10-21   The potential (Serber, 1949)

$$V = -g^2 \frac{e^{-kr}}{r} \tfrac{1}{2}(1 + P_M)$$

has been used. Show that it gives no force in odd states.

10-22   Show that for a spin $\tfrac{1}{2}$ particle scattering on a spinless center the spin component normal to the scattering plane cannot change. (Consider simultaneous conservation of parity and angular momentum.)

10-23   Show that in a time-reversible scattering of a spin $\tfrac{1}{2}$ particle on a spinless center,

$$p(-L+) = p(+L-)$$

10-24   Show that $D = 1$ for a nucleon on a spinless target. Show that $-1 + 2|P| \leqslant D \leqslant 1$.

10-25   Show that $R/R' = -\cot(\delta - \dot{\theta})$ and $A/A' = \tan(\gamma - \theta)$; θ is the scattering angle. Angles δ and γ should be found from Fig. 10-18.

10-26   Draw diagrams illustrating complete experiments to measure C_{pq} and C_{qp}.

10-27   Show that $C_{np} = C_{nq} = 0$.

10-28   Consider scattering between protons. Suitably define C_{ik} coefficients and show that $C_{pq} = -C_{qp}$.

10-29   Calculate the mean life of slow neutrons in water, assuming that they are absorbed only by radiative capture by free protons.

10-30   Show the equivalence of Eqs. (10-9.4), (10-9.5), and (10-9.6), (10-9.7).

10-31   Show that in Eq. 10-4.19

$$a = \tfrac{1}{2}(m_n + m_p)A + \langle TT_3|H_0|TT_3\rangle + E_c^{(0)}(A, T) - T(T + 1)E_c^{(2)}$$

$$b = (m_n - m_p) - E_c^{(1)}$$

$$c = 3E_c^{(2)}$$

H_0 is the charge-independent part of the nuclear hamiltonian.

10-32   In n–p scattering the experimental situation of a neutron impinging on a proton from a given direction gives a wave function, as far as i spin is concerned, of $\nu(1)\pi(2)$. This is the superposition of a state of i spin 1 and 0. The scattering amplitude is $\tfrac{1}{2}(f_1(\theta) + f_0(\theta))$. From this, derive

$$\sigma(\theta) = \tfrac{1}{4}|f_1(\theta) + f_0(\theta)|^2 \qquad \text{and} \qquad \sigma(\pi - \theta) = \tfrac{1}{4}|f_1(\theta) - f_0(\theta)|^2$$

and the inequalities

$$\left[\sigma_{np}(\pi - \theta)\right]^{1/2} + \left[\sigma_{np}(\theta)\right]^{1/2} \geqslant \left[\sigma_{pp}(\theta)\right]^{1/2}$$

$$4\sigma_{np}(90°) \geqslant \sigma_{pp}(90°)$$

ISBN 0-8053-8061-7

11

Nuclear Reactions

The study of nuclear reactions is one of the largest areas of nuclear and subnuclear physics. The threshold for forming pions is a suitable conventional energy boundary between the two fields. Below this threshold we deal only with nuclear reactions in the strict sense of the word; above this threshold mesic and particle phenomena become increasingly important until in the multi-GeV region, nuclear features become of secondary importance and the interplay between pions, single nucleons, and other particles are the prominent features, while the nuclear composition of targets becomes of secondary importance.

11-1   INTRODUCTION

At any energy, the conservation of energy and momentum imposes certain restrictions, improperly called "kinematic" restrictions, on the reactions. This very important part of the study is an application of mechanics, and is considered in part in this section and in Appendix D.

There are general formalisms, such as phase-shift analysis, the scattering-matrix theory, and the Breit–Wigner theory of resonance, which are broad enough to accommodate nuclear and subnuclear phenomena. These general

Emilio Segrè, Nuclei and Particles: An Introduction to Nuclear and Subnuclear Physics, Second Edition

501

methods and their nuclear applications are described in this chapter, whereas the nonnuclear applications are reserved for Chaps. 15 and 16. Actually, the general methods find application even in fields apparently remote from nuclear physics (e.g., electrical engineering), thus demonstrating the deep formal interrelations between different areas of physics.

From the experimental point of view it is possible to generate beams of neutrons, protons, deuterons, helium and heavier ions, electrons, and photons ranging over a tremendous energy interval. In the case of neutrons nuclear reactions that range from less than 10^{-3} to 10^{11} eV are produced. Charged particles cannot effectively react unless they have an energy comparable to the Coulomb barrier $zZe^2/R \cong zZ/A^{1/3}$ (in MeV) of the target, which sets the lower limit of usable energy. The upper energy attainable with accelerators is of the order of 10^{11} eV, but cosmic rays in rare cases give particles having as much as 10^{20} eV. Beams and targets can be polarized.

Reactions give information on many questions. In nuclear physics proper, for instance, they provide data on the assignment of quantum numbers to specific levels, on nuclear models, and on reaction mechanisms. Almost all the information in subnuclear physics comes from reactions.

Nuclear reactions at low energy are mostly of the type

$$A + a \rightarrow B + b + Q \qquad (11\text{-}1.1)$$

where A is a target nucleus, a the impinging particle, and B and b the products; b is usually a light nucleus or a gamma ray. The reaction represented in Eq. (11-1.1) is often described in a very convenient notation devised by Bothe:

$$A(a, b)B$$

where the first letter is the target and the last letter the final nucleus; those in parentheses are, first, the projectile and, second, the lighter escaping particle or particles. In this notation elastic proton scattering is represented by $A(p, p)A$, neutron capture followed by gamma-ray emission by $A(n, \gamma)B$, and so on.

The Q of the reaction is the rest-mass difference multiplied by c^2 of the left side minus the right side of the equation. If Q is positive, the reaction is exothermic; if Q is negative, the reaction is endothermic. In this case $|Q|$ is the minimum center-of-mass energy required for the reaction.

In a bombardment represented by Eq. (11-1.1), with A at rest in the laboratory, the threshold energy of incident particle a is larger than $|Q|$, because not all the kinetic energy of a is available, since the momentum of the center of mass is conserved.

Nonrelativistically, in the center-of-mass system, the velocities of a and A are related by

$$m_a v_a + m_A v_A = 0 \qquad (11\text{-}1.2)$$

ISBN 0-8053-8601-1

and for the minimum kinetic energy at which an endothermic reaction ($Q < 0$) occurs we have

$$\tfrac{1}{2}\left(m_a v_a^2 + m_A v_A^2\right) = |Q| \tag{11-1.3}$$

In the laboratory system (primed quantities)

$$v_A' = 0 \tag{11-1.4}$$

and the minimum velocity v_a' at which the reaction occurs is obviously

$$v_a' = v_a - v_A \tag{11-1.5}$$

Equations (11-1.2) to (11-1.5) give for the kinetic energy of a in the laboratory system

$$\tfrac{1}{2} m_a v_a'^2 = \tfrac{1}{2} m_a v_a^2 \left(1 + \frac{m_a}{m_A}\right)^2 = |Q| \frac{m_A + m_a}{m_A} \tag{11-1.6}$$

The threshold is thus obtained for a velocity v_a' such that

$$\frac{1}{2}\frac{m_a m_A}{m_A + m_a} v_a'^2 = |Q| \tag{11-1.7}$$

The particle must have a velocity v_a' such that a particle having the reduced mass $\mu = m_a m_A/(m_a + m_A)$ and traveling with velocity v_a' has kinetic energy $|Q|$.

In low-energy nuclear physics the nonrelativistic approximation is usually adequate. In high-energy nuclear physics, on the other hand, relativistic formulas are needed in almost every case (see Appendix D).

The experimental study of nuclear reactions constitutes in itself a large part of experimental nuclear physics. To give a general idea of the methods, we shall mention some of the experimental approaches, without entering into technical details.

By bombarding a nucleus, say, ^{27}Al, with protons, we may produce a nuclear reaction, and the product may be radioactive. In our example the reaction (p, γ) would give the stable nucleus ^{28}Si; the reaction (p, n) would give the positron emitter ^{27}Si; the reaction (p, d) would give the positron emitter ^{26}Al. A chemical or mass spectrographic separation of the products or an analysis of the radiations emitted by them will show the type and yield of the reactions. Using a proton beam of known intensity and a known target thickness, and measuring the yields, one obtains a reaction cross section from the relation

$$\varphi \sigma \mathfrak{N} = N \tag{11-1.8}$$

where φ is the number of projectiles crossing the target, $\mathfrak{N}$ is the number of target nuclei per unit area, σ is the cross section for the reaction considered, and N is the number of nuclei produced. This relation is valid for a target absorbing only a negligible fraction of the beam.

In our example φ could be measured by collecting the protons in a Faraday cage and obtaining their number from the charge; $\mathfrak{N}$ is obtained from the thickness of the aluminum foil and N from the radioactivity accumulated in the target, corrected for decay. By stacking a pile of foils and considering the energy loss of the protons, we may obtain σ as a function of energy, or "excitation function." The aluminum foils may also be coated (e.g., by evaporation) with a different substance, and comparison between the activity induced in the aluminum and that induced in the substance gives the excitation function of the substance, once that of aluminum is known. Figure 11-1 shows such an arrangement, and a typical excitation function obtained by a stacked-foil experiment is given in Fig. 11-30.

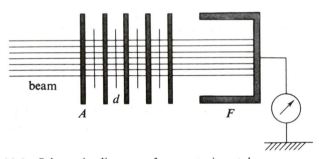

Figure 11-1 Schematic diagram of an experimental arrangement for an activation cross-section measurement. A, absorbers; d, detector foils; F, Faraday cage.

To observe a scattering cross section, we could detect the scattered protons with a counter, as in Fig. 11-2. The number scattered per unit solid angle as a function of θ gives the differential cross section $d\sigma/d\omega$. We may further measure the proton energy, in order to check whether the scattering is elastic or not. The energy may be obtained by measuring the range of the protons entering the counter, for instance, by putting over the counter a series of windows of increasing thickness. We thus obtain a differential cross section $d^2\sigma/d\omega\,dE$.

If we are interested in the polarization of the scattered protons, we must scatter them once more and look for a possible left–right asymmetry (Fig. 10-15). Similar measurements can be made for incident neutrons or other particles, and the flux is then measured by an integrating neutron counter such as the one described in Chap. 3 or by other suitable methods.

ISBN 0-805-3-8601-1

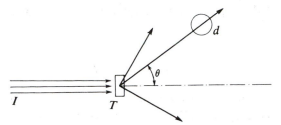

Figure 11-2 Measurement of the differential-scattering cross section by the detection of scattered neutrons. The principle is based on the assumption that the sample is thin, and hence does not attenuate the incident beam appreciably. I, incident beam; T, target; θ, scattering angle; d, detector.

We are often interested in the total cross section, determined by a measurement of the attenuation of the primary beam. If I is the intensity of the incident beam and $I - \Delta I$ the transmitted intensity, we have, for $\Delta I/I$ small,

$$\frac{\Delta I}{I} = \sigma \, \mathfrak{N} \tag{11-1.9}$$

Beams may contain particles of different velocities and the velocity dependence of the cross section is often of paramount interest. The primary beam must then be passed through a velocity selector, which may consist, for neutrons, of two shutters, separated by a known distance d, opening in succession at a time interval t. The neutrons threading the two shutters must then have the velocity d/t. For charged particles beams are mostly produced with a definite velocity.

11-2 GENERAL FEATURES OF CROSS SECTIONS

The following considerations apply in general only to cases not involving resonances, that is, where there is no single special nuclear level which plays a prominent role. Later we shall consider a treatment based on a generalization of the method of partial waves for collision processes (Appendix A). Here we shall be able to account for important characteristic features of some types of reactions by very simple considerations based on golden rule No. 2.

In the reaction described in Eq. (11-1.1) we start from a definite state, specified by the initial particles, momentum, angular momentum, and polarization. On the other hand, the final accessible states form a continuum, because the conservation theorems do not suffice to single out one accessible state only. A similar situation occurs, in atomic physics, in the emission of a photon by an excited atom: the photon may escape in any direction, and this alone gives a continuum of final states.

ISBN 0-8053-8601-7

Transitions of this type are best analyzed by using golden rule No. 2 (see, e.g., Ma 57):

$$w = \frac{2\pi}{\hbar} \langle |H_{if}|^2 \rangle \frac{dn}{dE} \qquad (11\text{-}2.1)$$

where w is the transition probability per unit time, dn/dE the density of the accessible final states, and $\langle |H_{if}|^2 \rangle$ an average value of the square of the matrix element connecting the initial and final states. We shall assume here that $|H_{if}|^2$ does not vary widely for the different accessible states. The density of the final states for a continuum is infinite, and at first sight Eq. (11-2.1) seems to give a divergent result. This is not the case, as can be verified by a limiting process. Enclose the system in a box of finite volume Ω. Then the levels of a free particle b, corresponding to the final state, do not form a continuum so long as Ω is finite; but we may still speak of a density of states, which is known to be

$$\frac{dn}{dE} = \frac{4\pi\Omega}{(2\pi\hbar)^3} \, p_b^2 \, \frac{dp_b}{dE} \qquad (11\text{-}2.2)$$

The matrix element H_{if} is

$$H_{if} = \int \psi_f^* \, U \psi_i \, d\tau \qquad (11\text{-}2.3)$$

where U is the interaction energy, which will be appreciable only in the region occupied by the nucleus. However, we shall need at present only some very simple qualitative features of H_{if}, which are obtainable by general considerations.

Although in general it is not possible to calculate H_{if} explicitly, since it depends on the nuclear structure, we may say something about its form. The wave functions ψ_f^*, ψ_i are normalized in the volume Ω. In the center-of-mass system and for the relative distance $r \gg$ range of nuclear forces (neglecting for a moment Coulomb forces) we can think of ψ_i as a product of a function $e^{ikr}/\Omega^{1/2}$ describing the relative motion of a and A and two functions, one representing the internal coordinates of a and the other the internal coordinates of A. A similar expression holds for ψ_f^*. The matrix element H_{if} is thus proportional to $1/\Omega$ and H_{if} of Eq. (11-2.1) goes to 0 as $1/\Omega$.

The contributions to H_{if} come only from the small region occupied by the nucleus. The remainder of the volume Ω does not contribute to the integral, because initial and final states are orthogonal. We can thus estimate H_{if} as

$$H_{if} \cong \frac{1}{\Omega} \int \varphi_f^* \, U \varphi_i \, d\tau \qquad (11\text{-}2.4)$$

ISBN 0-8053-8601-7

where the integral is extended over the nucleus and the φ are functions only of the internal coördinates. Approximately,

$$H_{if} \cong \frac{\langle U \rangle \times \text{(volume of the nucleus)}}{\Omega}$$

where $\langle U \rangle$ is an average over the nuclear volume.

For charged particles, ψ, at the nucleus, is reduced by the Coulomb repulsion. Thus, if particle a is positively charged, there will be in ψ_i a factor e^{-G_a}, where

$$G_a = \frac{1}{\hbar} \int_{R_1}^{R_2} [2m_a(V_a - E_a)]^{1/2} \, dr \cong \frac{\pi Z_A Z_a e^2}{\hbar v_a} \tag{11-2.5}$$

(see Chap. 7). This will give a factor e^{-2G_a} in $\langle |H_{if}|^2 \rangle$. Similarly, if particle b is charged, we shall have a reduction by a factor e^{-G_b} in ψ_f^*. We thus have, for charged particles,

$$H_{if} \cong \frac{\langle U \rangle \times \text{(vol. nucleus)}}{\Omega} \exp(-G_a - G_b) \tag{11-2.6}$$

To calculate a reaction cross section starting from w of Eq. (11-2.1) we note that

$$v_{aA} n_a \sigma_{A \to B} = w \tag{11-2.7}$$

where n_a is the number of particles a per unit volume and v_{aA} is the velocity of a relative to A. In fact, $v_{aA} n_a$ represents the number of particles of type a crossing a unit surface per unit time, or, in short, the flux of a. This quantity multiplied by the cross section $\sigma_{A \to B}$ for the reaction of Eq. (11-1.1) gives by definition the number of transitions per unit time w. With the normalization of one particle a in Ω, n_a is $1/\Omega$, and Eqs. (11-2.7), (11-2.1), and (11-2.2) give

$$\frac{1}{\Omega} v_{aA} \sigma_{A \to B} = \frac{2\pi}{\hbar} \langle |H_{if}|^2 \rangle \frac{4\pi}{(2\pi\hbar)^3} \Omega p_b^2 \frac{dp_b}{dE} \tag{11-2.8}$$

As it should, the volume Ω disappears from this formula because H_{if} is inversely proportional to Ω.

Next we replace dE by $v_b \, dp_b$, where for brevity v_{aA} and v_{bB} are designated v_a and v_b, respectively. Measuring p_b in the center-of-mass system and using the matrix element $\mathcal{H}_{if} = \Omega H_{if}$, which is independent of the normalization volume Ω, we obtain

$$\sigma_{A \to B} = \frac{1}{\pi \hbar^4} \langle |\mathcal{H}_{if}|^2 \rangle \frac{p_b^2}{v_a v_b} \tag{11-2.9}$$

ISBN 0-8053-8601-7

In practice one ordinarily wants the cross section averaged over all initial states of polarization and summed over all final states of polarization, such as would be found by starting with an unpolarized beam and observing outgoing particles with a detector insensitive to their polarization.

When the particles involved have spin, Eq. (11-2.9) must be slightly modified to take into account the statistical weights of the initial and final states. In our definition $\langle \mathcal{H}_{if} \rangle$ is an average over individual states. Taking into account the spin degeneracy, we have for the averaging, as far as spin is concerned,

$$\langle |\mathcal{H}_{if}|^2 \rangle = \frac{1}{g_i g_f} \sum_{if} |\mathcal{H}_{if}|^2 \tag{11-2.10}$$

where the sum is extended over all possible combinations of i and f generated by spin orientation. The statistical weights g_i, g_f are, respectively, $(2I_A + 1) \times (2I_a + 1) = g_i$ and $(2I_B + 1)(2I_b + 1) = g_f$.

Experimentally the initial system is a uniform assembly of different spin states, but it is so normalized as to have weight 1. In the final state, however, we do not discriminate different spin orientations and thus we must multiply $\langle |\mathcal{H}_{if}|^2 \rangle$ by g_f. We thus have

$$\sigma'_{A \to B} = \frac{1}{\pi \hbar^4} \langle |\mathcal{H}_{if}|^2 \rangle \frac{p_b^2}{v_a v_b} (2I_B + 1)(2I_b + 1) \tag{11-2.11}$$

where the prime specifies the particular type of cross section measured. Ordinarily we shall omit the prime, and unless otherwise specified σ indicates the cross section for an unpolarized beam on an unpolarized target.

Equation (11-2.11) permits several simple and interesting applications to specific cases, giving general features of the cross section versus energy for some types of reactions:

1.   *Elastic scattering* (n, n) (both particles uncharged). $v_a = v_b$, therefore, $p_b^2 / v_a v_b = (M_{\text{neutron}})^2$, a constant. At low energy $|\mathcal{H}_{if}|^2$ is approximately constant; therefore $\sigma \cong$ constant at low energy (Fig. 11-3a).

2.   *Exothermic reaction.* Low-energy *uncharged* bombarding particle as in (n, α), (n, p), (n, γ), (n, f) (neutron-induced fission) reactions. Q is usually positive and of the order of millions of electron volts, while neutron energy $\cong$ electron volts; therefore, $v_b \cong$ constant, and $p_b^2 / v_a v_b \propto 1/v_a$. Now $|\mathcal{H}_{if}|^2 \propto e^{-2(G_n + G_b)}$. However, G_n is 0 for a neutral particle and $e^{-G_b} \cong$ constant, since it depends on the almost constant energy of the outgoing particle. Therefore, $\sigma \sim 1/v_n$. The famous "$1/v$" law obtains (Fig. 11-3b).

3.   *Exothermic reaction; charged incoming particle* as in (p, n), (α, n), (α, γ), (p, γ) reactions. For incident energies $\ll Q$, the factor $p_b^2 / v_a v_b \propto 1/v_a$ and the barrier factor e^{-G_a} are operative and $\sigma \propto (1/v_a) e^{-2G_a}$ (Fig. 11-3c).

ISBN 0-8053-8601-7

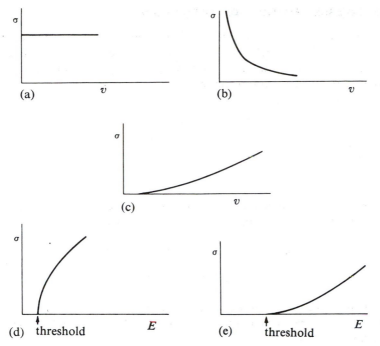

Figure 11-3 Behavior of cross sections at low energy as a function of the velocity v or energy E of the incident particle. (a) Neutron elastic scattering (n, n): $\sigma = $ constant. (b) Exothermic reactions, for example, (n, α), (n, γ): $\sigma \propto 1/v$. (c) Exothermic reaction with Coulomb barrier for projectile, for instance, (p, n), (α, n), (p, γ), (α, γ): $\sigma \propto (1/v)\exp[-2G\,(\text{incident})]$. (d) Behavior of cross sections near threshold. Endothermic inelastic scattering, for instance, (n, n'): $\sigma \propto (E - E_{\text{threshold}})^{1/2}$. (e) Behavior of cross sections near threshold. Endothermic reaction with Coulomb barrier for outgoing particle, for example, (n, α), (n, p): $\sigma \propto (E - E_{\text{threshold}})^{1/2} \exp[-2G\,(\text{outgoing})]$.

4. *Inelastic scattering*; (n, n') *reaction*. The nucleus is left in an excited state. The process is endothermic. Q is negative and $-Q$ is the excitation energy of the nucleus. For incident neutron energies slightly above the threshold, $v_n \cong$ constant, since the fractional change in incident energy is small. But $v_{n'}$ changes are great in this region: $v_{n'} \propto$ (excess of energy above threshold)$^{1/2}$. Therefore, $p_{n'}^2/v_n v_{n'} \propto$ (energy excess)$^{1/2}$. Hence near the threshold $\sigma \propto$ (energy excess)$^{1/2}$ (Fig. 11-3d).

5. *Endothermic reaction; charged outgoing particles* as in (n, α), (n, p) reactions. The reaction is exactly like that in case 4, except that the factor e^{-G_b} operates and is dominant. $\sigma \propto$ (energy excess)$^{1/2} \times e^{-2G_b}$ (Fig. 11-3e).

In all these reactions, no account has been taken of possible variations of $\mathcal{H}_{if}$ as shown, for instance, in resonance phenomena.

11-3   INVERSE REACTION—DETAILED BALANCE

Equation (11-2.11) gives a very important relation between the cross sections of reaction Eq. (11-1.1) when it proceeds to the right or to the left. Let us call the first cross section $\sigma_{A\to B}$ and the second $\sigma_{B\to A}$. Under very general assumptions H_{if} is hermitian, which means that

$$H_{if} = H_{fi}^* \tag{11-3.1}$$

and hence

$$\langle |H_{if}|^2 \rangle = \langle |H_{fi}|^2 \rangle \tag{11-3.2}$$

where the two matrix elements are obviously evaluated at the same energy in the center-of-mass system. We can then write, using Eq. (11-2.11),

$$\frac{\sigma_{A\to B}}{\sigma_{B\to A}} = \frac{p_b^2}{p_a^2} \frac{(2I_B + 1)}{(2I_A + 1)} \frac{(2I_b + 1)}{(2I_a + 1)} \tag{11-3.3}$$

The two reactions are considered in the center-of-mass system, and the velocities and momenta are measured in that system. Relation (11-3.3) is the principle of detailed balance.

The result of Eq. (11-3.3) may be obtained from statistical mechanics. Consider a box containing a mixture of A, a; B, b at equilibrium, the total energy of the system E being given. States of the system are described by giving the position and momenta of all particles and stating whether they are A, a or B, b. It is postulated that all states in the small energy interval ΔE near E have equal probability of being occupied. The justification of this postulate is one of the objectives of statistical mechanics and will not be considered here. Now divide the states into the Aa and Bb types. Our postulate asserts that the probability of an Aa-type state being occupied is the same as that of a Bb-type state being occupied; hence the ratio

$$\frac{\text{No. of } Aa \text{ states occupied}}{\text{No. of } Bb \text{ states occupied}} = \frac{\text{No. of possible } Aa \text{ states in } \Delta E}{\text{No. of possible } Bb \text{ states in } \Delta E} \tag{11-3.4}$$

Now the number of possible Aa states in the energy interval ΔE is equal to the maximum number $\mathcal{N}$ of Aa pairs (obtainable from the number of particles) multiplied by the number of possible states for one pair. This last quantity is the volume of phase space available to Aa pairs divided by $(2\pi\hbar)^3$, that is,

$$\frac{4\pi\Omega p_a^2}{(2\pi\hbar)^3} \frac{dp_a}{dE} \Delta E$$

ISBN 0-8053-8061-7

Thus the number of possible Aa states in ΔE is

$$\frac{\mathfrak{N} \, 4\pi\Omega}{(2\pi\hbar)^3} \frac{p_a^2}{v_a} \Delta E \qquad \text{because} \qquad \frac{dp_a}{dE} = \frac{1}{v_a} \qquad (11\text{-}3.5)$$

Similarly the number of possible Bb states in ΔE is

$$\mathfrak{N} \, \frac{4\pi\Omega}{(2\pi\hbar)^3} \frac{p_b^2}{v_b} \Delta E \qquad (11\text{-}3.6)$$

At equilibrium the number of $A \rightarrow B$ transitions per unit time must be equal to the number of $B \rightarrow A$ transitions per unit time. The first number is equal to the number of A states occupied times the transition probability for an A state, and similarly for a B state. Equations (11-2.7), (11-3.5), and (11-3.6) give

$$v_a \frac{p_a^2}{v_a} \sigma_{A \rightarrow B} = v_b \frac{p_b^2}{v_b} \sigma_{B \rightarrow A} \qquad \text{or} \qquad \frac{\sigma_{A \rightarrow B}}{\sigma_{B \rightarrow A}} = \frac{p_b^2}{p_a^2} \qquad (11\text{-}3.7)$$

We have omitted statistical weights. If we take them into account, we obtain Eq. (11-3.3) again. This equation is also valid for the differential cross section $d\sigma/d\omega$ (Fig. 11-4). To make it valid we need essentially only invariance under time reversal. The transition $A \rightarrow B$ must be the reverse of $B \rightarrow A$.

We have already applied the relation between direct and inverse reactions to the photodisintegration of the deuteron and to neutron capture by protons (Sec. 10-9).

Another important application is the determination of the spin of the pion by the reactions

$$p + p \rightleftarrows \pi^+ + D - 137 \text{ MeV} \qquad (11\text{-}3.8)$$

The statistical weight on the left side would be 4 if the protons were not identical, but Pauli's principle eliminates half the states and thus we put 2 as the statistical weight. On the right side the statistical weight is $3(2I_\pi + 1)$. Thus in the center-of-mass system we have

$$\frac{\sigma_{pp \rightarrow \pi D}}{\sigma_{\pi D \rightarrow pp}} = \frac{3(2I_\pi + 1)}{2} \frac{p_\pi^2}{p_p^2} \qquad (11\text{-}3.9)$$

The result from the measurements of the cross section is $I_{\pi^+} = 0$ (see Chap. 15).

Equation (11-3.7) applies if the beams involved are unpolarized. Otherwise more complicated relations must be used.

ISBN 0-8053-8061-7

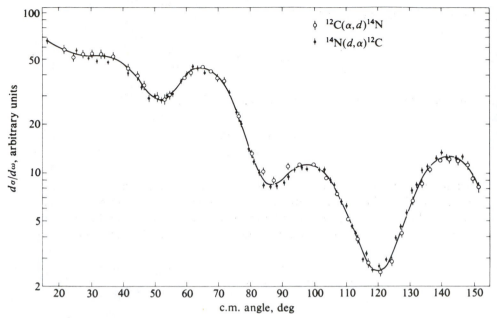

Figure 11-4 Angular distribution (center of mass) for the inverse reactions $^{12}C + \alpha \rightleftarrows {}^{14}N + d$ at matched energies: $E_\alpha = 41.7$ MeV, $E_d = 20.0$ MeV (lab). Probable errors on experimental points include statistical errors and estimated uncertainties in the subtraction of small background contributions to the measured counting rates. [From D. Bodansky, S. F. Eccles, G. W. Farwell, M. E. Rickey, and P. C. Robinson, *Phys. Rev. Letters*, **2**, 101 (1959).]

11-4   REACTION MECHANISMS. THE COMPOUND NUCLEUS (QUALITATIVE)

There are formal, rigorous theories based on quantum mechanics that are in principle adequate to describe nuclear reactions under very general assumptions (Feshbach, Kapur, Peierls, Wigner, and others). However, they are complicated and beyond the scope of this book.

Since the great variety of phenomena occurring in nuclear reactions precludes a unified treatment simple enough to be of practical use, it is necessary to use different models and approximations according to the nature and energy of the projectile and of the target.

Table 11-1 gives a survey of the most common reactions for intermediate ($30 \leqslant A < 90$) and heavy ($A \geqslant 90$) nuclei. Light nuclei have somewhat different reactions.

Following Weisskopf (Fig. 11-5) we shall first try to give an intuitive picture of a nuclear reaction, later describing in more detail special aspects of the phenomenon. We may consider the target nucleus as seen from an incoming particle as a region with a potential and an absorption coefficient. Remember, however, that this potential is due to single nucleons composing the

ISBN 0-8053-8601-7

TABLE 11-1   TABLE OF PARTICLES EMITTED IN TYPICAL NUCLEAR REACTIONS

Energy of incident particle	Intermediate nuclei (30 < A < 90)				Heavy nuclei (A > 90)			
	Incident particle							
	n	p	α	d	n	p	α	d
0–1 keV	n(el.) γ (res.)	No appreciable reaction	No appreciable reaction	No appreciable reaction	γ n(el.) (res.)	No appreciable reaction	No appreciable reaction	No appreciable reaction
1–500 keV	n(el.) γ (res.)	n γ α (res.)	n γ p (res.)	p n	n(el.) γ (res.)	Very small reaction cross section	Very small reaction cross section	Very small reaction cross section
0.5–10 MeV	n(el.) n(inel.) p α (res. for lower energies)	n p(inel.) α	n α(inel.) (res. for lower energies)	p n pn 2n	n(el.) n(inel.) p γ	n p(inel.) γ	n p γ	p n pn 2n
10–50 MeV	2n n(inel.) n(el.) p pn 2p α Three or more particles	2n n p(inel.) np 2p α(inel.) Three or more particles	2n n np 2p α(inel.) Three or more particles	p 2n pn 3n d(inel.) tritons Three or more particles	2n n(inel.) n(el.) p pn 2p α Three or more particles	2n n p(inel.) np 2p α Three or more particles	2n n p np 2p α(inel.) Three or more particles	p 2n np 3n d(inel.) tritons Three or more particles

ISBN 0-8053-8061-7

target nucleus. When the incoming particle hits the target it may be diffracted by the potential without losing any energy (elastic scattering). If the energy of the impinging particle is high and the absorption of the nucleus is such that it appears as a black sphere, we have the typical diffraction pattern exemplified in Fig. 11-6. If the energy of the incoming particle is low and can form standing waves in the target because it is absorbed only slightly, we shall find large scattering cross sections for the values of the nuclear radius (and correspondingly of A) that match the wavelength permitting the particle to form standing waves in the nucleus. The energies corresponding to these wavelengths are *resonance* energies.

If the incoming particle enters the nucleus, it may hit one nucleon and lift it to a higher energy state or even to an unbound state and still preserve enough energy to leave the nucleus. This process is called *direct interaction*. A typical example is given by collisions of high-energy protons in Li, where two protons emerge preferentially at 90 deg from each other and with energies roughly corresponding to those of a free proton-proton collision (Fig. 11-7).

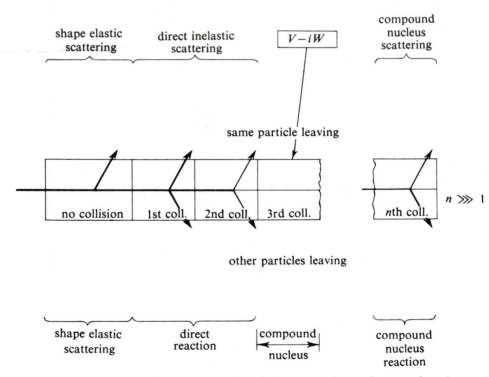

Figure 11-5 Graphic representation of the course of a nuclear reaction. As long as no collision takes place, only shape elastic scattering is possible. A first collision may produce direct reactions; later on, after many collisions, a compound nucleus is formed. [After V. F. Weisskopf, *Phys. Today*, **14**, 18 (1961).]

ISBN 0-8053-8061-1

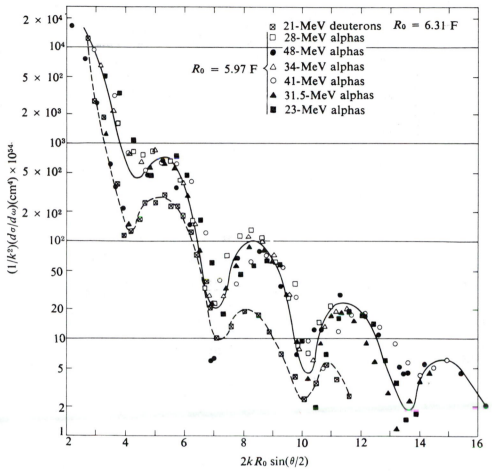

Figure 11-6   The observed elastic scattering cross section for alpha particles on magnesium divided by k^2, plotted as a function of $x = 2k\,R_0\sin(\theta/2)$. k is the reciprocal de Broglie wavelength and R_0 the nuclear radius. For a black disk $d\sigma/d\omega = (kR_0^2)^2[J_1(x)/x]^2$ [cf. Eq. (6-3.3)]. All data refer to center-of-mass system. [From J. S. Blair, G. W. Farwell, and D. K. McDaniel, *Nucl. Phys.*, **17**, 641 (1960).]

The incident particle may lose so much energy that it cannot escape the struck nucleus. If this energy is transferred to a nucleon leaving it bound, however, we have the case of formation of a compound nucleus. The energy is distributed among the nucleons and no nucleon can leave the nucleus until by further collisions the energy reconcentrates in one nucleon. The compound nucleus gives rise to a typical energy spectrum for the emitted particles having nearly a Maxwell distribution of velocities and a practically isotropic angular

ISBN 0-8053-8061-7

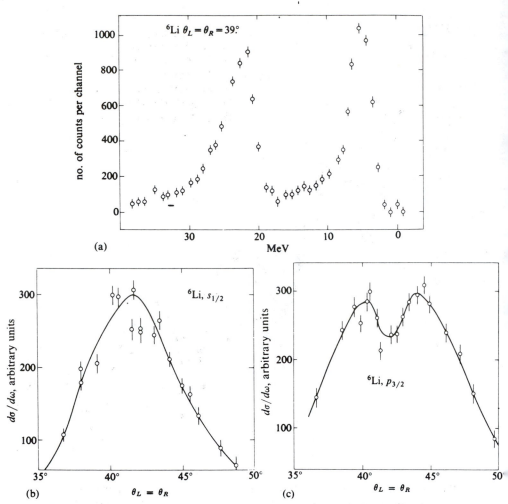

Figure   11-7   The energy carried away by each of two coincidence protons from (p, pp) reactions in ^{6}Li bombarded by 450-MeV protons has been measured. To the extent that the collision is quasielastic, the difference between the sum of these energies and the incident energy gives the binding energy of the struck proton in the nucleus. In part (a) the abscissa gives the difference between the sum of the energies of the escaping protons and the incoming proton. The two peaks correspond to the ejection of an $s_{1/2}$ and $p_{3/2}$ proton bound with 22.3 and 5.2 MeV, respectively. (b) and (c) show the angular correlation between the escaping protons. $\theta_L = \theta_R$ are deflections left and right. For protons at rest the curve would be a delta function at 45 deg. The broadening is due to the momentum of the struck proton. Note the difference between collisions with s and p protons and the minimum in the center of distribution for $p_{3/2}$ protons. Similar information is obtained from the $(e, e'p)$ reaction. [Courtesy H. Tyrén, S. Kullander, and G. Ramachandran.]

ISBN 0-805-8061-1

distribution. In contrast, the direct-interaction mechanism gives strong angular dependences, and characteristic maxima of the cross section as a function of energy (see Fig. 11-33). Between these two extreme cases there is also *preequilibrium emission*, in which a nucleon is emitted after some collisions, but before thermalization. The preequilibrium emission is recognizable from its energy and angular distribution; (p, α) and (n, α) reactions give also clear examples of preequilibrium reactions.

Diffraction scattering and direct interactions are expected to occur in periods of the order of 10^{-22} sec, the time a nucleon takes to travel the nuclear diameter. The compound nucleus, on the other hand, reaches thermal equilibrium in about five times this period. Excited levels have lives strongly dependent on the excitation energy. Their $\Gamma = \hbar/\tau$ goes from a small fraction of an eV for the levels reached by slow neutron capture to several MeV for the giant γ resonance (see Chap. 8).

Modern electronic techniques are still limited to times of 10^{-11} sec, but there is an ingenious method by which reaction times as low as 10^{-18} sec are observable. It uses the so-called *blocking effect*. Suppose the reacting atoms are exactly at the sites of a regular crystal, and one observes particles emitted in the exact direction of the crystalline plane. Then the crystalline plane shadows the particles emitted and one has a minimum of intensity in that specific direction. If on the other hand the emitting atoms move, before emission, to positions between the crystalline planes, the particles emitted in the direction of the plane are less inhibited. In a (p, p') reaction, the emitting nuclei have a velocity v due to the collision with the incident protons. The time of the reaction τ multiplied by v gives the distance the nucleus travels before emitting the proton. If this distance is short compared with the atomic radius, protons emitted in the direction of the crystalline plane will be blocked; if τ is such that $v\tau \approx d$, the distance between the crystal planes, the blocking effect gradually decreases. Since v is of the order of 10^{9} cm sec^{-1} and d of the order of 10^{-8} cm, reaction times $\tau \sim 10^{-17}$ sec are observable. Such experiments have made it possible to distinguish, in germanium, shape elastic (p, p) scattering from compound elastic scattering, which takes about 10^{-16} sec (see Fig. 11-8).

Some general results were obtained in Sec. 11-2 by a simplification of Eq. (11-2.1) on the hypothesis that $|\mathcal{H}_{if}|^2$ was approximately constant in the energy interval considered. However, this is not always the case. Thus, for the (n, γ) reaction produced by slow neutrons there are very important cross-section oscillations called resonances, as shown, for instance, in Fig. 11-9.

That the cross section is so large only in small energy intervals suggests the formation of a quantum state of the system in which the reaction occurs. This quantum state is not truly stationary. Actually, according to the uncertainty principle, the fact that the cross section is anomalously large in an

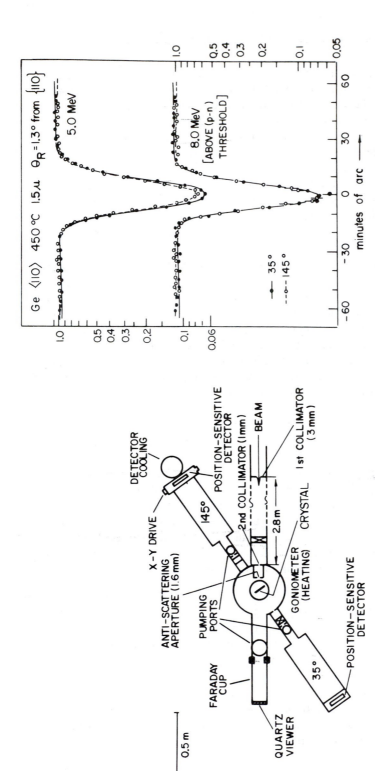

Figure 11-8 Left: experimental geometry used in the crystal blocking experiment to distinguish the shape (prompt) elastic from compound (delayed) elastic proton scattering. Right: experimental blocking dips in elastic-scattering yields from germanium observed at 35 deg (solid circles) and 145 deg (open circles); upper curves at 5.0-MeV proton bombarding energy clearly show delayed component at the larger angle (shallower dip). Lower curves obtained at 8.0-MeV proton energy (i.e., above neutron threshold) where the lifetime effect usually disappears, because the opening neutron channel widens the average compound-nuclear level beyond 1 keV (i.e., to a lifetime of less than 10^{-18} sec) where the blocking technique is no longer sensitive. Therefore, the two curves at large and small angles are seen to coincide, ruling out most sources of systematic error. [From G. M. Temmer et al., in J. de Boer and H. J. Mang (eds.), *Proceedings of the International Conference on Nuclear Physics, Munich* (Aug. 27–Sept. 1, 1973), Vol. I, North-Holland, Amsterdam, p. 512, 1973.]

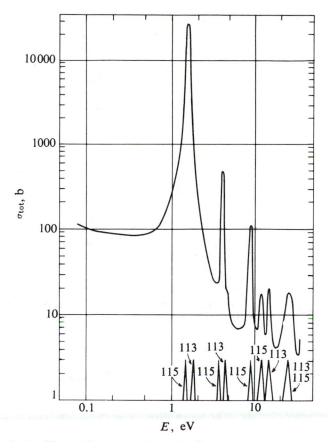

Figure   11-9   The total cross section of indium for neutrons, showing prominent resonances. The peaks are assigned to ^{113}In (4.23%) and ^{115}In (95.77%) by the triangles on the abscissa, which indicate also the resolution of the instrument. [V. L. Sailor and L. B. Borst, *Phys. Rev.*, **87**, 161 (1952).]

energy region of width ΔE indicates a mean life τ for the state, given by

$$\Delta E = \Gamma = \hbar/\tau = 6.58 \times 10^{-16}/\tau \qquad (11\text{-}4.1)$$

where ΔE is in eV and τ is in seconds. Such unstable states are called virtual states, and their energies are called resonance energies.

The discovery of these resonances gave rise to the physical model of the *compound nucleus* (Bohr, 1936) mentioned earlier. Bohr assumed that the nuclear reaction proceeds in two steps: (1) the formation from $A + a$ of a compound nucleus C, which survives a relatively long time and (2) later decomposes into $B + b$. It is assumed that the motions in the intermediate state C are very complicated and that the compound nucleus "forgets" how it was formed, except for its total energy, parity, and angular momentum.

With this simple picture we may split the collision into a first-phase formation of the compound nucleus for which we have a certain cross section σ_c, followed by a decomposition of the compound nucleus, with emission of the different particles n, p, γ, α, etc. Calling the probability of emission of each of them Γ_i, we may formally write the cross section for different processes as

$$\sigma_i = \sigma_c \frac{\Gamma_i}{\sum_i \Gamma_i} \tag{11-4.2}$$

The program implicit in this formula must be implemented by the evaluation of σ_c and of Γ_i, a difficult problem. However, in its great simplicity, the formula gives a useful insight.

As an immediate application of the general ideas involved we shall first give an example in which compound-nucleus formation and decay seem to be directly verified. Consider the bombardment $p + {}^{63}\text{Cu}$ or $\alpha + {}^{60}\text{Ni}$. The same compound nucleus ${}^{64}\text{Zn}$ is formed in both cases, and if the proton and the α have suitable energies, the compound nucleus also will have the same excitation energy. According to the compound-nucleus hypothesis, its further "destiny" must be independent of the mode of formation, and hence the ratio between the cross sections for emission of one neutron, one neutron and one proton, etc., must be constant (Fig. 11-10).

For the formation cross section σ_c in the region of resonances (if the virtual states are separated by energy intervals large compared with their width), Breit and Wigner (1936), on the basis of a theoretical calculation stressing the analogy with light absorption, suggested the following form:

$$\sigma_c = \frac{A/v}{(E - E_0)^2 + \Gamma^2/4} \tag{11-4.3}$$

where v is the velocity of the incident particle, E its energy, E_0 the resonance energy, Γ the level width, and A a constant depending on the specific reaction. This formula will be further discussed in Sec. 11-16.

In the case of light nuclei the levels of the compound nucleus are often widely separated, and their quantum numbers are known. The investigation of light nuclei is now well developed, and there are important summaries of this work (see the Bibliography).

We shall briefly review ${}^{8}\text{Be}$ as an example of an unstable light nucleus and its levels (Fig. 11-11). This nucleus may decay from its ground state, whose energy we shall assume to be 0, into two alpha particles according to

$$^{8}\text{Be} = {}^{4}\text{He} + {}^{4}\text{He} + 92 \text{ keV}$$

The mean life for such a decay is very short because, although the reaction is

ISBN 0-8053-8061-7

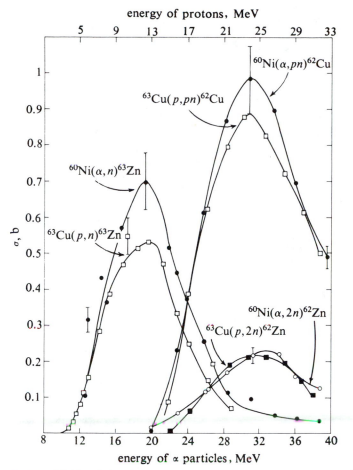

Figure 11-10 Yields of the decay products of the compound nucleus ^{64}Zn according to Ghoshal. The proton energy scale has been shifted 7 MeV to the right. [S. N. Ghoshal, *Phys. Rev.*, **80**, 939 (1950).]

barely exothermic, the Coulomb barrier is low. Calculation puts the mean life in the range of 10^{-15} to 10^{-16} sec and direct measurement gives 0.97×10^{-16} sec.

Information on the excited levels is obtained from reactions such as

$$^{7}Li + {}^{1}H \rightarrow {}^{8}Be \rightarrow 2\,{}^{4}He$$

$$^{7}Li + {}^{1}H \rightarrow {}^{7}Be + n$$

and from alpha–alpha scattering. For the last process the cross section should

ISBN 0-8053-8061-7

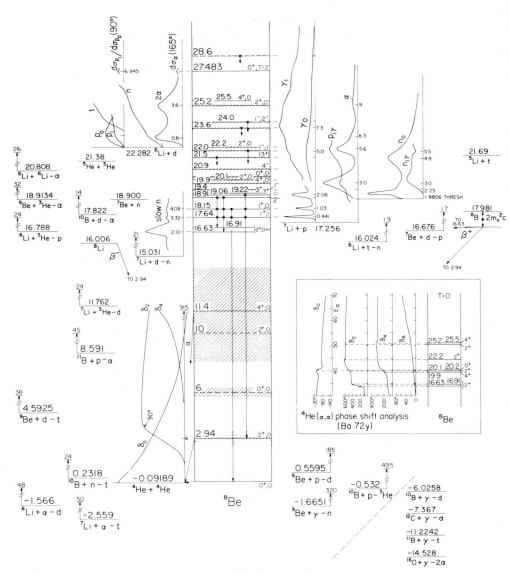

Figure 11-11 Energy levels of ^{8}Be with nuclear reactions leading to their formation. [Courtesy F. Ajzenberg-Selove.]

show a resonance. Precision experiments with a refined technique (Staub et al., 1967) have demonstrated this resonance at a laboratory energy of $2 \times$ 92.12 keV, and the measured width of the resonance gives the mean life quoted above. A second resonance is seen at 2.90 MeV (in the center-of-mass system). It is broad (1.2 MeV), corresponding to the short life of the excited state before it decomposes into two ^{4}He nuclei.

Another level corresponding to an alpha–alpha state in ^{8}Be is to be found at about 11.4 MeV (center of mass). It is several MeV wide. There are, however, levels in ^{8}Be that are stable with respect to disintegration into two ^{4}He. This is best understood by noting that, for a system of two identical particles, interchange of the particles is equivalent to reflection about the center of the line joining the particles. Alpha particles are bosons; hence interchange must leave the wave function unaltered. The corresponding states must thus be of even parity and have $I = 0, 2, 4, 6, 8$, etc. We conclude that only even-parity states of ^{8}Be, with I even, can decay into two ^{4}He. Odd-I states of ^{8}Be can decay into even- or odd-I states of ^{8}Be by gamma emission, according to the type of radiation: $E1$, $E2$, $M1$, etc.

The reaction ^{7}Li$(p, \gamma)^8$Be has a prominent narrow resonance at 441 keV proton energy, giving rise to a narrow line of gamma rays of 17.64 MeV and a broad one of ~ 14.7 MeV. The lines correspond to transitions from a state at 17.64 MeV, reached by proton capture in ^{7}Li. This state has $I = 1$, is even and decays to the ground and first excited state of 2.94 MeV by emission of gamma rays. It does not directly break up into two ^{4}He. ^{7}Li + H has other resonances, too; for instance, one at 1.03 MeV corresponding to 18.15 MeV excitation above the ground state of ^{8}Be associated with inelastic proton scattering. The reaction ^{7}Li$(p, \alpha)^4$He does not show sharp resonances but a broad maximum for a proton energy of 3 MeV, corresponding to a broad level at 19.9 MeV for ^{8}Be. The ^{7}Li$(p, n)^7$Be reaction shows a resonance at 2.25 MeV, corresponding to an odd-I state of ^{8}Be at 19.2 MeV. The ^{6}Li$(d, \alpha)^4$He reaction shows a peak corresponding to an even I level of ^{8}Be at 22.2 MeV (see Fig. 11-11).

Well-separated energy levels occur in heavy nuclei for slow neutron capture, in light nuclei for proton bombardment, and in many other cases. Often, however, there are many resonances close to each other (i.e., separated by distances comparable to or smaller than Γ). The cross section σ_c is then a smooth function of E. At high energy, where many levels are involved, σ_c may be described by the behavior of a wave in the presence of a potential well (see Sec. 11-7).

For an uncharged particle, we estimate σ_c by multiplying the geometric cross section by a penetration factor. The geometric cross section corresponds to the nuclear radius plus the de Broglie wavelength of the incoming particle because it is impossible to locate it with more accuracy; it is thus $\pi(R + \lambdabar)^2$. The penetration factor corresponding to a potential jump was calculated in

ISBN 0-8053-8061-7

Sec. 7-1 and it is

$$T = \frac{4kK}{(k + K)^2}$$

where k and K are the wave numbers of the particle outside and inside the nucleus respectively. We have thus

$$\sigma_c = \pi(R + \lambdabar)^2 \frac{4kK}{(k + K)^2} \tag{11-4.4}$$

Numerically, for nucleons impinging on heavy targets, $\pi\lambdabar^2 = 0.65 \times 10^{-18}E^{-1}(\text{eV})$ cm^2, $k = \lambdabar^{-1}$, is of the order of 10^{13} cm^{-1}.

In the case of a uniform potential well of depth V_0 we have

$$k^2 + K_0^2 = K^2 \tag{11-4.5}$$

where

$$\hbar^2 K_0^2/2M = V_0$$

Equation (11-4.4) gives, for $K_0 \gg k$ (slow particles), $\lambdabar \gg R$,

$$\sigma_c = \frac{4\pi\lambdabar}{K_0} \propto \frac{1}{v} \tag{11-4.6}$$

Equation (11-4.4) for high energy is not applicable because it refers only to the s wave and not to the total cross section for the formation of a compound nucleus (Fig. 11-12).

For charged particles Eq. (11-4.4) must be completed in order to take into account the Coulomb repulsion. This is obtained by multiplying the right side by a transmission factor $T \cong e^{-2G}$, where G is the Gamow barrier factor. Detailed numerical calculations for this are given by Blatt and Weisskopf in (BW 52) (see also Fig. 11-13).

For charged particles, the projectile, in order to touch the nucleus, must have (1) enough energy to overcome the potential barrier, and (2) a suitable impact parameter. We obtain by an elementary consideration, taking into account the conservation of angular momentum, that to reach the nuclear surface the impact parameter b must be smaller than $R\{1 - [V(R)/E]\}^{1/2}$, and from this follows

$$\sigma_c = \pi R^2\left(1 - \frac{V}{E}\right) \qquad \text{for} \quad E > V$$

$$= 0 \qquad \text{for} \quad E < V \tag{11-4.7}$$

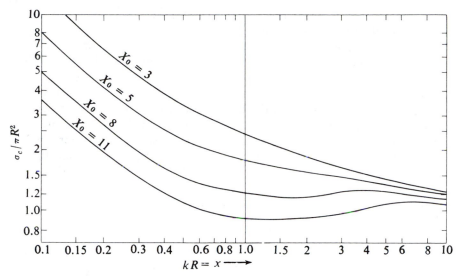

Figure 11-12 Cross section for the formation of the compound nucleus by neutrons. The abscissa is $x = kR = 0.218E^{1/2}$ if E is given in MeV and R in fermis. $X_0 = K_0R$ and is roughly equal to the nuclear radius in fermis. [(BW 52).]

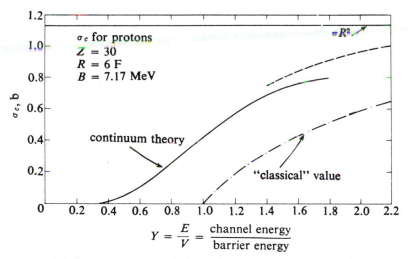

Figure 11-13 An example of the cross section for formation of the compound nucleus by protons. The solid curve represents the best theoretical value, the broken curve represents the approximation $\sigma_c \cong \pi(R + \lambdabar)^2\{1 - [V(R + \lambdabar)/E]\}$, and the dotted-and-dashed curve represents the classical value [Eq. (11.4.7)]. The asymptotic value, πR^2, is also shown. [(BW 52).]

ISBN 0-8053-8601-7

where $V = Zze^2/R$ is the Coulomb barrier and E is the kinetic energy of the projectile (in the center-of-mass system).

For further evaluation of Eq. (11-4.2) we note that in some cases the qualitative behavior of Γ_i may be easily understood. For instance, for a heavy nucleus and an incoming slow neutron of a few electron volts energy, it is clear that Γ_γ will not depend on the energy of the bombarding neutron. In fact, on capture there are several MeV of excitation, owing to the neutron binding energy, and the effect of the kinetic energy of the incoming neutron is negligible. On the other hand, for neutron reemission Γ_n is proportional to v, as can be seen by an application of Eq. (11-2.1), keeping in mind that in the circumstances $dn/dE \propto v_n$. Moreover in this physical situation $\Gamma_\gamma \gg \Gamma_n$.

The combination of Eq. (11-4.2) and Eq. (11-4.3) for σ_c and the evaluations given above provide a qualitative justification of the Breit–Wigner formulas,

$$\sigma(n, \gamma) = \frac{A/v}{(E - E_0)^2 + \Gamma^2/4} \tag{11-4.8}$$

$$\sigma(n, n') = \frac{A'}{(E - E_0)^2 + \Gamma^2/4} \tag{11-4.9}$$

which will be discussed more thoroughly later on.

However, the compound-nucleus picture is only part of the story, as is indicated by several experimental facts. For instance, in proton bombardment, (p, p) reactions are often favored over (p, n) reactions, even when the opposite should be true, because of the barrier effect on the outgoing particle. In inelastic scattering, small energy losses are favored. Angular distributions favor the forward direction in reactions such as (p, α), etc. These facts point to the importance of reaction mechanisms different from compound-nucleus formation.

11-5   FORMAL DEVELOPMENTS—SCATTERING MATRIX

To give a slightly more formal description of a large number of complex phenomena in a unified way, let us return to the theory of scattering of a spinless particle by a center of force (Appendix A) and generalize it so as to encompass the case where the impinging particle reacts with the scattering center.

Now we introduce the "channel" concept (Wigner), which is useful in considering nuclear reactions. Suppose that we have a particle a impinging on nucleus A, the whole system being completely specified from the quantum-mechanical point of view (i.e., in a definite quantum state). After this collision we may find the same particle and nucleus in a different quantum state, or even a different particle and a residual nucleus. The initial state is called the

ISBN 0-8053-8601-7

entrance channel, and any pair of possible residual nucleus and emerging particle, each in a definite quantum state, is called a *reaction channel*. A special channel is that of elastic collision.

To generalize collision theory (App. A) to the nonelastic case, consider again an incident plane wave, expressed asymptotically as

$$e^{ikz} \rightarrow \frac{1}{2kr} \sum_{l=0}^{\infty} (2l + 1)i^{l+1} \left\{ \exp\left[-i\left(kr - \frac{\pi l}{2} \right) \right] \right.$$

$$\left. - \exp\left[i\left(kr - \frac{\pi l}{2} \right) \right] \right\} P_l (\cos \theta) \quad (11\text{-}5.1)$$

where k represents $1/\lambdabar = Mv/\hbar = p/\hbar$, M being the reduced mass of the system and v the relative velocity of projectile and target, and where $z = r \cos \theta$.

The nuclear reaction changes expression [Eq. (11-5.1)] by acting on the outgoing wave only (i.e., on the coefficient of e^{ikr}/r). The asymptotic expression for the wave function is thus

$$\psi(r) = \frac{1}{2kr} \sum_{l} (2l + 1)i^{l+1} \left\{ \exp\left[-i\left(kr - \frac{\pi l}{2} \right) \right] \right.$$

$$\left. - \eta_l \exp\left[i\left(kr - \frac{\pi l}{2} \right) \right] \right\} P_l (\cos \theta) \quad (11\text{-}5.2)$$

The scattered wave is obtained by observing that

$$\psi(r) = e^{ikz} + \psi_{\text{scattered}} \quad (11\text{-}5.3)$$

and thus by subtracting Eq. (11-5.1) from Eq. (11-5.2) we have

$$\psi_s = \frac{1}{2kr} \sum_{l} (2l + 1)i^{(l+1)}(1 - \eta_l)e^{i(kr - \pi l/2)} P_l (\cos \theta) \quad (11\text{-}5.4)$$

which can also be written

$$\psi_s = \frac{e^{ikr}}{r} \sum_{l} \left(\frac{2l + 1}{2k} \right) i(1 - \eta_l) P_l (\cos \theta) \quad (11\text{-}5.5)$$

In the case of purely elastic scattering $\eta_l = e^{2i\delta_l}$, where δ_l is the usual real phase shift and hence $|\eta_l| = 1$. In the case of nonelastic scattering, $|\eta_l| < 1$, because some of the particles from the entrance channel are diverted to different channels. Then Eq. (11-5.5) is not a solution of a Schrödinger equation

ISBN 0-8053-8601-7

for a stationary state in the ordinary sense, implying the conservation of particles. If $|\eta_l| < 1$, we can easily find from Eq. (11-5.5) the elastic scattering cross section, but we shall also find that the flux removed by the target from the impinging wave is greater than the flux corresponding to the particles scattered elastically. The difference is due to particles that have disappeared following inelastic collisions, nuclear reactions, and, in general, diversion from the elastic scattering channel.

The scattering cross section is obtained by dividing the flux corresponding to ψ_s by the flux incident on the unit surface, which for the plane wave e^{ikz} is v.

The ingoing flux corresponding to ψ_s through a sphere of radius r_0 is, according to the rules of quantum mechanics,

$$\frac{\hbar r_0^2}{2iM} \int \left(\frac{\partial \psi_s}{\partial r} \psi_s^* - \frac{\partial \psi_s^*}{\partial r} \psi_s \right) d\omega \qquad (11\text{-}5.6)$$

By taking into account the relation

$$\int P_l P_{l'} \, d\omega = \frac{4\pi}{2l + 1} \delta_{ll'} \qquad (11\text{-}5.7)$$

Eqs. (11-5.5) to (11-5.7) yield, for the partial elastic scattering cross section of angular momentum l,

$$\sigma_s^l = \pi \lambdabar^2 (2l + 1) |1 - \eta_l|^2 \qquad (11\text{-}5.8)$$

The total flux entering a large sphere of radius r_0 may be computed from Eq. (11-5.6) now, by using $\psi(r)$, not ψ_s. We obtain a formula similar to Eq. (11-5.6), and on evaluation we find, for the wave of angular momentum l, a net ingoing flux of

$$\pi \lambdabar^2 v (2l + 1) \left(1 - |\eta_l|^2 \right) \qquad (11\text{-}5.9)$$

particles per second. These are lost to the incident beam and do not reappear in the scattered beam. They have "reacted" with the target, and the corresponding reaction cross section is

$$\sigma_r^l = \pi \lambdabar^2 (2l + 1) \left(1 - |\eta_l|^2 \right) \qquad (11\text{-}5.10)$$

Note that if $|\eta_l| = 1$, this cross section vanishes, because the scattering is purely elastic. Obviously $|\eta_l|^2 \leqslant 1$; otherwise there would be a negative reaction cross section or creation of particles.

ISBN 0-8053-8601-1

The total cross section is the sum of σ_r^l and σ_s^l and is given by

$$\sigma_t^l = 2\pi \lambdabar^2 (2l + 1)\left[1 - \mathrm{Re}\,\eta_l\right] \qquad (11\text{-}5.11)$$

All properties of a collision are specified by the η_l.

It is sometimes convenient to replace them by the "scattering amplitudes"

$$A_l = \frac{\eta_l - 1}{2ik} \qquad (11\text{-}5.12)$$

which for the particular elastic scattering case in which $|\eta_l| = 1$ gives

$$A_l = k^{-1}e^{i\delta_l} \sin \delta_l \qquad (11\text{-}5.13)$$

In a complex plane A_l is inside a circle of radius $\frac{1}{2}$ (see Fig. 11-14). These expressions give rise to interesting and important limiting cases (Figs. 11-14 and 11-15). For instance, to maximize the total cross section we must have $\eta_l = -1$; the cross section is then entirely due to elastic scattering and becomes

$$\sigma_s^l \mathrm{max} = 4\pi \lambdabar^2 (2l + 1) \qquad (11\text{-}5.14)$$

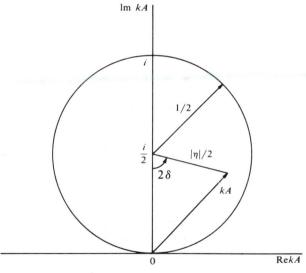

Figure 11-14 Plot of $f = kA = (\eta - 1)/2i$, (the complex scattering amplitude multiplied by k) in a complex plane. For $|\eta| = 1$ the vector representing f goes from the origin to a point on the circle of radius $\frac{1}{2}$ (unitary circle) and the scattering is purely elastic. For $|\eta| < 1$ the amplitude goes to a point inside the unitary circle. Resonance occurs when $\delta = \pi/2$.

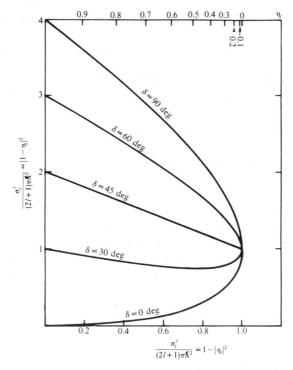

Figure 11-15 Relations between scattering cross section, reaction cross section, $|\eta|$, and δ.

For s waves

$$\sigma_s^0 \, \text{max} = 4\pi\lambdabar^2 \tag{11-5.15}$$

Equation (11-5.10) admits of a semiclassical interpretation. For angular momentum between l and $l+1$, particles in the incident beam must have an impact parameter between $b_l = l\hbar/p$ and $b_{l+1} = (l+1)\hbar/p$. The area between the two circles of radii b_l and b_{l+1} is

$$\pi \, \frac{\hbar^2(2l+1)}{p^2} = \pi\lambdabar^2(2l+1) \tag{11-5.16}$$

Complete removal of all the impinging particles from this ring gives the maximum reaction cross section for the partial wave and is obtained for $\eta_l = 0$. In fact, according to Eq. (11-5.10), $\eta_l = 0$ means that the nucleus absorbs the partial wave completely; in other words, it acts as a black disk. However, even in this case there is scattering. Equations (11-5.8) and (11-5.10) show that the elastic cross section is then equal to the reaction cross section.

If a nucleus acts as a black disk of radius R (i.e., for all particles of impact parameter $b \leqslant R$), the partial waves for which $R \geqslant \hbar l/p = \lambdabar l_{max}$ will give $\eta_l = 0$ and Eq. (11.5.11) gives

$$\sigma_l = 2\pi\lambdabar^2 \sum_{l=0}^{l_{max}=R/\lambdabar} (2l+1) = 2\pi\lambdabar^2(l_{max}+1)^2 \cong 2\pi\lambdabar^2\frac{R^2}{\lambdabar^2} = 2\pi R^2 \quad (11\text{-}5.17)$$

which is twice the geometric cross section. At first one would expect, thinking macroscopically, that a reaction cross section equal to πR^2 should not be

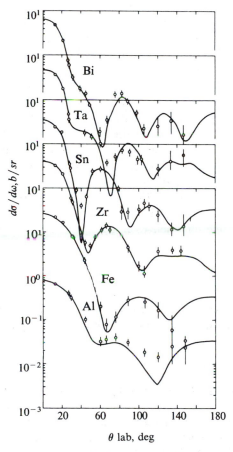

$d\sigma/d\omega, b/sr$

θ lab, deg

Figure 11-16 Experimental and theoretical differential elastic scattering cross sections for 7-MeV neutrons. The qualitative behavior is that of a diffraction curve from a sphere of radius $R = 1.4\ A^{1/3}$ F, but to obtain the detailed fit the sphere has been assumed not to be completely opaque and to have a fuzzy edge. [From S. Fernbach, *Rev. Mod. Phys.*, **30**, 414 (1958).]

accompanied by elastic scattering. However, from a wave point of view, in order to suppress completely the incoming wave behind the obstacle, we must place on it a source of the same amplitude, but opposite phase, uniformly spread over the obstacle. This source gives a beam, in a cone of angular aperture $\sim \lambdabar/R$, of intensity approximately equal to that intercepted by the obstacle. Hence, the true absorption giving rise to a cross section πR^2 is accompanied by a "diffraction scattering" corresponding to about the same cross section. The detectability of diffraction scattering clearly depends on the relative values of R and $\lambdabar$. It is observable when R is a few times $\lambdabar$, as in neutrons of a few MeV (Fig. 11-16). On the other hand, for macroscopic objects in the optical case, $\lambdabar/R$ is so small that the diffraction scattering is hardly observable.

At very high energies the wavelength of the projectile becomes small compared with R, and collisions occur with single nucleons in the nucleus. Between these two limiting cases, nuclei show a relative transparency for nucleons; they act as gray obstacles.

Scattering Matrix

● There are some formal features in the scattering problem that yield general conclusions important for their wide range of applicability, including atomic, nuclear, and subnuclear phenomena. We shall discuss here the barest principles of this approach; for an advanced treatment see (GW 64).

For the sake of simplicity we shall treat only s waves and spinless particles. Consider a system of sufficiently high energy so that it can decompose according to several channels α, β, γ, etc.; for simplicity also, we shall assume two particles in each channel. In a given channel, for instance, α, the wave function will be

$$\psi_\alpha = \psi(r_\alpha)\chi_\alpha \tag{11-5.18}$$

The first factor depends on the relative coordinate of the particles in channel α, and χ_α is a function describing the internal coordinates of the particles.

For distances r_α large compared with the range of nuclear forces, $\psi(r_\alpha)$ has the form

$$\psi(r_\alpha) = \left[A_\alpha \exp(-ik_\alpha r_\alpha) + A'_\alpha \exp(+ik_\alpha r_\alpha) \right] r_\alpha^{-1} (4\pi v_\alpha)^{-1/2} \tag{11-5.19}$$

where the first term represents an incoming wave and the second an outgoing wave. The wave numbers k_α and the relative velocity v_α depend on the total energy of the system and on the state α. For a given total energy they will thus be different in different channels. The last factor of Eq. (11-5.19) has the purpose of normalizing the wave to flux one when A_α, A'_α are one.

Consider a wave function containing an incoming wave only in channel α, not in the others (Fig. 11-17). It will have outgoing waves, not only in channel

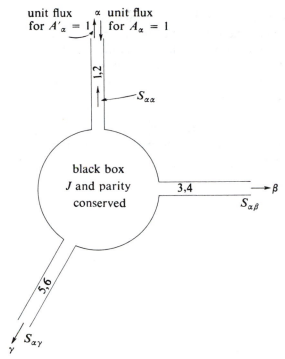

Figure 11-17 Symbolic figure and notations relative to scattering matrix. In channel α, particles 1 and 2; r_α their distance; χ_α their internal coordinates. In channel β, particles 3, 4; in channel γ, particles 5, 6.

α, but also in other channels. The outgoing wave in channel α corresponds to elastic scattering; the waves in other channels correspond to inelastic scattering and include reactions. The asymptotic behavior of the waves corresponding to reactions is obtainable by looking at ψ_α in the region of configuration space near channel β, γ, etc., and of course for relative distances r_β, r_γ, etc., large compared with nuclear dimensions; there

$$\psi_\alpha = \psi_{\alpha\beta}(r_\beta)\chi_\beta \qquad (11\text{-}5.20)$$

with

$$\psi_{\alpha\beta}(r_\beta) = -S_{\alpha\beta}\exp(ik_\beta r_\beta)(4\pi v_\beta)^{-1/2}r_\beta^{-1} \qquad (11\text{-}5.21)$$

where $S_{\alpha\beta}$ is a complex number, function of the energy.

At a given energy one may form as many ψ_α as there are entrance channels: call this number N. There are thus N ψ_α; we assume that they form a set of orthogonal, degenerate functions. The quantities $S_{\alpha\beta}$ are N^2 in number, and they form a matrix, called the S matrix. The S matrix completely determines the asymptotic behavior of the wave function. In this sense it represents the total information on the collision: the processes occurring at distances small compared with nuclear dimensions are unobservable, but the results of experiments are always observed at distances from the collision site large compared with nuclear dimensions, and the S matrix predicts the observations.

Exact calculation of the scattering matrix is in practice impossible. However, it has some very important properties, which are derivable from general principles. Although we cannot go into their derivation, the most important properties are as follows:

Unitarity is expressed by the relation

$$\sum_{\gamma} S_{\alpha\gamma}^* S_{\beta\gamma} = \delta_{\alpha\beta} \tag{11-5.22}$$

or $SS^{\dagger} = 1$ where $S^{\dagger}$ is the hermitian conjugate of S ($S_{ji}^* = S_{ij}^{\dagger}$) (note the order of the subscripts). In particular the relation

$$\sum_{\gamma} |S_{\alpha\gamma}|^2 = 1 \tag{11-5.23}$$

means that the sum of probabilities of ending in some channel is 1.

Time reversal. Consider in a process the incoming wave $\exp(-ik_{\alpha}r_{\alpha})$ $/(4\pi v_{\alpha})^{1/2}$ and the outgoing wave $S_{\alpha\beta} \exp(+ik_{\beta}r_{\beta})/(4\pi v_{\beta})^{1/2}$. In the time-reversed situation consider the normalized incoming wave $\exp(-ik_{\beta}r_{\beta})$ $/(4\pi v_{\beta})^{1/2}$ and the outgoing wave $S_{\beta\alpha} \exp(ik_{\alpha}r_{\alpha})/(4\pi v_{\alpha})^{1/2}$. If time reversal is permissible and there is no spin, the relation

$$S_{\alpha\beta} = S_{\beta\alpha} \tag{11-5.24}$$

obtains. If there are spins, one must be more careful in considering a situation and its time reversal. Without spin all we have to do to invert the direction of time is to invert all momenta. In the presence of spins, if we think of them as due to rotations, we must also invert the sense of rotations (i.e., the direction of the spin). If we define the channel $-\alpha - \beta$ by inverting all momenta as well as all spin directions, we have

$$S_{\alpha\beta} = S_{-\alpha-\beta} \tag{11-5.25}$$

The time-reversal relations are similar, but not identical, to the detailed balance conditions.

The conditions of unitarity and time-reversal invariance impose restrictions on the $2N^2$ real numbers entering into the scattering matrix. Other limitations come from information on the analytical properties of the $S_{\alpha\beta}$ considered as a function of energy.

ISBN 0-8053-8601-7

The diagonal elements of the scattering matrix are the $\eta = S_{\alpha\alpha}$ of this section. By arguments similar to those already used for η, one can show that the cross sections are given by

$$\sigma_{\alpha\beta} = \pi \lambda_\alpha^2 |\delta_{\alpha\beta} - S_{\alpha\beta}|^2 \qquad (11\text{-}5.26)$$

For $\alpha = \beta$ we have the elastic cross section. The reaction cross section is obtained by summing over all $\beta \neq \alpha$,

$$\sigma_s = \pi \lambda_\alpha^2 |1 - S_{\alpha\alpha}|^2 \qquad (11\text{-}5.27)$$

$$\sigma_r = \pi \lambda_\alpha^2 \sum_{\beta \neq \alpha} |S_{\alpha\beta}|^2 \qquad (11\text{-}5.28)$$

or, using the unitarity relation,

$$\sigma_r = \pi \lambda_\alpha^2 \left(1 - |S_{\alpha\alpha}|^2\right) \qquad (11\text{-}5.29)$$

identical to Eq. (11-5.10).

The "optical theorem" (Appendix A) can also be derived by use of the scattering matrix. The theorem relates the imaginary part of the forward-scattering amplitude $f(0)$ to the total cross section

$$\sigma_{\text{tot}} = \frac{4\pi}{k} \, \text{Im} \, f(0) \qquad (11\text{-}5.30)$$

where k is the wave number of the incident particle. ●

11-6 RESONANCES

To maintain a reasonable simplicity of treatment, we shall consider only s waves. This restriction does not affect the very extensive and important case of slow neutrons, where only the s waves are important. We shall also neglect spin effects. For a more general treatment we must refer to the book by Blatt and Weisskopf or to (GW 64). (From this point on, the subscript for the s partial wave will be omitted.) It is obvious that Eq. (11-5.8) for $l = 0$ and $\eta = e^{2i\delta}$ becomes

$$\sigma_s = \pi \lambda^2 |1 - \eta|^2 = 4\pi \lambda^2 \sin^2 \delta \qquad (11\text{-}6.1)$$

which is the equation used in treating low-energy neutron–proton scattering. Similarly,

$$\sigma_r = \pi \lambda^2 (1 - |\eta|^2) \qquad (11\text{-}6.2)$$

ISBN 0-8053-8601-7

and

$$\sigma_t = 2\pi\lambdabar^2(1 - \text{Re } \eta) \tag{11-6.3}$$

We see from these formulas that a knowledge of η is all that is required to find the three cross sections. Furthermore, a detailed knowledge of what happens inside the nucleus is not necessary in order to determine η. Assuming that the nucleus has a well-defined surface, η is determined by the value of

$$\left(\frac{u'}{u}\right)_{r=R} R = f \tag{11-6.4}$$

at the surface of the nucleus. Here $u = r\psi$, and R is the nuclear radius.

The wave function outside the nucleus, limited to s waves only, is

$$\psi_{\text{outside}} = \frac{i}{2kr}\left(e^{-ikr} - \eta e^{ikr}\right) \tag{11-6.5}$$

and the logarithmic derivative of u must be continuous everywhere, including R.

Everything happening inside the nucleus is transmitted to the outside by this logarithmic derivative u'/u. From Eqs. (11-6.4) and (11-6.5) we have

$$\left(\frac{u'}{u}\right)_{r=R} = \frac{f}{R} = \frac{-ike^{-ikR} - \eta ike^{ikR}}{e^{-ikR} - \eta e^{ikR}} = \frac{-ik(1 + \eta e^{2ikR})}{1 - \eta e^{2ikR}} \tag{11-6.6a}$$

or

$$\eta = \frac{f + ikR}{f - ikR}e^{-2ikR} \tag{11-6.6b}$$

We recall that kR for nucleons impinging on heavy nuclei is $3.2 \times 10^{-4}A^{1/3}[E(\text{eV})]^{1/2}$. By replacing η with this expression, we find that σ_r, σ_s take on interesting forms.

● If f is real, $|\eta| = 1$ and there is no reaction. In general f will be complex, and we indicate its real and imaginary parts by Re f and Im f,

$$f = \text{Re } f + i \text{ Im } f \tag{11-6.7}$$

Note that, to have $|\eta| < 1$, Im f must be negative.

After some reductions, substitution of Eq. (11-6.6b) in Eqs. (11-6.1) and (11-6.2) gives

$$\sigma_r = \pi\lambdabar^2\left[\frac{-4(\text{Im } f)kR}{(\text{Re } f)^2 + [kR - (\text{Im } f)]^2}\right] \tag{11-6.8}$$

ISBN 0-8053-8061-7

and

$$\sigma_s = 4\pi \lambdabar^2 \left| e^{ikR} \sin kR + \frac{kR}{i(kR - \operatorname{Im} f) - \operatorname{Re} f} \right|^2 \qquad (11\text{-}6.9)$$

We see again that for f real $\sigma_r = 0$ and σ_s takes a form that may be easily interpreted if $\sin kR$ is very large or very small compared with $|kR/(ikR - \operatorname{Re} f)|$.

In the first case

$$\sigma_s = 4\pi \lambdabar^2 \sin^2 kR$$

and, if also $kR \ll 1$,

$$\sigma_s = 4\pi R^2 \qquad (11\text{-}6.10)$$

which is four time the geometric cross section. This type of scattering is called potential scattering.

In the second case

$$\sigma_s = 4\pi \lambdabar^2 \frac{(kR)^2}{(\operatorname{Re} f)^2 + k^2 R^2} \qquad (11\text{-}6.11)$$

This scattering cross section has a maximum value $4\pi \lambdabar^2$ when $\operatorname{Re} f = 0$ (resonance) and depends critically on $\operatorname{Re} f$. Assume that, for a certain value E_0 of the energy of the incident particle, $\operatorname{Re} f = 0$ and call $\operatorname{Re} f'$ the value of the derivative of $\operatorname{Re} f$ at $E = E_0$. We may then write

$$\operatorname{Re} f(E) = (E - E_0)\operatorname{Re} f' \qquad (11\text{-}6.12)$$

It can be proved that $\operatorname{Re} f'$ is always negative. Substituting in Eq. (11-6.11) we have

$$\sigma_s = 4\pi \lambdabar^2 \frac{(kR)^2}{(E - E_0)^2(\operatorname{Re} f')^2 + (kR)^2} \qquad (11\text{-}6.13)$$

or calling

$$\Gamma_s = - \frac{2kR}{\operatorname{Re} f'} \qquad (11\text{-}6.14)$$

where the subscript s on Γ_s indicates "scattering," we obtain

$$\sigma_s = \pi \lambdabar^2 \frac{\Gamma_s^2}{(E - E_0)^2 + \Gamma_s^2/4} \qquad (11\text{-}6.15)$$

ISBN 0-8053-8061-7

This equation shows again that the maximum possible cross section is

$$\sigma_s \max = 4\pi\lambdabar^2$$

In a region near the energy E_0, the scattering cross section shows the typical resonance shape common to many phenomena, notably the refractive index of a medium near an absorption line. The physical meaning of Γ_s is the full energy width of the resonance peak at half its maximum value. More fundamentally, Γ_s is connected with the duration τ of the scattering process by $\hbar/\Gamma_s \approx \tau$, which, being finite, introduces an uncertainty in the resonance energy E_0. It is convenient to introduce Γ_s in the form of Eq. (11-6.14) in order to obtain Eq. (11-6.15) in the typical resonance form. Note that since Re f' is constant, Γ_s is proportional to k or to the velocity of the impinging neutrons. ●

A more refined treatment of scattering takes into account the full Eq. (11-6.9), which contains scattering coherent with the incident wave, caused by the nuclear surface. This coherent scattering is given by the term $e^{ikR} \sin kR$ of Eq. (11-6.9). The result is that the amplitudes of the scattered wave add, and Eq. (11-6.15) is replaced by

$$\sigma_s = 4\pi\lambdabar^2 \left| \frac{\Gamma_s/2}{(E - E_0) + i\Gamma_s/2} + \frac{R}{\lambdabar} \right|^2 \qquad (11\text{-}6.16)$$

when $R/\lambdabar \ll 1$. In the vicinity of the resonance, Eq. (11-6.16) reduces to Eq. (11-6.15), while at greater distances from the resonance, potential scattering prevails. We note that the potential and resonance scattering can interfere and give rise to characteristic dips in the cross section for energies just below resonance, as is shown in Fig. 11-18.

We must now extend our treatment to f complex in order to encompass the case of reactions besides elastic scattering. As pointed out earlier, we are no longer dealing with the ordinary stationary states of Schrödinger's equation. Left to itself, the nucleus decays, showing in the time dependence of the square of the modulus of its wave function the factor $e^{-\lambda t}$, where λ is the decay constant. In order to produce this time dependence, we formally introduce a complex value for the energy,

$$E = \epsilon - \tfrac{1}{2}i\Gamma_r \qquad (11\text{-}6.17)$$

where the subscript r on Γ reminds us that Γ_r refers to "reaction." The time-dependent factor of ψ, $e^{(-iEt/\hbar)}$, then becomes $\exp[(-i\epsilon t/\hbar) - (\Gamma_r t/2\hbar)]$, and the exponential decay of $\psi\psi^*$ follows the correct law if

$$\Gamma_r/\hbar = \lambda \qquad (11\text{-}6.18)$$

ISBN 0-8053-8061-7

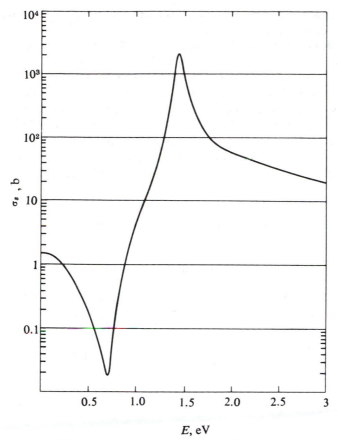

Figure 11-18 The scattering cross section for ^{115}In, an isotope with a resonance near thermal energy; the dip in cross section below the resonance corresponds to interference between the negative resonance amplitude and the positive potential amplitude. Compare also Fig. 11-9. Ordinate, logarithmic scale.

We have, numerically, for the lifetime (in seconds)

$$\tau = \frac{1}{\lambda} = \frac{6.58 \times 10^{-16}}{\Gamma_r(\text{eV})} \qquad (11\text{-}6.19)$$

This complex energy is analogous to the complex refractive index of optics, in which the imaginary part is connected with the absorption by the medium.

Thus we have a complex f which is a function of a complex variable E.

Assume that we have determined f by using a Schrödinger equation with real eigenvalues and that we have found

$$\text{Re } f(E_0) = 0$$

ISBN 0-8053-8061-7

When we use a complex eigenvalue, as required by the physical situation, we must introduce a complex f.

There will be a value of the complex variable near E_0 for which f will be zero. We call this value $E_r - \frac{1}{2} i\Gamma_r$ and we assume that both $(E_r - E_0)/E_0$ and $\Gamma_r/2E_0$ are $\ll 1$. The suffix r in E_r and Γ_r from now on, stands for "resonance." It is desired now to have f in the vicinity of its zero and for real values of the energy E. Expanding f in a power series and taking the first term only, we have, since $f(E_r - \frac{1}{2} i\Gamma_r) = 0$, $f(E) = (E - E_r + \frac{1}{2} i\Gamma_r) f'(E_r - \frac{1}{2} i\Gamma_r)$. The imaginary part of $f'(E_r - \frac{1}{2} i\Gamma_r)$ is small compared with the real part, as can be ascertained by using the definition Eq. (11-6.4), and to a sufficient approximation $f'(E_r - \frac{1}{2} i\Gamma_r)$ can be replaced by $\mathrm{Re}\, f'(E_r)$. Our expansion thus gives

$$f(E) = \left(E - E_r + \frac{i\Gamma_r}{2} \right) \mathrm{Re}\, f'(E_r) \tag{11-6.20}$$

Using Eq. (11-6.14) we have for the real and imaginary parts

$$\mathrm{Re}\, f(E) = -(E - E_r) \frac{2kR}{\Gamma_s} \tag{11-6.21}$$

$$\mathrm{Im}\, f(E) = \frac{-kR\Gamma_r}{\Gamma_s} \tag{11-6.22}$$

Inserting these values in Eqs. (11-6.8) and (11-6.9) we obtain in the vicinity of a resonance

$$\sigma_s = \pi \lambdabar^2 \frac{\Gamma_s^2}{(E - E_r)^2 + \Gamma^2/4} \tag{11-6.23}$$

and

$$\sigma_r = \pi \lambdabar^2 \frac{\Gamma_r \Gamma_s}{(E - E_r)^2 + \Gamma^2/4} \tag{11-6.24}$$

where $E_r \cong E_0$ and $\Gamma = \Gamma_r + \Gamma_s$.

Equations (11-6.23) and (11-6.24) should also contain a factor g on the right-hand side to take into account the spin of the target and of the projectile. For instance, for slow neutrons, where the incident particle has no orbital angular momentum but has spin $\frac{1}{2}$, the compound nucleus will have $2(I_A + \frac{1}{2}) + 1$ states if the neutron spin is parallel to the spin I_A of the target nucleus, and $2(I_A - \frac{1}{2}) + 1$ states if it is antiparallel. The total number of

ISBN 0-8053-8061-7

states is thus $4I_A + 2$. The probability of finding the spins parallel is

$$\frac{2(I_A + \frac{1}{2}) + 1}{4I_A + 2} = \frac{I_A + 1}{2I_A + 1} = g_+ \tag{11-6.25}$$

and the probability of finding them antiparallel is

$$\frac{2(I_A - \frac{1}{2}) + 1}{4I_A + 2} = \frac{I_A}{2I_A + 1} = g_- \tag{11-6.26}$$

The factor g has the value g_+ if the spins are parallel and g_- if they are antiparallel.

For slow neutrons Eqs. (11-6.23) and (11-6.24) may be further specified, noting that (with few exceptions, notable B and Li and fissionable nuclei) Γ_r corresponds to the emission of a gamma ray and can thus be called Γ_γ. It is essentially independent of the energy of the incoming slow neutron (Fig. 11-19). Γ_s corresponds to the reemission of a neutron and is proportional to the velocity of the impinging neutron, and we write for it Γ_n.

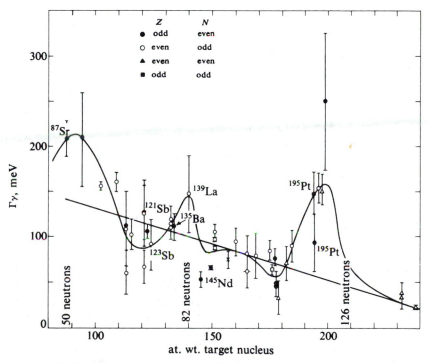

Figure 11-19 Measured radiation widths plotted against atomic weight, revealing the slow decrease of radiation width with atomic weight, as well as a sharp discontinuity at the closed nuclear shells near $A = 200$. [From D. J. Hughes, *Neutron Cross Sections*, Pergamon, London, 1957.]

ISBN 0-8053-8061-7

Consideration of Eqs. (11-6.23) and (11-6.24) and the foregoing remarks give the famous Breit–Wigner formulas, which are of great importance in both nuclear and particle physics.

$$\sigma(n, n) = \pi \lambdabar_r^2 \frac{\Gamma_n^{(r)2}}{\Gamma^2/4 + (E - E_r)^2} g \tag{11-6.27}$$

$$\sigma(n, \gamma) = \pi \lambdabar \lambdabar_r \frac{\Gamma_\gamma \Gamma_n^{(r)}}{\Gamma^2/4 + (E - E_r)^2} g \tag{11-6.28}$$

The Breit–Wigner formulas can be rewritten using dimensionless variables $\epsilon = 2(E_0 - E)/\Gamma$ and $x = \Gamma_s/\Gamma$ (called *elasticity*). We have then for the scattering amplitude

$$A/\lambdabar = x/(\epsilon - i) \tag{11-6.28a}$$

and

$$\sigma_s = 4\pi \lambdabar^2 x^2 / (\epsilon^2 + 1) \tag{11-6.28b}$$

$$\sigma_r = 4\pi \lambdabar^2 x(1 - x)/(\epsilon^2 + 1) \tag{11-6.28c}$$

$$\sigma_t = 4\pi \lambdabar^2 x/(\epsilon^2 + 1) \tag{11-6.28d}$$

The real and imaginary part of the scattering amplitude in a complex plane is given by Fig. 11-15, in which we also indicate η and δ.

If there is a nonresonant background to be added to a resonant one, the situation is more complex and we refer the reader to the literature.

Classical examples of the (n, γ) resonance are illustrated in Figs. 11-9, 11-18, 11-20, 11-21, 11-22, and 11-54. If there are several resonating levels near each other, the cross section takes a more complex form. Many examples of neutron cross sections are to be found in the atlas of Hughes (Hughes et al., 1960).

Equation (11-6.28) contains a $1/v$ factor, implicit in $\lambdabar$, multiplying a curve having the typical resonance shape. If the resonance is very broad with respect to the difference $E - E_r$, that is, if $\Gamma^2/4 \gg (E - E_r)^2$, or if $E_r \gg E$, so that variations of E do not affect the denominator, the resonance factor is almost constant and $\sigma(n, \gamma)$ obeys the famous $1/v$ law. This happens for instance in the reaction $^{10}B(n, \alpha)$, which is governed by a relation similar to Eq. (11-6.28), in which Γ_γ is replaced by Γ_α. Boron (normal isotopic composition 19.8% ^{10}B) shows a neutron-absorption cross section given by

$$\sigma(v) = \frac{\sigma(v_0)v_0}{v} = \frac{(760 \pm 2)(2200)}{v} b \tag{11-6.29}$$

ISBN 0-8053-8061-7

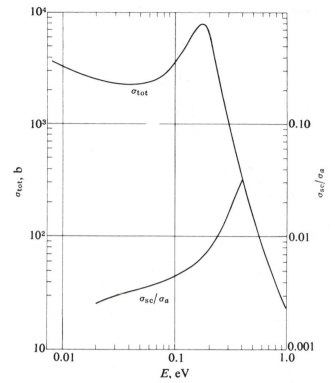

Figure 11-20 The energy dependence of the cadmium total cross section and the ratio between scattering and absorption cross sections for neutrons below 1 eV. Log-log scale.

where v is expressed in meters per second. This relation is valid for energies ranging from 10^{-3} to at least 10^2 eV. The absorption in boron is a good way of measuring the energy of neutrons.

As we have seen, the quantities η or f in Eqs. (11-5.8) to (11-5.11) and (11-6.6) completely describe scattering and reaction cross sections. One of the main problems in the study of nuclear reactions, therefore, is to calculate η from a model. At present we do not know how to analyze completely the nuclear dynamics (and thus obtain η). However, it is instructive to devise some extreme conditions under which η behaves in a simple, predictable fashion.

Consider two examples. Suppose first that the particle entering the nucleus moves inside it with a kinetic energy much larger than it had outside, and that it exchanges energy rapidly with the other nucleons. Under these assumptions an entering particle would have a very small probability of leaving the nucleus once it penetrated its surface. The wave function inside the nucleus

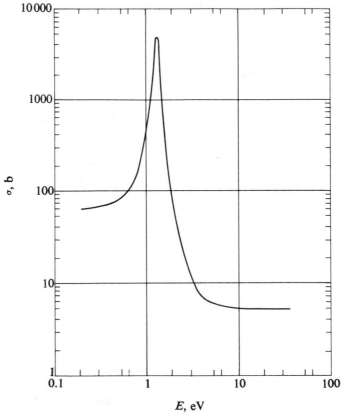

Figure 11-21 Total neutron cross section of Rh as a function of energy.
Log-log scale.

would have approximately the form

$$u_i \propto e^{-iKr} \qquad \text{for} \quad r < R \tag{11-6.30}$$

without an outgoing wave. From this expression we obtain immediately

$$f = -iKR \tag{11-6.31}$$

and from Eqs. (11-6.6) and (11-6.2) we obtain

$$\eta = \frac{K - k}{K + k} e^{-2ikR} \tag{11-6.32}$$

and

$$\sigma_r = \pi \lambdabar^2 \frac{4kK}{(K + k)^2} \tag{11-6.33}$$

ISBN 0-8053-8061-7

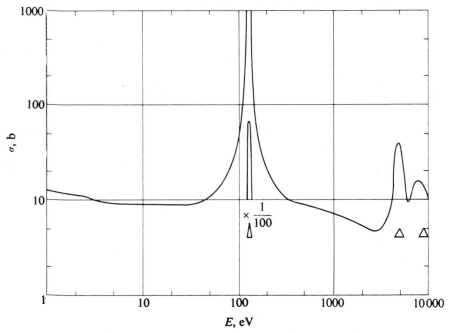

Figure 11-22 Total neutron cross section of cobalt as a function of energy. Log-log scale.

We have already found this expression (Eq. 11-4.4) as the product of the cross section $\pi \lambdabar^2$ multiplied by a transmission factor.

The analysis for $l > 0$ and for charged particles is more complicated and will not be discussed here. The situation described above applies to neutrons with energies of several hundred keV up to some MeV.

In our second example the energy of the incident particle (e.g., slow neutron) is such that no other channel except the entrance channel is open; the compound nucleus can reemit only the incident particle with the incident energy. The wave function inside the nuclear surface is thus

$$u_{in} \sim e^{-iKr} + e^{i(Kr+2\xi)} \tag{11-6.34}$$

with incoming and outgoing waves of the same amplitude but generally shifted in phase. We then have

$$f = -KR \tan(KR + \xi) \tag{11-6.35}$$

the phase ξ depending on the energy ϵ of the incident particle. At the values of ϵ for which f becomes 0, there is a sharp maximum for the scattering cross section. In the vicinity of this maximum the scattering cross section shows a

ISBN 0-8053-8061-7

resonance shape, and its width is

$$\Gamma = - \frac{2kR}{df/d\epsilon} \tag{11-6.36}$$

Actually the zeros of f correspond to virtual energy levels. The resonance levels occur according to Eq. (11-6.35) whenever

$$KR + \xi = Z(\epsilon) = n\pi \tag{11-6.37}$$

with n integral. If $dZ/d\epsilon$ were constant, the levels would be separated by an energy interval D^* such that

$$\frac{dZ}{d\epsilon} D^* = \pi \tag{11-6.38}$$

The true average distance between levels D is of the same order of magnitude as D^*, and from Eq. (11-6.36), observing that at resonance

$$-\frac{df}{d\epsilon} = \frac{KR}{d\epsilon}\frac{dZ}{d\epsilon} \tag{11-6.39}$$

we obtain the interesting relation

$$\Gamma = \frac{4k}{K} \frac{D^*}{2\pi} \tag{11-6.40}$$

which shows that the width of the levels is small compared with their distance as long as $k/K \ll 1$.

We can give a physical interpretation of Eq. (11-6.40) by noting that $\Gamma/\hbar$ is the reciprocal of the mean life of a level. Now our compound nucleus, in order to decay, must return to a suitable configuration, an event that occurs approximately periodically (period T). The probability of decay from this configuration is the transparency or transmission coefficient of the surface given by Eq. (11-6.33).

The periodic time T may be estimated, according to Weisskopf, from the following argument: Let us attempt to describe the motion of a particle semi-classically. For this purpose we shall construct a wave packet, and this can be done only by superimposing a number of states, say, N. The energy of the states will not be too far from E_0, and for the sake of argument we shall assume it to be

$$E_n = E_0 + nD \tag{11-6.41}$$

where n is an integral number and D is the distance between neighboring

ISBN 0-8053-86601-7

energy levels. The levels are thus assumed to be equally spaced, although this assumption is not essential.

The ψ of the system is then

$$\psi(t) = \sum_{n=0}^{N} a_n \varphi_n e^{-i(E_n t/\hbar)} = e^{-i(E_0 t/\hbar)} \sum_{n=0}^{N} a_n \varphi_n e^{-i(nDt/\hbar)} \qquad (11\text{-}6.42)$$

and $|\psi|^2$ has a period

$$T = \frac{2\pi\hbar}{D}$$

as can be verified immediately from Eq. (11-6.42).

The transparency is given by Eq. (11-6.33), which for $k/K \ll 1$ can be written $4k/K$, and hence

$$\frac{\Gamma}{\hbar} = \text{transparency} \times \frac{1}{T} = \frac{4k}{K}\frac{D}{2\pi\hbar} \qquad (11\text{-}6.43)$$

Resonance scattering governed by the considerations developed above is observed in light nuclei and at relatively low energies up to 1 MeV according to the A of the target or for very low-energy neutrons incident on intermediate nuclei. An example of the latter case is shown in Fig. 11-22, giving the resonance of cobalt at 132 eV. For this we have

$$\Gamma_\gamma \ll 1 \text{ eV} \qquad \Gamma_n \cong 4.9 \text{ eV}$$

Finally we have very important cases of resonance when radiative capture or fission competes successfully with the reemission of a slow neutron, that is, $\Gamma_\gamma \gg \Gamma_n$ or $\Gamma_f \gg \Gamma_n$.

11-7 OPTICAL MODEL

As a further step in developing nuclear models based on simple hypotheses we apply to nuclear reactions the same basic idea involved in the shell model. The impinging particle enters a potential well similar to the well used in the shell model. Because the projectile dissipates its energy we must add to the real part of the potential an imaginary one. We thereby obtain the "optical model" (Bethe, 1940; Feshbach, Porter, and Weisskopf, 1954). For example, for neutrons between 0 and 41 MeV, one can take

$$V(r) = V_0(r)(1 + i\xi) \qquad (11\text{-}7.1)$$

where $V(r)$ is assumed to be a rectangular well of the form

$$V_0(r) = -40 \text{ MeV} \qquad r < R$$

$$V_0(r) = 0 \qquad r > R$$

$$R = 1.45 A^{1/3} \times 10^{-13} \text{ cm}$$

$$\xi = 0.03$$

For numerical calculations a form (Woods and Saxon, 1954) often used is

$$V_0(r) = \frac{V_1}{1 + e^{+(r-R)/a}} \qquad (11\text{-}7.2)$$

with

$$V_1 = -60 \text{ to } -20 \text{ MeV}$$

$$\xi = 0.7 - 0.13$$

$$a = 0.5 - 0.7 \text{ F}$$

$$R = (1.15 - 1.35) A^{1/3} \text{F}$$

The constants of the model vary according to the nature and energy of the incoming particles.

The main feature of Eq. (11-7.2) is that it introduces a continuously changing potential, with a zone of rapid variation of thickness a at the nuclear surface ($r = R$). The imaginary part $\xi V_0(r)$ gives the absorption of the medium. This model has been called the *cloudy crystal ball*. Formal solution of the Schrödinger equation with the complex potential allows the evaluation of the logarithmic derivative of the wave function at the nuclear surface and hence the determination of the scattering and reaction cross sections.

The results of this model can be expected to give only the average behavior of the nucleus over many resonances; consequently the results do not reproduce the region of separate resonances. However, at higher energies where individual resonances are not resolved, the model gives good agreement with the experimental findings for total and elastic nuclear cross sections up to an energy of a few MeV (as shown in Fig. 11-23) where waves with $l \leqslant 5$ have been included.

Figure 11-23 (facing page) (a) Total neutron cross sections (calculated from the optical model) vs. $x^2 = (R/\lambda)^2$ and A for different choices of the imaginary part of the potential. (b) Measured total neutron cross sections, averaged over resonances, vs. E and A. (c) A similar plot using $x^2 = (R/\lambda)^2$ in place of E. [H. Feshbach, C. E. Porter, and V. F. Weisskopf, *Phys. Rev.*, **96**, 448 (1954).]

ISBN 0-8053-8601-7

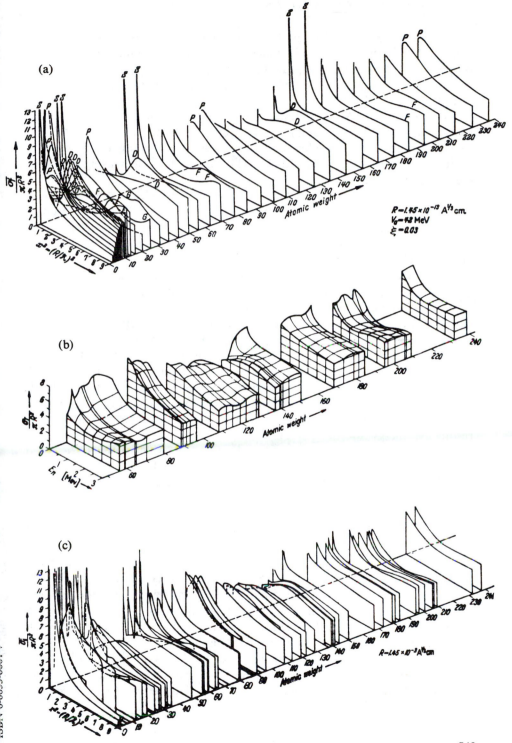

(a)

$R = 1.45 \times 10^{-13} A^{1/3}$ cm.
$V_0 = 42$ MeV
$\xi = 0.03$

(b)

(c)

$R = 1.45 \times 10^{-13} A^{1/3}$ cm.

The model may also be used to calculate angular distributions in neutron–nucleus scattering and it provides a good insight into more basic nuclear properties, such as the ratio Γ_n/D. Γ_n/D is approximately constant, but with the optical model we can take a further step and calculate it as a function of A and E_n (Fig. 11-24).

Above about 50 MeV the optical model can be used again in a simple fashion by considering Fraunhofer diffraction from the cloudy crystal ball (Fernbach, Serber, and Taylor, 1949), although the constants of the well are somewhat different from those of Eq. (11-7.2), with V_1 decreasing and ξ increasing with energy.

The optical model can be extended to charged particles, but we shall omit this complication. Once the potential is supplemented with a term proportional to $\boldsymbol{\sigma} \cdot \mathbf{l}$ (angular momentum and spin of the incident particle), it also gives a satisfactory explanation of the polarization phenomena observed in proton–nucleus scattering (cf. Chap. 10).

The cloudy crystal ball model can be related to the shell model by arguments of the following type, which, however, are not unambiguous.

If the imaginary part of the potential were absent, we should have an ordinary Schrödinger equation and corresponding stationary orbits. The shell model shows that such an approximation has considerable merit. One would then expect resonance levels of a compound nucleus to be determined by the characteristics of the well. With any of the well parameters ordinarily used, such resonances would be widely separated. Moreover, considering the levels available as a function of A, we should expect to find (for a given energy) a resonance for a narrow region of A values, because the energy considered is a resonance energy for the nuclear radius corresponding to a particular A. Suppose that we wanted to calculate V_0 from the shell model. We could observe that a certain level (e.g., the $3s$ level) is barely bound for a certain A; in this case $A \cong 150$. This means that for $\hbar k = (2mV_0)^{1/2}$ the $3s$ wave function at $r = R = r_0 A^{1/3}$ has the proper value of u'/u to ensure bare binding. From this, given R, we could find V_0. The agreement between V_0 calculated in this fashion and the value derived from scattering experiments is good if realistic forms of wells are used.

The resonance effects are indeed observed, however, not on single levels, but rather on groups of levels, clustering around certain energy regions because the single level obtainable from the shell model is highly degenerate. The main factor determining the position of the level is the well, but the residual interaction splits the one level into many. Experimentally, either the energy-resolving power of the instrument used to observe scattering is insufficient to resolve the single resonances or the levels actually overlap. The groups of resonances are then observed as relatively slow variations of the cross section as a function of energy. This is called the *giant resonance* phenomenon (Fig. 11-25).

ISBN 0-8053-8601-7

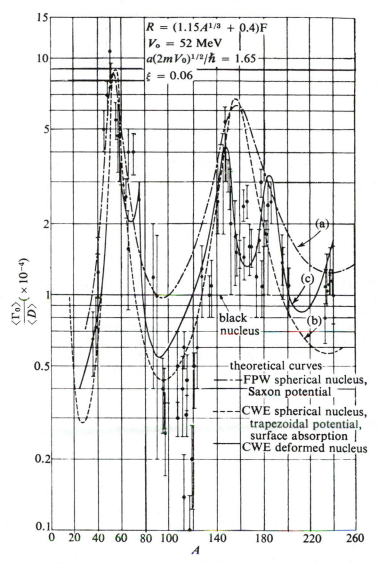

Figure 11-24 Neutron strength function at zero energy as a function of the mass number. Theoretical curve (a) corresponds to optical model with Saxon potential and constants indicated in the figure (FPW). Curve (b) corresponds to a spherical nucleus with trapezoidal potential and surface absorption (CWE). Curve (c) corresponds to a spheroidal nucleus according to Chase, Wilets, and Edmonds. [After J. A. Harvey, *Proc. Intern. Conf. Nucl. Structure*, Kingston, 1960.]

ISBN 0-8053-8601-7

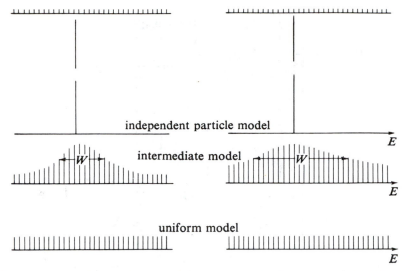

Figure   11-25   Figure from A. M. Lane, R. G. Thomas, and E. Wigner illustrating how the intermediate model is related to the strict independent-particle model, and to the uniform model. On the strict independent-particle model there is no coupling between individual particle states, so that the only resonances seen in neutron scattering are those for a single particle in a spherical potential well. In the uniform model the coupling between individual particle states is so strong that all configurations of the same spin and parity have, a priori, equal strengths for neutron-induced processes. In the intermediate model the coupling of many nuclear excited states to the incoming channel single nuclear excited state is such that the average strength function for nuclear levels of proper spin and parity displays a broad resonance behavior of width W peaked at the energy of the single-nucleon excited-state energy. Sum rules act to maintain the summed contribution of the strength function at a common value for all models.

A fundamental problem is how to relate the optical-model potential to the nucleon–nucleon interaction. Suffice it to say that Riesenfeld and Watson (1956) have found a connection between the forward-scattering amplitude in the nucleon–nucleon system and the constants of the optical model.

11-8 COMPOUND NUCLEUS—LEVEL DENSITY

The density of nuclear levels depends strongly on the energy of excitation and on A. Near the ground state of light nuclei the levels may be about 1 MeV apart, and in heavy nuclei about 50 keV apart (except for the magic numbers of neutrons or protons). This may be observed in scattering experiments.

ISBN 0-8053-8001-1

At an energy corresponding to the binding energy of the neutron (8 MeV), slow-neutron resonances for a medium-heavy element are a few electron volts apart. To illustrate the most important properties of the density of nuclear levels, assume that we have A nucleons to be assigned to equidistant single-particle levels, taking into account Pauli's principle. The total energy E of the system is given. We may put one nucleon or no nucleon in any level, provided the total energy obtained is E or in an energy interval ΔE near E. It is clear that there are many ways to accomplish this, and if there are Δn ways, $\Delta n / \Delta E = \rho(E)$ is the *density of nuclear levels* at energy E for that nucleus. As a first orientation assume the levels equally spaced by ϵ; then the problem is equivalent to that of decomposing the number $n = E/\epsilon$ into positive integral numbers $n_1, n_2, \ldots$. To see this, consider the example (Fig. 11-26) of a system with, say, 10 particles to be excited to an energy $4\epsilon = E$. Only one particle is permitted to enter in each substate (exclusion principle) and a state

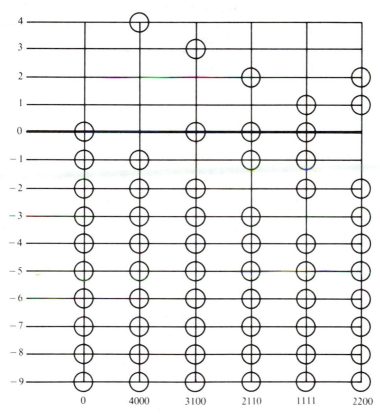

Figure 11-26 Occupied levels and vacancies for particles excited to an energy $E = 4\epsilon$.

ISBN 0-8053-8601-7

of the system is defined by its occupation numbers, 0 or 1. The zero energy corresponds to the state, when all low states are filled, as in (0). To reach energy $E = 4\epsilon$, I can push the highest particle from level 0 to level 4 as in (1); I can also push the highest particle to level 3 and the second highest particle to level 0 as in (2); I can also push the highest particle to level 2, the second highest to level 0, the third highest to level -1 as in (3), etc. I indicate the operations by writing the energy (in units of ϵ) given to the first particle first, that given to the second particle next, that given to the third particle in the third place, and so on, obtaining the numbers written under each column. All possibilities for energy 4 are shown in the figure and there is a one-to-one correspondence between each column and the expression of the number 4 as a sum of integer numbers. The problem is thus equivalent to that of finding the number of ways in which a given number can be expressed as a sum of positive integers (partitions). This problem has been solved by Ramanujan and Hardy, who give, for n large, $p(n) = (4\sqrt{3}\, n)^{-1} \exp[\pi(2n/3)^{1/2}]$, where $p(n)$ is the number of partitions possible. (The number A does not appear in the result until the energy is extremely large, beyond practical cases.)

The nuclear case differs from our example because there are neutrons and protons and two particles may go in the same level because of the spin. The formula for the level density then becomes, writing $a = \pi^2/6\epsilon$

$$\rho(E) = (\sqrt{\pi}\,/12)a^{-1/4}E^{-5/4} \exp\left[2(aE)^{1/2}\right] \tag{11-8.1}$$

See Fig. 11-27 as an example.

We may reach similar results by considering the properties of a Fermi gas and applying some concepts of statistical mechanics.

In Sec. 6-6 we described a nucleus in its ground state as a Fermi gas at absolute zero. If we excite the nucleus, it heats up and we may connect the excitation energy to its "temperature" using the specific heat formulas valid for a Fermi degenerate gas. [See, e.g., P. M. Morse, *Thermal Physics*, Benjamin, New York (1964) or (BM 69).]

For protons one has

$$(E_T - E_0)^{\text{prot}} = \frac{\pi^2}{4} Z \frac{(kT)^2}{E_F} \tag{11-8.2}$$

and by using for E_F, E_0 the expressions Eq. (6-6.5) we obtain

$$(E_T - E_0)^{\text{prot}} = \left(\frac{\pi^2}{72}\right)^{1/3} \frac{\Omega M}{\hbar^2} \left(\frac{Z}{\Omega}\right)^{1/3} (kT)^2 \tag{11-8.3}$$

with a similar expression for neutrons. The total energy is then, using $\Omega =$

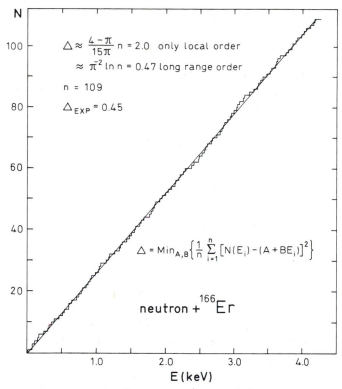

Figure 11-27 Number of levels N of spin and parity $I^{\pi} = \frac{1}{2}^{+}$ that have been observed for neutron energies up to a given value E [J. Rainwater et al., *Phys. Rev.*, **C5**, 974, (1972)]. The mean square deviation Δ from a uniform level spacing (corresponding to a straight line in the figure) is compared with the predictions based on two different models.

$(4/3)\pi r_0^3 A$, $r_0 = 1.3$ F, and the nucleon mass M (in MeV),

$$(E_T - E_0)^{\text{nucl}} = \left(\frac{\pi^2}{72} \right)^{1/3} \left(\tfrac{4}{3}\pi r_0^3 A \right)^{2/3} \frac{M}{\hbar^2} (Z^{1/3} + N^{1/3})(kT)^2$$

$$= 0.055 A^{2/3}(Z^{1/3} + N^{1/3})(kT)^2 = a(kT)^2 \quad (11\text{-}8.4)$$

If $A = 100$ and $Z = 44$, $E_T - E_0 = 8.7(kT)^2$. If $kT = 1$ MeV, the total excitation energy of the nucleus is only 8.7 MeV, as one would have for a classical gas of 9 degrees of freedom. Of the 300 degress of freedom, 29 are "frozen" by the degeneracy of the gas. Note that the energy for the degenerate gas depends quadratically on T, whereas for a classical gas E is known to be proportional to T.

The statistical definition of entropy is

$$S(T) = k \log[\omega(T)/\omega(0)] \tag{11-8.5}$$

where $\omega(T)$ is the number of quantum states available to the system at the specified temperature, and this number is proportional to $\rho(E)$. The thermodynamic definition of entropy,

$$S(T) = \int_0^T \frac{dE}{T} \tag{11-8.6}$$

in the specific case of a nucleus gives

$$S = 2ak^2 \int_0^T \frac{T \, dT}{T} = 2ak^2 T = 2(ak^2 E)^{1/2} \tag{11-8.7}$$

and we obtain from Eqs. (11-8.5) and (11-8.7)

$$\rho(E) = \rho(0)e^{2(aE)^{1/2}} \tag{11-8.8}$$

Equation (11-8.8) is qualitatively correct but the estimate of a given above gives too high a level density, because the hypotheses on which it is based are too crude. The constants $\rho(0)$ and a may be determined empirically, and Eq. (11-8.8) is then used as a fitting formula. The level density can be measured accurately at the energy corresponding to slow neutron capture by counting resonances, but one must then bear in mind that the only accessible states are those of spin $(I_0 \pm 1/2)$ where I_0 is the spin of the target nucleus. Figure 11-28 gives values of a from experimental data.

Equation (11-8.8) may be refined further considering the density of levels with a given I, and parity. The calculation for a Fermi gas gives

$$\rho(A, E, I, \pi) = \frac{2I+1}{24} a^{1/2} \left(\frac{\hbar^2}{2\mathcal{I}} \right)^{3/2} (E - E_r)^{-2} \exp\{2[a(E - E_r)]^{1/2}\} \tag{11-8.9}$$

where a [in $(\text{MeV})^{-1}$] is approximately $A/8$ and $E_r = (\hbar^2/2\mathcal{I})I(I+1)$ is the rotational energy of a *rigid* sphere of nuclear dimensions and mass.

The Breit–Wigner formula Eq. (11-6.28) integrated in an energy interval containing one resonance only gives

$$\int \sigma_r(E) \, dE = \pi g \int \frac{\lambdabar\lambdabar_r \Gamma_n \Gamma_\gamma \, dE}{(E - E_r)^2 + \Gamma^2/4} \cong 2\pi^2 g \lambdabar_r^2 \frac{\Gamma_n \Gamma_\gamma}{\Gamma} \tag{11-8.10}$$

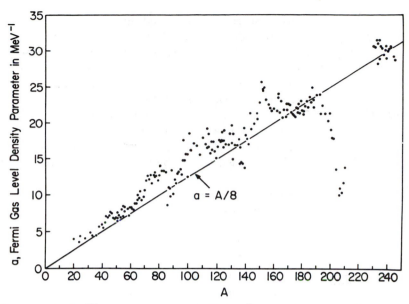

Figure 11-28 The parameter a [in $(MeV)^{-1}$] appearing in the Fermi gas level density formula determined empirically as a function of A from J. B. Garg in *Statistical Properties of Nuclei*, J. R. Huizenga, Ed., Plenum, New York, 1972.

To derive this formula, we have considered the isolated resonance as a sort of δ function having an area π and located at $E = E_r$. Consider now a neutron beam with an energy spread covering many resonances. We have, for the average capture cross section $\langle\sigma_r\rangle$ over a certain energy interval ΔE, the value given by Eq. (11-8.10) multiplied by the number of levels in the interval $\Delta E/D$ divided by the interval; that is,

$$\langle\sigma_r\rangle = 2\pi^2\lambdabar^2\langle g\rangle \frac{\langle\Gamma_n\rangle\langle\Gamma_\gamma\rangle}{\langle\Gamma\rangle D} \qquad (11\text{-}8.11)$$

For heavy elements at low energy $\langle\Gamma_\gamma\rangle \gg \langle\Gamma_n\rangle$, and, since Γ_n is proportional to $1/\lambdabar$,

$$\langle\sigma_r\rangle = 2\pi^2\lambdabar^2\langle g\rangle \frac{\langle\Gamma_n\rangle}{D} \propto \frac{1}{v} \qquad (11\text{-}8.12)$$

For higher energy, up to about 1 MeV, where $\Gamma_\gamma < \Gamma_n$,

$$\langle\sigma_r\rangle = 2\pi^2\lambdabar^2\langle g\rangle \frac{\langle\Gamma_\gamma\rangle}{D} \propto \frac{1}{E} \qquad (11\text{-}8.13)$$

Γ_γ depends little on energy and can be measured directly ($\Gamma_\gamma \sim 0.1$ eV). It is then possible to obtain D from Eq. (11-8.13) (see Fig. 11-29).

The thermodynamic approach can be extended by considering the emission of neutrons or of other particles by the excited compound nucleus as an evaporation process. The evaporated neutrons have an energy distribution corresponding to a Maxwellian distribution at the temperature of the residual nucleus. This has been demonstrated, at least qualitatively, by experiment (Gugelot, 1949). In the case of charged particles the Coulomb barrier prevents the evaporation of low-energy particles; it thus changes the energy distribution by multiplying the Maxwell distribution by the Coulomb barrier penetration factor.

If the initial excitation energy of the compound nucleus is sufficient, the evaporation of one particle leaves enough energy in the residual nucleus to

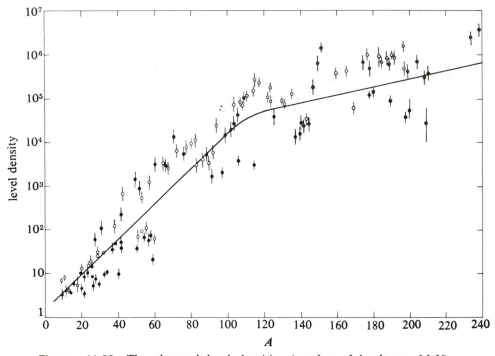

Figure   11-29   The observed level densities (number of levels per MeV) extrapolated to 7 MeV versus atomic number. The data are obtained from: the total level density of light nuclei ($A \leqslant 60$) the spacings of slow-neutron resonances ($A \geqslant 60$), and from the 1-MeV neutron cross sections, ● even-odd, odd-even, or even-even compound nuclei, ○ odd-odd compound nuclei. The full line in the density of levels at 7 MeV is given by Eq. (11-8.8). The errors are derived either from the number of levels used in determining the density or from errors involved in cross-section measurements. [From E. Vogt, BNL 331.]

ISBN 0-8053-8001-1

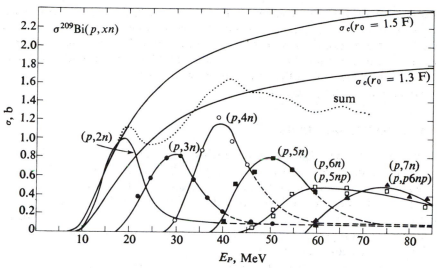

Figure 11-30 Excitation functions for the proton bombardment of bismuth, showing a statistical behavior of competing reactions. The calculated cross section for formation of the compound nucleus is indicated by σ_c. [R. E. Bell and H. M. Skarsgard, *Can. J. Phys.*, **34**, 745 (1956).]

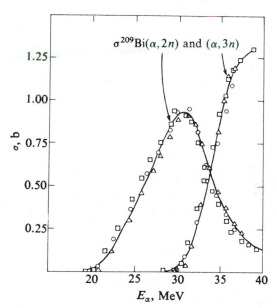

Figure 11-31 Excitation functions for alpha particle bombardment of bismuth [E. Kelly.]

permit the evaporation of a second particle, and so on, until the nucleus is left with so little excitation energy that only gamma emission is possible. We may thus have, for example (α, xn) reactions, with $x = 1, 2, 3$, etc., up to 6 or 7. The excitation functions for these processes and Figs. 11-30 and 11-31 give typical examples in which the competition of the different modes of decay of the compound nucleus is clearly seen.

11-9 DIRECT REACTIONS

In the qualitative description of nuclear reactions we have briefly mentioned direct interactions. An incoming particle hits a specific nucleon or a cluster of nucleons in the target and the remainder of the nucleus acts as a "spectator." If the nucleon is excited, the collision is inelastic. Such are, for instance, (p, p'), (n, n'), and (α, α') reactions. Characteristic for all of them, as opposed to compound nucleus formation, is the anisotropic angular distribution and the energy distribution.

The analysis of the angular distribution of the inelastically scattered particles gives important information about the quantum numbers of the initial and final states of the scattering nucleus. To see this qualitatively, write the vector angular momenta and spins of the particles and of the target. Referring to Fig. 11-32, which represents an inelastic collision, for instance (p, p'), we write

$$\mathbf{I}_i + \mathbf{l}_i + \mathbf{s}_i = \mathbf{I}_f + \mathbf{l}_f + \mathbf{s}_f \tag{11-9.1}$$

where $\mathbf{l}$ and $\mathbf{s}$ are the orbital and spin angular momentum of the proton and $\mathbf{I}$ is the total angular momentum of the target, in the initial and final states measured in units of $\hbar$.

From this we get, by constructing the extreme cases of orientation of the vectors,

$$I_i + I_f + 1 \geqslant |\mathbf{l}_f - \mathbf{l}_i| \geqslant |\mathbf{I}_f - \mathbf{I}_i - \mathbf{s}_i + \mathbf{s}_f|_{\min} \tag{11-9.2}$$

Moreover, $l_f - l_i \equiv \Delta l$ must be odd if initial and final states of the target have opposite parity or must be even if they have the same parity.

Now Δl is caused by the change of angular momentum of the impinging proton. If we can localize the collision, for instance by assuming it occurs at the nuclear surface at a distance R_0 from the center of mass of the system, we have

$$\Delta l = |\mathbf{k}_f - \mathbf{k}_i| R_0 \tag{11-9.3}$$

where $\hbar \mathbf{k}_i$, $\hbar \mathbf{k}_f$ are the initial and final linear momenta of the proton. Calling $\hbar \mathbf{k}$ the momentum transfer $\hbar |\mathbf{k}_i - \mathbf{k}_f|$, we have for Δl

$$\Delta l = k R_0 \tag{11-9.4}$$

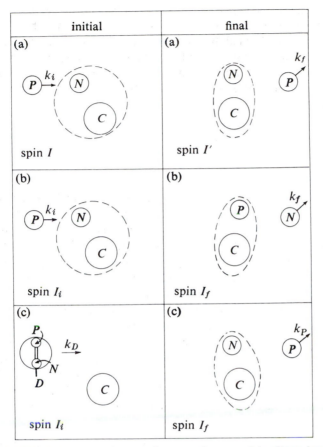

Figure 11-32 Schematic diagram for notations used in direct-interactions calculation. The nucleus NC has total angular momentum I or I' and its components N and C have orbital angular momentum l or l'. (a) Inelastic scattering, (b) knock-on reaction, (c) deuteron stripping.

Now from the cosine rule we have also

$$k^2 = k_i^2 + k_f^2 - 2k_ik_f \cos \theta \tag{11-9.5}$$

If Δl must have one or few values determined by the conditions of Eq. (11-9.2) there are only a few corresponding scattering angles θ that are preferred. For these the differential scattering cross section will show maxima. This qualitative semiclassical argument can be made into a theory of varying degrees of accuracy. To give an idea of the procedures used, we present the following simplified treatment of a special example.

ISBN 0-8053-8061-7

● We shall treat the case of inelastic alpha scattering in the Born plane waves approximation (see Tobocman, 1961). We assume that we may consider the target as formed by an alpha particle plus a nuclear core. The incident alpha particle has the reduced mass, with respect to the target, M_α and the center of mass momentum $\hbar k_i$. The ejected alpha particle has the momentum $\hbar k_f$. The differential cross section is, by the calculations of Sec. 11-2, or App. A,

$$\frac{d\sigma}{d\Omega} = \frac{M_\alpha^2}{(2\pi\hbar^2)^2}\frac{k_f}{k_i}|A_{if}|^2 \tag{11-9.6}$$

With reference to Fig. 11-32a, in which we now label the incident nucleus P with the symbol α and N is also an alpha particle, we find the the matrix element A_{if} is the sum of two integrals originating from the αN and αC interactions. We assume that these interactions are spherically symmetric and represent them by the functions $V_{\alpha N}(r_{\alpha N})$ and $V_{\alpha C}(r_{\alpha C})$, respectively. Here $r_{\alpha C}$ means the distance between the center of mass of the α particle and of body C, and a similar notation is used for other r's.

To treat a simple case we shall assume that the initial target nucleus of eigenfunction $\varphi_0(r_{NC})$ has orbital angular momentum $l = 0$ and total spin I. The final nucleus has an orbital angular momentum l' and a total angular momentum I'. Its eigenfunction is $\varphi_{I'}(r_{NC})$. The function $\varphi_0(r_{NC})$ is thus spherically symmetric; the function $\varphi_{I'}(r_{NC})$ has an angular dependence given by a linear combination of spherical harmonics $Y_{l'}^m(\theta, \varphi)$, with $m = l', l' - 1, \ldots, -l'$. The part of the matrix element due to $V_{\alpha N}(r_{\alpha N})$ is written

$$A_{\alpha\alpha'}(V_{\alpha N}) = \int \varphi_{I'}{}^*(r_{NC})\exp(-i\mathbf{k}_f \cdot \mathbf{r}_{\alpha I})V_{\alpha N}(r_{\alpha N})\exp(+i\mathbf{k}_i \cdot \mathbf{r}_{\alpha I})$$
$$\times \varphi_0(r_{NC})\,d\mathbf{r}_{\alpha I}\,d\mathbf{r}_{NC} \tag{11-9.7}$$

The integral Eq. (11-9.7) is simplified by a change of variables that permits us to decompose it into the product of two integrals. Replacing $r_{\alpha I}$ by its expression

$$\mathbf{r}_{\alpha I} = \mathbf{r}_{\alpha N} + \frac{M_C}{M_N + M_C}\mathbf{r}_{NC} \tag{11-9.8}$$

we have

$$A_{\alpha\alpha'}(V_{\alpha N}) = \int d\mathbf{r}_{\alpha N}\exp(-i\mathbf{k}\cdot\mathbf{r}_{\alpha N})V_{\alpha N}(r_{\alpha N})\int d\mathbf{r}_{NC}\exp(-i\mathbf{k}'\cdot\mathbf{r}_{NC})$$
$$\times \varphi_0(r_{NC})\varphi_{I'}{}^*(r_{NC}) \tag{11-9.9}$$

ISBN 0-8053-8601-7

where

$$\mathbf{k} = \mathbf{k}_f - \mathbf{k}_i \qquad (11\text{-}9.10)$$

and

$$\mathbf{k'} = \mathbf{k}\,\frac{M_C}{M_N + M_C} \qquad (11\text{-}9.11)$$

Now the calculation of each factor can be carried further. Consider first the integral in $d\mathbf{r}_{NC}$. Assume as the z axis the direction of $\mathbf{k'}$ and expand the plane wave in spherical harmonics according to the known relation (Appendix A)

$$e^{ikz} = \sum_L 4\pi(2L + 1)i^L j_L(kr) Y_L^0 (\cos \theta)$$

where $j_L(x)$ is the so-called *spherical Bessel function.*

The integration over angles gives zero unless $L = l'$ and $m = 0$, because of the orthogonality of the spherical harmonics, and the integral in $d\mathbf{r}_{NC}$ is proportional to

$$\int j_{l'}(k'r)\varphi_0(r)\varphi_{l'}{}^*(r)r^2\,dr \qquad (11\text{-}9.12)$$

where $k' = |\mathbf{k'}|$ and by $\varphi_{l'}{}^*(r)$ we mean only the radial part of the nuclear eigenfunction. Both $\varphi_0(r)$ and $\varphi_{l'}(r)$ decrease very rapidly beyond the nuclear surface corresponding to $r = R_I$. We may thus assume that the chief contribution to the integral comes to the region near the nuclear surface where the integrand is largest and conclude that the integral is proportional to $j_{l'}(k'R_I)$, which will thus appear as a factor in the final expression of $A_{\alpha\alpha'}$.

The evaluation of the other factor in the integral, Eq. (11-9.9), containing $V_{\alpha N}$, requires knowledge of this last function. One can give it different semi-empirical forms—for instance, the Yukawa form $V_{\alpha N}^0 e^{-r_{\alpha N}/\rho}/r$—and compute it. In view of the short range of the nuclear forces, we shall assume $V_{\alpha N} = V_{\alpha N}^0 \delta(r_{\alpha N})$, where $\delta(r)$ is the tridimensional Dirac delta function. It follows immediately that the integral reduces to $V_{\alpha N}^0$.

The integral corresponding to $V_{\alpha C}$ can now be written in analogy to Eq. (11-9.9) using the expression

$$\mathbf{r}_{\alpha I} = \mathbf{r}_{\alpha C} - \frac{M_N}{M_N + M_C}\,\mathbf{r}_{NC} \qquad (11\text{-}9.13)$$

However, when we write the expression corresponding to the second factor of Eq. (11-9.9) we find

$$\int d\mathbf{r}_{NC}\,\exp\{-i[M_N/(M_N + M_C)]\mathbf{k}\cdot\mathbf{r}_{NC}\}\varphi_0(r_{NC})\varphi_{l'}{}^*(r_{NC}) \qquad (11\text{-}9.14)$$

ISBN 0-8053-8601-7

Now the exponentional is nearly constant whenever $M_N \ll M_C$ (as we shall assume), and because of the orthogonality of φ_0 to $\varphi_{l'}$, the integral vanishes. For this reason we have neglected it.

In conclusion, we find that the matrix element A_{if} depends on θ only because k' appears in the Bessel function of Eq. (11-9.12). In fact we have

$$k^2 = k_i^2 + k_f^2 - 2k_i k_f \cos \theta = \left(\frac{M_N + M_C}{M_C} \right)^2 k'^2 \qquad (11\text{-}9.15)$$

where θ is the angle of scattering. The spherical Bessel functions are shown in Fig. A-1 of Appendix A. They resemble damped oscillations and this behavior determines the oscillations in the cross section.

It is apparent that if l had been different from zero there could have been many more terms in the matrix element, namely, those for which $|l - l'| \leqslant L \leqslant l + l'$, and correspondingly more Bessel functions. The formulas thus become complicated, but they preserve the structure.

$$A_{PP'} (V_{NP}) \approx \delta_{mm'} \sum_L B_{NP} (Ll'lm) j_L \left(\frac{M_C}{M_I} kR_I \right) \qquad (11\text{-}9.16)$$

with

$$\mathbf{k} = \mathbf{k}_f - \mathbf{k}_i \qquad (11\text{-}9.17)$$

The values of L over which the sum is to be taken are restricted by the conditions

$$|l - l'| \leqslant L \leqslant l + l' \qquad (11\text{-}9.18)$$

$$|I - I'| \leqslant L \leqslant I + I' \qquad (11\text{-}9.19)$$

$$\text{parity of } L = \text{parity of } l + l' \qquad (11\text{-}9.20)$$

These are often sufficient to leave in the sum only one or very few values of L (Fig. 11-33). For the case $l = 0$ explicitly treated, we have seen in fact only one value of L appearing in the sum. From the angular distribution of inelastic scattered particles we may thus extract information concerning the spin and the parity of the final state.

Similar to inelastic scattering are knock-on reactions, in which the incoming particle hits a particle of another kind and ejects it without forming a compound nucleus; for example, (α, p) (n, p) may go through this process (see Fig. 11-32b). An evaluation of the cross section in the Born approximation similar to that given above for inelastic scattering leads to a formula of the same structure as Eq. (11-9.16) and to the same selection rules, except that

ISBN 0-8053-8601-7

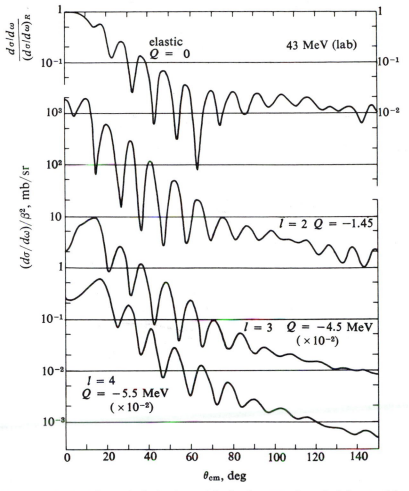

Figure 11-33 Theoretical elastic and inelastic scattering of alpha particles on ^{58}Ni. The angular distribution is calculated by direct-interaction theory and shows the influence of the angular momentum of the levels excited. Upper curve (elastic) gives the ratio of the differential scattering cross section to the Rutherford value. Lower curves correspond to excitations of rotational levels. For meaning of β^2, and nuclear deformation parameter, and parameters used in calculation, see original paper. Existing experimental data ($\theta < 60$ deg.) agree well with theory. [R. H. Bassel, G. R. Satchler, R. M. Drisko, and E. Rost, *Phys. Rev.*, **128**, 2693 (1963).]

ISBN 0-8053-8601-7

k in the Bessel function now means the magnitude of

$$\mathbf{k} = \mathbf{k}_i - \frac{M_I}{M_F} \mathbf{k}_f \qquad (11\text{-}9.21)$$

and that Eq. (11-10.19) is replaced by

$$L \leqslant I_P + I_N + I_i + I_f \qquad \bullet \qquad (11\text{-}9.22)$$

As an example we mention here an elegant application of direct interaction to the (p, n) reaction. Consider, for instance, the bombardment of ^{51}V in the reaction $^{51}V(p, n)^{51}Cr$. ^{51}V contains 23 protons and 28 neutrons. Outside the 20 neutron–20 proton core are 3 protons and 8 neutrons in $f_{7/2}$ orbits. In the final state there are 4 protons and 7 neutrons in the $f_{7/2}$ orbit of ^{51}Cr. The collision can be represented as changing a neutron into a proton without any change of orbits. The Q of the reaction must then be equal to the difference in Coulomb energy between ^{51}V and ^{51}Cr. In the spectrum of neutrons in this reaction one finds in fact a distinct peak corresponding to the incident proton energy minus the difference of the Coulomb energies (8.2 MeV calculated independently), emerging above the background of evaporation neutrons (see Fig. 11-34 and compare also with Fig. 6-13). In Fig. 6-13 the neutron or proton orbits, with the same quantum numbers, have the same energy measured from the bottom of their respective wells. This is a consequence of the charge independence of nuclear forces. The energy difference between a proton and a neutron orbit is then the same as the difference between the bottoms of the two wells.

Reactions initiated by deuterons show some interesting features attributable to surface effects and to the loose structure of the deuteron (see Chap. 10). These reactions often pass, not through an initial compound state in which the whole deuteron would be absorbed, but through a *stripping mechanism*, where the reaction may proceed at energies considerably below the Coulomb barrier (Oppenheimer and Phillips, 1936). The deuteron, without entering the target, is stripped of one of the nucleons (at very low energy, preferentially the neutron) and the other nucleon continues more or less along its initial trajectory. This process is predominant in the excitation function (d,p) of Fig. 11-35; a comparison of Figs. 11-34 and 11-35 shows the difference between the deuteron-initiated reaction and a reaction passing through the compound-nucleus process.

At low energy, but above the region in which Coulomb effects are important (3–14 MeV, depending on the Z of the target), (d, p) and (d, n) reactions behave quite similarly. Consider first (d, p) reactions, where the escaping protons show characteristic angular distributions, which can be explained by attributing them to deuterons that lose their neutron (stripping) without

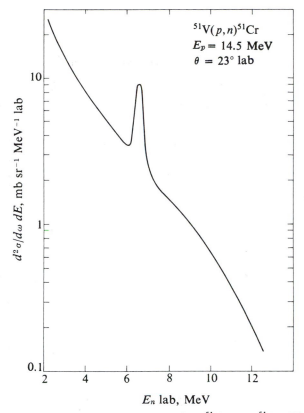

Figure 11-34 The proton–neutron reaction ^{51}V $(p, n)^{51}$Cr. The neutrons show an evaporation spectrum and superimposed on it a peak at energy 6.4 MeV. The energy left in the nucleus (14.5–8.1) MeV represents the energy difference between a neutron and a proton in the same type of orbit and corresponds to the energy difference between the two bottoms of the wells of Fig. 6-13; it is all of coulombic origin. [Courtesy J. D. Anderson, C. Wong, and J. W. McClure.]

the proton entering the field of the nucleus. The analysis of the angular distribution of these protons provides information about the angular momentum of the captured neutron and on the spin and parity of the final nucleus (Butler, 1951) (Fig. 11-36). The following qualitative argument is very similar to the argument given above for the (p, p') reaction.

In the stripping reaction at low energies, up to a few MeV, we must consider the initial and final states of the nucleus with angular momenta $\mathbf{I}_i$, $\mathbf{I}_f$, the orbital angular momentum of the deuteron with respect to the target nucleus $\mathbf{l}_d$ and the deuteron spin $\mathbf{s}_d$.

ISBN 0-8053-8601-1

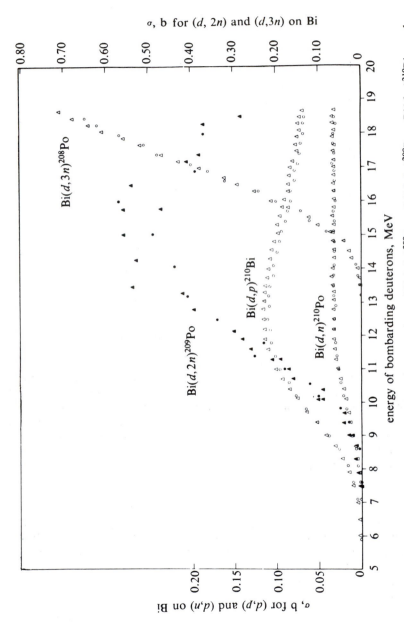

Figure 11-35 Excitation functions for the reactions $Bi(d, 2n)^{209}Po$, $Bi(d, 3n)^{208}Po$, $Bi(d, p)^{210}Bi$, and $Bi(d, n)^{210}Po$. [E. Kelly.]

ISBN 0-8053-8601-7

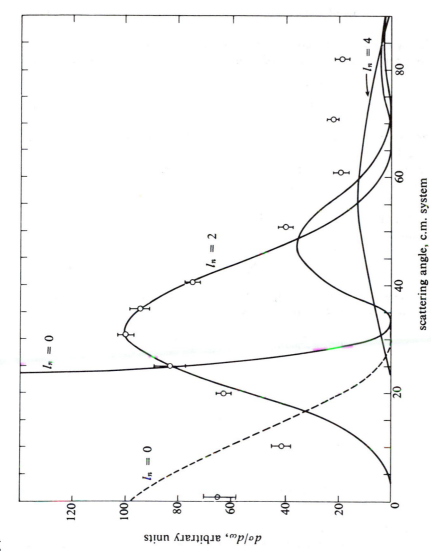

Figure 11-36 Angular distribution of protons from $^{35}Cl(d,p)^{36}Cl$. Curves are Butler curves for $l = 0$ and $l = 2$. The broken line is the curves for $l = 0$ reduced in scale by a factor of 20.

In the collision the neutron is captured, with orbital angular momentum $\mathbf{l}_n$, and the proton escapes at an angle θ with respect to the initial direction of motion of the deuteron. There is an interesting relation between $\mathbf{l}_n$ and the angle θ. To see this, we first note that

$$\mathbf{I}_f = \mathbf{I}_i + \mathbf{l}_d + \mathbf{s}_d - \mathbf{l}_p - \mathbf{s}_p = \mathbf{I}_i + \mathbf{l}_n + \mathbf{s}_n \qquad (11\text{-}9.23)$$

which simply expresses the conservation of angular momentum. The orbital angular momentum of the captured neutron $\mathbf{l}_n$ is thus restricted by the inequality

$$I_f + I_i + \tfrac{1}{2} \geqslant l_n \geqslant \left| |I_i - I_f| - \tfrac{1}{2} \right| \qquad (11\text{-}9.24)$$

as well as by the fact that, since the parity of the initial and final states is given, l_n must in some cases be even and in some cases odd, in order to satisfy parity conservation. For instance, starting from the ground state of ^{12}C, $I_i = 0$ (even); if we want to land in the ground state of ^{13}C, $I_f = \tfrac{1}{2}$ (odd), l_n must be $\leqslant 1$ and odd; that is, $l_n = 1$.

The linear momentum associated with the neutron and delivered to the capturing nucleus is

$$\hbar(\mathbf{k}_d - \mathbf{k}_p) = \hbar\mathbf{k}_n \qquad (11\text{-}9.25)$$

and this momentum gives rise to $\mathbf{l}_n$. Since the capture must occur at a radius not greater than the nuclear radius R, we have

$$|(\mathbf{k}_d - \mathbf{k}_p) \times \mathbf{R}| = |\mathbf{k}_n \times \mathbf{R}| \geqslant l_n \qquad (11\text{-}9.26\text{a})$$

or

$$(\mathbf{k}_d - \mathbf{k}_p)^2 \geqslant \frac{l_n^2}{R^2} \qquad (11\text{-}9.26\text{b})$$

or, squaring and rearranging,

$$2k_p k_d \cos\theta \leqslant k_p^2 + k_d^2 - \frac{l_n^2}{R^2} \qquad (11\text{-}9.27)$$

In a given transition k_p^2 and k_d^2 are constants, determined by the conservation of energy, and Eq. (11-9.27) shows that the preferred angle of scattering θ will increase with increasing l_n. Actually, this semiclassical argument is vastly oversimplified.

The calculation of stripping in the Born approximation may be done on lines similar to the calculation for inelastic scattering; however, it is appreciably more complicated, because it is not possible to factor the integrals in the

ISBN 0-8053-8061-7

matrix element by simple changes of variables and because the large size of the deuteron prevents the use of a zero-range potential.

The final result, however, is again very similar to that for the inelastic and knock-on reaction, but even simpler because the sum of Eq. (11-9.16) contains only one term. One finds

$$A_{DP}(V_{NP}) \approx \delta_{m0} D_{NP}(l) j_l(kR) \tag{11-9.28}$$

with

$$\mathbf{k} = \mathbf{k}_d - \frac{M_C}{M_F} \mathbf{k}_p \tag{11-9.29}$$

Here l is the orbital angular momentum of the final bound state. The inequality of Eq. (11-9.24) also obtains.

Stripping processes occur not only with deuterons but also with ^{3}H and ^{3}He. The (t, d) or $(^3He, d)$ reactions show many analogies with deuteron stripping. Stripping reactions are eminently suitable for determining the l and parity of nuclear levels from the angular distribution of the reaction products. In many cases, the results can be simply interpreted by the shell model. A different type of information concerning nuclear structure comes from the absolute value of the cross section.

At very high energies the stripping phenomenon may be described semiclassically (Serber, 1947) by considering one of the nucleons (e.g., the proton in the deuteron) as absorbed by a nucleus and the other as continuing along its trajectory almost unperturbed, with the initial velocity of the deuteron and, hence, half its energy. In this way, the extremely nonadiabatic collision gives rise to neutrons with an energy and angular distribution obtainable from the addition of the internal momentum of the deuteron to half the momentum of its center of mass. The energy spread and the angular spread of the neutron beam are

$$\Delta E = 1.5(\epsilon_d E_d)^{1/2} \tag{11-9.30}$$

$$\Delta \theta = 1.6 \left(\frac{\epsilon_d}{E_d} \right)^{1/2} \tag{11-9.31}$$

where ϵ_d is the binding energy of the deuteron and E_d its kinetic energy.

The converse of the stripping process is the "pick up" (n, d), (p, d) process in which, for example, a neutron entering a nucleus finds a proton with the right momentum and distance to form a deuteron, and the pair escapes as a bound deuteron. These reactions have been frequently observed and their cross sections are connected by detailed balancing with the stripping cross sections.

ISBN 0-8053-8061-7

11-10   THE FISSION PROCESS

Among nuclear reactions fission is very spectacular and of the greatest practical importance. A heavy nucleus such as uranium, under bombardment by a number of projectiles, splits into two large fragments such as ^{139}Ba and ^{97}Kr, which fly apart with an average energy of 170 MeV. The phenomenon is called fission (Hahn and Strassmann, 1939), and since it is accompanied by evaporation of neutrons which can start further fission, it offers the possibility of a chain reaction.

Occasionally, heavy nuclei undergo fission spontaneously, without outside stimulation (spontaneous fission; Flerov and Petrjak, 1940). Actually this mode of decay, which is negligible compared with alpha decay in nuclei having $Z \cong 92$, such as uranium, already shows for ^{252}Cf a branching of approximately 0.03 and for ^{254}Cf it becomes the main decay channel.

We shall now discuss the fission process without reference to whether it is spontaneous or stimulated. Consider first the mass formula of Chap. 6 or a table of masses. Let us inquire whether a nucleus of mass number A and atomic number Z, $M(A, Z)$, is stable with respect to its splitting into two equal parts of mass $M(A/2, Z/2)$. We have

$$M(A, Z) - 2M(A/2, Z/2) = 17.2A^{2/3}(1 - 2^{1/3})$$

$$+ 0.70 \frac{Z^2}{A^{1/3}}(1 - 2^{-2/3}) \quad \text{MeV} \quad (11\text{-}10.1)$$

This is positive for heavy nuclei and gives an energy of about 184 MeV for ^{236}U.

However, the original nucleus of mass $M(A, Z)$, assumed for the time being to be initially spherical, is stable for small deformations. In order to see this, we calculate the change in energy brought about by a small deformation. If this change is positive, the nucleus is stable. We try to change the nuclear shape from spherical to ellipsoidal, leaving the volume constant. The deformed ellipsoid has thus a major axis $a = R(1 + \epsilon)$ and minor axes $b = R(1 + \epsilon)^{-1/2}$. The volume is $V = (4/3)\pi ab^2$ and the surface is

$$S = 2\pi \left[b^2 + (ab \sin^{-1} e)/e \right] \approx 4\pi R^2 \left[1 + (2/5)\epsilon^2 \right]$$

with

$$e = \left[1 - (b^2/a^2) \right]^{1/2}.$$

The surface energy E_s is then, using the constants of Chap. 6, $E_s = 17.2A^{2/3}[1 + (2/5)\epsilon^2]$ in MeV. The electrostatic energy

$$E = \frac{1}{2} \int \rho(r_1)\rho(r_2) \frac{dv_1\, dv_2}{r_{12}}$$

ISBN 0-8053-8061-7

where dv_1, dv_2 are volume elements at a distance r_{12} and ρ is the charge density. On calculation this gives

$$E_c \approx \frac{3}{5} \frac{e^2 Z^2}{R} \left(1 - \frac{\epsilon^2}{5} \right)$$

The total change in energy is then

$$\Delta E = \Delta E_s + \Delta E_c = \epsilon^2 \left[\left(\tfrac{2}{5} \right) 17.2 A^{2/3} - \left(\tfrac{1}{5} \right) 0.70 Z^2 A^{-1/3} \right] \quad (11\text{-}10.2)$$

If ΔE is positive (i.e., for $Z^2/A < 49$), the spherical shape is stable.

The curve of the potential energy versus nuclear distance for two nuclei thus has the qualitative aspect of the dotted line in Fig. 11-37. The point for $\epsilon = 0$ is given by the nuclear mass; the region from $r = R_1 + R_2$ to infinity is a hyperbola of equation

$$V = \frac{Z_1 Z_2 e^2}{r} \quad (11\text{-}10.3)$$

as required by Coulomb's law. Here $R_1 + R_2$ is the distance between the electrical centers of the fission fragments at the moment of separation. If the nuclei at the moment of separation were spherical, $R_1 + R_2$ would be the sum

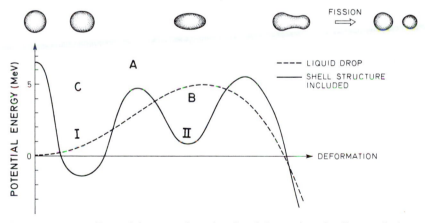

Figure   11-37   Potential energy function for deformations leading to fission. Shown are qualitative features in the structure of the fission barrier. The deformation parameter labels the path toward fission, as indicated by the shapes at the top of the figure. The shapes at the first (I) and second (II) minima correspond to those observed in the ground states and in the shape isomers in nuclei in the region of uranium. [From A. Bohr and B. Mottelson, *Ann. Rev. Nucl. Sci,* **23**, 363 (1973).]

ISBN 0-8053-8061-7

of the radii, but the deformation of the fragments is appreciable. The region between 0 and $r = R_1 + R_2$ is complicated and not entirely known but the difference between the energy at $r = 0$ and $r = \infty$ is obtainable from a table of masses, because it is the difference between the mass of the original nucleus and that of the two fragments. It is thus possible to trace the energy curve except for the region mentioned above. (We have neglected here the relatively small excitation energy of the two fragments.) The energy that must be supplied to the nucleus to provoke its fission, that is, the activation energy for the fission reaction, is thus the energy difference between the ground state of the nucleus at no deformation and at the lowest unstable configuration. This energy can be calculated with the liquid-drop model, corrected for shell effects (Fig. 11-38). The values obtained agree with the experimental results.

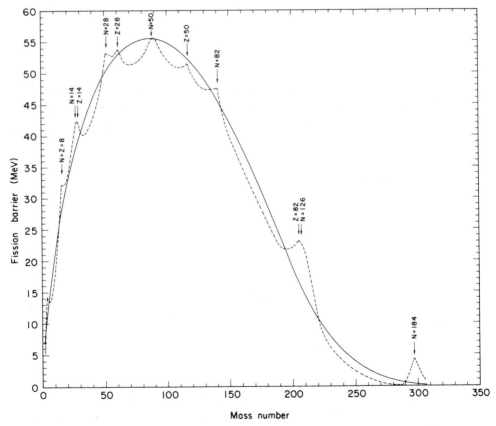

Figure   11-38   Fission barrier energy in MeV, along the line of beta stability. The smooth curve is the liquid drop result. The dotted curve is calculated from Myers and Swiatecki mass formula and shows shell effects. [From W. D. Myers and W. Swiatecki, Nuclear Physics **81**, 1, (1966).]

The tunnel effect of quantum mechanics has the consequence that there can be a small probability of fission even below the activation energy, owing to the zero-point energy of the nuclear motions. This is the reason for spontaneous fission. In Fig. 11-39 we have plotted the spontaneous-fission probability for heavy elements as a function of Z^2/A, the parameter that determines stability under small deformations. Although fission is a probable process for

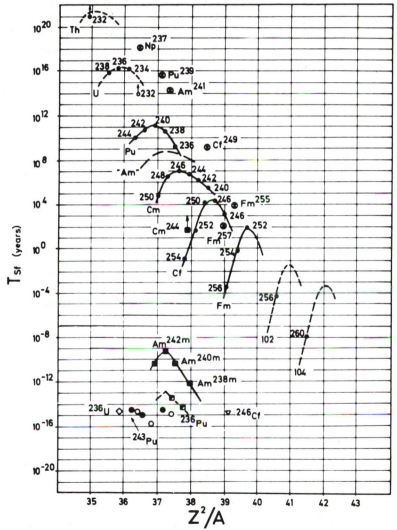

Figure 11-39 Reciprocals of spontaneous fission decay constants multiplied by $\log_e^2$, called spontaneous fission lifetimes. Measured lifetimes for the spontaneous fission isomers are also shown. [From V. M. Strutinsky and H. C. Paul, *Physics and Chemistry of Fission*, p. 157, IAEA, Vienna (1969).]

ISBN 0-8053-8061-7

a nucleus excited above the activation energy, its probability decreases very rapidly with decreasing energy. In a small energy interval the fission cross section becomes insignificant and one improperly speaks of a "fission threshold." See Figs. 11-40 and 11-41.

The binding energy of a neutron to a nucleus having an odd number of neutrons is greater by about 0.8 MeV than is to a nucleus having an even number of neutrons (see Chap. 6). Hence a slow neutron captured by a nucleus with N odd, such as ^{235}U or ^{239}Pu, gives rise to a compound nucleus that is more excited by about 0.8 MeV than the nuclei obtainable by slow-neutron capture in ^{238}U or other even-even nuclei. The excitation of about 7 MeV obtainable on capture of a slow neutron by an even-odd nucleus is sufficient to produce fission.

The fission fragments produced by slow-neutron fission in ^{235}U show a yield as a function of A, given in Fig. 11-42. It is most remarkable that fission

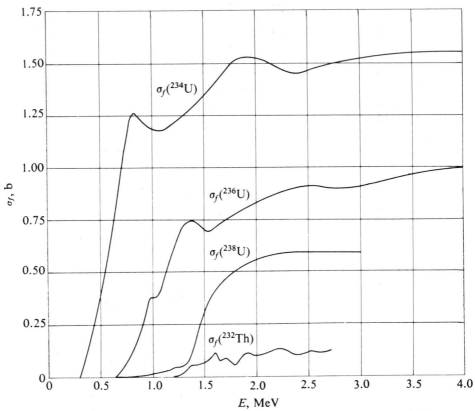

Figure   11-40   Neutron fission cross sections of ^{232}Th, ^{234}U, ^{236}U, and ^{238}U. [(WW 18).]

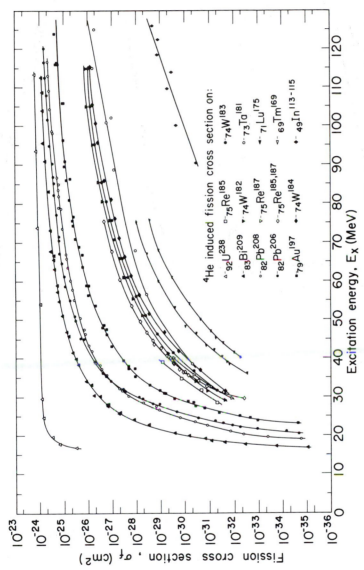

Figure 11-41 Examples of ^{4}He induced fission cross sections in ^{238}U and lighter elements. [From L. Moretto, in "Physics and Chemistry of Fission" IAEA, Vienna 1974.]

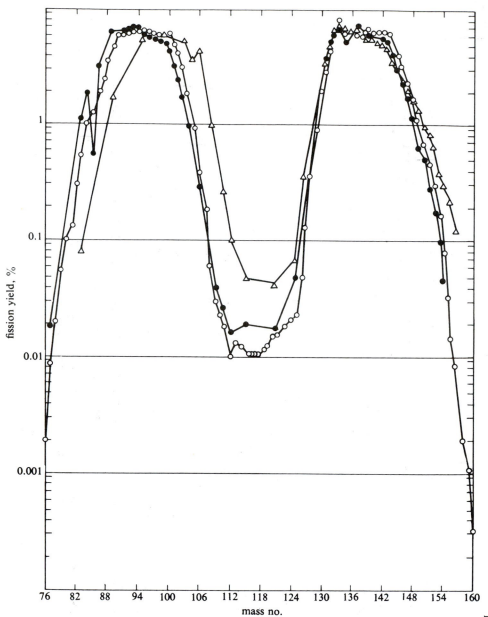

Figure   11-42   Abundance of different mass numbers in slow-neutron fission of ^{233}U (closed circles), ^{235}U (open circles), and ^{239}Pu (triangles). [(WW 58).]

ISBN 0-8053-8061-7

tends to occur with unequal fragments. The fragments have an excess of neutrons and revert to stability by a succession of beta decays, forming chains of isobars decaying into each other. As an example, we shall mention among the heavy fragments the famous chain of mass 140,

$$\underset{54}{}Xe \xrightarrow{(16s)} \underset{55}{}Cs \xrightarrow{(66s)} \underset{56}{}Ba \xrightarrow{(12.8d)} \underset{57}{}La \xrightarrow{(40h)} \underset{58}{}Ce \text{ (stable)} \quad (11\text{-}10.4)$$

in which fission was discovered. Among the light fragments, the chain of mass 99

$$\underset{41}{}Nb \xrightarrow{(2.4m)} \underset{42}{}Mo \xrightarrow{(67h)} \underset{43}{}Tc \xrightarrow{(2.12 \times 10^5 y)} \underset{44}{}Ru \text{ (stable)} \quad (11\text{-}10.5)$$

is an important source of the artificial element technetium. Both these chains are among the most probable modes of fission for ^{235}U bombarded by slow neutrons. Each occurs in about 6% of the fissions.

The asymmetry of fission (i.e., the difference in A of the fragments) has been the object of many investigations, but a completely satisfactory explanation has not yet been established.

Thus far we have considered small deformations of spherical nuclei. In reality the situation is more complicated. Often the ground state of the fissioning nucleus is deformed and gives rise to an average nonspherical potential. In this nonspherical potential one must further consider the effect of single-particle orbits. On deformation of the nucleus, the position of the levels moves, as shown, for example, in Fig. 6-42.

A given total number of neutrons (and protons) in a nucleus occupies the lowest possible orbits. Once the nucleons have been allocated, if there is an exceptionally large gap to the next orbit, we have a "closed shell" and a magic number. Because the levels move with the nuclear deformation, the magic numbers for a deformed nucleus may not be the same as for a spherical nucleus. The total energy of the nucleus varies with deformation for two types of reasons: the surface and electrostatic effects mentioned above give an average variation, a smooth function of A; in addition, the shell effects produce fluctuations (of the order of 1 MeV) around the smooth value (see Fig. 6-15). The energy is thus calculated as a sum of a smooth function of A and Z, which gives the main contribution, plus a smaller part containing the shell and other effects (Strutinski; Myers; Swiatecki).

Furthermore, the deformation must be described in more detail than a single parameter permits. The nuclear shapes for large deformations vary as shown in Fig. 11-43 and the description of these shapes pertaining to the elongation, the shape of the neck, possible asymmetries, etc. requires several parameters. For a qualitative understanding we restrict ourselves to three, one (c) giving the elongation, another (h) giving the neck, or neck radius, and a

ISBN 0-8053-8601-7

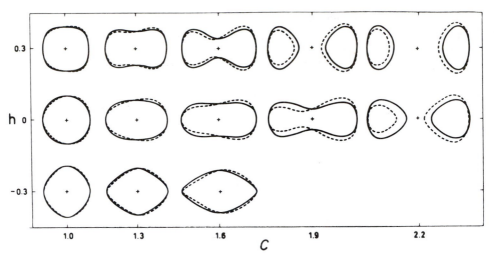

Figure 11-43 Some shapes in the $\{c, h\}$ parameterization. The solid lines show symmetric shapes ($\alpha = 0$); the dotted lines are shapes with an asymmetry parameter $\alpha = 0.2$. [From M. Brack et al., *Rev. Mod. Phys.*, **44**, 320 (1972).]

third (α) giving the asymmetry in the z direction. Figure 11-44 gives the total energy as a function of c and h and as a function of c and α for $h = 0$. Figure 11-45 shows surprising results: there are two minima as a function of h, and asymmetric shapes have less energy than symmetric ones. This gives some leads toward the solution of the puzzle of the asymmetry of fission near threshold, even if it falls short of a complete explanation. As we give more energy to the nucleus, the first occasion of fissioning appears in asymmetric configurations. On the other hand, at high energy the details of the potential surface become unimportant and fission tends to become symmetric. (See Fig. 11-46.)

The double humped barrier has more remarkable effects. There are three types of fission: (1) from states A in the continuum above the threshold, with decay times of the order of 10^{-16} to 10^{-20} sec; (2) from states B where only the second hump of the barrier is effective. These are the so-called fission isomers with decay times ranging from 10^{-2} (^{242}Am) to 10^{-9} (^{240}Pu) sec. About 30 nuclides exhibit this special form of isomerism, which was discovered in 1962 by S. M. Polikanov and co-workers. (3) Finally, spontaneous fission from states C is a much slower process, hindered by both humps of the barrier; its half-life ranges from hours to 10^9 yr or more. The isomeric states in Valley II are reached by (n, γ), $(\alpha, 2n)$, and other reactions. The passage from a state II to a state I by gamma transition is hindered because the nuclei in states II are much more elongated than those in states I and the Franck–Condon principle says that such transitions are improbable and compete relatively unfavorably with isomeric fission.

ISBN 0-8053-8601-7

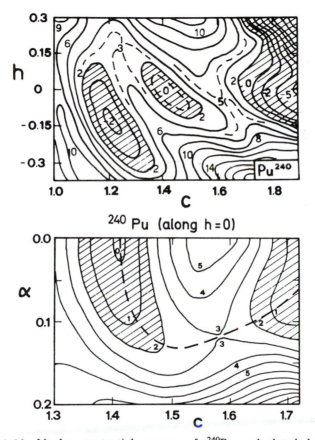

Figure 11-44 Nuclear potential energy of ^{240}Pu, calculated by Pauli, Ledergerber, and Brack. Top: contours of constant potential energy as functions of the fission coordinate c and necking coordinate h (interval between contours is variable). Bottom: potential-energy contours shown as functions of c and the mass-asymmetry coordinate α (interval between contours 1 MeV). The contours are labeled by the energy (in MeV) relative to the spherical liquid-drop energy. Regions where the potential energy is less than 2 MeV are hatched. The dashed curve in the lower part of the figure shows the location of the minimum in the potential energy with respect to α for constant values of c. The inclusion of mass-asymmetric deformations lowers the potential energy of the second saddle point by about 2.5 MeV. [From J. R. Nix, *Ann. Rev. Nucl. Sci.*, **22**, 65 (1972).]

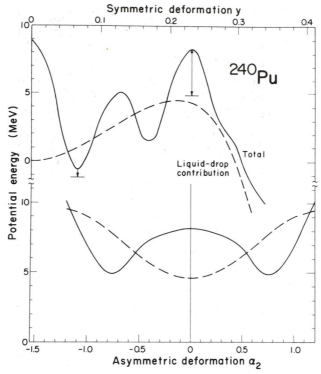

Figure 11-45 Nuclear potential energy of ^{240}Pu. Upper solid curve: dependence of the potential energy on the nuclear elongation; lower solid curve: its dependence on the mass-asymmetry coordinate

$$\alpha_2 = \frac{a_1^2 - a_2^2}{\frac{1}{2}\left(a_1^2 + a_2^2\right)}$$

where a_1 and a_2 are the transverse semiaxes of the two end spheroids forming the shape. Arrow at the ground-state minimum indicates the amount the potential energy is lowered when spheroidal deformations are included. At the second peak the solid point indicates the amount the potential energy is lowered when the remaining two symmetric coordinates are included, and the arrow the further lowering of 3.2 MeV associated with the asymmetric coordinate α_2. In the lower part of the figure, the three symmetric coordinates are held fixed at their values corresponding to the second symmetric saddle point (whose energy is given by the solid point). The dashed curves give the liquid-drop contributions to the total potential energies; all energies are relative to the spherical liquid-drop energy. [From J. R. Nix, *Ann. Rev. Nucl. Sci.*, **22**, 65, (1972).]

ISBN 0-8053-8601-7

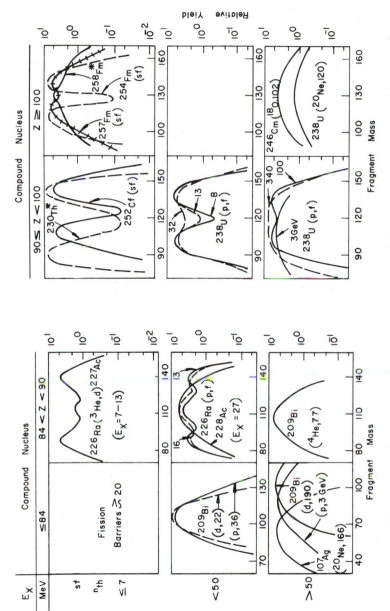

Figure 11-46   Summary of mass distributions from various types of fission. Target nucleus is given except for thermal-neutron capture where the compound nucleus is denoted by*. Projectile and energy in MeV (except as otherwise stated) are shown in parentheses. Yields in percentage. [From D. C. and M. M. Hoffman, *Ann. Rev. Nucl. Sci.*, **24**, 151, (1974).]

ISBN 0-8053-8601-7

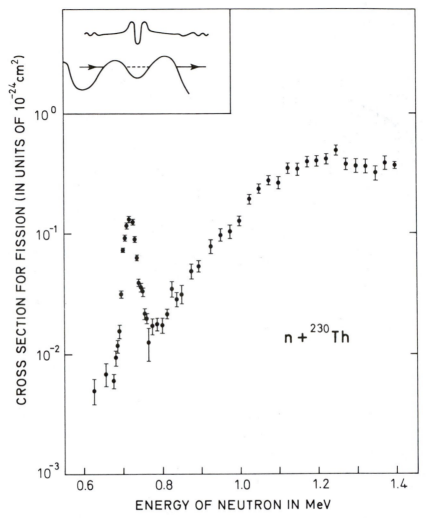

Figure 11-47   Resonance in threshold region for neutron induced fission of ^{231}Th. Shown is the cross section for fission as a function of the neutron energy. The insert illustrates the interpretation of the resonance phenomenon in terms of a semistationary state of vibration in the second minimum of the potential energy function. [From A. Bohr and B. Mottelson, *Ann. Rev. Nucl. Sci.*, **23**, 363 (1973).]

ISBN 0-8053-8601-7

Figure 11-47 shows a direct confirmation of this picture. Neutron bombardment of ^{230}Th produces fission. The cross section has, at about 720 keV, a resonance 40 keV wide. It corresponds to the formation of a quasi-stationary vibrational level of ^{231}Th associated with the second minimum of the potential energy.

In ^{240}Pu the (n, f) reaction as a function of energy shows the resonances given in Fig. 11-48. Between the (n, f) resonances there are many other (n, γ) resonances, corresponding to the level density at the energy of excitation of ^{241}Pu (not shown in the figure). The fissioning levels show a gross structure with an average separation of several hundred electron volts and a fine structure for each level of about 10 eV. This peculiar state of affairs is explained by assuming that the neutrons impinge on ^{240}Pu in the ground state configuration corresponding to Valley I and form excited states of ^{241}Pu which, below the fast fission threshold (E_f), decay by neutron or gamma emission. However, some of these states coincide in energy with the states corresponding to Valley II. In these cases the nucleus goes easily to Valley II, from which it then fissions. The levels in Valley II are spaced farther apart from each other than those of Valley I because they are lower with respect to the bottom of the valley. The gross spacing in the resonances corresponds thus to the level spacing to Valley II. These levels are also broad because fission starting from them is very rapid and the levels contain in their width several levels of Valley I. This gives rise to the structure.

An impressive confirmation of the double hump model is the observation (by H. J. Specht et al. in 1972) of rotational levels in ^{240}Pu corresponding to two moments of inertia, one for each deformation. One band is the normal

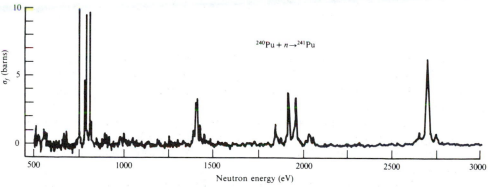

Figure 11-48 Cross section for the (n, f) reaction on ^{240}Pu. Resonance structure arising from a two-peaked fission barrier. The resonance groups, spaced at about 630 eV, occur when the excitation energy is close to a compound-nucleus level corresponding to a deformed nucleus, in the second well. The structure of the resonances (15 eV) corresponds to the distance between levels of the compound nucleus in the first well. [From E. Migneco and J. P. Theobald, *Nucl. Phys.*, **A112**, 603 (1968).]

ISBN 0-8053-8601-7

rotational band corresponding to the ground state of ^{240}Pu; the other obtains when the transition emitting the gamma rays occurs in the metastable nucleus deformed to the second minimum configuration. The time for the gamma emission in this second minimum is of the order of 10^{-11} sec and it is followed by fission with a lifetime of 4 nsec.

Refined studies of fission are relevant in the search for superheavy elements ($Z \cong 114$), where spontaneous fission is an important mode of decay. Such studies hint at the possibility of a zone of nuclear stability for $Z \cong 114$, $N \cong 178$.

We have already mentioned the difference in the distribution of A for fission fragments originating in fast or slow neutron fission. A considerable variety of distributions is obtained, as illustrated in Fig. 11-46.

Fission is accompanied by an instantaneous emission of neutrons, which are of decisive importance for a chain reaction. The average number of neutrons emitted per fission is call ν. It varies according to the excitation energy of the fissioning nucleus because each evaporation neutron carries away a certain amount of energy and enough must be left to permit the evaporation of the next neutron. The dependence of ν on the fissioning nucleus is primarily caused by its excitation energy on neutron capture. Approximate formulas for ν in the case of the important target nuclei ^{235}U and ^{239}Pu are $\nu = 2.432 + 0.066 E_n$ ($0 \leqslant E_n \leqslant 1$) and $\nu = 2.874 + 0.138 E_n$, respectively. The probability of the emission of n neutrons is given in a few examples in Fig. 11-49. The energy spectrum of fission neutrons is represented by the semi-empirical equation

$$f(E) = 2 \times 0.775 \left(\frac{0.775 E}{\pi} \right)^{1/2} e^{-0.775E} \qquad (11\text{-}10.6)$$

where $f(E)\,dE$ is the probability that a neutron has its energy in interval dE and E is measured in MeV (Fig. 11-50). It can be interpreted as an evaporation spectrum from the moving fragments, an interpretation confirmed by the study of the angular correlation of the direction of motion of the neutrons and of the fission fragments, at least for fission produced by slow neutrons. The light and heavy fragments emit neutrons and each mass has its own ν_L or ν_H, according to whether it is a light or heavy fragment. Figure 11-51 shows this correlation.

In addition to the neutrons promptly emitted in fission there are delayed neutrons associated only indirectly with the fission process. They are mentioned here for completeness. A fission fragment of atomic number Z, with a large neutron excess, may beta-decay to an excited state of an isobar of atomic number $Z + 1$. This state may be above the binding energy of the neutron and hence susceptible to emitting a neutron instantaneously. However, the emission of the neutron follows the beta decay of the preceding

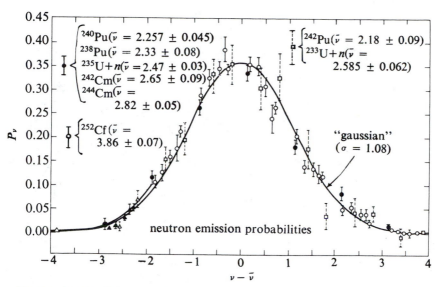

Figure   11-49   Experimental number of neutrons emitted per fission plotted to show the probability of emission of ν neutrons versus $\nu - \langle\nu\rangle$. [From J. Terrell, *Phys. Rev.*, **108**, 783 (1957).]

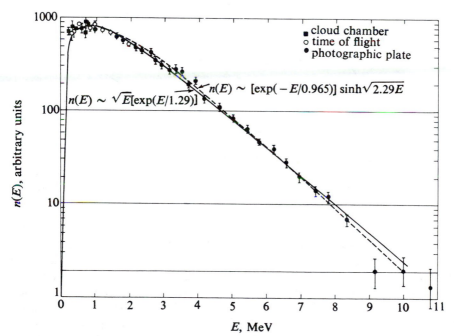

Figure   11-50   Fission neutron spectra from ^{235}U. [From (WW 58), after Leachman.]

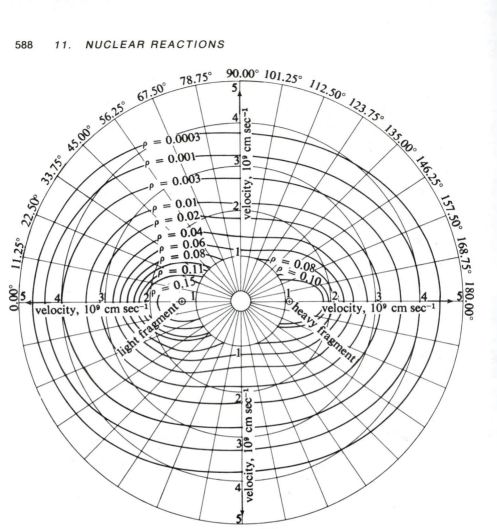

Figure   11-51 Spontaneous fission of ^{252}Cf. Contour diagram in polar
coordinates of observed neutron density distribution $\rho(V, \theta)$ as a function of
neutron velocity and angle for average asymmetry of mass division ($\langle V_L \rangle =$
1.37 cm nsec^{-1} $\langle V_H \rangle = 1.04$ cm nsec^{-1}, $\langle M_L \rangle = 108.5$, $\langle M_H \rangle = 143.5$). [From
H. R. Bowman, J. C. D. Milton, S. G. Thompson, and W. J. Swiatecki, *Phys.
Rev.*, **129**, 2133 (1963).]

nucleus. For this reason neutrons appear to be emitted according to an exponential decay law with a decay constant corresponding to the beta decay of
the preceding nucleus. Delayed neutrons are associated with the fission fragments and have the abundances indicated in Table 11-2. Outside the fission
fragments the same mechanism of emission of delayed neutrons is operative
in ^{17}N, which decays by beta emission into an excited ^{17}O*. This last nucleus
transforms instantaneously into ^{16}O by losing a neutron. A level scheme
showing the situation in two cases is reproduced in Fig. 11-52. A similar

ISBN 0-8053-8061-7

TABLE 11-2   DELAYED NEUTRONS[a]

Precursor	$T_{1/2}$ sec	E keV	Yield per 100 fissions		
			^{235}U	^{238}U	^{239}Pu
^{87}Br	54.5	250	0.052	0.054	0.021
^{137}I ^{88}Br	24.4 } 16.3 }	560	0.346	0.564	0.182
^{138}I ^{89}Br ^{93,94}Rb	6.3 4.4 } ~6 }	405	0.310	0.667	0.129
^{139}I Cs, Sb, Te ^{90,92}Br ^{93}Kr	2.0 1.6–2.4 { 1.6 1.5 }	450	0.624	1.60	0.199
^{140}I + Kr	0.5		0.182	0.93	0.052
Br, Rb, As	0.2		0.066	0.31	0.027
			1.58	4.12	0.61

[a] Adapted from G. R. Keepin, *Physics of Nuclear Kinetics,* Addison-Wesley, Reading, Mass., 1965.

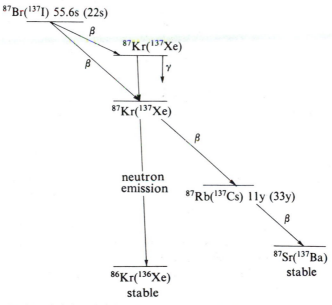

Figure 11-52 Origin of delayed neutrons in the fission chains of mass 87 and 137.

mechanism in the case of neutron-deficient isotopes leads to the emission of delayed protons. Thus, for instance, ^{25}Si by positron decay goes to an excited state of ^{25}Al which emits protons.

Occasionally fission is accompanied by the emission of long-range alpha particles (up to 20 MeV). In ^{235}U this occurs about once for every 500 fissions. The alpha particles show a definite tendency to escape in a direction perpendicular to that of the fragments (Fig. 11-53). Their energy spectrum is approximately Gaussian, with a maximum at 15 MeV and extending up to 25.

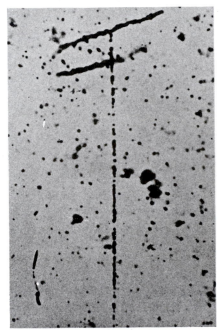

Figure   11-53   Fission tracks with a secondary alpha particle. [K. T. Titterton.]

Hundreds of times more rarely p, d, or t accompany fission, and occasionally even heavier fragments, such as ^{7}Li, ^{8}Li, ^{9}Li, ^{9}Be, and ^{10}Be appear.

Finally an emission of "prompt" gamma rays occurs at the moment of fission. The prompt gamma rays are emitted within 10^{-14} sec of fission, and thus mostly after neutron emission, as in all nuclear reactions. The gamma rays show a continuous spectrum with an average energy of 1 MeV; their total energy is about 7 MeV per fission. An empirical formula valid for $0.3 < E_\gamma < 7$ MeV for the spectrum is

$$dN(E_\gamma)/dE_\gamma = 6.7 \exp(-1.05E_\gamma) + 30 \exp(-38E_\gamma) \quad (11\text{-}10.7)$$

ISBN 0-8053-8001-7

It gives the number of photons per fission in an interval of 1 MeV around energy E_γ. The formula is approximately valid for ^{236}U, ^{240}Pu, ^{234}Th, and ^{252}Cf fission.

Table 11-4 gives the energy balance for an average fission produced in ^{235}U or ^{239}Pu by slow neutrons.

TABLE 11-3   ENERGY BALANCE FOR AN AVERAGE FISSION[a] (in MeV)

	^{235}U	^{239}Pu
Light fragments	99.8	101.8
Heavy fragments	68.4	73.2
Fission neutrons	4.8	5.8
Prompt gamma	7.5	~7
β of fission product	7.8	~8
γ of fission product	6.8	~6.2
Neutrinos (undetectable)	(~12)	(12)
Total detectable energy	195	202

[a] From G. R. Keepin, *Physics of Nuclear Kinetics,* Addison-Wesley, Reading, Mass., 1965.

Fission may be induced by many particles, and in general by any agent that can supply the activation energy to a heavy nucleus. In the practically important case of slow neutron-induced fission, we have for the fission cross-section formulas of the Breit–Wigner type:

$$\sigma_f = \pi \lambdabar^2 \frac{\Gamma_f \Gamma_n}{(E - E_r)^2 + \Gamma^2/4} \tag{11-10.8}$$

where Γ_f is the partial width for fission and $\Gamma = \Gamma_\gamma + \Gamma_f + \Gamma_n$. By direct measurement of the capture and fission cross sections, it is possible to obtain the practically important quantity,

$$\alpha = \frac{\sigma(n, \gamma)}{\sigma(n, f)} \tag{11-10.9}$$

In the immediate vicinity of a resonance this is the same as Γ_γ/Γ_f. Different resonances have different Γ_f, Γ_γ, and Γ_n; as an example, we give some data relative to ^{235}U in Fig. 11-54. The great complexity of the overlapping resonances is apparent. The value of Γ_f gives the decay constant for fission directly. It is through the measurement of the fission width that the value of 10^{-14} sec for the order of magnitude of the duration of the fission process was obtained.

We mention a few more miscellaneous items on fission. In neutron-produced fission, excitation curves such as those of Fig. 11-55 are obtained

ISBN 0-8053-8061-7

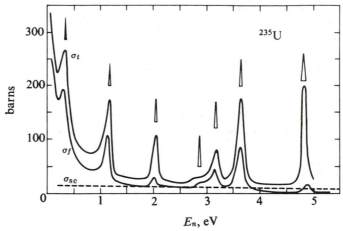

Figure   11-54   Observed total fission and scattering cross sections of ^{235}U from 1.0 to 5 eV. The fission cross section was normalized to 500 at 0.0253 eV. [From F. J. Shore and V. L. Sailor, *Proc. Intern. Conf. Peaceful Uses Atomic Energy*, Geneva, 1958.]

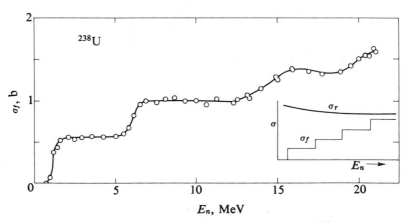

Figure   11-55   The fast-neutron fission cross section in ^{238}U according to Smith, Henkel, and Nobles. The inset is an idealization to show how the fission cross section σ_f increases suddenly at definite thresholds. By comparing σ_f to the total reaction cross section σ_r one can, in principle, determine how fissionability varies with excitation energy. [From I. Halpern, *Ann. Rev. Nucl. Sci.*, **9**, 245 (1959).]

ISBN 0-8053-8061-7

on increasing the energy of the neutrons. The striking steps observed originate at the energies at which new processes leading to fission become possible. Thus the first step is due to ^{238}U $(n, \gamma)^{239}$U when the neutrons have enough energy to leave ^{239}U with sufficient excitation to fission. The second step occurs when the neutrons have enough energy to excite ^{238}U by an (n, n') reaction to undergo fission. The next step occurs when the $(n, 2n)$ reaction leaves ^{237}U with enough energy to undergo fission, and so on.

With projectiles of sufficiently high energy it is possible to fission nuclei as light as Cu. Fission then competes with the process of spallation, the disintegration of the nucleus into a large number of free nucleons or very small fragments such as alpha particles, deuterons, etc.

When fission is induced by neutrons there are correlations between the direction of motion of the projectile and of the fission fragments (see Fig. 11-56) depending on the energy of the neutrons. For fission produced by

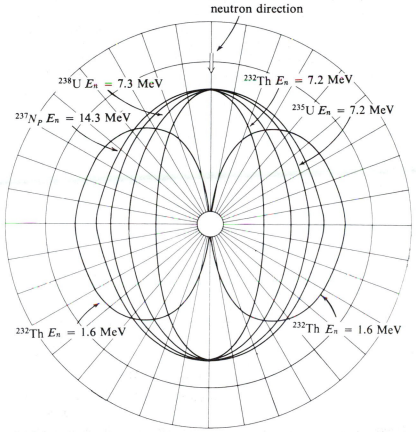

Figure   11-56   Angular distribution of fission fragments. [From R. L. Henkel and J. E. Brolley, *Phys. Rev.*, **103**, 1292 (1956).]

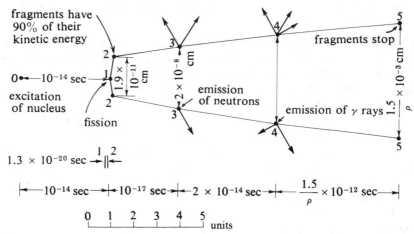

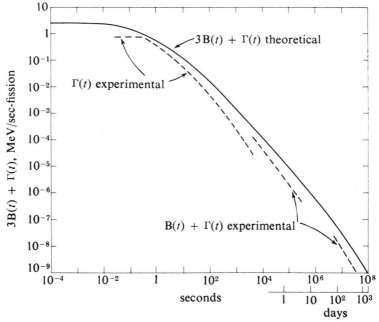

Figure 11-57 Graphic representation of the fission process. The events are: 0, excitation of nucleus; 1, fission; 2, fragments acquire 90% of their kinetic energy; 3, emission of neutrons; 4, emission of gamma rays; 5, fragments stop. The horizontal scale indicates the duration of the process: two events separated by the time t (sec.) are $20 + \log_{10} t$ units apart horizontally. The vertical scale indicates the distance of the fragments from each other: a distance r (cm) is indicated by $13 + \log_{10} r$ units; ρ is the density of the material in which the fragments travel, and the diagram is drawn for $\rho = 1$. [From (WW 58).]

Figure 11-58 Energy release from fission products. The neutrino energy is assumed to be twice the beta energy. $B(t)$ and $\Gamma(t)$ energy released as beta and gamma radiation, respectively. [From (WW 58).]

ISBN 0-8053-8061-7

gamma rays the directions of the gamma ray and of the fission fragment tend to be perpendicular. A graphic summary of the time sequence in the fission process is given in Fig. 11-57.

The total activity of all fission products is of interest in many practical problems. There are empirical formulas that give the beta-decay activity and the gamma activity of all fission products. If fission has occurred at time 0, the energy emitted as beta and gamma rays is given respectively by $B(t) = 1.26t^{-1.2}$ MeV sec^{-1} and $\Gamma(t) = 1.40t^{-1.2}$ MeV sec^{-1}, where $10^5 > t > 1$ sec (Fig. 11-58).

For an evaluation of the radiations emitted by the fission products long after fission, as required for studies of waste disposal, one must analyze the yield and radiations of the not many long life isotopes produced in fission and consider each of them separately.

11-11 HEAVY-ION NUCLEAR REACTIONS

The field of heavy-ion reactions is developing very rapidly, fostered by increasing experimental facilities. These include linear accelerators, cyclotrons, and even bevatrons that can give ions from a few MeV to about 2 GeV per nucleon. The ions accelerated range or will range all the way to uranium. For the detection, at relatively low energy, a combination of measurements of dE/dx and E obtained in a gas counter and a solid-state counter may give Z, A, and E of a product.

At sufficiently low energy the ions undergo Coulomb repulsion and thus show only Rutherford scattering. If, however, they touch the target nuclei, they react. Now, calling b the impact parameter and p the relative momentum, we have, by conservation of angular momentum,

$$pb = l\hbar = p'X \tag{11-11.1}$$

where $l\hbar$ is the angular momentum, X the minimum distance between the ions, and p' their relative momentum at that distance. We have also, by conservation of energy,

$$\frac{p^2}{2m} - E_B \frac{R}{X} = \frac{p'^2}{2m} \tag{11-11.2}$$

where m is the reduced mass and $E_B = Zze^2/R$ is the potential energy when the nuclei are in contact. From Eqs. (11-11.1) and (11-11.2), calling $p^2/2m = E$, we have as a condition for nuclear contact

$$2mR^2(E - E_B) \geq l_c^2\hbar^2 \tag{11-11.3}$$

The elastic scattering has been calculated (Blair, 1954) by taking the Coulomb scattering amplitude and subtracting from it all the partial waves

ISBN 0-8053-8061-7

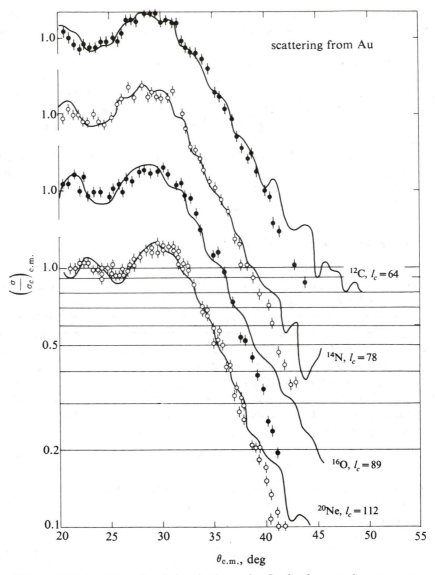

Figure 11-59 The ratio of the elastic to the Coulomb scattering cross section for ^{12}C, ^{14}N, ^{16}O, and ^{20}Ne scattered from gold at an energy of about 10 MeV per nucleon. The solid lines are the sharp cutoff calculations with the appropriate l_c indicated in the figure. [Data of Reynolds, Goldberg, and Kerlee, from A. Zucker, *Ann. Rev. Nucl. Sci.*, **10**, 27 (1960).]

ISBN 0-8053-8601-7

with $l < l_c$, assuming that these are absorbed in the collision. The only parameter available in this calculation is R, written in the form

$$R = r_0\left(A_1^{1/3} + A_2^{1/3}\right) \qquad (11\text{-}11.4)$$

where A_1, A_2 are the mass numbers of the two nuclei. Such a simple semiclassical method gives good results, as can be seen from Fig. 11-59. The value of r_0 thus obtained is near 1.45 F. Note the high values of l_c.

In distant collisions, where nuclear forces do not play a role, one often obtains nuclear excitation by the electromagnetic interaction (Coulomb excitation; cf. Sec. 8-7). In the case of heavy ions the Weizsäcker–Williams

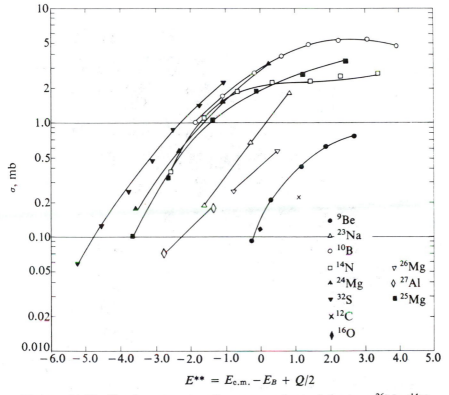

Figure 11-60 Total neutron-transfer cross sections of the type ^{26}Mg (^{14}N, ^{13}N) ^{27}Mg for 11 light elements plotted as a function of $E^{**} = E_{\text{c.m.}} - E_B + (Q/2)$, where $E_{\text{c.m.}}$ is the incident kinetic energy in the center-of-mass system, E_B is the barrier energy, and Q is the reaction energy. Note that $E^{**} + (Q/2)$ is the energy available to the two nuclei at the moment immediately after the transfer has occurred. [From D. E. Fisher, A. Zucker, and A. Gropp, *Phys. Rev.*, **113**, 542 (1959).]

method gives a good picture of the phenomenon. A characteristic effect occurring among the heavy ions is the excitation of high rotational levels. In even-even heavy nuclei, such as ^{238}U, highly developed rotational bands have been found, with only even levels $I = 0, 2, 4, 6, \ldots, 12$ (Fig. 6-48). All these have been excited by Coulomb interaction. The excitation process is multiple. The nucleus is excited by quadrupole transitions in successive jumps produced by the electromagnetic field of the *same* projectile.

What happens when the colliding nuclei touch each other? There are frequent simple reactions corresponding to the transfer of one nucleon, for

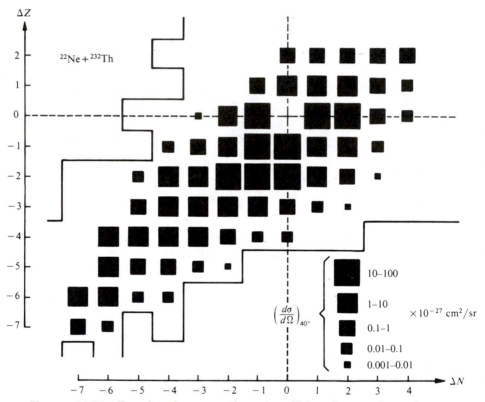

Figure   11-61   Transfer of many nucleons in collisions between heavy ions. Shown are the intensities of the reactions involving the transfer of ΔN neutrons and ΔZ protons from the target nucleus $^{232}_{90}$Th to the impinging $^{22}_{10}$Ne nucleus. The estimated limits of particle stability are indicated by the solid lines. The most probable transfers lead to an increase of the mass-to-charge ratio for the projectile, which can be understood in terms of a tendency toward a more uniform distribution of the neutron excess during the collision. [From A. Bohr and B. Mottelson, *Ann. Rev. Nucl. Sci* **23**, 363 (1973).]

ISBN 0-8053-8601-7

instance

$$^{26}\text{Mg}(^{14}\text{N}, \,^{13}\text{N})^{27}\text{Mg} \tag{11-11.5}$$

Such reactions occur in peripheral collisions without formation of a compound nucleus, with cross sections up to approximately 5 mb (Fig. 11-60). The bombardment of a heavy nucleus with another can produce many new nuclides, depending on how peripheral the collision is. As an example we give in Fig. 11-61 the results of a Ne bombardment of Th.

On formation of a compound nucleus, one observes a peculiar effect in the angular distribution of the evaporation particles. The compound nucleus frequently has very high angular momentum ($> 50\hbar$), necessarily directed perpendicular to the trajectory of the incoming particle. The evaporating particles then escape in a plane perpendicular to the angular momentum and, with equal probability, at an angle θ with the line of impact. As we rotate this

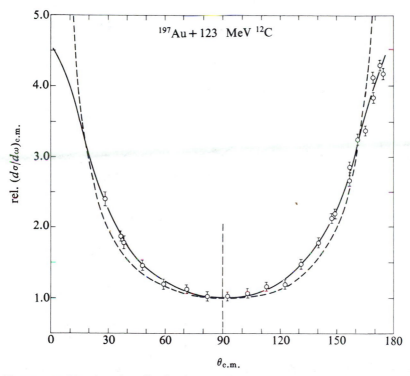

Figure 11-62 Angular distribution of fission fragments in the center-of-mass system for fission of ^{197}Au induced by ^{12}C ions of 123-MeV energy. Solid line, experimental curve; broken line, $1/\sin \theta$. [G. E. Gordon, A. E. Larsh, and T. Sikkeland, from E. Hyde, UCRL 9065.]

plane of escape of the particles, it is apparent that the angular distribution will have a $1/\sin\theta$ shape peaked in the forward and backward direction with respect to the line of impact. This has been observed in the alpha particles emitted by Ni bombarded by 160-MeV oxygen ions.

Similar effects appear in fission, where the fission fragments tend to escape in the line of collision for fission induced in ^{197}Au by 123-MeV ^{12}C ions (Fig. 11-62).

In bombardments of nuclei such as ^{107}Ag with ^{40}Ar of about 7 MeV per nucleon, one observes formation of fragments ranging from $A = 1$ to the sum of the two nuclei. In a good fraction of the cases the energy of the escaping fragment, say oxygen, in the c.m. system corresponds to the electrostatic energy of an oxygen nucleus in contact with one containing all the remaining nucleons. The initial kinetic energy is entirely or almost entirely distributed to many degrees of freedom (i.e., practically thermalized). We could think of

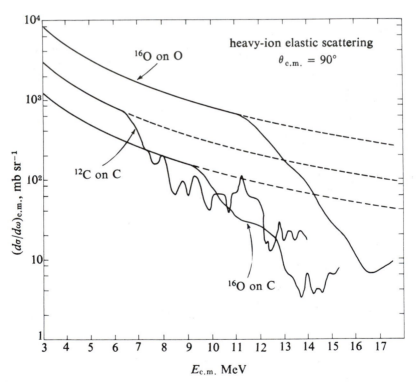

Figure 11-63 Differential cross sections at 90 deg c.m. for the elastic scattering of ^{16}O from oxygen, ^{12}C from carbon, and ^{16}O from carbon as a function of energy. The dashed lines are Rutherford scattering cross sections. [D. A. Bromley, J. A. Kuehner, and E. Almqvist, *Phys. Rev. Letters*, **4**, 365 (1960).]

ISBN 0-8053-8601-7

$^{40}_{18}Ar + ^{107}_{47}Ag$ combining to give a highly excited $^{147}_{65}Tb$ nucleus, which undergoes fission into $^{16}_{8}O$ and $^{131}_{57}La$. The nuclear matter in $^{147}_{65}Tb$ reaches thermal equilibrium before fission.

The angular distribution of the fragments in the c.m. system indicates a time for coming apart that is short with respect to the period of rotation of a compound nucleus which may have $I = 60\hbar$.

Another remarkable type of reaction has been observed in ^{12}C–^{12}C collisions at 90 deg (Bromley, Kuehner, and Almqvist, 1960). The cross section shows rapid fluctuations as a function of energy, and this has been attributed to the formation of a ^{12}C–$^{12}C(^{24}Mg)$ nucleus, strongly elongated and having rotational levels like a diatomic molecule (Fig. 11-63).

11-12 PHOTONUCLEAR REACTIONS

Gamma rays produce many reactions, for example, (γ, n), (γ, p), (γ, f), (γ, α), etc. These are observed by directing a beam of gamma rays at the target and detecting the emitted particles or the radioactivity of the product nucleus. Monochromatic sources of gamma rays are available from the rays emitted in some disintegrations, for instance, ThC′, ^{24}Na, and ^{60}Co, which however, are in the 1- to 3-MeV energy region. The $p + Li$ reaction gives a gamma ray of 17 MeV, which has been used extensively. Positron annihilation in flight is also a good gamma source. Electron accelerators give a continuum of gamma rays from the bremsstrahlung (see Chap. 2). The upper limit of the continuum is well determined, and it is possible to measure cross sections for photoreactions as a function of energy by using a bremsstrahlung spectrum and varying the energy of the electrons that produce it. However, this procedure is difficult and laborious.

The photodisintegration of the deuteron is the simplest photo-nuclear reaction. It has been treated in Chap. 10 for low energy region, for which a simple theory is adequate. Figure 11-64 shows an overall view of the photon absorption of an idealized nucleus.

A remarkable phenomenon in photonuclear absorption is the maximum shown by the cross section at energies around 20 MeV (Figs. 11-65 and 11-66). The shape of the curves and the positions of the maximum depend on A. A simple model accounting for this "giant resonance" has been proposed by Goldhaber and Teller. They consider a collective motion of all protons with respect to all neutrons in a nucleus. This motion gives rise to an electric dipole moment, which should account for the absorption. The frequency of the motion has been estimated by Jensen and others by means of a model in which neutrons and protons move in a fixed sphere in such a way that nuclear matter changes its composition from point to point, although the density of nucleons remains constant. The change of composition causes an increase in potential energy that is estimated from the nuclear-mass formula [Eq. (6-7.7)]. This energy produces a restoring force, and the consequent

ISBN 0-8053-8601-7

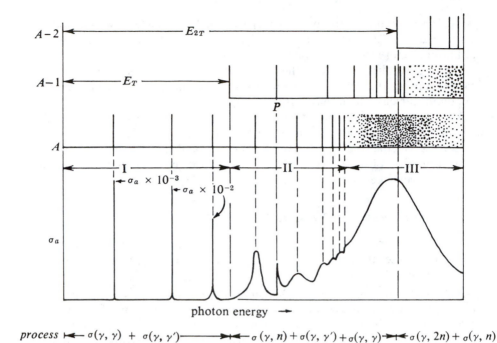

Figure   11-64   The photon absorption cross section for an idealized nucleus. Region I is that part of the energy scale below the particle thresholds where the absorption is into discrete energy levels. Region II is the energy range above the binding energy where structure may still exist in the absorption cross section. In region III the absorption cross section is smooth. The processes that can take place are indicated along the abscissa. The energy levels in the nuclei A, $A - 1$, and $A - 2$ are illustrated at the top of the diagram. The binding energies for one and two particles are designated by E_T and E_{2T}. The level P_1 in $A - 1$ represents a parent of the ground state of nucleus A. [From E. G. Fuller and E. Hayward, in (En 62).]

vibrations may be calculated hydrodynamically, yielding approximately the right frequency for light nuclei.

A theorem that is valid in the absence of exchange forces gives, for the integrated dipole photoabsorption cross section, the sum rule

$$\int_0^\infty \sigma_{\text{abs}}(E_\gamma)\, dE_\gamma = 2\pi^2 e^2\, \frac{\hbar}{mc}\, \frac{NZ}{A} = 0.058\, \frac{NZ}{A} \quad \text{MeV b} \quad (11\text{-}12.1)$$

where N, Z, and A are the number of neutrons, number of protons, and mass number of the nucleus, respectively; m is the mass of a nucleon. The giant resonance accounts for most of the integral, the high-energy cross section

ISBN 0-8053-8601-7

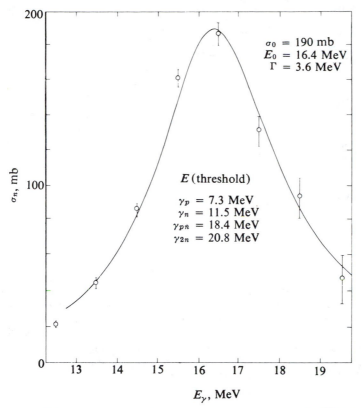

Figure 11-65 Cross sections for photoneutron production in ^{89}Y showing giant resonance. [P. F. Yergin and B. P. Fabricand, *Phys. Rev.*, **104**, 1334 (1956).]

contributing little to it. The sum rule expressed by Eq. (11-12.1) will now be proved from general principles.

In the case of $E1$ radiation (λ of the radiation $\gg$ dimensions of the nuclear system), the transition probability from the ground level i to a level f induced by an electric field $\mathcal{E} \cos \omega t$ directed in the z direction is

$$\int_{\text{line}} |a_f(t)|^2 \, d\omega = \mathcal{E}^2 e^2 \frac{\pi}{2\hbar^2} |z_{if}|^2 t \qquad (11\text{-}12.2)$$

where z_{if} is the matrix element of the coordinate z and $|a_f(t)|^2$ is the probability of finding the nucleus in state f at time t. This relation may be obtained by time-dependent perturbation theory.

The cross section for the photon absorption is by definition the transition probability per unit time divided by the photon flux per unit surface,

ISBN 0-8053-8601-7

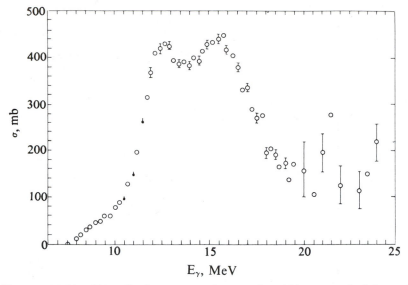

Figure   11-66   Giant dipole resonance for tantalum-181, a strongly deformed nucleus, showing two peaks. The separation of the two peaks reflects the quadrupole deformation of Ta ($Q_0 = 6.3$ b), while the relative integrated peak strengths yield a factor close to 2 in favor of the higher-energy peak which is associated with the smaller nuclear dimension, pointing to prolate deformation; axial symmetry is favored by the absence of a third peak. [E. G. Fuller and M. S. Weiss, *Phys. Rev.*, **112**, 560 (1958).]

$\mathscr{E}^2 c/8\pi\hbar\omega$. That is, limiting ourselves to a small frequency interval $\Delta\omega$ containing the absorption line of frequency

$$E_f - E_i = \hbar\omega_{fi} \tag{11-12.3}$$

$$\frac{(\mathscr{E}^2 e^2\pi/2\hbar^2)|z_{if}|^2}{(\mathscr{E}^2 c/8\pi)(1/\hbar\omega)} = \frac{4\pi^2 e^2}{\hbar c}\,\omega_{fi}|z_{if}|^2 = \langle\sigma\rangle\,\Delta\omega \tag{11-12.4}$$

Introducing the dimensionless quantity

$$f_{if} = \frac{2m}{\hbar}\,\omega_{fi}|z_{if}|^2 \tag{11-12.5}$$

called oscillator strength, we may rewrite Eq. (11-12.4) as

$$\langle\sigma\rangle\,\Delta\omega = \frac{2\pi^2 e^2}{mc}\,f_{if} \tag{11-12.6}$$

If we want the total absorption of our system, we shall integrate Eq. (11-12.6)

ISBN 0-8053-8061-7

on the left side and sum over all final states on the right:

$$\int \sigma \, d\omega = \frac{2\pi^2 e^2}{mc} \sum_f f_{if} \tag{11-12.7}$$

A remarkable theorem states that the

$$\sum_l f_{il}$$

of all oscillator strengths starting from one state i and ending in all other states is 1. Note that oscillator strengths may be positive or negative according to whether

$$E_i > E_l \quad \text{or} \quad E_i < E_l$$

This theorem is proved by starting from the fundamental commutation relation for conjugate variables,

$$pq - qp = \frac{\hbar}{i} \tag{11-12.8}$$

Writing this in matrix form for the z coordinate and its conjugate momentum $p_z = m\dot{z}$, we obtain

$$m \sum_l (\dot{z}_{jl} z_{lk} - z_{jl} \dot{z}_{lk}) = \frac{\hbar}{i} \delta_{jk} \tag{11-12.9}$$

Now, if we use a representation in which the energy is diagonal,

$$\dot{z}_{jl} = -\frac{i}{\hbar}(E_j - E_l)z_{jl} = i\omega_{lj} z_{jl} \tag{11-12.10}$$

Remembering that $z_{jl} = z_{lj}^*$ and substituting Eq. (11-12.10) into Eq. (11-12.9), we have

$$\sum_l \omega_{jl} |z_{jl}|^2 = \frac{\hbar}{2m} \tag{11-12.11}$$

Inserting Eq. (11-12.11) into the definition of f_{jl} [Eq. (11-12.5)] gives immediately

$$\sum_l f_{jl} = 1 \tag{11-12.12}$$

Equation (11-12.12) is the famous sum rule of Thomas and Kuhn (1924),

ISBN 0-8053-8061-7

which had great importance in the early development of quantum mechanics. In the present form this rule gives

$$\int \sigma \, d\omega = \frac{2\pi^2 e^2}{mc} \tag{11-12.13}$$

In order to apply this result to a nuclear system containing many nucleons, we must replace ez by

$$\sum_1^A e_i z_i$$

where e_i is the proton charge for the protons (i varying from 1 to Z) and 0 for the neutrons (i from $Z + 1$ to A). We write this sum as

$$\sum_1^A e_i z_i = e \left\{ \sum_1^Z z_i + \frac{Z}{A} \sum_{Z+1}^A z_i - \frac{Z}{A} \sum_{Z+1}^A z_i \right\}$$

$$= e \left\{ \frac{Z}{A} \sum_1^Z z_i + \frac{N}{A} \sum_1^Z z_i + \frac{Z}{A} \sum_{Z+1}^A z_i - \frac{Z}{A} \sum_{Z+1}^A z_i \right\} \tag{11-12.14}$$

The sum of the first and third terms on the right is the z of the center of mass of the system, $\langle z \rangle$, times the atomic number. We can thus write

$$\sum_1^A e_i z_i = e_p' \sum_1^Z z_i + e_n' \sum_{Z+1}^A z_i + eZ\langle z \rangle \tag{11-12.15}$$

where $e_p' = eN/A$, $e_n' = -eZ/A$ are called effective charges of the protons and neutrons, respectively.

Under the action of the external field the charges will move, and we may consider the motion as the sum of the motion of the center of mass, combined with a motion relative to the center of mass. It is this second motion that is important for nuclear reactions. The first motion gives rise to Thomson scattering of the nucleus as a whole. Applying the sum rule to the internal motion only, we have

$$\int \sigma \, d\omega = \frac{2\pi^2}{c} \frac{1}{m} \left(Ze_p'^2 + Ne_n'^2 \right) = \frac{2\pi^2 e^2}{mc} \frac{ZN}{A} \tag{11-12.16}$$

which is Eq. (11-12.1).

Exchange forces increase the value of the integral. More refined sum rules have been developed, and they are useful in analyzing reactions started by photons. Equation (11-12.16), when compared with the experimental results,

ISBN 0-8053-8061-7

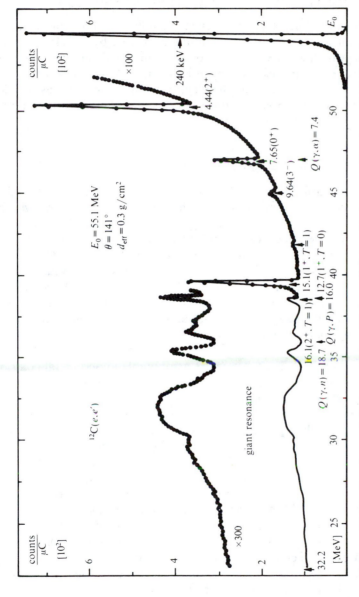

Figure 11-67 Spectrum of electrons scattered from a ^{12}C (graphite) target. For the peaks below the giant resonance region, excitation energy (MeV), spin, and parity of the respective ^{12}C levels are given. The three lowest particle thresholds are indicated. [From H. Theissen, *Springer Tracts in Modern Physics*, Vol. 65, p. 7 (1972).]

shows that the observed cross section in the giant resonance exhausts the integral. There is also evidence from the angular distribution of the photoneutrons and photoprotons that the transitions involved in the giant resonance are indeed electric dipole transitions (Figs. 11-64 and 11-65).

Electron bombardments produce effects similar to those produced by gamma rays, not through the bremsstrahlung generated by the electrons, but by a direct electron-nuclear interaction of electromagnetic nature. The effect of this interaction, as an order of magnitude, is $e^2/\hbar c = \alpha$ times that of a gamma ray of the same energy as the electrons. The cross sections for electron interaction are thus about 100 times smaller than those for photon interaction. (Compare Fig. 11-67 with Fig. 11-64.)

11-13 "INTERMEDIATE ENERGY" REACTIONS

We shall arbitrarily and conventionally set our limit for "intermediate energy" reactions at the pion formation threshold (135 MeV c.m.). An important problem for the study of these reactions is the calibration of the intensity of proton beams generated by accelerators. The cross section of the $^{12}C(p, pn)^{11}C$ reaction has been measured and is given in Fig. 11-68; together with the cross section given in Fig. 11-69, it may be used as a standard. These curves already show a form that is hardly reconcilable with compound-nucleus formation and there are many other signs pointing to direct interactions. For example, many more charged particles escape than can be accounted for by compound-nucleus formation.

Surface phenomena also occur; for example, the abundant high-energy inelastically scattered nucleons. These are attributable to peripheral collisions in which the incident particle chips off nucleons from the nuclear surface without compound-nucleus formation. The direct-reaction interpretation is corroborated by a study of the angular distribution of the inelastically scattered particles. Those of high energy are mostly scattered forward, as expected in a direct collision. Those of low energy are isotropically scattered, as expected in an evaporation process.

In a higher-energy collision ($E \cong 400$ MeV) a good picture is obtained by following the collisions of the impinging nucleus (nucleonic cascade). This has been done by the Monte Carlo method (Fig. 11-70), and part of the observed effects are thus accounted for. The struck nucleus after the nucleonic cascade is finished may be left in an excited state, from which an evaporation process takes place. However, this picture is certainly oversimplified. Figure 11-71 shows the mass yield obtained by proton bombardment of bismuth. At 40 MeV the reaction products have a mass number close to that of the target and are mainly accounted for by evaporation processes. At 480 MeV the yield curve shows two peaks. The peak around $A = 190$ is due to the primary process, which knocks off many nucleons or small aggregates of

ISBN 0-8053-8601-7

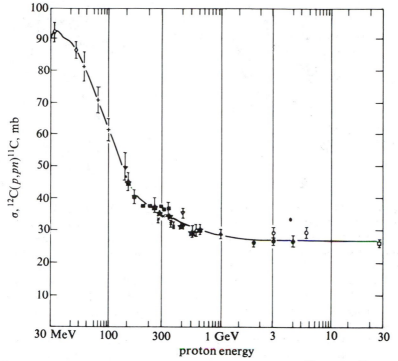

Figure 11-68 Excitation function for the reaction $^{12}C(p, pn)^{11}C$. The smooth curve is used as a basis for calculating cross sections. [From J. B. Cumming, *Ann. Rev. Nucl. Sci.*, **13**, 261 (1963).]

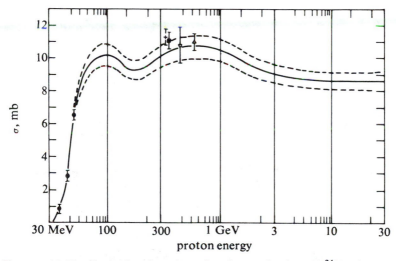

Figure 11-69 Excitation functions for the production of ^{24}Na, by proton bombardment of Al. The dashed curves indicate the estimated ($\pm 6.5\%$) standard deviation of the solid curve. [From J. B. Cumming, *Ann. Rev. Nucl. Sci.*, **13**, 261 (1963).]

ISBN 0-8053-8061-7

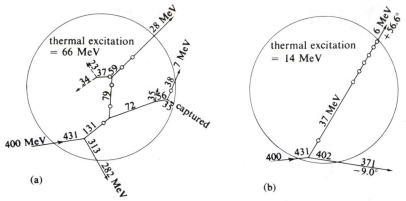

Figure 11-70 (a) A two-dimensional diagram of the Monte Carlo method, showing the development of a cascade. The entering nucleon has 400 MeV of kinetic energy; the numbers indicate the energy of the nucleons involved. The path of each nucleon is shown until it leaves the nucleus or until its energy becomes less than 35 MeV, indicated by a broken line and arrow. An open circle indicates the position at which a collision would have occurred but was forbidden by the Pauli exclusion principle. In this collision three cascade particles emerge in the forward hemisphere and the nucleus is left with a thermal excitation of 66 MeV. (taking into account all kinetic and binding energies). (b) A two-dimensional diagram for the Monte Carlo method, which illustrates a case where a quasi-elastic scattering interaction developed. [From G. Bernardini, E. T. Booth, and S. J. Lindenbaum, *Phys. Rev.*, **88**, 1017 (1952).]

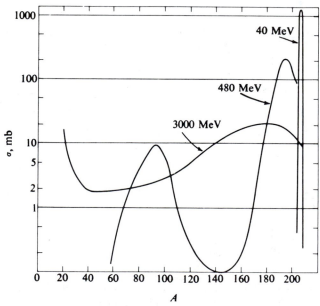

Figure 11-71 Mass-yield curves for the proton bombardment of bismuth. [From J. M. Miller and J. Hudis, *Ann. Rev. Nucl. Sci.*, **10**, 159 (1960).]

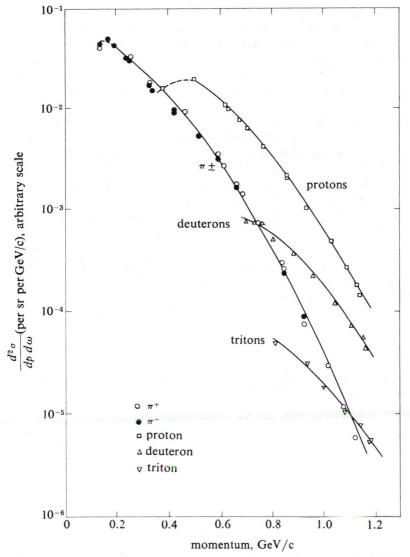

Figure 11-72 Momentum spectrum of various particles produced at 90 deg. by 30-GeV protons incident on a Be target. [V. L. Fitch, S. L. Meyer, and P. A. Piroué, *Phys. Rev.*, **126**, 1849 (1962).]

ISBN 0-8053-8061-7

nucleons (d, ^{4}He, etc.). This process is called *spallation*. It ends by leaving a variety of excited nuclei, some of which may have enough energy to undergo fission into two fragments having mass near 100; this is the origin of the second peak. At still higher energies of 3000 MeV, the yield curve changes character. Products with $6 < A < 30$ appear, which at lower energies are absent. These light nuclei are probably formed in the primary process (fragmentation).

On intermediate-A nuclei, protons from 10 to 300 GeV give approximately the same cross section for productions of lighter nuclei. Typical numbers for ^{59}Co bombarded by protons are (in millibarns): (p, p), 43; $(p, p4n)$, 0.7; $(p, 5p6n)$, 9.4; for formation of ^{7}Be 10.

At 10–20 GeV one sees a diffraction scattering of the nucleus as a whole and, superimposed on it, interactions with single nucleons.

Even at an energy of 30 GeV, pick-up reactions are manifest, as demonstrated in the bombardment of beryllium (Fig. 11-72).

At still higher energies, in the 100-GeV region, the phenomenon of jet formation appears. An impinging particle produces a narrow beam of secondaries in the forward direction (laboratory). The secondaries are mostly pions; sometimes the struck nucleus also evaporates nucleons or small fragments. These phenomena will be treated in Part III of this book.

BIBLIOGRAPHY

Ajzenberg-Selove, F. (ed), *Nuclear Spectroscopy*, Academic Press, New York, 1960.

Auerbach N., J. Hüfner, A. K. Kerman, and C. M. Shakin, "A Theory of Isobaric Analog Resonances," *Rev. Mod. Phys.* **44**, 48 (1972).

Bishop, G. R., and R. Wilson, "The Nuclear Photoeffect," in (Fl E), Vol. 42.

Blatt, J. M., and V. Weisskopf (BW 52).

Bodansky, D., "Compound Statistical Features in Nuclear Reactions," *Ann. Rev. Nucl. Sci.*, **12**, 79 (1962).

Brack, M., J. Damgaard, A. S. Jensen, H. C. Pauli, V. M. Strutinsky, and C. Y. Wong, "Funny Hills", *Rev. Mod. Phys.*, **44**, 320 (1972).

Burcham, W. E., "Nuclear Reactions, Levels, and Spectra of Light Nuclei," in (Fl E), Vol. 40.

Butler, S. T., and O. H. Hittmar, *Nuclear Stripping Reactions*, Wiley, New York, 1957.

Cohen, B. L., R. H. Fulmer, A. L. McCarthy, and P. Mukherjee, "Location of Neutron Single-Particle Levels from Stripping Reactions," *Rev. Mod. Phys.*, **35**, 332 (1963).

Cumming, J. B., "Monitor Reactions for High Energy Proton Beams," *Ann. Rev. Nucl. Sci.*, **13**, 261 (1963).

Danos, M., and E. G. Fuller, "Photonuclear Reactions," *Ann. Rev. Nucl. Sci.*, **15**, 29 (1965).

Eisberg, R. M., and C. E. Porter, "Scattering of Alpha Particles," *Rev. Mod. Phys.*, **33**, 190 (1961).

Endt, P. M., M. Demeur, and P. B. Smith (ed.), *Nuclear Reactions*, Wiley-Interscience, New York, 1959–1962.

Firk, F. W. K., "Low Energy Photonuclear Reactions," *Ann. Rev. Nucl. Sci.*, **20**, 39 (1970).

Glendenning, N. K., "Nuclear Stripping Reactions," *Ann. Rev. Nucl. Sci.*, **13**, 191 (1963).

Gol'danskii, V., and A. M. Baldin, *Kinematics of Nuclear Reactions*, Pergamon, London, 1961.

ISBN 0-8053-8061-7

Hayward, E., "Photodisintegration of Light Nuclei," *Rev. Mod. Phys.*, **35**, 324 (1963).

Hendricks, C. D., and J. M. Schneider, "Stability of a Conducting Droplet under the Influence of Surface Tension and Electrostatic Forces," *Am. J. Phys.*, **31**, 450 (1963).

Hodgson, P. E. "The Optical Model of the Nucleon–Nucleus Interaction," *Ann. Rev. Nucl. Sci.*, **17**, 1, (1967).

Hudis, J. , *Nuclear Chemistry*, Academic Press, New York, 1968.

Hughes, D. J., et al., *Neutron Cross Sections*, Brookhaven National Laboratory, Upton, New York, 1958-1960.

Huizenga, J. R., and L. Moretto, "Nuclear Level Densities," *Ann. Rev. Nucl. Sci.*, **22**, 165 (1972).

Jacob, G., and T. Maris, "Quasi-Free Scattering and Nuclear Structure," *Rev. Mod. Phys.*, **38**, 121 (1969); **45**, 6 (1973).

Lane, A. M., "Reduced Width of Individual Nuclear Energy Levels," *Rev. Mod. Phys.*, **32**, 519 (1960).

Levinger, J. S., *Nuclear Photodisintegration*, Oxford University Press, New York, 1960.

Macfarlane, M. H., and J. B. French, "Stripping Reactions and the Structure of Light and Intermediate Nuclei," *Rev. Mod. Phys.*, **32**, 567 (1960).

Milazzo-Colli L., and G. M. Braga Marazzan, "The Pre-Compound Emission Mechanism in Nuclear Reactions," *Riv. Nuovo Cim.* 3, 535 (1973).

Miller, J. M., and J. Hudis, "High Energy Nuclear Reactions," *Ann. Rev. Nucl. Sci.*, **9**, 159-202 (1959).

Nix, J. R., "Calculation of Fission Barriers for Heavy and Superheavy Nuclei," *Ann. Rev. Nucl. Sci.*, **22**, 65 (1972).

Rainwater, J., "Resonance Processes by Neutrons," in (Fl E), Vol 40.

Riesenfeld, W. B., and K. M. Watson, "Optical-Model Potential for Nucleons Scattered by Nuclei," *Phys. Rev.*, **102**, 1157 (1956).

Specht, H. J., "Nuclear Fission," *Rev. Mod. Phys.*, **46**, 773 (1974).

Talmi, I., and I. Unna, "Theoretical Interpretation of the Energy Levels of Light Nuclei," *Ann. Rev. Nucl. Sci.*, **10**, 353 (1960).

Thomas, T. D. "Compound Nuclear Reactions Induced by Heavy Ions," *Ann, Rev. Nucl. Sci.*, **18**, 343 (1968).

Tobocman, W., *Theory of Direct Nuclear Reactions*, Oxford University Press, New York, 1961.

Vandenbosch, R., and J. R. Huizenga, *Nuclear Fission* Academic Press, New York, 1973.

Wolfenstein, L., "Polarization of Fast Nucleons," *Ann. Rev. Nucl. Sci.*, **6**, 43-76 (1956).

There are many conferences on special types of reactions.

PROBLEMS

11-1 Find the threshold (in the laboratory system) for the reaction

$$^{23}\text{Na}(p, n)^{23}\text{Mg}$$

given that the upper limit of the positron spectrum emitted by ^{23}Mg is 3.0 MeV.

11-2 Fifty grams of copper containing a gold impurity bombarded with a thermal neutron flux of 10^8 neutrons sec^{-1} for 7 h gives 100 dis sec $^{-1}$ due to ^{198}Au in a certain apparatus. What is the percentage of gold in the sample, by weight? σ_{th} of Au = 96 b, $T_{1/2} = 2.7$ days. Plan the details of the experiment, including the calibrations and possible chemical operations required.

11-3 The reaction $^{14}\text{N}(d, \alpha)^{12}\text{C}$ has been studied in order to verify detailed balance [D. Bodansky, S. F. Eccles, G. W. Farwell, M. E. Rickey, and P. C. Robinson, *Phys. Rev. Letters*, **2**, 101 (1959)]. The deuterons used had 20 MeV of energy (lab); what was the corresponding lab energy of the helium ions? If in the direct experiment one observed

ISBN 0-8053-8061-7

helium ions escaping at 20 deg in the laboratory, at what angle did the corresponding deuterons in the inverse reaction escape? (See Fig. 11-4.)

11-4   The reaction

$$^3\text{He}(n, p)^3\text{H}$$

has a cross section in barns of $5000 \times (2000/v)$ (v in m sec^{-1}). Calculate the cross section for the inverse reaction and specify the energy of the incoming proton (neglect Coulomb barriers).

11-5   Prove classically that for a charged particle

$$\sigma_c = \pi R^2 (1 - V/E) \qquad \text{for } E > V$$

$$= 0 \qquad \text{for } E < V$$

where $V = Zze^2/R$ (R = nuclear radius).

11-6   If only elastic scattering is possible, the scattering matrix contains only one element, $S_{\alpha\alpha}$. Relate it to η.

11-7   Using Breit-Wigner formulas, show that the corresponding scattering matrix is for two channels (elastic scattering and reaction):

$$S = \frac{1}{\epsilon - i} \begin{pmatrix} x & [x(1-x)]^{1/2} \\ [x(1-x)]^{1/2} & 1 - x \end{pmatrix}$$

where $x = \Gamma_s/\Gamma$ and $\epsilon = (2/\Gamma)(E_r - E)$.

11-8   If there is only elastic scattering, the scattering matrix S_0 is diagonal, $S_{0,mm} = \exp(-2i\alpha_{mm})$. In the case of inelastic processes, weak compared to elastic, the scattering matrix has the form $S = S_0 + i\varepsilon$ and terms in ε^2 are negligible. Using unitarity and time reversal for S show that $\varepsilon S_0^\dagger - S_0\varepsilon^\dagger = 0; \varepsilon_{mn} = \varepsilon_{nm}$. From this derive $\varepsilon_{mn} = \rho_{mn} \exp[i(\alpha_m + \alpha_n)]$ with ρ_{mn} real and $\ll 1$. Finally

$$S_{mn} = \exp(-2i\alpha_m)\delta_{mn} + i\rho_{mn} \exp[i(\alpha_m + \alpha_n)]$$

For an application (see E. Fermi, *Collected Papers*, Vol. II, p. 1040) consider $\gamma + N \to \gamma + N$ (nuclear Compton scattering), $\pi + N \to \pi + N$ (pion–nucleon scattering), and the crosslink $\gamma + N \to \pi + N$.

11-9   Show that Fig. 11-14 is the correct representation of Eq. (11-5.12), including the geometric interpretation of η and δ. Plot Re A and Im A as a function of E. Assume $\eta = 1$, $\frac{1}{2}, \frac{1}{4}$.

11-10   The η_l of Eq. 11-5.2 may be averaged over a certain energy interval, giving $\bar{\eta}_l$. Show that the corresponding average elastic and total cross sections are

$$\bar{\sigma}_{\text{el}}^{(l)} = \pi \lambdabar^2(2l + 1)\left\{|1 - \bar{\eta}_l|^2 - |\bar{\eta}_l|^2 + \overline{|\eta_l|^2}\right\}$$

$$\bar{\sigma}_t^{(l)} = \pi \lambdabar^2(2l + 1)\left\{|1 - \bar{\eta}_l|^2 + 1 - |\bar{\eta}_l|^2\right\}$$

[H. Feshbach, C. E. Porter, and V. F. Weisskopf, *Phys. Rev.*, **96**, 448 (1954).]

11-11   Prove that $(r\psi)'/r\psi$ must be a continuous function of r.

11-12 ^{113}Cd has a capture cross section (n, γ) given in the following table:

E(eV)	0.01	0.02	0.03	0.05	0.08	0.10	0.15	0.20	0.30	0.40	0.50	0.70	1.00
$\sigma_{(n,\gamma)}$	3500	2700	2500	2500	2800	3600	7000	7000	1500	300	150	50	22

Find the constants for expressing it through a single-resonance Breit–Wigner formula.

ISBN 0-8053-8061-7

11-13 Using the following data referring to ^{135}Xe, plot its slow-neutron absorption cross section versus energy in the energy interval 10^{-2} to 1 eV. Use the Breit-Wigner formula with one resonance at

$$I = 3/2 \qquad E_0 = 0.082 \text{ eV} \qquad \Gamma_n^{(r)} = 24 \text{ meV} \qquad \Gamma_\gamma = 86 \text{ meV}$$

11-14 Show that the maximum possible scattering cross section due to a resonance is

$$\sigma_{sc} \text{ max} = 4g \frac{0.65 \times 10^6}{E_r \text{(eV)}} \text{ (barns)}$$

What relation is required between Γ_γ and Γ_n?

11-15 Shown that the maximum possible slow-neutron reaction cross section according to the Breit-Wigner formula is

$$\sigma_r \text{ max} = g \frac{0.65 \times 10^6}{E_r \text{(eV)}} \text{ (barns)}$$

What relation is required between Γ_γ and Γ_n? Apply this to ^{135}Xe ($I = 3/2$).

11-16 Show how one can find Γ_n/Γ_r by a measurement of σ_r at one resonance.

11-17 Calculate σ_s in the case $f = -iKR$.

11-18 Verify that if $f = -KR \tan(KR + \xi)$ the scattering cross section has a resonance shape in the vicinity of its maximum and calculate Γ.

11-19 Show that from Eq. (11-8.13) one obtains the numerical relation

$$\langle \sigma_r(E) \rangle = \frac{1.8 \times 10^{16} f}{E^{1/2}} \frac{\Gamma_r}{\Gamma_r + 0.44 \times 10^{10} f D E^{1/2}} \qquad \text{in } 10^{-24} \text{ cm}^2$$

E in eV, and $f = (\lambda/2)(\Gamma_n/\bar{D})$ in cm [(WW 58), p. 57].

11-20 Derive the Breit-Wigner formulas by a second-order perturbation calculation [see (Fe 50)].

11-21 Find the threshold energy for forming a π meson by bombarding proton with proton, carbon with proton, and carbon with helium. For the last two cases take the Fermi energy into account.

11-22 Shown that in a degenerate gas the average kinetic energy per particle is $\frac{3}{5}$ of the Fermi energy.

11-23 Plot the probable number of neutrons that a nucleus will emit as a function of excitation energy [see (Fe 50)].

11-24 Prove that for high-energy deuteron stripping the energy E of the escaping neutrons (lab) is $E_d/2$ with a spread

$$\Delta E = 1.5(\epsilon E_d)^{1/2}$$

and that they escape with an angular spread

$$\Delta \theta = 1.6(\epsilon/E_d)^{1/2}$$

where ϵ is the binding energy of the deuteron.

11-25 Estimate how much ^{99}Tc is produced in a natural uranium pile for every kilogram of ^{239}Pu produced.

11-26 Kr was found among gaseous fission products in quantities corresponding to its branching, whereas Xe was not. Can you guess why? (For chemists.)

11-27 Show that there are no levels of a given high (e.g., $I > 10$) angular momentum below a lowest energy, called "yrast" energy.

ISBN 0-8053-8061-7

CHAPTER
12

Neutrons

The neutron affords an interesting example of the complex path some-times followed in arriving at a major scientific discovery. Its existence had been hypothesized by Rutherford as early as 1920. His arguments, however, were purely speculative, and until 1932 no experimental evidence supporting them was available. In 1930, Bothe and Becker bombarded beryllium with alpha particles and observed a very penetrating radiation; they then showed that the penetrating radiation was composed of gamma rays. Further study by I. Curie and F. Joliot gave the surprising result (1932) that this "gamma radiation" also had a component that was capable of imparting energies of several MeV to protons. At first Curie and Joliot interpreted the observed energy transfer as a Compton effect on protons. The correct explanation, however, was soon provided by Chadwick, who showed that the recoil pro-tons observed by Curie and Joliot had been hit by a neutral particle of ap-proximately protonic mass, which he called the neutron. Chadwick bombarded with neutrons not only hydrogen but other light nuclei as well, and measured the range of the recoil particles. From the conservation of energy and momentum and range–energy relations he was able to determine the mass of the new particle "as very nearly the same as the mass of the proton."

Emilio Segrè, Nuclei and Particles: An Introduction to Nuclear and Subnuclear Physics, Second Edition

ISBN 0-8053-8601-7

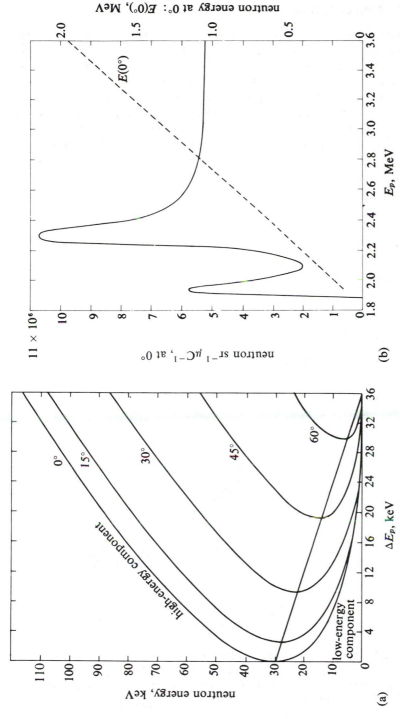

Figure 12-1 (a) Neutron energies from a thin lithium target as a function of the difference ΔE_p between the energy of the bombarding proton and the threshold energy, for various angles of emission in the laboratory system. (b) Yield of neutrons at 0° from the Li(p, n) reaction as a function of the proton energy. Target thickness 40 keV. [A. O. Hanson, R. F. Taschek, and W. S. C. Williams, *Rev. Mod. Phys.*, **21**, 635 (1949).]

2-1 NEUTRON SOURCES

The simplest neutron source is a mixture of a suitable radioactive substance and a light element such as beryllium or boron. Neutrons are generated by the (α, n) or (γ, n) process. These sources are weak compared with other types, but they are small and constant. For orientation we give two data: a Ra + Be source gives 1.35×10^7 neutrons $\text{sec}^{-1}\,\text{g}^{-1}$ of radium and a Po + Be source gives 73 neutrons for every 10^6 alpha particles. Spontaneous fission of some isotopes, such as ^{252}Cf, is also a practical neutron source.

Accelerators produce neutrons copiously. Moreover, by using a thin target and by selecting the neutrons emitted in a given direction, it is possible to obtain a fairly monochromatic beam of neutrons. For example, the reaction

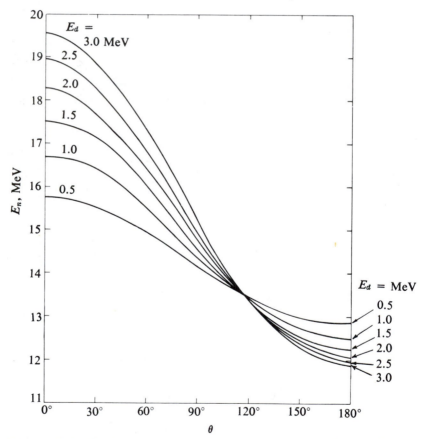

Figure 12-2 The energies of the neutrons emitted from the $^3\text{H}(d, n)^4\text{He}$ reaction for various values of E_d as a function of the angles of emission in the laboratory system. [A. O. Hanson, R. F. Taschek, and W. S. C. Williams, *Rev. Mod. Phys.*, **21**, 635 (1949).]

ISBN 0-8053-8601-7

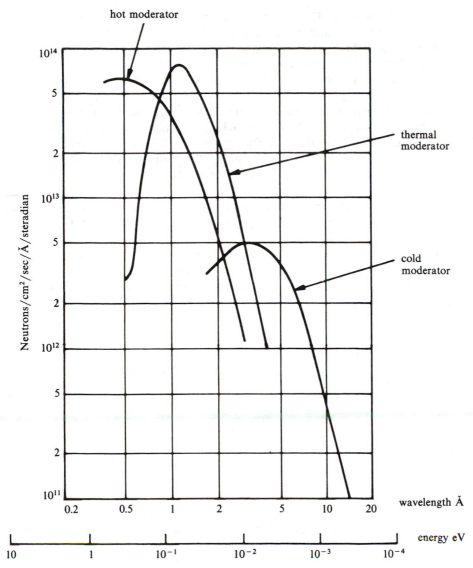

Figure 12-3 Flux inside the reactor in neutrons cm^{-2} sec^{-1}, per steradian and per angstrom of de Broglie wavelength of the neutrons, versus the energy or λ. Neutrons in the Institute M. v. Laue-P. Langevin reactor are moderated by 25 liters of liquid deuterium (cold), by the reactor itself (thermal), or by a graphite block at 2200°K.

$^7\text{Li} + {}^1\text{H} \rightarrow {}^7\text{Be} + n - 1.647$ MeV (endothermic) can produce neutrons with energies of a few keV and up, as shown in Fig. 12-1. The reactions $^3\text{H} + {}^2\text{H} = {}^4\text{He} + n + 17.6$ MeV, $^2\text{H} + {}^2\text{H} = {}^3\text{He} + n + 3.27$ MeV can give high-energy neutrons (Fig. 12-2). Photoneutrons are also fairly monochromatic. As a typical figure, 1 Ci of ^{24}Na gives 1.4×10^5 neutrons sec^{-1} of 0.9-MeV energy when it irradiates 1 g of Be at a distance of 1 cm.

Finally, nuclear reactors generate high neutron fluxes, 10^{15} neutrons cm^{-2} sec^{-1} being a typical neutron flux in a large machine. The total number of neutrons generated in a pile is approximately 6×10^{13} neutrons (sec kW)$^{-1}$. Neutrons from reactors can be thoroughly thermalized with a thermal column, and special energy regions can be selected by crystal reflection and other means.

A modern research reactor at the Institut M. v. Laue–P. Langevin in Grénoble, for instance, can give high fluxes at various energies and correspondingly intense collimated beams (Fig. 12-3).

Very high-energy neutrons (from 50 MeV on up) are best obtained by deuteron stripping or by charge-exchange collisions of protons with light nuclei. The collision of protons with deuterons yields neutrons of particularly well-defined energy (see Fig. 12-4).

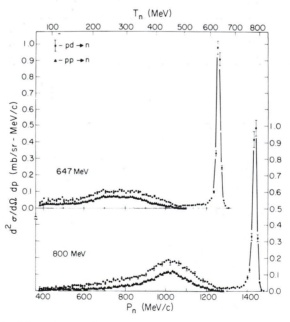

Figure 12-4 Monochromatic neutrons obtained from the p + d ⇒ n + 2p reaction in the forward direction, at incident energies of 647 and 800 MeV. For comparison the neutron spectrum obtained from the reaction p + p ⇒ p + n + π^+ is also shown. [C. W. Bjork et al., *Physics Letters* **63B**, (1976)]

ISBN 0-8053-8601-7

12-2 SLOWING DOWN OF NEUTRONS

Neutrons in reactors and in many nuclear processes are generated at energies of the order of 1 MeV. However, in penetrating matter they undergo a characteristic process of energy degradation (moderation). Whereas in the case of charged particles or of gamma rays the energy loss is mostly due to electromagnetic effects (e.g., ionization for charged particles or the Compton effect for gamma rays), neutrons are slowed down by nuclear collisions. These may be inelastic collisions, in which a nucleus is left in an excited state, or elastic collisions, in which the colliding nucleus acquires part of the energy of the neutron as kinetic energy. In the first instance the neutron must have enough kinetic energy (of the order of 1 MeV) to excite the collision partner. Below this limit only elastic collisions can slow down the neutrons, a process effective down to thermal energies (1/40 eV). At this stage the collisions, on the average, have no further effect on the neutron's energy: a state of thermal equilibrium has been reached (Fig. 12-5).

In order to analyze elastic collisions, consider a neutron (in the laboratory system) impinging on a nucleus of mass number A, and let the velocity of the neutron be V_1. In the center-of-mass system its velocity is then $\mathbf{v}_1 = [A/(A + 1)]\mathbf{V}_1$; the velocity of the target is $\mathbf{v}_2 = [-1/(A + 1)]\mathbf{V}_1$ (Fig. 12-6). After the collision the velocities in the center-of-mass system are unchanged in magnitude. However, in the laboratory system the neutron velocity $\mathbf{V}'_1$ is the vector difference of $\mathbf{v}'_1$ and the center-of-mass velocity of the target $[-1/(A + 1)]\mathbf{V}_1$. The law of cosines gives

$$V_1'^2 = \left(\frac{A}{A + 1} V_1 \right)^2 + \left(\frac{1}{A + 1} V_1 \right)^2 - 2V_1^2 A \left(\frac{1}{A + 1} \right)^2 \cos(\pi - \theta)$$

$$= \frac{V_1^2}{(A + 1)^2} (A^2 + 1 + 2A \cos \theta) \qquad (12\text{-}2.1)$$

where θ is the scattering angle in the center-of-mass system. The ratio of the neutron energies in the laboratory system after and before the collision is

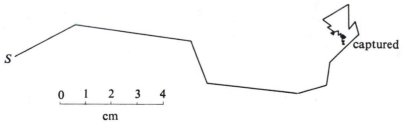

Figure 12-5 Schematic representation of the slowing down and diffusion trajectory of an initially fast neutron, from birth to capture, in a hydrogenous medium. The scale indicated corresponds approximately to the case of a water moderator.

ISBN 0-8053-8601-7

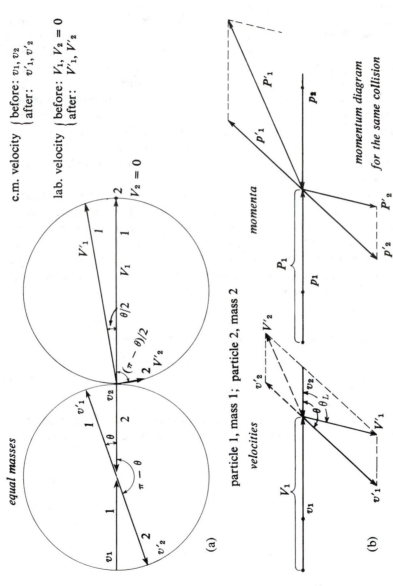

Figure 12-6 (a) Collision between particles of equal mass. At left, center of mass; at right, laboratory system (velocities and momentum). (b) Collision between particles of mass 1 and 2. At left a velocity diagram (in laboratory V_i, in center of mass v_i); at right, momentum diagram (in laboratory P_i, in center of mass p_i). Unprimed quantities are before the collision; primed quantities are after the collision.

ISBN 0-8053-8601-7

therefore

$$\frac{E}{E_0} = \frac{V_1'^2}{V_1^2} = \frac{1}{(A+1)^2}(A^2 + 1 + 2A\cos\theta) \tag{12-2.2}$$

which yields immediately the interesting inequality

$$\left(\frac{A-1}{A+1}\right)^2 \leqslant \frac{E}{E_0} \leqslant 1 \tag{12-2.3}$$

We can also calculate for future reference the relation between θ and θ_L, the laboratory scattering angle. We obtain from Fig. 12-6

$$V_1^2\left(\frac{A}{A+1}\right)^2 = V_1'^2 + \frac{1}{(A+1)^2}V_1^2 - \frac{2V_1V_1'}{A+1}\cos\theta_L$$

which gives, when combined with Eq. (12-2.2),

$$\cos\theta_L = \frac{A\cos\theta + 1}{(A^2 + 1 + 2A\cos\theta)^{1/2}} \tag{12-2.4}$$

For neutrons up to at least a few hundred keV energy, scattering is due only to the s wave and hence is spherically symmetric in the center-of-mass system. The probability dW that the scattering occurs within a solid angle $d\omega$ is thus proportional to $d\omega$, that is,

$$dW = \frac{d\omega}{4\pi} = \frac{\sin\theta\, d\theta}{2} = -\frac{d(\cos\theta)}{2} \tag{12-2.5}$$

Hence, after collision, we have equal probabilities of reaching equal intervals of $\cos\theta$ independently of θ. Differentiation of Eq. (12-2.2) gives

$$\frac{dE}{E_0} = \frac{2A}{(A+1)^2}d(\cos\theta) \tag{12-2.6}$$

which shows that equal intervals of $\cos\theta$ correspond to equal energy intervals. We conclude that after one collision the probability of reaching equal energy intervals anywhere between E_0 and $[(A-1)/(A+1)]^2 E_0 = \alpha E_0$ is the same (see Fig. 12-7). For $A = 1$, that is, collisions with hydrogen, any energy between E_0 and 0 is equally probable.

One can determine the average energy after n collisions, which in the case of hydrogen is $\langle E_n \rangle = (1/2^n)E_0$. However, this average arises from a distribution in which high energies are very rare. The median and the most probable energy are well below $\langle E_n \rangle$ (Fig. 12-8).

ISBN 0-8053-8601-7

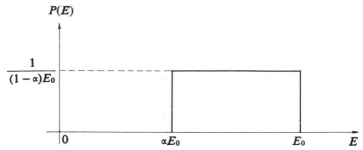

Figure 12-7   Distribution of the values of the neutron energy E after one collision in the case of isotropic scattering in the center-of-mass system for a neutron of initial energy E_0. The quantity $\alpha = (A - 1)^2/(A + 1)^2$.

In treating the slowing down of neutrons it is convenient to operate with the quantity $u = \log(E_0/E)$, where $E_0 = 10$ MeV, called the *lethargy*. This definition gives a different scale for measuring the energy (Fig. 12-9). We calculate first the average value of $\log(E_0/E_1)$ after one collision, a quantity which we shall call ξ. We have

$$\xi = \left\langle \log \frac{E_0}{E_1} \right\rangle = \frac{\int_{\alpha E_0}^{E_0} \log \frac{E_0}{E_1} \frac{dW_1}{dE_1} dE_1}{\int_{\alpha E_0}^{E_0} \frac{dW_1}{dE_1} dE_1}$$

$$= \frac{(A + 1)^2}{4AE_0} \int_{\alpha E_0}^{E_0} \log \frac{E_0}{E_1} dE_1 = 1 + \frac{(A - 1)^2}{2A} \log \frac{A - 1}{A + 1} \quad (12\text{-}2.7)$$

Each collision decreases the average value of $\log E$ by ξ, and hence, after n collisions we have

$$\langle \log E_n \rangle = \log E_0 - n\xi \quad (12\text{-}2.8)$$

The quantity ξ as a function of A has the values given in Table 12-1.

TABLE 12-1   AVERAGE VALUE OF THE DECREASE OF $\log E$ **PER COLLISION** (ξ)

$A = 1$	2	12	14	Large A
$\xi = 1$	0.725	0.158	0.136	$2/[A + (2/3)]$

Equation (12-2.8) can be used to calculate the number of collisions needed to thermalize neutrons of a given energy. For example, for $E_0 = 1$ MeV the

ISBN 0-8053-8601-7

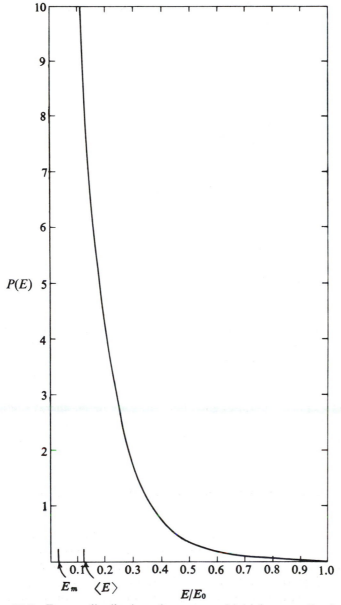

Figure 12-8 Energy distribution of neutrons of initial energy E_0 after four collisions with protons. Note the average energy $\langle E \rangle$ and the median energy E_m.

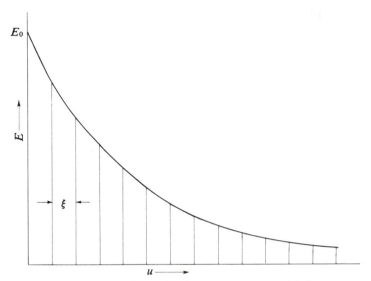

Figure   12-9   Relationship between energy and lethargy.

number required to reach thermal energy (1/40 eV) is 17.5 collisions in hydrogen or 110 collisions in carbon. Actually Eq. (12-2.8) breaks down for values of E_n near thermal energies because there, since the energy of the scattering nuclei is comparable to that of the neutrons, the collision may impart energy to the neutron instead of subtracting it. Moreover, for a molecular or crystalline moderating medium, on reaching an energy comparable to that of the chemical bond, the collision partner is no longer a free atom but a larger unit (molecule, lattice, etc.), and the laws governing energy loss become quite complicated.

12-3   ENERGY DISTRIBUTIONS OF NEUTRONS FROM A MONOENERGETIC SOURCE

In several problems of neutronology we shall use the cross sections per unit volume (i.e., the product of the nuclear cross section and the number of nuclei per unit volume), which we shall call Σ. The dimensions of Σ are $[L^{-1}]$. We may distinguish a scattering Σ or Σ_s and an absorption Σ, Σ_a. The scattering mean free path λ is clearly $1/\Sigma_s$, and the absorption mean free path is $\Lambda = 1/\Sigma_a$.

Let us now consider the total number of neutrons in a certain lethargy interval $du = -dE/E$ and the current of neutrons q along a u axis. The dimensions of q are $[T^{-1}]$. This is illustrated in Fig. 12-10. At a certain time each neutron is represented by a point on the logarithmic energy scale. Collisions move the points from left to right, and we obtain the "current" by

ISBN 0-8053-8601-7

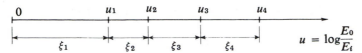

Figure 12-10 Representation of the slowing down of a neutron in a logarithmic scale (lethargy). ξ_i is $\log(E_{i-1}/E_i)$.

multiplying $n \equiv n(u)$, the number of neutrons per unit u, times their rate of change of u. This definition of q at once gives the relation

$$q = n(v/\lambda)\xi = nv\Sigma_s\xi \tag{12-3.1}$$

because v/λ is the number of collisions undergone by a neutron per unit time and ξ is the increase in lethargy per collision.

We now ask the question: What is the number of neutrons in a given energy interval in a neutron-moderating medium? Consider a steady-state situation. Let Q be the production rate of neutrons at the source; Q has the dimensions $[T^{-1}]$. We shall treat first the simpler case of no absorption. In a steady state, then, Q neutrons per second must pass any given lethargy, or

$$q(u) = Q = n(v/\lambda)\xi \tag{12-3.2}$$

and

$$n(u)\,du = -Q\frac{\lambda}{v\xi}\frac{dE}{E} \sim E^{-3/2}\,dE \tag{12-3.3}$$

if λ is energy independent.

In most practical cases one is interested in the neutron flux nv, rather than in the neutron density n in a certain lethargy interval. Using Eq. (12-3.3) we can write

$$-nv\,du = \frac{Q\lambda}{\xi}\frac{dE}{E} \tag{12-3.4}$$

which is the total neutron flux along the u axis of neutrons in the lethargy range du. This is, for example, approximately the energy distribution of epithermal neutrons in a reactor. Epithermal refers here to energy greater than $1/40$ eV and less than the first resonance of the moderator.

If absorption takes place, $q(u)$ is not constant, because neutrons disappear by absorption, and for an increase in the lethargy du the current q decreases according to

$$-dq(u) = \frac{nv}{\Lambda}\,du \tag{12-3.5}$$

ISBN 0-8053-8601-7

Remembering Eq. (12-3.2), which defines q, we may write

$$-\xi d(nv\Sigma_s) = n\frac{v}{\Lambda}\ du = nv\Sigma_a\ du \tag{12-3.6}$$

or

$$-\xi\frac{d(nv\Sigma_s)}{nv\Sigma_s} = \frac{\Sigma_a}{\Sigma_s}\ du \tag{12-3.7}$$

This equation can be integrated between the initial lethargy and an arbitrary lethargy value,

$$-\xi\log(nv\Sigma_s)\Big]_u^{u_0} = \int_u^{u_0}\frac{\Sigma_a}{\Sigma_s}\ du \tag{12-3.8}$$

and, by Eq. (12-3.2),

$$-\xi\log\left(\frac{Q}{\xi}\right) + \xi\log(nv\Sigma_s) = \int_u^{u_0}\frac{\Sigma_a}{\Sigma_s}\ du \tag{12-3.9}$$

from which

$$nv = \frac{Q}{\xi\Sigma_s}\ \exp\left(-\frac{1}{\xi}\int_{u_0}^u\frac{\Sigma_a}{\Sigma_s}\ du\right) \tag{12-3.10}$$

Implicit in Eq. (12-3.10) is the necessary hypothesis that Σ_a varies slowly with energy, or, better, $\Delta\Sigma_a/\Sigma_a \ll 1$ if $\Delta E \cong \xi E$.

12-4   MEAN DISTANCE FROM A POINT SOURCE VS. ENERGY

In this section we shall calculate $\langle R^2(E)\rangle$, the mean-square distance between a monoenergetic point source and all the neutrons of energy E generated by slowing down the neutrons from that source. If the mean free path were energy independent and the scattering one-dimensional, either left or right, we should argue as follows: In order to go from energy E_0 to energy E, we need N collisions, where

$$N = \frac{1}{\xi}\ \log\frac{E_0}{E} \tag{12-4.1}$$

We then have N steps of a random walk, each step of length λ. Hence

$$\langle R^2\rangle = N\lambda^2 = \frac{\lambda^2}{\xi}\ \log\frac{E_0}{E} \tag{12-4.2}$$

ISBN 0-8053-8061-7

In order to prove the last relation, that is, $\langle R^2 \rangle = N\lambda^2$ for a random walk in one dimension, consider the end points after $N = N_1 + N_2$ steps, where N_1 is the number of steps to the right and N_2 the number of steps to the left. The displacement to the right of the origin of the walk is

$$X_N = \lambda(N_1 - N_2) = \lambda(N - 2N_2) \tag{12-4.3}$$

Since the probability of a step to the right is equal to that of a step to the left, or $\langle N_1 \rangle = \langle N_2 \rangle = \frac{1}{2}N$, we have $\langle X_N \rangle = 0$. For N large, the end points of many samples will have a Gaussian density distribution around $\langle X_N \rangle = 0$, with a square of the standard deviation

$$\sigma^2 = \langle (X_N - \langle X_N \rangle)^2 \rangle = N\lambda^2 \tag{12-4.4}$$

as can be seen by noting that in Eq. (12-4.3) X_N gives rise to a standard deviation

$$\sigma^2(X_N) = \lambda^2 \sigma^2 (2N_2) \tag{12-4.5}$$

and that

$$\sigma^2(2N_2) = \langle 2N_2 \rangle = N \tag{12-4.6}$$

In three dimensions the displacement in the X direction is

$$X = \lambda \sum_1^N {}_i \cos \alpha_i \tag{12-4.7}$$

and

$$\langle (X - \langle X \rangle)^2 \rangle = \lambda^2 \sum_1^N {}_{ik} \cos \alpha_i \cos \alpha_k = \lambda^2 \frac{N}{3} \tag{12-4.8}$$

where α_i is the angle between the X axis and the direction of the ith step. The last step of Eq. (12-4.8) depends on the fact that $\Sigma_{i \neq k} \cos \alpha_i \cos \alpha_k$ averages to 0 and that the average value of $\cos^2 \alpha_i$ (on a sphere) is $\frac{1}{3}$. The same reasoning also applies to the Y and Z components of the displacement; and, since they are independent and lie in orthogonal directions, we have

$$\langle R^2 \rangle = \langle (X - \langle X \rangle)^2 \rangle + \langle (Y - \langle Y \rangle)^2 \rangle + \langle (Z - \langle Z \rangle)^2 \rangle = \langle \lambda^2 \rangle N \tag{12-4.9}$$

In Eq. (12-4.9) we have replaced λ^2 by $\langle \lambda^2 \rangle$ in order to take into account the variability of the free path between collisions.

We now compute $\langle \lambda^2 \rangle$ from λ. The probability of a free path of length between x and $x + dx$ is the probability of no collision between the origin and x, which is $e^{-x/\lambda}$, times the probability of a collision in dx, dx/λ. The

ISBN 0-8053-8061-7

average of the square of the mean free path is, therefore,

$$\int_0^\infty x^2 e^{-x/\lambda} \frac{dx}{\lambda} \bigg/ \int_0^\infty e^{-x/\lambda} \frac{dx}{\lambda} = 2\lambda^2 = \langle \lambda^2 \rangle \qquad (12\text{-}4.10)$$

We thus have from Eqs. (12-4.9) and (12-4.10)

$$\langle R^2 \rangle = \frac{2\lambda^2}{\xi} \log \frac{E_0}{E} \qquad (12\text{-}4.11)$$

This equation must be corrected to take into account the persistence of velocity (i.e., the fact that in the laboratory frame neutrons tend to continue to travel in their original direction after a collision). The persistence of velocity is inferred by averaging $\cos \theta_L$ as given by Eqs. (12-2.4). We have

$$\langle \cos \theta_L \rangle = \frac{1}{2} \int_{-1}^{+1} d(\cos \theta) \frac{A \cos \theta + 1}{(A^2 + 2A \cos \theta + 1)^{1/2}} = \frac{2}{3A} \qquad (12\text{-}4.12)$$

and not 0, as would be the case if there were no persistence of velocity. The effect increases $\langle R^2 \rangle$. Moreover, the mean free path is a function of energy, and Eq. (12-4.11) therefore is modified to

$$\langle R^2 \rangle = \frac{2}{[1 - (2/3A)]\xi} \int_E^{E_0} \lambda^2(E') \frac{dE'}{E'} \qquad (12\text{-}4.13)$$

Even Eq. (12-4.13) is only approximate, but it is useful in many cases and is simple to interpret. (See Fermi Coll. Papers, Vol. 2, p. 481).

An interesting application of Eq. (12-4.13) is the experimental determination of $\langle R^2 \rangle$ as a function of energy. A detector containing a thin indium foil sandwiched between two cadmium foils is very sensitive to neutrons of 1.44 eV energy, because indium has a sharp resonance at this energy and cadmium absorbs all neutrons of lower energy. When the detector is immersed in a tank of water at a distance R from a neutron source, it acquires a radioactivity by the (n, γ) reaction, proportional to the density of the 1.44-eV neutrons, and one obtains the curve of Fig. 12-11, from which one computes $\langle R^2 \rangle$ by graphical integration. Using a substance I with a different resonance energy, one obtains a curve different from that in Fig. 12-11. If λ does not vary with energy, as is the case for water and energies between 1 and 100 eV, we can write, from Eq. (12-4.13),

$$\langle R_I^2 \rangle - \langle R_{In}^2 \rangle = 6\lambda^2 \log \frac{E_{\text{res}}(In)}{E_{\text{res}}(I)} \qquad (12\text{-}4.14)$$

Such experiments were among the earliest performed by Fermi (1936) in order to determine resonance energies.

ISBN 0-8053-8061-7

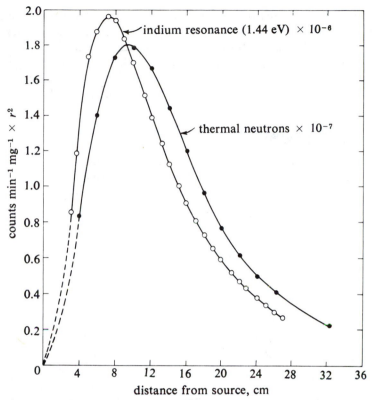

Figure 12-11 Distribution of slow neutrons from a Ra–α–Be source in water. The ordinate is the activity times r^2 of a thin indium foil. The curve labeled "indium resonance" represents the activity of a cadmium-covered foil; the curve labeled "thermal neutrons" represents the activity of a bare foil minus 1.07 times the activity of cadmium covered foil. The source intensity was 13.2×10^6 neutrons $\sec^{-1}$. The ordinate scale can be converted to a scale of thermal neutron flux $(nv) \times r^2$ by multiplying by 1.6×10^6. [From (Se 59).]

12-5 DIFFUSION THEORY—INTRODUCTION

We shall now treat the slowing down and diffusion of neutrons in an infinite medium (moderator). The process involves two steps. First the neutrons change their average energy and diffuse until they reach thermal energy. The density of the neutrons is thus a function of the coordinates and the energy. In a second phase the thermal neutrons diffuse without changing their average energy; the density is a function of the coordinates only. We shall consider steady-state problems, and the sources or sinks of neutrons will appear as boundary conditions. The elementary theory for the first phase is most applicable when $\xi \ll 1$, that is, when many collisions are needed to slow down the neutrons, and when the mean free path does not vary too much with energy. For these reasons the theory is a much better approximation for a graphite moderator than for a water moderator. In a water moderator 17 collisions only will slow a 1-MeV neutron to thermal energies, and the mean free path changes from about 4 cm to 0.4 cm. Note that this last value of the mean free path does not correspond to collisions with free protons, but is affected by the chemical bond of the hydrogen in the molecule.

12-6 THE AGE EQUATION

We shall now derive the neutron diffusion equation for energies above thermal. This procedure is known as the *Fermi age method*.

Let $n(\mathbf{r}, u)$ be the number of neutrons per unit volume per unit u. This is the same as the previous $n(u)$ except for the addition of "per unit volume." The dimensions of $n(\mathbf{r}, u)$ are $[L^{-3}]$. Let $q(\mathbf{r}, u)$ be the current density along the u axis. This is the previous $q(u)$ per unit volume. It is called the "slowing-down density" and is related to $n(\mathbf{r}, u)$ by

$$q(\mathbf{r}, u) = n(\mathbf{r}, u) \frac{v}{\lambda} \xi \qquad (12\text{-}6.1)$$

Its dimensions are $[L^{-3}T^{-1}]$.

Consider now a volume element of the medium and the neutrons with lethargy between u and $u + du$ contained in it. This number can vary for two reasons: (1) Neutrons of this lethargy diffuse into the volume element considered; changes due to this cause are indicated by ∂'. (2) Neutrons of lower lethargy reach the lethargy under consideration by collisions; changes due to this cause are indicated by ∂''.

For the first mechanism the fundamental diffusion relation

$$\mathbf{j} = - D \text{ grad } n \qquad (12\text{-}6.2)$$

(where $\mathbf{j}$ is the current density of neutrons and D the diffusion coefficient) gives, by a well-known argument,[1]

$$\frac{\partial' n(\mathbf{r}, u)}{\partial t} = D \nabla^2 n(\mathbf{r}, u) \qquad (12\text{-}6.3)$$

The diffusion coefficient D is defined by Eq. (12-6.2) and is related to λ and v, as kinetic theory shows, by

$$D = \frac{\lambda v}{3[1 - (2/3A)]} \qquad (12\text{-}6.4)$$

the quantity in brackets in the denominator originating from the persistence of velocity.

[1]See, e.g., P. M. Morse, *Thermal Physics*, Benjamin, New York, 1964.

ISBN 0-8053-8061-7

The second mechanism mentioned above gives another reason for a time change of n. Actually $q(u)$ neutrons enter the lethargy interval in question, but $q(u + du) = q(u) + (\partial q / \partial u)\, du$ leave it per unit time and per unit volume. Hence,

$$-\frac{\partial''n}{\partial t} = \frac{\partial q}{\partial u} \tag{12-6.5}$$

In the stationary state $\partial'n/\partial t + \partial''n/\partial t = 0$ and therefore

$$-\frac{\partial q}{\partial u} + D\,\nabla^2 n = 0 \tag{12-6.6}$$

It is now advantageous to introduce a new variable, called *Fermi age*, defined by the relation

$$\tau \equiv \int_{u_0}^{u} \frac{\lambda^2(u)}{3\xi[1 - (2/3A)]}\, du = \int_{t_0}^{t} D\, dt \tag{12-6.7}$$

the last equality deriving from Eq. (12-6.4) and $du = \xi(v/\lambda)\, dt$. For λ constant, Eq. (12-4.13) shows that

$$\tau = \frac{\lambda^2}{3\xi[1 - (2/3A)]}\,(u - u_0) = \tfrac{1}{6}\langle R^2 \rangle \tag{12-6.8}$$

where $\langle R^2 \rangle$ is the mean-square distance between the neutron and its source in passing from lethargy u_0 to lethargy u. From this relation it is clear that the dimensions of τ are $[L^2]$. Its physical significance is also evident—one-sixth the mean-square distance from the source. Introducing the new variable τ, we can write

$$\frac{\partial q}{\partial u} = \frac{\partial q}{\partial \tau}\frac{\partial \tau}{\partial u} = \frac{\partial q}{\partial \tau}\frac{\lambda^2}{3\xi[1 - (2/3A)]} \tag{12-6.9}$$

and Eq. (12-6.6) becomes

$$\frac{\lambda v}{3[1 - (2/3A)]}\,\nabla^2 n - \frac{\lambda^2}{3\xi[1 - (2/3A)]}\,\frac{\partial q}{\partial \tau} = 0 \tag{12-6.10}$$

which by remembering Eq. (12-6.1) yields immediately the age equation,

$$\nabla^2 q = \frac{\partial q}{\partial \tau} \tag{12-6.11}$$

ISBN 0-8053-8061-7

Mathematically this is the same as Fourier's heat-conduction equation,

$$\frac{k}{\rho c} \nabla^2 T = \frac{\partial T}{\partial t} \tag{12-6.12}$$

All the mathematical techniques developed for this equation are applicable to the age equation.

In particular a point source at the origin of the coordinates, emitting Q neutrons $\sec^{-1}$ of energy E_0 corresponding to $\tau = 0$ in an infinite medium, gives as a solution to Eq. (12-6.11),

$$q(r, \tau) = \frac{Q}{(4\pi\tau)^{3/2}} e^{-r^2/4\tau} \tag{12-6.13}$$

Equation (12-6.11) being linear, a sum of solutions is still a solution, and hence it is possible to construct a general solution of Eq. (12-6.11) by superposition of solutions of the type of Eq. (12-6.13). From the singular solution [Eq. (12-6.13)] we obtain the mean-square distance reached by neutrons of age τ,

$$\langle R^2 \rangle = \frac{\int_0^\infty r^2 e^{-r^2/4\tau} 4\pi r^2 \, dr}{\int_0^\infty e^{-r^2/4\tau} 4\pi r^2 \, dr} = 6\tau \tag{12-6.14}$$

which agrees with the result of Eq. (12-4.13).

Equation (12-6.11) can be solved by the classical methods of mathematical physics. As an example of practical importance we shall consider briefly a monoenergetic source at the center of the base of a square prism of base side b and indefinite length z. The variables in the age equation are first separated by putting

$$q = Q(\mathbf{r})T(\tau) \tag{12-6.15}$$

and obtaining

$$\frac{\nabla^2 Q}{Q} = \frac{1}{T}\frac{dT}{d\tau} = -k^2 \tag{12-6.16}$$

The boundary conditions for q are that it is 0 on the lateral faces of the prism, that it has a singularity at $x = y = b/2$, $z = 0$, and that it vanishes for $z \to \infty$. The conditions on the faces can be fulfilled only for a discrete set of numbers k^2, the eigenvalues of the problem. We have the general solution as a series of functions

$$q = \sum_n a_n Q_n(\mathbf{r})T_n(\tau) \tag{12-6.17}$$

ISBN 0-8053-8061-7

the constant coefficients a_n to be determined by the initial conditions, that is, by putting

$$\sum_n a_n Q_n (\mathbf{r}) T_n (0) = F(\mathbf{r}) \tag{12-6.18}$$

where $F(\mathbf{r})$ is the q of the source of neutrons at age 0. The development of this calculation is to be found in standard texts on heat conduction, the final result being

$$q(\mathbf{r}, \tau) = \frac{2Q}{b^2 (\pi \tau)^{1/2}} \exp\left(\frac{-z^2}{4\tau} \right) \sum_{m, n \text{ odd}} -(-1)^{(m+n)/2}$$

$$\times \sin \frac{m \pi x}{b} \sin \frac{n \pi y}{b} \exp\left[-\frac{\pi^2}{b^2} (m^2 + n^2) \tau \right] \tag{12-6.19}$$

where m, n are odd integral numbers and Q is the source strength. For a sufficiently large τ only the first harmonic corresponding to $m = n = 1$ is important.

The boundary condition assumed, $n(\mathbf{r}, \tau) = 0$ at the boundary of the moderator, seems reasonable, because free space should act as a perfect sink inasmuch as no neutron will ever be reflected by it. A more refined consideration, however, shows that the neutron density near the surface varies linearly with depth and that for a medium filling the half-space $x > 0$ it would vanish at $x = -\frac{2}{3} \lambda$ or, better, at $x = -0.71 \lambda$. The correct boundary condition, therefore, is that n vanishes at a distance of 0.71λ outside the moderator.

The particular example treated here was very important in the so-called sigma experiments that were used in testing materials for the first nuclear reactor. By using an indium sandwich detector to measure $q(0, 0, z, u_{\text{res}})$ one obtained $\tau(u_0, u_{\text{res}})$ for the source and medium under investigation. If the source is not monoenergetic, it may be considered as the sum of several monoenergetic sources (multigroup methods).

In the important case of fission neutrons the values of $\tau(1.4 \text{ eV})$ and of $\tau(\text{thermal})$ for several moderators are given in Table 12-2. These numbers are obtained by averaging over the fission neutron energy spectrum.

TABLE 12-2 FERMI AGE OF NEUTRONS OF 1.4 eV AND THERMAL IN VARIOUS MODERATORS

	$\tau(1.4 \text{ eV})$	τth
H_2O	31 cm^2	33 cm^2
D_2O	109	120
Beryllium	80	98
Graphite	311	350

ISBN 0-8053-8061-7

The diffusion theory gives a neutron density depending on r as $e^{-r^2/4\tau}$ for large r. However, neutrons arrive occasionally at large distances from the source without a collision, the probability of this event being $e^{-r/\lambda}$. It is clear that when $e^{-r/\lambda} \gg e^{-r^2/4\tau}$ there will be more neutrons than calculated by diffusion theory.

In these developments we have neglected absorption. In fact, absorption may be important, especially for moderators mixed with fissionable material showing strong resonances, as we shall see later.

12-7   DIFFUSION OF THERMAL NEUTRONS

Let $n(\mathbf{r})$ be the density of thermal neutrons, let

$$D = \lambda v / 3 \tag{12-7.1}$$

be their diffusion coefficient, and let Λ be their absorption mean free path, $1/\Sigma_a$. The mean life of a thermal neutron, T, is Λ/v. If Σ_a is proportional to $1/v$, as is often the case, T is independent of velocity. Thermal neutrons are generated in a medium when higher-energy neutrons are thermalized by a collision. Call $q_{th}(\mathbf{r})$ the number of neutrons per cubic centimeter becoming thermal per unit time. The quantity $q_{th}(\mathbf{r})$ is the slowing-down density considered previously for τ = thermal age.

We shall now establish a steady-state condition. We have, per unit volume, an increase in neutron density caused by

> Diffusion: $D \nabla^2 n$
> Slowing down: q_τ
> Absorption: $-n/T$

Summing these up, we have for the stationary state

$$\frac{\partial n}{\partial t} = 0 = D \nabla^2 n + q_\tau - \frac{n}{T} \tag{12-7.2}$$

or, using the expression for D [Eq. (12-7.1)],

$$\nabla^2 n - \frac{3}{\lambda \Lambda} n + \frac{3 q_\tau}{\lambda v} = 0 \tag{12-7.3}$$

The quantity

$$L = \left(\frac{\lambda \Lambda}{3} \right)^{1/2} \tag{12-7.4}$$

is called the *diffusion length*. Equation (12-7.3) must be solved with the

ISBN 0-8053-8061-7

boundary condition $n = 0$ on the surface of the moderator. The inhomogeneous term q_τ must be obtained by previously solving Eq. (12-6.11) with its sources and boundary conditions.

The mathematical formulation of the slowing-down and diffusion problems is thus complete in the approximation when diffusion theory is valid. Simultaneous solution of Eqs. (12-6.11) and (12-7.2) with the proper boundary conditions gives the complete answers. As an example of the solution of Eq. (12-7.2) we shall first give, as in the case of Eq. (12-6.11), the solution for a point source of thermal neutrons, located in the origin and emitting Q neutrons per second,

$$n(r) = \frac{Q}{4\pi Dr} e^{-r/L} \tag{12-7.5}$$

Note that this solution decreases less rapidly as a function of r than does the corresponding solution of Eq. (12-6.11).

As a second example we shall treat again the square column of side b in the x and y directions and of infinite height. With the boundary conditions $n = 0$ for $x = 0$ and $x = b$, $y = 0$ and $y = b$, the solution is

$$n(x, y, z) = \sum_{1 = j, k}^{\infty} n_{jk}(z) \sin \frac{\pi j x}{b} \sin \frac{\pi k y}{b} \tag{12-7.6}$$

and

$$n_{jk}(z) = c \exp\left[-\left(\frac{1}{L^2} + \pi^2 \frac{j^2 + k^2}{b^2} \right)^{1/2} z \right] \tag{12-7.7}$$

with j, k integral numbers and c proportional to the strength of the thermal neutron source located on the plane $z = 0$. The mode of slowest decay in the z direction is the one with $j = k = 1$, for which

$$n(x, y, z) = c \sin \frac{\pi x}{b} \sin \frac{\pi y}{b} \exp\left[-\left(\frac{1}{L^2} + \frac{2\pi^2}{b^2} \right)^{1/2} z \right] \tag{12-7.8}$$

The arrangement we are considering here, called the thermal column, can be realized in practice by putting a graphite column on a reactor. Essentially the same purpose is served by the sigma pile, in which a fast neutron source is located at the base of a graphite prism. Provided that one is far enough from the source ($z^2 \gg 4\tau_{th}$), the slow neutron density decays on the axis as

$$\exp\left[-\left(\frac{1}{L^2} + \frac{2\pi^2}{b^2} \right)^{1/2} z \right] \tag{12-7.9}$$

ISBN 0-8053-8061-7

and thus measurement of the slow neutron density gives L, the diffusion length. Naturally the measurements are sensitive to L^2 only if

$$\frac{2\pi^2}{b^2} \ll \frac{1}{L^2} \quad \text{or} \quad L \ll \frac{b}{(2\pi^2)^{1/2}} \tag{12-7.10}$$

Such measurements are suitable for determining the absorption coefficients of the weakly absorbing materials used in nuclear reactors and were of great importance in the development of the first pile: Fermi called them exponential experiments.

Table 12-3 gives data for important materials obtained by the combination of exponential experiments and measurements of total cross sections.

TABLE 12-3 DENSITY, DIFFUSION LENGTH, AND MEAN FREE PATH FOR SCATTERING AND ABSORPTION OF SLOW NEUTRONS IN IMPORTANT MATERIALS

	ρ, g cm^{-3}	$L = (\lambda\Lambda/3)^{1/2}$, cm	λ, cm	Λ, cm
H_2O	1.00	2.70	0.43	51.8
D_2O	1.10	102	2.4	1.34×10^4
Be	1.84	22.2	2.1	705
Graphite	1.62	47.2	2.7	2480

12-8 CHAIN-REACTING PILE

In Secs. 12-5 to 12-7 we have studied the moderation and diffusion of neutrons in an inert medium. If we add to this medium a fissionable material, for example, ^{235}U, neutrons are occasionally captured by it, fission results, and more neutrons are generated. The process may become divergent, giving rise to a chain reaction. A reactor is an apparatus in which fissionable material is accumulated (or piled, hence the name "pile") in such a way that a chain reaction develops (Figs. 12-12 and 12-13).

By a *generation* of neutrons we mean the processes from the production of a neutron to its final absorption. By *generation time* τ we mean the time required for this succession of processes. Naturally individual neutrons have different life histories. In a "generation" and "generation time" we consider

Figure 12-12 The research facility of the M. v. Laue–P. Langevin Research Institute (building diameter 62 m). The reactor proper is fueled by 8.5 kg of 93% ^{235}U mixed with Al and uses D_2O as moderator and neutron reflector. Power: 57 MW thermal; max. neutron flux about 10^{15} neutrons cm^{-2} sec^{-1}. [Courtesy M. v. Laue–P. Langevin Institute, Grénoble]

ISBN 0-8053-8061-7

ISBN 0-8053-8061-7

1 Reactor Fuel Element
2 Control Rod
3 Heavy Water Pump
4 Heat Exchanger D_2O - H_2O
5 Inclined Beam Tube
6 Hall for large Experiments
7 Laboratories
8 Neutron Guide Tubes Building
9 Fuel Element handling Device
10 Fuel Element handling Device
11 Conversion Electron Spectrometer
12 Cold Source Apparatus
13 Fission Product Spectrometer
14 Three axis Spectrometer
15 Capture Gamma Spectrometer
16 Rotating Crystal Spectrometer
17 Double Chopper
18 Small Angle Scattering
19 Multiple Diffraction Spectrometer "Hedgehog"
20 Three axis Spectrometer "Dancing floor"

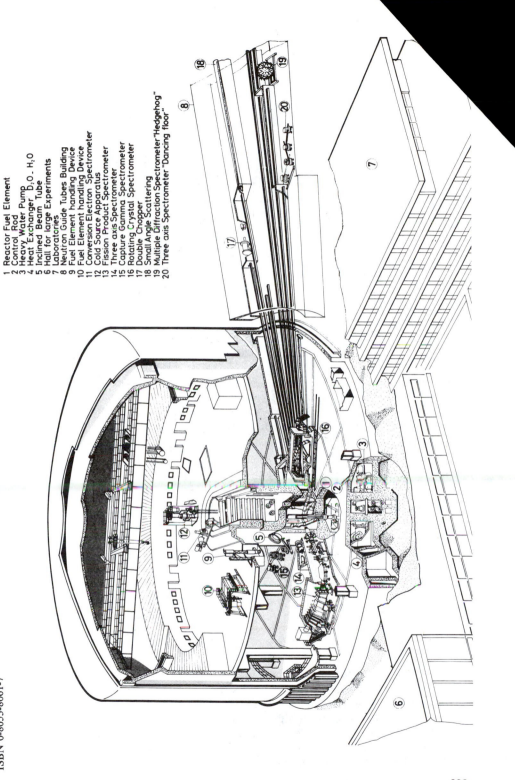

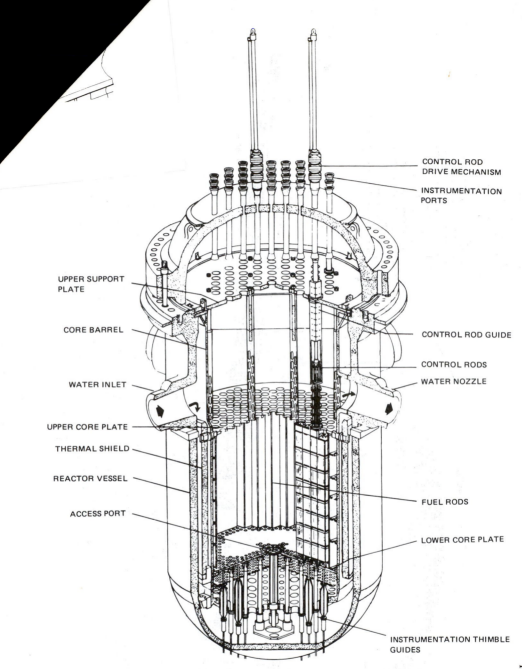

CONTROL ROD
DRIVE MECHANISM

INSTRUMENTATION
PORTS

UPPER SUPPORT
PLATE

CORE BARREL

CONTROL ROD GUIDE

CONTROL RODS

WATER INLET

WATER NOZZLE

UPPER CORE PLATE

THERMAL SHIELD

REACTOR VESSEL

FUEL RODS

ACCESS PORT

LOWER CORE PLATE

INSTRUMENTATION THIMBLE
GUIDES

Figure 12-13 A Westinghouse pressurized-water power reactor. The reactor
is used to produce steam, which generates electricity. Power: 3250 MW thermal, 1054 MW electric. Fuel 86 000 kg 3% ^{235}U as UO_2 rods clad in ZrNb
alloy. Moderator and coolant, ordinary water; core diameter 3.28 m; pressure
153 atm. Temperature: inlet 291°C, outlet 325°C. Average neutron flux 3.5 ×
10^{13} neutrons cm^{-2} sec^{-1}. Control rods of B_4C. [Westinghouse Preliminary
and Final Safety Analysis Reports; courtesy Diablo plant.]

suitable averages over all neutrons. We call k, the *multiplication constant*, the number of neutrons produced by a single neutron in a generation in an infinite medium, or the ratio between the number of neutrons produced and the number of neutrons absorbed per unit time in an infinite medium. For a chain reaction to occur it is necessary that $k > 1$. The neutron density in the medium then grows according to the equation

$$\frac{dn}{dt} = \frac{n(k-1)}{\tau} \tag{12-8.1}$$

whose solution is

$$n(t) = n(0)e^{(k-1)t/\tau} \tag{12-8.2}$$

In a reactor of finite size, some neutrons are lost by escape through the walls. To attain criticality, we can compensate for the losses by using a medium with a k sufficiently larger than 1. It is useful to define a number k_{eff} = average number of neutrons produced in reactor per unit time ÷ (average number of neutrons absorbed in reactor + average number of neutrons escaping from reactor per unit time). Clearly $k > k_{\text{eff}}$. The number k is a function of the medium only, and k_{eff} is a function of k and of the geometry of the reactor. The condition $k_{\text{eff}} > 1$ is the criticality condition for a finite reactor.

We shall first study k for a homogeneous mixture of uranium, possibly enriched in ^{235}U, and a moderator (graphite). We define the following:

$\sigma_r(U)$ = absorption cross section for thermal neutrons by all processes, excluding fission; mostly (n, γ) reaction
$\sigma_f(U)$ = fission cross section for thermal neutrons
$\sigma_f + \sigma_r = \sigma_a$ = total absorption cross section
N_C = number of atoms of species C per unit volume
$\sigma_r/\sigma_f = \alpha = (\sigma_a - \sigma_f)/\sigma_f$
ν = average number of neutrons produced per fission, is a nuclear constant depending slightly on the energy of the neutrons producing fission. Unless otherwise specified, ν refers to fission produced by thermal neutrons
η = average number of neutrons emitted per thermal neutron absorbed in fissionable material (fuel)

We have by definition

$$\eta = \nu \frac{\sigma_f(U)}{\sigma_f(U) + \sigma_r(U)} \tag{12-8.3}$$

Table 12-4 gives numerical values for the quantities mentioned above.

ISBN 0-8053-8061-7

TABLE 12-4   NEUTRON CROSS SECTIONS (IN BARNS) OF THE PRINCIPAL FISSILE NUCLIDES ^{233}U, ^{235}U, ^{239}Pu, AND ^{241}Pu[a]

	^{233}U	^{235}U	^{239}Pu	^{241}Pu
σ_a	578 ± 2	678 ± 2	1013 ± 4	1375 ± 9
σ_f	531 ± 2	580 ± 2	742 ± 3	1007 ± 7
σ_γ	47 ± 1	98 ± 1	271 ± 3	368 ± 8
α	0.089 ± 0.002	0.169 ± 0.002	0.366 ± 0.004	0.365 ± 0.009
η	2.284 ± 0.006	2.072 ± 0.006	2.109 ± 0.007	2.149 ± 0.014
ν	2.487 ± 0.007	2.423 ± 0.007	2.880 ± 0.009	2.934 ± 0.012

[a]G. C. Hanna, et al. *At. Energy Rev.*, **7**, 3–92 (1969). (Figures in the referenced article were all given to one additional significant figure.) (Neutron energy = 0.0252 eV, velocity = 2200 m sec^{-1}).

$\sigma_a = \sigma_f + \sigma_\gamma$; $\alpha = \sigma_\gamma / \sigma_f$; ν = neutrons per fission = $\eta(1 + \alpha)$.

The neutrons generated on fission are absorbed in part by the fuel, in part by the moderator. The fraction absorbed by the fuel will depend on the energy interval considered. For thermal neutrons it is called the *thermal utilization factor* and is indicated by the letter f. The quantity f is a function of the composition of the medium. Precisely, for a homogeneous medium

$$f = \frac{N_U \sigma_a(U)}{\sum_C N_C \sigma_a(C) + N_U \sigma_a(U)} \qquad (12\text{-}8.4)$$

If we could neglect all phenomena occurring before thermalization, the result would be that one neutron has a probability f of being absorbed in the fuel and will, in such a case, generate η neutrons. Hence

$$k = \eta f \qquad (12\text{-}8.5)$$

This relation would be approximately true in a reactor operating on a homogeneous mixture of ^{235}U and a moderator.

For natural or slightly enriched uranium we must consider the possibility of producing fission with fast-fission neutrons. By the "fast-fission factor" ϵ we mean the ratio of neutrons produced by fast and thermal fission to those produced by thermal fission only. In the case of lumped materials the number ϵ is appreciably different from 1. A fission neutron crossing the lump produces fission in about 1% of the cases, giving typical values of ϵ between 1.02 and 1.04.

In natural uranium a neutron has an appreciable probability of being captured by ^{238}U in the resonance region, before thermalization, without producing fission neutrons. We designate by p the probability that a neutron will escape resonance capture. Both ϵ and p depend on the composition and geometrical arrangement of the fuel and the moderator. The important invention

ISBN 0-8053-8061-7

of distributing the fuel on a lattice embedded in the moderator (lumping) has four effects: (1) It increases ϵ, which is advantageous. (2) ^{238}U has strong resonance absorption, and it is therefore desirable to accomplish the moderation in a uranium-free region; the lumping of uranium acts in this advantageous direction. (3) Resonance capture in uranium occurs essentially at the surface of a lump. Neutrons in the uranium resonance-energy region have a very large σ_r and do not penetrate the uranium lumps; hence the uranium inside a lump does not tend to lower p. Consequently we want large lumps in which the surface-layer volume is small compared with the total volume. (4) On the other hand, large-volume lumps tend to lower f, because the neutron density in or near the lumps tends to be smaller than in the moderator, favoring neutron absorption in the moderator, which is detrimental. It is clear that the requirements for optimum values of f and p conflict with each other and must be resolved by compromise. Of the four effects mentioned above, the most important is (3).

When we take into account fast neutron fission and resonance capture, k becomes

$$k = \eta f \epsilon p \tag{12-8.6}$$

a famous formula called the *four-factors formula*.

For the calculation p and ϵ we must refer to the special literature. Only for a diluted homogeneous mixture can p be approximated by

$$p(E) = \exp\left[-\frac{N_U}{\xi}\int_E^{E\,\text{source}}\lambda(E')\sigma_U(E')\,\frac{dE'}{E'}\right] \tag{12-8.7}$$

where λ is the mean free path in the medium and $\sigma_U(E)$ is the capture cross section of the fissionable material. For very dilute natural uranium λ may be considered energy independent and determined entirely by the moderator. It is thus convenient to evaluate p by using the "resonance integral"

$$I = \int_{0.3\,\text{eV}}^{E_0}\sigma_U\,\frac{dE}{E} = 240 \times 10^{-24}\ \text{cm}^2 \tag{12-8.8}$$

The lower limit of the integral corresponds to the absorption cut off of cadmium, and reflects the experimental technique by which the integral was measured. Then p is given by

$$p(0.3\ \text{eV}) = \exp\left(-\frac{N\lambda}{\xi}I\right) \tag{12-8.9}$$

The effect of lumping drastically reduces the resonance capture. For instance, the resonance integral for lumps of natural uranium is given empirically by

$$I = \int_{th}^{E_0} \sigma_{eff} \frac{dE}{E} = \left(9.25 + 24.7 \frac{S}{M}\right)(10^{-24}\ cm^2) \qquad (12\text{-}8.10)$$

where S is the surface of the lump in square centimeters and M its mass in grams. The term σ_{eff} indicates the cross section for the uranium atom, which gives a resonance integral, and this, when inserted in Eq. (12-8.9), gives the correct p.

We are now ready to consider a reactor of finite size, for which the effective multiplication factor k_{eff} may be written

$$k_{eff} = kP \qquad (12\text{-}8.11)$$

where P is the probability that a neutron will not leak out of the reactor. Table 12-5 shows diagrammatically the neutron economy in a finite reactor.

To determine the minimum critical dimensions for a reactor material of a given $k > 1$, and on the assumption (for the sake of simplicity) that the neutrons are thermal throughout their life, we return to Eq. (12-7.2) and consider the neutron balance, including multiplication. Equation (12-7.2) is then replaced by

$$D\ \nabla^2 n + \frac{(k-1)n}{T} + q_{th} = \frac{\partial n}{\partial t} \qquad (12\text{-}8.12)$$

and the criticality condition is characterized by a $\partial n/\partial t = 0$, even in the absence of sources q, that is, by

$$D\ \nabla^2 n + \frac{k-1}{T} n = 0 \qquad (12\text{-}8.13)$$

Introducing for D its value from Eq. (12-7.1) and remembering Eq. (12-7.4), we have

$$\nabla^2 n + \frac{k-1}{L^2} n = 0 \qquad (12\text{-}8.14)$$

This equation is to be solved with the boundary condition $n = 0$ on the surface of the reactor.

Equation (12-8.14) gives rise to an eigenvalue problem, and the lowest eigenvalue gives the criticality condition. Thus, for a cube of side a and with a corner in the origin of the coordinates,

$$n = C \sin \frac{\pi x}{a} \sin \frac{\pi y}{a} \sin \frac{\pi z}{a} \qquad (12\text{-}8.15)$$

ISBN 0-8053-8061-7

TABLE 12-5 NEUTRON ECONOMY IN NATURAL OR SLIGHTLY ENRICHED CHAIN REACTOR[a]

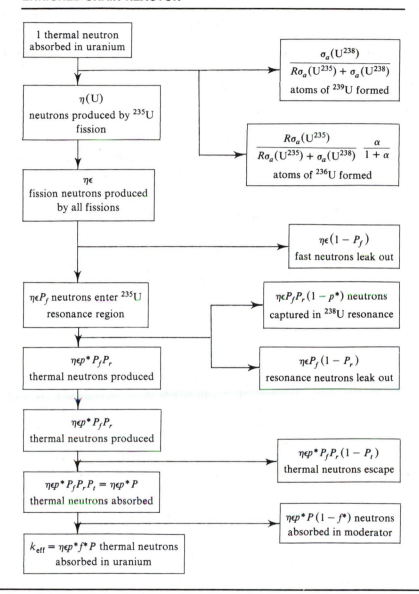

[a]From (WW 58). R is the atomic fraction of ^{235}U in uranium. For natural uranium $R = 0.007205$; p^*, f^* take into account the influence on the resonance escape probability and on the thermal utilization factor of the geometry of the reactor. In practice, $p^* \cong p$, $f^* \cong f$.

and

$$\frac{3\pi^2}{a^2} = \frac{k-1}{L^2} \tag{12-8.16}$$

In general the nuclear characteristics of the material determine $(k-1)/L^2$ (material buckling), and the geometry of the reactor determines $-(\nabla^2 n)/n$ (geometric buckling). The criticality condition is

$$-\frac{\nabla^2 n}{n} = B^2 = \frac{k-1}{L^2} \tag{12-8.17}$$

The considerations developed here are oversimplified and do not correspond to any real reactor, because fission neutrons are not generated at thermal energy.

In media to which the age theory is applicable we can still give a simple, fairly accurate theory.

Let us start again from Eq. (12-7.2); for thermal age

$$\frac{\lambda v}{3} \nabla^2 n - \frac{v}{\Lambda} n + q_{\text{th}} = \frac{\partial n}{\partial t}$$

and the age equation (12-6.11)

$$\nabla^2 q = \frac{\partial q}{\partial \tau}$$

We recall the meaning of q_{th} as the number of neutrons per unit time per unit volume becoming thermal at the position x, y, z. Appropriately, q is called the *density of nascent thermal neutrons*. Assume, as before, a cubical pile of side a. The boundary conditions require that the functions n, q vanish on the surface of the pile. Moreover, we assume that the neutrons are generated at an energy E (average energy of fission neutrons). We develop n, q in Fourier series and limit ourselves to the first harmonic, writing

$$n = n' \sin\frac{\pi}{a} x \sin\frac{\pi}{a} y \sin\frac{\pi}{a} z \tag{12-8.18}$$

$$q(x, y, z, \tau) = Q\Theta(\tau) \sin\frac{\pi}{a} x \sin\frac{\pi}{a} y \sin\frac{\pi}{a} z \tag{12-8.19}$$

Note that the space dependence of q is the same as that of n, irrespective of age. By Q we mean a constant to be determined, such that for $\tau = 0$, q becomes the q corresponding to the fission neutrons generated in the pile. From Eqs. (12-6.11) and (12-8.19) we have separating variables:

$$\frac{-3\pi^2}{a^2} \Theta = \frac{d\Theta}{d\tau} \tag{12-8.20}$$

ISBN 0-8053-8061-7

from which

$$\Theta(\tau) = \Theta(0)e^{-B^2\tau} \qquad (12\text{-}8.21)$$

with $-B^2 = -3\pi^2/a^2$ and $\Theta(0) = 1$, because for $\tau = 0$ the function q is $Q \sin(\pi/a)x \sin(\pi/a)y \sin(\pi/a)z$.

We obtain now from Eq. (12-7.3) by substitution

$$n\left(-\frac{\lambda v}{3}\frac{3\pi^2}{a^2} - \frac{v}{\Lambda} + \frac{Q}{n'}e^{-B^2\tau_{th}}\right) = \frac{\partial n}{\partial t} \qquad (12\text{-}8.22)$$

However, the ratio of Q/n' can be determined from the reproduction factor of the pile. In fact, if n is the density of slow neutrons, then nv/Λ neutrons per second and cubic centimeter are absorbed. Of these, the fraction f is absorbed in the fuel and each generates η fission neutrons. The fission neutrons are multiplied by ϵ in escaping from the lumps, but only the fraction p escapes resonance absorption.

In conclusion, remembering the four-factors formula [Eq. (12-8.6)], we find

$$\frac{kn'v}{\Lambda} = Q \qquad (12\text{-}8.23)$$

Substituting in Eq. (12-8.22) we have a criticality condition

$$-\frac{\lambda v}{3}\frac{3\pi^2}{a^2} - \frac{v}{\Lambda} + \frac{kv}{\Lambda}e^{-B^2\tau_{th}} = 0 \qquad (12\text{-}8.24)$$

or

$$-L^2B^2 - 1 + ke^{-B^2\tau_{th}} = 0 \qquad (12\text{-}8.25)$$

where

$$L^2 = \frac{\lambda\Lambda}{3} \qquad B^2 = \frac{3\pi^2}{a^2} = \text{geometric buckling} \qquad (12\text{-}8.26)$$

Solving for k we find

$$k = (1 + L^2B^2)e^{B^2\tau_{th}} \qquad (12\text{-}8.27)$$

which, as expected, for $\tau = 0$ gives the same result as Eq. (12-8.17). If, as usual, the exponential is near 1, we can write Eq. (12-8.27) as

$$k = 1 + (3\pi^2/a^2)(L^2 + \tau_{th}) \qquad (12\text{-}8.28)$$

ISBN 0-8053-8061-7

or

$$-\frac{\nabla^2 n}{n} = \frac{k-1}{\tau_{th} + L^2} \tag{12-8.29}$$

The similarity between Eqs. (12-8.29) and (12-8.17) is obvious: The diffusion length L has been replaced by the *migration length*,

$$\left(L^2 + \tau_{th}\right)^{1/2} = M \tag{12-8.30}$$

For a finite pile one often uses the *effective multiplication constant* k_{eff} defined by

$$k_{eff} = \frac{ke^{-B^2\tau_{th}}}{1 + B^2 L^2} \tag{12-8.31}$$

The criticality condition expressed through k_{eff} is then $k_{eff} > 1$.

Table 12-6 gives typical figures for a low-enrichment thermal reactor and affords a quantitative idea of the relative importance of the effects considered thus far.

TABLE 12-6   NEUTRON BALANCE—LARGE, LOW-ENRICHMENT THERMAL REACTORS

Virgin fast neutrons from thermal fissions	+ 100.0
Fast neutrons lost by fast absorption	− 0 to 3
Fast neutrons gained by fast fission	+ 0 to 8
Fast neutrons lost in leakage	− 2 to 10
Resonance capture in fertile material	− 15 to 25
Thermal neutrons lost in leakage	− 1 to 5
Thermal neutrons captured in structural materials, moderator, reflector, coolant, control	− 5 to 15
Thermal neutrons absorbed by fission product	− 2 to 5
Thermal-neutron capture in fertile material	− 10 to 20
Thermal neutrons captured parasitically in fissionable material	− 6 to 8
Thermal-neutron fission captures	− 40

To show the effect of distributing fissionable material in lumps, we note that for a homogeneous natural uranium—graphite mixture the maximum obtainable k is 0.78 for a ratio of 400 carbon atoms to 1 atom of uranium. It is therefore impossible to render such a mixture critical. On the other hand, for a lattice of uranium bars 2.5 cm in diameter, placed at a distance of 11 cm from axis to axis, in a graphite matrix,

$$\epsilon = 1.028 \quad p = 0.905 \quad f = 0.888 \quad \eta = 1.308 \quad k = 1.0625$$

$$\tau_{th} = 300 \text{ cm}^2 \quad L^2 = 350 \text{ cm}^2$$

and a "bare" cube measuring 5.55 m on a side would be critical.

ISBN 0-8053-8061-7

The reactors considered up to now are "bare." In practice most reactors are surrounded by a reflector. This is a layer of moderating material free of fissionable material. The reflector reduces the critical mass of the core, because neutrons that would escape from the bare core are reflected back into it by the surrounding material.

12-9 PILE KINETICS

Up to now we have considered the critical condition of a reactor without regard to the time dependence of the neutron flux. In practice a pile must be controlled; that is, the operator must be able to vary its k_{eff}. *Control rods* usually serve this purpose. These are rods of a strongly neutron-absorbent material, such as cadmium or boron-containing steel, which can be introduced into or removed from the pile structure. The introduction of the rod decreases k_{eff}, because neutrons are lost to the rod and because the neutron distribution is altered, thereby affecting the leakage from the pile.

We must study the behavior, with respect to time, of a pile when k is changed. For a neutron in a thermal pile most of the time is usually spent in diffusion after moderation; for example, moderation may require a few microseconds and diffusion a few milliseconds. It is therefore clear that in 1 sec there would be hundreds of generations, and, considering Eq. (12-8.1), we recognize that control of a pile would require prohibitively fast action. The situation is greatly eased by the delayed neutrons, as we shall see shortly. At first, however, let us neglect their presence. Equation (12-8.12) gives, for the time dependence of the neutron density, and taking Eq. (12-8.30) into consideration as well,

$$D \nabla^2 n + \frac{D(k-1)}{M^2} n = \frac{\partial n}{\partial t} \qquad (12\text{-}9.1)$$

which is valid if $M \ll$ pile dimensions. Note that, if k were 0, $M^2/D = \tau_0$ would represent the mean life for absorption of a neutron introduced into the pile [cf. Eq. (12-7.2)]. Separating the variables, we find from Eqs. (12-9.1) and (12-8.17) that

$$n = n(\mathbf{r}) e^{t/T} \qquad (12\text{-}9.2)$$

with

$$\frac{1}{T} = \frac{k - 1 - B^2 M^2}{M^2/D} = \frac{k_{eff} - 1}{\tau_0} \qquad (12\text{-}9.3)$$

where T is the pile *relaxation time* and B^2 is the geometric buckling for the fundamental mode. Actually a more accurate expression is obtained by using

ISBN 0-8053-8061-7

for τ_0 the expression

$$\tau_0 = \frac{L^2}{D_{\text{th}}} \tag{12-9.4}$$

In a graphite-moderated reactor τ_0 is of the order of 10^{-3} sec; hence even $k_{\text{eff}} = 1.001$ already gives a doubling time for the reactor flux of only 1 sec (approximately), which is too short for comfortable operation.

The delayed neutrons, however, change the situation radically, and we shall now consider their effect, although our treatment is oversimplified. The number of delayed neutrons per fission in ^{235}U amounts to about 0.016 neutron per fission. We shall call β the ratio of this number to ν; for ^{235}U $\beta = 0.0064$. The half-life of the delayed neutrons in ^{235}U ranges from 55.7 to 0.23 sec. Here we shall assume that all the neutrons have the same mean life τ_d.

It is clear that k may now be considered as the sum of a reproduction factor due to prompt neutrons only, k_p, and a reproduction factor due to delayed neutrons only, k_d. We thus write

$$k = k_p + k_d = k_p + \beta k \tag{12-9.5}$$

or

$$k_p = k(1 - \beta) \tag{12-9.6}$$

If the reactor is critical on the prompt neutrons only—that is, if $k_{\text{eff}} > 1$ or $k_{p,\,\text{eff}} > 1/(1 - \beta)$—the kinetics is practically unaffected by the delayed neutrons. On the other hand, if the pile is subcritical on prompt neutrons only, but supercritical on prompt plus delayed neutrons, the time constant of the pile is obtainable as follows: First Eq. (12-8.12) is replaced by

$$D \nabla^2 n + \frac{k_p - 1}{\tau_0} n + \frac{C}{\tau_d} = \frac{\partial n}{\partial t} \tag{12-9.7}$$

in which $C(\mathbf{r})$ is the density of the "pregnant" nuclei, as the delayed neutron emitters are called. The quantity C is simply related to n by

$$\frac{\partial C}{\partial t} = \frac{k_d}{\tau_0} n - \frac{C}{\tau_d} \tag{12-9.8}$$

One then solves the two coupled equations [(12-9.7) and (12-9.8)] by assuming $n(\mathbf{r}, t) = n_0 N(\mathbf{r})e^{t/\tau}$ and $C(\mathbf{r}, t) = C_0 N(\mathbf{r})e^{t/\tau}$, with $N(\mathbf{r})$, a function of the coordinates only, and n_0, C_0 constants, to be eliminated. From the coupled equations one obtains the connection between the period of the pile T and

ISBN 0-8053-8061-7

the excess reactivity ρ,

$$\rho = \frac{k_{\text{eff}} - 1}{k} = \frac{\tau_0}{kT} + \sum_i \frac{\beta_i \tau_i}{T + \tau_i} \qquad (12\text{-}9.9)$$

Equation (12-9.9) is called the *inhour* (inverse hour) equation. It relates the excess reactivity ρ to the parameters of the delayed neutron emitters and the period of the pile.

Substituting numerical values in Eq. (12-9.9) and defining as 1 inverse hour (inhr) the amount of excess reactivity that would give rise to a period of 1 hr, we find

$$(k_{\text{eff}} - 1) \text{ for 1 inhr} = 2.6 \pm 0.2 \times 10^{-5}$$

The relation between the excess reactivity in inhours and the period of the pile in seconds, as set forth in Eq. (12-9.9), is shown in Fig. 12-14.

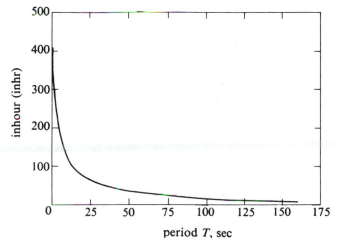

Figure 12-14 The inhour formula in graphical form. A measurement of the reactor period can be converted into an inhour of reactivity by referring to this graph.

An interesting physical application of Eq. (12-9.9) is the measurement of small-neutron-absorption cross section. A sample of the absorber is introduced into a pile, and the control rods are moved so as to compensate for the change in T of the pile produced by the absorber. The effect of the rod's position on ρ can be calibrated with known absorbers such as boron ($\sigma_a = 755 \times 10^{-24}$ cm^2 at 2200 m sec^{-1}) and by using Eq. (12-9.9) to interpolate or extrapolate the experimental data.

ISBN 0-8053-8061-7

12-10   BREEDING AND CONVERTING

In a uranium pile part of the neutrons are absorbed by ^{238}U and ultimately yield ^{239}Pu. The pile thus "converts" some ^{235}U, which is used to keep it going, into ^{239}Pu. The ratio of the desired nuclei generated to fuel nuclei destroyed is called the *conversion ratio* and is indicated by the letter C; the conversion gain G, or breeding, is $C - 1 = G$.

Indicating by l the neutrons lost by leakage or absorption in all but fissionable material we have

$$C = \eta - 1 - l \tag{12-10.1}$$

$$G = \eta - 2 - l \tag{12-10.2}$$

At thermal energies η is only slightly greater than 2 for all known nuclei except ^{233}U, and hence a thermal breeder is likely only with this substance. At higher energies conditions are more favorable, because η increases with energy, owing to the decrease of σ_a and to a lesser extent to the increase of ν. At 1 MeV η is estimated to reach 2.45 for ^{233}U, 2.3 for ^{235}U, and 2.7 for ^{239}Pu.

12-11   FUSION REACTIONS

We have seen how heavy elements may liberate large amounts of energy on fission. Light elements can do the same on "fusion," that is, by forming heavier nuclei through nuclear combinations of lighter ones. For example, if we could bind together two neutrons and two protons to form an alpha particle, we should set free 26 MeV.

Such fusion reactions are of colossal importance, because they are the source of solar energy. On the earth they occur in hydrogen bombs, and great effort is being devoted to produce them at a controlled rate.

We shall treat briefly some of the astrophysical reactions. Only binary reactions are of any practical importance; the collision of four particles cited above does not occur in actuality. However, the same result is achieved through a series of reactions which form a cycle, reconstituting some of the participating nuclei, which act only as catalysts.

There are two notable solar cycles. The first is

$$
\begin{array}{lll}
^{12}\text{C} + {^1}\text{H} \rightarrow {^{13}}\text{N} + \gamma & 1.93 \ \text{MeV} \\
^{13}\text{N} \quad\ \rightarrow {^{13}}\text{C} + e^+ + \nu + \gamma & 1.20 \ \text{MeV} \\
^{13}\text{C} + {^1}\text{H} \rightarrow {^{14}}\text{N} + \gamma & 7.60 \ \text{MeV} \\
^{14}\text{N} + {^1}\text{H} \rightarrow {^{15}}\text{O} + \gamma & 7.39 \ \text{MeV} \\
^{15}\text{O} \quad\ \rightarrow {^{15}}\text{N} + e^+ + \nu + \gamma & 1.71 \ \text{MeV} \\
^{15}\text{N} + {^1}\text{H} \rightarrow {^{12}}\text{C} + {^4}\text{He} & 4.99 \ \text{MeV}
\end{array}
\tag{12-11.1}
$$

ISBN 0-805-3-8001-1

This is called the *carbon cycle* (Fig. 12-15; Bethe, 1938), and its net effect is to combine four protons to form ^{4}He, gamma rays, and neutrinos (the two excess positrons will ultimately annihilate) and to liberate about 25 MeV of thermal energy, the neutrinos escaping. Note that the carbon is not destroyed in the process but acts only as a catalyst. The other cycle is

$$^1H + {}^1H \rightarrow {}^2H + e^+ + \nu \qquad 0.41 \text{ MeV}$$

$$^2H + {}^1H \rightarrow {}^3He + \gamma \qquad 5.51 \text{ MeV}$$

$$^3He + {}^3He \rightarrow {}^4He + 2{}^1H + \gamma \qquad 12.98 \text{ MeV} \qquad (12\text{-}11.2)$$

which also ultimately combines four protons to make an alpha particle.

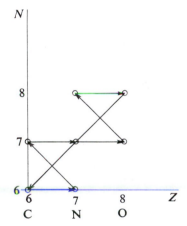

Figure   12-15   Carbon cycle on a Z–N diagram.

Temperatures at the interior of the stars are in the range of 10 to 20 million degrees centigrade, and the nuclear kinetic energies are of the order of 1 keV. All atoms are ionized, and there are no free neutrons.

The cross sections for the nuclear reactions between the charged particles involved show, at low energies, a very strong energy dependence on the Gamow factor (Fig. 12-16). Thus, in a gaseous mixture at a certain temperature T, the fastest particles in the Maxwellian distribution will react preferentially. The number of particles in a given energy interval, however, is a strongly decreasing function of energy, and the result is that the reactions will occur in a rather narrow energy interval around an optimum energy. More precisely, the Maxwellian distribution of velocities introduces into the reaction rate a factor $e^{-E/kT}$, and the effect of the Coulomb barrier gives the factor $e^{-Z_1 Z_2 e^2/\hbar v}$, where $Z_1 Z_2$ are the atomic numbers of the nuclei colliding

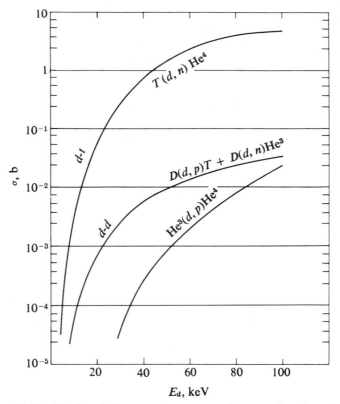

Figure 12-16   Nuclear-fusion reaction cross sections as a function of relative particle energy.

with relative velocity v. After integrating over all velocities, the reaction rate is approximately proportional to

$$\tau^2 e^{-\tau} \tag{12-11.3}$$

with $\tau = 3(\pi^2 m e^4 Z_1^2 Z_2^2 / 2\hbar^2 kT)^{1/3}$, where m is the reduced mass. The main contribution to this integral comes from the region where the function has a maximum which, under stellar conditions, corresponds to E between 3 and 50 keV (Fig. 12-17). The absolute rates in Table 12-7 are reported from a paper by Salpeter.

Fusion reactions on earth are used in the so-called hydrogen bomb and other nuclear explosions. For many years serious efforts have been directed at obtaining power from controlled thermonuclear reactions. Up to now these efforts have fallen short of their goal, but we will briefly describe the nature

ISBN 0-8053-8061-7

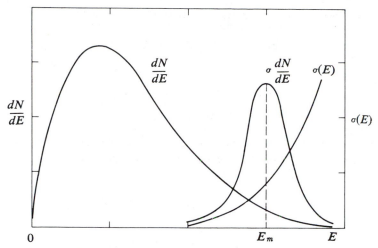

Figure 12-17 Effect of Maxwellian energy distribution on nuclear reaction rate. $\sigma(E)$, cross section; dN/dE, number of particles having relative energy E; the reaction rate is $\sigma\, dN/dE$.

TABLE 12-7   MEAN REACTION TIMES IN PROTON–PROTON CHAIN AND CARBON–NITROGEN CYCLE FOR TYPICAL STELLAR CONDITIONS[a]

Reaction	t_r
$^1H + {}^1H \rightarrow {}^2H + e^+ + \nu$	7×10^9 y
$^2H + {}^1H \rightarrow {}^3He + \gamma$	4 sec
$^3He + {}^3He \rightarrow {}^4He + 2(^1H)$	4×10^5 y
$^{12}C + {}^1H \rightarrow {}^{13}N + \gamma$	10^6 y
$^{13}N \rightarrow {}^{13}C + e^+ + \nu$	10 min
$^{13}C + {}^1H \rightarrow {}^{14}N + \gamma$	2×10^5 y
$^{14}N + {}^1H \rightarrow {}^{15}O + \gamma$	$< 3 \times 10^7$ y
$^{15}O \rightarrow {}^{15}N + e^+ + \nu$	2 min
$^{15}N + {}^1H \rightarrow {}^{12}C + {}^4He$	10^4 y

[a]Mean reaction time t_r for a temperature T of $15 \times 10^{6\circ}$K, density of 125 g cm^{-3}, hydrogen concentration x_H of 0.8, and total concentration x_{CN} of C and N of 0.01 (by mass).
From E. E. Salpeter, *Ann. Rev. Nucl. Sci.,* **2**, 41, (1952).

ISBN 0-8053-8061-7

of the problem. The reactions most commonly considered are

$$^2H + {}^2H \rightarrow {}^3H + {}^1H + 3.25 \text{ MeV}$$

$$^2H + {}^2H \rightarrow {}^3He + n + 4.0 \text{ MeV}$$

$$^3H + {}^2H \rightarrow {}^4He + n + 17.6 \text{ MeV}$$

$$^3H + {}^2H \rightarrow {}^4He + {}^1H + 18.3 \text{ MeV}$$

In order to obtain power from them, they must occur at an appreciable rate in a controllable, possibly steady-state, situation. The power density produced by any binary reaction is

$$p_{12} = n_1 n_2 \langle \sigma v \rangle_{12} W_{12} \qquad (12\text{-}11.4)$$

where p_{12} is the power density, n_1, n_2 the nuclear densities, $\langle \sigma v \rangle_{12}$ the average of the product of the cross section times the velocity, and W_{12} the energy released in one nuclear reaction. Figure 12-18 shows $\langle \sigma v \rangle_{12}$ for the reactions, assuming Maxwellian velocity distributions.

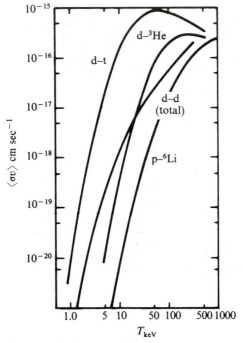

Figure   12-18   Plots of fusion reaction rate parameters $\langle \sigma v \rangle$ for D–T, D–^{3}He, D–D (total), and p–^{6}Li reactions as obtained by averaging over a Maxwellian distribution. 1 keV kinetic temperature $= 1.16 \times 10^7{}^\circ$K. [From R. F. Post, *Ann. Rev. Nucl. Sci.*, **20**, 518 (1970).]

ISBN 0-8053-8061-7

The central problem in fusion reactions is to obtain a plasma, a neutral mixture of electrons and ions, at a temperature such that the desired reaction occurs at a sufficient rate *and to maintain this condition*. The temperature required is of the order of 10 keV ($1.16 \times 10^{8}\,°K$) and all materials vaporize at such temperatures. To contain the plasma one uses magnetic fields, which in suitable configurations act as an impenetrable wall for charged particles. One can say that the magnetic field exerts a pressure of $B^2/8\pi$ on the plasma. The plasma stays confined for a certain containment time τ until instabilities develop and the plasma leaves the confinement region. The plasma must also be heated to bring it to the ignition temperature and the power developed must compensate energy losses which occur mainly by bremsstrahlung of the plasma electrons. The minimum requirement that the plasma heat up to the ignition temperature during the confinement time gives the Lawson criterion

$$p_{12}\tau = \tfrac{3}{2}(n_1 + n_2)kT_i \qquad (12\text{-}11.5)$$

or, assuming $n_1 = n_2 = n/2$

$$6kT_i/\langle \sigma v \rangle W = n\tau \qquad (12\text{-}11.6)$$

On the left-hand side we have quantities depending on nuclear properties; $n\tau$ is a controllable parameter. In order to have a thermonuclear reaction, $n\tau$ must actually exceed 10^{14}.

Table 12-8 gives examples of desired operating conditions for thermonuclear reactions according to Post (1970).

TABLE 12-8   EXAMPLE OF OPERATING CONDITIONS FOR STEADY-STATE AND PULSED FUSION REACTORS[a]

	Steady state		Pulsed
	D–T	D–^{3}He	D–T
Fuel cycle			
Fuel density (cm^{-3})	3×10^{14}	2×10^{14}	2.5×10^{16}
Temperature (keV)	25	300	10
Required confinement time (sec)	0.5	1.0	.025
Power density (W cm^{-3})	25	8	18 000
Pressure (ions only, atm)	12	96	400
Minimum confining field (kG)	25	60	140

[a] From R. F. Post, *Ann. Rev. Nucl. Sci.*, **20**, 516 (1970).

12-12   EFFECT OF CHEMICAL BINDING OF HYDROGEN SCATTERER

In the remainder of this chapter we shall discuss some properties of slow neutrons and some typical problems that have been investigated by slow neutron techniques. The chain-reacting pile has been the preferred neutron source for the investigations we shall report here.

ISBN 0-8053-8061-7

When a neutron of energy much greater than the molecular binding energy strikes a hydrogen nucleus in a molecule, it knocks this proton out of the molecule; and it loses, on the average, one half its energy to the proton (assuming elastic s-wave scattering). However, if the energy of the neutron is less than the $\hbar\omega$ of the molecular vibration, it cannot lose any energy to vibration or to freeing of the hydrogen. Consequently the proton acts as if it had the mass of the whole molecule. This makes it hard for the slow neutron to lose energy. Thus, as the thermal region is approached, neutrons find it increasingly difficult to lose energy. For these reasons, at energies below $\hbar\omega$ the reduced mass of a neutron and proton approaches $\mu = M$ rather than $\mu = M/2$. As a consequence the scattering cross section increases by a factor of 4 in passing from large energies to energies small compared with that of the chemical bond.

By the Born approximation (see Appendix A), the differential scattering cross section is

$$\sigma(\theta) = \frac{\mu^2}{4\pi^2\hbar^4} \left| \int \psi_f^* U\psi_i \, d\tau \right|^2 \tag{12-12.1}$$

where ψ_f and ψ_i are plane waves normalized for unit volume. Thus, in Eq. (12-12.1) $\sigma_{bound} = 4\sigma_{free}$, because μ_{bound} is twice as large as μ_{free}. Although the ordinary criterion for the applicability of the Born approximation is not satisfied for thermal energies, it can be shown that Eq. (12-12.1) is still valid.

That the total cross section for a bound proton is four times larger than for a free proton may also be proved by a simple argument developed by Blatt and Weisskopf. The incident slow neutron has a wavelength λ large compared with the dimensions of the molecule in which the proton is bound, and hence we have spherical symmetric scattering (s wave only) in the center-of-mass system, which practically coincides with the system in which the molecule is at rest,

$$\frac{d\sigma}{d\Omega} = c \tag{12-12.2}$$

In scattering by a free proton, the scattering is also spherically symmetric in the center-of-mass system, which, however, moves with half the velocity of the neutron with respect to the laboratory system. If we call a the n–p incoherent scattering length $[a = (\frac{3}{4} a_t^2 + \frac{1}{4} a_s^2)^{1/2}$; see Sec. 10-3], we know from Eq. (10-2.8) that in the center-of-mass system

$$\frac{d\sigma}{d\omega} = a^2 \tag{12-12.3}$$

where $d\omega$ is the center-of-mass solid angle. In the laboratory with Θ the scattering angle, this gives

$$\frac{d\sigma}{d\Omega} = \frac{d\sigma}{d\omega} \frac{d\omega}{d\Omega} = a^2 \frac{d\cos 2\Theta}{d\cos\Theta} = 4a^2 \cos\Theta \tag{12-12.4}$$

ISBN 0-8053-8061-7

In the forward direction ($\cos \Theta = 1$) the two cross sections must be equal, because the collision does not impart any momentum to the proton and thus cannot be affected by the proton being free or bound. Hence

$$c = 4a^2 \tag{12-12.5}$$

and the total cross section

$$\sigma_{\text{tot bound}} = 4\pi c = 16\pi a^2 \tag{12-12.6}$$

is four times as large as

$$\sigma_{\text{tot free}} = 4\pi a^2 \tag{12-12.7}$$

To estimate the energy at which the cross section increases because of the chemical bond, note that for the carbon–hydrogen bond in paraffin the longitudinal vibration is 3,000 cm^{-1}, or 1/3 eV. The transverse vibration is 600 cm^{-1}, or 1/15 eV.

This type of argument can be extended from molecules to crystal lattices, and the results are closely related to the recoilless processes of gamma emission and absorption (Sec. 8-8; Lamb, 1939).

12-13 LOW-ENERGY SCATTERING FROM COMPLEX NUCLEI

Let us first consider the scattering of a slow neutron from a free isolated nucleus. Only s-wave scattering will be important, because the de Broglie wavelength of the neutron is very large compared with the range of nuclear forces. The asymptotic expression of the wave function of the system (see Sec. 11-5) is

$$\psi \sim e^{ikz} + C_0 (e^{ikr}/r) \tag{12-13.1}$$

and

$$k = 1/\lambdabar = p/\hbar \tag{12-13.2}$$

The constant C_0 is $(i/2k)(1 - \eta_0)$ of Eq. (11-5.5). It is related to the s-phase shift δ by

$$C_0 = \frac{\lambdabar}{2i} (e^{2i\delta} - 1) = \lambdabar e^{i\delta} \sin \delta \cong \lambdabar \delta \text{ (for } \delta \text{ small)} \tag{12-13.3}$$

For pure elastic scattering, δ is real, and we have for the total scattering cross section

$$\sigma_s = 4\pi \lambdabar^2 \sin^2 \delta \tag{12-13.4}$$

We now use the scattering length (see Sec. 10-2) defined by

$$\lambdabar \sin \delta = -a \tag{12-13.5}$$

The scattering length in the case of

$$\left|\frac{a}{\lambda}\right| = |\sin \delta| \ll 1 \tag{12-13.6}$$

represents the intercept of the tangent to $r\psi$ at $r = R$ (nuclear radius) with the r axis. The total elastic cross section is then

$$\sigma_s = 4\pi|a|^2 \tag{12-13.7}$$

The order of magnitude of a is usually 10^{-12} cm. If there is a reaction besides scattering, we can use a complex δ and a complex a defined by Eq. (12-13.5).

However, for slow neutrons the real part of a is large compared with the imaginary. On the assumption of a real, the scattered wave according to Eqs. (12-13.1), (12-13.3), and (12-13.5) may be written $-ae^{ikr}/r$. It has no phase shift with respect to the incident wave if a is negative and a phase shift of π if a is positive. In most cases, a is positive, but there are a few exceptions, notably n–p scattering in the singlet state.

12-14   DETERMINATION OF SCATTERING LENGTHS

Clearly the scattering length for slow neutrons is a function of the scatter and of the mutual orientation of the neutron and scatterer spin if the latter is not 0.

Experimental determination of the scattering length rests on the scattering by material systems composed of many nuclei. The result of scattering by bulk matter depends not only on the scattering lengths of the nuclei involved but also on the arrangement of nuclei in the scatterer, because there are interference phenomena between waves scattered by different nuclei. These interference phenomena determine many important physical effects, such as Bragg reflection, scattering from polyatomic molecules, and index of refraction. They are also essential for determining the relative sign of the scattering length of different nuclei.

The fundamental idea is expressed mathematically by considering an incoming plane wave traveling in the z direction and scattering centers having the scattering length a_j. The distance of point j from the point of observation, at a great distance from the scatterer, is called r_j. The sum of incident and scattered waves has an amplitude, with multiple scattering neglected,

$$A = e^{ikz} - \sum_j \frac{a_j}{r_j} e^{ik(z_j + r_j)} \tag{12-14.1}$$

and the scattered intensity is obtained by taking

$$I = \left|\sum_j \frac{a_j}{r_j} e^{ik(z_j + r_j)}\right|^2 \tag{12-14.2}$$

where z_j is the z coordinate of the j scattering center.

In the scattering of slow neutrons (a similar situation occurs with X rays), often only part of the amplitudes scattered by different centers interfere with one another and with the incident wave. The corresponding cross section is called the "coherent cross section." The scattered waves that do not interfere either with one another or with the incident wave form the *incoherent scattering*. Note that in the forward direction the scattering is always coherent, because $r_j + z_j$ in Eq. (12-14.1) is the same for all centers.

To visualize coherent scattering, we may think of centers set in vibration in phase with the incident wave. For incoherent scattering we may think of centers that frequently and suddenly change phase with respect to the incident wave.

In a gas the interference terms due to waves scattered from different molecules cancel out, and the scattered intensity is the sum of the intensities scattered by single molecules. Thus, in the case of a simple monatomic monoisotopic gas the scattering is spherically symmetric in the center-of-mass system, and the scattering cross section is

$$\sigma = 4\pi|a|^2 \tag{12-14.3}$$

Note that a simple scattering experiment is insufficient to determine the sign of a.

Even in this case we must apply a small correction depending on whether we want a for a free or a bound nucleus. In the first case the reduced mass is given by $1/\mu_f = 1 + (1/A)$. In the latter case the reduced mass for the target nucleus bound to a molecule of mass M is given by $1/\mu_b = 1 + (1/M)$, and using, for example, the Born approximation, we see that

$$\sigma_{\text{bound}} = \left[\frac{1 + (1/A)}{1 + (1/M)} \right]^2 \sigma_{\text{free}} \tag{12-14.4}$$

or

$$a_{\text{bound}} = \frac{M}{A} \left(\frac{A + 1}{M + 1} \right) a_{\text{free}} \tag{12-14.5}$$

A nucleus is to be regarded as bound when the energy transferred in the collision is small compared with the binding energy of the molecule. For nuclei bound in solids or liquids M tends to infinity.

Let us now consider a gas of diatomic molecules of a substance having two spinless isotopes and assume that λ of the neutron is very large compared with the interatomic distance in the molecule, so that for scattering purposes we may sum the amplitudes scattered by the two nuclei of each molecule. Let the relative abundance of the two nuclei be p_1, p_2 ($p_1 + p_2 = 1$), and their scattering lengths a_1 and a_2, respectively. In the gas there are molecules having both nuclei of species 1, both nuclei of species 2, or mixed nuclei.

Their relative abundance is

$$p_1^2, p_2^2, 2p_1 p_2$$

The average cross section for scattering per molecule of the gas is then given by

$$\sigma = 4\pi \left[p_1^2 (2a_1)^2 + p_2^2 (2a_2)^2 + 2p_1 p_2 (a_1 + a_2)^2 \right]$$

$$= 4\pi \left[4(p_1 a_1 + p_2 a_2)^2 + 2p_1 p_2 (a_1 - a_2)^2 \right] \tag{12-14.6}$$

The scattering is thus the same as that produced by two equal nuclei scattering coherently, each with scattering length $a = p_1 a_1 + p_2 a_2$, and at the same time scattering incoherently, with a cross section

$$\sigma_{\text{inc}} = 4\pi p_1 p_2 (a_1 - a_2)^2 \tag{12-14.7}$$

If we define

$$\sigma_{\text{coh}} = 4\pi (p_1 a_1 + p_2 a_2)^2 \tag{12-14.8}$$

the scattering from the molecule takes the form

$$\sigma = 4\sigma_{\text{coh}} + 2\sigma_{\text{inc}} \tag{12-14.9}$$

the factors 4 and 2 pointing to the fact that for coherent scattering we have summed the amplitudes and for incoherent scattering the intensities. Note also that

$$\sigma_{\text{coh}} + \sigma_{\text{inc}} = 4\pi \left(p_1 a_1^2 + p_2 a_2^2 \right) \tag{12-14.10}$$

These results may be generalized. If we have N nuclides randomly mixed with relative abundance p, the scattering length

$$a = \sum_j p_j a_j \tag{12-14.11}$$

determines the interference properties of the substance. In the absence of nuclear polarization one isotope with spin I acts as a mixture of two isotopes with relative abundance $(I + 1)/(2I + 1)$ and scattering length a_+, relative abundance $I/(2I + 1)$ and scattering length a_-, respectively. Here a_+ and a_- are the scattering lengths for parallel or antiparallel neutron and nuclear spin and the I factors express the statistical probabilities of the corresponding spin orientations.

ISBN 0-8053-8061-7

From what we have said we discover, at least in principle, methods for determining the relative sign of the scattering amplitude to two nuclear species. Measuring a phenomenon where the waves produced by the two scatterers interfere, we can measure $(a_1 + a_2)^2$. Measuring the scattering by the two species separately, we have a_1^2 and a_2^2. From this, one obtains the relative sign of a_1 and a_2.

A notable property of scattering is that when neutrons scatter flipping their spin, their waves cannot interfere; the spin-flip scattering is incoherent. To see this, note that if we scatter a neutron from a system of nuclei in such a way as to produce interference, we cannot specify on which nucleus the scattering has occurred. Interference can arise only when scattering occurs on more than one nucleus. On the other hand, if we have spin flip in the neutron, we can specify the scattering nucleus by observing which nucleus has changed its I_z. To treat this situation, we must generalize Eq. (12-14.1) in order to take into account the wave function of the scatterer as well.

To discuss this question in mathematical terms, consider an incident plane neutron wave with spin up, traveling in the z direction, and the scatterer's initial wave function φ. The system is then represented by

$$ e^{ikz} \begin{pmatrix} 1 \\ 0 \end{pmatrix} \varphi $$

Call φ_j the wave function of the scatterer when the j nucleus has flipped its spin and denote by a', a the scattering length with or without spin flip. We look for an asymptotic solution of the form

$$ e^{ikz} \begin{pmatrix} 1 \\ 0 \end{pmatrix} \varphi - \sum_j \frac{a_j}{r_j} e^{ik(z_j + r_j)} \begin{pmatrix} 1 \\ 0 \end{pmatrix} \varphi - \sum_j \frac{a_j'}{r_j} e^{ik(z_j + r_j)} \begin{pmatrix} 0 \\ 1 \end{pmatrix} \varphi_j $$

The scattered intensity is obtained by taking the modulus square of the scattered part of the wave, integrated over the scatterer coordinates. By definition,

$$ \int \varphi_i^* \varphi_k \, dq = \delta_{ik} \quad (\delta_{ik} \text{ is the Dirac delta function}) \quad (12\text{-}14.12) $$

and this eliminates all terms of the sum containing $\varphi\varphi_j^*$, leaving those with $|\varphi|^2$ and $|\varphi_j|^2$. The scattered intensity has the form

$$ \left| \sum_j \frac{a_j}{r_j} e^{ik(r_j + z_j)} \right|^2 + \sum_j \frac{|a_j'|^2}{r_j^2} $$

The first term is the scattering without spin flip and shows interference between scattering from different centers. The second is spin-flip scattering and is incoherent.

ISBN 0-8053-8061-7

12-15   SCATTERING IN ORTHO- AND PARAHYDROGEN

Scattering from ortho- and parahydrogen is another important example of the application of the ideas of Sec. 12-14. We shall consider only scattering at very low temperatures and for neutron wavelengths large compared with the internuclear distance (0.74Å). We cannot apply the formulas of Sec. 12-14 directly, because at very low temperatures, in the presence of a catalyst, all hydrogen molecules are in the zero rotational state, which has antiparallel nuclear spins. The spins are thus not distributed at random. We shall calculate the scattering cross section for such a molecule and for a molecule in which the spins of the two hydrogens are parallel (orthohydrogen).

To do this we write the neutron–proton scattering length formally as

$$a = a_s \pi_s + a_t \pi_t \qquad (12\text{-}15.1)$$

where a_s and a_t are the ordinary singlet and triplet scattering length corrected for the mass effect

$$a_{s,\,t} = \tfrac{4}{3} a_{s,\,t\,\text{free}} \qquad (12\text{-}15.2)$$

and π_s is an operator which has the value 1 for singlet states, 0 for triplet states. The operator π_t has instead the value 0 for singlet states, 1 for triplet states. Such operators (projection operators) are, as we can derive immediately from the calculation at the beginning of Sec. 10-5

$$\pi_s = \tfrac{1}{4}\left(1 - \boldsymbol{\sigma}_n \cdot \boldsymbol{\sigma}_p\right) \qquad (12\text{-}15.3)$$

and

$$\pi_t = \tfrac{1}{4}\left(3 + \boldsymbol{\sigma}_n \cdot \boldsymbol{\sigma}_p\right) \qquad (12\text{-}15.4)$$

with $\boldsymbol{\sigma} \equiv$ Pauli spin operator. To compute the scattering from our molecule, we write the scattering length of each atom as a_1, a_2, using Eq. (12-15.1). The scattering length for the molecule is

$$a_1 + a_2 = a_s \pi_{s1} + a_t \pi_{t1} + a_s \pi_{s2} + a_t \pi_{t2}$$

$$= a_s\left[\tfrac{1}{2} - \tfrac{1}{4}\boldsymbol{\sigma}_n \cdot (\boldsymbol{\sigma}_{p1} + \boldsymbol{\sigma}_{p2})\right] +$$

$$a_t\left[\tfrac{3}{2} + \tfrac{1}{4}\boldsymbol{\sigma}_n \cdot (\boldsymbol{\sigma}_{p1} + \boldsymbol{\sigma}_{p2})\right] \qquad (12\text{-}15.5)$$

Now $\tfrac{1}{2}(\boldsymbol{\sigma}_{p1} + \boldsymbol{\sigma}_{p2})$ is the spin $\mathbf{S}$ of the two hydrogen nuclei; for orthohydrogen $S = 1$, and for parahydrogen $S = 0$. Substituting in Eq. (12-15.5) and re-

ISBN 0-8053-8061-7

arranging, we have

$$a = \left(\tfrac{3}{2}a_t + \tfrac{1}{2}a_s\right) + \left(\tfrac{1}{2}a_t - \tfrac{1}{2}a_s\right)(\boldsymbol{\sigma}_n \cdot \mathbf{S}) \qquad (12\text{-}15.6)$$

For parahydrogen the scattering cross section is thus

$$\sigma_{\text{para}} = 4\pi\left(\tfrac{3}{2}a_t + \tfrac{1}{2}a_s\right)^2 \qquad (12\text{-}15.7)$$

For orthohydrogen Eq. (12-15.6) gives, on squaring, linear terms in $\boldsymbol{\sigma}_n \cdot \mathbf{S}$ which for unpolarized beams and target cancel. The term containing $(\boldsymbol{\sigma}_n \cdot \mathbf{S})^2$ averaged gives $S(S + 1) = 2$. We thus obtain

$$\sigma_{\text{ortho}} = 4\pi\left[\left(\tfrac{3}{2}a_t + \tfrac{1}{2}a_s\right)^2 + \tfrac{1}{2}\left(a_t - a_s\right)^2\right] \qquad (12\text{-}15.8)$$

The cross section of parahydrogen depends very strongly on the relative sign of a_s, a_t, as was pointed out by Teller (1937). According to experiment, $\sigma_{\text{ortho}} = 125$ b, $\sigma_{\text{para}} = 4$ b; this result is explainable only if a_s, a_t have opposite signs. The quantitative interpretation (i.e., derivation of the scattering lengths from the scattering by molecular hydrogen) requires numerous refinements that are beyond the scope of this book.

12-16 INTERFERENCE PHENOMENA IN CRYSTALS

Since the slow-neutron de Broglie wavelength is

$$\lambda = 2.86 \times 10^{-9}/E^{1/2}\ cm \qquad (E \text{ in eV}) \qquad (12\text{-}16.1)$$

it reaches 1.81×10^{-8} cm for thermal values. It is then comparable to the interatomic distances in solid lattices, and the neutrons will show many of the phenomena characteristic of the interaction of X rays with crystals, such as Laue and Bragg diffraction, a small refractive index, and others.

The experimental techniques employed are similar, and crystal spectrographs and similar instruments are used for both radiations. Figure 12-19 shows the similarity of the photographs obtained by diffraction with neutrons and X rays. Strong monochromatic neutron sources are usually obtained by a mechanical velocity selector (Fig. 12-20), by crystalline reflection, or by some other device. However, there are important differences between the two types of radiation. In many respects they supplement each other very usefully in solid-state investigations. For instance, hydrogen, which is most difficult to detect by X-ray diffraction, has a large a and is easily observable by means of neutrons. We shall treat briefly some of the most important topics of this aspect of neutron physics.

We shall begin with the scattering by a crystalline lattice, giving rise through interference phenomena to Bragg reflection. Consider a set of parallel planes having a distance d between successive planes and containing

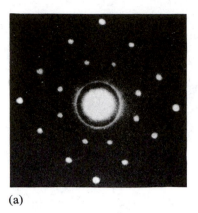

(a)

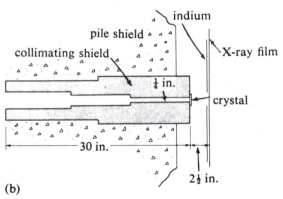

(b)

Figure 12-19   (a) Laue photograph showing neutron diffraction by a NaCl crystal. [From Wollan, Shull, and Marney, *Phys. Rev.*, **73**, 527 (1948).] (b) Schematic diagram of the Laue camera used to obtain the neutron diffraction pattern.

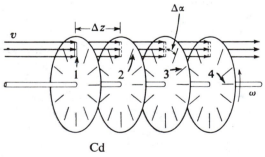

Cd

Figure 12-20   An example illustrating the principle of a mechanical monochromator for slow neutrons. The selector is transparent to neutrons of velocity $v = (\Delta z / \Delta \alpha)\omega$, where ω is the angular velocity of the selector, Δz the distance between the disks, and $\Delta \alpha$ is the angular shift of the window from one disk to the next. Neutrons of this velocity will thread, for instance, slots 1,2,3,4.

ISBN 0-8053-8061-7

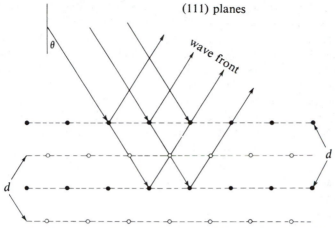

(111) planes

wave front

d

d

Figure   12-21   Neutron diffraction from the (111) planes of a cubic crystal.

identical spinless scattering nuclei of scattering length a. The scattered neutrons, originating at depth md will give rise to an amplitude (see Fig. 12-21).

$$A = \sum_{m=1}^{m=N} ae^{2ikmd \cos \theta} \qquad (12\text{-}16.2)$$

where N is the number of layers involved. Performing the sum, which is a geometric progression, we find

$$A = a \left\{ \frac{1 - e^{2ikNd \cos \theta}}{1 - e^{2ikd \cos \theta}} \right\} e^{2ikd \cos \theta} \qquad (12\text{-}16.3)$$

The intensity is proportional to $|A|^2$, where

$$|A|^2 = |a|^2 \frac{\sin^2(Nkd \cos \theta)}{\sin^2(kd \cos \theta)} \qquad (12\text{-}16.4)$$

is large for those values of θ for which the denominator vanishes, that is, if

$$kd \cos \theta = n\pi$$

or

$$n\lambda = 2d \cos \theta$$

which is the famous Bragg relation (W. Bragg and L. Bragg, 1913). If the

ISBN 0-8053-8061-7

crystal contains several isotopes, we must use in Eq. (12-16.4) the weighted sum of the scattering lengths in order to determine the effective scattering length for coherent scattering. Naturally such a crystal is not a "perfect" crystal, because the isotopes are randomly distributed in the lattice and, in addition to the coherent scattering, incoherent scattering must be present.

Consider now the case of atoms of two kinds, as in NaCl. The (111) planes are equidistant and consist alternately of sodium and chlorine. In the first-order Bragg reflection the optical path for reflection from sodium planes differs by $\lambda/2$ from the path for reflection from chlorine planes. Consequently, if sodium and chlorine nuclei cause the same change in phase of the scattered neutron wave, their contributions will subtract and the order will have low intensity; the contrary will happen if they scatter with opposite phase change.

The situation for the second order is reversed. High intensity here is connected with scattering by sodium and chlorine when both produce the same change in phase. On these effects is superimposed a continuous decrease of intensity from order to order, as in X rays.

In more complicated cases the intensity of the various orders is determined, as for X rays, by the form factor, defined as the effective coherent scattering amplitude per unit cell of the crystal,

$$F = \left| \sum_{j} a_j \exp\left(\frac{2\pi i n \delta_j}{d} \right) \right|$$ (12-16.5)

where d is the spacing of the lattice planes, n is the order of the Bragg reflection, and δ_j is the perpendicular distance of the jth nucleus to one of the planes of reflection assumed as reference. The sum is extended to all nuclei in the unit cell. It is clear that F depends strongly on the signs of a_j, and an analysis of the intensities for the different orders may lead to the determination (Fig. 12-21; NaCl) of the relative sign of the scattering lengths for the different nuclei of the lattice.

An interesting application of the Bragg condition is the preparation of beams of very slow neutrons. By passing a beam of neutrons of different velocities through a polycrystalline medium, the neutrons that satisfy the Bragg relation for some crystal are reflected out, and soon only those of $\lambda > 2d$, which can never satisfy the Bragg condition, are left in the beam. Thus we have a cutoff at high neutron velocity. The intensity falls off at low neutron velocity because of the original velocity distribution in the beam (Fig. 12-22).

Bragg reflection can also be used to investigate the magnetic properties of a material. We shall see that the scattering length is affected by the orientation of a neutron with respect to the electronic magnetic moment of an atom.

ISBN 0-8053-8061-7

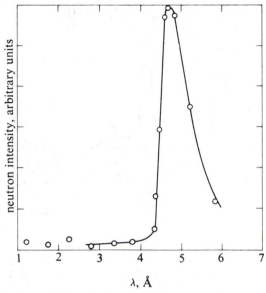

Figure 12-22 Spectral distribution of cold neutrons from a beryllium oxide filter (Fermi and Marshall).

In a saturated ferromagnetic material there will be a strong dependence of the reflecting power on the neutron polarization, and thus Bragg reflection can be used to polarize neutrons.

In a different application the Laue patterns for neutrons, produced by ferromagnetic substances, depend on the magnetization of the scatterer. Finally, we have the interesting case of antiferromagnetic compounds. These are solids in which the spins of the atoms are aligned, but always in pairs with opposite directions, as in two saturated ferromagnetic lattices with equal and opposite magnetizations, superimposed on each other. These substances do not show macroscopic properties such as ferromagnetism but reveal their structure in neutron diffraction analysis (Fig. 12-23).

Another interesting application of very slow neutrons has been in the investigation of the properties of liquid helium II. A very slow neutron ($\lambda > 4 \, \overset{\circ}{A}$) scatters, producing a phonon in the liquid. The energy and momentum of the phonon must be equal by the conservation laws to the measurable changes of energy and momentum of the neutron. One thus obtains the relation between E and p for the phonon (dispersion relation), which is of great importance for the theory of He II. The applications of neutrons to molecular and solid-state problems are rapidly increasing. The examples cited should give the reader an idea of the possibilities of this most elegant tool of investigation.

ISBN 0-8053-8061-7

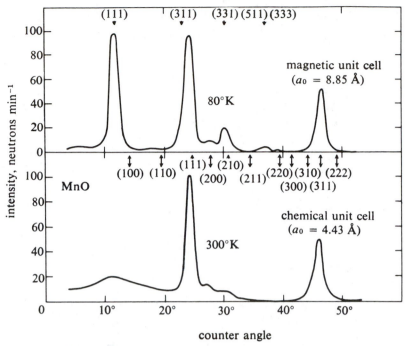

Figure   12-23   Powder diffraction patterns for crystalline manganese oxide at 80°K and at room temperature, due to Shull and Smart [*Phys. Rev.*, **76**, 1256 (1949)]. The additional peaks, in the low-temperature pattern, are characteristic of a magnetic unit cell of twice the dimensions of the chemical unit cell, clearly indicating the antiferromagnetic structure of manganese oxide.

12-17   INDEX OF REFRACTION

Another application of the theory of coherent scattering is the atomic explanation of the refractive index. The forward scattering of neutrons, light, X rays, etc., is always coherent, and its interference with the incoming beam gives rise to the refractive index of the scattering material.

The connection between scattering length and refractive index is obtained by the following calculation: Assume for the sake of simplicity that the scattering is spherically symmetric, and consider a thin slab of material of thickness T, containing N nuclei per unit volume. A plane wave $\psi_0 = e^{ikz}$ impinges normally on the slab, and we calculate the amplitude of the wave at the point P (Fig. 12-24),

$$\psi_P = e^{ikz} - 2\pi NT \int_0^\infty \frac{a}{r} e^{ikr} x \, dx \qquad (12\text{-}17.1)$$

where we have neglected the absorption in the slab. Now $x^2 + z^2 = r^2$,

ISBN 0-8053-8061-7

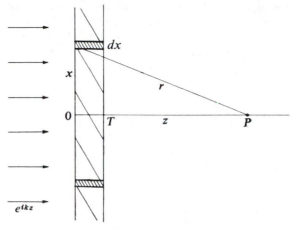

Figure 12-24 Plane wave of slow neutrons impinging (from the left) on an infinite slab of scattering material. The symbols are those used in the calculation of the relation between the index of refraction and the scattering length.

whence $x\,dx = r\,dr$ and

$$\psi_P = e^{ikz} - 2\pi NTa \int_z^\infty e^{ikr}\,dr \qquad (12\text{-}17.2)$$

The last integral can be evaluated either by using the artifice of the Fresnel zones or by adding a small positive imaginary part to k and passing to the limit for this small part vanishing, the result being that the value of the integral at the upper limit vanishes. We thus obtain

$$\psi_P = \left(1 - \frac{2\pi i NTa}{k}\right)e^{ikz} \qquad (12\text{-}17.3)$$

Now consider the effect of attributing a refractive index n to the slab. The propagation vector in the material is nk versus k in vacuum; hence

$$\psi_P \cong e^{ik(z-T)+inkT} = e^{ikz}e^{ik(n-1)T} \qquad (12\text{-}17.4)$$

and for $k(n-1)T \ll 1$

$$\psi_P \cong e^{ikz}\left[1 + ik(n-1)T\right] \qquad (12\text{-}17.5)$$

Comparison of Eqs. (12-17.3) and (12-17.5) gives

$$n = 1 - \frac{2\pi Na}{k^2} = 1 - \frac{\lambda^2}{2\pi}Na \qquad (12\text{-}17.6)$$

ISBN 0-8053-8061-7

For positive a the index of refraction is thus smaller than 1. Its value is very near 1 for slow neutrons; in fact, typical values would be, for example, $\lambda = 10^{-8}$ cm, $N = 10^{23}$ cm^{-3}, $a = 10^{-12}$ cm, $n - 1 = \pm 2 \times 10^{-6}$. In the case $n < 1$ total reflection from a surface is possible; it occurs whenever $\sin i_c \geqslant n$. Since $n \cong 1$, $i_c \cong 90$ deg, and calling $(90 \text{ deg} - i_c) = \epsilon$, we have $\sin i_c = \cos \epsilon \cong 1 - (\epsilon^2/2)$, or

$$-\frac{\epsilon^2}{2} \geqslant n - 1 \tag{12-17.7}$$

Remembering Eq. (12-17.6) we thus have the limiting angle

$$\epsilon = \left(\frac{N\lambda^2}{\pi} a \right)^{1/2} = \lambda \left(\frac{N}{\pi} \right)^{1/2} \left(\frac{\sigma_{\text{tot}}}{4\pi} \right)^{1/4} \tag{12-17.8}$$

Clearly total reflection occurs at a given angle only for λ greater than a critical λ_c defined by Eq. (12-17.8). We therefore have a way of obtaining a neutron velocity selector (Fig. 12-25).

Even more interesting is the reflection on magnetized mirrors. It can be shown (see Sec. 12-18) that the refractive index of a neutron for a ferromag-

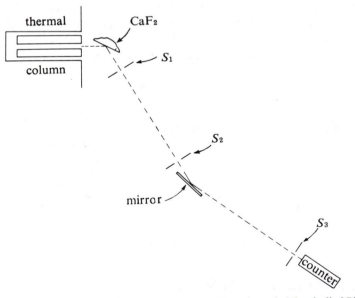

Figure 12-25 Experimental arrangement of Fermi and Marshall [*Phys. Rev.*, **71**, 666 (1947)] for the determination of the critical angle for reflection of monoenergetic neutrons from various mirrors.

ISBN 0-8053-8061-7

netic substance depends on the relative orientation of the magnetization and the neutron spin. More precisely,

$$n_{\pm} = 1 - \frac{\lambda^2 Na}{2\pi} \mp \frac{\mu_n B}{2E_n} \qquad (12\text{-}17.9)$$

where the $\pm$ refers to the orientation of the field with respect to the neutron spin, μ_n is the magnetic moment of the neutron, E_n its energy, B the magnetic flux density in the mirror in the direction of the mirror plane, and a the scattering length of the mirror material. Hence from Eq. (12-17.8) it can be seen that neutrons of opposite spin direction may have different limiting angles. Total reflection from a magnetized mirror then affords a method of polarizing neutrons.

Finally, total reflection from hydrocarbons of different C/H ratio provides an important means of measuring the ratio of the scattering amplitude of carbon to that of hydrogen, or, more precisely, to that of

$$\tfrac{3}{4}a_3 + \tfrac{1}{4}a_1 = a_H \qquad (12\text{-}17.10)$$

the weighted average of singlet and triplet scattering lengths (Hughes, Burgy, and Ringo, 1950). The carbon scattering length (practically mono isotopic, spinless ^{12}C) can be measured directly from free carbon cross-section measurements.

12-18   POLARIZATION OF SLOW-NEUTRON BEAMS

Since the scattering lengths a_+ and a_- are not generally equal, a crystal with all its nuclei oriented in the same direction will have different cross sections for the two directions of the spin of the incident neutrons. If $a_+ > a_-$, more of the neutrons with spin parallel to that of the scatterer (spin up) will be scattered away, leaving a transmitted beam which is predominantly "spin down." The polarization would approach 100% if the path through the crystal were made sufficiently long. However, this method is not very practical, since one has to line up the nuclear spins, which is generally difficult.

In ferromagnetic materials, the atomic magnetic moments can be lined up. We shall now show how this can give different cross sections for the two possible neutron orientations, by considering the scattering using the Born approximation for the amplitude.

Let $U = b\delta(r) \mp \mu_N B_z$ represent the interaction potential between the neutron and the Fe atom in its lattice position. The upper sign obtains for the magnetic moment of the neutron parallel to B, the lower sign for the magnetic moment of the neutron antiparallel to B. (Remember that spin and

ISBN 0-8053-8061-7

magnetic moment in the neutron are antiparallel.) Using the Born approximation, one obtains

$$\sigma(\theta) = \frac{m^2}{4\pi^2\hbar^4} \left| \int U e^{i(\mathbf{k}'-\mathbf{k})\cdot\mathbf{r}} \, d\tau \right|^2$$

$$= \frac{m^2}{4\pi^2\hbar^4} \left| b \mp \mu_N \int B_z(r) e^{i(\mathbf{k}'-\mathbf{k})\cdot\mathbf{r}} \, d\tau \right|^2 \tag{12-18.1}$$

where the region of integration is that part of the lattice belonging to one Fe
atom. For slow neutrons λ is of the order of the lattice distance, and the
exponential is $\cong 1$ over the region of integration. Thus

$$\sigma(\theta) = \frac{m^2}{4\pi^2\hbar^4} \left| b \mp \mu_N \int B_z(r) \, d\tau \right|^2 \tag{12-18.2}$$

where the integral is extended to the lattice volume occupied by an Fe atom.
To give an idea of the order of magnitude of the magnetic field contribution,
we shall take b as 0 in the following estimate. Let $b = 0$; $\lambda \gg$ atomic dimen-
sions; $B_z = 23\,000$ G (saturated Fe); $\int B_z \, d\tau = 2.7 \times 10^{-19}$ G $\cdot$ cm^3. Then $\sigma = 4.9 \times 10^{-24}$ cm^2.

The experimental result for thermal neutrons on magnetized iron is $\sigma = (12 + 3.15)$ b or $\sigma = (12 - 3.15)$ b; the first value, $\sigma = 15.15$ b, obtains for
the neutron spin parallel to B; the second value, $\sigma = 8.85$ b, corresponds
to spin antiparallel to B. Naturally, if we filter neutrons through magnetiz-
ed iron, we shall obtain a polarized beam with an orientation such that $\sigma = 8.85$ b, the opposite orientation being preferentially removed from the beam.

A disadvantage of this method is that it happens to be very sensitive to
incomplete polarization of the iron. Iron is completely polarized in each
domain, but in the small number of domains that are not properly oriented,
the neutrons will precess about B and undo most of the work done by the
previous "good" domains.

Let us call $\sigma_0 \pm \sigma_m$ the cross sections for neutrons with spin parallel or
antiparallel to the magnetization, and let us assume complete saturation of
the iron. If we pass a beam of incident unpolarized intensity I_0 through a slab
of thickness d containing N atoms of iron per unit volume, the emerging
beam will have intensities I_+ polarized with spin parallel to the magnetization, and I_- polarized antiparallel.

The following relations are immediately derived:

$$I_+ = \tfrac{1}{2} I_0 e^{-(\sigma_0 + \sigma_m)Nd} \tag{12-18.3}$$

$$I_- = \tfrac{1}{2} I_0 e^{-(\sigma_0 - \sigma_m)Nd} \tag{12-18.4}$$

$$P = \frac{I_+ - I_-}{I_+ + I_-} = -\tanh N\sigma_m d \cong -N\sigma_m d \tag{12-18.5}$$

ISBN 0-8053-8061-7

A second slab of magnetized iron may then be used as an analyzer; depending on the direction of its magnetization, we have, for the intensity passing through both slabs,

$$I = I_0 e^{-N\sigma_0(d_1 + d_2)} \cosh N\sigma_m (d_1 \pm d_2)   \quad\quad (12\text{-}18.6)$$

where the upper sign denotes parallel, and the lower sign denotes opposite, magnetization of the slabs.

BIBLIOGRAPHY

Amaldi, E., "The Production and Slowing Down of Neutrons," in (Fl E), Vol. 38.2.
Bacon, G. E., *Neutron Diffraction*, Oxford University Press, New York, 1962.
Bethe, H. A., Energy Production in Stars (Nobel lecture), *Science*, **161**, 541 (1968).
Blatt, J. M., and V. Weisskopf (BW 52).
Cole, T. E., and A. M. Weinberg, "Technology of Research Reactors," *Ann. Rev. Nucl. Sci.*, **12**, 221 (1962).
Etherington, H. (ed.), *Nuclear Engineering Handbook,*, McGraw-Hill, New York, 1958.
Feld, B. T., "The Neutron," in (Se 59), Vol. II.
Fermi, E. (Fe 50).
Fermi, E., *Collected Papers of Enrico Fermi*, University of Chicago Press, Chicago, Vol. 1, 1962; Vol. 2, 1966.
Glasstone, S., and M. C. Edlund (GE 52).
Goldberger, M., and F. Seitz, "Theory of the Refraction and the Diffraction of Neutrons by Crystals," *Phys. Rev.,* **71**, 294 (1947).
Hughes, D. J. (Hu 53).
Hughes, D. J., *Neutron Optics*, Wiley-Interscience, New York, 1954.
Hughes, D. J., "Reactor Techniques," in (Fl E), Vol. 44.
Keepin, G. R., *Physics of Nuclear Kinetics*, Addison-Wesley, Reading, Mass, 1965.
Lamarsch, J. R. (La 66).
Marion, J. B., and J. L. Fowler (eds.), *Fast Neutron Physics*, Wiley-Interscience, New York, 1960–1963.
Post, R. F., "Controlled Fusion, Research and High Temperature Plasmas," *Ann. Rev. Nucl. Sci.*, **20**, 509 (1970).
Ribe F. L. "Fusion Reactor Systems," *Rev. Mod. Phys.* **47**, 71, (1975).
Schwinger, J., and E. Teller, "The Scattering of Neutrons by Ortho- and Parahydrogen," *Phys. Rev.*, **52**, 286 (1937).
Weinberg, A., and E. Wigner (WW 58).
Wilkinson, M. K., E. O. Wollan, and W. C. Koehler, "Neutron Diffraction," *Ann. Rev. Nucl. Sci.*, **11**, 303 (1961).

PROBLEMS

12-1 Devise methods for producing neutron beams within an energy range from 10^{-3} to 10^8 eV and for measuring their energy and intensity.

12-2 How many neutrons per second are emitted by 1 g of Ra enclosed in a Be sphere having an internal cavity of 0.75 cm radius and an external radius of 2.5 cm?

12-3 Prove that for large A the average lethargy increase per collision tends to $2/[A + (2/3)]$.

12-4 Find the number of collisions with D, He, Be, U necessary to reduce the energy of a neutron from 10^6 to 1 eV.

ISBN 0-8053-8061-7

12-5   Show that the energy distribution of neutrons of initial energy E_0, after, n collisions in hydrogen, is given by

$$P_n(E)\, dE = \frac{[\log(E/E_0)]^{n-1}}{(n-1)!} \frac{dE}{E_0}$$

12-6   A neutron of initial energy E_0 has reached by elastic collisions in hydrogen the energy E. Show that the probability $p_n(E)$ of the number of collisions suffered being n is given by

$$p_n(E) = \frac{[\log(E_0/E)]^{n-1}}{(n-1)!} \frac{E}{E_0}$$

12-7   Give the physical interpretation and explanation of the different factors of Eqs. (12-4.1) and (12-4.2).

12-8   Show that the diffusion coefficient is given by

$$\frac{\lambda v}{3[1 - (2/3A)]}$$

12-9   Assuming λ constant, find the relationship between the "age" of the neutron and the period of time that has passed since its birth. The neutron diffuses and passes from an initial energy E_0 to an energy E.

12-10   High-energy neutrons are produced in the upper atmosphere. In the high-energy range they are slowed down by both elastic and inelastic scattering. Call Q the number crossing the 10^5-eV energy per second. How many reach thermal energy? Assume $\sigma_s = 11$ b for N, 4.2 b for O, and σ_r for N = 1.78 b $\times (2200/v)$, where v is in meters per second. [See (Fe 50), p. 184.]

12-11   Develop Eq. (12-6.19). [See (Fe 50), p. 190.] The source is a δ function.

12-12   Calculate the diffusion length in graphite if $\sigma_s = 4.8$ b and $\sigma_a = 4.5$ mb. Compare this with the actual distance traveled by a thermal neutron before capture.

12-13   An isotropic source of thermal neutrons is placed at the center of a beryllium sphere of 10 cm radius. Calculate the flux distribution if the source emits 100 neutrons sec^{-1}, assuming the sphere is surrounded with cadmium so that all neutrons leaving the sphere are lost by radiative capture. What fraction of the source neutrons will be captured in the cadmium? Suppose that in the case of beryllium Σ_a were equal to zero. What fraction of the source neutrons would leak out under these conditions? Explain the difference.

12-14   How much fuel would be consumed by a nuclear-powered submarine traveling at an average power of 30 MW for 1 month?

12-15   Calculate the critical size of a sphere of pure ^{235}U.

12-16   Calculate the critical thickness of a slab of material having a certain k. Repeat the calculation for the slab sandwiched between two layers of material having the diffusion coefficient D and thickness T.

12-17   Calculate the critical mass of a pile having a given k and a cylindrical form. Find the most favorable ratio between height and radius.

12-18   What is the minimum concentration of uranyl nitrate in heavy water which can go critical? Assume pure ^{235}U.

12-19   A reaction of current interest in the application of thermonuclear processes to power production is the following:

$$D + D = {}^3He + n$$

(a) If the deuterium nuclei interact at rest, calculate the total final kinetic energy of the neutron and of the helium nucleus (in MeV). (b) Calculate the momentum of each (in

ISBN 0-8053-8061-7

MeV/c). (c) If the deuterium nuclei must come within 10^{-11} cm of each other, what energy must be supplied to overcome the electrostatic repulsion? (d) If this energy is supplied by heating the deuterium to a very high temperature, what order of magnitude of temperature is required?

12-20 Show that if neutron scattering is due to the effect of a single Breit–Wigner resonance level, then the scattering length is given by

$$- a = \frac{\lambda_R \Gamma_n}{E - E_R + i\Gamma}$$

where Γ_n and Γ are the neutron and the total half-width at half-maximum, and λ_R and E_R refer to the resonance.

12-21 Represent a heavy nucleus by a potential well of radius $R = 1.2 \times 10^{-13} A^{1/3}$ cm and a depth of 30 MeV. Thermal neutrons will have several waves of $r\psi$ inside the well. As a consequence (draw a figure) it is unlikely that the intercept of the tangent to $r\psi$ at R cuts the x axis on the negative side. Moreover, the scattering length is likely to be comparable to the nuclear radius.

12-22 Show that for a complex scattering length $a = a_r + ia_i$ with $a_r \gg a_i$ one has $a_r / a_i = \lambda(\sigma_s)^{1/2}/\pi^{1/2} \sigma_a$.

12-23 Show that the probability of spin flip per scattering is given by

$$Q = \frac{2}{3} \frac{\sigma_{\text{inc}}}{3\sigma_s}$$

[G. C. Wick, *Phys. Z*, **38**, 403, 689 (1937).]

12-24 Find the probability that an n–p scattering is accompanied by spin flip.

12-25 Show that on the scattering of neutrons in liquid He II the energy and momentum of the phonon are given by

$$E = \frac{\hbar^2 (\lambda_i^{-2} - \lambda_f^{-2})}{2m_N}$$

$$p^2 = \hbar^2 (\lambda_i^{-2} + \lambda_f^{-2} - 2\lambda_i^{-1}\lambda_f^{-1} \cos \varphi)$$

assuming $2\pi\lambda_i = 4$ Å and $E = vp(v = 240 \text{ m sec}^{-1})$. Plan an experiment to verify this relation [Yarnell and others, *Phys. Rev.*, **113**, 1379 (1959).]

12-26 For NaCl on (001) planes make a table of the wavelengths reflected in first and second order at angles 1, 5, 10, 20, and 40 deg. What is the relative intensity of the first- and second-order beam assuming a distribution $\sim v^3 e^{-mv^2/2kT}$ for the neutrons?

12-27 Derive the relation between refractive index (complex) and scattering length for a complex.

12-28 Derive the relation between refractive index and scattering for a scattered wave amplitude $f(\theta)$.

12-29 Calculate the polarization of a slow-neutron beam from the intensities measured in a double absorption experiment in magnetized iron.

ISBN 0-8053-8061-7

CHAPTER 13

Introduction to Particle Physics

13-1 GENERAL IDEAS, NOMENCLATURE, AND CATALOGUE OF PARTICLES

In the third part of this book we shall treat particle physics. The study of subnuclear entities, which we shall loosely call *particles*, takes us one step beyond the nucleus in analyzing the structure of matter. The situation is similar in some respects to the step from atomic to nuclear physics. To study the atom it was sufficient to assume the validity of Coulomb's law and the existence of the nucleus and peripheral electrons. The structure of the nucleus could be neglected to a large extent. We needed to know its charge and its approximate size, but little else was necessary except for very detailed investigations. In the next step, the study of the nucleus, it was sufficient to know that its constituents are neutrons and protons and to have an idea of the range and strength of nuclear forces in order to derive a fairly complete picture. Beta decay introduces new problems and a new type of force: the so-called weak interaction or Fermi force. However, until one asks detailed questions about nuclear forces, it is sufficient in describing nuclear phenomena to consider only neutrons, protons, electrons, neutrinos, and gamma quanta.

From the theoretical point of view, a deeper study of beta decay and of nuclear forces was the starting point of the next development. A parallel

Emilio Segrè, Nuclei and Particles: An Introduction to Nuclear and Subnuclear Physics, Second Edition

ISBN 0-8053-8061-7

development was the experimental discovery of numerous particles, many of them first detected in cosmic rays. The technical development of large accelerating machines made it possible to generate these particles in the laboratory and thus to study them in detail.

The whole concept of "elementary particles," including its definition, is not clearly settled. Attempts have been made to classify particles in various ways, attributing to some of them a more fundamental role than to others. There have also been attempts to put them all on equal footing, introducing a single field which should describe them all. This field obeys a nonlinear equation which contains a fundamental length as a parameter. The different particles and their properties should be derivable from this field (Heisenberg). Other approaches consider the observable collision phenomena in terms of a *scattering matrix*. This matrix becomes the main object of the theory, and its study should provide the ultimate information. Until now, however, none of these ambitious theories has been able to make enough predictions to establish itself solidly.

We shall limit ourselves to a semiphenomenological approach, describing the experimental facts, their immediate consequences, and the theoretical interpretations which seem solidly established.

Many of the relevant concepts have already been introduced in previous chapters. For instance, spin and statistics are treated in Chap. 6 and i spin is treated in Chap. 10. The whole chapter on beta decay has many arguments in common with the following chapters. For the convenience of the reader we shall first give some general nomenclature and classifications.

One may consider four families of particles in order of increasing rest mass: the first contains only one member, the *photon*, a boson of spin 1. The second family, *leptons*, contains fermions of spin $\frac{1}{2}$, lighter than the proton (i.e., two types of neutrinos, electrons, and muons). Leptons are subject to electromagnetic and Fermi interactions only, not to the strong interactions. The third family, *mesons*, comprises bosons of integral spin. These are heavier than the leptons and subject to all three types of interaction. The fourth family, *baryons*, comprises the proton and heavier fermions. Baryons are subject to all three types of interaction; those heavier than the neutron are called *hyperons*. All particles subject to strong interactions are *hadrons*.

The nomenclature is further refined by taking into account the i spin, hypercharge, and strangeness of a particle and, when applicable, g parity (see Sec. 13-4). Table 13-1 gives the conventions. The mass is occasionally added, in parentheses, to the letter: $\Delta(1236)$. When there is more than one particle with the same quantum numbers, primes are used. For the proton and neutron one still uses p and n, respectively. The reader should note, however, that the nomenclature is still variable [e.g., what in the standard scheme is called $\Sigma'(1385)$ is often written as $Y^*(1385)$].

TABLE 13-1   PARTICLE NAME CONVENTIONS[a, b]

Name	T		Y	S	G
		Mesons			
η	0		0	0	$+$
ω or ϕ^c	0		0	0	$-$
ρ	1		0	0	$+$
π	1		0	0	$-$
K^+, K^0	$\frac{1}{2}$		$+1$	$+1$	
$K^-, \overline{K}^0$	$\frac{1}{2}$		-1	-1	
		Baryons			
N	$\frac{1}{2}$		$+1$	0	
Δ	$\frac{3}{2}$		$+1$	0	
Z_0, Z_1	0, 1		$+2$	$+1$	
Λ	0		0	-1	
Σ	1		0	-1	
Ξ	$\frac{1}{2}$		-1	-2	
Ω	0		-2	-3	

[a]Isotopic spin is indicated by the symbol T or by the symbol I. Nuclear physicists tend to prefer T, particle physicists I, but there are many exceptions. In this book we tend to prefer T, but we are not consistent. We hope the reader will not be confused.

[b]Adapted from Particle Data Group, "Review of Particle Properties," *Phys. Letters*, **50B**, No. 1 (1974). See also *Rev. Mod. Phys.*, **48**, S1, (1976).

[c]The ω is used for those $T^G = 0^-$ mesons that decay mainly into 3π [$\omega(784)$, and $\omega(1675)$]; the ϕ is reserved for $\phi(1019)$ and possible future higher-mass $T^G = 0^-$ mesons that decay mainly into K–$\overline{K}$.

The discovery and classification of these particles is a major achievement of the decade 1950–1960. During this period there originated a complicated nomenclature that is no longer used. (The study of natural radioactivity also left us with a complicated nomenclature for the natural radioactive substances.) Table 13-2 lists the most important particles and their principal characteristics as they are known today, with the pertinent modern nomenclature.

Measurement of the mass of the different particles is made by observations of reactions in which they transform into other particles of known mass, as will be seen later. The assignment of spin is for the most part indirect; in the arguments used, high spins generally are ruled out a priori. From the decay modes of the particles we know with certainty whether they are fermions or bosons and hence whether they have integral or half-integral spins.

Particles that cannot decay by strong interaction and thus have mean lives long compared with the characteristic nuclear time 10^{-22} sec are designated

TABLE 13-2 PARTICLE PROPERTIES[a]

Stable Particles

Particle	$I^G(J^P)C_n$	Mass (MeV) Mass² (GeV)²	Mean Life (sec) cτ (cm)	Partial decay mode — Mode	Partial decay mode — Fraction[b]	p or p_{max}[c] (MeV/c)
γ	$0, 1(1^-)^-$	$0(<2)10^{-21}$	stable	stable		
ν_e		$0(<60\text{eV})$	stable	stable		
ν_μ	$J=\frac{1}{2}$	$0(<1.2)$				
e	$J=\frac{1}{2}$	0.5110034 ±.0000014	stable ($>2\times10^{21}$y)	stable		
μ	$J=\frac{1}{2}$	105.65948 ±.00035 $m^2=0.01116$ $m_\mu-m_{\pi\pm}=-33.909$ ±.006	2.197134×10^{-6} ±.00008 cτ $=6.587\times10^4$	$e\nu\bar\nu$ $e\gamma\gamma$ $3e$ $e\gamma$	100% $(<1.6)10^{-5}$ $(<6)10^{-9}$ $(<2.2)10^{-8}$	53 53 53 53
π±	$1^-(0^-)$	139.5688 ±.0064 $m^2=0.0195$	2.6030×10^{-8} ±.0023 cτ $=780.4$ $(\tau^+-\tau^-)/\bar\tau=$ $(0.05\pm0.07)\%$ (test of CPT)	$\mu\nu$ $e\nu$ $\mu\nu\gamma$ $\pi^0e\nu$ $e\nu\gamma$ $e\nu e^+e^-$	100% $(1.24\pm0.03)10^{-4}$ $(1.24\pm0.25)10^{-4}$ $(1.02\pm0.07)10^{-8}$ $(3.0\pm0.5)10^{-8}$ $(<3.4)10^{-8}$	30 70 30 5 70 70
π⁰	$1^-(0^-)^+$	134.9645 ±.0074 $m^2=0.0182$ $m_{\pi\pm}-m_{\pi^0}=4.6043$ ±.0037	0.84×10^{-16} ±.10 cτ $=2.5\times10^{-6}$	$\gamma\gamma$ γe^+e^- $\gamma\gamma\gamma$ $e^+e^-e^+e^-$ $\gamma\gamma\gamma\gamma$	$(98.83\pm0.05)\%$ $(1.17\pm0.05)\%$ $(<5)10^{-6}$ $(3.47)10^{-5}$ $(<6.1)10^{-5}$	67 67 67 67 67

ISBN 0-8053-8061-7

ISBN 0-8053-8061-7

$K^\pm$ $\frac{1}{2}(0^-)$ 493.707 ± 0.037 $m^2 = 0.244$

1.2371×10^{-8}
±.0026
$c\tau = 370.8$
$(\tau^+ - \tau^-)/\bar{\tau} = (.11 \pm .09)\%$
(test of CPT)

$m_{K^\pm} - m_{K^0} = -3.99 \pm 0.13$

Mode	Branching ratio	
$\mu\nu$	$(63.54 \pm 0.19)\%$	236
$\pi\pi^0$	$(21.12 \pm 0.17)\%$	205
$\pi\pi^-\pi^+$	$(5.59 \pm 0.03)\%$	125
$\pi\pi^0\pi^0$	$(1.73 \pm 0.05)\%$	133
$\mu\pi^0\nu$	$(3.20 \pm 0.09)\%$	215
$e\pi^0\nu$	$(4.82 \pm 0.05)\%$	228
$e\pi^0\pi^0\nu$	$\left(1.8 \begin{smallmatrix}+2.4\\-0.6\end{smallmatrix}\right) 10^{-5}$	207
$\pi\pi^\mp e^\pm\nu$	$(3.7 \pm 0.2) 10^{-5}$	203
$\pi\pi^\pm e^\mp\nu$	$(<5) 10^{-7}$	203
$\pi\pi^\mp\mu^\pm\nu$	$(0.9 \pm 0.4) 10^{-5}$	151
$\pi\pi^\pm\mu^\mp\nu$	$(<3) 10^{-6}$	151
$e\nu$	$(1.38 \pm 0.20) 10^{-5}$	247
$e\nu\gamma$	$(<7) 10^{-5}$	247
$\pi\pi^0\gamma$	$(2.71 \pm 0.19) 10^{-4}$	205
$\pi\pi^+\pi^-\gamma$	$(10 \pm 4) 10^{-5}$	125
$\mu\pi^0\nu\gamma$	$(<6) 10^{-5}$	215
$e\pi^0\nu\gamma$	$(3.7 \pm 1.4) 10^{-4}$	228
πe^+e^-	$(<0.26) 10^{-6}$	227
$\pi^\mp e^\pm e^\pm,$	$(<1.5) 10^{-5}$	227
$\pi\mu^+\mu^-$	$(<2.4) 10^{-6}$	172
$\pi\gamma\gamma$	$(<3.5) 10^{-5}$	227
$\pi\gamma\gamma\gamma$	$(<3) 10^{-4}$	227
$\pi\nu\bar{\nu}$	$(<0.6) 10^{-6}$	227
$\pi\gamma$	$(<4) 10^{-6}$	227
$e\pi^\mp\mu^\pm$	$(<3) 10^{-8}$	214
$e\pi^\pm\mu^\mp$	$(<1.4) 10^{-8}$	214
$\mu\nu\bar{\nu}$	$(<6) 10^{-6}$	236

K^0 $\frac{1}{2}(0^-)$ 497.70 ± 0.13 $m^2 = 0.248$

50% K_{Short},   50% K_{Long}

TABLE 13-2   (continued)
Stable Particles

Particle	$I^G(J^P)C_n$	Mass (MeV) Mass² (GeV)²	Mean Life (sec) $c\tau$ (cm)	Partial decay mode Mode	Fraction[b]	p or p_{max}^c (MeV/c)
K_S^0	$\frac{1}{2}(0^-)$		0.886×10^{-10} ±.0023 $c\tau = 2.68$	$\pi^+\pi^-$	(68.77)% ± 0.26	206
				$\pi^0\pi^0$	(31.23)%	209
				$\mu^+\mu^-$	(<0.3) 10^{-6}	225
				e^+e^-	(<35) 10^{-5}	249
				$\pi^+\pi^-\gamma$	(2.0 ± 0.4) 10^{-3}	206
				$\gamma\gamma$	(<0.4) 10^{-3}	249
K_L^0	$\frac{1}{2}(0^-)$		5.179×10^{-8} ± 0.040 $c\tau = 1553$	$\pi^0\pi^0\pi^0$	(21.3 ± 0.6)%	139
				$\pi^+\pi^-\pi^0$	(11.9 ± 0.4)%	133
		$m_{K_L} - m_{K_S}$ = $0.5349 \times 10^{10}\,\hbar$ sec^{-1} ± 0.0022		$\pi\mu\nu$	(27.5 ± 0.5)%	216
				$\pi e\nu$	(39.0 ± 0.6)%	229
				$\pi e\nu\gamma$	(1.3 ± 0.8)%	229
				$\pi^+\pi^-$	(0.201 ± 0.006)%	206
				$\pi^0\pi^0$	(0.093 ± 0.019)%	209
				$\pi^+\pi^-\gamma$	(<0.6) 10^{-4}	206
				$\pi^0\gamma\gamma$	(<2.4) 10^{-4}	231
				$\gamma\gamma$	(4.9 ± 0.4) 10^{-4}	249
				$e\mu$	(<1.6) 10^{-9}	238
				$\mu^+\mu^-$	(<1.6) 10^{-8}	225
				e^+e^-	(<1.6) 10^{-9}	249
				$e^+e^-\gamma$	(<2.8) 10^{-5}	249
η	$0^+(0^-)^+$	548.8 ± 0.6	$^e\Gamma = (0.85 \pm 0.12)$keV Neutral decays 71.1%	$\begin{cases} \gamma\gamma \\ \pi^0\gamma\gamma \\ 3\pi^0 \end{cases}$	(38.0 ± 1.0)% (3.1 ± 1.1)% (30.0 ± 1.1)%	274 258 180

$m^2 = 0.301$

		p (MeV/c)
$\pi^+\pi^-\pi^0$	$(23.9 \pm 0.6)\%$	175
$\pi^+\pi^-\gamma$	$(5.0 \pm 0.1)\%$	236
$\pi^0 e^+ e^-$	$(<0.04)\%$	258
$\pi^+\pi^-$	$(<0.15)\%$	236
$\pi^+\pi^- e^+ e^-$	$(0.1 \pm 0.1)\%$	236
$\pi^+\pi^-\pi^0\gamma$	$(<6)\,10^{-4}$	175
$\pi^+\pi^-\gamma\gamma$	$(<0.2)\%$	236
$\mu^+\mu^-$	$(2.2 \pm 0.8)\,10^{-5}$	253
$\mu^+\mu^-\pi^0$	$(<5)\,10^{-4}$	211

Charged decays 28.9%

p $\quad \frac{1}{2}(\frac{1}{2}^+)$ $\quad$ 938.2796 ± 0.0027 $\quad m^2 = 0.8804$

stable $\quad (>2 \times 10^{28}\text{y})$

	100%	1

n $\quad \frac{1}{2}(\frac{1}{2}^+)$ $\quad$ 939.5731 ± 0.0027 $\quad m^2 = 0.8828$

$m_p - m_n = -1.29344 \pm 0.00007$

918 ± 14 $\quad c\tau = 2.75 \times 10^{13}$

$pe^-\nu$	100%	1

Λ $\quad 0(\frac{1}{2}^+)$ $\quad$ 1115.60 ± 0.05 $\quad m^2 = 1.245$

2.578×10^{-10} $\pm .021$ $\quad c\tau = 7.73$

		p
$p\pi^-$	$(64.2)\% \pm 0.5$	100
$n\pi^0$	$(35.8)\%$	104
$pe^-\nu$	$(8.13 \pm 0.29)\,10^{-4}$	163
$p\mu^-\nu$	$(1.57 \pm 0.35)\,10^{-4}$	131
$p\pi^-\gamma$	$(0.85 \pm 0.14)\,10^{-3}$	100

Σ^+ $\quad 1(\frac{1}{2}^+)$ $\quad$ 1189.37 ± 0.06 $\quad m^2 = 1.415$

$m_{\Sigma^+} - m_{\Sigma^-} = -7.99 \pm .08$

0.800×10^{-10} $\pm .006$ $\quad c\tau = 2.40$

$\dfrac{\Gamma(\Sigma^+ \to l^+ n\nu)}{\Gamma(\Sigma^- \to l^- n\nu)} < .035$

		p
$p\pi^0$	$(51.6)\% \pm 0.7$	189
$n\pi^+$	$(48.4)\%$	185
$p\gamma$	$(1.24 \pm 0.18)\,10^{-3}$	225
$n\pi^+\gamma$	$(0.93 \pm 0.10)\,10^{-3}$	185
$\Lambda e^+\nu$	$(2.02 \pm 0.47)\,10^{-5}$	72
$\leftarrow \begin{cases} n\mu^+\nu \\ ne^+\nu \end{cases}$	$(<2.4)\,10^{-5}$	202
	$(<1.0)\,10^{-5}$	224
pe^+e^-	$(<7)\,10^{-6}$	225

Σ^0 $\quad 1(\frac{1}{2}^+)$ $\quad$ 1192.48 ± 0.08 $\quad m^2 = 1.422$

$<1.0 \times 10^{-14}$ $\quad c\tau < 3 \times 10^{-4}$

		p
$\Lambda\gamma$	100%	74
$\Lambda e^+ e^-$	$(5.45)\,10^{-3}$	74

ISBN 0-8053-8061-7

TABLE 13-2 (continued)
Stable Particles

Particle	$I^G(J^P)C_n$	Mass (MeV) Mass² (GeV)²	Mean Life (sec) cτ (cm)	Mode	Fraction[b]	p or p_{max}^c (MeV/c)
Σ^-	$1(\frac{1}{2}^+)$	1197.35 ± 0.06	1.482×10^{-10} ±.017 cτ = 4.44	$n\pi^-$	100%	193
				$ne^-\nu$	$(1.08 \pm 0.04)\,10^{-3}$	230
		$m^2 = 1.434$		$n\mu^-\nu$	$(0.45 \pm 0.04)\,10^{-3}$	210
	$m_{\Sigma^0} - m_{\Sigma^-}$	$= -4.87$ ±.06		$\Lambda e^-\nu$	$(0.60 \pm 0.06)\,10^{-4}$	79
				$n\pi^-\gamma$	$(1.0 \pm 0.2)\,10^{-4}$	193
Ξ^0	$\frac{1}{2}(\frac{1}{2}^+)^h$	1314.9 ± 0.6	2.96×10^{-10} ±.12 cτ = 8.93	$\Lambda\pi^0$	100%	135
				$p\pi^-$	$(< 3.6)\,10^{-5}$	299
		$m^2 = 1.729$		$pe^-\nu$	$(< 1.3)\,10^{-3}$	323
	$m_{\Xi^0} - m_{\Xi^-}$	$= -6.4$ ±.6		$\Sigma^+ e^-\nu$	$(< 1.5)\,10^{-3}$	119
				$\Sigma^- e^+\nu$	$(< 1.5)\,10^{-3}$	112
				$\Sigma^+ \mu^-\nu$	$(< 1.5)\,10^{-3}$	64
				$\Sigma^- \mu^+\nu$	$(< 1.5)\,10^{-3}$	49
				$p\mu^-\nu$	$(< 1.3)\,10^{-3}$	309
Ξ^-	$\frac{1}{2}(\frac{1}{2}^+)^h$	1321.29 ± 0.14	1.652×10^{-10} ±.023 cτ = 4.95	$\Lambda\pi^-$	100%	139
				$\Lambda e^-\nu$	$(0.70 \pm 0.21)\,10^{-3}$	190
		$m^2 = 1.746$		$\Sigma^0 e^-\nu$	$(< 0.5)\,10^{-3}$	123
				$\Lambda\mu^-\nu$	$(< 1.3)\,10^{-3}$	163
				$\Sigma^0 \mu^-\nu$	$(< 0.8)\,10^{-3}$	70
				$n\pi^-$	$(< 1.1)\,10^{-3}$	303
				$ne^-\nu$	$(< 3.2)\,10^{-3}$	327
Ω^-	$0(\frac{3}{2}^+)^h$	1672.2 ± .4 $m^2 = 2.797$	$1.3 ^{+0.3}_{-0.2} \times 10^{-10}$ cτ = 3.9	$\left.\begin{array}{l}\Xi^0\pi^-\\ \Xi^-\pi^0\\ \Lambda K^-\end{array}\right\}$	Total of 41 events seen	293 290

[a] Adapted from Particle Data Group, "Review of Particle Properties," *Phys. Letters,* **50B**, No. 1 (1974). See also *Rev. Mod. Phys.,* **48**, Part II, April 1976. The *i*-spin denoted by *I* in these tables is designated *T* elsewhere in the

ISBN 0-8053-8091-L

ISBN 0-8053-8061-7

ADDITIONAL PARAMETERS FOR STABLE PARTICLES

	Magnetic moment		μ Decay parameters[j]		
e	1.001 159 6567	$\dfrac{e\hbar}{2m_e c}$			
	$\pm$.000 000 0035				
μ	1.001 166 897	$\dfrac{e\hbar}{2m_\mu c}$	$\rho = 0.752 \pm 0.003$ $\eta = -0.12 \pm 0.21$ $h = 1.00 \pm 0.13$		
	$\pm$.000 000 027		$\xi = 0.972 \pm 0.013$ $\delta = 0.755 \pm 0.009$		
			$|g_A/g_V| = 0.86 {\,}^{+0.33}_{-0.11}$ $\phi = 180° \pm 15°$		

$K^{\pm}$ Mode	Partial rate (sec^{-1})	$\Delta I = \tfrac{1}{2}$ rule for $K \to 3\pi^k$
$\mu\nu$	$(51.36 \pm 0.19)10^6$	
$\pi\pi^0$	$(17.07 \pm 0.15)10^6$	$K^+ \to \pi^+\pi^+\pi^-$ $g = -.214 \pm .005$
$\pi\pi^+\pi^-$	$(4.52 \pm 0.02)10^6$	$K^- \to \pi^-\pi^+\pi^-$ $g = -.214 \pm .007$
$\pi\pi^0\pi^0$	$(1.40 \pm 0.04)10^6$	$K^{\pm} \to \pi^{\pm}\pi^0\pi^0$ $g = .522 \pm .020$
$\mu\pi^0\nu$	$(2.58 \pm 0.07)10^6$	$K^0_L \to \pi^0\pi^-\pi^+$ $g = .610 \pm .021$
$e\pi^0\nu$	$(3.90 \pm 0.04)10^6$	

K^0_S	Mode		Form factors for K_{l3} decays
	$\pi^+\pi^-$	$^l(0.776 \pm .006)10^{10}$	K^+_{e3} $\lambda^e_+ = .029 \pm .004$ K^0_{e3} $\lambda^e_+ = .026 \pm .004$
	$\pi^0\pi^0$	$^l(0.352 \pm .004)10^{10}$	

K^0_L	Mode		CP violation parameters^{l,n}
	$\pi^0\pi^0\pi^0$	$(4.11 \pm 0.13)10^6$	$|\eta_{+-}| = (2.17 \pm .07)10^{-3}$ $|\eta_{00}| = (2.25 \pm .09)10^{-3}$
	$\pi^+\pi^-\pi^0$	$(2.31 \pm 0.07)10^6$	$\phi_{+-} = (46.6 \pm 2.5)°$ $\phi_{00} = (48 \pm 13)°$
	$\pi\mu\nu$	$(5.31 \pm 0.11)10^6$	
	$\pi e\nu$	$(7.53 \pm 0.12)10^6$	
	$\pi^+\pi^-$	$^l(3.42 \pm 0.36)10^4$	$|\eta_{+-0}|^2 < 0.12$ $|\eta_{000}|^2 < 0.28$ $\delta = (.34 \pm .01)10^{-2}$
	$\pi^0\pi^0$	$^l(1.80 \pm 0.36)10^4$	$\Delta S = -\Delta Q$
			$Rex = .008 \pm .022$ $Imx = .012 \pm .030$

η	Mode	Asymmetry parameter
	$\pi^+\pi^-\pi^0$	$(0.12 \pm .17)\%$
	$\pi^+\pi^-\gamma$	$(0.88 \pm .40)\%$

ADDITIONAL PARAMETERS FOR STABLE PARTICLES

	Magnetic moment $(e\hbar/2m_pc)$	Decay mode	Decay parameters[n] Measured α	$\phi(degree)$	γ	Derived $\Delta(degree)$	g_A/g_V	g_V/g_A
p	2.7928456 ±.0000011							
n	−1.913148 ±.000066	$pe^-\nu$					-1.250 ± 0.009 $\delta = (181.1 \pm 1.3)°$	
Λ	−0.67 ±.06	$p\pi^-$	0.647 ± 0.013	$(-6.5 \pm 3.5)°$	0.76	$\left(7.6 {+4.0 \atop -4.1}\right)°$		
		$n\pi^0$	0.651 ± 0.045					
		pev					-0.66 ± 0.05	
Σ^+	2.62 ±.41	$p\pi^0$	-0.979 ± 0.016	$(36 \pm 34)°$	0.17	$(187 \pm 6)°$		
		$n\pi^+$	$+0.066 \pm 0.016$	$(167 \pm 20)°$	−0.97	$\left(-73 {+136 \atop -10}\right)°$		
		$p\gamma$	$-1.03 {+.52 \atop -.42}$					
Σ^-	−1.6 to 0.8	$n\pi^-$	-0.069 ± 0.008	$(10 \pm 15)°$	0.98	$\left(249 {+12 \atop -115}\right)°$		
		$ne^-\nu$ $\Lambda e^-\nu$					$\pm (0.435 \pm 0.035)$	0.24 ± 0.20
Ξ^0		$\Lambda\pi^0$	-0.44 ± 0.08	$(21 \pm 12)°$	0.84	$\left(216 {+13 \atop -19}\right)°$		
Ξ^-	−1.93 ±.75	$\Lambda\pi^-$	-0.393 ± 0.023	$(2 \pm 7)°$	0.92	$(184 \pm 15)°$		

[b]Quoted upper limits correspond to a 90% confidence level.

ISBN 0-8053-8061-

ISBN 0-8053-8061-7

[c] In decays with more than two bodies, p_{max} is the maximum momentum that any particle can have.

[e] Theoretical value.

[f] The direct emission branching ratio is $(1.56 \pm .35) \times 10^{-5}$.

[h] P for Ξ and J^P for Ω^- not yet measured. Values reported are $SU(3)$ predictions.

[i] Assumes rate for $\Xi^- \rightarrow \Sigma^0 e^- \nu$ small compared with $\Xi^- \rightarrow \Lambda e^- \nu$.

[j] $|g_A/g_V|$ defined by $g_A^2 = |C_A|^2 + |C'_A|^2$, $g_V^2 = |C_V|^2 + |C'_V|^2$, and $\Sigma \langle e|\bar{\partial}|\Gamma_i|\mu\rangle\langle\nu|\Gamma_i(C_i + C'_i\gamma_5)|\nu\rangle$; ϕ defined by $\cos\phi = -Re(C_A^* C_V + C'_A C'^*_V)/g_A g_V$.

[k] The definition of the slope parameter of the Dalitz plot is

$$|M|^2 = 1 + g\left(\frac{s_3 - s_0}{m_{\pi^+}^2}\right)$$

[l] The $K_S^0 \rightarrow \pi\pi$ and $K_L^0 \rightarrow \pi\pi$ rates (and branching fractions) are from independent fits and do not include results of $K_L^0 - K_S^0$ interference experiments. The $|\eta_{+-}|$ and $|\eta_{00}|$ values given are these rates combined with the $|\eta_{+-}|$ and $|\eta_{00}|$ results from interference experiments.

[m] The definition for the CP violation parameters is as follows.

$$\delta = \frac{\Gamma(K_L^0 \rightarrow l^+) - \Gamma(K_L^0 \rightarrow l^-)}{\Gamma(K_L^0 \rightarrow l^+) + \Gamma(K_L^0 \rightarrow l^-)}$$

$$|\eta_{+-0}|^2 = \frac{\Gamma(K_S^0 \rightarrow \pi^+\pi^-\pi^0)}{\Gamma(K_L^0 \rightarrow \pi^+\pi^-\pi^0)}$$

$$|\eta_{000}|^2 = \frac{\Gamma(K_S^0 \rightarrow \pi^0\pi^0\pi^0)}{\Gamma(K_L^0 \rightarrow \pi^0\pi^0\pi^0)}$$

[n] The definition of these quantities is as follows.

$$\alpha = \frac{2|s||p|\cos\Delta}{|s|^2 + |p|^2}$$

$$\beta = \frac{-2|s||p|\sin\Delta}{|s|^2 + |p|^2}$$

$$\beta = (1 - \alpha^2)^{1/2}\sin\phi$$

$$\gamma = (1 - \alpha^2)^{1/2}\cos\phi$$

g_A/g_V defined by $\langle B_f|\gamma_\lambda(g_V - g_A\gamma_5)|B_i\rangle$

δ defined by $g_A/g_V = |g_A/g_V|e^{i\delta}$

MESON PROPERTIES

Nonstrange (Y = 0)						Strange (|Y| = 1)	
Entry	$I^G(J^P)C_n$	Entry	$I^G(J^P)C_n$	Entry	$I^G(J^P)C_n$	Entry	$I(J^P)$
π (140)	$1^-(0^-)+$	→η_N (1080)	$0^+(N)+$	ρ' (1600)	$1^+(1^-)-$	K (494)	$1/2(0^-)$
η (549)	$0^+(0^-)+$	A_1 (1100)	$1^-(1^+)+$	A_3 (1640)	$1^-(2^-)+$	K^* (892)	$1/2(1^-)$
		→M (1150)		ω (1675)	$0^-(N)-$	κ	$1/2(0^+)$
ρ (770)	$1^+(1^-)-$	→$A_{1.5}$ (1173)		g (1680)	$1^+(3^-)-$	Q	$1/2(1^+)$
ω (783)	$0^-(1^-)-$	B (1235)	$1^+(1^+)-$			K^* (1423)	$1/2(2^+)$
→M (940)		→ρ' (1250)	$1^+(1^-)-$			→K_N (1660)	$1/2$
→M (953)		f (1270)	$0^+(2^+)+$			→K_N (1760)	$1/2$
η' (958)	$0^+(0^-)+$	D (1285)	$0^+(A)+$			L(1770)	$1/2(A)$
δ (970)	$1^-(0^+)+$	A_2 (1310)	$1^-(2^+)+$				
→H (990)		E (1420)	$0^+(A)+$				
S^* (993)	$0^+(0^+)+$	→X (1430)	0				
ϕ (1019)	$0^-(1^-)-$	→X (1440)	1				
→M (1033)		f' (1514)	$0^+(2^+)+$				
→B_1 (1040)		F_1 (1540)	$1(A)$			→Exotics	

N means $J^P = 0^+$, 1^-, etc. A means $J^P = 0^-$, 1^+

ISBN 0-8053-8061-7

BARYON PROPERTIES[a]

Given are the name, nominal mass, quantum numbers (when known), and status of the baryon states. For the quantum numbers the symbols $P11$ mean that the baryon *considered as a nucleon-pion state* has orbital angular momentum $L=1$, $2T=1$, $2J=1$. Thus for the proton $P11$ means $L=1$, $T=1/2$, $J=1/2$, parity even (because of the negative intrinsic parity of the pion). The proton has $TJ^P = 1/2\ 1/2^+$. For other unstable particles L is the orbital angular momentum in pionic decay, followed by $2T$ and $2J$.

N	Δ	Λ	Σ	Ξ / Ω
$N(939)$ $P11$ ****	$\Delta(1232)$ $P33$ ****	$\Lambda(1116)$ $P01$ ****	$\Sigma(1193)$ $P11$ ****	$\Xi(1317)$ $P11$ ****
$N(1470)$ $P11$ ****	$\Delta(1650)$ $S31$ ****	$\Lambda(1330)$ Dead	$\Sigma(1385)$ $P13$ ****	$\Xi(1530)$ $P13$ ****
$N(1520)$ $D13$ ****	$\Delta(1670)$ $D33$ ***	$\Lambda(1405)$ $S01$ ****	$\Sigma(1440)$ Dead	$\Xi(1630)$ **
$N(1535)$ $S11$ ****	$\Delta(1690)$ $P33$ *	$\Lambda(1520)$ $D03$ ****	$\Sigma(1480)$ *	$\Xi(1820)$ ***
$N(1670)$ $D15$ ****	$\Delta(1890)$ $F35$ ***	$\Lambda(1670)$ $S01$ ****	$\Sigma(1620)$ $S11$ **	$\Xi(1940)$ ***
$N(1688)$ $F15$ ****	$\Delta(1900)$ $S31$ *	$\Lambda(1690)$ $D03$ ****	$\Sigma(1660)$ $P11$ **	$\Xi(2030)$ **
$N(1700)$ $S11$ ****	$\Delta(1910)$ $P31$ ****	$\Lambda(1750)$ $P01$ **	$\Sigma(1670)$ $D13$ ***	$\Xi(2250)$ *
$N(1700)$ $D13$ **	$\Delta(1950)$ $F37$ ****	$\Lambda(1815)$ $F05$ ****	$\Sigma(1690)$ **	$\Xi(2500)$ **
$N(1780)$ $P11$ ***	$\Delta(1960)$ $D35$ **	$\Lambda(1830)$ $D05$ ***	$\Sigma(1750)$ $S11$ ***	$\Omega(1672)$ $P03$ ****
$N(1810)$ $P13$ ***	$\Delta(2160)$ $P33$ **	$\Lambda(1860)$ $P03$ **	$\Sigma(1765)$ $D15$ ****	
$N(1990)$ $F17$ **	$\Delta(2420)$ $H311$ ***	$\Lambda(1870)$ $S01$ **	$\Sigma(1840)$ $P13$ *	
$N(2000)$ $F15$ **	$\Delta(2850)$ ***	$\Lambda(2010)$ $D03$ **	$\Sigma(1880)$ $P11$ **	
$N(2040)$ $D13$ **	$\Delta(3250)$ ***	$\Lambda(2020)$ $F07$ **	$\Sigma(1915)$ $F15$ ****	
$N(2100)$ $S11$ *		$\Lambda(2100)$ $G07$ ****	$\Sigma(1940)$ $D13$ ***	
$N(2100)$ $D15$ *		$\Lambda(2110)$ $F05$ *	$\Sigma(2000)$ $S11$ *	
$N(2190)$ $G17$ ***		$\Lambda(2350)$ ****	$\Sigma(2030)$ $F17$ ****	
$N(2220)$ $H19$ ***		$\Lambda(2585)$ ***	$\Sigma(2070)$ $F15$ *	
$N(2650)$ ***			$\Sigma(2080)$ $P13$ **	
$N(3030)$ ***			$\Sigma(2100)$ $G17$ *	
$N(3245)$ *			$\Sigma(2250)$ ****	
$N(3690)$ *			$\Sigma(2455)$ ***	
$N(3755)$ *			$\Sigma(2620)$ ***	
			$\Sigma(3000)$ **	

**** Good, clear, and unmistakable.   *** Good, but in need of clarification or not absolutely certain.   ** Needs confirmation.   * Weak.

ISBN 0-8053-8061-7

stable. For particles that can decay by strong interaction only a summary of the information available is given. Table 13-2 is based on the table of particle properties updated and published annually by the Particle Data Group (RPP 74).

● Each particle has its own antiparticle. For the electron and for fermions of spin $\frac{1}{2}$ this derives from Dirac's equation. For an electron in an electromagnetic field

$$\left(\frac{\partial}{\partial t} + \boldsymbol{\alpha} \cdot \boldsymbol{\nabla} + im\beta \right)\psi = ie(\boldsymbol{\alpha} \cdot \mathbf{A} - V)\psi \tag{13-1.1}$$

where $\hbar = c = 1$, $\mathbf{A}$ and V are the vector and scalar electromagnetic potentials, and $\boldsymbol{\alpha}$ and β are Dirac's matrices, obeying the relations

$$\alpha_\mu \alpha_\nu + \alpha_\nu \alpha_\mu = 2\delta_{\mu\nu} \qquad (\mu, \nu = 1, 2, 3, 4)$$

$$\alpha_4 \equiv \beta \tag{13-1.2}$$

It is possible to choose $\boldsymbol{\alpha}$ and β such that all the matrix elements of $\boldsymbol{\alpha}$ are real and those of β are purely imaginary. The ψ has four components.

If we take the complex conjugate on both sides of Eq. (13-1.1) we obtain

$$\left(\frac{\partial}{\partial t} + \boldsymbol{\alpha} \cdot \boldsymbol{\nabla} + im\beta \right)\psi^* = -ie(\boldsymbol{\alpha} \cdot \mathbf{A} - V)\psi^* \tag{13-1.3}$$

That is, ψ^* obeys the same equation as ψ except that the charge e has been changed into $-e$. If Eq. (13-1.1) has a positive energy solution $\psi \sim e^{-iEt}$ ($E > 0$), then Eq. (13-1.3) has a negative energy solution, and vice versa. ●

It is clear that the negative energy solutions of Eq. (13-1.3), whose meaning may at first seem questionable, correspond to positive energy solutions of Eq. (13-1.1), provided the sign of charge is reversed. They correspond to the motion of *positrons.*

Dirac interpreted the positrons as "holes" in a distribution, with almost all states of negative energy filled, but here Dirac's theory is formulated in a perfectly symmetric form in which the electron and positron are treated on an equal footing. This is important because it shows that the symmetry between electron and positron is not related to their statistics, as one might think from the "hole" picture, in which the exclusion principle plays a fundamental role.

The concept of charge conjugation can be extended to bosons. One finds that for each charged particle there exists an antiparticle that has the same mass and spin as the particle. Their electric charge and magnetic moment are equal in magnitude but opposite in sign.

For neutral particles, if they have electromagnetic properties (e.g., a magnetic moment, as in the case of the neutron) the same relations apply as for

ISBN 0-8053-8601-7

charged particles. Thus for instance a neutron and an antineutron with parallel spins have equal and opposite magnetic moment.

If they have no magnetic or electric moments, the particles may be self-conjugate like the π^0 or they may have distinct antiparticles like the neutrino. Whether one case or the other occurs can be decided by experiment; for instance, in the case of the neutrino, by observation of the inverse beta decay. The relations between particle and antiparticle are summarized in Table 13-3.

TABLE 13-3   RELATION BETWEEN PARTICLE AND ANTIPARTICLE

	Particle	*Antiparticle*	
Mass		Same	
Spin		Same	
Charge	$+q$		$-q$
Magnetic moment	μ		$-\mu$
Mean life (free decay)		Same	
Annihilation	In pairs	For fermions only	
Generation	In pairs	For fermions only	
Isospin		Same	
Third comp. of *i* spin	T_3		$-T_3$
Intrinsic parity		Same for bosons;	
(see Sec. 15-3)		opposite for fermions	

In the case of fermions, particle and antiparticle must be generated or annihilated in pairs. This can be looked upon as a consequence of the hole theory of Dirac and it has been experimentally verified in every case. It is also required by the principle of conservation of leptons and nucleons.

Annihilation occurs in a variety of forms. For electron–positron pairs annihilation is mostly electromagnetic, and light quanta are emitted. In the case of nucleons–antinucleons, pi mesons are the most frequent annihilation product (Fig. 13-1) but occasionally K mesons are also emitted. Electromagnetic annihilation into gamma rays or electron–positron pairs occurs only seldom in comparison with pion-producing annihilation. Particle and antiparticle have the same i spin but opposite T_3.

Particle–antiparticle symmetry is not limited to particles, but because of the identity of the forces extends to nuclei, atoms, and worlds. In the laboratory on earth physicists have created antideuterons and antitritons; there is no doubt that in suitable circumstances all other antinuclei could be formed and surrounded by positrons to boot. Whether antiworlds exist, however, is not known.

There are important empirical laws stating that

1. The number of baryons is conserved. For instance, $T_{1/2}$ of the proton is larger than 10^{28} yr, some 10^{19} times the age of the universe! It is the conservation of baryons that prevents a reaction such as $^3H \rightarrow p + \gamma$ or $p + \pi^0$.

ISBN 0-8053-8601-7

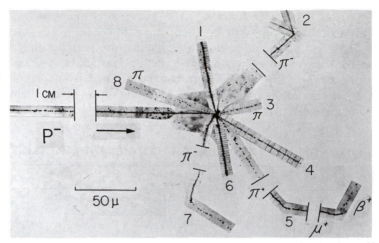

Figure   13-1   Proton–antiproton annihilation in a photographic emulsion. The annihilating proton is bound in a nucleus. The nucleus is disrupted and tracks 1, 4, and 6 are fragments of the nucleus. The other tracks are charged pions coming from the annihilation. Neutral pions are invisible. The total visible energy released is 1300 MeV > 938 MeV (the mass of the incoming antiproton), proving that annihilation has taken place. A study of the primary antiproton track permits the determinations of the $\bar{p}$ mass with an accuracy of a few percent.

In these conservation laws an antiparticle is considered as having the baryon number -1; thus, $p + \bar{p} \rightarrow \pi^{+} + \pi^{-}$ does not violate the conservation of baryons.

2.   The number of leptons is conserved or, more precisely the number of electronic leptons (e, ν_e) and muonic leptons (μ, ν_μ) are conserved separately. (See Chap. 9.)

13-2   ASSOCIATED PRODUCTION; STRANGENESS

In 1947 Rochester and Butler obtained cloud-chamber photographs of two cosmic-ray events in which the tracks had the configuration of a V (Fig. 13-2). A careful analysis indicated that the tracks originated from the decay of two new species of particles, now called K^0 and K^+. Subsequently emulsion photographs of other cosmic-ray events showed particles decaying into three pions (Brown, Camerini, Fowler, Muirhead, Powell, and Ritson, 1949) (Fig. 13-3). More new particles appeared in cosmic-ray events during the next few years.

When accelerators reached an energy sufficient for producing these particles in the laboratory, it became apparent that there was a discrepancy

ISBN 0-8053-8601-7

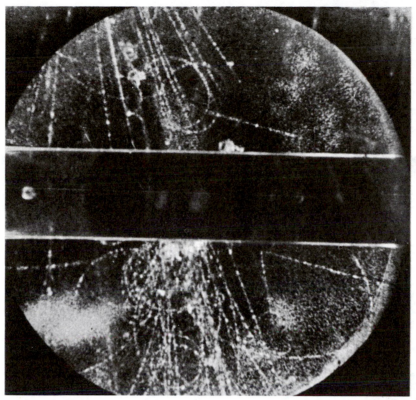

Figure 13-2 First picture showing a neutral particle decaying into 2 pions. Such a neutral particle is now called a K^0 particle. Originally the authors called them V particles. [From G. D. Rochester and C. C. Butler, *Nature*, **160**, 855 (1947).]

between the relatively large cross section for production and the long decay period. This led to the discovery of associated production and to the introduction of the "strangeness" quantum number (Gell-Mann and Nishijima, 1953).

If we examine the decay constant of the hyperons (Table 13-2), we find that they are all of the order of 10^{10} sec^{-1}, except for Σ^0, which, however, decays by gamma emission and transforms into another hyperon, the Λ. K mesons have similar or smaller decay constants.

Production cross sections for hyperons and K mesons at about 1.5 GeV are about one tenth the nucleon–pion scattering cross section. On this basis we could try to estimate the decay constant of the hyperons, attributing the decay to an interaction

$$\Lambda \Rightarrow p + \pi \tag{13-2.1}$$

ISBN 0-8053-8601-7

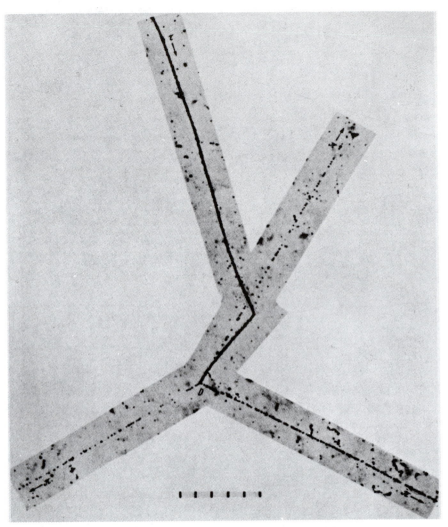

Figure 13-3 Decay of a τ meson (in modern notation $K^{\pm}$) into three pions. [Hodgson, 1951, from (PFP 59).]

Its matrix element could be approximately calculated from production if the production reaction were

$$\pi + p \Rightarrow \Lambda + \pi \qquad (13\text{-}2.2)$$

(which it is not). This, however, leads to a discrepancy, showing that the decay is very strongly forbidden.

From the production cross section we conclude that the matrix element for Λ production is of the order of $10^{-1/2}$ or one third that of the Yukawa

ISBN 0-8053-8601-7

interaction. If we use this datum to estimate the decay constant of the Λ, we arrive at 10^{22} sec^{-1}, a number 10^{12} times larger than the experimental one. A similar contradiction is obtained if we compare the reactions

$$K^0 \Rightarrow 2\pi^0 \tag{13-2.3}$$

and

$$p + \pi^- \Rightarrow n + K^0 \tag{13-2.4}$$

In other words, the decay constant is much too small if we assume that the production occurs according to Eq. (13-2.2) or (13-2.4). In Eqs. (13-2.1) to (13-2.4) we have used (for purposes of argument) the strong-interaction symbol $\Rightarrow$. Actually all these reactions are forbidden for strong interactions, as we shall see shortly.

Several hypotheses have been put forward in order to resolve this apparent discrepancy. However, the fact is that the reactions by which Λ and K are formed are not those of Eqs. (13-2.2) and (13-2.4) but

$$\pi^- + p \Rightarrow \Lambda^0 + K^0 \tag{13-2.5}$$

or similar reactions, in which, starting from pions and nucleons, two particles of the Λ, K, Σ, and Ξ families are formed simultaneously. The large cross sections for production occur only for "associated production" of a Λ^0 and K^0, starting from $\pi^- + p$ (Figs. 13-4 and 13-5). Which particles are to be associated will be seen shortly.

The decay and production processes thus involve different particles and cannot be simply related (Nambu, Nishijima, and Yamaguchi, 1951; Pais, 1952; Fowler, 1954).

This whole subject is systematized by introducing a new additive quantum number called *strangeness* (S) (Gell-Mann and Nishijima, 1953) and by postulating that, in strong interactions, strangeness must be conserved. At present, these concepts are unrelated to other ideas in physics, and we have no profound "explanation" of their meaning, but shall employ them as very useful semiempirical rules.

The strangeness S of a particle is defined by the relation

$$Q = T_3 + \frac{N}{2} + \frac{S}{2} \tag{13-2.6}$$

where Q is the electric charge in units of $+e$ (proton charge), T_3 is the third component of the i-spin, and N is the baryon number (the number of nucleons that ultimately appear in the decay or in the final products of a chain of decays of the particle). The quantity $N + S = Y$ is called *hypercharge*

ISBN 0-8053-8601-7

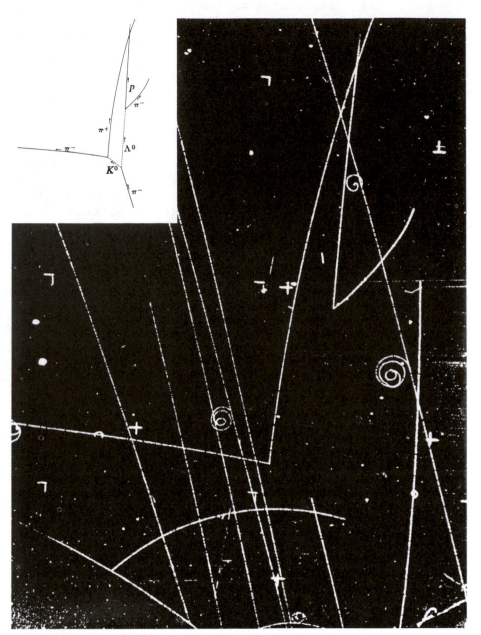

Figure 13-4 Bubble-chamber picture showing the following reactions:

$$\pi^- + p \Rightarrow \Lambda^0 + K^0$$

$$\Lambda^0 \to \pi^- + p$$

$$K^0 \to \pi^- + \pi^+$$

[Courtesy Lawrence Berkeley Laboratory, Berkeley, Calif.]

700

ISBN 0-8053-8601-7

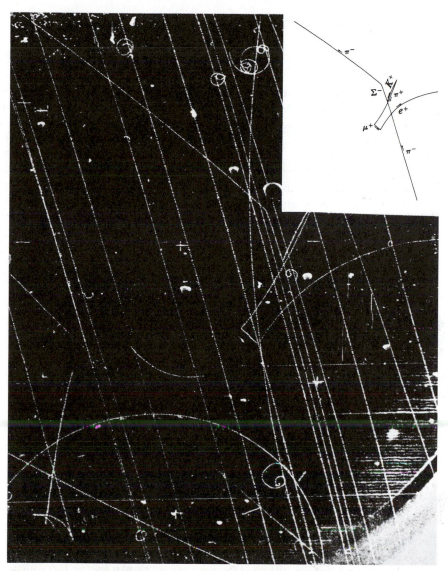

Figure 13-5 Bubble-chamber picture showing the following reactions:

$$\pi^- + p \Rightarrow \Sigma^- + K^+$$
$$\Sigma^- \to \pi^- + n$$
$$K^+ \to \pi^+ + \pi^0 + \pi^0$$
$$\hookrightarrow \mu^+ + \nu$$
$$\hookrightarrow e^+ + \nu + \bar{\nu}$$

[Courtesy Lawrence Berkeley Laboratory,]

ISBN 0-8053-8601-7

and one may also write Eq. (13-2.6) in the form

$$Q = T_3 + \frac{Y}{2} \qquad (13\text{-}2.6a)$$

In order to assign the strangeness to a particle, we must know its T_3. Since particles appear in multiplets—three pions, two K mesons, two nucleons, one Λ, three Σ, etc.—it is possible, from the very multiplicity of these particle multiplets, to assign a T, because $2T + 1$ particles appear in a multiplet, although this criterion is not always sufficient. (Thus, both pions and K mesons appear in three charge states, positive, neutral, and negative, but pions have i-spin 1, whereas K mesons form two doublets of i-spin $\frac{1}{2}$.) The strangeness of antiparticles is assigned in a similar manner (considering that one antinucleon is counted as -1 nucleon) and is equal and opposite to the strangeness of the corresponding particle. Strangeness cannot be defined for particles which do not have strong interactions, because i-spin is not defined for them.

Analysis of the empirical material yields the classification of Fig. 13-6. The assignments for the "stable" particles were made as follows:

1. Λ^0 appears in only one charge state, neutral; hence, it is an i-spin singlet; $T = 0 = T_3$. It decays, producing a proton; hence, $N = 1$. Equation (13-2.6) then gives $S = -1$.

2. Σ^+, Σ^0, Σ^-. Here $T = 1$ because we have a charge triplet with $T_3 = 1, 0, -1$. Σ particles decay into one nucleon plus bosons; hence, $N = 1$. Equation (13-2.6) then gives $S = -1$.

3. Ξ^0 and Ξ^- have been observed, indicating that $T = \frac{1}{2}$, $T_3 = \frac{1}{2}, -\frac{1}{2}$. The decay gives rise to one nucleon, and Eq. (13-2.6) gives $S = -2$. This is confirmed, once the strangeness of the K is assigned, by the observed production reaction (Alvarez, Eberhard, Good, Graziano, Ticho, and Wojcicki, 1959), for example:

$$p + K^- \Rightarrow \Xi^- + K^+$$

4. K mesons. From their decay in pions only, we conclude that $N = 0$.

We know three charged states of K, and we could think that $T = 1$. However, the charged K have different strangeness, as shown by the forbiddenness of $\pi^- + p \Rightarrow K^- + \Sigma^+$, whereas $\pi^- + p \Rightarrow K^+ + \Sigma^-$ is allowed. Furthermore there are two kinds of K^0 (see Sec. 19-5). The K^+, K^-, and the two K^0, respectively, are then interpreted as charge conjugates of each other. We have thus two i-spin doublets: K^+ and K^0 with $T = \frac{1}{2}$, $T_3 = \frac{1}{2}, -\frac{1}{2}$ respectively, and hence by Eq. (13-2.6), $S = 1$, and their charge conjugates K^- and $\overline{K^0}$ with $T = \frac{1}{2}$, $T_3 = -\frac{1}{2}, +\frac{1}{2}$, respectively, and $S = -1$.

5.   Pions with $T = 1$, $N = 0$ have $S = 0$.

6.   Nucleons with $T = \frac{1}{2}$, $N = 1$ have $S = 0$.

ISBN 0-8053-8601-7

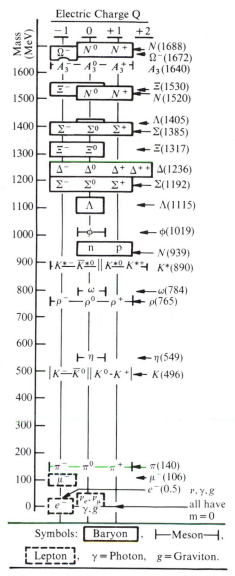

Figure 13-6 Particles of mass < 1700 MeV. Charge, mass, and nucleon number are obvious from the figure. Number of particles in a multiplet (box) is $2T + 1$. Strangeness is the same for all particles in a box: $S = 2(Q - T_3) - N$ or $Y = 2\langle Q \rangle$ for all particles in a box. Antiparticles would give a similar diagram with $Q' = -Q$, $M' = M$, $T' = T$, $T_3' = -T_3$, $S' = -S$, $Y' = -Y$.

ISBN 0-8053-8601-7

7. The Ω^- formed by the reaction $K^- + p \Rightarrow \Omega^- + K^+ + K^0$ has strangeness -3, $N = 1$, and hence $T_3 = 0$. It is expected (see Sec. 17-2) to be an i-spin singlet.

The arguments may be extended to other particles.

The conservation of strangeness immediately gives selection rules that forbid Eq. (13-2.2) and allow associated production. This was the starting point for the establishment of this quantum number. However, it also gives additional selection rules that have been verified experimentally. Thus reactions such as

$$\pi + N \Rightarrow \overline{K} + \Sigma \qquad \text{or} \qquad N + N \Rightarrow \Lambda + \Lambda \qquad (13\text{-}2.7)$$

do not occur, while reactions such as

$$\pi + N \Rightarrow K + \Sigma \qquad \text{or} \qquad \pi + N \Rightarrow K + \Lambda \qquad (13\text{-}2.8)$$

are observed.

13-3 INTERACTIONS; CONSERVATION LAWS

As important as the acquaintance with particles and relative nomenclature is a knowledge and classification of the interactions to which they are subject. Although this topic has been touched upon in other chapters we shall again summarize its results.

The interactions between elementary particles can be classified in four groups:

1. *Strong interactions.* This group comprises the Yukawa interaction and its derivatives, such as the nucleon–nucleon force. It is characterized by a coupling constant f. The dimensionless constant $f^2/\hbar c$ is found to be of the order of 1.
2. *Electromagnetic interactions.* These are characterized by the coupling constant $e^2/\hbar c = 1/\text{``137''}$.
3. *Weak interactions* such as the one causing beta decay or the decay of the K meson into three pions. Their coupling constant (see Chap. 9) may be expressed in a dimensionless form by the Fermi constant of beta decay g divided by the rest energy of a particle and the cube of its Compton wavelength. Its value is $G = (g/mc^2)(\hbar/mc)^{-3} = 1.03 \times 10^{-5}$ if for m we take the protonic mass. (See Ch. 9.)
4. *Gravitational interaction.* A dimensionless coupling constant is $Km^2/\hbar c \cong 5.88 \times 10^{-39}$, where $K (K = 6.67 \times 10^{-8}$ dyne cm^{2}g^{-2}) is Newton's constant and for m we take the mass of the proton. This interaction is much weaker than (3); it will not be considered here.

ISBN 0-8053-8601-7

To understand the meaning of the numerical values of the dimensionless coupling constants, first consider the electromagnetic interactions. In estimating the cross section for a process, (e.g., Compton scattering), one finds orders of magnitude by taking a dimensionally correct natural unit and multiplying it by $e^2/\hbar c$ as many times as the order of the process. Thus, in the case of Compton scattering one could take $(\hbar/m_e c)^2(e^2/\hbar c)^2$. The last factor is squared because the photon is absorbed and reemitted. In the electron–positron annihilation one would take $\hbar/m_e c^2)(e^2/\hbar c)^2$ as the mean life for two-quanta annihilation, etc. Naturally these crude estimates have to be taken with great caution and are valid only for allowed processes.

For the strong interactions the constant $f^2/\hbar c$ can be estimated from the pion–nucleon cross section. If we take the cross section to be $(\hbar/mc)^2(f^2/\hbar c)^2$, with m equal to the nucleon mass, we have $f^2/\hbar c \cong 15$. However, this order of magnitude depends on the specific form of the interaction. The theory is often formulated in such a way that the fundamental coupling constant to be considered is $(f^2/\hbar c)(m_\pi/2m_p)^2 = 0.08$. The value 1 quoted above is a sort of compromise between the different definitions. According to some modern ideas the strong interactions might even be "as strong as possible." By this it is meant that if one had a complete theory considering all strong interactions between all particles, the scattering matrix elements would be as large as permitted by the unitarity and other analytic properties of the matrix. According to these ideas no special constant may be attached to the strong interactions, and the f here considered originates from our considering only a part of the total scattering matrix. The constant f is an empirical parameter describing the neglected interactions. In a complete theory it would vanish. However, $f^2/\hbar c$ is, in any reasonable definition, considerably larger than $e^2/\hbar c$.

The constant g of weak interactions is dimensionally different from e and f. To derive a dimensionless constant from g one must introduce a mass as explained above and in Ch. 9-4.

In conclusion: Although in establishing the numerical values of the coupling constants for the different types of interactions, there is a certain amount of arbitrariness, depending on the specific form given to the theory, nevertheless, the strengths of the interactions are so different that the classification is useful. A sharper and more precise criterion of classification is given by the conservation laws of Table 13-4. Whenever necessary we shall indicate strong interactions by $\Rightarrow$, electromagnetic interactions by$\rightsquigarrow$, and weak interactions by $\rightarrow$.

The conservation laws listed in Table 13-4 have all been verified empirically. The conservation of energy, momentum, and angular momentum are fundamentally related to the invariance properties of space-time. The conservation of charge is related to the so-called gauge invariance of the electromagnetic field [see (Ja 75)]. Strangeness has been discussed in Sec. 13-2. For

ISBN 0-8053-8601-7

TABLE 13-4   CONSERVATION LAWS AND THEIR VALIDITY

Conservation of	Strong	Electromagnetic	Weak	Character[a]
Energy and momentum	Yes	Yes	Yes	A
Angular momentum	Yes	Yes	Yes	A
Parity	Yes	Yes	No	M
Isospin	Yes	No	No	A
Strangeness	Yes	Yes	No	A
Number of nucleons	Yes	Yes	Yes	A
Number of Leptons (separate e and μ)	Yes	Yes	Yes	A
Electric charge	Yes	Yes	Yes	A
Invariance under charge conjugation	Yes	Yes	No	M
Invariance under time reversal	Yes	Yes	Yes[b]	M

[a] Quantum numbers are either additive (A) or multiplicative (M).
[b] Except for the special case of K decay; see Sec. 19-6.

other conservation properties we have no "theoretical" justification. This is especially true of the conservation of nucleons and of leptons.

The quantities conserved and the relative quantum numbers behave in two radically different ways when one considers a system formed by the combination of two other systems. For instance, the angular momentum of the whole is the sum of the angular momenta of the components. The corresponding quantum numbers are called *additive* quantum numbers. On the other hand, the parity of a compound system is the product of the parities of the components. The corresponding quantum numbers are called *multiplicative* quantum numbers. The last column of Table 13-4 indicates the character of the quantum numbers by A (additive) or M (multiplicative). The character of the conservation laws is ultimately an experimental question. As an example, note that, if strangeness were a multiplicative quantum number, it could not account for the difference between reactions (13-2.7) and (13-2.8).

Experiment shows that the electromagnetic interaction is strictly proportional to the charges. There are no electromagnetic interactions besides the usual ones between charges and currents. Anomalous magnetic moments and such phenomena are not be be attributed to special couplings but can be reduced to interactions produced by an electric current. Gell-Mann expresses this by saying that there is a "minimal electromagnetic interaction."

For electromagnetic interactions, on the assumption of "minimality," one also finds a selection rule for i-spin. The charge may be written as

$$Q = T_3 + \frac{Y}{2} \tag{13-3.1}$$

It is therefore a linear function of the third component of the i-spin vector

ISBN 0-8053-8601-7

and in a vector interaction one obtains the selection rule

$$\Delta T = \pm 1, 0 \qquad (13\text{-}3.2)$$

as we remember from the angular momentum selection rules for electromagnetic radiation.

Table 13.4 can be further refined by distinguishing a weak and a very weak interaction. The weak interaction obeys the CP and T symmetries separately; the very weak interaction does not. Similarly, there is a very strong interaction obeying the $SU(3)$ symmetry and a strong interaction that violates it. The ratio between weak and very weak interaction is of the order of 10^3; that between very strong and strong interaction is of the order of 10.

13-4   SOME NEW SYMMETRIES AND SELECTION RULES

We give here some additional selection rules for strong and electromagnetic interactions mainly as a summary of results. To be complete, the explanations given here, admittedly inadequate, would require theoretical knowledge beyond the level of this book. The relevant materials can be found in more advanced books (Kä 64, W 71, Kae 65). However, the results can be understood to a certain extent without proofs and they are practically useful.

For convenience, we shall use the Dirac notation and indicate by $|K\rangle$ the eigenfunction (ket) corresponding to the quantum number K, in particular to a particle K, and by $\langle K|$ its complex conjugate (bra). The symbol $\langle K'|K\rangle$ means $\int \psi_{K'}^{*}\psi_{K}\,d\tau$. It is thus the Fourier coefficient of the development of ψ_K in $\psi_{K'}$ eigenfunctions. An operator $\mathcal{O}$ operating on $|K\rangle$ gives as a result the function $\mathcal{O}|K\rangle$ and the symbol $\langle K'|\mathcal{O}|K\rangle$ stands for the matrix element of the operator $\mathcal{O}$ when one uses as base the functions $|K\rangle$. In different notation

$$\langle K'|\mathcal{O}|K\rangle = \int \psi_{K'}^{*}\mathcal{O}\psi_{K}\,d\tau \qquad (13\text{-}4.1)$$

As is known from quantum mechanics, there is an equivalent notation in which a ket is indicated by a one-column matrix and a bra by a one-row matrix. Thus, by choosing a suitable basis (i.e., a suitable set of quantum numbers), we can represent the three pions by

$$|\pi^{+}\rangle = \begin{vmatrix} 1 \\ 0 \\ 0 \end{vmatrix} \qquad |\pi^{0}\rangle = \begin{vmatrix} 0 \\ 1 \\ 0 \end{vmatrix} \qquad |\pi^{-}\rangle = \begin{vmatrix} 0 \\ 0 \\ -1 \end{vmatrix} \qquad (13\text{-}4.3)$$

The minus sign for the π^{-} is chosen for convenience. Using as quantum numbers $T = 1$ and $T_3 = \pm 1, 0$, we can also denote the three pions as

$$|\pi^{+}\rangle = |1\ 1\rangle \qquad |\pi^{0}\rangle = |1\ 0\rangle \qquad |\pi^{-}\rangle = |1\ -1\rangle$$

ISBN 0-8053-8601-7

The electric-charge operator in matrix form is then

$$T_3 = \begin{vmatrix} 1 & 0 & 0 \\ 0 & 0 & 0 \\ 0 & 0 & -1 \end{vmatrix} \tag{13-4.4}$$

and the $|\pi^+\rangle$, etc. are eigenfunctions with eigenvalues $\pm 1, 0$. The operators T_1 and T_2 are analogous to the angular momentum operators and in matrix form they are

$$T_1 = \frac{1}{\sqrt{2}} \begin{vmatrix} 0 & 1 & 0 \\ 1 & 0 & 1 \\ 0 & 1 & 0 \end{vmatrix} \quad T_2 = \frac{1}{\sqrt{2}} \begin{vmatrix} 0 & -i & 0 \\ i & 0 & -i \\ 0 & i & 0 \end{vmatrix} \tag{13-4.5}$$

The parity operator P applied to a pion changes the sign of the ket:

$$P|\pi^+\rangle = -|\pi^+\rangle \quad P|\pi^0\rangle = -|\pi^0\rangle \quad P|\pi^-\rangle = -|\pi^-\rangle \tag{13-4.6}$$

This fact is expressed by saying that a pion has a negative intrinsic parity. The parity operator applied to a wave function labeled according to angular momentum gives

$$P|l\rangle = (-1)^l|l\rangle \tag{13-4.7}$$

For a pion of angular momentum l, $P|\pi, l\rangle = (-1)^{l+1}|\pi\rangle$ because the negative intrinsic parity introduces an additional factor of -1.

For bosons, the particle and antiparticle have the same intrinsic parity; for fermions, they have opposite intrinsic parity.

The charge conjugation operator for nucleons gives, by definition,

$$C|p\rangle = |\bar{p}\rangle \quad C|n\rangle = |\bar{n}\rangle \quad C|\bar{p}\rangle = |p\rangle \quad C|\bar{n}\rangle = |n\rangle \tag{13-4.8}$$

For pions C gives

$$C|\pi^0\rangle = |\pi^0\rangle \tag{13-4.9}$$

indicating that the π^0 is self-conjugate and that $|\pi^0\rangle$ is an eigenstate of C with eigenvalue $+1$. On the other hand, $|\pi^+\rangle$ and $|\pi^-\rangle$ are not eigenstates of C, because

$$C|\pi^{\mp}\rangle = |\pi^{\pm}\rangle \tag{13-4.10}$$

The result follows from the fact that C changes π^+ into π^-, and that $C^2 = 1$. The sign in front of $|\pi^{\pm}\rangle$ is to some extent arbitrary: we shall assume $C|\pi^{\pm}\rangle = |\pi^{\mp}\rangle$.

ISBN 0-8053-8601-7

The operator C changes the signs of all charges and hence the direction of the electric field $\mathbf{E}$. This is expressed by saying that for one light quantum $C|\gamma\rangle = -|\gamma\rangle$ and for n quanta the eigenvalue of C is $(-1)^n$.

The operation of charge conjugation gives interesting results for systems containing the same number of particles and their respective antiparticles plus any number of self-conjugate particles. Such systems are necessarily electrically neutral and have baryon number 0. These systems are eigenstates of C with eigenvalue $(-1)^{l+S}$ where S is the total spin and l the total orbital angular momentum. To make this plausible, note that the C operation may be considered equivalent to the exchange of all particles with their antiparticles. In the simple case of two bosons (e.g., π^+, π^-) this exchange brings about a factor of $(-1)^l$ where l is the orbital angular momentum. If the particles have spin, necessarily integral for bosons, they combine their spins $s_1 = s_2$ to form the total spin S. On exchange of the particles, the wavefunction is invariant if S is even or changes sign if S is odd. This statement derives from properties of the Clebsch–Gordan coefficients. In conclusion, the wavefunction gets multiplied by $(-1)^{l+S}$. For fermion–antifermion pairs one can make a similar argument. On exchange, a factor of $(-1)^l$ comes from the orbital angular momentum. The opposite intrinsic parity of fermion and antifermion introduces another factor of -1 and the combination of two equal half-integral spins produces an S symmetric with respect to the exchange of s_1 and s_2 for S odd and antisymmetric for S even, as can be recognized, for instance, for two spins $\frac{1}{2}$. In conclusion, we again find the factor $(-1)^{l+S}$. This factor appears as C_n in Table 13-2 for neutral nonstrange particles.

As an example consider positronium in the $^1S_0$ state or in the $^3S_1$ state. For the former C has the eigenvalue $(-1)^{0+0} = 1$, for the latter C has the eigenvalue $(-1)^{0+1} = -1$. Hence, because the electromagnetic interactions conserve C the $^1S_0$ state can annihilate in an even number of quanta, including two, but the $^3S_1$ state can annihilate only in an odd number of quanta (larger than one because with one it is impossible to conserve energy and momentum).

In strong interactions both C and T are conserved. The two operators do not commute, however; therefore, the simultaneous conservation of both gives new results that go beyond those obtainable from the conservation of each separately. As an example consider nucleons and their antiparticles. Starting from the relation $Q = T_3 + (N/2)$ we can find the T, T_3 assignments for the four particles[1]

$$|p\rangle = |T\, T_3\rangle = |\tfrac{1}{2}\ \tfrac{1}{2}\rangle \qquad |n\rangle = |\tfrac{1}{2}\ -\tfrac{1}{2}\rangle \qquad |\bar{p}\rangle = -|\tfrac{1}{2}\ -\tfrac{1}{2}\rangle \qquad |\bar{n}\rangle = |\tfrac{1}{2}\ \tfrac{1}{2}\rangle$$

$$(13\text{-}4.11)$$

[1] The minus sign for $|\bar{p}\rangle$ has been chosen to represent the nucleon–antinucleon states as four-component vectors in a product space of baryon number (in which C operates), and T. In this space baryon number N, Q, T, and T_3 all commute. From the sign of the antiproton kets follow the unusual signs in $|1\ 0\rangle$, $|0\ 0\rangle$.

ISBN 0-8053-8601-7

For a nucleon–antinucleon system one has, in a T, T_3 representation, for $T = 1$,

$$|1\ 1\rangle = |p\rangle|\bar{n}\rangle$$

$$|1\ 0\rangle = \frac{1}{\sqrt{2}}\,(|p\rangle|\bar{p}\rangle - |n\rangle|\bar{n}\rangle) \qquad (13\text{-}4.12)$$

$$|1\ -1\rangle = |\bar{p}\rangle|n\rangle$$

and for $T = 0$

$$|0\ 0\rangle = \frac{1}{\sqrt{2}}\,(|p\rangle|\bar{p}\rangle + |n\rangle|\bar{n}\rangle) \qquad (13\text{-}4.13)$$

Now the eigenstates of C are $|p\rangle|\bar{p}\rangle$ and $|n\rangle|\bar{n}\rangle$ but not their linear combinations, thus T and C do not have common eigenstates and do not commute. In strong interactions, however, both are conserved and this gives important physical consequences. In particular, we shall consider the operator (G parity)

$$G = C e^{i\pi T_2} \qquad (13\text{-}4.14)$$

product of a rotation in i-spin space and charge conjugation. This operator has eigenvalues $\eta_G = \pm 1$ and states having one value of η_G must conserve it in strong interactions. The operator $e^{i\pi T_2}$ is similar to the operator $e^{i\theta J \cdot \mathbf{n}}$ representing a rotation in ordinary space by an angle θ around an axis $\mathbf{n}$. The operator $e^{i\pi T_2}$ is thus a rotation by π around the second axis in i-spin space. To see its effect on nucleons, write

$$e^{i\pi T_2} = 1 + i\pi T_2 - \frac{\pi^2}{2!}\,T_2^2 + \cdots \qquad (13\text{-}4.15)$$

and for T_2 use the expression valid for $T = \frac{1}{2}$, $T_2 = \frac{1}{2}\tau_2$ with

$$\tau_2 = \begin{vmatrix} 0 & -i \\ i & 0 \end{vmatrix}$$

By direct calculation, remembering that $\tau_2^2 = 1$, one has

$$e^{i\pi T_2} = 1 \cos\frac{\pi}{2} + 2iT_2 \sin\frac{\pi}{2} = 2iT_2 = \begin{vmatrix} 0 & 1 \\ -1 & 0 \end{vmatrix}$$

and

$$G|p\rangle = -|\bar{n}\rangle \qquad G|\bar{p}\rangle = -|n\rangle \qquad G|n\rangle = |\bar{p}\rangle \qquad G|\bar{n}\rangle = |p\rangle \qquad (13\text{-}4.16)$$

The single-particle states are not eigenstates of G, but states containing equal

ISBN 0-8053-8601-7

numbers of nucleons and antinucleons *and* eigenstates of T are also eigenstates of G. For such systems the eigenvalue of C, as we have seen, is $(-1)^{l+S}$; another factor $(-1)^T$ comes from the i-spin and thus

$$\eta_G = (-1)^{l+S+T} \tag{3-4.17}$$

For pions the effect of G can be calculated in a similar fashion. We know the effect of C. The effect of $e^{i\pi T_2}$ is found by calculating Eq. (13-4.15) for $T = 1$. In this case we verify directly that $T_2^{2n} = T_2^2$ for $n \geqslant 1$ and $T_2^{2n+1} = T_2$ for $n \geqslant 0$. Then

$$e^{i\pi T_2} = 1 + iT_2 \sin \pi + T_2^2 (\cos \pi - 1)$$

$$= \begin{vmatrix} 1 & 0 & 0 \\ 0 & 1 & 0 \\ 0 & 0 & 1 \end{vmatrix} - \begin{vmatrix} 1 & 0 & -1 \\ 0 & 2 & 0 \\ -1 & 0 & 1 \end{vmatrix} = \begin{vmatrix} 0 & 0 & 1 \\ 0 & -1 & 0 \\ 1 & 0 & 0 \end{vmatrix} \tag{13-4.18}$$

hence, writing

$$|\pi^+\rangle = \begin{vmatrix} 1 \\ 0 \\ 0 \end{vmatrix} \qquad |\pi^0\rangle = \begin{vmatrix} 0 \\ 1 \\ 0 \end{vmatrix} \qquad |\pi^-\rangle = \begin{vmatrix} 0 \\ 0 \\ -1 \end{vmatrix}$$

we obtain

$$e^{i\pi T_2}|\pi^+\rangle = -|\pi^-\rangle \qquad e^{i\pi T_2}|\pi^0\rangle = -|\pi^0\rangle \qquad e^{i\pi T_2}|\pi^-\rangle = -|\pi^+\rangle \tag{13-4.19}$$

and the eigenvalue of G is thus -1 for any pion. For a system with n pions

$$\eta_G = (-1)^n \tag{13-4.20}$$

It follows that a system having $\eta_G = 1 (-1)$ can give rise by strong interaction only to a system containing an even (odd) number of pions.

For instance, a proton–antineutron system in the $^3S_1$ state has $T = 1, l = 0$, $S = 1, \eta_G = +1$, and it cannot annihilate into three pions, for which $\eta_G = -1$. A p–$\bar{p}$ or an n–$\bar{n}$ system is not an eigenstate of T but a mixture of states with $T = 0, 1$ and its G parity is not determined unless one knows T by specifying in detail the superposition. Table 13-5 summarizes some of the results of the preceding discussion.

TABLE 13-5   RESULTS OF P, C, G OPERATIONS

Operator	$|\gamma\rangle$	$|p\rangle$	$|n\rangle$	$|\bar{p}\rangle$	$|\bar{n}\rangle$	$|\pi^+\rangle$	$|\pi^0\rangle$	$|\pi^-\rangle$
P	$(-1)^l \Pi_{\text{int}}^a$	$|p\rangle$	$|n\rangle$	$-|\bar{p}\rangle$	$-|\bar{n}\rangle$	$-|\pi^+\rangle$	$-|\pi^0\rangle$	$-|\pi^-\rangle$
C	$-|\gamma\rangle$	$|\bar{p}\rangle$	$|\bar{n}\rangle$	$|p\rangle$	$|n\rangle$	$|\pi^-\rangle$	$|\pi^0\rangle$	$|\pi^+\rangle$
G		$-|\bar{n}\rangle$	$|\bar{p}\rangle$	$-|n\rangle$	$|p\rangle$	$-|\pi^+\rangle$	$-|\pi^0\rangle$	$-|\pi^-\rangle$

$^a\Pi_{\text{int}} = 1$ for E radiation, $= -1$ for M radiation.

13-5   *CPT* THEOREM

Another important theoretical result is the so-called *CPT* theorem, which in a simplified form states that any hermitian interaction, relativistically invariant, commutes with all products of the three operators *C* (charge conjugation), *P* (parity inversion), and *T* (time reversal) in any order. This theorem (Lüders, 1954; Pauli, 1956) is proved under broad assumptions.

It has important practical consequences because even if an interaction is not invariant under one of the three factors (an example is the beta interaction, which does not conserve parity), it must be invariant under *CPT*. In the case of beta interactions the evidence is that the interaction is invariant under *T* and *CP* separately.

An important consequence of the *CPT* theorem is the identity of the mass and lifetime of particle and antiparticle. *CPT* applied to a free particle in motion gives the antiparticle with the same energy and momentum and hence the same rest mass. If we consider the decay constant as an imaginary part of the mass, as it would appear in a Schrödinger equation, one has also the identity of the mean lives. *CPT* also prescribes the identity of the gyromagnetic ratios.

Experimentally there are several measurements relevant to the *CPT* theorem. They are summarized in Table 13-6. The K_0, $\overline{K}_0$ case is especially impressive, because it is possible to infer that the masses differ by less than one part in 10^{15}.

TABLE 13-6   EXPERIMENTAL CHECKS OF *CPT* INVARIANCE[a]

Quantity measured	*Result*
	Equality of masses
$m(\pi^+)/m(\pi^-)$	$\begin{cases} 1.0002 & \pm 0.0004 \\ 1.0002 & \pm 0.0005 \end{cases}$
$m(\bar{p})/m(p)$	$1.008 \quad \pm 0.005$
$m(e^-)/m(e^+)$	1.000101 ± 0.000185
$1 - m(\overline{K^0})/m(K^0)$	$\leqslant 2 \times 10^{-16}$
	Equality of gyromagnetic ratios
$\frac{1}{2}(g_e^+ - g_e^-)$	$(\ 1.5 \ \pm 2)\alpha^2/\pi^2$
$\frac{1}{2}(g_\mu^+ - g_\mu^-)$	$(-0.09 \pm 0.14)\alpha^2/\pi^2$
	Equality of total lifetimes
$\tau(\mu^+)/\tau(\mu^-)$	$1.000 \quad \pm 0.001$
$\tau(K^+)/\tau(K^-)$	0.99910 ± 0.00078
$\tau(\pi^+)/\tau(\pi^-)$	1.00055 ± 0.00071

[a]From D. S. Ayres et al., *Phys. Rev.*, **3D**, 1051 (1971).

ISBN 0-8053-8601-7

There is still a puzzle related to *CPT*, namely, the violation of *CP* in K_l decay. It is possible that this violation is compensated by a violation of *T*, restoring *CPT* conservation. (See Sec. 19-6.)

13-6   CROSSING RELATIONS

Important relations of great generality derive from a property of the scattering matrix. Its elements are, according to theory, analytic functions of the kinematic variables appearing in them. This property has been connected with causality, but the proof is not completely free from objection; field theory has also been used for the same purpose. Ignoring the difficult question of a rigorous proof of analyticity, we shall assume it to be true and shall exploit it to find relations between various reactions by using the crossing principle, to be presently explained. In order to simplify our discussion, let us assume that we have the reaction

$$a(p_a) + b(p_b) \rightarrow c(p_c) + d(p_d) \tag{13-6.1}$$

neglecting spin and *i*-spin; *a, b, c, d* are the names of the particles; they have four momenta

$$p_a = (\mathbf{p}_a, iE) \qquad p_a^2 = -m_a^2 \text{ etc.} \qquad (c = 1) \tag{13-6.2}$$

[Note that many authors use a different convention for four vectors: $p = (-E, \mathbf{p})$.] The conservation of energy and momentum gives

$$p_a + p_b = p_c + p_d \tag{13-6.3}$$

In a collision p_a and p_b are known; the eight unknown components of p_c, p_d are subject to two conditions, $p_c^2 = -m_c^2$, $p_d^2 = -m_d^2$, and to four equations (13-6.3), and the remaining degrees of freedom are the scattering angle θ and the inessential arbitrary azimuth of the collision plane.

The 16 variables p_a, p_b, p_c, p_d are redundant in describing a collision. Physically it is clear that the collision is described by two numbers only. For example, consider m_b at rest; then the magnitude of $\mathbf{p}_a$ and the scattering angle give all the essential information. Relativistically appropriate variables (Mandelstam, 1959) are

$$-s = (p_a + p_b)^2 = (p_c + p_d)^2 \tag{13-6.4}$$

$$-t = (p_a - p_c)^2 = (p_d - p_b)^2 \tag{13-6.5}$$

$$-u = (p_a - p_d)^2 = (p_c - p_b)^2 \tag{13-6.6}$$

ISBN 0-8053-8601-7

Of these three variables only two are independent because

$$-s - t - u = p_a^2 + p_b^2 + p_c^2 + p_d^2 + 2(p_a^2 + p_a p_b - p_a p_c - p_a p_d)$$

$$= -m_a^2 - m_b^2 - m_c^2 - m_d^2 + 2p_a(p_a + p_b - p_c - p_d)$$

where the last term in parentheses is zero. Hence

$$s + t + u = m_a^2 + m_b^2 + m_c^2 + m_d^2 \qquad (13\text{-}6.7)$$

In the laboratory system $p_b = (0, im_b)$, $p_a = (\mathbf{p}_a, iE_a)$

$$-s = (\mathbf{p}_a, i(m_b + E_a))^2 \qquad (13\text{-}6.8)$$

$$-s = \mathbf{p}_a^2 - m_b^2 - E_a^2 - 2m_b E_a, \quad = -m_a^2 - m_b^2 - 2m_b E_a \qquad (13\text{-}6.9)$$

In the c.m. system, $-s = (0, i(E_a + E_b))^2 = -(E_a + E_b)^2$ that is, s is the total c.m. energy squared; in the same system

$$t = m_a^2 + m_c^2 - 2(E_a E_c - |\mathbf{p}_a||\mathbf{p}_c| \cos \theta) \qquad (13\text{-}6.10)$$

which for the special case $m_a = m_b = m_c = m_d$ gives

$$-t = 2\mathbf{p}^2(1 - \cos \theta) \qquad (13\text{-}6.11)$$

where θ is the scattering angle and $\mathbf{p}$ the momentum. In a physical situation $s \geqslant (m_a + m_b)^2$, $t \leqslant 0$, because $\mathbf{p}^2$ is positive and $|\cos \theta| \leqslant 1$.

The scattering matrix element for reaction Eq. (13-6.1) may now be expressed as $f_{ab \to cd}(p_a, p_b, p_c, p_d)$. We keep the redundant variable for symmetry. Now consider the reaction

$$a + \bar{d} \to c + \bar{b} \qquad (13\text{-}6.12)$$

in which $\bar{b}$ stands for the antiparticle of b and $\bar{d}$ for the antiparticle of d. The crossing principle says that the matrix element corresponding to the values p_a, p_b, p_c, p_d for reaction Eq. (13-6.1), $a + b \to c + d$, is the same as for reaction Eq. (13-6.12), $a + \bar{d} = c + \bar{b}$, corresponding to the p values p_a, $-p_d, p_c, -p_b$. In other words, if a particle is moved from the right-hand side to the left-hand side of the reaction equation, and its four momentum is changed in sign, the scattering matrix element does not change. This principle includes the identity of the scattering matrix for $a + b \to c + d$ and $\bar{a} + \bar{b} \to \bar{c} + \bar{d}$. The crossing operation can be repeated and several reactions may be related by crossing; for example

$$p + n \to p + n \qquad p + \bar{n} \to p + \bar{n} \qquad p + \bar{p} \to \bar{n} + n \qquad \bar{p} + \bar{n} \to \bar{p} + \bar{n}$$

ISBN 0-8053-8601-1

Rather than use the four momenta p directly, it is convenient to express the scattering matrix as a function of the Mandelstam variables. Define s, t, and u according to Eqs. (13-6.4) to (13-6.6) for the direct reaction; then for the crossed reaction Eq. (13-6.12) $s' = -(p_a - p_d)^2 = u$, $t' = -(p_a - p_c)^2 = t$, and $u' = -(p_a + p_b)^2 = s$. Thus if for $a + b \to c + d$ the scattering matrix is given by $f(s, t, u)$, for $a + \bar{d} \to c + \bar{b}$ the scattering matrix is given by $f(u, t, s)$. However, if $f(s, t, u)$ corresponds to a physical situation for the direct reaction, there are no physical values of the momenta for the crossed reaction corresponding to those values of s, t, and u. Physical values of the momenta for the crossed reaction correspond to unphysical ones for the direct reaction, making impossible a direct experimental verification of the crossing principle. This does not negate its usefulness because f is an analytic function that can be continued to the whole plane, passing from physical regions of the direct reaction to physical regions of the crossed one. Up to now this is only a theoretical possibility.

Table 13-7 shows the crossing symmetry for the case of the four reactions

$$\text{I} \quad p + n \to p + n$$

$$\text{II} \quad p + \bar{p} \to \bar{n} + n$$

$$\text{III} \quad \bar{p} + n \to \bar{p} + n$$

$$\text{IV} \quad \bar{p} + \bar{n} \to \bar{p} + \bar{n}$$

For reaction I, call the four p_i of the particles p_1, p_2, p_3, and p_4. We then have the respective crossing symmetries shown in the table. We have used $p_1 + p_2 = p_3 + p_4$. The limitations $s \geqslant 4m^2$, $t \leqslant 0$ and $u \leqslant 0$ confine the physical region for reaction I and similar limitations obtain for reactions II and III. This is illustrated in Fig. 13-7. The physical regions are hatched and marked by the reaction to which they refer. It is common to name a reaction according to the variable representing the energy. Thus reaction I is the s channel, reaction II is the t channel, and reaction III is the u channel. To bring to fruition the crossing relations one needs a deep study of the analytic character of the function $f(s, t, u)$, particularly of its singularities, but such a study is beyond the scope of this book.

ISBN 0-8053-8601-7

TABLE 13-7   CROSSING AND VARIABLES FOR $a + b \Rightarrow c + d$ REACTION

I	$-s_{\text{I}} = (p_1 + p_2)^2$	$-t_{\text{I}} = (p_1 - p_3)^2$	$-u_{\text{I}} = (p_1 - p_4)^2$
II	$-s_{\text{II}} = (p_1 - p_3)^2 = -t_{\text{I}}$	$-t_{\text{II}} = (p_1 + p_2)^2 = -s_{\text{I}}$	$-u_{\text{II}} = (p_1 - p_4)^2 = -u_{\text{I}}$
III	$-s_{\text{III}} = (p_2 - p_3)^2 = -u_{\text{I}}$	$-t_{\text{III}} = (p_3 - p_1)^2 = -t_{\text{I}}$	$-u_{\text{III}} = (p_1 + p_2)^2 = -s_{\text{I}}$
IV	$s_{\text{IV}} = s_{\text{I}}$	$t_{\text{IV}} = t_{\text{I}}$	$u_{\text{IV}} = u_{\text{I}}$

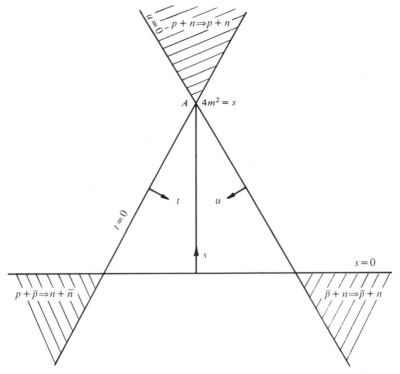

Figure 13-7 Mandelstam variables in neutron proton scattering and crossed reactions. Shaded regions are physical for the reactions indicated. Mass of the proton assumed equal to that of the neutron. Vertex A has $s = 4m^2$

13-7 EXPERIMENTS ON MASS, LIFE, AND OTHER PARTICLE PROPERTIES

The first step for a study of particles is the determination of their mass, mean life, charge, spin, i-spin, and other properties. Considerable experimental effort is devoted to this purpose and a great deal of theoretical effort is expended on the classification and prediction of the experimental results.

Rather than discuss every case separately, we will give here enough typical examples to provide an idea of the techniques used.

The electric charge is assumed to be strictly quantized, and thus it is relatively easily measured.

A variety of methods exist for measuring the mass. Some derive, with adaptations, from mass spectroscopy; others are entirely different. The example of the muon will illustrate some of the methods available for the determination of the mass of particles. The instability of the muon and the conditions for its production bring about peculiarities worth discussing.

The fundamental idea is always to measure two quantities, such as energy and momentum, on the same particle, and to derive the mass. This method

ISBN 0-8053-8601-7

was applied to the muons found in cosmic rays and gave a first measurement accurate to within about 5% of the muon mass. The quantities measured were momentum (from the curvature of a cloud-chamber track in a magnetic field) and the specific ionization (from the number of droplets condensed along the track) (see Fig. 13.8). In photographic emulsions one can count the number of grains from the end of the range, thereby obtaining the energy of the particle, and the grain density, thus obtaining the velocity. From these data one finds the mass.

When particles are produced in the laboratory more precise methods become feasible. These methods measure the ratio of the mass of a particle to that of some standard, for example, the proton. They are based on the observation that many quantities (e.g., momentum, energy, range, and total ionization) have the form

$$p = m\pi(v) \qquad E = m\epsilon(v) \qquad R = m\rho(v)/z^2 \quad \text{etc.} \qquad (13\text{-}7.1)$$

where m is the rest mass of the particle, and π, ϵ, ρ are functions of the velocity only (cf. Sec. 2-3).

If two different kinds of particles 1, 2, *having the same velocity*, are stopped in a photographic emulsion, then the momenta are in the same ratio as their ranges and their masses. This relation derives from the equations

$$\frac{p_1}{p_2} = \frac{m_1}{m_2}\frac{\pi(v_1)}{\pi(v_2)} \qquad\qquad \frac{R_1}{R_2} = \frac{m_1}{m_2}\frac{\rho(v_1)}{\rho(v_2)}$$

which are immediate consequences of Eq. (13-7.1).

Experimentally Barkas and co-workers (1956) measured the ratio of the pion–muon mass and of the pion–proton mass. The variables measured were the range (in emulsion) and momentum. These measurements give the masses with an accuracy of about one part in a thousand.

A completely different way of obtaining the muon mass is offered by the study of radiative transitions in mu-mesic atoms, already discussed in Sec. 6-4. The mass of the muon is determined from the energy of the muonic hydrogenic spectral lines. By a fortunate accident in the case of muonic X rays one can use the very precise method of critical absorption (see Chap. 2). In particular, careful work on the muonic $3D_{3/2, 5/2} - 2P_{3/2, 1/2}$ multiplet in phosphorus has shown that the strongest line ($D_{5/2} - P_{3/2}$) has an energy of $88{,}014 \pm 10$ eV. This gives a precise value of $m_\mu/m_e = 206.76$, accurate to one part in 10^4.

The best value of the mass of the muon, given in Table 13-2, is 105.65948 ± 0.00035 MeV and is obtained by a combination of the measurement mentioned above with others on the magnetic moment and $g - 2$ of the muon (See Chap. 14). Note that $m_\mu/m_e = 206.7686 \pm 0.0003$.

ISBN 0-8053-8601-7

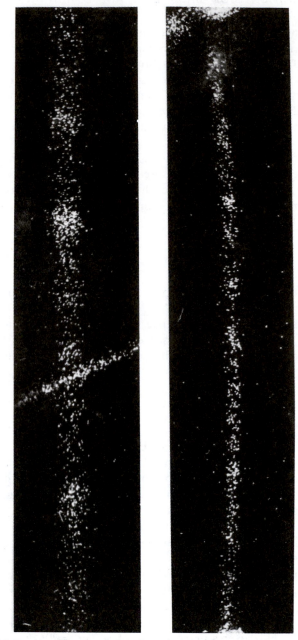

Figure 13-8 An early cloud chamber picture showing the velocity (from the ionization, i.e., the number of droplets per unit path length) and momentum (from the curvature) of a meson. Pressure in cloud chamber 1.5 atm of N_2. Magnetic field 800 G. Expansion occurred 0.5 sec after particle had crossed the chamber to give time to the ions to diffuse and make droplets countable. [Courtesy R. B. Brode and D. R. Corson.]

ISBN 0-8053-8601-7

In this section we have assumed that the charge of the muon is exactly equal to that of the electron. This is hardly debatable, and direct experiments have confirmed the equality of charge to one part in 20 000 or better.

When a pion decays at rest according to $\pi \rightarrow \mu + \nu$, conservation of energy and momentum gives a relation between the p of the escaping particles and their rest mass,

$$m_\pi^2 - m_\mu^2 - m_\nu^2 = 2\frac{p^2}{c^2}\left[1 + \left(1 + \frac{m_\mu^2 c^2}{p^2}\right)^{1/2}\left(1 + \frac{m_\nu^2 c^2}{p^2}\right)^{1/2}\right] \quad (13\text{-}7.2)$$

The momentum p has been accurately measured, yielding 29.789 MeV/c or 4.119 MeV for the energy of the muon. Inserting the independently measured m_π, m_μ one finds $m_\nu \leqslant 0.8$ MeV. Conversely, the relation assuming $m_\nu = 0$ gives $m_\pi - m_\mu = 33.908 \pm 0.006$ MeV.

In other cases the mass of a particle is obtained from energy measurements in reactions. For instance, although the mass of charged pions may be measured by methods similar to those used for muons, including mesic atoms, the neutral pion presents new problems compounded by its very short life.

The neutral pion decays spontaneously electromagnetically into two gamma rays. The observed reactions

$$\pi^- + p \rightarrow \pi^0 + n \qquad \pi^0 \rightsquigarrow 2\gamma \qquad (13\text{-}7.3)$$

serve to determine its mass. Assume that the π^- is captured at rest. The π^0 and n acquire a relative kinetic energy equal to $(m_{\pi^-} + m_p - m_{\pi^0} - m_n)c^2$. This energy can be directly measured by the observation of the gamma spectrum of the decaying π^0. The two-decay gamma rays have equal and opposite momentum in the coordinate system in which the π^0 is at rest. In the laboratory system they appear at an angle to each other, and their frequency shows the Doppler shift due to the motion of the π^0. The spectral "line" of the decay radiation is transformed into a band with a maximum and minimum frequency (Fig. 13.9). The mass difference between π^- and π^0 is thus obtainable. The experiment was first carried out by Panofsky, Steller, and Steinberger (1951). They stopped π^- in hydrogen and observed the spectrum of the gamma rays emitted, using a pair spectrometer. Two reactions occur,

$$\pi^- + p \rightarrow n + \pi^0 \qquad (13\text{-}7.4)$$

$$\pi^- + p \rightarrow n + \gamma \qquad (13\text{-}7.5)$$

The second gives rise to a monochromatic gamma ray and the first to a continuum from which one can derive the mass of π^0, as explained above. The probability of reaction (13-7.4) relative to reaction (13-7.5) is 1.56 ± 0.02 (Panofsky ratio).

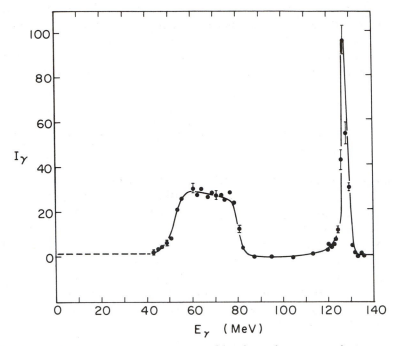

Figure 13-9 Gamma-ray spectrum resulting from the capture of π^- mesons in hydrogen. The reaction $\pi^- + p \to n + \gamma$ leads to the spike at 130 MeV. The square pulse at 60 MeV is due to $\pi^- + p \to n + \pi^0$, $\pi^0 \to 2\gamma$. [Crowe and Phillips, *Phys. Rev.*, **96**, 470 (1954).]

More direct, but experimentally difficult, is the measurement of the neutron velocity in reaction (13-7.4) by a time-of-flight method (Crowe). The result is

$$m_{\pi^0} - m_{\pi^\pm} = 4.6043 \pm 0.0037 \text{ MeV} \qquad (13\text{-}7.6)$$

From the evaluation of all experiments there results the value of the mass of the π^0

$$m_{\pi^0} = 134.9645 \pm 0.0074 \text{ MeV} \qquad (13\text{-}7.7)$$

For a particle at rest the mean life is by definition the average time it lasts, starting from an arbitrary initial time. For instance, a muon arriving in a scintillator gives a signal on arrival while stopping and a second ·signal on decay. The mean time elapsed between the two signals is the mean life. Figure 13-10 shows an apparatus for this determination; Figure 13-11 illustrates a similar method for π^+.

ISBN 0-8053-8601-7

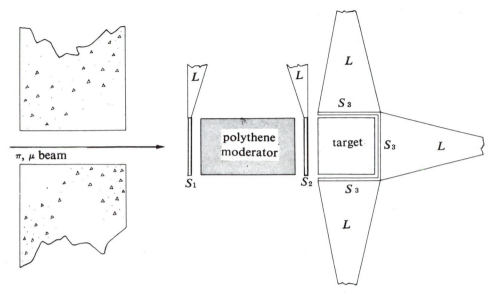

Figure   13-10 A typical apparatus used in measuring the μ^+ lifetime. A mixed beam of pions and muons enters the polythene moderator, which stops the pions. The arrival of a muon is signaled by an S_1, S_2 coincidence. The successive decay of a muon stopped in the target is detected by S_3. The average time interval between arrival and decay gives the mean life of the muon; S, scintillators; L, light guides. [After Astbury, Hattersley, Hussain, Kemp, and Muirhead (RoC 60).]

For particles in flight the mean life τ in their rest frame is different from that in the laboratory frame τ'

$$\tau = \tau'(1 - \beta^2)^{1/2} = \tau'/\gamma \tag{13-7.8}$$

A beam in the laboratory is attenuated by decay according to

$$I/I_0 = \exp(-x'/\upsilon\tau') = \exp(-x'/\beta\gamma c\tau) = \exp(-x'm/pc\tau) \tag{13-7.9}$$

where x' is the beam length, υ the velocity, p the momentum, and m the rest mass, all measured in the laboratory. The mean life can thus be measured from the attenuation of beams of known p between stations at a known distance.

Direct methods or time-of-flight methods may go as low as 10^{-10} sec; beyond this limit they become impractical. For the neutral pion decaying into two gammas with a half-life of the order of 10^{-16} sec, ingenious methods using photographic emulsions have made possible a direct measurement. One

ISBN 0-8053-8601-7

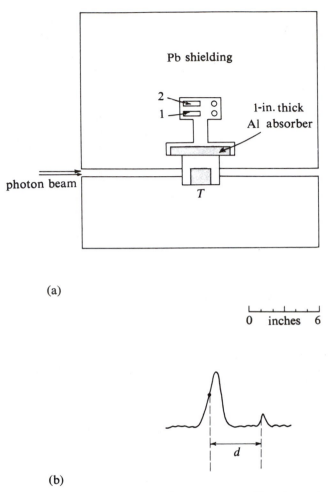

(a)

0   inches   6

(b)

Figure 13-11 Measurement of the π^+ mean life. (a) The photon beam generates π^+ in the polyethylene target T. These mesons are stopped in target No. 2 after crossing 1; the resulting 1, 2 coincidence signal starts the sweep of the oscilloscope. This signal appears in (b) as the large pulse. The $\pi \rightarrow \mu$ decay gives the succeeding pulse. The average of time interval d is the mean life of the π^+. [From C. E. Wiegand, *Phys. Rev.*, **83**, 1085 (1951).]

method is as follows: occasionally, one of the decay gamma rays is replaced by a pair of electrons of opposite sign forming a pair (Dalitz pair, 1951; Fig. 13-12). The probability of this event occurring is approximately $e^2/\hbar c$. Suppose that a neutral pion emerges from a star in the photographic emulsion and decays into a Dalitz pair and an invisible gamma ray. It is possible to infer the time interval between emission and decay of the neutral pion from

ISBN 0-8053-8601-7

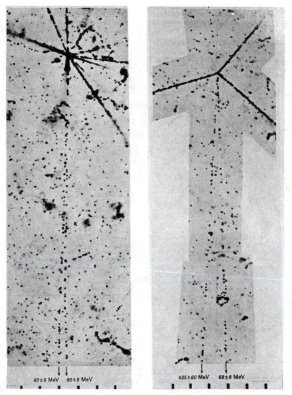

Figure 13-12 Electron pairs, apparently emerging directly from nuclear disintegrations attributed to the decay of the neutral π-meson into a Dalitz pair and an invisible gamma quantum. [From (PFP 59).]

the distance between the star and the origin of the Dalitz pair and from the momentum of the decaying pion. This momentum is also derived from a study of the pair (see Fig. 13-13).

A different approach was used by von Dardel and others (1963). An energetic π^0 shows an appreciable, although short, free path before decaying ($1\,\mu$m for $p = 5$ GeV/c); the absorption of the gamma rays produced on decay shows a transition effect in a Pt foil a few microns thick and from this one can infer the mean life (see Prob. 13-7).

Application of detailed balance to the reaction

$$\pi^0 \rightleftarrows \gamma + \gamma$$

also gives the mean life of π^0. The inverse reaction, or its equivalent, is obtained by using high-energy photons to produce π^0 in the electric field near

ISBN 0-8053-8601-7

Figure 13-13 Dalitz pair seen in a hydrogen bubble chamber. A π^- makes a charge-exchange collision with a proton generating a π^0 which decays into an invisible gamma ray and a Dalitz pair. [Courtesy Lawrence Berkeley Laboratory.]

the nucleus of a heavy element. The cross section for production is proportional to the decay constant of the pion, and experiment again gives a mean life of approximately 2 10^{-16} sec.

The relation between the mass width of a particle and its decay constant

$$1/\tau = \Delta E/\hbar = \Gamma \tag{13-7.10}$$

gives a general method for determining the mean life. However, for $\Delta E = 1$ MeV, $\tau = 6.58 \times 10^{-22}$ sec, way beyond anything directly measurable. The measurement of the full width at half maximum ΔE in a reaction cross section (e.g., production of the 3/2, 3/2 resonance in p–π^+) is a good example of this method.

ISBN 0-8053-8601-7

BIBLIOGRAPHY

Bethe, H. A., and F. de Hoffmann (BH 55).
Fermi, E. (Fe 51).
Feynman, R. (Fey 61).
Frazer, W. M. (Fr 66).
Källén, G. (Kä 64).
Kämpffer, F. A. (Kae 65).
Leon, M. (Le 73).
Marshak, R. E., and E. C. G. Sudarshan, *Introduction to Elementary Particle Physics*, Wiley-Interscience, New York, 1961.
Nishijima, K., *Fundamental Particles*, Benjamin, New York, 1963.
Omnès, R. (O 71).
Particle Data Group (RPP 74).
Perkins, D. (Pe 72).
Powell, C. F., P. H. Fowler, and D. H. Perkins (PFP 59).
Wick, G. C., "Invariance Principles of Nuclear Physics," *Ann. Rev. Nucl. Sci.*, **8**, 1 (1958).
Williams, W. S. C. (W 71).

PROBLEMS

13-1 Of the following decay processes, state which are strictly forbidden, which are weak, which are electromagnetic, and which are strong; give the reasons for your decision.

(a) $p \rightarrow e^+ + \gamma$

(b) $^8_3\text{Li} \rightarrow ^8_4\text{Be} + e^+ + \nu$

(c) $\pi^+ \rightarrow \mu^+ + \gamma$

(d) $\gamma \rightarrow e^+ + e^-$ (in vacuum)

(e) $p \rightarrow n + e^+ + \nu$ (for free proton)

(f) $K^+ \rightarrow \pi^+ + \gamma$

(g) $\Lambda^0 \rightarrow p + e^- + \nu$

(h) $\Lambda^0 \rightarrow n + \gamma$

(i) $\Xi^- \rightarrow \Sigma^0 + K^-$

(j) $e^+ + e^- \rightarrow p + \bar{p}$

(k) $e^+ + e^- \rightarrow n + n$

(l) $e^+ + e^- \rightarrow \nu + \nu$ (specify neutrinos).

13-2 If the spin of the K^0 meson were 1, the decay mode $K^0 \rightarrow 2\pi^0$ would not be allowed. Why not?

13-3 Consider a system composed of a nucleon and its antinucleon in the $^1S_0$, $^3S_1$, $^1P_1$, $^3P_{0,1,2}$ states. Find their possible parity, *i*-spin, G quantum number, and charge-conjugation quantum number.

13-4 Try to prove as much as possible of Table 13-8.

13-5 The conservation laws of Table 13-4 give rise to constants of the motions. These are quantized with discrete values, except energy and momentum. Why? Note that if the system is enclosed in a box, energy and momentum also have only discrete values.

13-6 Devise a method for measuring the pion mass solely from observation of photographic emulsions. (Protons have been stopped in the same emulsion.) See (PFP 59).

13-7 Assume that protons impinging on a Pt foil of thickness t produce π^0 uniformly, all traveling in the same direction as the primary beam. Denote by λ the mean free path for free decay of the π^0, by X the radiation length of Pt for gamma rays produced by the π^0 decay, and by B the branching ratio to Dalitz pairs. Show that the number of positrons emerging from the foil is

$$R(t) = Kt \left\{ B + \frac{1}{X} \left[\frac{t}{2} - \lambda + \frac{\lambda^2}{t} (1 - e^{-t/\lambda}) \right] \right\}$$

ISBN 0-8053-8601-7

TABLE 13-8 STRONG-INTERACTION DECAY PROPERTIES OF $N-\bar{N}$ COMBINATIONS

$N-\bar{N}$ state	J^P	T	C	G	2π	3π	π + γ's	$(n\gamma)_{min}$	$K-\bar{K}$
$^1S_0$	$0-$	0	+	+	No (P)^a	No (G)^a also no 4π	$\pi^0 + 2\gamma$ $\pi^+ + \pi^- + \gamma$	2	No (P)
		1	+	−	No (P, G)	Yes $\left\lvert \dfrac{\pi^\pm + \pi^- + \pi^-}{\pi^\mp 2\pi^0} = 4 \;;\; \dfrac{\pi^+ + \pi^- - \pi^0}{3\pi^0} = \tfrac{2}{3} \right\rvert$	$\pi + 2\gamma$ $2\pi + \gamma$ (no $2\pi^0$)		No (P)
$^3S_1$	$1-$	0	−	−	No (G)	Yes(no $3\pi^0$)	$\pi^0 + \gamma$	3	Yes $\left(\dfrac{K^+ + K^-}{K^0 + \bar{K}^0} = 1 \right)$
		1	−	+	Yes (no $2\pi^0$)	No (G)	$\pi + \gamma$		Yes $\left(\dfrac{K^+ + K^-}{K^0 + \bar{K}^0} = 1 \right)$
$^1P_1$	$1+$	0	−	−	No (P, G)	Yes (no $3\pi^0$)	$\pi^0 + \gamma$	3	No (P)

Possible decays

ISBN 0-8053-8601-7

ISBN 0-8053-8601-7

	1	−	+	No (P)	No (G)	$\pi + \gamma$		No (P)
$^3P_0$ 0+	0	+	+	Yes $\left(\dfrac{\pi^+\pi^-}{2\pi^0} = 2\right)$	No (P, G)	$\pi^0 + 2\gamma$	2	Yes [no $K\overline{K} + \pi$] (P)
	1	+	−	No (G)	No (P)	$\pi + 2\gamma$ $2\pi + \gamma$		Yes [no $K\overline{K} + \pi$] (P)
$^3P_1$ 1+	0	+	+	No (P)	No (G)	$\pi^0 + 2\gamma$		No (P, G)
	1	+	−	No (P, G)	Yes (see $^1S_0$)	$\pi + 2\gamma$ $\pi^\pm + \gamma$ $2\pi + \gamma$ (no $2\pi^0$)	4	No (P, G)
$^3P_2$ 2+	0	+	+	Yes (see $^3P_0$)	No (G)	$\pi^0 + 2\gamma$		Yes
	1	+	−	No (G)	Yes (see $^1S_0$)	$\pi^\pm + \gamma$ $\pi^0 + 2\gamma$ $\pi^+ + \pi^- + \gamma$	2	Yes

^{a}(P), (G) means forbidden by parity or G-parity conservation.

13-8   The reaction $d + d = \text{He} + \pi^0$ has never been observed. Why?

13-9   Show that isospin invariance for nuclear forces is sufficient to ensure the equality of p–p, n–n, and n–p interaction in all states antisymmetric in spin and space.

13-10   Show that for pion–pion states symmetric in space, isospin invariance does not require that the interaction be the same in all states. Show that there are two independent interactions.

13-11   Show that the interaction of the electromagnetic field with a system of neutrons and protons can be written, nonrelativistically, as

$$H = \frac{e}{mc} \sum_i \tfrac{1}{2} \mathbf{p}_i \cdot \mathbf{A}(x_i)(1 + 2\tau_3^{(i)})$$

$$+ \sum_i \left\{ \tfrac{1}{2} \mu_n (1 - 2\tau_3^{(i)}) + \tfrac{1}{2} \mu_p (1 + 2\tau_3^{(i)}) \right\} \boldsymbol{\sigma}^{(i)} \cdot \nabla \times \mathbf{A}(x_i)$$

where $\mathbf{p}_i$ is the momentum of the ith nucleon, $\mathbf{A}(x_i)$ the vector potential at its position, $\boldsymbol{\sigma}^{(i)}$ its Pauli spin vector, and $\tau^{(i)}$ its i-spin.

From this show that the selection rules $\Delta T = 0, \pm 1$ obtain. (*Hint:* The interaction may be divided into a part H_s independent of i spin and a part H_v proportional to τ_3.)

13-12   Given that the total cross section for π^+–p reactions at energy E_0 is σ_0, what specific prediction can you make purely on the basis of charge symmetry?

13-13   Prove that η cannot decay strongly into $\pi^0 + \gamma$.

13-14   Prove that ω cannot decay strongly or electromagnetically into $\eta + \pi$.

13-15   Prove that ρ cannot decay strongly into $\eta + \pi$.

13-16   Calculate the threshold energy for the process

$$\gamma + p \rightarrow \pi^0 + p$$

13-17   Consider the reaction $\pi^- + p \rightarrow n + \pi^0$, where the initial π^- and proton are at rest. Calculate (*a*) the velocity of the emitted π^0; (*b*) the maximum and minimum energies of the gamma rays emitted in the π^0 decay; (*c*) the spectrum of the γ emitted.

13-18   Plan an experiment to determine the mean life of π^0, using the decay

$$K^+ \rightarrow \pi^+ + \pi^0$$

and the formation of Dalitz pairs from the π^0.

13-19   Develop the crossing relations for the reaction $\pi^+ + p^+ \rightarrow \pi^+ + p^+$ and find the physical regions for s, t, u in the various channels.

13-20   What will be the absolute minimum energy necessary to form K mesons by bombarding hydrogen with protons? Consider separately strong and weak interactions.

13-21   How could one produce a Ξ^- in the laboratory? Find the threshold. Find the threshold for the production of K^0 and $\bar{K}^0$ in π^-–proton collisions.

13-22   Suppose that you are studying reactions produced by high-energy protons incident on a hydrogen target. Write a strong reaction which can lead to each of the following products: (*a*) a Λ^0, (*b*) a K^-, (*c*) a Ξ^-, (*d*) an anti-Λ^0.

13-23   Show that the reactions

$$
\begin{array}{lll}
\text{(a) } K^- + {}^4\text{He} & \Rightarrow & \Sigma^- + {}^3\text{He} \\
\text{(b) } & \Rightarrow & \Sigma^0 + {}^3\text{H} \\
\text{(a) } K^- + {}^4\text{He} & \Rightarrow & \Lambda^0 + \pi^- + {}^3\text{He} \\
\text{(b) } & \Rightarrow & \Lambda^0 + \pi^0 + {}^3\text{H}
\end{array}
$$

have branching ratios $a : b = 2 : 1$.

13-24   Develop the argument showing that the absence of the decay $K^+ \Rightarrow \pi^+ + \gamma$ indicates a spin 0 for K^+.

ISBN 0-8053-8601-7

13-25 Prove that three pions cannot be in a state $J^P = 0^+$.

13-26 Show that the i-spin eigenfunction of a system of three pions with $T = 0$ is antisymmetric with respect to the exchange of the coordinates of any two pions. Show that as a consequence the coordinate eigenfunction must also be antisymmetric with respect to the exchange of pions.

13-27 Show that Dirac's matrices α_k transformed by $U\alpha_k U^{-1} = \alpha_k'$ with $U = 2^{-1/2} \times (\alpha_2 + \beta) = U^{-1}$ have all real elements and β' has all imaginary elements.

ISBN 0-8053-8601-7

CHAPTER

14

Leptons

In this chapter we discuss the neutrinos, electrons, and muons. They form the lepton family, which is characterized by the absence of strong interactions. These particles all have spin $\frac{1}{2}$ and charge $\pm e$, 0. Their masses are 0 for the neutrinos, 0.511 MeV for the electrons, and 105.66 MeV for the muons. What produces these mass differences and what produces the difference between electronic and muonic neutrinos are major unsolved problems.

The absence of strong interactions, and for neutrinos, of electromagnetic interactions as well, is demonstrated by the ability of these particles to penetrate great thicknesses of matter without attenuation. In astrophysics the unimpeded escape of neutrinos from the interior of stars seems to be required to account for the phenomena of stellar evolution.

The interaction cross section of neutrinos with nucleons is of the order of $10^{-44}p^2$ cm^2, where p is the neutrino momentum in MeV/c, up to about 1 GeV/c, after which it increases to $10^{-38} p$ cm^2, p in GeV/c. The mean free path of a 1-MeV neutrino is 10^{21} g cm^{-2}, whereas star radii range from 10^9 to 10^{13} g cm^{-2}.

Similarly, cosmic-ray muons penetrate the surface of the earth and the only energy loss observed is accounted for by the electromagnetic interaction.

Emilio Segrè, Nuclei and Particles: An Introduction to Nuclear and Subnuclear Physics, Second Edition

ISBN 0-8053-8601-7

In proton–muon scattering, as well as in proton–electron scattering, nothing has been detected, even at the highest energies investigated thus far, that goes beyond the electromagnetic interaction. At low energy, as we shall see in some detail, high-precision measurements of magnetic moments agree with pure electrodynamics. The same is true of investigations of electron and μ pair production by gamma rays.

14-1 NEUTRINOS

Neutrinos are produced in beta decay and positron emission. We discussed their properties in Chap. 8; the results of that discussion are summarized here:

ν_e mass < 60 eV;   charge, magnetic moment $= 0$

ν_μ mass < 0.8 MeV;   charge, magnetic moment $= 0$

Both have spin $\frac{1}{2}$. The simplest description is that they are two-component Dirac particles, thus have fixed helicity, -1 for the particles, $+1$ for the antiparticles. The upper limit to the mass of $\bar{\nu}_e$ comes from beta decay, of ν_μ from the decay $K_L \to \pi^\pm + \mu^\mp + \nu_\mu$ or $\bar{\nu}_\mu$; the decays $\pi^+ \to \mu^+ + \nu_\mu$ and $\mu^+ \to e^+ + \nu_e + \bar{\nu}_\mu$ (see Table 13-2) give slightly higher limits. Neutrinos are produced in beta decay, as well as in many particle reactions such as, $\pi^+ \to \mu^+ + \nu_\mu$, $\pi^- \to \mu^- + \bar{\nu}_\mu$. The neutrino appearing in the last reactions is different from the neutrino of beta decay. This is demonstrated by bombarding nuclei with ν_μ. One occasionally gets muons ($\sigma \sim 10^{-38}$ cm^2 at 1 GeV) which are interpreted as the product of the reactions

$$\nu_\mu + n \to \mu^- + p \tag{14-1.1}$$

$$\bar{\nu}_\mu + p \to \mu^+ + n \tag{14-1.2}$$

However, the reactions

$$\nu_e + n \to e^- + p \tag{14-1.3}$$

$$\bar{\nu}_e + p \to e^+ + n \tag{14-1.4}$$

TABLE 14-1 TABLE OF LEPTON HELICITIES

	e^-	e^+	μ^-	μ^+	ν_e	$\bar{\nu}_e$	ν_μ	$\bar{\nu}_\mu$
Leptonic number	1	-1	1	-1	1	-1	1	-1
Helicity[a]	$(-\beta)$	$(+\beta)$	$(-\beta)$	$(+\beta)$	-1	$+1$	-1	$+1$
Muonic number	0	0	1	-1	0	0	1	-1

[a] $\beta = v/c$.

(the inverse of beta decay) are known and under the conditions of the bombardment one would also expect to produce electrons, provided the two kinds of neutrinos were identical. In fact, no electrons have been found, leading to the conclusion that there are two different kinds of neutrinos. By this, we do not mean neutrinos ν, and antineutrinos $\bar{\nu}$, but two totally different kinds of particles, which we have indicated by ν_e and ν_μ. The antiparticles of ν_e and ν_μ exist and are denoted by $\bar{\nu}_e$ and $\bar{\nu}_\mu$, respectively. Other forbidden reactions pointing to the existence of two types of leptons, including neutrinos, separately conserved, are

$$
\begin{aligned}
\mu^+ &\rightarrow e^+ + e^+ + e^- && \text{branching} < 6 \times 10^{-9} \\
\mu^+ + e^- &\rightarrow 2\gamma && \text{branching} < 5 \times 10^{-6} \\
\mu^\pm &\rightarrow e^\pm + \gamma && \text{branching} < 2 \times 10^{-8} \\
\mu^+ &\rightarrow e^+ + 2\gamma && \text{branching} < 4 \times 10^{-6}
\end{aligned}
$$

Apart from the phenomenological distinction implied by the experiments described, we do not know what the difference between the two neutrinos is.

14-2 MUON PRODUCTION AND DECAY

Muons were first discovered in cosmic rays about 1936 by C. D. Anderson, S. H. Neddermeyer, and others. Previously, particles of approximately the muon mass had been postulated by Yukawa (1935) as quanta of the field responsible for nuclear forces. It turned out, however, that the particles of intermediate mass (mesons) first seen in cosmic rays were *not* the particles postulated by Yukawa; for, although they have about the right mass, they do not interact appreciably with nucleons (Conversi, Pancini, and Piccioni, 1947). These particles, now called muons, in many ways resemble heavy electrons. They belong to the lepton family, which comprises particles that are not subject to specific nuclear interactions, but only to electromagnetic and weak interactions. For historical reasons the muon is sometimes called, improperly, the mu meson. The word meson should be reserved for bosons.

The charge of muons (positive and negative) is equal in magnitude to the charge of the electron. There are no neutral muons. Muons are usually obtained as a decay product of the pion according to the reactions

$$\pi^+ \rightarrow \mu^+ + \nu_\mu \tag{14-2.1}$$

$$\pi^- \rightarrow \mu^- + \bar{\nu}_\mu \tag{14-2.2}$$

Although the commonest source of muons is pion decay, muons appear as

ISBN 0-8053-8601-7

branchings in other decays, such as

$$\Sigma^{\pm} \rightarrow \mu^{\pm} + \nu_{\mu} + n \qquad (14\text{-}2.3)$$

$$\Lambda \rightarrow \mu^{-} + \bar{\nu}_{\mu} + p \qquad (14\text{-}2.4)$$

$$K \rightarrow \mu + \nu_{\mu}, \quad \text{etc.} \qquad (14\text{-}2.5)$$

Muons can also be generated in pairs by photoproduction. The cross section for this process is similar to that for electron pair production, except that the mass of the muon replaces the mass of the electron in the formula.

Free muons decay into electrons and neutral particles. The electron spectrum is a continuum; thus more than one invisible neutral partner is required to conserve energy and momentum. The form of the spectrum agrees with the predictions of a reaction

$$\mu^{-} \rightarrow e^{-} + \nu_{\mu} + \bar{\nu}_{e} \qquad (14\text{-}2.6)$$

proceeding by weak interaction and completely similar to beta decay. Of the two neutrinos, one is believed to be of the kind found in beta decay, and the other of the kind found in pion decay. Direct experiment failed to reveal any sizable amount of gamma radiation associated with muon decay. The maximum momentum of the decay electrons is given by $m_{\mu}c^2 = (p_e^2 c^2 + m_e^2 c^4)^{1/2} + p_e c$ and it is 52.827 MeV/c.

The continuous energy spectrum of decay electrons has the form indicated in Fig. 14-1. The energy of the electrons is greatest when the two neutrinos escape in the same direction, opposite to that of the electron. The probability of this occurrence, in the case of identical neutrinos, is 0. The qualitative reason for this is related to Pauli's principle, which prevents two identical neutrinos from moving into the same quantum state. Thus, if the two neutrinos were identical, there would be no electrons with the maximum possible energy, and the energy spectrum would look like the line $\rho = 0$ in Fig. 14-1. We conclude, therefore, that in muon decay two different neutrinos are emitted. They belong to the two different species of neutrinos. This in turn gives two possibilities for the neutrino pair: it could consist, for the μ^- decay, of an electron antineutrino $\bar{\nu}_e$ and a muon neutrino ν_{μ} or of a ν_e and a $\bar{\nu}_{\mu}$; in any case their combined helicity is zero. The choice $\bar{\nu}_e + \nu_{\mu}$ is preferred because it associates the e^- with the $\bar{\nu}_e$ as in beta decay, and, of course, e^+ with ν_e.

The decay of the muon by weak interaction according to Eq. (14-2.6) and a similar relation for μ^+ is particularly important because none of the participating particles is subject to strong interactions (which affect, e.g., beta decay). One has a clean case on which to test any weak-interaction theory.

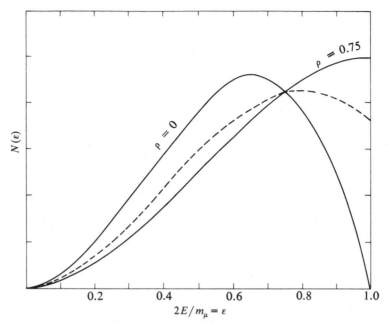

Figure 14-1 Energy spectrum of electrons emitted in the decay of the muon. Maximum energy is 52.8 MeV. The equation of the curve is $N(\varepsilon) = 4\varepsilon^2[3(1 - \varepsilon) + \frac{2}{3}\rho(4\varepsilon - 3)]$. Curves for the Michel parameter $\rho = 0$ and $\rho = 0.75$ are shown, as well as a dashed curve representing the statistical phase-space distribution. Experiments indicate a spectrum with $\rho = 0.7518 \pm 0.0026$.

The simplest assumes a weak interaction with a matrix element of the form $g(C_V V - C_A A)$ that has been successful in beta decay. Note that we have not assumed the coefficients of V and A to be equal in magnitude and that g is not *of necessity* the same as in beta decay.

Given the matrix element, one calculates everything relating to the decay by using golden rule No. 2 or relativistic modifications of it. The calculation requires techniques beyond the scope of this book [see (Com 73)], but the result is simple. Neglecting the mass of the electron m compared to the mass of the muon m_μ and calling **n** the unit vector in the direction of the electron momentum, in the rest frame of the muon, one has for the probability of emission of the electron of energy ε in the solid angle element $d\omega$ at an angle θ

$$dW = \frac{g^2 m_\mu^5 c^4}{3 \cdot 2^6 \cdot \pi^3 \hbar^7} \left[2\varepsilon^2(3 - 2\varepsilon) \right]\left[1 + \frac{1 - 2\varepsilon}{3 - 2\varepsilon} \cos \theta \right]\left[\frac{1 - \mathbf{n} \cdot \mathbf{s}_e}{2} \right] \frac{d\varepsilon \, d\omega}{4\pi}$$

$$(14\text{-}2.7)$$

ISBN 0-8053-8601-7

where $\varepsilon = E/E_{max} = 2E/m_\mu c^2$; s_μ, s_e are unit vectors in the direction of the muon and electron spin, respectively, and $\cos \theta = s_\mu \cdot n$.

This formula is written so that the first bracket gives the electron spectrum, the second the asymmetry of the electron emission with respect to the μ spin, and the third the electron helicity.

For more general interactions containing scalar, tensor, etc. terms, the spectrum Eq. (14-2.7) acquires different forms. Limiting ourselves to the most important parameter, we have that the first bracket in Eq. (14-2.7) is replaced by

$$2\varepsilon^2 \left[(6 - 4\rho) - \left(6 - \tfrac{16}{3}\rho\right)\varepsilon \right] \qquad (14\text{-}2.8)$$

The Michel parameter ρ depends on the interaction and takes the value $3/4$ for $V - A$. Experiment singles out this value (Fig. 14-2).

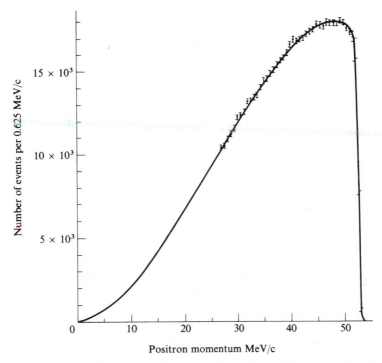

Figure 14-2 Results of experiment to determine ρ. Shown are the experimental points, along with a theoretical curve for $\rho = 3/4$ corrected for radiative effects and ionization loss. [From M. Bardon et al., *Phys. Rev. Letters*, **14**, 449 (1965).]

ISBN 0-8053-8601-7

The decay constant is

$$\frac{1}{\tau} = \frac{g^2 m_\mu^5 c^4}{192\pi^3 \hbar^7} = \frac{1}{(2.199 \pm 0.001)\ \mu\text{sec}} \tag{14-2.9}$$

multiplied by a radiative correction of the order of $1/\text{``}137\text{''}$. With this correction $g = (1.43506 \pm 0.00026) \times 10^{-49}$ erg cm^3, very close to the g of beta decay! Let us use the same symbol for both.

The second bracket of Eq. 14-2.7 shows that $(1 - 2\varepsilon)/(3 - 2\varepsilon) = \alpha$, the asymmetry factor, varies from $1/3$ for $\varepsilon = 0$ to -1 for $\varepsilon = 1$. We see this by considering the component of the angular momentum in the direction of motion of the electron. For electrons having the maximum possible energy, the neutrino pair has zero angular momentum in that direction, and the electron must have the same angular momentum as the decaying muon. The angular momentum of the positron has actually been measured in the case of positive muons and found to be parallel to the direction of motion, which indicates that the spin of the muon itself is preferentially parallel to the direction of emission of the high-energy positrons (see Fig. 14-3).

helicities in

$\pi \to \mu \to e$ decay

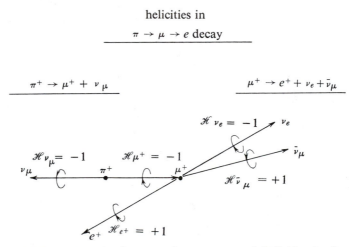

Figure 14-3 Schematic diagram of momenta and helicities in the decays $\pi^+ \to \mu^+ + \nu_\mu$ and $\mu^+ \to e^+ + \bar{\nu}_\mu + \nu_e$.

The orientations of all spins are thus consistent with the assignments of decay,

$$\pi^- \to \mu^- + \bar{\nu}_\mu \qquad \pi^+ \to \mu^+ + \nu_\mu \tag{14-2.10}$$

$$\mu^- \to e^- + \bar{\nu}_e + \nu_\mu \qquad \mu^+ \to e^+ + \nu_e + \bar{\nu}_\mu \tag{14-2.11}$$

ISBN 0-8053-8601-7

which in turn are equivalent to the conservation of leptons and of muonic number, provided that one defines e^- as a lepton and assigns the leptonic and muonic numbers as shown in Table 14-1.

The average value of the asymmetry given by the second bracket of Eq. 14-2.7 is $-1/3$; this has been verified by observing the decay of muons stopped in a photographic emulsion in a very strong magnetic field. The third bracket shows that the helicity of the electron is -1, independent of the energy, in the $m = 0$ approximation. The helicity of the electrons or positrons has been verified by Møller ($e^- - e^-$) or Bhabha ($e^+ - e^-$) scattering of the decay electrons in magnetized iron in which the $4d$ orbital electrons are polarized. These and other experiments observing the circular polarization of the bremsstrahlung produced by muonic electrons or the annihilation radiation of positrons in magnetized iron have given helicities of -0.9 ± 0.3 for electrons and 1.0 ± 0.15 for positrons.

The value of g as derived from muon decay is nearly the same as that for beta decay. On the other hand, C_A/C_V for muon decay is 1, whereas for beta decay it is 1.23. The identity of g points to the identity of the weak interaction operating in beta and muon decay (universal Fermi interaction), but why is not C_A the same for both interactions when C_V is the same? This important question will be treated in Chap. 19.

14-3   MUON CAPTURE

Negative muons stopped in matter may decay freely, but they may also interact with nuclei of the stopping substance and disappear by capture. In very light elements they show about the same mean life as positive muons. This important fact, discovered by Conversi, Pancini, and Piccioni in 1947, demonstrated that they interact only weakly with nuclei and laid the foundation for the distinction between pions and muons.

When negative muons are stopped in substances of a given Z, their mean life is

$$\frac{1}{\tau} = \frac{1}{\tau_\mu} + \frac{1}{\tau_c} \qquad (14\text{-}3.1)$$

The first term on the right corresponds to the free decay and the second to the disappearance caused by interaction with the nucleus of the stopping medium.

For $Z = 11$ the two terms are almost equal. This fact throws an interesting light on the capture process. We try to estimate $1/\tau_c$, assuming an interaction that produces the process

$$p + \mu^- \rightarrow n + \nu_\mu$$

(cf. Chap. 9).

ISBN 0-8053-8601-7

Take as matrix element for this interaction a characteristic constant $g_{\mu N}$ divided by a normalization volume Ω. The rate for the process is calculated by golden rule No. 2 as

$$\text{rate} = \frac{2\pi}{\hbar} \left(\frac{g_{\mu N}}{\Omega} \right)^2 \rho \tag{14-3.2}$$

The density of final states ρ in this case is, remembering that $dE = (c + v_N)dp$,

$$\rho = \frac{\Omega p^2}{2\pi^2 \hbar^3} \frac{1}{(c + v_N)} \tag{14-3.3}$$

where p is the momentum of the neutrino and $v_N \sim 0.1c$ the velocity of the recoil neutron, which is given by

$$m_\mu c^2 = cp + \frac{p^2}{2M_N} \tag{14-3.4}$$

and

$$p = m_N v_N$$

For a muon bound to a nucleus of low Z such that the muonic orbit is external to the nucleus, the normalization volume is replaced by the volume of the mesic atom. We may think of the muon as enclosed in a box of volume $\pi a^3 / Z^3$, where $a = \hbar^2 / e^2 m_\mu$ is the radius of the muonic hydrogen Bohr orbit. We multiply the result by Z, in order to take into account that there are Z protons in the nucleus, obtaining

$$\frac{1}{\tau_c} = \frac{1}{\pi^2} g_{\mu N}^2 \frac{E_\nu^2 Z}{\hbar^4 c^2 (c + v_N)} \left(\frac{Z}{a} \right)^3 \tag{14-3.5}$$

or, neglecting v_N compared with c and putting $E_\nu = m_\mu c^2$,

$$\frac{1}{\tau_c} = \frac{g_{\mu N}^2}{\pi^2} \frac{m_\mu^5 c^4}{\hbar^7} \alpha^3 Z^4 \tag{14-3.6}$$

The ratio of τ_c to τ given by Eq. (14-2.9) is

$$\frac{\tau_c}{\tau} = \frac{g^2}{g_{\mu N}^2} \frac{\text{“}137\text{”}^3}{192\pi} Z^{-4} \tag{14-3.7}$$

showing that if $g = g_{\mu N}$, the probability of capture equals that of free decay for $Z = 8$, in fair agreement with the experimental result.

ISBN 0-8053-8601-7

Equation (14-3.6) is not intended to be quantitative. An important correction is introduced by the exclusion principle, which rules out transitions in which the neutron would go into an occupied orbit. In order to obtain quantitative results, one must start from the appropriate interactions and take into account all the complications introduced by the nuclear structure of the capturing nucleus. An exact calculation is possible only for proton capture; for a few of the lightest nuclei (^{3}H, ^{4}He) the approximations are very good and theory and experiment agree if $g_{\mu N}$ is put equal to g, indicating the identity of the interaction producing beta decay, muon decay, and muon capture (Fig. 14-4).

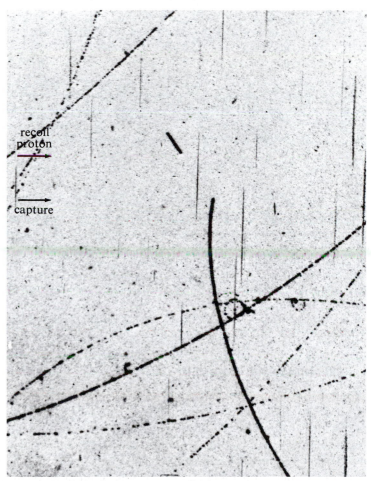

recoil proton

capture

Figure 14-4 About one in every 10^3 muons stopped in liquid hydrogen is captured according to the reaction $\mu^- + p \rightarrow n + \bar{\nu}$. The neutron recoils with an energy of 5.2 MeV and collides with a proton giving a track about 2.5 mm long (in the original). [Courtesy R. Hildebrand.]

ISBN 0-8053-8601-7

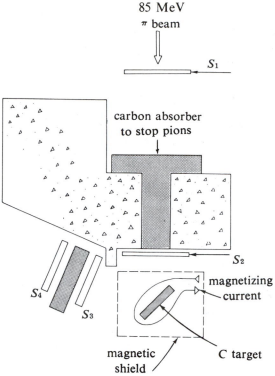

85 MeV
π beam

S_1

carbon absorber
to stop pions

S_2

magnetizing
current

S_4

S_3

magnetic
shield

C target

Figure   14-5   A beam of pions and muons passes scintillator S_1. The pions but not the muons are stopped before reaching scintillator S_2. Thus, 1 and 2 coincidences signal the arrival of a muon which will be stopped in the carbon target. The positrons from the muon decay are revealed by a coincidence between counters 3 and 4. The magnetic field perpendicular to the figure, acting in the target area, produces a precession of the muon spin. The correlation between spin direction and preferred direction for the positron emission is shown in Fig. 14-6. [After R. L. Garwin, L. M. Lederman, and M. Weinrich, *Phys. Rev.*, **105**, 1415 (1957).]

14-4   SPIN AND MAGNETIC MOMENT OF MUONS

The spin of the muon is $\frac{1}{2}$. The proof came from experiments which also showed the nonconservation of parity in muon decay. Assume that a π^+ comes to rest and decays, emitting a neutrino and a μ^+ in opposite directions. The pion has no angular momentum, and hence the μ^+ has an angular momentum equal and opposite to that of the neutrino. If the neutrino is polarized with the spin antiparallel to the direction of motion, so is the muon. This has been verified by direct experiment (Hyams et al., 1961). The muon comes to rest in a very short time compared with its half-life; if the polarization is not destroyed in the moderation process, we have the emission of a

ISBN 0-8053-8601-7

positron and two particles of zero rest mass from a polarized source. This process shows a marked asymmetry in the sense that the positrons tend to travel parallel to the spin direction (Figs. 14-5 and 14-6) and Eq. 14-2.7.

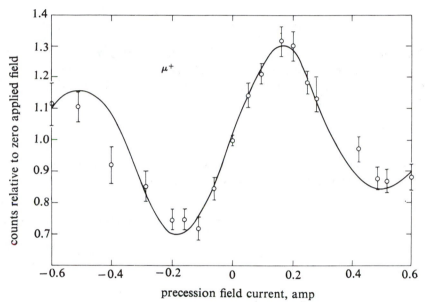

Figure 14-6 Variation of gated 3–4 counting rate with magnetizing current. The solid curve is computed from an assumed electron angular distribution $1 - \frac{1}{3} \cos \theta$, with counter and gate-width resolution folded in.

Suppose that we apply a magnetic field to the muon at rest. The field produces a rotation of the direction of polarization with an angular velocity

$$\omega = g \frac{eH}{2m_\mu c} \tag{14-4.1}$$

where $g = \mu/I$ is the gyromagnetic ratio, that is, the magnetic moment in natural units $e\hbar/2m_\mu c$, divided by angular momentum in units $\hbar$. This rotation is observed as a change of the direction of preferential emission of the positrons. These phenomena were observed in experiments by Friedman and Telegdi and by Garwin, Lederman, and Weinrich, who found g to be very close to the g value of the electron, approximately 2. This result is a strong, if not binding, argument for a spin $\frac{1}{2}$ of the muon. Indeed, if this spin were $\frac{3}{2}$ or larger, the g value would be very odd indeed.

Another argument comes from mu-mesic atoms, whose spectra show structures consistent with the assignment of spin $\frac{1}{2}$ and $g = 2$ for the muons.

ISBN 0-8053-8601-7

Precision measurements on muons have yielded important results in quantum electrodynamics and have helped to establish universal constants.

In quantum electrodynamics the simple Dirac theory gives, for the magnetic moment of the muon,

$$\mu = \frac{e\hbar}{2m_\mu c} = g_\mu \frac{e}{2m_\mu c} I \quad \text{with} \quad g_\mu = 2 \tag{14-4.2}$$

More refined theories that take into account renormalization effects and radiative corrections have revealed some departures from the simple theory, such as the Lamb shift in the hydrogen spectrum and the departure from 2 for the value of g_e. In particular, the quantity

$$a = \tfrac{1}{2}(g - 2) \tag{14-4.3}$$

can be calculated from first principles on the basis of electrodynamics alone. The result expressed as a power series in $\alpha = e^2/\hbar c$ is

$$a_e = \frac{\alpha}{2\pi} - 0.32848 \frac{\alpha^2}{\pi^2} + (1.49 \pm 0.2) \frac{\alpha^3}{\pi^3} \tag{14-4.4}$$

$$a_\mu - a_e = 1.09426 \frac{\alpha^2}{\pi^2} + (19.4 \pm 1.3) \frac{\alpha^3}{\pi^3} \tag{14-4.5}$$

The difference betweeen a_e and a_μ indicates that some of the higher-order corrections, such as vacuum polarization, are mass dependent and differ for the two particles.

Extremely accurate experiments on electrons, in which polarized electrons precess for a known time in a known magnetic field (Wesley and Rich, 1970), show agreement with the theoretical a_e.

In similar experiments with muons a polarized muon with momentum initially parallel to spin is made to circle in a magnetic field B perpendicular to the plane of the orbit. The angular velocity of the particle and of its momentum is $\omega = eB/mc\gamma$ where $\gamma = (1 - \beta^2)^{-1/2}$ and m is the rest mass of the particle. The spin rotates in the laboratory system with the angular velocity

$$\omega_s = \omega(1 + a\gamma) \tag{14-4.6}$$

After a certain time, spin and velocity make an angle, in the rest frame of the muon,

$$\varphi = a\gamma\omega t \tag{14-4.7}$$

Thus the spin precesses with respect to the velocity. Figures 14-7 and 14-8 illustrate an experiment conducted at CERN. The muons rotate in the

ISBN 0-8053-8601-7

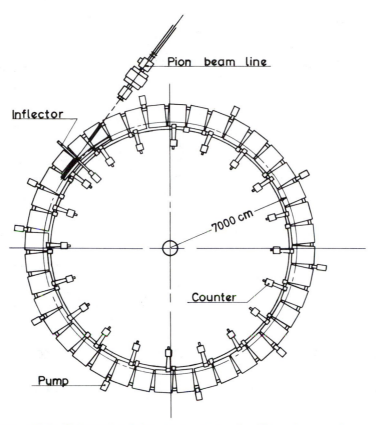

Figure 14-7 Schematic of the muon storage ring. Pions decay and generate muons, which revolve in the ring [From J. Bailey et al., *Phys. Letters*, **55B**, 420 (1975).] with a period of 0.15 μsec while their spin precesses with respect to their momentum with a period of 4.3 μsec. The counters count only high-energy decay electrons emitted forward in the muon rest frame.

chamber. Only decay electrons with maximum energy are detectable by a biased detector. The precession of the spin modulates the intensity of such electrons and from the magnetic field prevailing in the chamber and the frequency of the modulation one finds

$$a_\mu = (11658.95 \pm 0.27) \times 10^{-7} \tag{14-4.8}$$

which differs from the theoretical value by 13 ± 29 ppm.

The absence of an "anomalous" part in the muon magnetic moment is an argument to indicate that the muon is subject only to weak and electromagnetic interactions, and confirms the high accuracy of quantum electrodynamics. Knowledge of a gives g with extreme precision, and this yields

ISBN 0-8053-8601-7

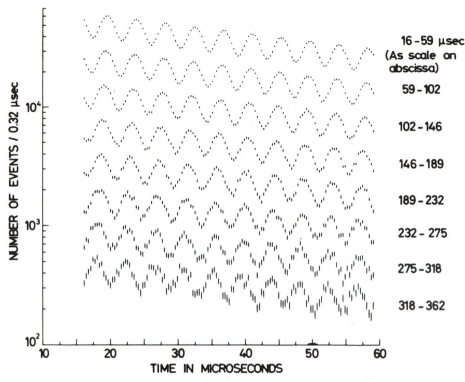

Figure   14-8   In the apparatus Fig. 14-7, the counting rate is given by

$$N(t) = N(0) \exp(-t/\tau)\left[1 - A\,\cos(\omega_a t - \psi)\right]$$

where τ is the muon mean life in the lab, $\omega_a = eaB/m_\mu c$, and A is determined in the experiment. [From J. Bailey et al., *Phys. Letters*, **55B**, 420 (1975).]

the ratio between the masses of the particles for which g is known. Experiments of the Garwin–Lederman–Weinrich type have been repeated with great care, calibrating the magnetic field directly on the proton precession frequency, with the result that the ratio of the precession frequencies $\omega_\mu/\omega_p = \mu_\mu/\mu_p = 3.1833467(82)$. The ratio μ_p/μ_e is also known with great accuracy. From these one obtains

$$\frac{m_\mu}{m_e} = \frac{g_\mu \mu_e \mu_p}{g_e \mu_p \mu_\mu} = 206.7683(6) \tag{14-4.9}$$

The study of muonium, a system composed of a positive muon and an electron, resembling a hydrogen atom, gives important information on the fine structure constant α. Muonium is a hydrogenic system calculable with extreme precision entirely from first principles; in contrast, the hydrogen

ISBN 0-8053-8601-7

atom contains corrections due to the structure of the proton and its magnetic moment which are not calculable from pure electrodynamics.

Muonium is obtained (Hughes et al., 1960) by stopping muons in argon. To reveal it, one observes the direction of emission of positrons coming from the decay of the muon. The muonium atom in a magnetic field has four levels (illustrated in Fig. 14-9). These levels are unequally populated because the muons are initially polarized; hence, on decay the muons emit positrons preferentially in one direction. Applying both a constant magnetic field and an oscillating field, one induces transitions between the levels (see Figs. 14-9 and 14-10) which alter the intensity of the β^+ emitted in one direction. From the resonance frequency and the applied field one finds $\Delta\nu$, which to a first approximation is

$$\Delta\nu = \frac{8}{3}\,\alpha^4\,\frac{m_e c^3}{2\pi\hbar}\,\frac{\mu_\mu}{\mu_e} \qquad \text{(Hertz)} \qquad (14\text{-}4.10)$$

Telegdi and co-workers (1971) attained even higher precision in the measurement of $\Delta\nu$ by an ingenious refinement of the resonance method. Using a

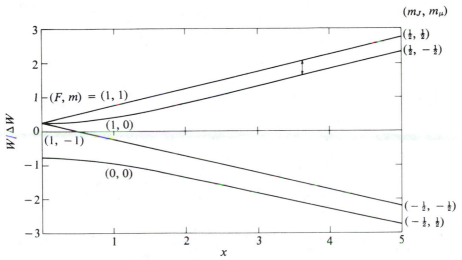

Figure 14-9 Energy-level diagram for the ground state of muonium in a magnetic field H obtained from the Breit-Rabi equation:

$$W_{F=\frac{1}{2}\pm\frac{1}{2},\,m} = -\frac{\Delta W}{4} + \mu_0 g_\mu Hm \pm \frac{\Delta W}{2}\left(1 + 2mx + x^2\right)^{1/2}$$

ΔW is the zero-field hfs separation between the $F=1$ and $F=0$ states, $x = (g_J - g_\mu)\,\mu_0 H/\Delta W$, g_J and g_μ are the electron and muon g values in units in which $g_J \cong 2$, $\mu_0 = $ Bohr magneton, and $\Delta\nu = \Delta W/h = 4463.301$ MHz. The levels are designated by both their weak-field quantum numbers (F, m) and their strong-field quantum numbers (m_J, m_μ). The transition observed is indicated by the arrow at $x \cong 3.6$.

ISBN 0-8053-8601-7

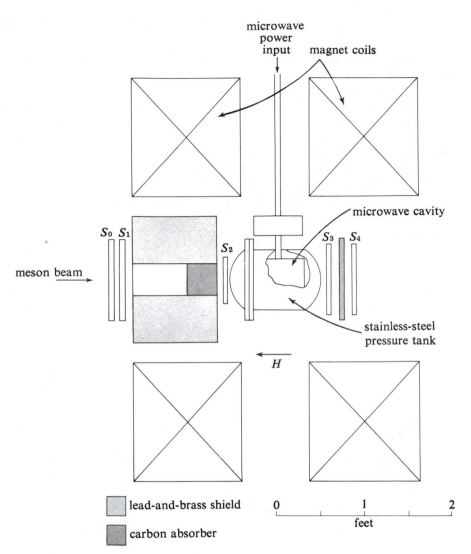

Figure 14-10 Apparatus used by Hughes for the detection of muonium. A muon is stopped in purified argon at high pressure and is subject to microwave magnetic field. A stopped muon is indicated by a $12\bar{3}$ count; a decay positron is indicated by a $34\bar{2}$ count and recorded as an event if it occurs in the time interval between 0.1 μsec and 3.3 μsec after the $12\bar{3}$ count. The number of events and the number of $12\bar{3}$ counts are measured, with the microwave field off and with the microwave field on, as a function of the static magnetic field. The ratio R is then computed, where

$$R = \frac{(\text{events}/12\bar{3})_{\text{microwaves on}}}{(\text{events}/12\bar{3})_{\text{microwaves off}}}$$

At resonance the transition should be observed as an increase of R to a value greater than 1. [From V. W. Hughes, *Ann. Rev. Nucl. Sci.*, **16**, 445 (1966).]

ISBN 0-8053-8601-7

formula including all higher-order corrections, one can find from the experimental $\Delta\nu$ that the value of $1/\alpha = 137.03638(19)$, in agreement with other determinations based on very different phenomena, such as the Josephson effect.

BIBLIOGRAPHY

Bargmann, V., L. Michel, and V. Telegdi, "Precession of the Polarization of Particles Moving in a Homogeneous Electromagnetic Field," *Phys. Rev. Letters*, **2**, 435 (1959).

Bethe, H., and F. de Hoffmann (BH 55).

Combley, F., and E. Picasso, "The Muon ($g - 2$) Precession Experiments, Past, Present, Future," *Phys. Repts.*, **14**, 1 (1974).

Commins, E. (Com 73).

De Rújula A., H. Georgi, S. L. Glashow, H. R. Quinn, "Fact and Fancy in Neutrino Physics," *Rev. Mod. Phys.*, **46**, 391 (1974).

Feinberg, G., and L. M. Lederman, "The Physics of Muons and Muonic Neutrinos," *Ann. Rev. Nucl. Sci.*, **13**, 431 (1963).

Fermi, E. (Fe 51).

Feynman, R. P., *The Theory of Fundamental Processes*, Benjamin, New York, 1961.

Hughes, V. W. "Muonium," *Ann. Rev. Nucl. Sci.*, **16**, 445 (1966).

Lederman, L. "Neutrino physics" in E. H. S. Burhop (ed.), *High-Energy Physics*, Academic Press, London, 1969.

Morita, M., *Beta Decay and Muon Capture*, Benjamin, New York, 1973.

Primakoff, H., "Theory of Muon Capture," *Rev. Mod. Phys.*, **31**, 802 (1959).

Rich, A., and J. C. Wesley, "Current Status of the Lepton g-factors," *Rev. Mod. Phys.*, **44**, 250 (1972).

Wu, C. S., and L. Wilets, "Muonic Atoms and Nuclear Structure," *Ann. Rev. Nucl. Sci.*, **19**, 527 (1969).

PROBLEMS

14-1 (a) Calculate the maximum energy of the electrons emitted in muon decay. (b) Calculate the muon energy in the decay $\pi^+ \to \mu^+ + \nu$ and the electron energy in the decay $\pi^+ \to e^+ + \nu$.

14-2 Show by comparing the energy released in beta decay of the neutron and in muon decay on the one hand, and the mean life for the two processes on the other, that their coupling constants are of the same order of magnitude.

14-3 Estimate the mean life for annihilation of positronium and compare it with the experimental value. Try to do the same for muonium ($\mu^- + e^+$). Experiment shows that the mean life for annihilation is much larger than the mean life for spontaneous decay. Try to find an explanation.

14-4 Molecular ions containing a deuteron and a proton bound by one muon, corresponding to the ordinary molecular ion HD^+, are occasionally formed by stopping μ^- in liquid hydrogen containing a small amount of deuterium. Estimate the internuclear distance in the ion and show that the nuclei may react to form ^3He. In this case what happens to the muon? Discuss, in qualitative terms, possible life histories of muons in liquid hydrogen containing small amounts of deuterium in solution (see Feinberg and Lederman, 1963).

14-5 Calculate the energy of X rays emitted in muonic transitions in O and S and plan an experiment to measure them. Consider especially any doublet due to the muon magnetic moment.

14-6 In muon capture there is emission not only of X rays but also of Auger electrons. Estimate the fluorescent yield in a specific case.

ISBN 0-8053-8601-7

CHAPTER
15

Pions and Other Bosons

This chapter is devoted to pions and other strongly interacting bosons. We have discussed before how Yukawa postulated, as the quantum of nuclear forces, a particle with mass approximately $m = \hbar/rc$, where r is the range of nuclear forces, and how these particles were found in cosmic rays by Lattes, Occhialini, and Powell in 1947 (see Fig. 1-6). Soon thereafter, when the Berkeley cyclotron reached sufficient energy to produce pions (Lattes and Gardner, 1948), beams of charged pions became available in the laboratory. The neutral pion has such a short life that it is practically impossible to obtain neutral-pion beams. However, its decay into two gamma rays of about 70 MeV in the center-of-mass system is so characteristic that from this radiation it was possible to infer the existence of the π^0 and to determine several of its properties (Björklund, Crandall, Moyer, and York, 1950). Pions are bosons of spin 0.

15-1 THE YUKAWA INTERACTION

In 1935 Yukawa tried to develop a theory of nuclear forces. The most important experimental feature of these forces is that they have a *range*. That is, nuclear forces decrease extremely rapidly when the interacting particles are more than about 10^{-13} cm apart. Experiment shows that in practice there is a

Emilio Segrè, Nuclei and Particles: An Introduction to Nuclear and Subnuclear Physics, Second Edition

ISBN 0-8053-8601-7

critical length beyond which the interaction does not extend. In this respect a nuclear force differs fundamentally from a $1/r^2$ force such as the electric force. We must thus expect that if there is a nuclear potential, it will contain a parameter with the dimensions of a length. Actually such a potential was first proposed by Yukawa in the form

$$V(r) = g\,\frac{e^{-kr}}{r} \tag{15-1.1}$$

where k is the reciprocal of the length which can be assumed to represent the range of nuclear forces.

Although field theory is definitely beyond the scope of this book, it is necessary at this point to introduce, at least qualitatively, several principles of field theory. The electromagnetic field is classically described by Maxwell's equations. Planck's constant does not appear in the equations and they do not describe any of the quantum phenomena. However, it is possible to subject the expressions of the fields or of the potentials to a process of quantization somewhat similar to that by which Schrödinger's equation is obtained from a classical hamiltonian. One finds that the energy of the field is quantized; that is, it has the expression

$$E = \sum_s \hbar\omega_s\left(n_s + \tfrac{1}{2}\right) \tag{15-1.2}$$

where the ω_s are the classical eigenfrequencies of the radiation field and the n_s are integral numbers $\geqslant 0$. This expression for the energy is interpreted as a manifestation of light quanta of energy

$$\epsilon_s = \hbar\omega_s \tag{15-1.3}$$

The energy of the electromagnetic field is the sum of the energies of its quanta. A momentum is also associated with a light quantum, its direction is that of the propagation of the light and its magnitude is

$$p = \frac{\hbar\omega}{c} = \frac{\hbar}{\lambdabar} \tag{15-1.4}$$

Any field in general shows quantum properties, and the process of quantization connects the wave with the corpuscular aspects.

When we try to apply this type of idea to nuclear forces, the first problem to arise, by analogy with the electromagnetic force, is that of finding the properties of the quanta of a field that is associated with the Yukawa potential of Eq. (15-1.1). Table 15-1 shows a very suggestive correspondence. In the first line we note that for the electromagnetic field the expression of the static potential does not contain any characteristic length, whereas the opposite is true in the case of nuclear forces. The second line gives the well-known

ISBN 0-8053-8601-7

TABLE 15-1 COMPARISON BETWEEN ELECTROMAGNETIC AND YUKAWA FIELDS

	Electromagnetic field	*Nuclear field*
Static potential	$V = g/r$	$V = g(e^{-kr}/r)$
Eq. of propagation for the potential	$\nabla^2 V - (1/c^2)(\partial^2 V/\partial t^2) = 0$	$\nabla^2 V - k^2 V - (1/c^2)(\partial^2 V/\partial t^2) = 0$
Plane-wave solution	$V = V_0 \exp\{-i[(z/\lambdabar - \omega t)]\}$	$V = V_0 \exp\{-i[(z/\lambdabar - \omega t)]\}$
Relation between $\lambdabar$ and ω	$\dfrac{-1}{\lambdabar^2} + \dfrac{\omega^2}{c^2} = 0$	$\dfrac{-1}{\lambdabar^2} + \dfrac{\omega^2}{c^2} - k^2 = 0$
Relation between p and E	$\dfrac{-p^2}{\hbar^2} + \dfrac{E^2}{\hbar^2 c^2} = 0$	$\dfrac{-p^2}{\hbar^2} + \dfrac{E^2}{\hbar^2 c^2} - k^2 = 0$
or	$E = cp$	$(c^2 p^2 + \hbar^2 k^2 c^2)^{1/2} = E$

wave-propagation equation in the case of an electromagnetic field and for the nuclear forces the natural generalization (Klein–Gordon equation). The third line gives the relation of $\omega, \lambdabar$, and k to one another in the plane wave solution of the preceding equation.

Note that according to de Broglie we have the relation

$$\lambdabar = \hbar/p \tag{15-1.5}$$

between the momentum of a particle and the de Broglie wavelength of its associated wave. The relation $\hbar\omega = E$ is also valid. With the help of these expressions we write line 6. The relation

$$c^2 p^2 + \hbar^2 k^2 c^2 = E^2 \tag{15-1.6}$$

immediately suggests a connection between $\hbar^2 k^2 c^2$ and a rest mass m for the quantum of the nuclear force. In fact, if

$$k = mc/\hbar \tag{15-1.7}$$

the expression for the energy becomes

$$E = \left(m^2 c^4 + c^2 p^2\right)^{1/2} \tag{15-1.8}$$

the well-known expression for the total energy of a particle of rest mass m and momentum p.

We are thus led to expect a connection between the range of nuclear forces and the rest mass of their quanta. Numerically, if we assume that

ISBN 0-8053-8601-7

$k = 10^{13}$ cm^{-1}, we find that $m = 0.3 \times 10^{-24}$ g, approximately 300 times the mass of the electron, in very good qualitative agreement with the pion mass ($273m_e$).

From a different point of view an argument due to Wick (1938) justifies Eq. (15-1.7). The interaction between two nucleons is transmitted by the field of nuclear forces and is accompanied by the emission and absorption of the corresponding quanta. If a nucleon emits a quantum of a certain rest mass m, the energy of the system is increased at least by mc^2; and if we could measure it between the time of emission of the quantum and its absorption by the other nucleon, we should find a creation of energy in the amount of at least mc^2. However, in order to perform this measurement, according to the uncertainty principle, we require a time

$$t \cong \frac{\hbar}{E} = \frac{\hbar}{mc^2} \qquad (15\text{-}1.9)$$

Unless the quantum lasts a time comparable to t, no violation of the conservation of energy can be ascertained. During such a time, the quantum may travel a distance of up to

$$ct = r_0 = \hbar/mc \qquad (15\text{-}1.10)$$

and thus interacting particles must be closer than r_0, which is then, by definition, the range of nuclear forces.

The existence of exchange forces for which the interaction brings about the transformation of a neutron into a proton, and vice versa (Chap. 10), leads to the expectation that quanta of nuclear force may carry an electric charge of magnitude e. On the other hand, there must also be neutral quanta in order to explain ordinary forces, and in fact there are pions of charge $\pm e$ and 0.

What we have said thus far brings us to a consideration of the process by which a nucleon (N) transforms into a nucleon plus a pion,

$$N \rightarrow N' + \pi \qquad (15\text{-}1.11)$$

as the elementary step of the nuclear interaction, similar to that of the emission of a quantum by an electric charge; this is called the Yukawa process. Nuclear interactions come about by a succession of Yukawa processes in which a nucleon emits a pion, thereby changing its momentum; the pion is then absorbed by another nucleon. In the process there is overall conservation of energy, momentum, and electric charge.

The nucleon–nucleon interaction is then a process of the second order according to the scheme

$$N + N \rightarrow N' + \pi + N \rightarrow N' + N'' \qquad (15\text{-}1.12)$$

where the primes on a symbol indicate its charge and momentum.

The matrix element corresponding to the process of Eq. (15-1.11) can be established as an order of magnitude from simple considerations based on the elements of field theory [see (Fe 51)]. It is

$$g \frac{\hbar c}{(2\Omega w_s)^{1/2}} \tag{15-1.13}$$

where g is a constant characteristic of the Yukawa interaction. Its value must be derived from experiment; w_s is the total energy of the pion that has been created or destroyed in the Yukawa process. In this connection we note that the Yukawa process does not necessarily consider the creation or destruction of *real* pions. The process may be a "virtual" one, describable formally as a real process but with any value of w_s, even those incompatible, for the single step, with the conservation laws. Ω is a normalization volume which disappears from the final formulas.

The process represented by Eq. (15-1.12) is a second-order process, and the matrix elements corresponding to a transition between the initial and final state are calculable by second-order perturbation theory.

For a second-order perturbation the effective matrix element has the form

$$H'_{if} = \sum_m \frac{H_{im} H_{mf}}{W_i - W_m} \tag{15-1.14}$$

where the H_{im}, H_{mf} are the first-order matrix elements relating the initial state of energy W_i to an intermediate state of energy W_m. In the specific case of the Yukawa interaction, the intermediate state is the one in which a pion is created, and in which $W_i - W_m = -w_s = -(m_\pi^2 c^4 + c^2 p^2)^{1/2}$, at least approximately. Substituting the first-order matrix elements in Eq. (15-1.14), we obtain

$$H'_{if} = -\frac{1}{w_s}\left[\frac{g\hbar c}{(2\Omega w_s)^{1/2}}\right]^2 = -\frac{g^2\hbar^2 c^2}{2\Omega(m_\pi^2 c^4 + c^2 p^2)} \tag{15-1.15}$$

The momentum p is equal to the momentum of the pion emitted by the first nucleon because in the intermediate states momentum (not necessarily energy) is conserved. When the pion is absorbed by the second nucleon, this momentum is transferred to it.

Appendix A shows that the matrix element of the transition from a state of momentum p' to a state of momentum p'' is the Fourier transform of the

ISBN 0-8053-8601-7

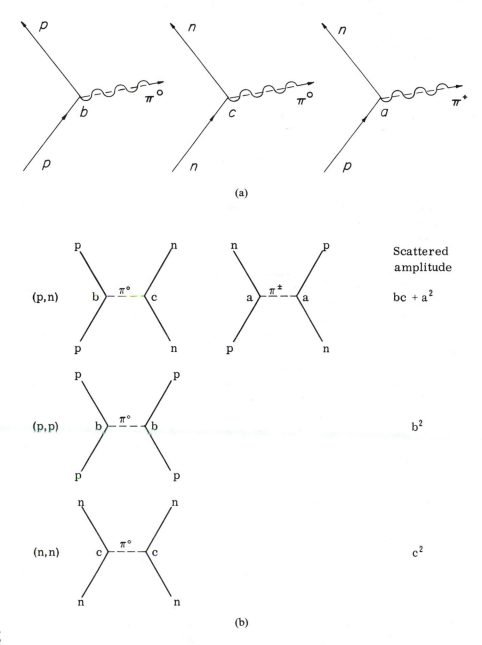

Figure 15-1 (a) The diagrams corresponding to the fundamental pion-nucleon interactions. Charge independence requires that the coupling constants a, b, c be connected by $a = (2b)^{1/2}$ and $b = -c$. (From Fey 61) (b) The combined diagrams giving all cases of nucleon-nucleon scattering. The scattered amplitude for (p, n) is $bc + a^2$, for (p, p) it is b^2, for (n, n) it is c^2.

potential $U(r)$. We must then have

$$-\frac{g^2\hbar^2c^2}{2\Omega(m_\pi^2c^4 + c^2p^2)} = \frac{1}{\Omega}\int U(\mathbf{r})e^{-i\mathbf{p}\cdot\mathbf{r}/\hbar}\, d\mathbf{r} \qquad (15\text{-}1.16)$$

where $\mathbf{p}$ is the momentum transfer. This equation can be solved to yield

$$U(\mathbf{r}) = \frac{-g^2}{16\pi^3\hbar}\int \frac{e^{i\mathbf{p}\cdot\mathbf{r}/\hbar}}{m_\pi^2c^2 + p^2}\, d\mathbf{p} = \frac{-g^2}{8\pi r}e^{-m_\pi cr/\hbar} \qquad (15\text{-}1.17)$$

This potential is then equivalent to the Yukawa interaction. Experimentally a collision process may be interpreted either as a transition produced by the Yukawa interaction or as a deflection resulting from a nucleon–nucleon potential. Thus we again find the relation among nuclear forces, their range, and the mass of the pion from which we started.

According to Feynman the fundamental processes Eq. (15-1.1) may be symbolically represented by the diagrams Fig. 15-1a. By combining them as in Fig. 15-1b one may obtain the scattering amplitude for all cases of nucleon–nucleon scattering. The diagrammatic form is the starting point for a powerful calculation techniques that goes beyond the scope of this book (see, e.g., Fey 61 or Le 73).

All the considerations discussed in this section are qualitative only.

Their quantitative development is mathematically complicated and not yet complete. In particular, it is not legitimate to use perturbation theory for strong interactions. Furthermore, pions are not the only particles involved in the nucleon–nucleon force. Heavier particles, corresponding to shorter ra1ii of action, enter the picture. For instance, in the case of the repulsive core of the nucleon, heavier mesons play a role. (See Sec. 15-7.)

15-2   SPIN OF THE PIONS

The neutral pion decays into two gamma rays. This fact shows that it is of integral spin and a boson.

Yang (1950) developed an argument excluding spin 1. Both the result and the method are important and may be applied to all particles decaying into two gammas.

Consider a π^0 at rest at the origin of the coordinates and sending, on decay, two gamma rays in the direction of the positive and negative z axis, as in Fig. 15-2. Analyze the photons into circularly polarized components of helicity $+1$ or -1. We denote the photon wave functions by $|+R\rangle$, $|+L\rangle$, $|-R\rangle$, $|-L\rangle$ where $+$ indicates a photon traveling in the positive z direction and $-$ a photon traveling in the negative z direction, R a right-handed photon and L a left-handed one. The photons of wave function $|+R\rangle$ and $|-L\rangle$ have m, the z component of angular momentum, equal to $+1$; those of

ISBN 0-8053-8061-7

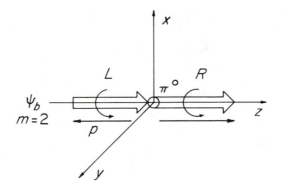

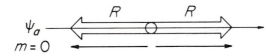

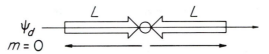

Figure 15-2 Analysis of possible polarizations of two gamma rays generated by the decay of a neutral pion.

wave function $|+L\rangle$, $|-R\rangle$ have $m = -1$. Considering the two quanta together, we have four possible wave functions:

$$
\begin{aligned}
\psi_a &= |+R\rangle\,|-R\rangle \quad m = 0 \\
\psi_b &= |+R\rangle\,|-L\rangle \quad m = 2 \\
\psi_c &= |+L\rangle\,|-R\rangle \quad m = -2 \\
\psi_d &= |+L\rangle\,|-L\rangle \quad m = 0
\end{aligned}
\tag{15-2.1}
$$

We now rotate the system by 180 deg around the y axis and describe the same event. The z axis has changed direction but the handedness of the photons remains the same. The four possibilities in the new system are

$$
\begin{aligned}
\psi_a' &= |-R\rangle\,|+R\rangle = \psi_a \\
\psi_b' &= |-R\rangle\,|+L\rangle = \psi_c \\
\psi_c' &= |-L\rangle\,|+R\rangle = \psi_b \\
\psi_d' &= |-L\rangle\,|+L\rangle = \psi_d
\end{aligned}
\tag{15-2.2}
$$

Thus only ψ_a and ψ_d are eigenstates of this rotation, with eigenvalues 1.

If the neutral pion had spin 1, obviously it could decay into $m = 0$ states, but not into $m = 2$ states. This rules out decay into ψ_b or ψ_c. Furthermore, under rotation it would behave like a spherical harmonic of order one, and having $m = 0$ only Y_{10} could be considered. Now Y_{10} is proportional to $z/r = \cos \theta$ and under our rotation $\theta \rightarrow \pi - \theta$ and the wave function changes sign. Thus it cannot be represented by ψ_a or ψ_d, which are invariant under this rotation. It follows that no particle of spin 1 can decay into two photons, and π^0 cannot have spin 1. Spin 0 is the favorite value, by analogy with π^+.

This type of argument can be extended by considering other rotations and inversion. (See Prob. 15-18.) The photons may also be analyzed into linearly polarized components, leading to important results concerning the correlation of their polarizations.

For the π^+ meson there is direct experimental evidence for spin 0. Consider the reaction

$$p + p \rightleftarrows \pi^+ + d \tag{15-2.3}$$

and its inverse. Going from left to right, the reaction is the production of π^+ and d in p–p collisions; proceeding in the opposite direction, it is the absorption of π^+ in deuterium. We may apply the principle of detailed balance (see Sec. 11-3) to the two reactions and obtain the relation, valid for an unpolarized beam on an unpolarized target,

$$\frac{\sigma(E_1)_{1 \rightarrow 2}}{\sigma(E_2)_{2 \rightarrow 1}} = \frac{g_2}{g_1} \frac{p_\pi^2}{p_p^2} \tag{15-2.4}$$

where the cross sections and the momenta are measured in the center-of-mass system and g_2 and g_1 are the statistical weights of the final states

$$g_2 = (2I_\pi + 1)(2I_d + 1) = (2I_\pi + 1)3 \tag{15-2.5}$$

because the spin of the deuteron is 1. On the other hand,

$$g_1 = (\tfrac{1}{2})(2I_p + 1)^2 = 2 \tag{15-2.6}$$

because the spin of the proton is $\tfrac{1}{2}$. The factor $\tfrac{1}{2}$ in g_1 originates from the fact that we have identical protons; and the states obtained by interchanging them, in a classical sense, are not distinguishable. This argument has been mentioned in a different form in Sec. 11-3. The cross sections for the reactions of Eq. (15-2.3) may be either total or differential; in the latter case the center-of-mass scattering angle must be the same in both the direct and the inverse process. The energies E_1 and E_2 must also be the same in the center-of-mass system. Thus to the collision of a slow π^+ (kinetic energy 0) there

ISBN 0-8053-8061-7

corresponds a kinetic energy in the p–p system given by the mass difference $\pi^+ + d - 2p = 138$ MeV.

Evaluation of the experimental results gives $I_{\pi^+} = 0$. The spin of the π^- is also 0, as shown, for instance, by the absence of structure in the spectra of pionic atoms.

In conclusion, all pions have spin 0, are bosons, and have no magnetic moment.

15-3 INTRINSIC PARITY

A full treatment of intrinsic parity is beyond the scope of this book. We shall limit ourselves to the most elementary notions, which are, however, indispensable even for an introduction to the subject.

The principle of conservation of parity may be stated crudely by saying that the mirror image of a natural phenomenon represents another possible physical phenomenon. We can rearrange our apparatus, substances, etc., in such a way as to realize exactly what we see in the mirror. Then, performing an experiment with the original apparatus or with the rearranged apparatus, we obtain as the result of our observations certain numbers (e.g., differential cross sections). If the numbers obtained are identical, we say that the phenomenon obeys the principle of conservation of parity, or that it is reflection invariant. This invariance is true only if we limit ourselves to strong and electromagnetic interactions. For weak interactions, the invariance holds only if the reflection is accompanied by the simultaneous replacement of each particle by its own antiparticle (Fig. 15-3 and see Chap. 9).

Instead of the operation of mirroring, we may consider the operation of changing the sign of all coordinates, or the reflection on a point (Fig. 15-4). In both mirroring and reflection, the right hand changes into the left hand. Again the two physical situations must be equivalent as far as strong and electromagnetic interactions are concerned.

The requirement just mentioned has important physical consequences, such as selection rules forbidding certain reactions that, at first sight, would seem possible. For example, consider slow neutron capture in ^{7}Be according to the equation

$$^7\text{Be} + n \Rightarrow 2(^4\text{He}) \tag{15-3.1}$$

On the right hand the system of two identical bosons must have l even, and hence positive parity, because a spherical harmonic of even l does not change sign on reflection. On the left hand ^{7}Be, according to the shell model, has an odd neutron in a p state. Hence, the overall parity of ^{7}Be in the ground state is the same as that of a neutron in a p state, that is, negative. A slow neutron is necessarily in an s state. Therefore the total parity on the left hand is negative. We conclude that the reaction is forbidden by the conservation of

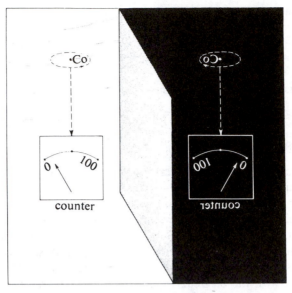

Figure 15-3 Symbolic drawing (C. N. Yang) of the beta decay of matter and antimatter. The operation of reflection and charge conjugation is illustrated by representing antimatter by white lines against a black background. In this figure the reading of the meter on the left is L, that of the meter on the right is R. Experiments show that these two readings are the same.

parity, and indeed it has never been observed experimentally. Note the assumption that neutrons or protons in a state of orbital angular momentum l have a parity $(-1)^l$. The conservation of parity imposes conditions on many nuclear reactions. These conditions are always satisfied when the forces involved are nuclear or electromagnetic.

A special consideration is needed when we extend the rule to processes in which new particles are created.

When a particle is emitted or absorbed (e.g., when a nucleon emits a pion), we must study the influence of this absorption or emission on the parity of the system. For some particles, such as pions, we find that the parity of a system composed of a nucleon and a pion in an even orbital angular momentum state is the opposite of the nucleon parity; and we then say that the pion has negative, or odd, intrinsic parity. If the parity of a nucleon is not changed by emission of the particle in the even angular momentum state, the emitted particle is said to have positive, or even, intrinsic parity. In other words, in a reaction involving strong interactions, we postulate that parity is conserved. We check the parity on the two sides of the equation by measuring the usual orbital angular momentum, we assign positive parity, by definition,

ISBN 0-8053-8061-7

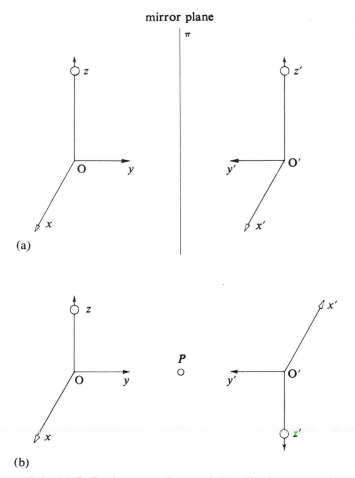

mirror plane

(a)

(b)

Figure 15-4 (a) Reflection on a plane and (b) reflection on a point *P*. Part (b) may be obtained by reflection on a plane, as in part (a), followed by a rotation of 180 deg around the *y* axis.

to nucleons, and if the final result seems to violate the conservation of parity, a case occurring only when new particles such as pions are created, we ascribe to the pion an "intrinsic parity," thereby reestablishing the conservation of parity rule. The procedure has a physical significance—it is not a play on words, as one might think. First, the result is always the same for a given particle, no matter which reaction is used. Furthermore it prevents the superposition of even and odd states by strong interactions and thus establishes special symmetries in angular distributions, gives selection rules, and so forth.

As an example of determination of intrinsic parity, consider the emission of a pion by colliding nucleons: Is the parity of the system changed when the pion is emitted in a $l = 0, 2, 4$ state or when it is emitted in a $l = 1, 3, 5$ state, etc? The question is answered by considering the reverse process, the capture of π^- in deuterium. Here we have the reaction

$$\pi^- + d \Rightarrow n + n \tag{15-3.2}$$

The π^- slow down by ionization until they are practically at rest; an analysis of the capture process shows that it occurs from an s orbit ($l = 0$).

The total angular momentum on the left is thus given by the spin of the deuteron, because the pion is spinless. Hence, the total angular momentum has the value 1. On the right hand, the two neutrons can only be in states $^1S_0$, $^3P_{012}$, $^1D_2$, etc., other states being ruled out by the exclusion principle. The only allowed state of spin 1 is thus $^3P_1$, which has parity -1. We must conclude that a pion plus a deuteron in an s state has negative parity. This state of affairs is then expressed by saying that the pion has a negative intrinsic parity, assuming conventionally that a nucleon has a positive intrinsic parity.

The experimental evidence for reaction (15-3.2) is based on the branching ratio between the reactions

$$\pi^- + d \Rightarrow 2n \tag{15-3.3}$$

$$\pi^- + d \Rightarrow 2n + \gamma \tag{15-3.4}$$

$$\pi^- + d \Rightarrow 2n + \pi^0 \tag{15-3.5}$$

For pions at rest one finds

$$\frac{\text{probability of } \pi^- + d \to 2n}{\text{probability of } \pi^- + d \to 2n + \gamma} = 2.97 \pm 0.07 \tag{15-3.6}$$

and $5.6 \ 10^{-4}$ for the ratio of reaction, 15-3.5 to 15-3.4.

The weakness of reaction (15-3.5) confirms s-state capture. In fact, when capture occurs in the s state, J on the left side of (15-3.5) is 1. On the right side there is very little available energy and the orbital angular momentum of the two neutrons is bound to be zero, which means that the two neutrons are in the $^1S_0$ state. The pion also has insufficient energy to go into a state of angular momentum 1 in the center of mass of the system. Consequently there is no possibility of balancing angular momenta.

The experiment is accomplished by observing the gamma rays emitted when the π^- are stopped in deuterium. Only gamma rays corresponding to reaction (15-3.4)—that is, with energy approximately equal to the rest energy of the π^-—are found. The $2n$ and $2n + \pi^0$ reactions are difficult to observe directly.

ISBN 0-8053-8061-7

The qualitative fact of the high probability of reaction (15-3.3) is sufficient to prove that it is allowed and hence to establish the negative intrinsic parity of the π^-.

In the case of electromagnetic radiation, the parity of a system changes on emission of $E1$, $E3$, El(odd), etc., or $M2$, $M4$, Ml(even), etc., radiation. It does not change on emission of $E2$, $E4$, El(even), etc., or $M1$, $M3$, Ml(odd), etc., radiation. The angular momentum carried away is always l. We may summarize the results by saying that the parity of the radiation is

$$(-1)^l \Pi_{int}$$

where Π_{int} is 1 for all electric radiation, -1 for all magnetic radiation (cf. Sec. 8-2 and Table 13.5).

Some of these assignments, such as the intrinsic parity of $E1$ radiation and the intrinsic parity of the proton, have the character of convention. Other assignments of intrinsic parity then follow from experimental facts and the initial conventions. For instance, it can be shown, by a development of the Yang argument given at the beginning of Sec. 15-2, that a system of spin 0, decaying into two gamma rays, gives rise to quanta with parallel polarization planes if the original system has even parity, or to quanta with perpendicular polarization planes if the original system has odd parity. In the case of positronium in an $^1S_0$ state, which annihilates into two quanta (cf. Chap. 2), experiment shows that the system has odd parity. We conclude that the electron and the positron have opposite intrinsic parities. For the case of the π^0 the experiment mentioned above has not been performed directly, but the decay into Dalitz pairs gives a direct determination of the intrinsic parity. If the planes of two Dalitz pairs originating from the same pion tend to be parallel, the intrinsic parity of the pion is positive; if they tend to be perpendicular, the intrinsic parity is negative. Enough cases of this rare type of decay have been observed to show that the intrinsic parity of π^0 is negative. In fact the intrinsic parity of all pions is negative.

Note an interesting consequence of this property for the Yukawa interaction

$$N \Rightarrow N' + \pi \tag{15-3.7}$$

This interaction conserves parity and angular momentum. On the left hand, the parity is even by definition, and the angular momentum is the spin $\frac{1}{2}$ of the nucleon. On the right hand, in order to have angular momentum $\frac{1}{2}$, we may consider only the $l=0$ or $l=1$ state, where l is the orbital angular momentum in the center-of-mass system of the pion and nucleon. But only the $l=1$ state is odd; hence in the Yukawa interaction the pion is emitted (virtually) in a p state.

ISBN 0-8053-8061-7

Finally we wish to comment briefly on the intrinsic parity of strange particles. Remembering the definition (Sec. 13-2) and the strangeness property, we see that in strong interactions only particles with the same strangeness participate in a reaction such as

$$p + K^- \Rightarrow \Lambda + \pi^0 \tag{15-3.8}$$

Even if we knew all the angular momenta on both sides of the equation, we could conclude, by an argument similar to the one used for pions, only that the Λ and K have the same intrinsic parity or opposite intrinsic parity. In other words, the parity of the (K, Λ) system with respect to the nucleon is defined, but not that of each separate particle. The same is true for the (Σ, K) system and for the (Σ, Λ) system. These three parities, however, are not independent, and two of them determine the third. Thus, if (Σ, K) and (Σ, Λ) have the same parity, (Λ, K) must have parity $+1$.

The case of the Ξ is simpler. Its strangeness, -2, permits reactions such as

$$\Xi^- + p \Rightarrow \Lambda + \Lambda \tag{15-3.9}$$

The intrinsic parity on the right side is obviously positive; hence the Ξ^- intrinsic parity with respect to the nucleon is in principle measurable. The practical methods for determining intrinsic parities are rather indirect. The results are

(Λ, K) system   odd

(Σ, K) system   odd

(Σ, Λ) system   even

The Ξ parity has not yet been directly measured, but according to $SU(3)$ systematics is expected to be $+$.

As an example of methods for determining the relative parity, consider the (Λ, K) system. The reaction $K^- + {}^4\text{He} \Rightarrow {}^4_\Lambda\text{He} + \pi^-$ leading to the formation of the ${}^4_\Lambda\text{He}$ hypernucleus (see Sec. 16-3) has been observed. Now assume that the spin of ${}^4_\Lambda\text{He}$ is 0. Then since the orbital angular momentum on both sides of the formation equation must be the same, the K^- and Λ have opposite parity. In other words, the (K, Λ) system is odd with respect to the nucleon. The weakness of the argument lies in the assumption that ${}^4_\Lambda\text{He}$ has spin 0 and that it is formed in the ground state. Although both assumptions are very probable, they are not proved. The parity of the (Σ, K) system with respect to the proton follows from a study of the reaction $\pi^+ + p \Rightarrow \Sigma^+ + K^+$ using polarized protons, and the parity of the (Σ, Λ) system derives from the electromagnetic decay $\Sigma^0 \rightsquigarrow \Lambda^0 + (e^+ + e^-)$.

A remarkable relation between the intrinsic parity of a particle and its antiparticle, provable in the same sense as the relation between spin and statistics, is that such parity is the same for bosons and opposite for fermions.

ISBN 0-8053-8061-7

15-4   ISOTOPIC SPIN OF PIONS

We have already introduced the concept of isotopic spin for the nucleon (Chap. 10). The same concept is extended to pions by assigning them a coordinate susceptible of three values only, similar to the z component of a spin 1 vector. We shall consider this new variable as the third component of a vector $\mathbf{t}$ in *isotopic spin space*. The physical interpretation of t_3 is given by its relation to the charge Q of the meson,

$$Q/e = t_3 \tag{15-4.1}$$

The complete eigenfunction of a meson will thus be the product of a part depending on the space coordinates and of a part depending on i-spin. The i-spin obeys the same commutation relation as ordinary angular momentum and for spin 1 is easily represented by eigenfunctions having the properties of spherical harmonics of order 1.

These formal considerations are important because, as discussed in Chap. 10, nuclear forces are charge independent or, more precisely, the behavior of nuclear forces is determined by the total i-spin of the system, not by its third component. Moreover, strong interactions conserve i-spin; that is, the initial and final states of a system undergoing a transition produced by strong interactions have the same i-spin. These statements are based on experimental facts, as we shall illustrate by examples. The conservation of i-spin is formally similar to that of angular momentum for an isolated system, except that it is only approximate when there are interactions other than strong ones, because neither electromagnetic nor weak interactions conserve i-spin. Hence, there can be transitions between states of different i-spin, but their probability is in general much smaller than that of transitions due to interactions involving nuclear forces.

The eigenfunction of a particle will contain a factor depending on i-spin. We shall indicate this part symbolically for a proton as p and for a neutron as n, with a similar notation for other light nuclei. This notation is analogous to the α, β notation for ordinary spin. For pions we use a superscript $+$, 0, $-$, according to the charge of the pion. Thus, p^+ means proton and positive pion, dp means deuteron and proton, etc. For a system of two or more particles we define a total i-spin.

$$\mathbf{T} = \mathbf{t}^{(1)} + \mathbf{t}^{(2)} + \cdots \tag{15-4.2}$$

and

$$T_3 = t_3^{(1)} + t_3^{(2)} + \cdots \tag{15-4.3}$$

The eigenfunctions corresponding to a physical system such as p^0 are *not* always eigenfunctions of $\mathbf{T}^2$ and T_3. Generally we have to make linear combinations with appropriate coefficients (Clebsch–Gordan coefficients), in perfect analogy with the composition of angular momenta (Table 15-2; see also

TABLE 15-2   EIGENFUNCTIONS FOR THE NUCLEON–PION SYSTEM

$T = \frac{3}{2}$	$T = \frac{1}{2}$
T_3	T_3
$\frac{3}{2}\chi_3^3 = p^+$	
$\frac{1}{2}\chi_3^1 = \sqrt{(\frac{2}{3})}p^0 + \sqrt{(\frac{1}{3})}n^+$	$\frac{1}{2}\chi_1^1 = -\sqrt{(\frac{1}{3})}p^0 + \sqrt{(\frac{2}{3})}n^+$
$-\frac{1}{2}\chi_3^{-1} = \sqrt{(\frac{2}{3})}n^0 + \sqrt{(\frac{1}{3})}p^-$	$-\frac{1}{2}\chi_1^{-1} = \sqrt{(\frac{1}{3})}n^0 - \sqrt{(\frac{2}{3})}p^-$
$-\frac{3}{2}\chi_3^{-3} = n^-$	

Appendix E). Remember that one often calls, for brevity, an eigenfunction of $\mathbf{T}^2$ with eigenvalue $K(K+1)$ an eigenfunction corresponding to a state with $T = K$. We shall write, according to the usual conventions, the eigenfunctions of $\mathbf{T}^2$ and T_3 as $\chi_T^{T_3}$, where for brevity the subscript is twice the i-spin and the superscript twice its third component. Thus

$$p^+ = \chi_3^3 \qquad (15\text{-}4.4)$$

meaning that the physical situation p^+ corresponds to an eigenstate of $\mathbf{T}^2$ and T_3 with eigenvalues $\frac{3}{2}(\frac{3}{2}+1)$ and $+\frac{3}{2}$.

Using a table of Clebsch–Gordan coefficients, we can write Table 15-2 for the nucleon–pion system. The relations can be solved by expressing the six functions p^+, p^0, p^-, n^+, n^0, and n^- through the χ's. Note that we can experimentally realize the pure $T = \frac{3}{2}$ state by bombarding protons with positive pions or, at least in principle, by bombarding neutrons with negative pions. The pure $T = \frac{1}{2}$ state, however, cannot be realized in a nucleon–pion bombardment.

We list the following examples of experimental facts demonstrating the conservation of i-spin.

1.   The equality between p–p and n–p interactions was discussed in Chap. 10.

2.   Consider the reactions

$$p + p \Rightarrow d + \pi^+$$
$$n + p \Rightarrow d + \pi^0 \qquad (15\text{-}4.5)$$

The i-spin of the deuteron is 0, hence on the right-hand side of the equation the i-spin is 1, due to the pion. On the left-hand side, $p + p$ has i-spin 1, but $n + p$ is a superposition of i-spins 1 and 0 with equal amplitudes. From this it follows that the differential cross section for the first reaction must be

ISBN 0-8053-8061-7

twice that for the second. Figure 15-5 shows a result of the measurements of the angular distributions; they agree as they should.

3. Consider the reaction

$$p + d \Rightarrow {}^3\mathrm{He} + \pi^0$$

$$\Rightarrow {}^3\mathrm{H} + \pi^+ \qquad (15\text{-}4.6)$$

The left side of this equation has i-spin $\frac{1}{2}$ and $T_3 = \frac{1}{2}$, because the i-spin of the deuteron is 0 and that of the proton is $\frac{1}{2}$ with $T_3 = \frac{1}{2}$. The i-spins of ${}^3\mathrm{H}$ and ${}^3\mathrm{He}$ are $\frac{1}{2}$, as can be seen by considering them as a deuteron plus a neutron or a proton. The third component of the i-spin is $-\frac{1}{2}$ and $+\frac{1}{2}$ for ${}^3\mathrm{H}$ and ${}^3\mathrm{He}$, respectively. An eigenstate of T and T_3 with eigenvalues $\frac{1}{2}$ and $+\frac{1}{2}$, respectively, is obtained by combining the eigenfunctions of the π and of ${}^3\mathrm{He}$ or ${}^3\mathrm{H}$, respectively.

$$\chi_1^1 = -\sqrt{\left(\tfrac{1}{3}\right)}he^0 + \sqrt{\left(\tfrac{2}{3}\right)}h^+ \qquad (15\text{-}4.6a)$$

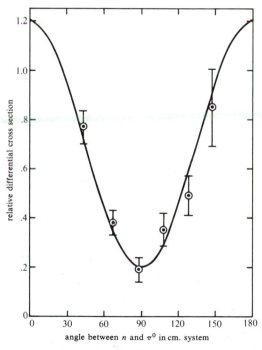

Figure 15-5 Angular distribution of π^0 formed in $n + p \Rightarrow d + \pi^0$. The solid line corresponds to π^+ production in $p + p \Rightarrow d + \pi^+$. [R. Hildebrand, *Phys. Rev.*, **89**, 1090 (1953).]

ISBN 0-8053-8061-7

as can be seen by analogy with the $T = \frac{1}{2}$, $T_3 = \frac{1}{2}$ state of Table 15-2. Here *he* and *h* replace *p* and *n*, respectively.

Now the *i*-spin part of the *pd* state is unchanged by collision and is χ_1^1 (Table 15-2). The amplitudes with which he^0 and h^+ are represented in this state are in the ratio 1 : $\sqrt{2}$. The probabilities of ending in the states h^+ and he^0 are proportional to the square of these amplitudes and hence the differential cross sections for the reactions in Eq. (15-4.5) must be in the ratio 1 : 2. This has been well verified (Fig. 15-6). A similar example, involving strange particles is

$$K^- + d \Rightarrow \Lambda + n + \pi^0 \tag{15-4.7}$$

$$\Rightarrow \Lambda + p + \pi^- \tag{15-4.8}$$

the *i*-spin on the left is $\frac{1}{2}$. The eigenstate of $T = \frac{1}{2}$, $T_3 = -\frac{1}{2}$, on the right, is

$$\chi_1^{-1} = \sqrt{\left(\tfrac{1}{3}\right)}\Lambda n^0 - \sqrt{\left(\tfrac{2}{3}\right)}\Lambda p^- \tag{15-4.9}$$

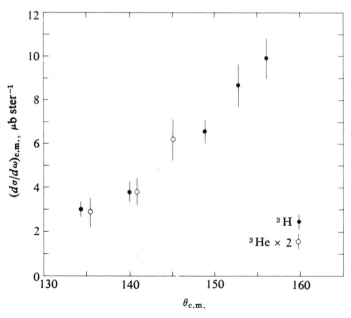

Figure   15-6   The differential cross sections for the reactions $p + d \Rightarrow \pi^+ + $ ^{3}H and $p + d \Rightarrow \pi^0 + {}^3$He with 450-MeV protons. The measured cross sections for the second reaction have been multiplied by a factor of 2 to facilitate a direct comparison. The cross sections are given in arbitrary units, which are, however, approximately microbarns per steradian. [From A. V. Crewe, E. Garwin, B. Ledley, E. Lillethun, R. March, and S. Marcowitz, *Phys. Rev. Letters*, **2**, 269 (1959).]

ISBN 0-8053-8061-7

(Compare Table 15-2.) Hence, the cross section for the second process is twice as large as for the first.

4.   In the reaction $d + d \Rightarrow {}^4\mathrm{He} + \pi^0$, i-spin would not be conserved; the reaction does not occur.

5.   The conservation of i-spin gives selection rules for photonuclear reactions that are verified experimentally.

6.   Light nuclei show energy levels that are well explained on the charge-independence hypothesis (cf. Chap. 10).

7.   One of the most striking instances of charge independence is nucleon–pion scattering, which we shall discuss in the next section.

15-5   PION–NUCLEON SCATTERING AND RESONANCES

There are the following observable scattering processes between nucleons and pions:

$$\text{(a)} \quad \pi^+ + p \Rightarrow \pi^+ + p$$
$$\text{(b)} \quad \pi^- + p \Rightarrow \pi^- + p \qquad\qquad (15\text{-}5.1)$$
$$\text{(c)} \quad \pi^- + p \Rightarrow \pi^0 + n$$

In addition there are the processes

$$\pi^0 + p \Rightarrow \pi^0 + p$$

and the processes obtained from Eqs. (15-5.1) by replacing the proton with a neutron and the π^+ with a π^-. Because neutron targets are not available, one proceeds indirectly, measuring the pion (or other particle)–deuteron cross section and subtracting from it the pion–proton cross section. This procedure would be correct if the deuterons contained independent protons and neutrons at great distances from one another. This is not the case and a crude approximation is obtained by assuming that the σ_{xn}, σ_{xd}, and σ_{xp} are related by Eq. (15-5.2), which can be justified on a simple geometric argument (*Glauber correction*).

$$\sigma_{xn} = \sigma_{xd} - \sigma_{xp} + \frac{\sigma_{xp}\sigma_{xn}}{4\pi}\langle r^{-2}\rangle \qquad\qquad (15\text{-}5.2)$$

The mean inverse square distance between neutron and proton $\langle r^{-2}\rangle$ may be derived by the deuteron eigenfunction and is approximately 0.03 mb^{-1}, at 15 GeV/c but depends somewhat on the energy.

Let us now introduce the scattering amplitudes f_1, f_3 for the isotopic spin states $T = \frac{1}{2}$, $\frac{3}{2}$, respectively. An incoming plane wave described by

$$e^{ikz}\chi_1 \qquad\qquad (15\text{-}5.3)$$

ISBN 0-8053-8061-7

gives rise to a scattered wave $f_1\chi_1$, and an incoming plane wave $e^{ikz}\chi_3$ gives rise to a scattered wave $f_3\chi_3$. In these statements we have expressed implicitly the i-spin conservation, because, according to the statements, scattering does not change χ_1 into χ_3, or vice versa. The scattering amplitudes f_1 and f_3 are functions of the energy and of θ, the scattering angle.

In the scattering of positive pions on protons, p^+, only the i-spin $\frac{3}{2}$ state is involved, and hence the scattering amplitude is f_3. For the case of scattering of negative pions on protons, p^-, the initial state contains a mixture of χ_1^{-1} and χ_3^{-1}, as can be seen immediately.

Solving the equations in Table 15-2, we have the physical state

$$p^- = \sqrt{(\tfrac{1}{3})}\chi_3^{-1} - \sqrt{(\tfrac{2}{3})}\chi_1^{-1} \tag{15-5.4}$$

Similarly we have

$$n^0 = \sqrt{(\tfrac{2}{3})}\chi_3^{-1} + \sqrt{(\tfrac{1}{3})}\chi_1^{-1} \tag{15-5.5}$$

The p^- scattering is thus described by

$$p^- e^{ikz} = \left[\sqrt{(\tfrac{1}{3})}\chi_3^{-1} - \sqrt{(\tfrac{2}{3})}\chi_1^{-1}\right]e^{ikz} \rightarrow \sqrt{(\tfrac{1}{3})}f_3\chi_3^{-1} - \sqrt{(\tfrac{2}{3})}f_1\chi_1^{-1} \tag{15-5.6}$$

We now express χ_3^{-1}, χ_1^{-1} through physical states n^0, p^- and obtain

$$p^- e^{ikz} \rightarrow \sqrt{(\tfrac{1}{3})}\left[\sqrt{(\tfrac{2}{3})}n^0 + \sqrt{(\tfrac{1}{3})}p^-\right]f_3 + \sqrt{(\tfrac{2}{3})}\left[-\sqrt{(\tfrac{1}{3})}n^0 + \sqrt{(\tfrac{2}{3})}p^-\right]f_1$$

$$= n^0\sqrt{2}(f_3 - f_1)\tfrac{1}{3} + p^-(f_3 + 2f_1)\tfrac{1}{3} \tag{15-5.7}$$

We see that $(f_3 + 2f_1)/3$ is the amplitude for elastic scattering (p^- goes into p^-) and $\sqrt{2}(f_3 - f_1)/3$ is the amplitude for charge-exchange scattering (p^- goes into n^0).

The processes of Eq. (15-5.1) must have cross sections in the ratio of the squares of their amplitudes. It has been found that at energies up to about 200 MeV $f_3 \gg f_1$. Then if we neglect f_1 compared with f_3, we have, for the ratios of the cross sections, total as well as differential, $p^+ \Rightarrow p^+ : p^- \Rightarrow n^0 : p^- \Rightarrow p^- = 9 : 2 : 1$. In fact the measurement of these ratios led to the conclusion, theoretically predicted, that the important scattering amplitude for energies below 200 MeV corresponds to $T = \frac{3}{2}$ and verified in an impressive way i-spin conservation.

The measurements of cross section and angular distribution for the p^+, p^- scatterings [Eq. 15-5.1] show a resonance process (Fig. 15-7), and the question arises what the total angular momentum of the wave involved in the resonance is. Several arguments indicate that it is $J = \frac{3}{2}$. First, the maximum

ISBN 0-8053-8061-7

value of the scattering cross section for a state of given J and l is

$$\sigma_{\max} = \frac{4\pi \lambda^2 (2J + 1)}{(2s_1 + 1)(2s_2 + 1)}$$

(15-5.8)

where $2\pi\lambda$ is the de Broglie wavelength of the pion in the center-of-mass system and s_1, s_2 the spins of the colliding particles. This formula extends Eq. (11-5.12) to the case of particles with spin. See (BW 52). The value of $\sigma_{\max}$ for $J = \frac{1}{2}$ would definitely be smaller than the experimental cross section at 195

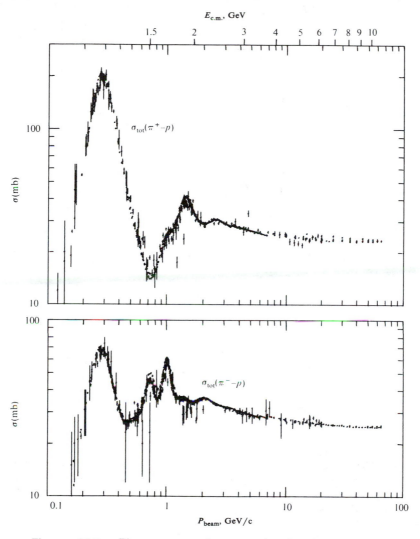

Figure 15-7a Pion proton total cross section data from RPP 1974.

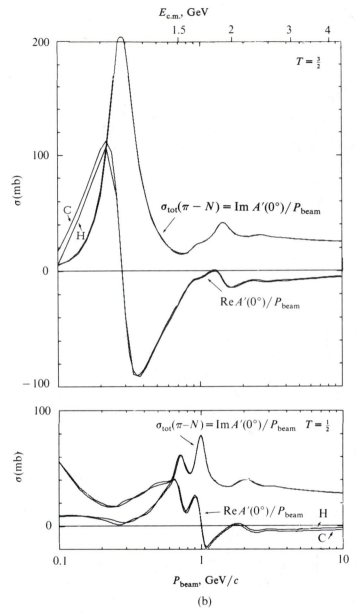

(b)

Figure   15-7b A smooth interpolation of the pion nucleon cross section for $T = \frac{3}{2}$ and $T = \frac{1}{2}$ and the real parts of the forward scattering amplitudes. The normalization of the curves for each value of T is such that the sum of their squares divided by 19.6 gives $d\sigma/dt$ at $0°$ in $mb/(GeV/c)^2$. From RPP 1974.

ISBN 0-8053-8061-7

MeV (laboratory), which is 195×10^{-27} cm². This value is very close to $8\pi \lambdabar^2$, which is the maximum obtainable, according to Eq. (15-5.8), for $J = \frac{3}{2}$. Moreover, the angular distribution in p^+ elastic scattering, $\sigma(\theta) \sim 1 + 3 \cos^2 \theta$ (Fig. 15-8), shows that the scattering occurs in a $J = \frac{3}{2}$ state, and in view of the energy this is a $p_{3/2}$ and not a $d_{3/2}$ state; this is further confirmed by data using polarization techniques.

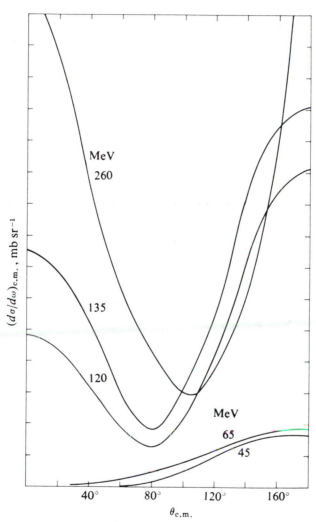

Figure   15-8   The angular distribution for the elastic π^+–proton scattering in the center-of-mass system. Differential cross section and scattering angle in the c.m. system. Curves are labeled according to the laboratory kinetic energy of the pion.

ISBN 0-8053-8061-7

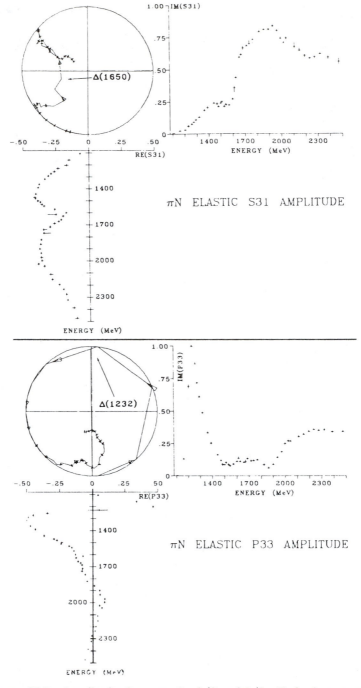

Figure 15-9 Amplitudes for some $I = 3/2$ and $1/2$ πN elastic scattering in the $J = 1/2$ and $J = 3/2$ waves. The energy dependence of each amplitude is displayed by plotting its real and imaginary parts vs. energy, in alignment with

772

ISBN 0-8053-8061-7

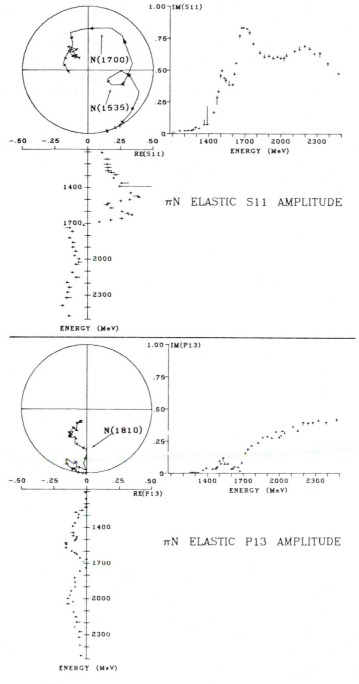

πN ELASTIC S11 AMPLITUDE

πN ELASTIC P13 AMPLITUDE

the corresponding Argand plot. In addition, arrows are plotted on the Argand plots with bases positioned at integer multiples of 50 MeV and a base-to-tip length of 5 MeV. All the energy axes run from elastic threshold to 2500 MeV.

ISBN 0-8053-8061-7

To see this, take as axis of quantization the direction of motion of a pion impinging on a proton at rest. The system then has $m_J = \frac{1}{2}$, $m_l = 0$, and if the orbital momentum l is 1, then J, the total angular momentum, is $\frac{1}{2}$ or $\frac{3}{2}$. Assume it is $\frac{3}{2}$; then by Clebsch–Gordan coefficients the wave function

$$\psi\left(J = \tfrac{3}{2}, \, m_J = \tfrac{1}{2}\right) = \sqrt{\left(\tfrac{1}{3}\right)}Y_{11}\beta + \sqrt{\left(\tfrac{2}{3}\right)}Y_{10}\alpha \qquad (15\text{-}5.9)$$

where α and β are the usual spin-up and spin-down wave functions. The angular distribution of the decay products is given by

$$I(\theta) = \psi\psi^* = \tfrac{1}{3}|Y_{11}|^2 + \tfrac{2}{3}|Y_{10}|^2 \qquad (15\text{-}5.10)$$

where we have used the orthogonality properties of α, β. Replacing the spherical harmonics by their expressions, we have

$$I(\theta) = \frac{3}{4\pi}\left(\tfrac{1}{6}\sin^2\theta + \tfrac{2}{3}\cos^2\theta\right) \qquad (15\text{-}5.11)$$

or $I(\theta)$ proportional to $1 + 3\cos^2\theta$. The reader may verify that $\psi(\frac{3}{2}, -\frac{1}{2})$ gives the same result and that $\psi(\frac{1}{2}, \pm\frac{1}{2})$ gives an isotropic distribution.

Now $J = l \pm \frac{1}{2}$ for the pion–nucleon system, and we conclude that the resonance corresponds to a state with $T = \frac{3}{2}$, $l = 1$, $J = \frac{3}{2}$. Another experimental argument confirming this assignment is the energy dependence, at low energies, of the total cross section on the relative momentum, $\sigma_t \sim p^4$, a dependence characteristic of p-wave scattering.

The total scattering cross section for the $T = \frac{3}{2}$ state may be described by a single-level resonance formula (Brueckner, 1952).

$$\sigma = 2\pi\lambdabar^2 \frac{\Gamma^2}{(E - E_0)^2 + \Gamma^2/4} \qquad (15\text{-}5.12)$$

where $E_0 = 1236$ MeV is the total c.m. energy and Γ is 110 MeV.

A more refined analysis of the nucleon–pion interaction involves the measurement and calculation of the phase shifts of the scattered partial waves. Great effort has been spent on this important task. At low energy only s and p waves are important, the p waves dominating, as mentioned above. The states are now classified according to T, J, and l. The usual notation for the phase shifts and scattering amplitudes is $2T$ and $2J$ as first and second subscripts. Figure 15-9 gives the plots in the complex Argand plane of the scattering amplitudes of some of the low partial waves. A resonance obtains when the phase of the scattering amplitude is $\pi/2$.

Note that phase shifts near threshold depend on the $(2l + 1)$th power of momentum. This dependence is required by theory if the interaction has a range small compared with the de Broglie wavelength of the pions.

ISBN 0-8053-8601-7

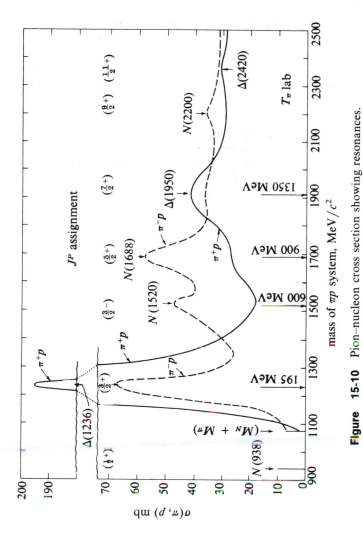

Figure 15-10 Pion–nucleon cross section showing resonances.

The $T = \frac{3}{2}, J = \frac{3}{2}$ is the most prominent and best-studied resonance of the pion–nucleon system. However, there are other resonances at higher energies. Their characteristic quantum numbers are derived from angular distributions and, if necessary, by the study of the polarization of the recoil nucleon, or by the use of polarized proton targets. The results available are summarized in (RPP 76) and in Fig. 15-10.

The phase shifts should be derivable from meson theory, provided that one knows the correct form of the meson–nucleon interaction and can solve the mathematical problem involved. Approximate forms of the interaction are known, and calculations of phase shifts based on them have been developed in close analogy with the effective range expressions in n–p scattering. For instance, for the δ_{33} phase shifts one has approximately

$$\frac{q^3 \cot \delta_{33}}{\omega^*} = \frac{3}{4f^2} (1 - \omega^* r_{33}) \tag{15-5.13}$$

where q and $\omega = \omega^* - (q^2/2M)$ (M of the proton) are the momentum and total energy of the pion in the center-of-mass system, in units of $m_\pi c$ and

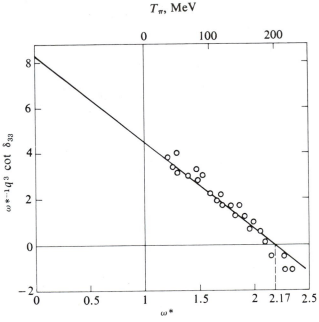

Figure 15-11   Effective range plot for the 3, 3 phase shift. $(q^3 \cot \delta_{33})/\omega^*$ is plotted as a function of $\omega^* = \omega + q^2/2M$. The intercept at $\omega^* = 0$ gives the coupling constant f^2. [From S. W. Barnes, B. Rose, G. Giacomelli, J. Ring, K. Miyake, and K. Kinsey, *Phys. Rev.*, **117**, 226 (1960).]

ISBN 0-8053-8061-7

$m_\pi c^2$; r_{33} is an "effective range" for the $N-\pi$ interaction in the $T = \frac{3}{2}$, $J = \frac{3}{2}$ state, in units of $\hbar / m_\pi c$, and f is the fundamental pion–nucleon coupling constant (Fig. 15-11).

From an experimental plot of this formula one derives f^2 and r_{33}. The data give $f^2 = 0.080 \pm 0.002$; $r_{33} = 0.65$ F.

The same value of the constant f^2 is obtained from several other phenomena (e.g., other phase shifts, pion photoproduction, and nucleon–nucleon scattering).

Note that dimensionally $[f^2] = [e^2]$, where e is the electric charge. In the same units, $e^2 = (1/4\pi)(1/'137')$.

15-6 NUCLEAR-COLLISION PRODUCTION AND PHOTOPRODUCTION OF PIONS

Pions may be produced by nucleon–nucleon collisions. The simplest case is when only one pion is produced according to the reaction

$$N + N \Rightarrow 2N + \pi \tag{15-6.1}$$

This reaction can assume many forms,

$$p + p \Rightarrow p + p + \pi^0$$
$$\Rightarrow p + n + \pi^+ \tag{15-6.2}$$

$$n + p \Rightarrow p + p + \pi^-$$
$$\Rightarrow n + p + \pi^0$$
$$\Rightarrow n + n + \pi^+ \tag{15-6.3}$$

etc. Moreover, the neutron and proton at the right may escape as a bound deuteron. The conservation of i-spin allows us to express all the cross sections of Eq. (15-6.2 and 15-6.3) as a function of only three of them (Rosenfeld, 1954). The experimental verification of these relations is another confirmation of the conservation of i-spin.

As regards pion photoproduction, processes such as

$$\gamma + p \rightsquigarrow p + \pi^0$$
$$\gamma + p \rightsquigarrow n + \pi^+ \tag{15-6.4}$$

have been observed.

The source of gamma rays is generally the bremsstrahlung of electrons accelerated by machines. The spectrum is thus continuous. However, determination of the momentum vector of the pion or nucleon in one of the reactions of Eq. (15-6.4) is sufficient to reconstruct completely the kinematics of the process and thus to determine a unique gamma-ray energy.

ISBN 0-8053-8061-7

TABLE 15-3   CLASSIFICATION OF THE PHOTOPRODUCTION OF PIONS

Type of radiation absorbed	Parity of final state	l of pion	J of final state	Momentum dependence of σ
M1	+	1	$\frac{1}{2}$	q^3
M1	+	1	$\frac{3}{2}$	q^3
E1	−	0	$\frac{1}{2}$	q
E1	−	2	$\frac{3}{2}$	q^5
E2	+	1	$\frac{3}{2}$	q^3
E2	+	3	$\frac{5}{2}$	q^7

Near the threshold only the lowest angular momentum states are involved, and we can analyze the reaction according to Table 15-3 (we must remember that the nucleon has $J = \frac{1}{2}$ and even parity and the pion has odd intrinsic parity). The momentum dependence of the cross section at threshold is q^{2l+1}, and hence at low energy $E1$ is the dominating process, when it occurs. When it does occur, photoproduction depends on the electric dipole moment of the nucleon–pion system, and we see from Fig. 15-12 that for the same distance the dipole moments for a nucleon and a pion are in the ratios

$$\text{dipole } 1 + \frac{m_\pi}{M} \; : \; 1 \; : \; \frac{m_\pi}{M} \; : \; 0$$

$$\text{for } \pi^- p; \quad \pi^\pm n; \quad \pi^0 p; \quad \pi^0 n$$

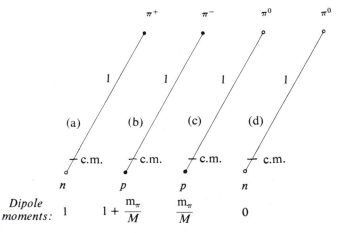

Figure 15-12 Electric dipole moments for a nucleon–pion system: (a) $(\pi^\pm, n)$, (b) (π^-, p), (c) (π^0, p), (d) (π^0, n). The moments are in the ratios $1 : 1 + (m_\pi/M) : m_\pi/M : 0$.

ISBN 0-8053-8061-7

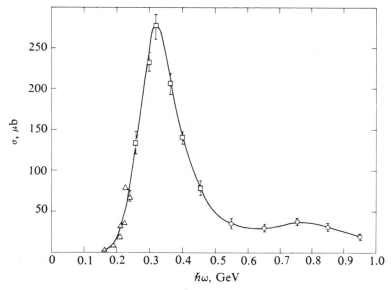

Figure   15-13   Reaction $\gamma + p \leadsto \pi^0 + p$; total cross section as a function of energy.

Near threshold we expect the cross section for photoproduction to be in the same ratio as the square of the dipole moments, and this is approximately verified.

The $\pi^- + p \leadsto n + \gamma$ process, the inverse of photoproduction, is observed by capturing negative pions in hydrogen; it then competes with the $\pi^- + p \Rightarrow \pi^0 + n$ process. The *Panofsky ratio* of the second to the first is 1.55 ± 0.01. At first sight this number seems small, because the numerator contains a strong interaction and the denominator an electromagnetic one, and the matrix elements are expected to be in a ratio of 137. Taking into account the phase space, we could expect a ratio of about 6. The experimental result shows that the $n-\pi^0$ final state is inhibited. This state has $J = \frac{1}{2}$ and negative parity because the captured π^- is in an s state; the electromagnetic radiation emitted is $E1$ and the nucleus must flip its spin. We conclude that the strong interaction in an s state is ineffective and this is borne out by the s-wave shifts found in scattering experiments. (See Fig. 15-18.)

The resonance of the $\Delta(1235)$ state so prominent in pion nucleon scattering appears also in photoproduction (Fig. 15-13).

15-7   THE ρ, ω, AND OTHER STRONGLY DECAYING BOSONS

Pions are the lightest of an extended family of strongly interacting bosons. The next are the K, η, ρ, and ω, in order of increasing mass. Kaons are strange and thus cannot decay by strong interaction into pions; they will be

ISBN 0-8053-8061-7

treated in Sec. 19-5. The η decays only electromagnetically into pions or gammas and thus is relatively stable. Other bosons are extremely short-lived because they decay by strong interaction.

Production

One observes these strongly decaying bosons in reactions in which they are produced (*production*) by measuring the invariant mass of the reaction products. An example will explain the method. In the reaction

$$\pi^+ + p \Rightarrow \pi^+ + \pi^0 + p \tag{15-7.1}$$

at fixed energy, plot the frequency of occurrence of the invariant mass of $\pi^+\pi^0$, that is, the quantity $(E_1 + E_2)^2 - (\mathbf{p}_1 + \mathbf{p}_2)^2 = M^2$ of the two pions, versus M. The diagram will look like that in Fig. 15-14a, with a continuous background but a peak at 770 MeV. This energy corresponds to the mass of the "resonance" or particle. The reaction may be thought of as proceeding in two steps. The first is the production of the proton and a particle R (resonance); in the second, the particle R decays into two pions. The particle R lasts a long time on a nuclear scale and when it decays it may be at a distance large compared with nuclear dimensions from the spot at which it was formed. The energy or mass of R is indicated by the maximum of the spike in Fig. 15-14a and its mean life by the width of the spike according to the uncertainty relation. The production method is not limited to particles decaying into two pions. For instance, at higher energy the $\pi^+ - p$ collision may generate three pions; the invariant mass of these three pions is shown in Fig. 15-14c. A resonance at 784 MeV is evident and it corresponds to the ω meson.

The ω just mentioned was discovered by Maglic, Alvarez, Rosenfeld, and Stevenson (1961) by production in a different context. They examined a great number of proton–antiproton annihilation stars, corresponding to five pions. For each star they combined the pions three at a time and measured the invariant mass of each triplet. This mass was found to vary between a minimum of $3m_\pi$ and a maximum, for annihilation at rest, of $2(m_p - m_\pi)$. The probability of finding an intermediate value contains a continuously varying phase-space factor, but the experiment (Fig. 15-15) showed a characteristic spike corresponding to the mass of the ω.

Formation

The other system for finding resonances, called *formation*, requires observing the cross section for a given reaction as a function of the energy of the incident particle. For instance, in the reaction

$$\pi^+ + p \Rightarrow \pi^+ + p \tag{15-7.2}$$

there is a maximum at 1230 MeV; its width as indicated in Fig. 15-7, is 115 MeV, and this corresponds to the "formation" of the particle. An interesting

ISBN 0-8053-8061-7

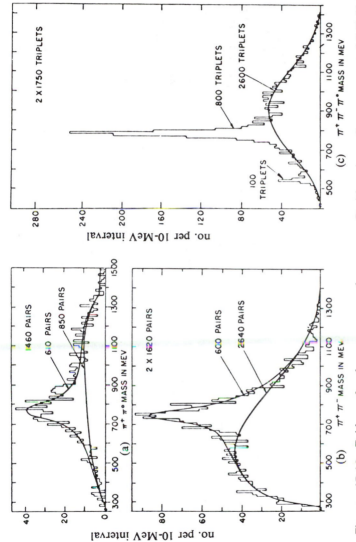

Figure 15-14 Evidence for the ρ mesons and ω mesons. Positive pions of momentum ∼2.9 GeV/*c* impinge on protons, generating two, three, or four pions. The pions produced are often associated as ρ, η, or ω mesons. This is shown in (a), where the mass distribution (in their own center of mass) for $\pi^+\pi^0$ pairs from $\pi^+ + p \Rightarrow \pi^+ + \pi^0 + p$ is given. The peak corresponds to ρ^+. In (b) the mass of $\pi^+\pi^-$ from the reaction $\pi^+ + p \Rightarrow \pi^+ + \pi^+ + \pi^- + p$ is plotted. The peak corresponds to ρ^0. In (c) the mass of $\pi^+\pi^0\pi^-$ triplets originating from $\pi^+ + p \Rightarrow \pi^+ + p + \pi^+ + \pi^- + \pi^0$ is given. The peaks correspond to η and ω. The smooth curves correspond to phase space. [From C. Alff et al. (Columbia and Rutgers Universities), *Phys. Rev. Letters,* **9**, 322 (1962).]

ISBN 0-8053-8061-7

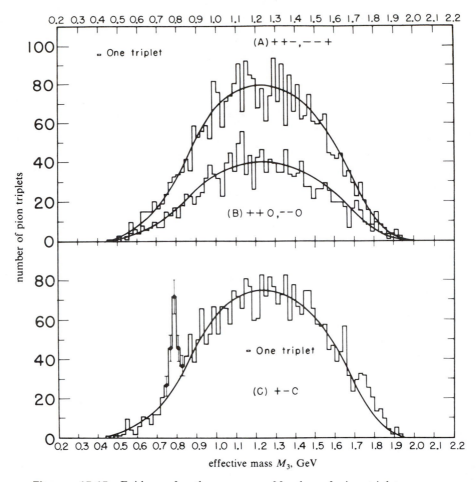

Figure 15-15 Evidence for the ω meson. Number of pion triplets versus effective mass (M_3) of the triplets for reaction $\bar{p} + p \Rightarrow 2\pi^+ + 2\pi^- + \pi^0$. ($A$) is the distribution for the combinations of charge 1; (B) for the combinations of charge 2; and (C) for the neutral combinations. [From B. Maglich, L. Alvarez, A. Rosenfeld, and L. Stevenson, *Phys. Rev. Letters*, **7**, 178 (1961).]

example of a formation experiment is the reaction

$$e^+ + e^- \rightsquigarrow \rho^0 \Rightarrow \pi^+ + \pi^- \qquad (15\text{-}7.3)$$

obtainable with colliding beams. (See Fig. 18-8.) This ρ is of course the same as the one found in reaction Eq. (15-7.5) and in $p\bar{p}$ annihilation.

Resonances have definite quantum numbers: T, J, parity, etc.; and these may sometimes be found by relatively simple observations. Let us consider

ISBN 0-8053-8061-7

TABLE 15-4   QUANTUM NUMBERS FOR TWO-PION SYSTEMS

System	Q	l	$P = (-1)^l = C$	T	G
(π^+, π^-)	0	Even	1	0 or 2	+
	0	Odd	-1	1	+
(π^0, π^0)	0	Even	1	0 or 2	+
$(\pi^\pm, \pi^0)$	± 1	Even	1	2	+
	± 1	Odd	-1	1	+
$(\pi^\pm, \pi^\pm)$	± 2	Even	1	2	+

bosons decaying into two or three pions. The possible quantum numbers for two-pion decay are given in Table 15-4. The reader is invited to prove the statements of the table using the material in Sec. 13-4 and the fact that pions are bosons. If the decay of the resonance is by strong interaction, the resonance must have the same angular momentum, parity, and i-spin as the two-pion system. If the decay is by weak interaction, only angular momentum is certainly conserved. The width of the resonance indicates unambiguously whether the decay is by strong interaction.

As an example, in the reactions

$$\pi^+ + p \Rightarrow \rho^+ + p$$
$$\hookrightarrow \pi^0 + \pi^+ \tag{15-7.4}$$

$$\pi^- + p \Rightarrow \rho^0 + n$$
$$\hookrightarrow \pi^+ + \pi^- \tag{15-7.5}$$

one forms a resonance (ρ meson) (Fig. 15-14a), decaying into two pions of 770-MeV invariant mass and with a width of 152 MeV. No resonance decaying into $\pi^+ \pi^+ n$ or $\pi^0 \pi^0 n$ is observed. This alone establishes that the resonance decays by strong interaction and has $T = 1$ with L odd. To find the angular momentum $J = L$, note that the two pions must have an eigenfunction symmetric with respect to their exchange (including i-spin). The i-spin part of the eigenfunction is antisymmetric because $T = 1$, thus L must be odd, as given in Table 15-4. The lowest value is 1 and this is proved correct by the angular distribution of the line of flight of the two pions with respect to the line of flight of their center of mass. Choose this second line as the axis of quantization; the angular momentum then has $L_z = 0$ and the two decay pions must have in their eigenfunction the factor Y_{L0} which produces an angular distribution $|Y_{L0}(\theta, \varphi)|^2$ or, for $L = 1$, $\cos^2 \theta$. The actual observation gives a distribution $A + B \cos \theta + C \cos^2 \theta$. This is explained by the interference of a spherically symmetric background amplitude with the resonance, but in any case the absence of higher powers of $\cos \theta$ suffices to assign $J = 1$ and negative parity. In conclusion, for ρ, $T^G(J^P) = 1^+(1^-)$.

ISBN 0-8053-8061-7

For the resonance at 1264 MeV (f meson), one has not observed two positive pions but only (π^+, π^-) or (π^0, π^0). This shows that f has $T = 0$ and hence L even. The angular distribution of the decay pions indicates that $J = 2$. In conclusion, for f, $T^G(J^P)$ is $0^+(2^+)$.

For a three-pion system the assignment of quantum numbers is more complicated. The possible i-spin values are 3, 2, 1, 0. If the resonance is observed only when the decaying system has charge zero, although there is no constraint forcing this condition by charge conservation alone, the i-spin is 0. This is the case of the ω (see Fig. 15-15). Similarly, if only 0, ± 1 charges appear, the i-spin is 1, and so forth.

For the determination of angular momentum and parity one uses Dalitz plots, described in the next section.

15-8   DALITZ PLOTS

In 1953 Dalitz and Fabri devised a very ingenious method for obtaining information on the spin and parity of a particle decaying into three particles. If the decay is by strong interaction, as in the ω, their method will give information on both spin and parity; if the decay occurs by weak interaction, as in the K^+, the information is limited to the spin.

Consider, for example, a K^+ meson at rest decaying into one negative and two positive pions. A statistical study of many such decays in which one measures the energies of the pions and the angles between their trajectories leads to several important conclusions.

The three particles emitted may have any momentum and energy permitted by the conservation laws. The largest energy values are found when the decay occurs with all the pions traveling on the same line of flight. It is easy to see, for instance, that the maximum energy the negative pion may have is $\frac{2}{3}Q = E_0$ where $Q = (m_K - 3m_\pi)c^2$. It is also obvious that $E_1 + E_2 + E_3 = m_K c^2$ in the system in which the K is at rest.

The decay is governed by golden rule No. 2

$$w = \frac{2\pi}{\hbar} |M_{if}|^2 \, \delta(P_i - P_f) \, \frac{dN}{dW} \qquad (15\text{-}8.1)$$

where the matrix element may be a function of the momenta of the decay pions, the four-momenta delta function ensures energy and momentum conservation, and the last factor, density of final states, may be calculated separately. This last factor is proportional to $dE_1 \, dE_2$ (E_3 is determined once E_1 and E_2 are given) for all the accessible regions of phase space. If one thus plots the events using E_1 and E_2 as coordinates, one obtains a diagram (a Dalitz plot) in which the accessible region is contained in a closed curve and the points representing the events are uniformly distributed within the curve if the matrix element is constant. Any departure from uniformity is to be

ISBN 0-8053-8061-7

attributed to the dependence of the matrix element on the momenta of the pions.

In the special case of three pions, the equality of the masses makes it expedient to use a triangular plot. The distances of a representative point from the three sides of the equilateral triangle are chosen proportional to the kinetic energy of the pions and the distance from the base is attributed to the unequal pion. The sum of these three distances is constant, in agreement with the conservation of energy. Nonrelativistically the conservation of momentum restricts the representative points to the circle inscribed in the triangle (see Prob. 15-10); in the extreme-relativistic case they are restricted to an inscribed triangle. We put the origin of a coordinate system in the center of the triangle, and plot $(T_-/Q) - \frac{1}{3}$ on the y axis, and $(T'_+ - T_+)/\sqrt{3}\,Q$ on the x axis. With these coordinates $dE_1\,dE_2$ is proportional to $dx\,dy$; thus for a constant matrix element the representative points have uniform density in the occupied region. In the nonrelativistic case the area between $E = T_-$ and $E + dE$ is then

$$E^{1/2}(E_0 - E)^{1/2}\,dE \tag{15-8.2}$$

as can be seen from the geometry of Fig. 15-16, and this area is proportional to the probability of finding the π^- in that energy interval.

In order to analyze the departures from uniformity and gain information on the spin and parity of the three-pion system, we study the decay in the center-of-mass system of two equally charged pions; the three momenta of the pions appear as in Fig. 15-17. The pions have negative intrinsic parity and no spin. The system of the two identical pions, these being bosons, must have

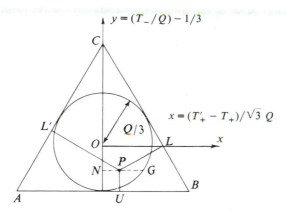

$y = (T_-/Q) - 1/3$

$x = (T'_+ - T_+)/\sqrt{3}\,Q$

Figure 15-16 Dalitz plot. PL', PL, kinetic energies of $\pi^+ = T'_+$, T_+ (T_+ has the lower energy). PU, kinetic energy of π^-. Circle contains zone permitted by energy–momentum conservation. The coordinates of a representative point referred to the axes marked in the figure are $x = (T'_+ - T_+)/\sqrt{3}\,Q$ and $y = (T_-/Q) - \frac{1}{3}$.

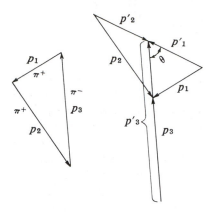

Figure 15-17 Momentum vectors and definitions of l, L and θ. The laboratory momenta p_1, p_2, p_3 belong to π^+, π^+, π^-; p'_1, p'_2, p'_3 are the momenta of π^+, π^+, π^- in a system in which the center of mass of π^+, π^+ is at rest. p'_1, p'_2 give angular momentum L; p'_3 gives angular momentum l.

an even angular momentum L. Assume that the third (negative) pion has an angular momentum l with respect to the center of mass of the identical pions. The total angular momentum of the system J is limited by

$$L + l \geqslant J \geqslant |L - l|$$

Its intrinsic parity is $(-1)^{l+1}$, because the two identical pions necessarily have positive parity; and the odd pion, with negative intrinsic parity, determines the parity of the system.

If we limit ourselves to the smallest possible values of L and l, we have the possibilities given in Table 15-5 of realizing a given J and parity, with given L and l. Note that a 0^+ state is impossible because it would require $L = l$ and then the parity would be odd.

TABLE 15-5   DECOMPOSITION OF THE TOTAL ANGULAR MOMENTUM IN THREE-BODY DECAY

J^P	L	l
0^-	0	0
1^+	0	1
1^-	2	2
2^+	2	1

By not too difficult methods based on symmetry and group theory one obtains the general results shown in Fig. 15-18, but even simpler methods yield considerable information provided the linear dimensions of the region

ISBN 0-8053-8061-7

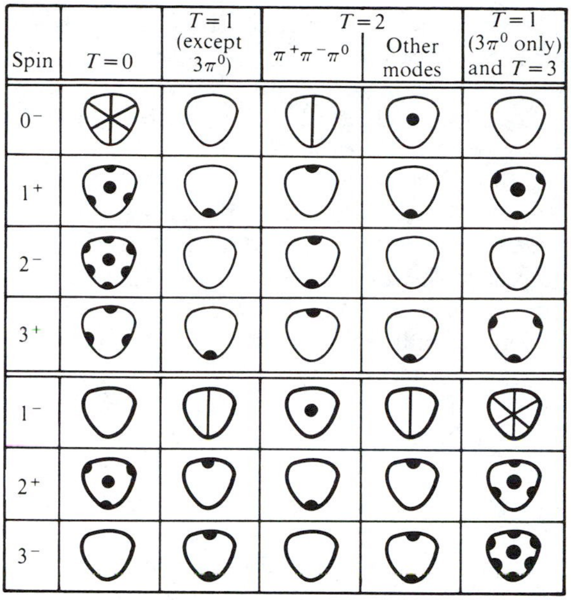

Figure 15-18 Regions of the three-pion Dalitz plot where the density must vanish because of symmetry requirements are shown in black. The vanishing is of higher order (stronger) where black lines and dots overlap. In each isospin and parity state, the pattern for a spin of J and $J + 2n$ are identical, provided $J \geqslant 2$. (Exception: vanishing at the center is not required for $J \geqslant 4$.) [From C. Zemach, *Phys. Rev.*, **133B**, 1201 (1964).]

in which the interaction takes place are small compared with the $\lambdabar$ of the escaping pions. Then the eigenfunctions of the latter, near the interaction region, vary as $p_{\pi^+}^L$, $p_{\pi^-}^l$, where the p_{π^+}, p_{π^-} are the momenta of the pions. The transition probability then contains factors $p_{\pi^-}^{2l}$, $p_{\pi^+}^{2L}$ from the matrix element. In particular, for a particle having $J = 0$, l and L are zero, and the matrix element is constant, and the transition probability depends on E only through the density of final states.

For other J^P, the probability depends on E and on θ, as indicated in Table 15-6.

TABLE 15-6   PROBABILITY DISTRIBUTION AS A FUNCTION OF ENERGY AND ANGLE FOR THREE-BODY DECAY

Spin parity	$w(E, \theta)$
0^-	$E^{1/2}(E_0 - E)^{1/2}$
1^+	$E^{3/2}(E_0 - E)^{1/2}$
1^-	$E^{5/2}(E_0 - E)^{5/2} \sin^2 \theta \cos^2 \theta$
2^+	$E^{3/2}(E_0 - E)^{5/2} \sin^2 \theta$

ISBN 0-8053-8061-7

Experimental realization of this analysis is obtained by plotting the triangular Dalitz plot. Note that in this plot $\cos \theta = PN/NG$ of Fig. 15-16. Furthermore, the plot necessarily has three symmetry axes, the three heights of the triangle; thus all representative points may be plotted in a sextant only. Differences between sextants imply violations of charge conjugation that are not expected for strong interactions.

It is sometimes possible to understand the behavior of a Dalitz plot qualitatively, without having to resort to calculations. For instance, the probability of emission of a pion with vanishing kinetic energy goes to 0 for a system of odd spin–odd parity or even spin–even parity. By this we mean that the matrix element vanishes independently of the phase-space factor. In fact, the pion of small momentum must be produced in an s state of odd parity, and then the two other pions must have the same angular momentum as, but opposite parity to that of the initial system. For instance, if the initial system is 1^- or 2^+, the two pions should be 1^+ or 2^-, respectively. But these states are not possible for two pions, and hence the transition is forbidden for low-energy pions.

As an example, Figs. 15-19 and 15-20 give the density of points expected for particles having $J^P = 0^-$ or 1^- and the experimental result based on the observation of the ω meson (see Sec. 15-8) showing that it is a 1^- meson.

In these considerations we have used a nonrelativistic approach for the sake of simplicity. Relativistic corrections are small if the energies involved are small compared with the rest mass of the pions.

When the decay products have different masses, the triangular plot is not suitable and one uses other plots that give similar information. For example, consider the reaction

$$K^- + p \rightarrow (1) + (2) + (3) \tag{15-8.3}$$

Call E the total energy in the c.m. system and E_1, E_2, E_3 the energies of the three particles (1), (2), (3) formed. Plot, on rectangular coordinates, E_1 against E_2. Lines of constant E_3 are inclined at 45 deg to the axes because $E_1 + E_2 = E - E_3$. On a line $E_1 = $ constant particles (2) and (3) are constrained to have, in their own center-of-mass system, a given rest mass and a given momentum $|\mathbf{p}|$. The only free variable is the angle between $\mathbf{p}$ and the direction of flight of the center of mass of particles (2) and (3).

The region in the E_1–E_2 plane containing possible representative points for a given E is limited by a curve. On its boundary the three particles move on the same straight line. Pure phase space gives a uniform density within the boundary. If two of the three particles, for instance (2) and (3), tend to emerge in a state (resonance) of fixed energy in their center of mass, the representative points line up on lines of constant E_1.

Often one uses as coordinates the square of the masses (in their center-of-mass system) of couples of particles. For instance, in the example of Eq.

ISBN 0-8053-8061-7

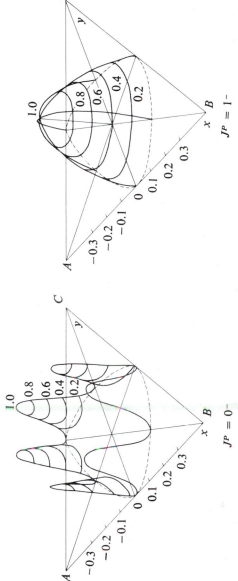

Figure 15-19 Density of points expected in a Dalitz plot for a $J^P = 0^-$ or 1^- system decaying into three pions.

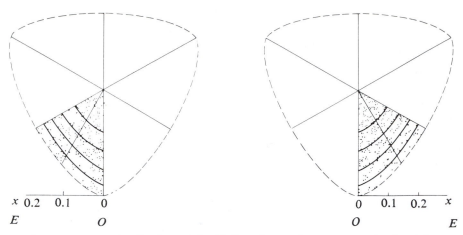

Figure 15-20 Application of the Dalitz plot to the ω meson. At the right only points at the resonant energy have been used, at the left the background. The figures correspond to those in Fig. 15-19, case 1^- seen from above. Line OE corresponds to AB.

(15-8.3) the square of the mass $M^2(2, 3)$ and $M^2(1, 3)$ may be used instead of E_1 and E_2. The two sets of variables are linearly related because, as in our example, one calculates that

$$M^2(2, 3) = E^2 + M^2(1) - 2EE_1 \quad (c = 1) \qquad (15\text{-}8.4)$$

which for E constant shows the linear relation between E_1 and $M^2(2, 3)$. Figure 15-21 is an example of this type of plot for the reaction

$$K^+ + p \Rightarrow K^0 + p + \pi^+$$

For each event measure the invariant mass (i.e., the mass in their c.m. system) of the π^+-p system and the $K^0-\pi^+$ system, plotting one against the other. We obtain the graphs shown in Fig. 15-21. There are clearly two preferred energies for the two systems, one around 1235 MeV, corresponding to the production of the $\Delta(1235)$ resonance Δ^{++}, and one corresponding to the K^{*+} at 892 MeV, a meson of short life.

15-9   THE η AND K MESONS

These two bosons cannot decay by strong interaction and thus are semi-stable. Both have among their modes of decay three pions.

The η can be generated in several ways. It was discovered in the reaction

$$\pi^+ + d \Rightarrow p + p + \pi^+ + \pi^- + \pi^0$$

ISBN 0-8053-8061-7

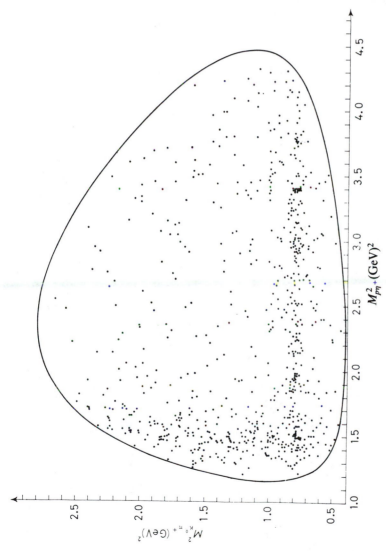

Figure 15-21 A Dalitz plot for the $K^+p \Rightarrow K^0p\pi^+$ events, $P_{K^+} = 3.0$ GeV/c, 747 events. [M. Ferro Luzzi et al., *Nuovo Cimento* **36**, 1101 (1965).]

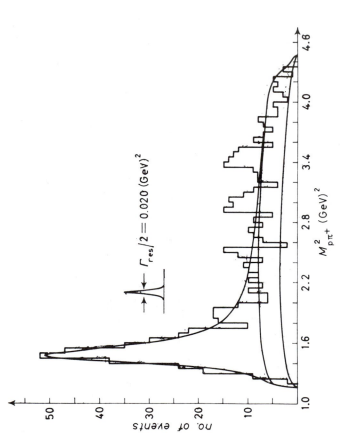

Figure 15-21a The projection of the Dalitz plot in Fig. 15-21 onto the $M^2_{p\pi^+}$ axis. The top curve represents the best fit to the data. The bottom curve is the contribution of the general background. The middle curve represents all events other than the Δ^{++} events.

ISBN 0-8053-8061-7

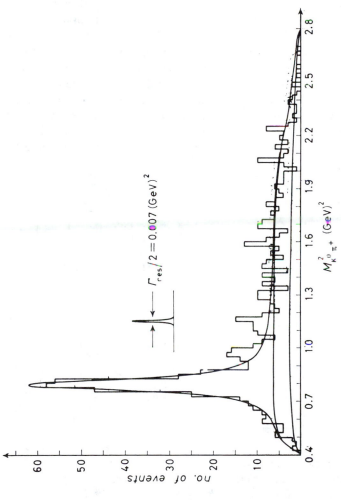

Figure 15-21b Projection of the Dalitz plot of Fig. 15-21 onto the $M^2_{K^0\eta^+}$ axis. The curves are as defined in Fig. 15-21a, except that the middle curve excludes the K^{*+} events.

which showed a peak for the three pions at 549 MeV (Pevsner et al., 1961). Other methods of production are

$$K^- + p \Rightarrow \Lambda + \eta \qquad (15\text{-}9.1)$$

(Bastien et al., 1962) and

$$\gamma + p \rightsquigarrow p + \eta \qquad (15\text{-}9.2)$$

(Mencuccini et al., 1962). In both cases the three pions π^+, π^-, π^0 or $3\pi^0$ show a resonance at 549 MeV. This resonance appears only for the neutral system of three pions. Attempts to produce a charged η by the reaction $\pi^+ + n \Rightarrow n + \eta^+$ were unsuccessful, whereas the reaction $\pi^+ + n \Rightarrow p + \eta$ has an appreciable yield. We conclude that η has $T = 0$. The decay into pions only, means that η is a boson and hence of integral spin. Looking at (RPP 74), we find that the η also has a decay into 2γ. Here the final state has $C = +1$ (see Sec. 13-4) and this information together with $T = 0$ implies that $G = +1$. On the other hand, the final state of the three-pion decay obviously has $G = -1$. We reconcile the two facts by observing that electromagnetic decay conserves neither G nor T. Thus even if the initial state, the η, is an eigenstate of T with $T = 0$, the final three-pion state is not necessarily an eigenstate of T. This explains the departure from the sixfold symmetry of the triangular Dalitz plot, easily observed in Fig. 15-22. To assign J^P to the η, we note that, of the

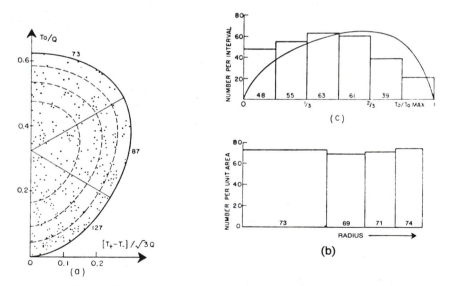

Figure 15-22 Dalitz plot for a sample of the $\eta \rightarrow \pi^+ + \pi^- + \pi^0$ decay. (a) Distribution of points; (b) radial density; (c) projection of the points on the π^0 axis. [From C. Alff *et al., Phys. Rev. Letters,* **9**, 325 (1962).]

ISBN 0-8053-8061-7

possible simple states, 0^+ would not allow the observed three-pion decay, and 1^+ or 1^- are incompatible with the 2γ decay. We thus conclude that for the $\eta T^G(J^P)$ are $0^+(0^-)$.

Other important bosons are the $K^\pm$, K^0, $\overline{K^0}$. They are all strange and thus cannot, by strong interaction, decay into pions only. Typical modes of production (associated) are

$$\pi^- + p \Rightarrow K^+ + \Sigma^-$$
$$\pi^- + n \Rightarrow K^0 + K^- + n \tag{15-9.3}$$

$$\pi^- + p \Rightarrow K^0 + \Sigma^0 \text{ or } \Lambda^0$$
$$\pi^+ + n \Rightarrow \overline{K^0} + K^+ + n \tag{15-9.4}$$

From these modes of production, since Λ has $T = 0$ and the nucleon has $T = \frac{1}{2}$, it follows that K must have $T = \frac{1}{2}$ and the formula in Eq. (13-2.6) gives strangeness $+1$ to K^+, K^0 and -1 to K^-, $\overline{K^0}$. This assignment is confirmed by comparing the reactions

$$K^- + d \Rightarrow \Lambda + n + \pi^0$$
$$\Rightarrow \Lambda + p + \pi^- \tag{15-9.5}$$

The i-spin at the left is $\frac{1}{2}$; at the right the eigenstate of $T = \frac{1}{2}$, $T_3 = -\frac{1}{2}$ (see Table 15-2) is $3^{-1/2}(\Lambda, n)\pi^0 - (\frac{2}{3})^{1/2}(\Lambda, p)\pi^-$; thus the branching ratio of the reactions is $1 : 2$, in agreement with experiment. The decay of the K into pions indicates a boson and the absence of any structure in K-mesic spectra and the Dalitz plot suggest spin 0. This value of J is consistent with the absence of $K^\pm \to \pi^\pm + \gamma$, with the presence of the $K^0 \to 2\pi^0$ decay, and with the similarity of the polarization in $K^+ \to \mu^+ + \nu_\mu$ and $\pi^+ \to \mu^+ + \nu_\mu$ decay. We conclude that the J of the K is zero. For the parity, as pointed out in Sec. 15-3, we know only the (K, Λ) parity and if we assume by convention a positive parity for Λ, then the parity of K is negative.

The decay of K mesons by weak interactions has a very interesting and complex phenomenology. There are decays into pions alone, pions and leptons, and leptons only. The particle tables (RPP 74) give branching ratios. In particular, the decay of neutral K exhibits a small violation of CP. This will be treated in Chap. 19.

15-10   PERIPHERAL COLLISIONS

Formation reactions often involve so-called peripheral collisions, similar to the direct reactions of Chap. 11. Consider the reaction

$$\pi^- + p \Rightarrow \pi^0 + \pi^- + p \tag{15-10.1}$$

in the limiting case in which the final nucleon receives a very small momentum. We may interpret this reaction as a collision of the initial π^- with a meson in the mesonic cloud around the proton, with the latter acting as a "spectator." If the spectator receives a very small momentum, the collision almost corresponds to that between free pions. Such collisions are said to be peripheral. The intuitive, qualitative argument given here can be made precise (Goebel; Chew and Low, 1959) and quantitative. Studies of reactions of the type of Eq. (15-10.1) gave the first experimental indications of resonances in the pion–pion system. (Anderson, Bang, Burke, Carmony, and Schmitz, at Berkeley). The method of peripheral collision was used to establish the ρ meson (Erwin, March, Walker, and West, at Wisconsin, 1961; Stonehill, Baltay, Courant, Fickinger, Fowler, Kraybill, Sandweiss, Sanford, and Taft, at Yale, 1961) in the reactions

$$\pi^- + p \Rightarrow \rho^- + p$$
$$\;\;\llcorner_{\rightarrow\; \pi^- + \pi^0} \tag{15-10.2}$$

$$\pi^- + p \Rightarrow \rho^0 + n$$
$$\;\llcorner_{\rightarrow\; \pi^+ + \pi^-} \tag{15-10.3}$$

The peripheral character of the $\pi^- + p \Rightarrow \rho^0 + n$ and $\pi^- + p \Rightarrow \rho^- + p$ reactions can be verified by measuring the angle φ between the plane of the incident π^- and the scattered neutron or proton and that of the two pions forming the ρ. An isotropic distribution of φ (Treiman–Yang angle, 1962) is evidence that the proton is only a spectator and that the collision is peripheral. In this case the ratio between the yield of ρ^0 and ρ^- can be calculated (see Prob. 15-9) and is predicted to be 1 : 2, in agreement with experiment.

The peripheral collision $\pi^- + p \Rightarrow \rho^0 + n$ may be examined in a reference system in which the line of flight of the incident pion is the z axis and the π^- collides head-on with a pion of the cloud surrounding the nucleon. The system is such that the two pions have equal and opposite velocity. If in this system the momentum of the initial proton and the final neutron is small, the collision is peripheral, and when the energy of the two pions equals the mass of the ρ, the cross section is especially large, decreasing on increasing momentum transfer. This is demonstrated in Fig. 15-23. Call θ the angle between the initial and final lines of flight of the negative pion, in the system described above. The object formed, the ρ in our case, has $J_z = 0$ and it decays, giving rise to an angular distribution $|Y_{10}(\theta)|^2$. It is thus possible to find the angular momentum of the resonance. When applied to the ρ, the method shows $J = 1$.

The simple qualitative idea of peripheral collision has been elaborated by Chew and Low (1959) into a quantitative formulation. Call t the square of the momentum transfer to the nucleon and call ω the center-of-mass energy of the two escaping pions. If the collision is extremely peripheral, $t \ll m_\pi^2$ and

ISBN 0-8053-8061-7

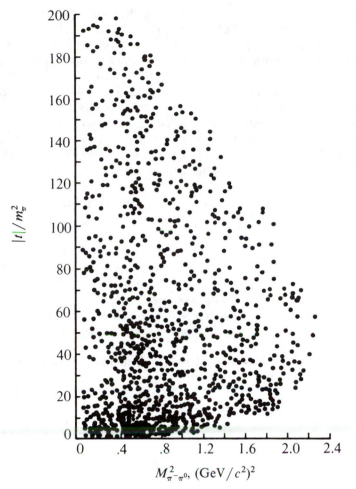

Figure   15-23   Chew-Low plot of four-momentum transfer squared $(-t)$ to the $\pi^-\pi^0$ system versus $M^2_{\pi^-\pi^0}$ from the final state $\pi^-\pi^0 p$. A Chew–Low plot of events in the reaction $\pi^- + p \Rightarrow \pi^0 + \pi^- + p$. The ordinate is the square of the momentum transferred to the proton and M^2 is the square of the mass of the $\pi^- - \pi^0$ system. Peripheral collisions require t small and the graph shows the peripheral production of ρ of $M^2 = 0.585$ (GeV)^2. [From D. H. Miller et al., *Phys. Rev.* **153**, 1423 (1967).]

one has the formula

$$\frac{d^2\sigma}{dt\,d\omega^2} = \frac{f^2}{2\pi m_\pi^2} \frac{t}{\left(t - m_\pi^2\right)^2} \frac{\omega}{P^2} \left(\frac{\omega^2}{4} - m_\pi^2\right)^{1/2} \sigma_{\pi\pi}(\omega) \quad (15\text{-}10.4)$$

with f the pion–nucleon coupling constant and P the incident momentum in

ISBN 0-8053-8061-7

the laboratory. This formula is valid only in the limit as $t - m_\pi^2 \to 0$ but it is also applied if $t - m_\pi^2$ is small compared to m_π^2. It can then be used to find $\sigma_{\pi\pi}$ from directly measurable quantities and it reveals resonances in the $\pi-\pi$ cross section at energies corresponding to the ρ or other two-pion particles.

Pion–pion interactions appear in still other phenomena. In fact, the first indication of two-pion resonances occurred in the analysis of nucleon form factors (Chap. 18). Qualitatively these are interpreted as a consequence of the presence of a meson cloud around a central bare nucleon. The size of the cloud is directly measured by experiment and is related, according to the argument of Sec. 15-1, to the mass of the particles present in the cloud; the cloud is so large that these particles cannot be anything heavier than pions. The anomalous part of the magnetic moment of the nucleons has the same origin. From the measurement of the magnetic moment and charge form factors it has been possible to conclude that they are associated with the virtual emission or absorption of pion pairs. However, the probability of emission or absorption of virtual pions is calculable from the value of the constant f of Sec. 15-5, and is insufficient to account for the experimental magnetic properties of the nucleons if we consider the pions as independent. To resolve this difficulty a strong pion–pion interaction was postulated and the energy and quantum numbers of a resonance of the $\pi-\pi$ system were assigned, in fair agreement with those of the not yet discovered ρ meson (Frazer and Fulco, 1960).

As a further example of pion–pion interaction Abashian, Booth, and Crowe (1960) studied the reactions

$$p + d \Rightarrow 2\pi^0 + {}^3\text{He}$$

$$\Rightarrow \pi^+ + \pi^- + {}^3\text{He}$$

$$\Rightarrow \pi^+ + \pi^0 + {}^3\text{H} \tag{15-10.5}$$

They found that in the first reaction ${}^3\text{He}$ tends to recoil as if the two pions escaped in a bound state. This effect is prominent near the threshold for two-pion production. No such effect is found in the second and third reactions. The authors concluded from this that the matrix element for two-pion production is unusually large near the threshold for $T = 0$ states. A possible interpretation is that the two-pion system in the $T = 0, J = 0$ state has a strong interaction that can be represented by a scattering length (see Chap. 10) of about $2\hbar/m_\pi c$. This is relatively large. The situation is in this respect similar to that of the $n-p$ system in the singlet state.

BIBLIOGRAPHY

Barkas, W. H., W. Birnbaum, and F. M. Smith, "Mass Ratio Method Applied to the Measurement of L-Meson Masses and the Energy Balance in Pion Decay," *Phys. Rev.*, **101**, 778 (1956).

ISBN 0-8053-8061-7

Butterworth, I., "Boson Resonances," *Ann. Rev. Nucl. Sci.*, **19**, 179 (1969).

Chew, G., *S-Matrix Theory of Strong Interactions*, Benjamin, New York, 1961.

De Franceschi, G., A. Reale, and G. Salvini, "The η and η' Particles in the Pseudoscalar Nonet," *Ann. Rev. Nucl. Sci.*, **21**, 1 (1971).

Fermi, E. (Fe 51).

Fermi, E., "Lectures on Pions and Nucleons," *Nuovo Cimento, Suppl.*, **11**, 17 (1955).

Frazer, W. R. (Fr 66).

Gell-Mann, M., and K. M. Watson, "The Interactions between π-Mesons and Nucleons," *Ann. Rev. Nucl. Sci.*, **4**, 219 (1954).

Glauber, R. J., "High Energy Collision Theory" Lectures in Theor. Phys., v.1. Interscience, New York, 1959.

Lattes, C. M. G., Occhialini, G., and C. F. Powell, "Observations on the Tracks of Slow Mesons in Photographic Emulsions," *Nature*, **160**, 453 (1947).

Leon, M. (Le 73).

Moorhouse, R. G., "Pion–Nucleon Interactions," *Ann. Rev. Nucl. Sci.*, **19**, 301 (1969).

Nishijima, K., *Fundamental Particles*, Benjamin, New York, 1963.

Omnès, R., and M. Froissart, *Mandelstam Theory and Regge Poles*, Benjamin, New York, 1964.

Quigg, C., and C. J. Joachain, "Multiple scattering expansions in several particle dynamics," *Rev. Mod. Phys.*, **46**, 279 (1974).

Salvini, G., "The η Particle," *Riv. Nuovo Cimento*, **1**, 57 (1969).

Tripp, R. D., "Spin and Parity Determination of Elementary Particles," *Ann. Rev. Nucl. Sci.*, **15**, 325 (1965).

Williams, S. C. (W 71).

Yukawa, H., "Interaction of Elementary Particles," *Proc. Phys. Math. Soc. of Japan*, **17**, 48 (1935).

PROBLEMS

15-1 Show by an argument of the type developed by Wick [*Nature*, **142**, 993 (1938)] that the quanta corresponding to Coulomb's law have zero mass.

15-2 Assume that a neutron and a proton have an interaction

$$V = \pm g e^{-kr}/r$$

Using the Born approximation, calculate the total and differential scattering cross section. Do the same for a potential

$$V' = -(V/2)(1 + P)$$

where P is the neutron–proton exchange operator.

15-3 Calculate the energy for a π beam impinging on d which is to be used to study detailed balance in the reaction $p + p \Rightarrow d + \pi$ at $T_p = 450$ MeV (laboratory system).

15-4 Why does the reaction $\pi^- + d \Rightarrow 2n + \pi^0$ not occur for pions at rest?

15-5 Find the i-spin eigenfunctions of all three-pion systems.

15-6 What would be the ratio between the cross sections of the following reactions if $f_3 = f_1$ or $f_3 = 0, f_1 \neq 0$?

$$\pi^+ + p \Rightarrow \pi^+ + p$$
$$\pi^- + p \Rightarrow \pi^- + p$$
$$\pi^- + p \Rightarrow \pi^0 + n$$

15-7 Prove that the phase shift on a pion–nucleon collision at low energy is proportional to p^{2l+1}, where p is the momentum and the angular momentum of the colliding particles is $\hbar l$.

ISBN 0-8053-8061-7

15-8   An ω particle (three-pion state) of kinetic energy 100 MeV is generated in a $p-\bar{p}$ annihilation process. (*a*) How far does it travel before decomposing into three pions? (Use the resonance width to find the mean life.) Compare this distance with nuclear dimensions. (*b*) Show that the fact that only neutral ω particles are known indicates that ω has *i*-spin 0. (*c*) What is the maximum momentum in the laboratory for an ω meson generated in proton–antiproton annihilation at rest? (*d*) What is the maximum momentum, in the laboratory, of one of its decay pions?

15-9   Calculate the branching ratio between reactions (15-10.2) and (15-10.3): $\pi^- + p \Rightarrow \rho^- + p$ and $\pi^- + p \Rightarrow \rho^0 + n$. Hint: Use *i*-spin and remember that the coupling at the vertexes $p \Rightarrow p' + \pi^0$ and $p \Rightarrow n + \pi^+$ are in the ratio 1 : 2 (see Sec. 15-1).

15-10   Show that in Fig. 15-16 the distance $r = OP$ between the center of the circle and P is given by $r^2 = \frac{4}{9}[Q^2 - 3(T_1 T_2 + T_1 T_3 + T_2 T_3)]$ where the T_i are the kinetic energies of the pions. Furthermore, nonrelativistically $|\mathbf{p}_1 - \mathbf{p}_2| \leqslant |\mathbf{p}_3| \leqslant |\mathbf{p}_1 + \mathbf{p}_2|$ and from this and $T_i = p_i^2/2m$ obtain $T_- - T_+ - T'_+ \leqslant 2(T_+ T'_+)^{1/2}$ and $r^2 < Q^2/9$, showing that the representative points are within the circle of the figure.

15-11   Show that in the Dalitz plot in Fig. 15-16, $PN/NG = \cos\theta$.

15-12   Show that the representative points in a Dalitz plot for extremely relativistic particles are confined to an equilateral triangle inscribed in the Dalitz triangle.

15-13   Prove that a two-pion system with $T = 1$ must have odd angular momentum and odd parity and that two pions with $T = 0$ must have even angular momentum and even parity.

15-14   Find the *i*-spin eigenfunctions of all two-pion systems.

15-15   The reaction $^9\text{Be}(d, t\alpha)^4\text{He}$ at a deuteron laboratory energy of 26 MeV gives a three-particle final state. One may apply to it the Dalitz plot technique. Predict results on the basis of level diagrams of ^8Be and ^7Li (Figs. 10-6 and 11-10). [See M. A. A. Sonnemans et al., *Phys. Rev. Letters*, **31**, 1359 (1973).]

15-16   On the basis of phase-space and energy considerations alone, calculate the branching ratio of the decay

$$\Sigma^0 \Rightarrow \Lambda^0 + \gamma$$

$$\Rightarrow n + \gamma$$

Compare your results with the experimental data.

15-17   Find the methods and threshold energies for the production of K^+ and K^-.

15-18   Using the notation in Sec. 15-2, show that the three states $2^{-1/2}(\psi_a + \psi_d)$, ψ_b, and ψ_c, corresponding to a two-photon system, are eigenstates of the parity operator with eigenvalues $+1$ and that the $2^{-1/2}(\psi_a - \psi_d)$ state is an eigenstate with eigenvalue -1.

15-19   By bombarding nucleons with pions, K mesons, or $\bar{K}$ mesons, it is possible to produce resonances with a given hypercharge and *i*-spin. Make a table for the various possibilities, specifying Y and T for the different cases. Investigate energy thresholds for the different cases.

ISBN 0-8053-8061-7

CHAPTER

16

Baryons

Baryons are strongly interacting particles "containing" nucleons. By this we mean that they are protons or, if unstable, include protons among their ultimate decay products. We have treated some of them—protons, neutrons, and the (33) pion resonance—in previous chapters. Nine of the baryons (p, n; Λ; Σ^+, Σ^-, Σ^0; Ξ^-, Ξ^0; Ω^-) are stable with respect to strong interactions. The proton is absolutely stable; the free neutron beta-decays into a proton. The other "stable" baryons are strange, or contain "charmed quarks" (See Ch. 17), decaying by weak interaction, and have mean lives of the order of 10^{-10} sec (except Σ^0, which decays electromagnetically in $< 10^{-14}$ sec). All other baryons are unstable and decay by strong interaction.

16-1 BARYON GENERATION

With modern large accelerators it is possible to generate beams of unstable particles. Usually a target of beryllium or platinum, according to whether a light or heavy element is preferred, is bombarded with a primary proton beam. A great variety of nuclear reactions occur in the target, either directly or through the formation of intermediate pions. The target becomes a source of particles of different momenta, mass, and charge. Production of a beam of

Emilio Segrè, Nuclei and Particles: An Introduction to Nuclear and Subnuclear Physics, Second Edition

ISBN 0-8053-8061-7

specific particles requires systems of magnets and electrostatic fields (see Sec. 4-10).

Such beams are used in experiments. If the particles are unstable, they decay in flight and one has spontaneous attenuation in space, with a mean flight distance $\bar{l}$,

$$\bar{l} = \tau c \beta \gamma = \tau p / m_0 \tag{16-1.1}$$

where τ is the mean life in the rest system. The length τc amounts most often to a few centimeters; thus, to have practical beams one requires fairly high γ.

In order to obtain strange baryons by strong interaction either one uses associated production or one must start with strange particles. Of these, charged K and K_L a linear combination of K^0 and $\overline{K^0}$, are available. At low energy the simplest reactions are those of K^+ with protons and neutrons,

$$
\begin{aligned}
K^+ + p &\Rightarrow K^+ + p \quad \text{elastic scattering} \\
K^+ + n &\Rightarrow K^+ + n \quad \text{elastic scattering} \\
K^+ + n &\Rightarrow K^0 + p \quad \text{charge exchange}
\end{aligned}
\tag{16-1.2}
$$

This is in agreement with the assignment to the K^+ of $+1$ strangeness. No other particle of strangeness $+1$ can be formed except K^0.

On the other hand, the K^-, in addition to elastic and charge-exchange scattering with formation of $\overline{K^0}$, gives rise to the following exothermic reactions:

$$
K^- + p \Rightarrow
\begin{cases}
\Sigma^+ + \pi^- & + 103.00 \text{ MeV} \\
\Sigma^0 + \pi^0 & + 104.53 \text{ MeV} \\
\Sigma^- + \pi^+ & + 95.06 \text{ MeV} \\
\Lambda^0 + \pi^0 & + 181.42 \text{ MeV} \\
\Lambda^0 + \pi^+ + \pi^- & + 37.24 \text{ MeV} \\
\Lambda^0 + \pi^0 + \pi^0 & + 46.46 \text{ MeV}
\end{cases}
$$

With a neutron, the corresponding reactions are

$$
K^- + n \Rightarrow
\begin{cases}
\Sigma^- + \pi^0 & + 100.97 \text{ MeV} \\
\Sigma^0 + \pi^- & + 101.22 \text{ MeV} \\
\Lambda^0 + \pi^- & + 178.11 \text{ MeV} \\
\Lambda^0 + \pi^- + \pi^0 & + 43.15 \text{ MeV}
\end{cases}
$$

Neutral particles obviously cannot be deflected electromagnetically. They are often formed by charge-exchange processes. For example, antineutrons are formed by the process (Fig. 16-1)

$$p + \bar{p} \Rightarrow n + \bar{n} \tag{16-1.3}$$

and Λ–$\overline{\Lambda}$ pairs have been formed similarly (Fig. 16-2).

ISBN 0-8053-8061-7

Figure 16-1 Formation of an antineutron (by charge exchange) observed in a propane bubble chamber. The antiproton undergoes charge exchange $p + \bar{p} \Rightarrow n + \bar{n}$ at the spot indicated by the larger arrow. The annihilation star of the $\bar{n}$ is also indicated. [Courtesy Lawrence Berkeley Laboratory.]

ISBN 0-8053-8061-7

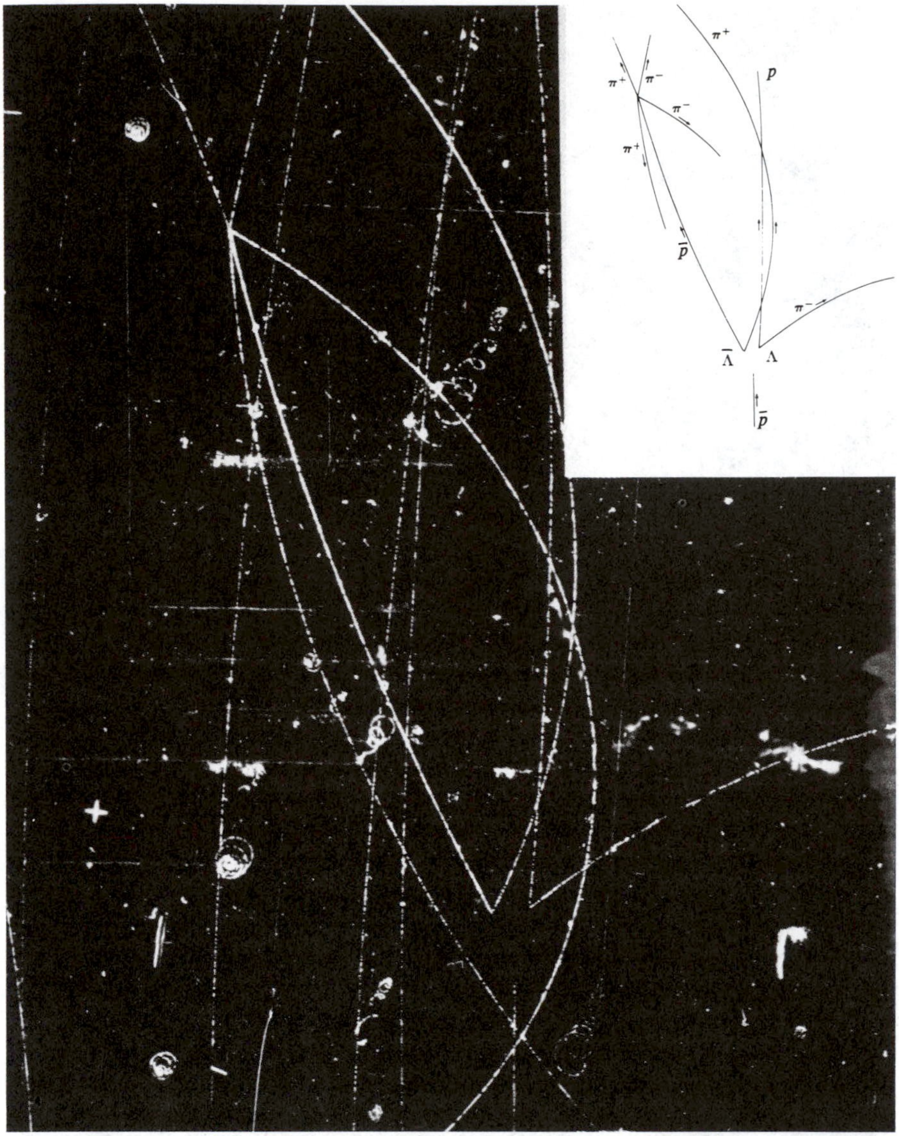

Figure 16-2 Production and decay of neutral lambda and antilambda pair in a hydrogen bubble chamber. The $\overline{\Lambda}$ was produced with a momentum of 720 MeV/c (laboratory). It went backwards in the center-of-mass system. The $\bar{p}$ from the decay annihilates, forming four charged pions. [Courtesy Lawrence Berkeley Laboratory.]

ISBN 0-8053-8061-7

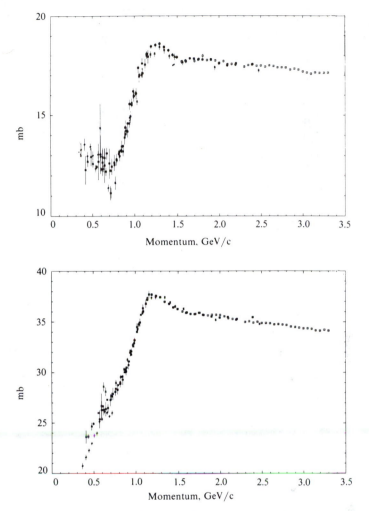

Figure **16-3** Compilation of recent K^+p (top) and K^+d (bottom) total cross-section measurements. [From (RPP 74).]

The particles of the beams are, for the most part, made to interact with protons or deuterons, often in the form of liquid H_2 or D_2. The latter is used to study the interaction with neutrons, since free neutrons are not available. Observations are made by means of bubble chambers, counter arrays, photographic emulsions, spark chambers, etc.

Resonances are also observed in the case of baryons either by *production* or by *formation* (See Ch. 15-7).

By increasing the K energy, one finds numerous and complicated resonances (Fig. 16-3, 4a and b). They are studied in ways similar to the $\Delta(1235)$

ISBN 0-8053-8061-7

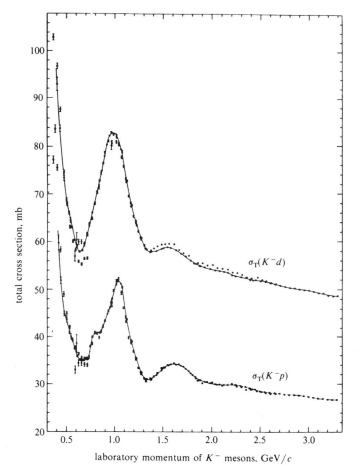

total cross section, mb

laboratory momentum of K^- mesons, GeV/c

Figure   16-4a K^-p and K^-d total cross-section data. [From (RPP 74).]

resonance, and much that has been said for it is applicable, with suitable changes, to other resonances. Pertinent information is given in (RPP 74); for more detail, consult (W 71). At zero kinetic energy the resonances appear in production, as demonstrated, for instance, in the reaction

$$K^- + p \Rightarrow \Lambda + \pi^+ + \pi^- \tag{16-1.4}$$

If this is a genuine three-body reaction, the kinetic energies of the two pions plotted in a Dalitz plot of T_{π^+} versus T_{π^-} will uniformly occupy the kinematically permissible region, assuming that the matrix element is constant and the probabilities of the different final states are proportional to the corresponding phase space available. If, however, the reaction proceeds in two steps through

ISBN 0-8053-8061-7

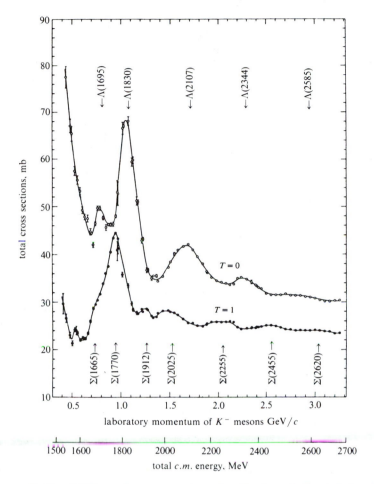

Figure **16-4b** K^-N total cross sections for $T = 0$ and $T = 1$ below 3.3 GeV/c. [From (RPP 74).]

the formation of an excited hyperon Σ' of short life,

$$K^- + p \Rightarrow \Sigma'^{\pm} + \pi^{\mp}$$
$$\Downarrow$$
$$\Lambda^0 + \pi^{\pm} \qquad\qquad (16\text{-}1.5)$$

the first pion will have a fixed energy in the K^-p center-of-mass system, because it results from a two-body reaction. The second pion will also have a fixed energy in the center of mass of Σ', and this will give an energy band in the K^-p center-of-mass system. The band width depends on the Q of the Σ' disintegration. The short life of Σ' produces an uncertainty in its total energy.

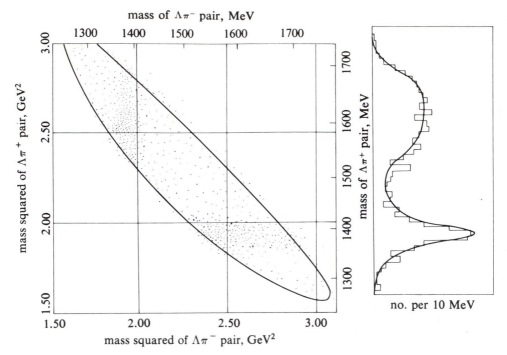

mass of $\Lambda\pi^-$ pair, MeV

mass squared of $\Lambda\pi^+$ pair, GeV^2

mass squared of $\Lambda\pi^-$ pair, GeV^2

mass of $\Lambda\pi^+$ pair, MeV

no. per 10 MeV

Figure 16-5 Dalitz plot of $\Lambda\pi^+\pi^-$ events from K^-p interactions at 1.22 GeV/c laboratory. The square of $\Lambda\pi^+$ effective mass is plotted against the square of $\Lambda\pi^-$ effective mass. Scales giving the masses in MeV are also shown. If the matrix element for the reaction were constant, the area within the oval would be uniformly populated. Projection of the events onto the $\Lambda\pi^+$ mass axis is displayed to the right of the figure. The curve represents the fitting of Breit-Wigner resonance expressions to the (Λ, π^+) and (Λ, π^-) systems. [From J. Shafer, J. J. Murray, and D. O. Huwe, *Phys. Rev. Letters*, **10**, 179 (1963).]

Figure 16-5 gives a plot of the kinetic energies of the two combinations (Λ, π^+) and (Λ, π^-) in Eq. (16-1.4) when the K^- have a momentum of 1220 MeV/c lab. It is apparent from the figure the (Λ, π^+) or (Λ, π^-) system has preferentially the mass 1383 ± 1 MeV in its own c.m. system.

We have evidence, therefore, of a resonance in the $(\Lambda, \pi^\pm)$ system with a total energy of 1383 ± 1 MeV. The Σ' of energy 1383 has been observed not only positively and negatively charged, but also neutral. The 'next step is to assign quantum numbers to the Σ'. The i-spin is 1, as can be seen from the three possible charges. The spin J is shown to be $\frac{3}{2}$ by a consideration of the angular distribution of the Σ' production in the K^-p center-of-mass system.

ISBN 0-8053-8061-7

Similar consideration of the reaction

$$K^- + p \Rightarrow \Lambda' + \pi^0$$
$$\Downarrow$$
$$\Sigma + \pi \qquad\qquad (16\text{-}1.6)$$

indicates the existence of an excited hyperon of i-spin 0 at mass 1405 MeV. The results obtained thus far are recorded in (RPP 74) and Table 13-2.

16-2 BARYON SPIN MEASUREMENTS

The determination of spin, intrinsic parity and magnetic moment is one of the important tasks in the study of particles. To the methods indicated in previous sections we add one (Fig. 16-6), due to Adair (1955), suitable for short-lived particles. Consider the reaction

$$\pi + p \Rightarrow K + \Lambda$$

and assume that the spin of K is 0. In the center-of-mass system consider those cases in which the line of flight of K and Λ is parallel to that of π and p. The component of the angular momentum in this common direction m_J is $\pm \frac{1}{2}$, coming—on the left side of the equation—from the spin of the proton

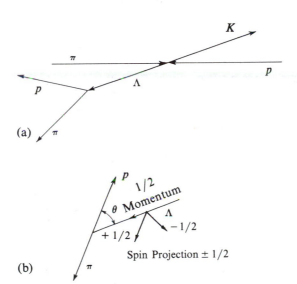

Figure 16-6 Momentum vectors in Adair's model for spin measurement. (a) Center-of-mass-system momenta of reaction $\pi + p \Rightarrow K + \Lambda$. (b) Rest system of Λ. Angle of decay θ.

ISBN 0-8053-8061-7

and—on the right side of the equation—from the component of the Λ spin in the direction of the line of flight. Orbital angular momenta are directed perpendicularly to the line of flight and thus do not contribute to m_J. The Λ decays into a proton and a pion. Designate by θ the angle between the proton and the line of flight of the Λ in the rest system of the Λ. If the Λ has spin $\frac{1}{2}$, we shall have an isotropic angular distribution of the decay proton. (An anisotropic angular distribution would indicate a spin $> \frac{1}{2}$.) The experiment indicates spin $\frac{1}{2}$. Essentially the method furnishes a sample of aligned (not polarized) Λ and examines their decay. No alignment without polarization is possible if $J = \frac{1}{2}$. In practice, it is necessary to examine Λ escaping at a small angle from the line of flight $\pi-p$. This introduces a certain probability of observing Λ with $|m_J| \neq \frac{1}{2}$, but it is possible to correct for this effect.

Particles with a spin different from zero may have a magnetic moment. They can be made to precess in magnetic fields and from the change of orientation of the spin as revealed by the details of the decay, one finds the gyromagnetic ratio and hence the magnetic moment. Mesic atoms or exotic atoms in general can also give the magnetic moment from the structure of their X-ray lines. Results are given in (RPP 74).

16-3 HYPERFRAGMENTS

In 1953 Danysz and Pniewski found a fragment, emerging from a high-energy nuclear star formed by cosmic rays, which they interpreted as a nucleus in which a neutron had been replaced by a Λ^0 (Fig. 16-7). Many similar fragments were found later, and the name hyperfragments or hypernuclei has been adopted for such particles, together with the notation $^4_\Lambda \mathrm{He}$, etc., indicating a $^4\mathrm{He}$ nucleus in which a neutron has been replaced by a Λ^0. Hyperfragments are formed on capture of strange particles by nuclei. For instance, capture of K^- at rest by He gives rise in a few percent of the cases to $^4_\Lambda \mathrm{He}$ or $^4_\Lambda \mathrm{H}$. Even two neutrons have, in rare cases, been replaced by two Λ.

Hyperfragments are unstable and decay, emitting either nucleons and a pion (mesonic decay) (Fig. 16-8) or only nucleons (nonmesonic decay) (Fig. 16-7). The mesonic modes of decay can be thought of as the normal decay of the Λ inside the nucleus according to the elementary process $\Lambda \to p + \pi^-$ or $\Lambda \to n + \pi^0$. The nonmesonic decay corresponds to the elementary processes $\Lambda + p \to p + n$ or $\Lambda + n \to n + n$. Examples of mesonic decays are $^4_\Lambda \mathrm{H} \to \pi^- + {}^4\mathrm{He}$; $\pi^- + n + {}^3\mathrm{He}$; $\pi^- + {}^2\mathrm{H} + {}^2\mathrm{H}$; etc. Examples of nonmesonic decay are $^8_\Lambda \mathrm{Be} \to {}^3\mathrm{He} + {}^4\mathrm{He} + n$.

The mean life of hyperfragments has yet to be measured accurately. For $^4_\Lambda \mathrm{H}$ there is a measurement giving $(1.2 \pm 0.6) \times 10^{-10}$ sec, comparable to the mean life of the free Λ^0; the same order of magnitude is found for other hyperfragments.

ISBN 0-8053-8061-7

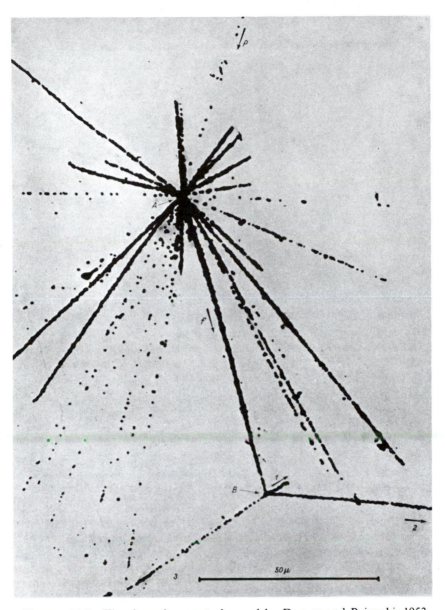

Figure 16-7 First hyperfragment observed by Danysz and Pniewski, 1953. The fragment *f* was produced in star *A* and decayed nonmesonically at point *B* into three particles: 1, 2, 3.

ISBN 0-8053-8061-7

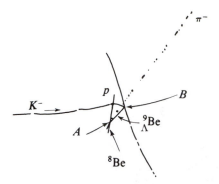

Figure   16-8   A $^9_\Lambda$Be hyperfragment was produced by the nuclear absorption of a negative K meson at rest. The star produced by the K^- meson absorption is indicated by arrow B. The $^9_\Lambda$Be hyperfragment decayed (at the point indicated by arrow A) into a π^- meson of 26.6 ± 0.6 MeV, a proton, and ^{8}Be. The two alpha tracks of ^{8}Be begin to separate near their end. [W. F. Fry, J. Schneps, and M. S. Swami, *Phys. Rev.*, **101**, 1526 (1956)].

The binding energy of the Λ^0 in the nucleus can be measured by suitable energy cycles (Fig. 16-9). For instance, we observe a $^9_\Lambda$Be hyperfragment disintegrating into two ^{4}He, a proton, and a π^-, releasing the total kinetic energy of 30.92 ± 0.5 MeV. The energy released by Λ^0 in the $p-\pi$ disintegration is 37.56 MeV. Starting from two ^{4}He, a proton, and a negative pion, reassemble the Λ^0 from the proton and π^- spending this energy. We reassemble the

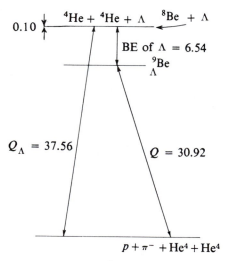

Figure   16-9   Energy balance of the $^9_\Lambda$Be hyperfragment disintegration shown in Fig. 16-8. Energies in MeV.

ISBN 0-8053-8061-7

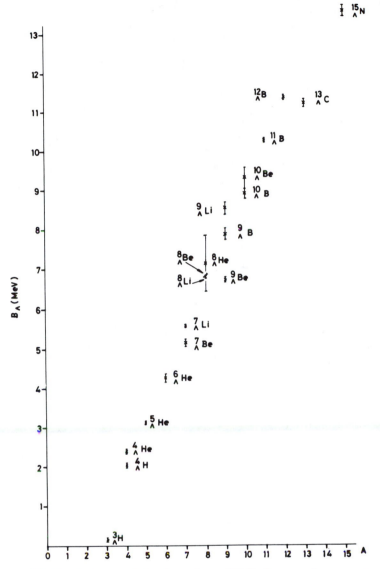

Figure 16-10 Variation of the B_Λ values with the hypernuclear mass numbers. [From M. Jurić et al., *Nuc. Phys.*, **B52**, 1 (1973).]

ISBN 0-8053-8061-7

two ^{4}He nuclei to form ^{8}Be, gaining 0.10 MeV. We now have a Λ^0 and ^{8}Be and have spent, altogether, 37.46 MeV. On binding the Λ^0 to ^{8}Be, we must obtain $37.46 - 30.92 = 6.54$ MeV, which is the binding energy of Λ^0 to ^{8}Be, to form $^9_\Lambda$Be.

The results of an analysis of the available experimental data are summarized in Fig. 16-10, which gives the binding energy of the Λ^0 in several light nuclei. It is clear that B_Λ is approximately proportional to A. This is in striking contrast to the binding energy of the nucleons, which fluctuates considerably in the case of light nuclei, owing to the exclusion principle. The single Λ^0 is not subject to such special limitations. Thus, for example, whereas ordinary nuclei with $A = 5$ are unbound, there are hyperfragments of mass 5. On the assumption that the Λ^0–nucleon forces are charge independent, it has been possible to analyze the experimental data, the main conclusions being that the Λ^0–nucleon force is comparable in magnitude to the nucleon–nucleon force and is spin–dependent. The evidence also favors the value of $\frac{1}{2}$ for the Λ^0 spin.

The ratio between mesonic and nonmesonic decay gives another argument for this spin assignment (Cheston and Primakoff, 1953; Ruderman and Karplus, 1956). The mesonic decay of a hyperfragment is considered similar to gamma emission by an excited nucleus. The nonmesonic decay represents an additional channel in the same way that the emission of conversion electrons represents another mode of deexcitation for nuclei. The ratio between mesonic and nonmesonic decays is, then, analogous to the internal conversion coefficient. The calculation indicates a strong dependence of the ratio on the orbital angular momentum l with which the pion is emitted and points to $l = 0$ and to a spin $\frac{1}{2}$ for the Λ^0.

BIBLIOGRAPHY

Danysz, M., and J. Pniewski, "Delayed Disintegration of a Heavy Nuclear Fragment," *Phil. Mag.*, **44**, 348 (1953).

Jurić, M., et al., "Hypernuclei Binding Energies," *Nucl. Phys.*, **B52**, 1 (1973).

Leon, M., (Le 73).

Particle Data Group (RPP 74).

Tripp, R. D., "Spin and Parity Determination of Elementary Particles," *Ann. Rev. Nucl. Sci.*, **15**, 325 (1965).

Williams, W. S. C. (W 71).

PROBLEMS

16-1 Show that the decay $K^+ \leadsto \pi^+ + \gamma$ is forbidden if K^+ has spin 0. In fact, it is not observed, giving another argument for spin 0 of the K^+.

16-2 What isotopic spin states are involved in the reactions:

$$(a) \quad \pi^- + p \Rightarrow \Lambda^0 + K^0$$
$$(b) \quad \pi^+ + p \Rightarrow \Sigma^+ + K^+$$
$$(c) \quad \pi^- + p \Rightarrow \Sigma^- + K^+$$
$$(d) \quad K^- + d \Rightarrow \Lambda^0 + p + \pi^-$$

ISBN 0-8053-8601-7

16-3 Consider the $\Sigma-\pi$ interaction. There are nine possible charge combinations. How many independent interactions may exist under the isospin invariance hypothesis?

16-4 An excited hyperon state Σ' has been found which decays into Λ^0 and π. It has a total energy of 1380 MeV with a full width of half-maximum of about 30 MeV. Calculate (*a*) the mean life of the Σ' and (*b*) its strangeness.

ISBN 0-8053-8601-7

CHAPTER 17

Classification of Hadrons, Quarks, and $SU(3)$

The great number of particles contained in the particle tables, (RPP 74) a major result of physics after World War II, cries for ordering principles, for classification. The periodic system of the elements stands as an example of a great synthesis achieved semiempirically at a time when its theoretical basis was still lacking. Maybe something similar can be achieved for elementary particles.

There have been many attempts to develop a classification of particles on the assumption that some are more "elementary" than others. For instance, as far back as 1949, Fermi and Yang tried to consider the pion as composed of an $N–\overline{N}$ pair. More recently, Sakata (1956) assumed the neutron, proton, and Λ as the fundamental particles and tried to build a model in which all particles are formed by these three and their antiparticles.

17-1 SAKATA'S MODEL; QUARKS

Sakata's choice was based on esoteric ideas but the result is that each of the three "elementary" particles is responsible for one of the three additive conserved quantities: baryon number (n), charge (p), and strangeness (Λ). Thus, for instance, a π^0 could be obtained by an $n–\bar{n}$ or $p–\bar{p}$ combination, a K^+ by a $\overline{\Lambda}–p$ combination, and so on. This attempt, however, does not agree

Emilio Segrè, Nuclei and Particles: An Introduction to Nuclear and Subnuclear Physics, Second Edition

ISBN 0-8053-8601-7

with experiment because it predicts objects that are not found and that are difficult to rule out a priori. For instance, one knows a particle corresponding to $p\bar{n}\Lambda$ but not one corresponding to $pn\Lambda$.

Other attempts have been made to discover symmetries among particles. It has been remarked by S. Sakata, Y. Ne'eman, Gell-Mann, and others that it is possible to collect all presently known particles and resonances in families containing either 1, or 8, or 10, or 27 members.

These numbers form the basis of a classification based on the hypothesis of the invariance of the interaction with respect to two newly introduced operators, U and V, similar to the known i-spin invariance of strong interactions.

The i-spin T gives rise to multiplets containing $2T + 1$ members differing in T_3. They have approximately the same mass and we assume that they would have exactly the same mass in the absence of electromagnetic interaction. The operators T_+ and T_-, raising and lowering operators, transform a particle into another in the same multiplet by raising or lowering its T_3. The T_i matrices give a representation of the rotation group in three-dimensional space, called in mathematical terminology $SU(2)$ (special unitary group in two complex dimensions). The two new operators U and V have commutation properties similar to those of T and to those of ordinary angular momentum.

In the case of the ordinary angular momentum [$SU(2)$ group] the simplest representation is obtained by the Pauli matrices and their commutation rules. Similarly for the three operators T, U, and V there is a fundamental representation by nine 3×3 matrices (including the unit matrix); commutation rules, however, are more complicated than those for ordinary angular momentum. The group properties of these operators are called $SU(3)$ (special unitary group in three complex dimensions) or the eightfold way. The numbers 1, 8, and 10 for the number of members in a multiplet emerge naturally from the properties of $SU(3)$. The third components, of T, U, V have simple physical meanings expressible through the charge and hypercharge;

$$T_3 = Q - \tfrac{1}{2}Y \qquad U_3 = -\tfrac{1}{2}Q + Y \qquad V_3 = -\tfrac{1}{2}(Q + Y) \quad (17\text{-}1.1)$$

Note that $T_3 + U_3 + V_3 = 0$.

The mathematical study of $SU(3)$ is beyond the scope of this book, but it is possible to obtain an idea of some of its results in a less abstract manner by formulating a model based on some new "fundamental" primary entities which would be the building stones of the observed particles. This of course is a new form of the old idea of "elementary" particles. We have already seen that the Sakata attempt using p, n, and Λ does not work. Gell-Mann and Zweig proposed a model postulating new, *thus far unobserved*, particles called *quarks*. Quarks are supposed to have fractional charges. Obviously, their discovery would be a major accomplishment and physicists have devoted great

ISBN 0-8053-8601-7

effort to finding them, either in cosmic rays or embedded in ordinary matter, but to no avail.

It could be that quarks are permanently trapped in the real particles and manifest themselves only in combinations of integral charge. The question of their "existence" then becomes one of defining the word "existence." The forces between quarks may have their own quanta, sometimes called *gluons*.

The quark concept contains $SU(3)$, but goes farther. It predicts more than the simple $SU(3)$ and on the whole the additional predictions are correct.

In the simplest version there are three species of quarks and corresponding antiquarks. They carry baryon number, electric charge, and strangeness, as in the Sakata model, but in $1/3$ amounts, as indicated in Table 17-1. Quarks have spin $\frac{1}{2}$ and as such should have Fermi statistics; quarks and antiquarks have opposite parity. The symbols for the three fundamental quarks can be written by using T_3, U_3, and V_3 as quantum numbers:

$$|u\rangle = |\tfrac{1}{2}, 0, -\tfrac{1}{2}\rangle \qquad |d\rangle = |-\tfrac{1}{2}, \tfrac{1}{2}, 0\rangle \qquad |s\rangle = |0, -\tfrac{1}{2}, \tfrac{1}{2}\rangle \quad (17\text{-}1.2)$$

The letters u, d and s stand for (isotopic spin) *up*, *down* and for *strange*. The corresponding quarks are given in the first rows of Table 17-1. Three quarks were sufficient for describing all particles known to the end of 1974. New particles have been discovered since and they require a fourth quark c (for *charmed*) and a new quantum number C (*Charm*). The group corresponding to four quarks is $SU(4)$. The last row and column of Table 17-1 contain the fourth quark of $SU(4)$. We will give some information on particles containing charmed quarks in Sec. 17-6.

Note that all quarks have B (baryon number) $= 1/3$ and spin $1/2$. For antiquarks reverse the signs of Q, T_3, B, S and C.

Although free quarks have never been seen, one may speak of their mass in the sense that a number of features of the real particles may be described by non relativistic motions of bound quarks. This approximation requires binding energy of the quarks small compared with the mass of the particles. In this approximation one has $m_u = m_d = 0.336\text{GeV}$; $m_s = 0.538$ GeV and $m_c = 1.65$ GeV.

The notation for quarks vary. One finds: q_1, q_2, q_3, q_4 or p, n, λ, p' and several others.

TABLE 17-1   QUARKS AND THEIR QUANTUM NUMBERS

Symbol	Q	T	T_3	B	S	C
u	2/3	1/2	1/2	1/3	0	0
d	−1/3	1/2	−1/2	1/3	0	0
s	−1/3	0	0	1/3	−1	0
c	2/3	0	0	1/3	0	1

ISBN 0-8053-8601-7

17-2 COMBINATION OF QUARKS

Most known particles do not contain charmed quarks and they follow the rules of $SU(3)$ represented by the combinations of the u, d, s quarks. The group structure $SU(3)$ is approximate, but the approximation is better than for the $SU(4)$ group. It is thus expedient to treat first particles containing only u, d, s, quarks.

All these particles, including resonances, are formed by $(q\bar{q})$ combinations if they are bosons or by (qqq) or $(\bar{q}\bar{q}\bar{q})$ combinations if they are baryons. Besides reproducing the $SU(3)$ multiplets, the quark model excludes certain quantum numbers. For three quarks, Table 17-1 shows that it is impossible to have $S > 0$; that for $S = 0$ it is impossible to have $T > \frac{3}{2}$; that for $S = -1$ it is impossible to have $Q = 2$; etc. These *exotic* particles or resonances have never been positively identified, despite considerable effort.

To give the quantum numbers of quark combinations and classify them into multiplets we use a graphic method justifiable on $SU(3)$ properties.

For the simpler case of $SU(2)$, suppose we want to combine two spins $\frac{1}{2}$. We represent them graphically on an S_z axis as two points of abscissa $\pm\frac{1}{2}$. We indicate the first spin by an open circle and the second by a dot. We superimpose to the left-hand dot and to the right-hand circle the centers of two copies of the original diagram obtaining the diagram in Fig. 17-1a. There are then two states with $S_z = 0$ and one each with $S_z = \pm 1$. They form the well-known $S_z = 0, \pm 1$ states of a spin 0 singlet and spin 1 triplet (Fig. 17-1b).

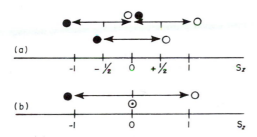

Figure 17-1 Two spins $\frac{1}{2}$ combine to give spin 1 or 0. (a) Origin of second spin is brought onto the ends of the first spin (dot or open circle). Ends of resultant spins, marked by arrows, are at $S_z = \pm 1, 0$. The point at 0 is double. (b) Resulting spins 1 and 0.

For quarks the situation is more complicated. A quark is represented in a two-dimensional diagram having T_3 as the abscissa and Y as the ordinate. The unit of length of the ordinate is, however, $\sqrt{3}/2$ times that of the abscissa. The three fundamental quarks and antiquarks are represented in Fig. 17-2.

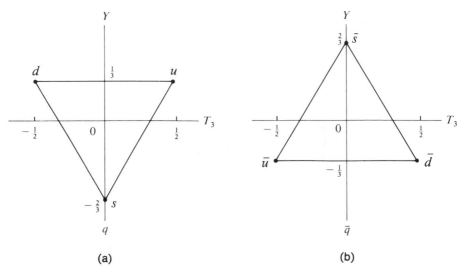

(a)   (b)

Figure 17-2 Fundamental quarks (a) and antiquarks (b) in a T_3–Y diagram. Quarks are indicated by the points marked d, u, s; antiquarks by the points marked $\bar{d}$, $\bar{s}$, $\bar{u}$.

If real particles are combinations of quarks, the (qq) combination does not occur, because it would give fractional electric charges, contrary to experience. The $(q\bar{q})$ combination is more interesting. To combine a quark with an antiquark, bring the origin of the antiquark diagram in Fig. 17-2b on u and mark the coordinates in the quark plane of the points representing $\bar{s}$, $\bar{u}$, $\bar{d}$. These points give the T_3 and Y of the combinations $u\bar{s}$, $u\bar{u}$, $u\bar{d}$, where the first letter comes from the quark at the center of the antiquark triangle and the second from the antiquarks at the vertices of the antiquark triangle. Repeating this operation by bringing successively the origin of Fig. 17-2b to the places marked u, d, s, we cause the vertices of the antiquark triangles to land on the vertices of a hexagon at the places marked $u\bar{s}$, $u\bar{d}$, $d\bar{s}$, $s\bar{u}$, $d\bar{u}$, $s\bar{d}$, and also on the center of the hexagon marked $u\bar{u}$, $d\bar{d}$, $s\bar{s}$ (see Fig. 17-3). We have thus $3 \times 3 = 9$ wave functions. These split into an $SU(3)$ octet and an $SU(3)$ singlet, a fact that is denoted symbolically as

$$\textcircled{\scriptsize 3}\times\textcircled{\scriptsize $\bar{3}$}=\textcircled{\scriptsize 8}+\textcircled{\scriptsize 1}$$

The octet wave functions correspond to the quark combinations on the periphery of the hexagon and to two linear combinations of the wave functions indicated at the center of the hexagon. Of these, one is part of an i-spin triplet, the other is an i-spin singlet, and a third is not in the $SU(3)$ octet, but is an $SU(3)$ singlet. The quark contents are given in Table 17-2, but a proof is beyond the scope of this book.

ISBN 0-8053-8601-7

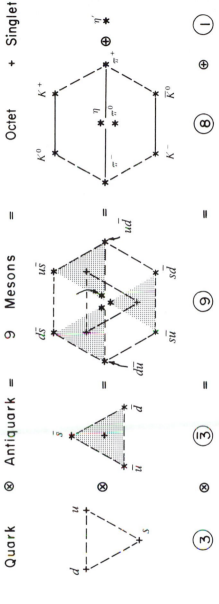

Figure 17-3 Quark–antiquark combination in a $T_3 - Y$ plane. Antiquark triangles are shaded. The three central points have $d\bar{d}$, $u\bar{u}$, $s\bar{s}$.

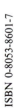

TABLE 17-2   QUARK COMPOSITION OF REAL BOSONS

Y	T	Quark composition		$J^P = 0^-$	$J^P = 1^-$
0	1	$\bar{u}d\;(\bar{u}u - \bar{d}d)/\sqrt{2}$	$\bar{d}u$	$\pi^-\;\pi^0\;\pi^+$	$\rho^-\;\rho^0\;\rho^+$
1	$\frac{1}{2}$	$\bar{s}d$	$\bar{s}u$	$K^0\;K^+$	$K^{*0}\;K^{*+}$
-1	$\frac{1}{2}$	$\bar{u}s$	$\bar{d}s$	$K^-\;\overline{K^0}$	$K^{*-}\;\overline{K^{*0}}$
0	0	$(\bar{u}u + \bar{d}d - 2\bar{s}s)/\sqrt{6}$		η	
0	0	$(\bar{u}u + \bar{d}d + s\bar{s})/\sqrt{3}$		η'	
0	0	$(\bar{u}u + \bar{d}d)/\sqrt{2}$			ω
0	0	$\bar{s}s$			φ

As far as ordinary spin and parity are concerned, a $q\bar{q}$ system without any orbital angular momentum is in an $^1S_0$ or $^3S_1$ state with $J^P = 0^-$ or 1^-, respectively. (Remember that q and $\bar{q}$ have opposite intrinsic parity.) The $q\bar{q}$ combinations with 0^- and 1^- have been identified as in Fig. 17-4. States with orbital angular momentum, such as $^1P_1$ and $^3P_{2,1,0}$, have also been tentatively assigned to some higher resonances.

Next we consider the combination of three quarks (qqq) containing $(3) \times (3) \times (3) = (10) + (8) + (8) + (1) = 27$ real particles.

To obtain the combinations of three quarks, we proceed in a manner similar to the one used for two quarks. We start from a one-quark diagram (Fig. 17-2) and we superimpose the origin of a second quark diagram on each quark of the first. We mark the points thus obtained with the name of the two quarks that make up the combination. We then superimpose the origin of a third quark diagram on each point containing two quarks, and we mark with

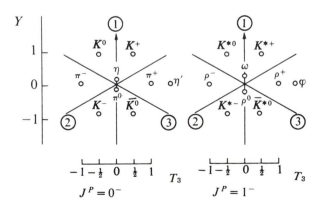

Figure 17-4   The quark–antiquark combinations $(3) \times (\bar{3}) = (8) + (1)$ for $J^P = 0^-$ and 1^-. The first is the same as Fig. 17-3. Note that a * or a prime indicate an excited state.

ISBN 0-8053-8601-7

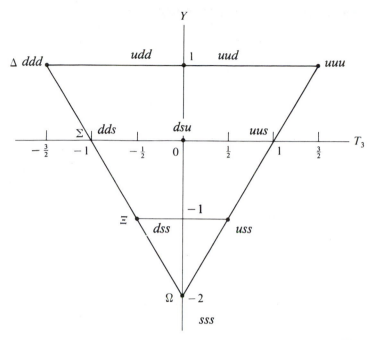

Figure 17-5 Combination of three quarks. The combination gives ③×③ ×③ = ⑩ + ⑧ + ⑧ + ① real particles, as shown in Fig. 17-6. One of the octets is completely known; the other is only partially known.

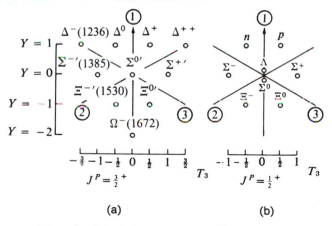

(a)   (b)

Figure 17-6 The $J^P = \frac{3}{2}^+$ decimet and the $J^P = \frac{1}{2}^+$ octet extracted from Fig. 17-5.

ISBN 0-8053-8601-7

TABLE 17-3   QUARK COMPOSITION OF REAL FERMIONS

Y	T	Quark composition		$J^P = \frac{1}{2}^+$ (octet)			$J^P = \frac{3}{2}^+$ (decimet)				Mass
1	$\frac{1}{2}$	ddu	uud	n	p						939
0	0	$s(du{-}ud)/\sqrt{2}$			Λ						1116
0	1	sdd $s(du+ud)/\sqrt{2}$ suu		Σ^-	Σ^0	Σ^+					1193
-1	$\frac{1}{2}$	ssd	ssu	Ξ^-	Ξ^0						1317
1	$\frac{3}{2}$	ddd ddu duu uuu					Δ^-	Δ^0	Δ^+	Δ^{++}	1235
0	1	sdd sdu suu					$\Sigma^{-\prime}$	$\Sigma^{0\prime}$	$\Sigma^{+\prime}$		1385
-1	$\frac{1}{2}$	ssd ssu					$\Xi^{-\prime}$	$\Xi^{0\prime}$			1530
-2	0	sss					Ω^-				1672

three quarks the points obtained from the vertices of the third quark. The result is shown in Fig. 17-5. The places with *ddd*, *sss*, and *uuu* are occupied only once; those with *udd*, *duu*, *sdd*, etc. are occupied three times and the central spot *sdu* is occupied four times. All places in the triangle form a multiplet ($J^P = \frac{3}{2}^+$) of 10 (a decimet), which we identify with the particles in Fig. 17-6a. The remaining combinations give two octets and a singlet. One of the octets ($J^P = \frac{1}{2}^+$) is identified in Fig. 17-6b; the other is not yet completely known. The singlet is the $\Lambda'(1405)$ with a small admixture of $\Lambda'(1675)$. Table 17-3 gives the quark composition.

To obtain the wavefunctions of the quark systems corresponding to real particles one has to choose the proper linear combinations of the individual quark wave functions, including spin. We refer to Feld's book quoted in the bibliography for a detailed treatment of this subject.

17-3   MASS FORMULAS

We may examine the multiplets using as coordinates U_3 and Q in lieu of T_3 and Y, obtaining the same diagrams, rotated by an angle of 120 deg, as those in Fig. 17-7. For the peripheral points there is no ambiguity, but the central point of Fig. 17-6b, containing Λ and Σ^0, eigenstates of T_3, now has a linear combination of these two which is an eigenstate of U_3. It is important to identify this linear combination $\Sigma_U = \alpha\Sigma^0 + \beta\Lambda$ where Σ^0 and Λ are the ordinary particles, eigenstates of T_3. To obtain orthonormal linear combinations with real coefficients, we must have $\alpha^2 + \beta^2 = 1$. To obtain another relation between α and β, start from the neutron that has $U = 1$, $U_3 = 1$ and apply to it the operator U_- (lowering operator), obtaining $U_-|n\rangle = \sqrt{2}\,|\alpha\Sigma^0 + \beta\Lambda\rangle$ where the ket on the right-hand side is an eigenstate of U, U_3 with

ISBN 0-8053-8601-7

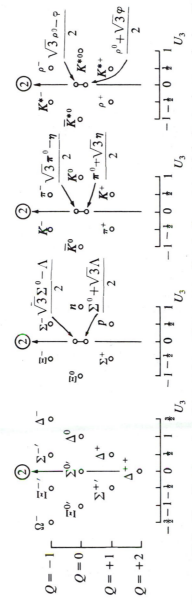

Figure 17-7 Classification of particles and resonances according to the octet and decimet scheme—abscissa U_3, ordinate Q. All particles in a multiplet have the same J^P. These are the same figures as in Fig. 17-6 rotated by 120 deg.

$U = 1$, $U_3 = 0$. The coefficient $\sqrt{2}$ comes from the formulas

$$U_-|U, U_3\rangle = \left[U(U+1) - U_3(U_3 - 1)\right]^{1/2}|U, U_3 - 1\rangle$$

$$U_+|U, U_3\rangle = \left[U(U+1) - U_3(U_3 + 1)\right]^{1/2}|U, U_3 + 1\rangle$$

(Similar formulas obtain for the T_- and T_+ operators.) Apply T_+ to the result obtaining $T_+ U_-|n\rangle = 2\alpha|\Sigma^+\rangle$. Proceeding in the opposite sequence, one has $T_+|n\rangle = |p\rangle$ and $U_- T_+|n\rangle = U_-|p\rangle = |\Sigma^+\rangle$. The two results must be equal because U_- and T_+ commute and comparing them one has $\alpha = 1/2$, $\beta = \sqrt{3}/2$. We thus conclude that $\Sigma_U^0 = \frac{1}{2}(\Sigma^0 + \sqrt{3}\,\Lambda)$. The state Λ_U corresponding to the Λ is orthogonal to Σ_U^0 and is $\frac{1}{2}(\sqrt{3}\,\Sigma^0 - \Lambda)$.

If U-spin were exactly conserved in strong interactions, as i-spin is, all particles in the same U multiplet, except for electromagnetic effects, would have the same mass. This is not so and one must assume that U-spin invariance of the strong interactions is only approximately valid. An indication of the accuracy of inaccuracy of the approximation is given by the relative mass difference of particles in the same multiplet differing in Y. This difference is of the order of 0.1 of the total mass. We hypothesize a medium-strong interaction that conserves i-spin but not U-spin. This interaction produces the mass differences in a U multiplet. In the quark model this means that u and d must have the same mass; s is supposed to be heavier.

The interaction that breaks the U symmetry has an analogy with the electromagnetic interaction, which is not invariant with respect to i-spin rotation (a change in T_3 means a change in charge, even when T is constant). As a hypothesis it has been assumed that this symmetry-breaking interaction bears the same relation to U-spin that the electromagnetic interaction bears to i-spin. The simplest form of this interaction makes it proportional to U_3.

The mass in a U multiplet would then be given by the expression

$$M = a + bU_3$$

with a and b constants. In particular, in the decimet the masses of the U quartet Δ^- (1235), $\Sigma^{-\prime}$ (1385), $\Xi^{-\prime}$ (1530), and Ω^- (1672) should be and are equally spaced. This relation was used for predicting the mass of Ω^-, subsequently found in the Brookhaven bubble chamber. The discovery of the Ω^- was one of the major successes of $SU(3)$ theory. Since this original discovery, several Ω^- have been observed and some have also been identified in older unexplained observations. The mass of Ω^- precludes its strangeness-conserving decay because the lightest combination of strangeness -3 and $B = 1$ is K^-, Ξ^0 with mass 1805. Hence, the Ω^- can decay only by weak interaction; several channels are open and have been observed (Fig. 17-8).

ISBN 0-8053-8601-7

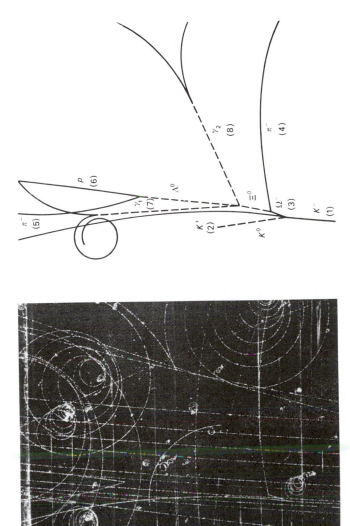

Figure 17-8 The first Ω^- event (Barnes et al., BNL 1964). The drawing illustrates the events:

$$K^-(1) + p \Rightarrow K^+(2) + K^0 + \Omega^-(3);$$

$$\Omega^- \to \Xi^0 + \pi^-(4); \quad \Xi^0 \to \pi^0 + \Lambda; \quad \Lambda \to \pi^-(5) + p(6)$$

$$\pi^0 \leadsto \gamma_1 + \gamma_2 \leadsto (e^+ + e^-)(7) + (e^+ + e^-)(8).$$

The numbers refer to visible tracks (solid lines). Neutrals (broken lines) are invisible.

 ISBN 0-8053-8601-7

The U triplet containing $|n\rangle$, $|\frac{1}{2}(\Sigma^0 + \sqrt{3}\,\Lambda)\rangle$, and $|\Xi^0\rangle$ should also have equally spaced masses, and the masses of n, $\frac{1}{4}(\Sigma^0 + 3\Lambda)$, and Ξ^0 are 939.55, 1134.81, and 1314.9, respectively, with mass differences of 195.3 and 180.1 MeV.

Okubo and Gell-Mann (1962) have further generalized the mass formula, obtaining for hyperons

$$M = M_1 + M_2 Y + M_3\left[T(T+1) - \frac{Y^2}{4}\right] \tag{17-3.1}$$

where M_1, M_2, M_3 are constant in one multiplet. For bosons they gave

$$\mu^2 = \mu_0^2 + \mu_1^2\left[T(T+1) - \frac{Y^2}{4}\right] \tag{17-3.2}$$

where μ_0^2, μ_1^2 are again constant in one multiplet.

The electromagnetic interaction conserves U exactly, but does not conserve i-spin. It follows that the electromagnetic properties of the members of the same U multiplet are equal. Thus the magnetic moments of p and Σ^+, n and Ξ^0, Σ^- and Ξ^- are expected to be equal.

We may also calculate relations between electromagnetic mass differences between members of the same i-spin multiplet. For instance, in the baryon decimet (Fig. 17-6a), assume that the symmetry is broken first by the medium-strong interaction and then by the electromagnetic, without any cross term. Then the mass differences in an i-spin multiplet are equally spaced and we have

$$\Delta^- - \Delta^0 = \Sigma^{-\prime} - \Sigma^{0\prime} = \Xi^{-\prime} - \Xi^{0\prime}$$
$$\Delta^0 - \Delta^+ = \Sigma^{0\prime} - \Sigma^{+\prime}$$

where the particle symbol stands for its mass. A more elaborate procedure taking into account both medium-strong and electromagnetic interactions gives relations such as

$$\Xi^- - \Xi^0 + n - p = \Sigma^- - \Sigma^+$$

involving all six particles on the periphery of the octet diagram. Such a relation gives 7.8 MeV for both sides which is well within the experimental error of about 0.2 MeV.

The quark model also predicts magnetic moments on the assumption that for each quark the magnetic moment is proportional to the electric charge. One can then calculate the ratio of the magnetic moments of baryons, obtaining the ratios indicated in Table 17-4. Without detailed models it is however ambiguous whether each magnetic moment contains in the denominator the mass of the particle. For n and p this prediction is very well satisfied; for other particles the experimental data are insufficient for us to know.

ISBN 0-8053-8061-7

TABLE 17-4 BARYON MAGNETIC MOMENTS PREDICTED BY QUARKS

Baryon	p	n	Σ^+	Σ^0	Σ^-	Λ^0	Ξ^0	Ξ^-
μ	1	$-\frac{2}{3}$	1	$\frac{1}{3}$	$-\frac{1}{3}$	$-\frac{1}{3}$	$-\frac{2}{3}$	$-\frac{1}{3}$

17-4 CROSS-SECTION PREDICTIONS BY $SU(3)$ AND QUARKS

Conservation of U-spin in strong interactions predicts the ratio of cross sections or branching ratios in decays in a manner similar to what obtains from i-spin conservation. The results are in fair agreement with experiment. For this topic see (W 71).

The quark model predicts cross sections with moderate success. The underlying idea is simple. Assume for a moment that all qq cross sections are equal, irrespective of the nature of the quarks. In the forward direction the scattering amplitudes add and for a pion–proton collision there are $2 \times 3 = 6$ combinations of each quark of the pion and of the proton. For the proton –proton collision there are $3 \times 3 = 9$ such combinations. The amplitudes should thus be in the ratio of $6/9$. The optical theorem, however, says that σ_T is proportional to Im $f(0)$; thus we could expect a ratio $\sigma(\pi p)/\sigma(pp) = 2/3$, neglecting the small real part of $f(0)$. This relation is not too far from the truth. By refining the argument and introducing less sweeping hypotheses, one obtains much better agreement with experiment. For instance, it is reasonable to assume that at high energy $\sigma(qq) = \sigma(q\bar{q})$, according to a theorem due to Pomeranchuk, and, by i-spin invariance, that $\sigma(du) = \sigma(uu) = \sigma(dd)$ and $\sigma(ds) = \sigma(us)$. With these assumptions and the method indicated above we find

$$\sigma(\pi^+ p) + \sigma(\pi^- p) + \tfrac{1}{3}\left[\sigma(K^+ p) + \sigma(K^- p) - \sigma(K^+ n) - \sigma(K^- n)\right]$$

$$= \tfrac{2}{3}\left[\sigma(pp) + \sigma(\bar{p}p)\right]$$

and also

$$\sigma(\Lambda p) = \sigma(pp) + \sigma(K^- n) - \sigma(\pi^+ p)$$

in good agreement with experiment.

17-5 REGGE RECURRENCES

Another set of hypotheses, associated with the name of T. Regge, predicts the "recurrence" of a particle if, all other quantum numbers being unchanged, one or two units of angular momentum are added. The Regge recurrences have an analogy to the levels of an atomic system, having the same radial quantum number but increasing l. Think of a hydrogen atom (neglecting spin); assume as zero the energy of the electron and proton at a

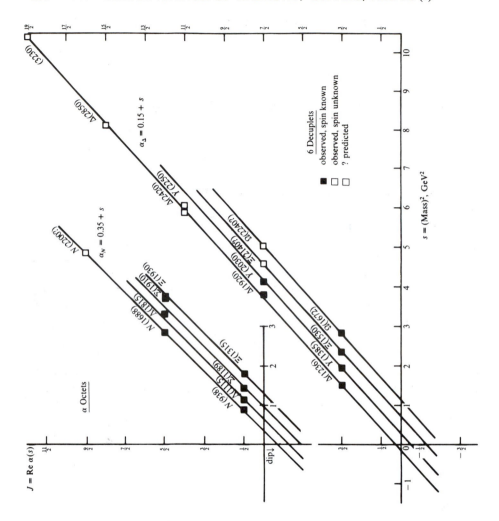

ISBN 0-8053-8061-7

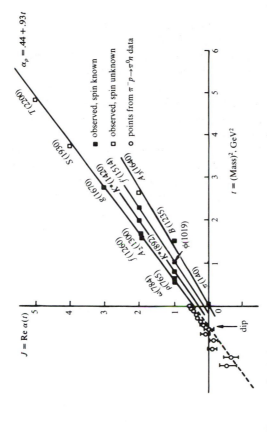

Figure 17-9a and b Regge trajectories. (a) Baryon trajectories. Note the $\Delta J = 2$ separation. (b) Meson trajectories. Note the $\Delta J = 1$ separation. The trajectory contains even and odd status. This situation is called "exchange degeneracy". [From C. B. Chiu, *Ann. Rev. Nucl. Sci.*, **22**, 255 (1972).]

great distance from each other and at rest, then the atomic energy is given by $-R/(n_r + l)^2$, where R is the Rydberg constant. It is possible for an atomic system to generalize the energy formula and to define the energy as an analytic function of l, considered as a complex variable. However, physical states correspond only to l real and integral. All levels having the same n_r lie on a *Regge trajectory* and are called *recurrences* of the first. A Regge trajectory may contain stable (negative energy) and unstable points. The latter appear physically as resonances.

In baryons or mesons addition of 1 or 2 units of angular momentum respectively give a recurrence. This is related to the behavior of exchange forces which change sign for even or odd angular momenta (Sec. 10-7). The recurrences of an octet should then form a new octet.

A family of particles in the Regge classification is thus formed by particles having the same i-spin and strangeness but increasing angular momentum. In such families the mass squared is related to the angular momentum by the formula

$$J = \alpha' m^2 + \alpha_0 \qquad (17\text{-}5.1)$$

where α' is about $1(\text{GeV}/c^2)^{-2}$. Notable examples of such families are shown in Fig. 17-9a and b.

17-6   CHARM

Thus far we have seen the success of the $SU(3)$ group and its quark representation. Many generalizations have been prompted by mathematical motivations. Experimental evidence supports an approximate $SU(4)$ symmetry and the introduction of a fourth quark. The first experimental evidence that called for a fourth quark was the absence of a strangeness changing neutral current in weak interactions (See Ch.19) explained by S. L. Glashow, J. Iliopoulos and Maiani as far back as 1970 by the existence of a fourth quark. Later the value of R (see Sect. 17-7), the discovery of the J/ψ particle (See Sect. 18-4) and the physics of states obtained by e^+-e^- collisions has corroborated the existence of a fourth quark to a practical certainty. Table 17-1 gives the quantum numbers of this fourth, charmed, quark.

Charm is a new additive quantum number in many ways similar to strangeness. Charm is conserved in strong interactions and enters in a generalized Gell–Mann Nishijima formula

$$Q = T_3 + \tfrac{1}{2}(B + S + C)$$

With four quarks the multiplets' graphical construction requires three dimensions and is much more complicated. For instance the $q\bar{q}$ multiplet contains 16 members, 9 of which are in the old $SU(3)$ meson octet and singlet. Table 17-5 gives the $q\bar{q}$ system of $J^P = 0^-$. The part that does not contain any

ISBN 0-8053-8601-7

TABLE 17-5 0⁻ MESONS

$\bar{q}$ \ q	u	d	s	c
$\bar{u}$	$\dfrac{\pi^0}{\sqrt{2}} + \dfrac{\eta}{\sqrt{6}} + \dfrac{\eta'}{\sqrt{3}}$	π^-	K^-	D^0
$\bar{d}$	π^+	$-\dfrac{\pi^0}{\sqrt{2}} + \dfrac{\eta}{\sqrt{6}} + \dfrac{\eta'}{\sqrt{3}}$	$\overline{K^0}$	D^+
$\bar{s}$	K^+	K^0	$-\sqrt{\tfrac{2}{3}}\,\eta + \dfrac{\eta'}{\sqrt{3}}$	F^+
$\bar{c}$	$\overline{D^0}$	D^-	F^-	η_c

c quarks is the old $SU(3)$ octet and singlet. The new charmed mesons are contained in the last row and column with the nomenclature that is at present most usual. Table 17-6 presents the same $q\bar{q}$ multiplet for $J^P = 1^-$. This multiplet contains mesons that can be directly reached in e^+e^- collisions. Note that the $c\bar{c}$ mesons called J/ψ have $C = 0$.

Figure 17-9 shows the generalization of Fig. 17-4 for $SU(4)$. For 3 quarks one obtains baryons. Polyhaedra containing the spin $\frac{1}{2}$ octet and the spin 3/2 decimet are shown in Fig. 17-10. The $SU(3)$ multiplets are contained in the planes with $C = 0$.

The introduction of charm, conserved in strong interactions requires associated production of charmed particles in the same manner as strangeness requires associated production of strange particles. The neutral K mesons K^0 and $\overline{K^0}$ have opposite strangeness and give rise to interesting phenomena in their weak decay. Analogously there must be neutral charmed mesons D^0 and $\overline{D^0}$ that may show in their weak decay effects similar to those shown by the K^0 and $\overline{K^0}$. (See Ch. 19). However the fast decay of D^0, $\overline{D^0}$ makes it difficult to observe the mixing effect.

Experimentally one has observed a particle formed by $c\bar{c}$ quarks. It exhibits no charm although composed of two charmed quarks. This particle and its excited states are similar to positronium composed of e^+ and e^- and exhibiting no charge. The particle is called charmonium and it is formed in electron–positron collisions. It is treated more in detail in Ch. 18-4.

TABLE 17-6 1⁻ MESONS

$\bar{q}$ \ q	u	d	s	c
$\bar{u}$	$(\omega + \rho^0)/\sqrt{2}$	ρ^-	K^{*-}	D^{*0}
$\bar{d}$	ρ^+	$(\omega - \rho^0)/\sqrt{2}$	$\overline{K^{*0}}$	D^{*+}
$\bar{s}$	K^{*+}	K^{*0}	φ	F^{*+}
$\bar{c}$	$\overline{D^{*0}}$	D^{*-}	F^{*-}	J/ψ

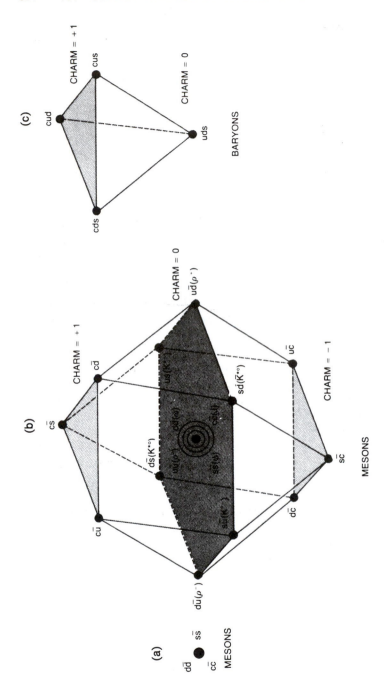

Figure 17-10 Supermultiplets of hadrons that include the predicted charmed particles can be arranged as polyhedrons. Each supermultiplet consists of particles with the same value of J^P. Within each supermultiplet the particles are assigned positions according to three quantum numbers: positions on the shaded planes are determined by i-spin and strangeness; the planes themselves indicate values of charm. The mesons are represented by a point (a) and by a solid called a cuboctahedron (b), which comprises 15 particles, including six charmed ones. The mesons shown are those with $J^P = 1^-$, but all mesons fit the same point and cuboctahedron representations.

ISBN 0-1093-8553-7

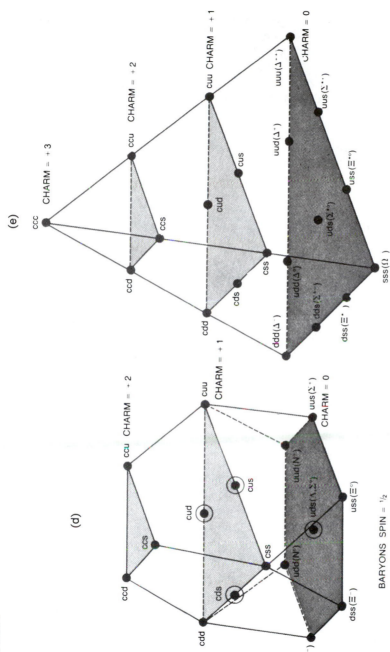

BARYONS SPIN = ³/₂

BARYONS SPIN = ½

The baryons form a small regular tetrahedron (*c*) of four particles, a truncated tetrahedron (*d*) of 20 particles and a larger regular tetrahedron (*e*), also of 20 particles. Both mesons and baryons are identified by their quark constitution, and for those particles that have been observed the established symbol is also given. Each figure contains one plane of uncharmed particles identical with Figs. 17-4 and 17-6, from L. Glashow *Scientific American* Oct. 1975.

Particles exhibiting charm have also been observed. In e^+e^- collisions at SLAC when the beam reaches an energy of 4 GeV one finds a neutral state at about 1865 MeV decaying into $K^{\pm}\pi^{\mp}$ or $K^{\pm}\pi^{\mp}\pi^{\pm}\pi^{\mp}$. The evidence for $C \neq 0$, i.e. for charm, is 1) the high production threshold shows associated production and this is confirmed by an analysis of the kinematics of the decay products, that show the recoil from an about equally massive partner. 2) The identity of the invariant mass of the different decay channels shows the presence of one type of particle only. 3) The narrowness (< 40 MeV) of the level shows a decay forbidden for strong interaction. 4) The presence of a K is important because in the hadronic weak decay of a charmed quark c one obtains prevalently $s, u, \bar{d}$ quarks and one has the approximate selection rule $\Delta Q = \Delta C = \Delta S$ as will be discussed in Ch. 19. A D^0 charmed meson containing $c\bar{u}$ quarks goes to $s u d \bar{u}$. The $s\bar{u}$ form K^- and $u\bar{d}$ form π^+. Any number of $\pi^+\pi^-$ pairs compatible with the energy available may be added. We have thus the formation of $K^-\pi^+$ (and $\pi^+\pi^-$ pairs), but not of $K^+\pi^-$. 5) Parity is violated in the decay showing a decay by weak interaction. In the actual experiment D^0 and $\overline{D^0}$ are formed. The latter decays into $K^+\pi^-$ (and $\pi^+\pi^-$ pairs).

Charmed baryons have also been obtained. For instance physicists at Fermilab have obtained by photoproduction a narrow state at 2.26 GeV decaying into $\overline{\Lambda}\pi^-\pi^-\pi^+$ but not into $\overline{\Lambda}\pi^+\pi^+\pi^-$. These particles are attributed to the decay of a charmed baryon containing the $\bar{c}\bar{u}\bar{d}$ quarks, and called $\overline{\Lambda_c^+}$. The decay of $\bar{c}$ gives most probably $\bar{s}\bar{u}\bar{d}$ and one has the quarks $\bar{s}\bar{u}\bar{d}\bar{u}\bar{d}$. We combine the $\bar{s}\bar{u}\bar{d}$ into a $\overline{\Lambda}$ and $\bar{u}\bar{d}$ into a π^-. To this we add $\pi^+\pi^-$ pairs. We see that in this way one obtains $\overline{\Lambda}\pi^-(\pi^+\pi^-)$. One can see that there is no charmed antibaryon leading to a $\overline{\Lambda}\pi^+(\pi^+\pi^-)$ state.

It is apparent that many charmed particles will soon be discovered and that a whole new particle spectroscopy is in the offing.

17-7   COLOR

The quantum numbers thus far considered (*i*-spin, strangeness, charm) are collectively called "flavors". Quarks seem to need a further qualification called "color". Color comes in three varieties and form another $SU(3)$ group, however actually observed particles are always strictly color singlets. With the introduction of color the fundamental quarks become 12 because each of the four quarks u, d, s, c may have three colors. The antiquarks have anticolors.

What first prompted the introduction of color (O. W. Greenberg 1964) was a serious difficulty with the spin statistics connection of colorless quarks. Particles containing three quarks of spin $\frac{1}{2}$ must be fermions and as such must have a wave function antisymmetric with respect to the exchange of any two quarks. Careful analysis of specific examples, for instance $\Delta^{++}(1236)$, gives a wavefunction symmetric with respect to the exchange of any two quarks. In the Δ^{++} the three quarks are in an s state with respect to each other and this gives symmetry on exchange. Attempts to introduce other an-

ISBN 0-8053-8601-7

gular momenta lead to insurmountable difficulties. One thus faces a paradox. Color solves it because the singlet state in color is always antisymmetric with respect to the exchange of 2 quarks (as the expression $a \cdot (b \times c)$ is antisymmetric on the exchange of any two factors) and multiplying the symmetric wavefunction by the wavefunction corresponding to color singlet one obtains the desired antisymmetry. If this were the only reason for introducing color it would be only an ad hoc artifice of little value. There are, however, more independent corroborations from three other sources.

The rate of $\pi^0 \rightarrow \gamma\gamma$ decay can be calculated with good agreement with experiment. Without color one would be off by a factor of 9.

The ratio

$$R = \frac{\sigma(e^+e^- \rightarrow \text{hadrons})}{\sigma(e^+e^- \rightarrow \mu^+\mu^-)}$$

is comparatively easy to measure and significant because at high energy the cross section for $e^-e^+ \rightarrow$ hadrons, according to most theories, should tend to the cross section for $e^+e^- \rightarrow \bar{q}q$;. This assumes that the primary phenomenon is the electromagnetic formation of quark antiquark pairs that immediately turn into hadrons with unit probability. The diagram Fig. 17-11 compares μ and quark production in e^+e^- collisions. What is important for R are the first two vertices that, for point like quarks and point like muons are the same except for the charges $\pm 1/3$ or $\pm 2/3$ for quarks and ± 1 for muons. The quantity R is expected to be energy independent. Under these circumstances the ratio R is given by

$$R = \Sigma Q_j^2$$

where Q_j is the charge of the j kind of quark. The measurement of R gives thus a means of counting the species of quarks. In the case of $SU(3)$ for 3 quarks we have $\Sigma Q_j^2 = (1/3)^2 + (1/3)^2 + (2/3)^2$ and $R = 2/3$. For $SU(4)$ $R = 10/9$. However the experimental R (Fig. 17-12) shows much larger values than 1.11 or 10/9 indicating that there are more varieties of quarks than in

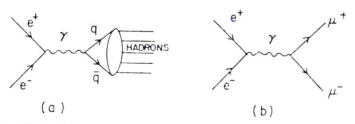

(a)   **(b)**

Figure 17-11 Diagrams for $e^+e^- \rightarrow$ hadrons and $e^+e^- \rightarrow \mu^+\mu^-$. The first two vertexes are the same in both cases. The hadron formation in part (a) occurs with probability one.

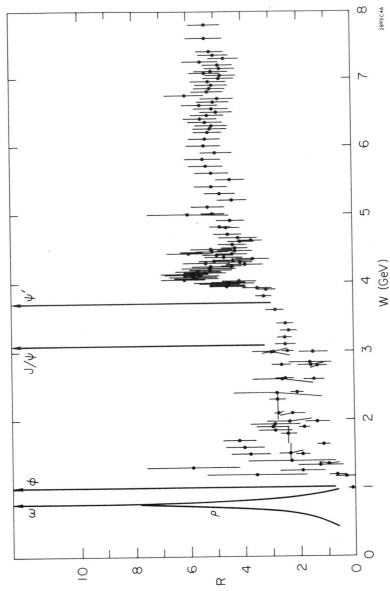

Figure 17-12 Values of $R = \sigma(e^+e^-) \rightarrow \text{hadrons}/\sigma(e^+e^-) \rightarrow \mu^+\mu^-$ as a function of energy. The resonances correspond to specific particles. The plateaus below 4 GeV is at a level of approximately 2. Beyond 4 GeV charmed quarks raise R, almost to a level of 5. (R. F. Schwitters, and K. Strauch, *Ann. Rev. Nucl. Sci.* **26**, 89 (1976)).

ISBN 0-8053-8601-7

$SU(3)$ or even in $SU(4)$. Color helps the situation by multiplying by three the number of quarks. The value of R thus obtained, 3.33, is in indifferent agreement with observation. The low energy value 2 corresponds to the energy domain in which charmed quarks are yet inaccessible. Beyond the threshold for creation of charmed quarks the R value reaches about 5. However the theoretical value 3.33 is lower than the experimental value of R, arousing the suspicion that there may be more types of quarks. An alternative hypothesis is the existence of leptons much heavier than muons, that would be counted in hadronic production.

Renormalizable gauge theories of weak interactions require that the sum of the lepton charges neutralizes the quark charges. The four leptons have charge -2; the quarks (Table 17-1) have charge 2/3 without color, but 3 colors reestablish the required neutrality!

If a heavy charged lepton should be established with certainty, renormalizable theories would require at least two more quarks as well as one more neutral lepton, a third neutrino, not necessarily massless.

The whole quark theory has to face the problem of the confinement of quarks and why they do not appear free. There are many theoretical speculations for justifying the confinement. For instance it is contemplated that the energy necessary for separating quarks increases infinitely with distance, as if the quarks were bound by a force independent of distance. Quark pairs would then be created as the distance is increased, causing multiple production of particles before the quarks are separated. If the strong interaction between quarks is mediated by massless mesons of i-spin one which themselves have color (gluons) the force likely to be realized is proportional to distance as in a classical harmonic oscillator.

BIBLIOGRAPHY

Chiu, C. B., "Evidence for Regge Poles, etc.," *Ann. Rev. Nucl. Sci.*, **22**, 255 (1972).
Feld, B. T., *Models of Elementary Particles*, Blaisdell Pub. Co., Waltham, Mass., 1969.
Gaillard, M. K., B. W. Lee, J. L. Rosner, "Search for Charm" *Rev. Mod. Phys.* **47**, 277 (1975).
Gasiorowicz, S., *Elementary Particle Physics*, Wiley, New York, 1966.
Gell-Mann, M., and Y. Ne'eman, *The Eightfold Way*, Benjamin, New York, 1964.
Lipkin, H. J., *Lie Groups for Pedestrians*, Wiley, New York, 1965.
Lipkin, H. J., "Quarks for Pedestrians," *Phys. Repts.*, **8C**, 175 (1973).
Morpurgo, G., "A Short Guide to the Quark Model," *Ann. Rev. Nucl. Sci.*, **20**, 105 (1970).
Nishijima, K., "Charge Independence Theory of V-Particles," *Progr. Theoret. Phys. (Kyoto)*, **13**, 285 (1955).
Nishijima, K., *Fundamental Particles*, Benjamin, New York, 1963.
Samios, N. P., M. Goldberg, and B. T. Meadows, "Hadrons and $SU(3)$," *Rev. Mod. Phys.*, **46**, 49 (1974).
Zakharov, V. I., Ioffe, B. L., and L. B. Okun, "New Elementary Particles," *Sov. Phys. Usp.* **18**, 75 (1975).

ISBN 0-8053-8601-7

PROBLEMS

17-1 Using a graphic method similar to that of Fig. 17-1, combine spins 1 and $\frac{1}{2}$, obtaining $\frac{3}{2}$ and $\frac{1}{2}$; do the same for two spins 1, obtaining spins 2, 1, and 0.

17-2 Show that according to the quark model, for mesons we must have $|S| < 1$, $|T| < 1$, and $|Q| < 1$ (S is the strangeness). Furthermore, indicating by L the orbital angular momentum and by s the spin, we have

$$P = (-1)^{L+1} \qquad C = (-1)^{L+s} \qquad G = C(-1)^T$$

17-3 Construct diagrams similar to Fig. 17-5 for qq, $qq\bar{q}$, and $qqqq$. Find the multiplets they contain. These quark multiplets are not found in nature; what charges would they have?

17-4 Show that the quark model predicts

$$\sigma(\Sigma^- p) = \sigma(pp) + \sigma(K^- p) - \sigma(\pi^- p) + 2[\sigma(K^+ n) - \sigma(K^+ p)]$$

$$\sigma(\Sigma^- n) = \sigma(pp) + \sigma(K^- p) - \sigma(\pi^- p)$$

At high energy, the first is 35 mb and theory agrees with experiment. For the second, the agreement is moderate (34 mb).

17-5 Consider the decays

$$\Sigma^-(1328) \Rightarrow \Lambda + \pi^- \qquad \text{and} \qquad \Delta^-(1236) \Rightarrow n + \pi^-$$

The initial state has $U = \frac{3}{2}$. For $\Lambda + \pi^-$ the final state is not an eigenstate of U but can be expressed as a superposition of two eigenstates. $n + \pi^-$ has $U = \frac{3}{2}$. Assuming U invariance, calculate the ratio of the matrix elements for the two decays. To compare with experiment you also need the phase-space factor.

17-6 Study Fig. 17-8 and give reasons why the decays are labeled weak and electromagnetic, respectively. Prove that the strangeness of Ω^- must be -3. Find alternate modes for the decay of Ω^-. Compare other Ω^- in the literature. [Consult current (RPP) or L. W. Alvarez, *Phys. Rev.*, **D8**, 702 (1973).]

17-7 As a very crude approximation we could consider the mass of a particle as the sum of the mass of the constituent quarks. Denoting the mass of a particle by its symbol we then have for the baryon decimet:

$$\Delta^{++} = 3u \quad \Sigma^{+\prime} = 2u + s \quad \Xi^{0\prime} = u + 2s \quad \Omega^- = 3s$$

$$\Sigma^{+\prime} - \Delta^{++} = \Xi^{0\prime} - \Sigma^{+\prime} = \Omega^- - \Xi^{0\prime} = s - u$$

Compare the results with the masses given in Table 13-2.

17-8 Do the same as in the preceding problem for the meson octet comparing K^-, K^+, K^0, $\bar{K}^0$, π and $\eta = \sqrt{\frac{1}{6}}\,(u\bar{u} + d\bar{d}) - \sqrt{\frac{2}{3}}\,s\bar{s}$. Obtain $3\eta + \pi = 4K$ and compare with the experimental results.

17-9 Try to do the same for the baryon octet. You would find $\Lambda = \Sigma = 2u + s$. However Λ and Σ have slightly different masses. The mass difference is connected to the *i*-spin. Compare the results of the Gell-Mann Okubo mass formula, Eq. 17-3.1.

17-10 Show that the decays $D^+ \to K^+ \pi^0$ or $D^+ \to K^+ \pi^+ \pi^-$ are forbidden, but that $D^+ \to K^- \pi^+ \pi^+$ is allowed.

ISBN 0-8053-8001-1

C H A P T E R

18

Form Factors and e^+e^- collisions

18-1 FORM FACTORS FOR NUCLEONS

High-energy electrons scattered on nucleons give information about the distribution of the electric charge and magnetism in the nuclei. This information is best conveyed by using form factors, which are a generalization of those used in Sec. 6-3. Inelastic scattering, with the generation of new particles reveals further subnuclear structures that are not completely understood.

To treat these phenomena it is first necessary to understand the kinematics, which in all interesting cases is relativistic. For brevity we shall use units of $\hbar = c = 1$ and often neglect the energy corresponding to the electron rest mass.

Consider first a collision of an electron with a proton and represent it by the diagram in Fig. 18-1. In order to treat elastic and inelastic collisions together, we consider the possibility of several escaping particles besides the scattered electron. Call M' their total mass and P' their total four momentum. In the elastic case obviously $M' = M$, the mass of the proton.

The incident electron has four momentum p, the scattered electron p', the proton before collision P, and the proton after collision (or all other particles formed in the case of inelastic collision) P'. We take four vectors $p = (\mathbf{p}, iE)$; thus $p^2 = \mathbf{p}^2 - E^2 = -m^2$ for a real particle. The four momentum transferred by the electron in the collision is $p - p' = q$. All four vectors squared are invariants in a Lorentz transformation and four momentum is conserved in collisions. Let us calculate

$$q^2 = (\mathbf{p} - \mathbf{p}')^2 - (E - E')^2 = (\mathbf{p}^2 - E^2) + (\mathbf{p}'^2 - E'^2) - 2\mathbf{p}\cdot\mathbf{p}' + 2EE'$$

$$= -2m^2 - 2|\mathbf{p}\,||\,\mathbf{p}'|\cos\theta + 2EE' \tag{18-1.1}$$

Emilio Segrè, Nuclei and Particles: An Introduction to Nuclear and Subnuclear Physics, Second Edition

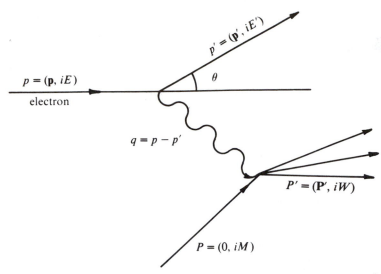

Figure 18-1   Diagram of an electron colliding with a proton $e + p \rightarrow e' + X$ where X may be a proton (elastic collision) or any particle or group of particles allowed by conservation laws.

If we neglect the electron mass, the last term is $2|\mathbf{p}|\,|\mathbf{p}'|$ and we have

$$q^2 = 2|\mathbf{p}||\mathbf{p}'|(1 - \cos\theta) = 4|\mathbf{p}||\mathbf{p}'|\sin^2(\theta/2) \qquad (18\text{-}1.2)$$

Here θ is the electron deflection.

We can also calculate q^2 from the energy imparted to the proton, supposed originally to be at rest in the laboratory. For the proton or other particles formed

$$P' = P + q, \qquad P = (0, iM) \quad \text{and} \quad W^2 = M'^2 + \mathbf{P}'^2 = (M' + T)^2 \quad (18\text{-}1.3)$$

with W the total and T the kinetic energy of the nucleon (and possibly other particles) after the collision. Hence

$$q^2 = (P' - P)^2 = -M^2 - M'^2 + 2MW$$

$$= M^2 - M'^2 + 2M\nu \qquad (18\text{-}1.4)$$

where $\nu = -q \cdot P/M$ is a relativistic invariant that, in the laboratory system, takes the value $\nu = E - E'$.

It is apparent from Eq. (18-1.4) that q^2 is positive in a real collision. Vectors with positive lengths squared are called *spacelike*. If we transfer the momentum with a real particle, q^2 must be negative, because $q^2 = -m^2$ of

ISBN 0-8053-8601-7

the particle transferring the momentum. For a real photon $q^2 = 0$; for a "virtual photon" q^2 may be positive or negative.

Note that given q^2 and ν, M' is determined. The two scalars ν and q^2 completely determine the collision unless one distinguishes polarizations, and/or analyzes further the particles of total mass M'.

We shall treat the elastic case first. Assume that a Dirac electron scatters on a Dirac proton (i.e., on a proton obeying Dirac's equation for the electron, except for the mass). Such a proton would have the magnetic moment $e\hbar/2M_p c$. The scattering of a Dirac electron by a spinless point charge of mass M, with θ and ω measured in the laboratory system and calling $e^2/mc^2 = r_0 = 2.82 \times 10^{-13}$ cm, is given in Eq. (6-3.20);

$$\frac{d\sigma}{d\omega_M} = r_0^2 \left(\frac{m}{2E} \right)^2 \frac{1}{\sin^4(\theta/2)} \times \frac{\cos^2(\theta/2)}{1 + (2E/M)\sin^2(\theta/2)} \tag{18-1.5}$$

where the first factor is the usual Rutherford scattering and the second factor takes into account electron spin and proton recoil. This formula is due to Mott (1930). The proton spin gives another factor, $1 + (q/2M)^2 2[\tan^2(\theta/2)]$. However, the proton magnetic moment is anomalous and this modifies the magnetic factor. Rosenbluth (1950) obtained for a point nucleus

$$\frac{d\sigma}{d\omega_R} = \frac{d\sigma}{d\omega_M} \left\{ 1 + \frac{q^2}{4M^2} \left[2(1+\kappa)^2 \tan^2\left(\frac{\theta}{2} \right) + \kappa^2 \right] \right\} \tag{18-1.6}$$

where κ, the anomalous part of the magnetic moment, is 1.79 for the proton and -1.91 for the neutron. This formula is generalized to extended nucleons by introducing form factors: an electric form factor $F_1(q^2)$ and a magnetic form factor $F_2(q^2)$. They are normalized to give $F_1^p(0) = F_2^p(0) = F_2^n(0) = 1$; $F_1^n(0) = 0$, where p and n denote proton and neutron respectively. In a nonrelativistic approximation they are Fourier transforms of the charge density and of the magnetization density due to the anomalous magnetic moment only. With these form factors one obtains (Rosenbluth, 1950).

$$\frac{d\sigma}{d\omega_R} = \frac{d\sigma}{d\omega_M} \left\{ F_1^2(q^2) + \frac{q^2}{4M^2} \left[2(F_1(q^2) \right. \right.$$

$$\left. \left. + \kappa F_2(q^2))^2 \tan^2\left(\frac{\theta}{2} \right) + \kappa^2 F_2^2(q^2) \right] \right\} \tag{18-1.7}$$

It is, however, more convenient to use the linear combinations

$$G_M^{p,n} = F_1^{p,n} + \kappa F_2^{p,n} \quad \text{and} \quad G_E^{p,n} = F_1^{p,n} - \frac{q^2}{4M^2} \kappa F_2^{p,n} \tag{18-1.8}$$

ISBN 0-8053-8601-7

which are normalized to give

$$G_M^{p,\,n}\,(0) = \mu_{p,\,n} \qquad G_E^p\,(0) = 1 \qquad G_E^n\,(0) = 0 \qquad (18\text{-}1.9)$$

where p and n stand for proton and neutron, respectively, and $\mu_{p,\,n}$ is in units of the nuclear magnetons: $\mu_p = 2.79$, $\mu_n = -1.91$. With these form factors, Rosenbluth's formula becomes

$$\frac{d\sigma}{d\omega_R} = \frac{d\sigma}{d\omega_M}\left[\frac{G_E^2 + (q^2/4M^2)G_M^2}{1+(q^2/4M^2)} + \frac{q^2}{4M^2}\,2G_M^2\,\tan^2\!\left(\frac{\theta}{2}\right)\right] \qquad (18\text{-}1.10)$$

The $G(q^2)$ can be measured experimentally (Hofstadter, 1960) by measuring $R = (d\sigma/d\omega)_R/(d\sigma/d\omega)_M$ at the same q^2, but different θ; this permits us to separate $G_E(q^2)$ from $G_M(q^2)$. Under these conditions R should be a linear function of $\tan^2(\theta/2)$, as has been verified (see Fig. 18-2). Note that for $q^2 \ll 4M^2$, $G_E(q^2)$ is the dominant term, but as q^2 increases G_M becomes more important and for high q^2 it overshadows the electric scattering.

For neutrons the best one can do, since a target of free neutrons is unavailable, is to use deuterons and subtract the proton effect. This procedure needs correction (Glauber, 1953) because the two nucleons partially shield

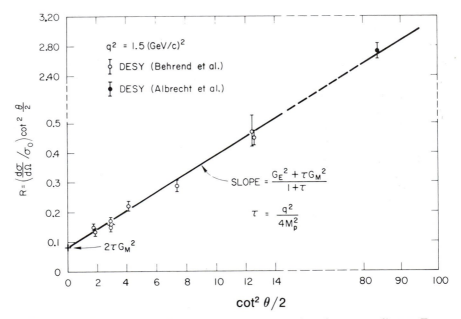

Figure 18-2 Analysis of electron–proton scattering data according to Eq. (18-1.10); $(d\sigma/d\omega)_R/(d\sigma/d\omega)_M\,\cot^2(\theta/2)$ at constant q^2 is a linear function of $\cot^2(\theta/2)$. Intercept and slope give the quantities indicated in the figure.

ISBN 0-8053-8601-7

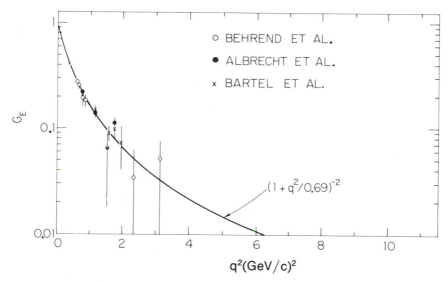

Figure   18-3   Electric form factor for $e-p$ elastic scattering as a function of q^2.

each other. The data are formally analyzed with Eq. (18-1.10). The results are shown in Figs. 18-3 and 18-4.

The form factors obey the simple empirical scaling law

$$G(q^2) = G_E^p(q^2) = \frac{G_M^p(q^2)}{|\mu_p|} = \frac{G_M^n(q^2)}{|\mu_n|} \qquad G_E^n(q^2) = 0 \quad (18\text{-}1.11)$$

which tells us essentially that the electric density and the density of magnetization are equally distributed in the proton, that the same distribution obtains for magnetism in the neutron, and that the neutron is neutral everywhere.

The function $G(q^2)$ is well reproduced, in the zone where it has been measured, by the formula

$$G(q^2) = \left(1 + q^2/M_q^2\right)^{-2} \qquad (18\text{-}1.12)$$

with $M_q^2 = 0.71$ in $(GeV)^2$. This is the Fourier transform of an exponential charge density $\rho(r) \propto \exp(-M_q r)$ A density proportional to a Yukawa-type potential $V(r) = g \exp(-M_q r)/r$, due to a quantum of mass M_q, would give

$$G(q^2) \propto \left(1 + q^2/M_q^2\right)^{-1}$$

ISBN 0-8053-8601-7

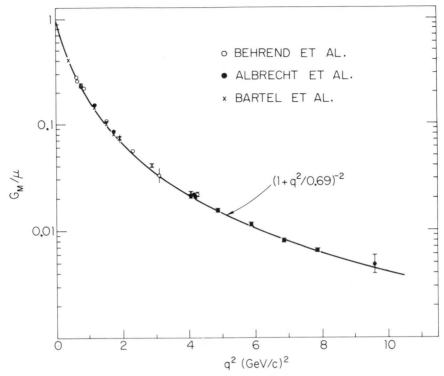

Figure 18-4   Magnetic form factor for *e−p* elastic scattering as a function of q^2.

without the square in the denominator, and it is attractive to analyze $G(q^2)$ as due to a sum of Yukawa-type terms. The interaction would be represented by a diagram in which the unknown particle must have the same quantum numbers as the γ, that is, $J^{PC} = 1^{--}$, strangeness and charge zero. Candidates of this description are the ω, $ρ^0$, and φ; furthermore two particles are needed, one of *i*-spin 0 (the ω) and one of *i*-spin 1 (the $ρ^0$) because the γ (electromagnetic interaction) may change *i*-spin by one unit. In fact, such particles were first hypothesized by Frazer, Fulco, and Chew (1960) to account for the form factors.

18-2   ELECTRON–PROTON INELASTIC SCATTERING

Electron–nucleon scattering may be also inelastic, with the creation of one or more particles, as indicated in Fig. 18-1. For one particle this inelastic scattering leads simply to the excitation of the proton by collision. The excited state could be, for instance, Δ′(1236) or other pion–nucleon resonances. This experiment is, in a different context, similar to a Franck–Hertz experi-

ISBN 0-8053-8601-7

ment. The mass M' of the final state, be it one or more particles, depends on q^2 and v, according to Eq. (18-1.4), and one must consider a doubly differential cross section. Calculation of this cross section separating the known electrodynamic part from the unknown strong interaction part described phenomenologically by form factors gives

$$\frac{d^2\sigma}{d\omega\, dE'} = \frac{d\sigma}{d\omega_{nr}} \left[W_2(v, q^2) + 2W_1(v, q^2) \tan^2(\theta/2) \right] \quad (18\text{-}2.1)$$

where $d\sigma/d\omega_{nr}$ is the Mott cross section calculated for an infinite mass (no recoil) and $d\omega$ is the solid angle in which one measures the electrons of energy E'. One can measure W_1, W_2 by measuring $d\sigma/d\omega$ at constant q^2 and v, and at various θ. At small θ, W_2 is the important factor, and in any case W_1 is found experimentally to be small compared to W_2.

Graphs of $d^2\sigma/d\omega\, dE'$ at constant θ and constant E are given in Figs. 18-5 to 18-7 as a function of the mass M' of the particles not measured. For $M' = M$ one would have a large elastic peak not shown in the figure. The maxima at 1.24 GeV, etc. correspond to excited states of the nucleon. Above them there is a continuum. As one goes to higher incident energies, the bumps disappear but the continuum persists. The bumps decrease about as fast as the elastic scattering, showing that the whole nucleon is involved.

In Fig. 18-7 we give $2MW_1$ and $vW_2(q^2)$ plotted as a function of *one* variable, $2Mv/q^2 = \omega$ (Not to be confused with solid angle). At low ω (in a region not contained in the figure), one clearly sees the dependence on v and q^2 separately, but for $\omega > 1$, W_2 is a function of ω, not of v and q^2 separately. This remarkable phenomenon (Bjørken and Paschos, 1969), is called *scaling*; it occurs for both v and q large, but in a constant ratio.

Scaling may be interpreted as indicating that the scattering is due to point-like particles (*partons*) contained in the nucleon. The argument qualitatively is as follows. If the electron collides with a pointlike particle of mass m, Eq. (18-1.4) gives $q^2 = 2mv$. The proton contains i pointlike partons (possibly related to quarks) of charge Q_i; they share the longitudinal momentum $\mathbf{P}$ of the proton, so that each has momentum $x_i\mathbf{P}$. Because of the relativistic space contraction and time dilation, they appear to the electron as slowly moving, transversally, in a disk corresponding to the relativistically flattened proton. One assumes that individual collisions with free particles are a good approximation.

For free point particles Eq. (18-2.1) must reduce to Mott scattering, and calculation shows that this requires that the form factor W_2 take the form $M\delta(q \cdot P - \frac{1}{2}q^2)$, which can also be written as $\delta[v + (q^2/2M)]$ with M the mass of the proton. W_1 is negligible. Now consider an aggregate of partons of mass m_i and charge Q_i in a frame in which the proton has a huge longitudinal momentum with respect to the electron. In particular, the longitudinal momentum is large compared with the transverse momentum of the partons.

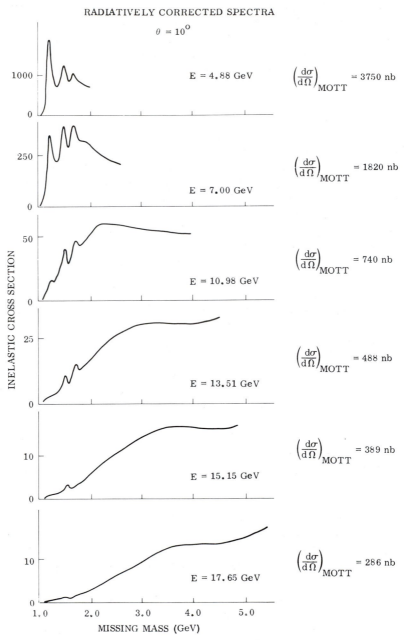

Figure 18-5 Visual fits to spectra showing the scattering of electrons from hydrogen at 10 deg for primary energies E from 4.88 to 17.65 GeV.. The elastic peaks have been subtracted and radiative corrections applied. The cross sections are expressed in nanobarns per GeV per steradian. [From J. I. Friedman and H. W. Kendall, *Ann. Rev. Nucl. Sci.,* **22,** 203 (1972).]

ISBN 0-8053-8601-7

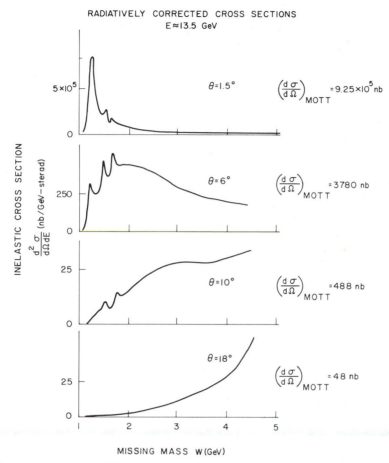

Figure 18-6 Visual fits to spectra showing the scattering of electrons from hydrogen at a primary energy E of approximately 13.5 GeV, for scattering angles from 1.5 to 18 deg. [From J. I. Friedman and H. W. Kendall, *Ann. Rev. Nucl. Sci.*, **22**, 203 (1972).]

Call **P** the total longitudinal momentum, and $x_i P$ the longitudinal four momentum of the ith parton. The contribution to W_2 of each parton is then

$$W_2^{(i)} = x_i Q_i^2 M\delta\left(x_i(q\cdot P) - \tfrac{1}{2}q^2\right) = Q_i^2 M\delta\left(q\cdot P - \frac{q^2}{2x_i}\right) = -Q_i^2\delta\left(\nu + \frac{q^2}{2Mx_i}\right)$$

(18-2.2)

(These relations use $\delta(\alpha x) = \alpha^{-1}\delta(x)$).

ISBN 0-8053-8601-7

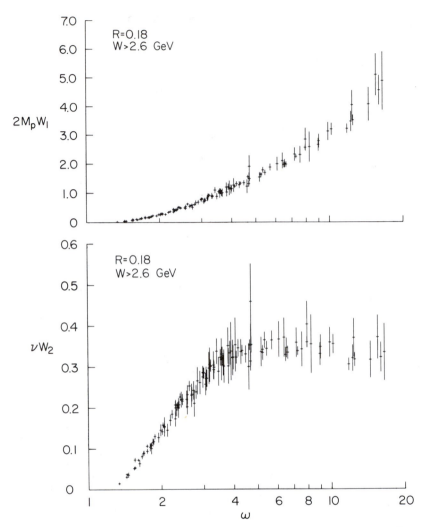

Figure 18-7 $2MW_1$ and νW_2 for the proton as functions of $\omega = 2M\nu/q^2$ for $W > 2.6$ GeV, $q^2 > 1$ $(GeV/c)^2$, and $R = 0.18$. The scaling property is implicit in plotting the variables as functions of ω only. [From J. I. Friedman and H. W. Kendall, *Ann. Rev. Nucl. Sci.*, **22**, 203 (1972).]

For a distribution of N partons one has the sum of all $W_2^{(i)}$ because the partons scatter incoherently.

$$W_2(\nu, q^2) = -\sum_N P(N)\left\langle \sum_i Q_i^2 \right\rangle_N \int_0^1 dx\, f_N(x)\delta\left(\nu + \frac{q^2}{2xM}\right) \quad (18\text{-}2.3)$$

ISBN 0-8053-8601-7

where $P(N)$ is the probability of finding N partons and $f_N(x)$ is their distribution in x, that is, in the longitudinal momentum; $\langle \Sigma_i Q_i^2 \rangle_N$ equals the average value of $\Sigma_i Q_i^2$ for such a distribution of partons. From this equation, by integrating over x one obtains

$$\nu W_2(\nu, q^2) = \sum_N P(N) \left\langle \sum_i Q_i^2 \right\rangle_N \xi f_N(-\xi) = F(\omega) \qquad (18\text{-}2.4)$$

where ξ is the value of x that nullifies the variable in the delta function, $\xi = q^2/2M\nu = 1/\omega$.

The parton model has also been used for describing proton–proton elastic scattering. Consider both protons as parton clouds and assume for a moment that the electron–proton form factor squared gives the probability that the cloud does not go to pieces under the impact of a particle (no matter which), transferring the momentum q to the cloud. For two colliding protons we should then expect a cross section proportional to the fourth power of the form factor, because both colliding particles must survive the collision. In addition to this type of p–p scattering, there is also diffraction scattering (see Chap. 20) and other phenomena. The observed p–p scattering is thus due to the combination of several mechanisms. Which is most important depends on the c.m. energy of the colliding protons and on the momentum transferred. If the parton–parton model were accurate, we should have

$$\left(\frac{d\sigma(q^2)}{dq^2} \bigg/ \frac{d\sigma(0)}{dq^2} \right)_{pp} = \left(\frac{d\sigma(q^2)}{dq^2} \bigg/ \frac{d\sigma(0)}{dq^2} \right)_{ep}^2$$

or

$$\left(\frac{d\sigma(q^2)}{dq^2} \right)_{pp} \propto \left[G(q^2) \right]_{ep}^4 \qquad (18\text{-}2.5)$$

Figure 20-7 shows the extent to which this relation obtains.

18-3   ELECTRON-POSITRON COLLISIONS

Electron–positron collisions can be studied up to c. m. energies of about 10 GeV by colliding beams. Techniques and accelerators are improving rapidly and luminosities of 10^{32} cm^{-2} sec^{-1} are within reach. Also, $e^- - e^-$ at comparable energies are under study.

The electromagnetic interaction between electron and electron or positron may be calculated with extreme accuracy by modern quantum electrodynamics. For instance, one has the $e^- - e^-$ scattering cross section (Møller,

1932), which for $E \gg m$ is given by

$$\frac{d\sigma}{d\omega} = \frac{r_0^2}{8} \left(\frac{m}{E} \right)^2 \left[\frac{1 + \cos^4(\theta/2)}{\sin^4(\theta/2)} + \frac{2}{\sin^2(\theta/2) \cos^2(\theta/2)} + \frac{1 + \sin^4(\theta/2)}{\cos^4(\theta/2)} \right]$$

(18-3.1)

and the $e^- - e^+$ scattering cross section (Bhabha, 1936)

$$\frac{d\sigma}{d\omega} = \frac{r_0^2}{8} \left(\frac{m}{E} \right)^2 \left[\frac{1 + \cos^4(\theta/2)}{\sin^4(\theta/2)} - 2 \frac{\cos^4(\theta/2)}{\sin^2(\theta/2)} + \frac{1 + \cos^2\theta}{2} \right]$$

(18-3.2)

where $r_0 = e^2/mc^2$. In both formulas energy, solid angle, and scattering angle are in the c. m. system, very much like what would be observed in colliding beams. These formulas need corrections for radiative processes and other electromagnetic processes. All of these can be calculated for any specific experimental situation. Other electromagnetic reactions are

$$e^+ + e^- \rightsquigarrow \mu^+ + \mu^-$$
$$e^+ + e^- \rightsquigarrow \gamma + \gamma$$
$$e^+ + e^- \rightsquigarrow e^+ + e^- + \gamma \qquad (18\text{-}3.3)$$

and so on. No leptons heavier than the muons have been seen with certainty, although some experiments indicate them. The outstanding result of this work is that quantum electrodynamics consistently gives the right answer within an experimental error of a few percent.

All the resulting formulas calculated for point charges have the form

$$\frac{d\sigma}{d\omega} = r_0^2 \left(\frac{m}{E} \right)^2 |f(\theta)|^2 \qquad (18\text{-}3.4)$$

where θ is a suitable angle, such as the scattering angle. One may generalize the formulas by introducing form factors function of the four-momentum transfer squared q^2. These form factors represent the effect of interactions other than the electromagnetic one, or they represent departures of the electromagnetic interaction from its standard form. The form factors in Eq. (18-1.7) show how they would appear in a typical example. The form factors may be parameterized in the form

$$F(q^2) = 1/\left[1 + (q^2/\Lambda^2) \right] \qquad (18\text{-}3.5)$$

Experiment agrees with all form factors $= 1$ when only leptons and photons

ISBN 0-8053-8601-7

participate in the reaction and the accuracy of the experiment sets a limit to Λ^2, usually of the order of 100 $(\text{GeV}/c)^2$. The length $\hbar/\Lambda < 2 \times 10^{-15}$ cm may also be used as a suitable parameter.

It is noteworthy that electrodynamics is valid over at least 27 powers of 10 in linear dimensions. As a practical consequence purely electrodynamic phenomena may be used for calibrating beams, luminosities, etc.

The situation is much more complicated when hadrons are involved. Collisions between e^+ and e^- may create hadrons and leptons, provided there is sufficient energy and the internal quantum numbers allow it. This limits the systems to $Q = S = 0 = N_{\text{nucl}} = N_{\mu \text{ leptons}} = N_{\text{e leptons}}$. Furthermore, considering $e^+ - e^-$ as equivalent to a photon, $J^{PC} = 1^{--}$. (See Chap. 13.) In particular one may have

$$e^+ + e^- \leadsto \pi^+ + \pi^-$$

$$e^+ + e^- \leadsto K^+ + K^- \quad \text{etc.} \tag{18-3.6}$$

Cross sections for such processes are of the form

$$\frac{d\sigma}{d\omega} = \frac{1}{8} r_0^2 \left(\frac{m}{E} \right)^2 \beta_\pi^3 \sin^2 \theta |F(q^2)|^2 \tag{18-3.7}$$

where the form factors contain the strong interaction. Note that $q^2 \leqslant -4m_\pi^2$. The reactions are often dominated by an intermediate particle, such as the ρ, ω or φ meson. Figure 18-8 shows, for example, the $e^+ - e^-$ cross section in the region of the φ resonance.

Similarly one has the reaction

$$e^- + e^+ \leadsto p + \bar{p} \tag{18-3.8}$$

This reaction is related to elastic electron–proton scattering by "crossing." The cross section is

$$\frac{d\sigma}{d\omega} = \frac{1}{4} r_0^2 \left(\frac{m}{E} \right)^2 \beta_p \left[|F_1(q^2) + \kappa F_2(q^2)|^2 (1 + \cos^2 \theta) \right.$$

$$\left. + \left| \frac{M_p}{E} F_1 + \frac{E}{M_p} \kappa F_2 \right|^2 \sin^2 \theta \right]$$

using c. m. coordinates. According to crossing, the form factors are the same analytic functions of q^2 as those of Eq. (18-3.7) but for different values of q^2. In scattering $q^2 \geqslant 0$; in annihilation $q^2 < -4M^2$, as can be easily recognized in the c. m. system. In this formula β_p is the proton velocity and κ the anomalous part of the magnetic moment of the proton.

Studies on the neutrino–proton collisions have been analyzed and yield corresponding form factors. They are similar to those obtained with electrons.

ISBN 0-8053-8601-7

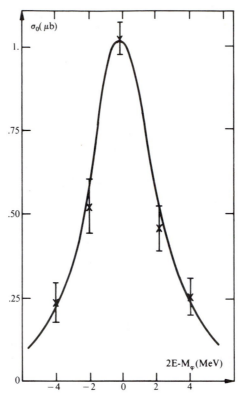

Figure 18-8 Excitation curve for the formation of φ mesons by e^+e^- collisions. The φ has a 34% branching into $K_S^0 K_L^0$. In the experiment the K_S^0 are detected by their decay into two pions. The curve is analyzed using a Breit Wigner formula with $M_\varphi = 1020$ MeV and $\Gamma_\varphi = 3.8$ MeV. The factor $|F(q^2)|^2$ of equation 18-3.7 is given by the resonance. The beam is calibrated on e^+e^- elastic scattering [From G. Cosme *et al.*, *Phys. Letters*, **48B**, 159 (1974).]

18-4   THE ψ-PARTICLES

In November 1974 the discovery of a new type of particles was simultaneously announced by two groups of physicists from The Massachusetts Institute of Technology–BNL and from SLAC–LBL. The MIT–BNL group bombarded Be with 30 GeV protons and found $e^+ - e^-$ pairs having a fixed energy in their center of mass. They interpreted the pairs as due to the decay of a neutral particle of mass 3.1 GeV called now J/ψ. The same resonance appeared very conspicuously at SLAC in $e^+ - e^-$ collisions. The finding was immediately confirmed and extended at Frascati (ADONE) and other laboratories.

In these experiments one finds a maximum for the cross section at 3.095 and at 3.684 GeV and a broad enhancement at 4.1 GeV. The resonances

ISBN 0-8053-8601-7

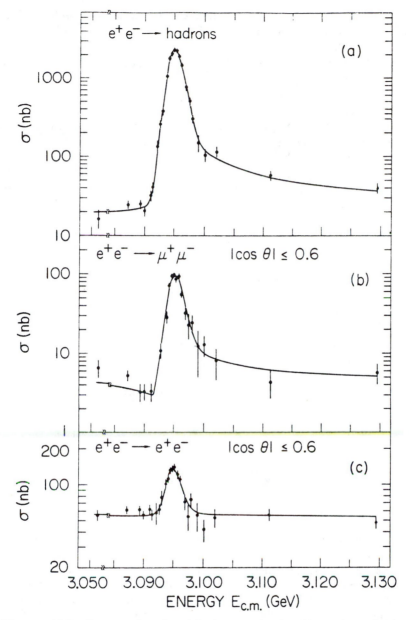

Figure 18-9 Cross sections for a) hadron production, b) μ pair production, and c) e pair production and scattering in the vicinity of ψ(3095). The notation cosθ ⩽ |0.6| refers to the angle observed. Courtesy SLAC and LBL.

ISBN 0-8053-8601-7

decay in e^+–e^- pairs, and μ^+–μ^- pairs, as well as hadronically (π, K, etc.). The $\psi'(3.7)$ resonance decays about half of the time into $J/\psi(3.1)$ and two pions (Fig. 18-10).

The width of the resonances is too small for direct measurement. The resolving power of the colliding beams is about 1.5 MeV and the Γ of $J/\psi(3.1)$ is only 70 keV. The Γ is inferred by assuming a Breit–Wigner formula for the resonance, subtracting the cross sections given by quantum electrodynamics, and measuring the branching ratios of the resonant part for the two processes mentioned above and for hadronic decay. The total area under the peak combined with the branching ratios mentioned above give the results summarized in Table 18-1.

The unusual sharpness of the resonances poses the problem of the decay mechanism. A pure weak decay would not be sufficient to give the observed

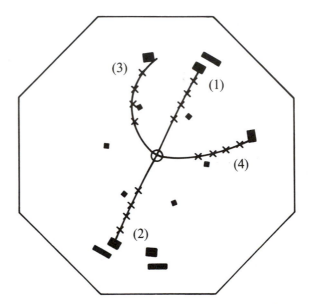

Figure   18-10   A spark chamber observation of the $\psi'(3.7)$ decaying into $J/\psi(3.1) + \pi^+ + \pi^-$. The lighter $J/\psi(3.1)$ decays into $e^+ + e^-$. The primary electrons and positrons generating the $\psi'(3.7)$ move perpendicularly to the plane of the figure. Parallel to them there is a 4000 G magnetic field sufficient to deflect the π^+ and π^- (3)(4) that have a kinetic energy of about 150 MeV each, but insufficient to visibly deflect the decay electrons (1), (2) that have an energy of about 1550 MeV each. They form the stem of the Ψ configuration visible in the figure. The curved part of the Ψ are the pions. The figure is a computer reconstruction of the tracks determined by the sparks located at the crosses. The black spots are triggering counters. The spots on the regular hexagon are mechanical supports. [From G. S. Abrams *et al.*, *Phys. Rev. Letters*, **34**, 1181, (1975).]

ISBN 0-8053-8601-7

TABLE 18-1   PROPERTIES OF $J/\psi(3.1)$ AND $\psi'(3.7)$

m	$3.095 \pm .004$ GeV	$3.684 \pm .005$ GeV
J^{PC}	1^{--}	1^{--}
$\Gamma_e = \Gamma_\mu$	$4.8 \pm .6$ keV	$2.1 \pm .3$ keV
Γ_h	59 ± 14 keV	224 ± 56 keV
Γ	69 ± 15 keV	228 ± 56 keV
Γ_e/Γ	$0.069 \pm .009$	0.0093 ± 0.0016
Γ_h/Γ	$0.86 \pm .02$	0.981 ± 0.003
$\Gamma_{\gamma h}/\Gamma$	0.17 ± 0.03	0.029 ± 0.004

width. The scarcity of gamma rays in the branching and the abundance of hadronic products negate a purely electromagnetic decay. A highly forbidden strong decay is the most likely explanation.

Concerning the nature of J/ψ, the e^+-e^- production shows that it is not a lepton. If J/ψ is formed in a one-quantum process, as is likely, its quantum numbers should be the same as those of the photon, that is, $J^P = 1^-$. There is direct confirmation of this assignment from interference between electron and muon pairs produced electromagnetically and by J/ψ decay. When $J/\psi(3.1)$ decays into pions only, these are in an odd number, indicating that J/ψ has a negative G parity and pointing also to strong decay, conserving G. Decay into $\Lambda\bar{\Lambda}$, and other modes too, assigns to both ψ the i-spin zero.

The possibility that J/ψ is a heavy boson mediating weak interactions is unlikely because both ψ's can be photoproduced by high-energy (20 GeV) gamma rays with cross sections appropriate to a hadron; for these and other valid reasons not mentioned here the heavy boson assignment is unacceptable.

All these arguments point to a hadronic nature for ψ's, which raises the question of what forbids the decay.

A satisfactory explanation interprets the $J/\psi(3.095)$ as a $c\bar{c}$ quark combination, i.e. as the ground state of charmonium, $J^{PC} = 1^{--}$. Charm is conserved in strong and electromagnetic interactions and a charmed particle can not decay strongly into a non charmed one, but charmonium states although composed of charmed quarks have $C = 0$ and can decay strongly into non charmed particles. This form of decay, however is hindered by a rule found by S. Okubo, G. Zweig and J. Iizuka, often called the Zweig rule. It says that a transition that changes the nature of the initial quarks is inhibited. The rule is essentially empirical and is exemplified in the $\varphi(1020)$ decay. The φ may decay into 3π with $\Gamma_{3\pi} = 670$ keV and into $2K$ with $\Gamma_{2k} = 3$ MeV. Phase space would favor the pionic decay, furthermore if one scales up from the analogous $\omega(783)$ decay one would obtain a 3π of about 100 MeV versus the 670 keV observed. The diagram relevant to the φ decay is given in Fig. 18-11 and it shows that for the φ decay the s quarks must change nature in going into 3 pions, but not in going into $2K$. The decays of the ω that contains the

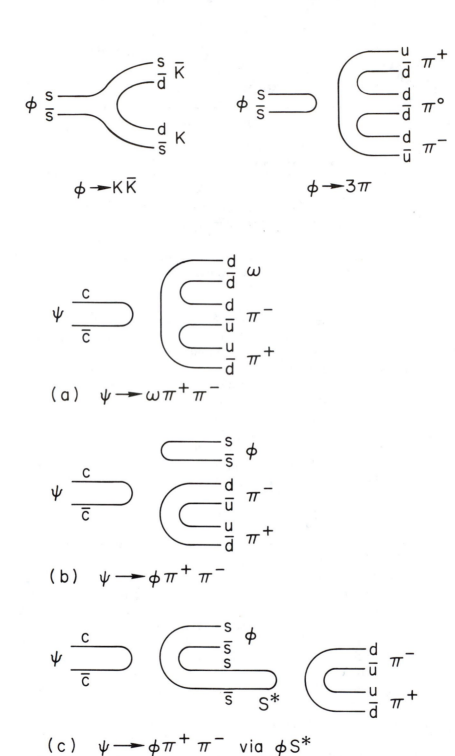

Figure 18-11 Quark diagrams illustrating Zweig's rule in φ and ψ decays.

ISBN 0-8053-8601-7

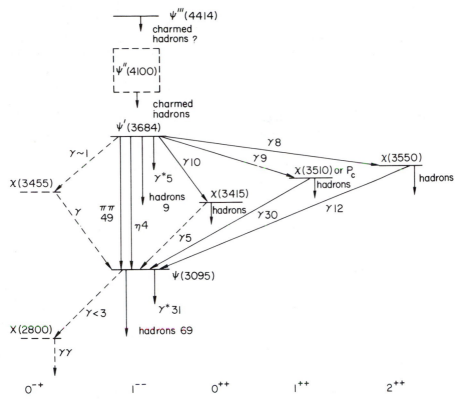

Figure 18-12 Levels of charmonium and their decays. The numbers denote the percentage of the branchings. The first column contains levels $^1S_0$, the second $^3S_1$, the third, fourth and fifth $^3P_0$, $^3P_1$, and $^3P_2$ respectively. (Courtesy SLAC-LBL)

$u\bar{u} + d\bar{d}$ quarks is not subject to the hindrance provided by the Zweig rule. One can formulate Zweig's rule by defining "*disconnected diagrams*" as those in which one or more particles can be isolated by drawing a line which does not cross any quark lines. Zweig's rule says that disconnected diagrams are suppressed compared to connected ones. See Fig. 18-11.

The ψ which decays into ordinary particles that do not contain c quarks are all hindered by this rule. States containing charmed quarks are energetically inaccessible.

The hypothesis that J/ψ (3.1) and $\psi'(3.7)$ are composed of $c\bar{c}$ quarks is supported by the observation of 2 or more intermediate states (χ states) to which the ψ' decays by gamma emission. The intermediate states decay in turn either by gamma emission to $J/\psi(3.1)$ or into hadrons. The intermediate states are not formed directly in $e^+ - e^-$ collisions probably because they do not have the quantum numbers $J^P = 1^-$. A possible explanation in terms of

ISBN 0-8053-8061-7

charmed quarks would assign $J/\psi(3.1)$ to a 1 $^3S_1$ (1 is the principal quantum number $n = l + n_r$) state of charmonium, a compound of $c\bar{c}$ in the same way as positronium is formed of e^+e^-. The χ states would be 2 $^3P_{012}$ states of charmonium with $J^{PC} = 0^{++}$, 1^{++}, 2^{++} and $\psi'(3.7)$ would be a 2 $^3S_1$ state, $J^{PC} = 0^{++}$, as shown in Fig. 18-12. To obtain the χ states at the mass values observed (Fig. 18-12) one assumes a potential between charmed quarks that is coulombian at very short distances and depends linearly on r at large distance. Such a potential produces automatically the confinement of quarks.

It is obvious that the investigations of this field are in a very preliminary stage and that we have seen only the tip of the iceberg.

BIBLIOGRAPHY

Bjørken, F. D., and E. A. Paschos, "Inelastic Electron–Proton Scattering," *Phys. Rev.*, **185**, 1975 (1969).

Brodsky, S. J., and S. D. Drell, "The Present Status of Quantum Electrodynamics," *Ann. Rev. Nucl. Sci.*, **20**, 147 (1970).

Friedman, J. I., and H. W. Kendall, "Deep Inelastic Electron Scattering," *Ann. Rev. Nucl. Sci.*, **22**, 203 (1972).

Gaillard, M. K., B. W. Lee, and J. L. Rosner, "Search for Charm," *Rev. Mod. Phys.*, **47**, 277 (1975).

Gatto, R., "Theoretical Aspects of Colliding Beam Experiments," *Ergeb. exakt. Naturw.*, **39**, (1965).

Gilman, F. J., "Photoproduction and Electroproduction," *Phys. Repts.*, **4**, 95 (1972).

Hofstadter, R. (ed.), *Nuclear and Nucleon Structure*, Benjamin, New York, 1963.

Kogut, J., and D. Susskind, "The Parton Model," *Phys. Repts.*, **8**, 75 (1973).

Schwitters, R. F. and K. Strauch, "The Physics of $e^+ - e^-$ Collisions", *Ann. Rev. Nucl. Sci.*, **26**, 89 (1976).

SLAC, Summer Institute on Particle Physics, 1976, SLAC, Stanford, Ca. 1976.

PROBLEMS

18-1   In the second graph in Fig. 18-6, there is a peak at about 1.5 GeV missing mass. From the data in the figure, find the value of E' corresponding to this peak.

18-2   In order to obtain an overview of electron–nucleon scattering, use a W^2–q^2 diagram with W^2 as abscissa. Plot the lines $\omega =$ constant ($\omega = 2M\nu/q^2$). On this diagram locate points corresponding to the curves of Fig. 18-6. [See J. I. Friedman and H. W. Kendall, *Ann. Rev. Nucl. Sci.*, **22**, 220 (1972).]

18-3   Compare the Bhabha, Møller, and Compton scattering formulas and plot $d\sigma/d\omega$ in typical cases.

18-4   Show that $-\nu$ of Eq. (18-1.4) is a scalar $= q \cdot P/M$ and that in a real collision $q^2 > 0$. Show that for $W_2 = -Q^2 \cdot \delta[\nu + (q^2/2M)]$, Eq. (18-2.1) reduces to point scattering.

18-5   In order to find $F_1^2(q^2)$ and $F_2^2(q^2)$ [Eq. (18-1.7)], measurements of $d\sigma/d\omega$ at different values of θ, but at the same value of q, are useful. One procedure is to measure the quantity in the braces of Eq. (18-1.7), called R, and plot curves with F_1 as abscissa and F_2 as ordinate for constant q, giving the number R. Vary θ, at constant q, and consider a new curve for the new measured R. From these measurements F_1 and F_2 are found (with some ambiguities). Discuss the algebra of the problem.

18-6   Calculate the first energy levels for a $c\bar{c}$ system assuming an harmonic oscillator potential, a potential linear in r or a finite potential well and compare the results with the levels of Fig. 18-12.

ISBN 0-8053-8061-7

CHAPTER

19

Weak Interactions Revisited

In Chaps. 9 and 14 we treated two important manifestations of the weak interaction: beta decay and the decay of the muon. There are many more phenomena that depend on weak interactions and we shall try here to generalize and unify the picture, first basing it on the $V - A$ interaction and then introducing the so-called conserved current and the current–current interaction.

Weak decays are conveniently classified as (a) leptonic (e.g., the muon decay into an electron and two neutrinos; only leptons intervene in these); (b) semileptonic (e.g., ordinary beta decay, the decay of a pion into a muon and a neutrino; leptons and hadrons intervene in these); (c) hadronic (e.g., decay of a Λ into a proton and a pion, in which only hadrons intervene).

Leptonic decays are the simplest because there are no secondary effects due to strong interactions. In this sense they represent the purest example of weak interaction. We pointed out in Chap. 14 that the muon decay is fully explained by the $V - A$ interaction with a value $g_\mu = 1.43 \times 10^{-49}$ erg cm³ of the coupling constant and $C_V = - C_A = 1$.

Emilio Segrè, Nuclei and Particles: An Introduction to Nuclear and Subnuclear Physics, Second Edition

ISBN 0-8053-8061-7

19-1   CONSERVED CURRENT

The striking coincidence of the g values for beta decay and for muon decay, coupled with the fact that for the A part of the interaction one has $C_A = -1.23 C_V$ for beta decay, whereas $C_A = -C_V$ for the muon, led to the formulation of the *conserved vector current* (CVC) theory by Gershtein and Zeldovich and independently by Feynman and Gell-Mann. We will try to give an idea of this theory. As we have mentioned repeatedly, the beta-decay theory has analogies with electromagnetism that were discovered and used as a guide by Fermi as early as 1934. Electromagnetism involves the charge conservation principle expressed by the equation

$$\frac{\partial \rho}{\partial t} = \text{div } \mathbf{j} \tag{19-1.1}$$

or in four-dimensional notation

$$\sum_\mu \frac{\partial j_\mu}{\partial x_\mu} = 0 \tag{19-1.2}$$

When we consider the electromagnetic interaction of a proton, its charge appears constant even if part of the time the proton is dissociated, for instance, into a neutron and a positive pion. The charge appearing on the pion exactly compensates the one destroyed on the proton passing into a neutron. On the other hand, the magnetic moment of the proton is affected by the dissociation because there is no principle of conservation of the magnetic moment (see Sec. 6–12).

One might think that a bare nucleon has a weak interaction very different from that of a nucleon surrounded by pions, and that thus neutron decay and muon decay should be very different. The latter particle is a lepton not subject to strong interactions and as such its decay, according to $\mu^- \to e^- + \bar{\nu}_e + \nu_\mu$, represents a pure case of weak interaction. On the other hand, neutron decay should be affected by the dissociation and give rise to a different apparent strength of the weak interaction. We may write the two decays, each as resulting from the steps indicated in Eq. (19-1.3).

$$\begin{array}{ll} \mu^- \to e^- + \bar{\nu}_e + \nu_\mu \quad \text{or} & \mu^- \to \nu_\mu, \, \nu_e \to e^- \\ n \to p + \bar{\nu}_e + e^- & n \to p, \, \nu_e \to e^- \end{array} \tag{19-1.3}$$

●Only the steps containing leptons are analogous, but those containing the neutron and proton could be very different from the corresponding one containing μ^- and ν. The experimental fact shows that this is the case for the A part of the interaction, but not for the V part. To account for this, we rewrite the electric current for bare nucleons using the i-spin formalism. For a bare

ISBN 0-8053-8061-7

proton we have, according to Dirac's formulas,

$$j_\mu = e\bar{\psi}_p \gamma_\mu \psi_p \tag{19-1.4}$$

For a neutron there is no current, and both cases can be described by

$$j_\mu = e\bar{\psi}_N \gamma_\mu \left(\tfrac{1}{2} + t_3\right)\psi_N \tag{19-1.5}$$

where ψ_N refers to a nucleon, proton, or neutron, and contains the dichotomic variable on which the i-spin operator t_3 operates. The eigenvalues of t_3 are of course $\pm\tfrac{1}{2}$, as explained in Chap. 10. The total current can thus be decomposed into a scalar part in i-spin space and a vector part in the same space.

$$j_\mu = j_\mu^S + j_\mu^V = (e/2)\bar{\psi}_N \gamma_\mu \psi_N + e\bar{\psi}_N \gamma_\mu t_3 \psi_N \tag{19-1.6}$$

The first isoscalar part obeys the continuity equation $\sum_\mu \partial j_\mu^S / \partial x_\mu = 0$, by virtue of Dirac's equation. The isovector part, on the other hand, is not conserved. If a proton becomes a neutron, the charge is destroyed but reappears in a virtual pion; thus j_μ^V is conserved only if we also include another current due to pions.

For the weak interaction a typical term of the vector interaction is $\bar{\psi}_p \gamma_\mu \psi_n$, which can be written as

$$j_\mu = \bar{\psi}_N \gamma_\mu t_- \psi_N \tag{19-1.7}$$

using the i-spin formalism in which the operator $t_+ = (t_1 + it_2)$ changes a neutron into a proton, and the operator $t_- = (t_1 - it_2)$ changes a proton into a neutron. The operators t_1, t_2, and t_3 are components of a vector in i-spin space. In the CVC theory this vector part of the current is conserved and thus in beta decay too, the possible dissociation of a proton has no influence on the decay. ●

Symbolically a neutron can decay as shown in the two diagrams in Figs. 19-1 and 19-2. In Fig. 19-1 the neutron decays directly; in Fig. 19-2 it dissociates first, emitting a π^- and becoming a proton. The probability of e-ν emission by the pion offsets the fact that a proton cannot emit an electron. The weak-interaction current is thus conserved in the same way as the electric current. The two currents, electromagnetic and weak interaction, are the t_+, t_-, and t_3 of the same conserved i-spin vector current.

There are several direct experimental tests substantiating the CVC theory:

1. The agreement of the g constant in beta and muon decay. The agreement is not perfect but the discrepancy can be accounted for by Cabibbo's theory (see Sec. 19-2) and the residual discrepancy by small electromagnetic corrections.

ISBN 0-8053-8061-7

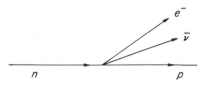

Figure 19-1 Feynman diagram for neutron beta decay in the direct case.

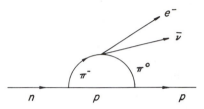

Figure 19-2 Feynman diagram for neutron beta decay through virtual negative pion formation.

2. The ^{12}B, ^{12}C, ^{12}N experiments. These three nuclei have excited states with $T = 1$, $T_3 = \pm 1, 0$. They decay to the ^{12}C ground state, which has $T = 0$, by β^-, γ, β^+ emission, respectively. According to the CVC theory the matrix elements are connected to each other; furthermore, the shapes of the β^+, β^- spectra depart slightly from the allowed shape in a characteristic and predictable fashion. All this has been confirmed by experiment with satisfactory accuracy.

3. The $\pi^+ \to \pi^0 + e^+ + \nu_e$ reaction is analogous to ordinary 0^+–0^+ beta decay and the CVC prescribes that the coupling constant be g, as in beta decay. The decay constant for this process is given by the beta-decay formula Eq. (9-5.32)

$$\lambda = g^2 \frac{m^5 c^4}{2\pi^3 \hbar^7} |M|^2 f = g^2 \frac{m^5 c^4}{\pi^3 \hbar^7} \frac{\epsilon_0^5}{30} \tag{19-1.8}$$

where ϵ_0 is the upper limit of the beta spectrum in units of mc^2. In writing the last equation we have used $|M|^2 = 2$ and $f = \epsilon_0^5/30$, as in Eq. (9-5.29).

The resulting λ is 0.43 sec^{-1}. The more frequent decay $\pi^+ \to \mu^+ + \nu_\mu$ has $\lambda = 3.84 \times 10^7$ sec^{-1}; hence the branching ratio should be 1.07×10^{-8}. This agrees with the experimental result within an error of about 10%. The result is significant because the CVC theory predicts this decay quantitatively, whereas other theories give numbers anywhere between 10^{-8} and 5×10^{-9}.

4. Alpha–beta angular correlations in ^{8}Li, ^{8}B, and ^{8}Be. ^{8}Li, ^{8}Be, and ^{8}B have three excited states with $T = 1$, $T_3 = \pm 1, 0$. The ground state of ^{8}Be has $T = 0$; another $T = 0$ state of ^{8}Be, at 2.90 MeV above the ground state, decays instantly into two alphas. (See Fig. 11-11.) ^{8}Li and ^{8}Be decay to this excited

ISBN 0-8053-8061-7

state, and the CVC theory predicts an angular correlation between the directions of the beta and alpha emission, in fair agreement with experiment.

In conclusion, the current–current interaction of the form

$$\frac{g}{\sqrt{2}} \, \mathcal{J}_\alpha \mathcal{J}_\alpha^\dagger \tag{19-1.9}$$

with

$$\mathcal{J}_\alpha = \bar{\mu}\gamma_\alpha(1+\gamma_5)\nu_\mu + \bar{e}\gamma_\alpha(1+\gamma_5)\nu_e + \mathcal{J}_\alpha^{\Delta S=0} + \mathcal{J}_\alpha^{\Delta S=-1} \tag{19-1.10}$$

$$\mathcal{J}_\alpha^\dagger = \bar{\nu}_\mu\gamma_\alpha(1+\gamma_5)\mu + \bar{\nu}_e\gamma_\alpha(1+\gamma_5)e + \left(\mathcal{J}_\alpha^{\Delta S=0}\right)^\dagger + \left(\mathcal{J}_\alpha^{\Delta S=-1}\right)^\dagger \tag{19-1.11}$$

accounts for the weak interaction. We have written the leptonic part of the current explicitly. The hadronic terms are written symbolically, keeping separate strangeness-constant and strangeness-changing parts because they are found experimentally to be different.

A particle symbol such as e, stands for ψ_e, and to avoid confusion we use $\alpha = 1, 2, 3, 4$ as a component index, whereas in Eqs. (19-1.4) to (19-1.7) we used the familiar μ. The electronic neutrino is denoted by ν_e and the muonic neutrino by ν_μ. Particles with an overbar are created, those without an overbar are destroyed. The operator $(1 + \gamma_5)$ is a projection operator that gives the handedness of the interaction. It singles out left-handed particles and thus does not conserve parity. A complete understanding of the interaction requires relativistic quantum mechanics and destruction and creation operators. See, for example, (Com 73).

●That the axial vector current is not conserved in weak interactions is demonstrated by the difference between C_A and C_V in beta decay. It is possible, however, to establish a useful theoretical relation (the Goldberger–Treiman relation) between the axial vector coupling constant of beta decay, the coupling constant $g_{NN\pi}$ of strong interactions (as obtainable, e.g., from pion–nucleon scattering), and the rate of decay of the pion into a muon plus a neutrino.

In units of $\hbar = c = 1$ the transition probability per unit time for the $\pi \rightarrow \mu + \nu_\mu$ reaction may be written as

$$w = \frac{G^2 f_\pi^2}{8\pi} m_\mu^2 m_\pi \left[1 - \frac{m_\mu^2}{m_\pi^2}\right]^2 \tag{19-1.12}$$

where G is the weak-interaction constant and f_π is a constant with the dimensions of mass. Its numerical value cannot be obtained from weak-interaction theory alone, because it can be thought of as related to the potential dissocia-

ISBN 0-8053-8061-7

tion of the pion; thus f_π is to be regarded as a parameter in the theory to be obtained by identifying the decay constant of the pion with w. Equation (19-1.12) is derived under the assumption that the axial vector portion of the weak hadronic current is operative in pion decay. One can prove that the vector part of the current cannot contribute. The numerical value of f_π derived from Eq. (19-1.12) and the decay constant of the pion is $f_\pi = 0.93 m_\pi$.

The Goldberger–Treiman relation says that

$$-C_A = f_\pi g_{NN\pi} \sqrt{2} \, / \, (m_p + m_n) \qquad (19\text{-}1.13)$$

and is in agreement with actual values of C_A, f_π, and $g_{NN\pi}$ to within about 10%.

An important assumption in the derivation of Eq. (19-1.13) is the so-called partially conserved axial current (PCAC), according to which the axial vector hadronic current would be conserved in a fictitious world in which the pion mass was zero. The Goldberger–Treiman relation is approximately correct because the mass of a virtual pion emitted by a neutron while it beta-decays, as in the Feynman diagram in Fig. 19-2, may be neglected in comparison to the mass of the neutron itself.

S. Adler and independently W. Weisberger carried this kind of argument one step further in 1965 and used the PCAC hypothesis and the techniques of "current algebra" (involving speculative assumptions about the algebraic properties of the hadronic weak currents under $SU(3)$ transformations) to calculate the numerical value of C_A from pion–nucleon scattering cross-section data. The result they obtained is in excellent agreement with experiment: $C_A = -1.24$; and the assumptions on which their arguments are based are justified if quarks are fundamental constituents of all strongly interacting particles. ●

19-2   SELECTION RULES IN STRANGE DECAYS; CABIBBO'S THEORY

The closeness of the values of the coupling constant in beta decay, muon decay, and other weak decays has prompted the idea that the Fermi coupling is "universal." This universality has taken various forms; initially, before the study of strange particles, it led us to consider a "Puppi triangle" (Fig. 19-3) where the coupling constants for particles located at the vertices are the same. The idea was later extended, yielding a tetrahedron, at the fourth vertex of which strange particles were placed; however, if we compare the $K^+ \to \mu^+ + \nu_\mu$ and $\pi^+ \to \mu^+ + \nu_\mu$ decays, we find that the decay rates, which from phase space alone should be in the ratio

$$\frac{M_K \left(1 - M_\mu^2 / M_K^2\right)^2}{M_\pi \left(1 - M_\mu^2 / M_\pi^2\right)^2} = 17.7 \qquad (19\text{-}2.1)$$

ISBN 0-8053-8061-7

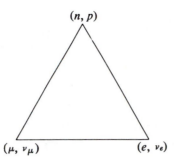

Figure 19-3 Puppi's triangle. Fermions at the vertices of the triangle interact with the universal Fermi interaction. The pairs $(\mu, \nu_\mu)(e, \nu_e)$ correspond to the mu decay; the pairs $(\mu, \nu_\mu)(n, p)$ correspond to the capture of a mu meson in a nucleus; the pairs $(e, \nu_e)(n, p)$ to the beta decay. The real situation is more complicated than what the figure shows.

are in the ratio 1.34. Similarly, comparing the experimental semileptonic decay rates for the strange particles ($\Lambda \rightarrow pe\nu$; $\Sigma \rightarrow ne^-\nu$; $\Xi \rightarrow \Lambda e\nu$) with a calculation in strict analogy with the neutron beta decay, one finds that the experimental result is about 20 times too small.

Furthermore, the following empirical selection rules obtain.

1. $\Delta S = \pm 1, 0$. Bigger changes of strangeness, such as would occur in the decay $\Xi \rightarrow N + \pi$, $\Xi \rightarrow N + e + \nu$ or $\Omega^- \rightarrow N + \pi$, have never been seen and they occur, if at all, in less than one in a thousand of the more common decays.

2.   For $\Delta S = 0$, $\Delta T_3 = \pm 1$. The i-spin refers only to the hadrons. The isovector current automatically produces $\Delta T_3 = \pm 1$.

3.   For $\Delta S = \pm 1$, $\Delta S = \Delta Q$, *in semileptonic decays.* This means that if the strangeness increases or decreases by one unit in the transformation, the electric charge *of the hadrons only* does the same. A striking example of this rule is that whereas hundreds of examples of $K^+ \rightarrow \pi^+ + \pi^- + e^+ + \nu_e$ are known, $K^+ \rightarrow \pi^+ + \pi^+ + e^- + \bar{\nu}_e$ has never been seen. There might, however, be rare exceptions to this rule.

4.   For $\Delta S = \pm 1$, $|\Delta T| = \frac{1}{2}$. The relation Eq. (13-2.6) $Q = T_3 + \frac{1}{2}(N + S)$, combined with $\Delta S = \Delta Q$ and conservation of nucleons $\Delta N = 0$, gives $\Delta T_3 = \frac{1}{2} \Delta S = \pm \frac{1}{2}$. In any case Q must be an integral number and this forces $|\Delta T| = \frac{1}{2}, \frac{3}{2}, \frac{5}{2}$, etc. The rule says that of these possible values only $|\Delta T| = \frac{1}{2}$ is realized. The rule $|\Delta T| = \frac{1}{2}$ is stronger than $\Delta S = \Delta Q$ because a transformation obeying the latter rule does not necessarily have $|\Delta T| = \frac{1}{2}$, but could have $|\Delta T| = \frac{3}{2}$, etc.

As an application of this selection rule, consider the branching ratio in the decay

$$\Lambda^0 \rightarrow p + \pi^-$$
$$\rightarrow n + \pi^0 \tag{19-2.2}$$

ISBN 0-8053-8061-7

The i spin of the Λ^0 is 0. On the right the (p, π^-) system and (n, π^0) system are described, as in Eqs. (15-5.4) and (15-5.5), by wave functions containing $T = \frac{1}{2}$ and $T = \frac{3}{2}$ components. Only the former counts, for in the latter case $\Delta T = \frac{3}{2}$, and transitions to them are forbidden. The amplitudes of states with $T = \frac{1}{2}$ in $p + \pi^-$ and $n + \pi^0$ are in the ratio $\sqrt{2} : 1$. Hence, the branching ratio must be $2 : 1$, in agreement with the experimental result.

A comprehensive clarification of the phenomenology and quantitative predictions of the strange-particle weak decays was offered by Cabibbo (1963).

Cabibbo's theory concerns real physical particles and uses $SU(3)$. It was established independently of the quark model. For the sake of simplicity, however, it is advantageous to use the quark model when presenting Cabibbo's theory. Thus in lieu of operating with real particles containing quarks, such as $K^+ = u\bar{s}$ or $\pi^+ = u\bar{d}$, we will consider the constituent quarks.

First, note that in the absence of medium-strong interactions, members of the same U_3 multiplet within each $SU(3)$ family, such as π^+, K^+ or Σ^+, p, would have the same mass. The s and d quarks (eigenquarks for strong interactions) have the same charge and could not be distinguished from each other because an $SU(3)$ rotation would change one into the other. The medium-strong interaction breaks the degeneracy and then a definite linear combination of s and d quarks is operative for weak interactions. The specific linear combination of d and s quarks (eigenquarks for weak interactions) that makes a weak current with u is

$$d' = d \cos \theta + s \sin \theta \qquad (19\text{-}2.3)$$

and when we write the current

$$\bar{u}\gamma_\alpha(1 + \gamma_5)d' = \bar{u}\gamma_\alpha(1 + \gamma_5)(d \cos \theta + s \sin \theta) \qquad (19\text{-}2.4)$$

we see that the coupling $d\bar{u}$ that does not change strangeness ($\Delta S = 0$) has a factor $\cos \theta$, while the coupling $s\bar{u}$ that changes strangeness ($\Delta S = 1$) has a factor $\sin \theta$.

The fourth charmed quark c is treated in a way completely similar to u inasmuch as it is coupled to a linear combination of d and s. The linear combination chosen is

$$s' = -d \sin \theta + s \cos \theta \qquad (19\text{-}2.5)$$

with the same angle θ as in Eq. (19-2.3). The two linear combinations d', s' are thus orthogonal, and $s'\bar{s}' + d'\bar{d}' = s\bar{s} + d\bar{d}$ independently of θ, an important relation that has the physical consequence of suppressing the extremely rare $K_L \rightarrow \mu^+ + \bar{\mu}$, as we shall see later.

ISBN 0-8053-8061-7

In order to determine the angle θ, consider the reactions $K^+ \to \mu^+ + \nu_\mu$ and $\pi^+ \to \mu^+ + \nu_\mu$. They are completely similar except that one involves an s quark and the other a d quark because K^+ is an $\bar{s}u$ combination and π^+ is a $d\bar{u}$ combination. The ratio of their rates corrected for the phase-space factor thus gives

$$\frac{\text{rate}(K^+ \to \mu^+ + \nu)}{\text{rate}(\pi^+ \to \mu^+ + \nu)} = \frac{m_K\left[1 - (m_\mu^2/m_K^2)\right]^2}{m_\pi\left[1 - (m_\mu^2/m_\pi^2)\right]^2} \tan^2\theta \quad (19\text{-}2.6)$$

and inserting numbers one obtains $\tan\theta = 0.275$.

Strictly speaking, one should distinguish between the V and A parts of the interaction and give to each a Cabibbo angle. However, the two θ seem to be the same within experimental error. Cabibbo's angle may be derived with consistent results from many decays and this is the principal empirical test of the theory. The best value of θ is 0.235 ± 0.006 rad; $\cos\theta = 0.973$; $\sin\theta = 0.233$. If θ were 0, it is apparent that $\Delta S = \pm 1$ decays would be forbidden; if θ were $\pi/2$, there would be no $\Delta S = 0$ decays, and if $\theta = \pi/4$, there would be a "naive universality" in which ΔS would not influence weak interactions.

The expression for the complete charged current is thus

$$\mathcal{G}_\alpha^C = \bar{\nu}_\mu \gamma_\alpha (1 + \gamma_5) \mu + \bar{\nu}_e \gamma_\alpha (1 + \gamma_5) e + \bar{u} \gamma_\alpha (1 + \gamma_5) d' + \bar{c} \gamma_\alpha (1 + \gamma_5) s'$$

$$(19\text{-}2.7)$$

and its hermitian conjugate

$$\mathcal{G}_\alpha^{C\dagger} = \bar{\mu} \gamma_\alpha (1 + \gamma_5) \nu_\mu + \bar{e} \gamma_\alpha (1 + \gamma_5) \nu_e + \bar{d}' \gamma_\alpha (1 + \gamma_5) u + \bar{s}' \gamma_\alpha (1 + \gamma_5) c$$

$$(19\text{-}2.8)$$

In the current $\mathcal{G}_\alpha^C$, the term $\bar{\nu}_e \gamma_\alpha (1 + \gamma_5) e$ causes the transformations

$$e^- \to \nu_e \qquad \bar{\nu}_e \to e^+$$

with $\Delta Q = +1$. In the hermitian conjugate current $\mathcal{G}_\alpha^{C\dagger}$, the term $\bar{e} \gamma_\alpha (1 + \gamma_5) \nu_e$ causes the transformations

$$\nu_e \to e^- \qquad e^+ \to \bar{\nu}_e$$

with $\Delta Q = -1$. The action of the term $\bar{\nu}_\mu \gamma_\alpha (1 + \gamma_5) \mu$ is similar.

In the hadronic part of the current $\mathcal{G}_\alpha^C$, using as objects of the transformation the eigenquarks of strong interactions u, d, s, c, we find terms containing $\cos\theta$ or $\sin\theta$ as a factor in the amplitudes. Thus we have, in the transformations produced by $\bar{u}\gamma_\alpha(1 + \gamma_5)d'$, the terms originating from

$$d \to u \qquad \bar{u} \to \bar{d} \qquad \text{ampl } \cos\theta$$
$$s \to u \qquad \bar{u} \to \bar{s} \qquad \text{ampl } \sin\theta$$

They both have $\Delta Q = 1$. The hermitian conjugate term produces

$$u \to d \qquad \bar{d} \to \bar{u} \qquad \text{ampl } \cos\theta$$
$$u \to s \qquad \bar{s} \to \bar{u} \qquad \text{ampl } \sin\theta$$

with $\Delta Q = -1$.

Symbolically this is summarized in the highly symmetrical scheme

$$\begin{pmatrix} \nu_\mu \\ \mu \end{pmatrix}_L \quad \begin{pmatrix} \nu_e \\ e \end{pmatrix}_L \quad \begin{pmatrix} u \\ d' \end{pmatrix}_L \quad \begin{pmatrix} c \\ s' \end{pmatrix}_L$$

where L denotes that left-handed components are operative.

In real hadrons the $\mathcal{J}_\alpha^C$ current can make changes as in the following examples.

In a π^-, written in term of quarks as $\bar{u}d$, the $\bar{u}\gamma_\alpha(1 + \gamma_5)d$ part of the current (with amplitude $\cos\theta$) transforms the d quark of the pion into a u quark, yielding the combination $\bar{u}u$, equivalent, as quantum numbers, to the vacuum. Thus that part of the current can destroy a negative pion.

The same part of the current can transform a neutron into a proton. In terms of quarks, the neutron is ddu; by changing a d quark into a u quark one obtains uud, a proton.

The complete interaction, by associating the current with its hermitian conjugate in terms of the form $\mathcal{J}_\alpha^C \mathcal{J}_\alpha^{C\dagger}$, conserves charge and hadronic and leptonic numbers.

As examples, consider the leptonic decay

$$\mu^- \to e^- + \bar{\nu}_e + \nu_\mu$$

which can also be written as $\mu^- + \nu_e \to e^- + \nu_\mu$. It is produced by the term $\bar{\nu}_\mu\gamma_\alpha(1 + \gamma_5)\mu[\bar{\nu}_e\gamma_\alpha(1 + \gamma_5)e]^\dagger$.

In the semileptonic beta decay

$$n \to p + \bar{\nu}_e + e^-$$

the $\bar{u}\gamma_\alpha(1 + \gamma_5)d$ term in the current transforms a neutron into a proton as described above and the $[\bar{\nu}_e\gamma_\alpha(1 + \gamma_5)e]^\dagger$ term generates the electron and antineutrino.

The semileptonic decay

$$\pi^- \to \mu^- + \bar{\nu}_\mu$$

ISBN 0-8053-8061-7

is obtained from the $\bar{u}\gamma_\alpha(1 + \gamma_5)d$ term in the current, which destroys a π^-, and the $\left[\bar{\nu}_\mu\gamma_\alpha(1 + \gamma_5)\mu\right]^\dagger$ term, which creates a μ^- and a $\bar{\nu}_\mu$.

In the hadronic decay

$$K^+ \rightarrow \pi^+ + \pi^0$$

the quark content of K^+ is $u\bar{s}$, the quark content of π^+ is $u\bar{d}$, and that of π^0 is the same as the vacuum. The part of $\mathcal{J}_\alpha^{C\dagger}$ that changes $\bar{s}$ into $\bar{u}$ is $\left[\bar{u}\gamma_\alpha(1 + \gamma_5)\bar{s}'\right]^\dagger$ (sin θ as amplitude) and the part that changes $\bar{u}$ into $\bar{d}$ is $u\gamma_\alpha(1 + \gamma_5)d'$ (cos θ as amplitude). Together they change K^+ into π^+, and we add $u\bar{u}$, $d\bar{d}$ for the π^0. This last step is permissible if there is available energy.

In addition to conservation of charge, leptons (separately), and nucleons, other selection rules emerge from the structure of the charged current–current interaction.

In the charged hadronic current that does not contain c we see that the destruction of an s quark is accompanied by the creation of a u quark or vice versa, and we read from Table 17-1 that this requires

$$\Delta S = \Delta Q \qquad \Delta T = \frac{1}{2} \qquad (19\text{-}2.9)$$

The charm-changing part of the charged current gives approximate selection rules. Large terms containing cos θ in amplitude (Cabibbo allowed) give

$$\Delta Q = \Delta S = \Delta C \qquad \Delta T = 0 \qquad (19\text{-}2.10)$$

Small terms containing sin θ in amplitude (Cabibbo forbidden) give

$$\Delta Q = \Delta C \qquad \Delta S = 0 \qquad \Delta T = \frac{1}{2} \qquad (19\text{-}2.11)$$

A c quark transforms predominantly into the $su\bar{d}$ combination, as can be seen by considering the products of the currents

$$\bar{s}\gamma_\alpha(1 + \gamma_5)c\left[d\gamma_\alpha(1 + \gamma_5)u\right]^\dagger \quad (\text{ampl } \cos^2\theta)$$

It follows that in real particles we expect a strange particle among the decay products of a charmed particle. This is important for the identification of charmed particles. As an example, a D^0 formed by $\bar{u}c$ can go into $su\bar{d}u$, and this gives $K^-\pi^+$ or $\bar{K}^0\pi^0$ but not $K^+\pi^-$ or $K^0\pi^0$.

19-3 NEUTRAL CURRENTS; UNIFICATION OF ELECTROMAGNETISM AND WEAK INTERACTIONS

In addition to the charged currents (called charged because they change the electric charge, e.g., by destroying a ν_μ and creating a μ^-, there are neutral currents, so called because they destroy a particle and create another one

ISBN 0-8053-8061-7

of the same charge. Such currents are demonstrated in high-energy neutrino events. To observe them, one produces high-energy neutrino beams by first forming high-energy pion or kaon beams. The mesons decay in flight and produce neutrinos in a poorly collimated beam of energy extending from 20 to about 300 GeV. By suitable choice of the primary beam, one can have ν_e, $\bar{\nu}_e$, ν_μ, $\bar{\nu}_\mu$. The neutrino beam is filtered through tens or hundreds of meters of solid material and then enters a sensitive region, such as a bubble chamber, where tons of material may be observed. The reactions

$$\bar{\nu}_\mu + p \rightarrow \mu^+ + n \qquad \nu_\mu + n \rightarrow \mu^- + p \qquad (19\text{-}3.1)$$

have been ascertained. For them the cross section in square centimeters is of the order of $10^{-38}E$ (GeV). Also, their counterparts with ν_e and electrons obtain. Notable are the rare events

$$\bar{\nu}_\mu + e^- \rightarrow \bar{\nu}_\mu + e^- \qquad \nu_\mu + e^- \rightarrow \nu_\mu + e^- \qquad \nu_\mu + N \rightarrow \nu_\mu + \text{hadrons}$$

$$\bar{\nu}_\mu + p \rightarrow \bar{\nu}_\mu + p \qquad \nu_\mu + p \rightarrow \nu_\mu + p \qquad (19\text{-}3.2)$$

These events have withstood the stringent tests necessary to identify them. They demonstrate the existence of neutral currents because the current connects only muonic lepton with muonic lepton, electronic lepton with electronic lepton, and hadron with hadron. Both muonic leptons have the same charge (zero) and thus the current must be neutral.

The new gauge theories that try to unify electromagnetic and weak interactions (S. Weinberg, 1967, A. Salam, 1968) require neutral currents and their discovery lends support to these theories.

According to these theories neutral weak currents are formed by a linear combination of two neutral currents, one transforming like t_3 and the other scalar, in a way similar to the electromagnetic current [as described in Eq. (19-1.6)]. The scalar and vector parts of the leptonic and hadronic currents are denoted by $\mathcal{J}_{lS}^0$, $\mathcal{J}_{lV}^0$, $\mathcal{J}_{hS}^0$, $\mathcal{J}_{hV}^0$. How they combine to form a neutral weak current is determined by a new parameter called the Weinberg angle. By developing the calculation, one finds that the Cabibbo angle enters only in the charged currents but not in the neutral currents. The neutral weak currents do not change flavors (T_3, S, and C) because the orthogonality of s' and d' cancels flavor-changing terms, such as $s\bar{u}$, in the expression $d'\bar{d}' + s'\bar{s}'$. An immediate consequence of this is that there is no strangeness-changing neutral current, and this explains the absence of the decay $K_L \rightarrow \mu^+ + \mu^-$ at the level of the first-order weak interaction. The minute branching observed (10^{-8}) can be accounted for by higher-order corrections. The introduction of the c quark is essential to avoid the strangeness-changing neutral weak current. This cancellation mechanism was first suggested by Glashow, Iliopoulos,

ISBN 0-8053-8061-7

and Maiani in 1970 and the absence of the $K_L \rightarrow 2\mu$ decay was the first experimental evidence, although indirect, of charm.

Summing up, we have the currents

$$\text{charged leptonic} \quad \mathcal{G}_l^C = \bar{\nu}_e e + \bar{\nu}_\mu \mu$$

$$\text{neutral leptonic} \quad \mathcal{G}_l^0 = \bar{\nu}_e \nu_e + \bar{\nu}_\mu \nu_\mu - \bar{e}e - \bar{\mu}\mu$$

$$\text{charged hadronic} \quad \mathcal{G}_h^C = \cos\theta \, (\bar{u}d + \bar{c}s) + \sin\theta \, (\bar{u}s - \bar{c}d) \tag{19-3.3}$$

$$\text{neutral hadronic} \quad \mathcal{G}_h^0 = \bar{u}u + \bar{c}c + \bar{d}d + \bar{s}s$$

where for brevity we have suppressed the operators $\gamma_\alpha(1 + \gamma_5)$ in the charged currents and similar ones in the neutral currents.

The current–current interaction has the form

$$2^{-1/2}G\,\mathcal{G}_\alpha^C \mathcal{G}_\alpha^{C\dagger} + 2^{-1/2}G'\,\mathcal{G}_\alpha^0 \mathcal{G}_\alpha^{0\dagger} \tag{19-3.4}$$

with

$$\mathcal{G}_\alpha^C = \mathcal{G}_{l,\,\alpha}^C + \mathcal{G}_{h,\,\alpha}^C \qquad \mathcal{G}_\alpha^0 = \mathcal{G}_{l,\,\alpha}^0 + \mathcal{G}_{h,\,\alpha}^0$$

where the coupling constant G' for the neutral current interaction is generally different from G, the difference being related to the mass difference of the hypothetical neutral intermediate boson Z^0 and the equally hypothetical charged intermediate boson $W^\pm$. (See Sec. 19-3.)

The detailed quantitative development of these concepts permits the prediction of many decay constants, but is beyond the scope of this book. To give an idea of the results, we show in Table 19-1 the theoretical and experimental results for semileptonic decays. Note also that C_V for beta decay should not be exactly 1, but $\cos^2\theta = 0.9435$. This value improves the agreement with muon decay. The residual small discrepancy may be accounted for by electromagnetic corrections.

TABLE 19-1 CABIBBO'S THEORY PREDICTIONS AND OBSERVED RESULTS

	Branching ratio	
Decay	*Exp.* $\cdot 10^4$	*Theory* $\cdot 10^4$
$\Sigma^- \rightarrow \Lambda e^- \bar{\nu}_e$	0.60 ± 0.06	0.62
$\Sigma^+ \rightarrow \Lambda e^+ \nu_e$	0.20 ± 0.04	0.19
$\Lambda \rightarrow p e^- \bar{\nu}_e$	8.13 ± 0.29	8.6
$\Lambda \rightarrow p \mu^- \bar{\nu}_\mu$	1.57 ± 0.35	1.41
$\Sigma^- \rightarrow n e^- \bar{\nu}_e$	10.8 ± 0.4	10.1
$\Sigma^- \rightarrow n \mu^- \bar{\nu}_\mu$	4.5 ± 0.4	4.8
$\Xi^- \rightarrow \Lambda e^- \nu_e$	6.9 ± 1.8	5.4
$\Xi^- \rightarrow \Sigma^0 e^- \bar{\nu}_e$	< 5	0.8

ISBN 0-8053-8601-7

19-4　INTERMEDIATE BOSONS

Most of the facts of weak-interaction physics suggest the existence of new particles that have not yet been seen. They would be the quanta of the weak-interaction field. Because the weak interactions are described as contact interactions, that is, the currents coupling with each other are taken at the same space–time point, the quanta should be very massive. In fact, the Yukawa relation $\hbar/mc$ = the range of the force gives an infinite mass for range zero. The quanta should also be bosons of spin 1 and charge $\pm e$, as suggested by the analogy between weak interactions and electromagnetism and the presence of charged and neutral currents. These particles are called *intermediate bosons* and we denote them by the symbols $W^\pm, Z$. Intermediate bosons should be able to decay by weak interaction into leptons + neutrinos; pions; K + pions; etc., like the transformations induced by the weak currents. For instance, the decay $W^+ \to$ lepton + ν has a calculated rate

$$\frac{gM_W^3}{6\pi\sqrt{2}}\ \frac{c^3}{\hbar^4} = 2 \times 10^{20}\ \text{sec}^{-1}$$

for $M_W = 10$ GeV/c^2.

The best chance of observing W seems at present to be by forming it through the reaction

$$\nu_\mu + p \to W^+ + \mu^- + p \tag{19-4.1}$$

The cross section is a rapidly rising function of E_ν and decreases with M_W. Up to now, however, intermediate bosons have not been seen. Experiments set a lower limit to the mass at about 7.5 GeV.

In modern attempts to unify weak and electromagnetic interactions by *gauge theories*, the neutral intermediate boson Z is related to the photon. In a situation of higher symmetry than the real world there are four massless vector mesons. The symmetry is spontaneously broken and three of the mesons of charge $\pm e$, 0 become the massive $W^\pm, Z^0$; the fourth remains massless and is identified with the photon. the model requires that

$$M_W = \left(\frac{\pi\alpha}{\sqrt{2}\ G}\right)^{1/2} \frac{m_p}{\sin \xi} \geqslant 37.3\ \text{GeV}/c^2 \tag{19-4.2}$$

$$M_Z = \frac{m_W}{\cos \xi} = \frac{74.6\ \text{GeV}/c^2}{\sin 2\xi} \geqslant 74.6\ \text{GeV}/c^2 \tag{19-4.3}$$

where $\tan \xi = G'/G$ with G and G' as defined in Eq. (19-3.4); m_p is the mass of the proton; G the dimensionless Fermi constant; and $\alpha = 1/\text{'}137\text{'}$.

The unification program is still far from its goal, but its complete success would be a substantial advance in natural philosophy, as the example of the unification of electricity and magnetism suggests.

ISBN 0-8053-8061-7

19-5 SOME FURTHER EXAMPLES OF WEAK DECAYS

We have already treated beta decay and muon decay (in Chap. 9 and Sec. 14-2). No change is required for muon decay. For the vector part of beta decay, g is replaced by $g \cos \theta$, as pointed out in Sec. 19-2. Here we discuss some other examples of weak decay, starting with pion decay, a semileptonic case.

Pion Decay

The most common decay is

$$\pi^+ \rightarrow \mu^+ + \nu_\mu \tag{19-5.1}$$

with a similar formula for π^-. These decays are briefly designated $\pi_{\mu 2}$ decay, where the symbols mean that the decaying particle, π, decays into a charged μ and a total of two particles. This decay may be thought of as a two-step process involving an intermediate nucleon–antinucleon state:

$$\pi^+ \rightarrow p + \bar{n} \rightarrow \mu^+ + \nu_\mu \tag{19-5.2}$$

In this model the matrix element of the transition will contain a factor related to the virtual dissociation of the pion and a factor $\bar{\psi}_\nu \gamma_\alpha (1 + \gamma_5) \psi_\mu$ characteristic of the $V - A$ interaction. Because of the first factor the decay constant is not calculable from first principles (see, however, the Goldberger–Treiman relation), but it is possible to calculate the branching ratio between the $\pi_{\mu 2}$ decay and the analogous π_{e2} decay. This is explicitly written as

$$\pi^+ \rightarrow p + \bar{n} \rightarrow e^+ + \nu_e \tag{19-5.3}$$

In the branching ratio there is a common factor relevant to the dissociation of the pion that cancels, a phase-space factor favoring the π_{e2} mode, and the factor coming from the expressions $\bar{\nu}_e \gamma_\alpha (1 + \gamma_5) e$ and $\bar{\nu}_\mu \gamma_\alpha (1 + \gamma_5) \mu$, which strongly favors the $\pi_{\mu 2}$ mode. The phase-space factor for the $\pi_{\mu 2}$ mode is

$$\frac{dN}{dE} = \frac{4\pi p^2}{(2\pi\hbar)^3} \frac{dp}{dE} = \frac{1}{2\pi^2 \hbar^3} \frac{p^2}{c+v} \tag{19-5.4}$$

where v is the velocity of the μ. Considered alone it would give for the ratio of the decay constants λ_e / λ_μ:

$$\frac{p_e^2 (1 + v/c)_\mu}{p_\mu^2 (1 + v/c)_e} = \frac{(m_\pi^2 + m_e^2)(m_\pi^2 - m_e^2)^2}{(m_\pi^2 + m_\mu^2)(m_\pi^2 - m_\mu^2)^2} = 3.3 \tag{19-5.5}$$

In the $V - A$ theory the neutrino has helicity -1 and the electron or muon has helicity v/c. In our decay the pion has no angular momentum; hence, the electron and neutrino emitted in opposite directions must have opposite spins; thus they must have the same helicity. The heavy particle, muon or electron, has the "wrong" helicity and the process is only possible because its velocity is not c. For the electron, the velocity is higher than for the muon and hence the inhibition of the decay is greater. In $V - A$ theory the helicity introduces a factor $1 - (v/c)$, and taking into account this factor and expressing everything through the masses, one finds

$$R = \frac{\lambda_e}{\lambda_\mu} = \frac{p_e^2(1 + \beta_\mu)(1 - \beta_e)}{p_\mu^2(1 + \beta_e)(1 - \beta_\mu)}$$

$$= \left(\frac{m_e}{m_\mu} \right)^2 \left[\frac{m_\pi^2 - m_e^2}{m_\pi^2 - m_\mu^2} \right]^2 = 1.3 \times 10^{-4} \tag{19-5.6}$$

If some radiative corrections are taken into account, R becomes 1.24×10^{-4}, in excellent agreement with the experimental result $(1.25 \pm 0.03) \times 10^{-4}$.

A similar calculation for the K decays $K_{\mu 2}$ and K_{e2} also gives results in agreement with experiment.

The rare decay $\pi^+ \to \pi^0 + e^+ + \nu_e$ has been discussed in Sec. 19-1.

Λ Decay

This is an example of hadronic decay, with a change of strangeness. The decay constant agrees with the prediction of Cabibbo's theory.

The angular distribution of the decay products for Λ particles polarized in the z direction and at rest at the origin of the coordinate system is of interest. The spin of the Λ is $\frac{1}{2}$. The decay products $p + \pi$ must be in a state of angular momentum $\frac{1}{2}$, that is, in either an $S_{1/2}$ or $P_{1/2}$ state. Also, a combination of the two is permissible if, and only if, parity is not conserved in the decay.

● Call f_s and f_p the complex amplitudes of the s- and p-wave functions in the decay products and call θ and φ the polar angles of escape of the proton. The $S_{1/2}$, $P_{1/2}$ wave functions, using Clebsch–Gordan coefficients for $J_z = +\frac{1}{2}$, are

$$Y_0^0 \begin{vmatrix} 1 \\ 0 \end{vmatrix}$$

and

$$\sqrt{\tfrac{2}{3}}\, Y_1^1 \begin{vmatrix} 0 \\ 1 \end{vmatrix} - \sqrt{\tfrac{1}{3}}\, Y_1^0 \begin{vmatrix} 1 \\ 0 \end{vmatrix}$$

ISBN 0-8053-8061-7

and similar expressions obtain for $J_z = -\frac{1}{2}$. Since the initial Λ has $J_z = \frac{1}{2}$, the wave function after decay is

$$\psi = f_s Y_0^0 \begin{vmatrix} 1 \\ 0 \end{vmatrix} + f_p \left[\sqrt{\tfrac{2}{3}} \; Y_1^1 \begin{vmatrix} 0 \\ 1 \end{vmatrix} - \sqrt{\tfrac{1}{3}} \; Y_1^0 \begin{vmatrix} 1 \\ 0 \end{vmatrix} \right] \tag{19-5.7}$$

The complex amplitudes f_s, f_p contain four real numbers; however, one phase is arbitrary. These numbers may be found by measurements on the angular distribution and polarization of the decay proton.

To find the angular distribution of the decay protons, one takes $|\psi|^2$. Taking into consideration the orthogonality relations for spin functions, and introducing the explicit expressions of the spherical harmonics, one obtains

$$4\pi|\psi|^2 = |f_s - f_p \cos\theta|^2 + |-f_p \sin\theta|^2$$

$$= |f_s|^2 + |f_p|^2 - 2 \operatorname{Re} f_s^* f_p \cos\theta \tag{19-5.8}$$

Defining

$$\alpha = \frac{2 \operatorname{Re} f_s^* f_p}{|f_s|^2 + |f_p|^2} \tag{19-5.9}$$

we have, for the angular distribution,

$$dw(\theta) = \tfrac{1}{2}(1 - \alpha \cos\theta)\, d(\cos\theta). \tag{19-5.10}$$

It is clear that an asymmetry of the decay occurs only if f_s and f_p are *both* different from 0, which means that parity is not conserved. ●

In practice one does not have polarized Λ at rest. The experiment is then carried out as follows: Λ^0 are produced in the reaction

$$\pi^- + p \Rightarrow \Lambda + K^0 \qquad \Lambda \to \pi^- + p \tag{19-5.11}$$

The conservation of momentum restricts the trajectories of the incident π, Λ, and K^0 in a single plane; and the Λ may be, and actually is, formed with its spin preferentially oriented perpendicular to this plane. Actually this is the only direction in which the Λ can be polarized by the process of Eq. (19-5.11), because parity is conserved in strong interactions and the argument given in Sec. 10-7 on the direction of the polarization applies here as well. On decay of the Λ the proton produced may escape preferentially up or down with respect to the K–Λ plane if, and only if, parity is not conserved in the Λ decay (Fig. 19-4).

ISBN 0-8053-8601-7

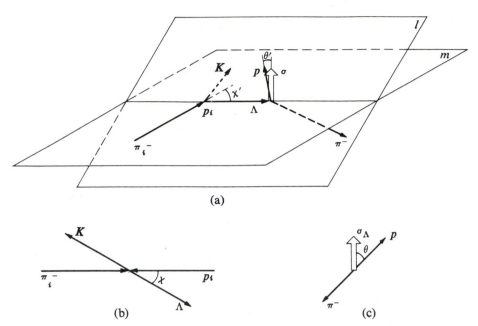

Figure 19-4 Schematic diagram showing the momentum vectors important in the polarization and decay of a Λ. (a) Laboratory system. Note that the momentum of π_i^-, K, and Λ are in plane m (production plane). Spin σ is perpendicular to plane m. The momenta of Λ, p, π^- are in plane l (decay plane). (b) The production in the c.m. system of π_i^- and p_i. (c) The Λ decay in a system in which the Λ is at rest. The angles χ, θ correspond to angles χ', θ' in the lab.

Looking from a system in which the Λ^0 is at rest, we find that it is partially polarized and that the direction of emission of the proton is correlated to the direction of polarization, in a way similar to that which occurs in beta decay. The experiment measures the product of the polarization of the Λ and the asymmetry coefficient α. If the Λ has been formed by the reaction in Eq. (19-5.11), escaping at an angle χ (c.m.) from the π–p direction, it has a polarization $p(\chi)$. Now let θ designate, as before, the angle between the momentum of the decay proton and the perpendicular to the K–Λ plane in the rest system of the Λ. The probability for the direction of escape is given by

$$w(\chi, \cos \theta) \, d \cos \theta = A\left[1 - p(\chi)\alpha \cos \theta\right] d \cos \theta \qquad (19\text{-}5.12)$$

We can interpret $p(\chi)$ as a measure of the polarizing efficiency of the production process and α as a measure of the analyzing efficiency of the decay process. The experiment on anisotropy of decay measures $p\alpha$. Protons

ISBN 0-8053-8061-7

tend to escape parallel to the spin of the Λ ($p\alpha < 0$). Measuring the polarization of the decay proton, for instance by scattering, gives α separately, by an argument that we do not reproduce here. One finds $\alpha = 0.65 \pm 0.01$.

Charged K Mesons

K mesons show a great variety of decay modes. These have been observed in photographic emulsions and in cloud or bubble chambers. The decay modes of the charged K, with their branching ratios, are given in Table 12-2.

The decay constant of $K^\pm$ is 0.8083×10^8 sec^{-1}, corresponding to a mean life of $1.2371 \pm 0.0026 \times 10^{-8}$ sec. It has been measured in photographic emulsions and with counters. That all modes of decay have the same mean life has been verified, indicating that they are due to the disintegration of the same particle. The magnitude of the decay constant indicates that it is due to weak interactions.

Other conceivable modes of decay, such as

$$K^+ \to \pi^+ + \gamma$$
$$K^+ \to \nu + \bar\nu + \pi^+ \qquad (19\text{-}5.13)$$

have never been observed. Their absence is important in order to establish some of the properties of $K^\pm$ mesons. For example, the absence of the decay

$$K^+ \to \pi^+ + \gamma \qquad (19\text{-}5.14)$$

is a strong argument for the assignment of spin 0 to K^+.

The decays of K^- are more difficult to observe than those of K^+ because the K^- interact strongly with nuclei. For K^- the Coulomb attraction favors nuclear interactions at the expense of free decay, whereas the free decay of the K^+ is favored by the Coulomb repulsion, which prevents them from approaching nuclei; furthermore, strangeness conservation prevents many K reactions. However, the free K^- mesons decay with decay constants identical to those of K^+ in the charge conjugate modes; for example,

$$K^- \to \pi^- + \pi^0$$
$$\to \pi^- + \pi^- + \pi^+, \text{ etc.} \qquad (19\text{-}5.15)$$

have the same decay constants as

$$K^+ \to \pi^+ + \pi^0$$
$$\to \pi^+ + \pi^+ + \pi^-, \text{ etc.} \qquad (19\text{-}5.16)$$

The case of K^0 is more complicated and will be discussed separately.

ISBN 0-8053-8601-7

19-6   THE K^0–$\overline{K^0}$ DOUBLET

The spinless K^0, $\overline{K^0}$ particles are generated by strong interactions in which strangeness is conserved, and hence they must have a well-defined strangeness. Typical production reactions are

$$
\begin{array}{ll}
K^- + p \Rightarrow \overline{K^0} + n & S = -1 \\
\pi^+ + p \Rightarrow K^+ + \overline{K^0} + p & S = 0 \\
\pi^- + p \Rightarrow K^0 + \Lambda^0 & S = 0 \\
K^+ + n \Rightarrow K^0 + p & S = 1
\end{array}
\qquad (19\text{-}6.1)
$$

Note that $\overline{K^0}$ can be produced only by charge exchange or in association with K^+ or K^0.

The strangeness of $\overline{K^0}$ is -1, that of K^0 is $+1$, and because of this they act differently in subsequent strong interactions. For instance, K^0 may interact according to the reaction

$$
K^0 + p \Rightarrow K^+ + n \qquad (19\text{-}6.2)
$$

but not according to the reaction

$$
K^0 + p \Rightarrow \Sigma^+ + \pi^0 \quad \text{or} \quad K^0 + n \Rightarrow K^- + p \qquad (19\text{-}6.3)
$$

whereas $\overline{K^0}$ may react according to

$$
\overline{K^0} + p \Rightarrow \Sigma^+ + \pi^0 \quad \text{or} \quad \overline{K^0} + n \Rightarrow K^- + p \qquad (19\text{-}6.4)
$$

but not according to

$$
\overline{K^0} + p \Rightarrow K^+ + n \qquad (19\text{-}6.5)
$$

We may thus distinguish a K^0 from a $\overline{K^0}$ by a strong interaction (Fig. 19-5).

In the decay determined by weak interactions the important conservation law is the CP law (charge conjugation times parity transformation). Now the operator C changes $|K^0\rangle$ into $|\overline{K^0}\rangle$, and P leaves it unchanged, except for a possible change of sign. In any case, C^2 and P^2 are identity operators. We may define the $|K^0\rangle$ state as giving

$$
CP|K^0\rangle = |\overline{K^0}\rangle \qquad (19\text{-}6.6a)
$$

and we have

$$
CP|\overline{K^0}\rangle = |K^0\rangle \qquad (19\text{-}6.6b)
$$

ISBN 0-8053-8601-7

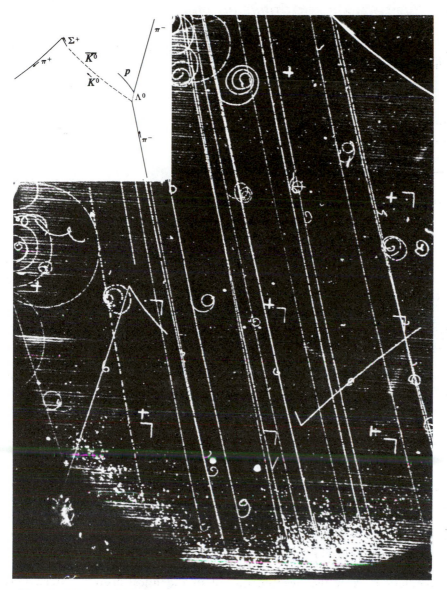

Figure 19-5 Bubble-chamber picture showing the reactions

$$\pi^- + p \Rightarrow \Lambda^0 + K^0$$
$$\Lambda^0 \rightarrow \pi^- + p$$
$$K^0 \rightarrow \overline{K^0}$$
$$\overline{K^0} + p \Rightarrow \Sigma^+ + \pi^0$$
$$\Sigma^+ \rightarrow \pi^+ + n$$

Note the invisible transformation $K^0 \rightarrow \overline{K^0}$ evidenced by the subsequent formation of a Σ^+ in a $\overline{K^0}$ collision.

ISBN 0-8053-8061-7

881

It is hence clear that K^0 and $\overline{K^0}$ are not eigenstates of CP. However, the linear combinations

$$|K_1\rangle = 2^{-1/2}\big(|K^0\rangle + |\overline{K^0}\rangle\big) \qquad |K_2\rangle = 2^{-1/2}\big(|K^0\rangle - |\overline{K^0}\rangle\big) \quad (19\text{-}6.7)$$

are eigenstates of CP, as can be seen by applying Eq. (19-6.6a, b) to them, and precisely

$$CP|K_1\rangle = |K_1\rangle \qquad CP|K_2\rangle = -|K_2\rangle \qquad\qquad (19\text{-}6.8)$$

We can also express $|K^0\rangle$ and $|\overline{K^0}\rangle$ through $|K_1\rangle$, $|K_2\rangle$ as

$$|K^0\rangle = 2^{-1/2}\big(|K_1\rangle + |K_2\rangle\big) \qquad |\overline{K^0}\rangle = 2^{-1/2}\big(|K_1\rangle - |K_2\rangle\big) \quad (19\text{-}6.9)$$

We can then say that the production of a K^0 is equivalent to the production of a K_1 and a K_2 with equal amplitude and a prescribed phase relation (Gell-Mann and Pais, 1955).

The phenomenon is similar to that of the superposition with a given phase relation of light linearly polarized in perpendicular directions. If the phase difference is $\pm\pi/2$, the result for equal amplitudes is circularly polarized light (left or right). Conversely, the superposition of left and right circularly polarized light gives rise to linearly polarized light, the direction of the plane of polarization depending on the phase difference of the two rays of equal amplitude. In general the results discussed in this section closely parallel the properties of polarized light, and it is possible to develop optical analogies for these phenomena. An important difference, however, appears in the spontaneous decay of the K^0 mesons, for which there is no analogy in light quanta.

From the fact that K^0, $\overline{K^0}$ are eigenstates of strangeness and K_1, K_2 are eigenstates of CP it follows that we should consider K^0 and $\overline{K^0}$ when we study strong interactions, and K_1, K_2 when we consider weak interactions. In the decay produced by weak interactions K_1 may go only in states for which the eigenvalue of CP is $+1$ and K_2 only in states for which the eigenvalue of CP is -1. We now apply this to the decay into pions.

The (π^+, π^-) or (π^0, π^0) systems have $CP = +1$. In the latter case this is immediate, because the eigenfunction of two identical bosons is unchanged when the particles are interchanged, an operation which, in the center-of-mass system, is clearly equivalent to P. Moreover, the π^0 is unaffected by charge conjugation; hence $CP = 1$. For the (π^+, π^-) system the same result obtains. To see this, note that operation C interchanges the two pions, replacing π^+ with π^- and vice versa, but operation P interchanges the two momenta and the product of the two is the identity. For three pions $\pi^+\pi^-\pi^0$ or $\pi^0\pi^0\pi^0$, CP can be -1. Consider first two pions either neutral or of opposite charge; for arguments given above CP is $+1$. Suppose they have an angular momentum l in their center of mass. The third, neutral, pion has in

ISBN 0-8053-8601-7

the same system an angular momentum $l' = -l$ because all three pions together have spin 0. Operation P on the third pion alone gives a factor $(-1)^{l'+1}$ because of the negative intrinsic parity of the pion. Operation CP then gives $(-1)^{l+1}$ and $CP = \pm 1$ according to the l value. From this it is clear that K_1 can decay into $\pi^0 + \pi^0$ or $\pi^+ + \pi^-$, but K_2 cannot. The decay constant of K_1 is 1.129×10^{10} sec^{-1}. The K_2 decays in three pions or by leptonic decays, and its decay constant is 1.934×10^7 sec^{-1}. The difference is mainly due to the phase-space difference for two- or three-pion decay.

If we form a beam of K^0 by strong interaction [Eq. (19-6.1)], after a few tenths of a nanosecond the K_1 component has disappeared and we are left with a beam of half the initial intensity consisting solely of K_2. This can be proved by verifying that the beam composed of "old" particles can now produce strong interactions characteristic of both $\overline{K}^0$ and K^0, which was impossible for the "young" beam composed only of K^0. Moreover, in passing through matter, the beam of K_2 changes the relative amplitudes and phases of its K^0, $\overline{K}^0$ components (because K^0 and $\overline{K}^0$ behave differently with respect to strong interactions) and emerges with a K_1 regenerated component, which appears as a short-lived component decaying into two pions (Pais and Piccioni, 1955; Fig. 19-6).

The phenomenon is similar to the change of polarization suffered by a beam of linearly polarized light propagating in a birefringent medium.

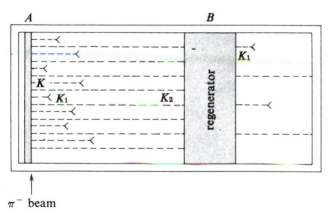

Figure 19-6 Schematic diagram showing the regeneration of K_1 events in a multiplate cloud chamber. The symbol $\langle$ indicates the decay

$$K_1 \rightarrow \pi^+ + \pi^-$$

In target A, pions generate K^0. The K_1 component decays fast and only K_2 reach the regenerator B. In crossing it some of them undergo strong interactions and emerge as K^0 having a regenerated K_1 component. [From A. Pais and O. Piccioni, *Phys. Rev.*, **100**, 1487 (1955).]

ISBN 0-8053-8061-7

We shall treat a simplified case more quantitatively to illustrate the phenomenology. (We use units of $\hbar = c = 1$.) Suppose that a beam of K_2 enters a substance, and that the K_0 and $\overline{K_0}$ making up the K_2 have a different strong interaction with the material; after such an interaction, the phases and amplitudes of the beam are changed and its state is now represented by

$$|\psi\rangle = 2^{-1/2}\left(a|K^0\rangle - b|\overline{K^0}\rangle\right) \qquad (19\text{-}6.10a)$$

with a, b complex. If we look in the direction of propagation of the beam, a and b are proportional to the forward scattering amplitude. The $|\psi\rangle$ after the interaction may be rewritten as

$$|\psi\rangle = \tfrac{1}{2}(a - b)|K_1\rangle + \tfrac{1}{2}(a + b)|K_2\rangle \qquad (19\text{-}6.10b)$$

If $a \neq b$, the probability of finding a K_1 is finite, that is, the beam has been regenerated.

In a more realistic case, consider a slab of material of thickness L and a beam of K_2 entering from the left, directed toward the right. Consider the probability f_{21}, proportional to $a - b$, of regenerating a K_1 in an interaction. The amplitude at the right edge of the slab produced by a K_1 interacting at depth x in the slab is

$$A = \exp(ik_2 x) f_{21} \exp\left[ik_1(L - x)\right] \qquad (19\text{-}6.11)$$

where k_1 is the wave number of the impinging K_1 and k_2 that of the K_2 that is produced. The two are slightly different because of the mass difference between K_1 and K_2 [see Eq. (19-6.24)]. The expression has to be further corrected for the decay in flight of K_1; we can neglect the decay of the long-lived K_2. This correction introduces a factor $\exp[-(L - x)/2v\gamma\tau]$ where v is the velocity of K_1 and τ its mean life in its rest system. The factor 2 is present because A is an amplitude. Taking into account relativity, we have

$$\frac{m_1}{k_1} = \frac{1}{v\gamma} \qquad (19\text{-}6.12)$$

where $\gamma = [1 - (v^2/c^2)]^{-1/2}$. Introducing in Eq. (19-6.11), one has

$$A = \exp(ik_2 x) f_{21} \exp\left[ik_1(L - x) - \frac{m_1}{2k_1\tau}(L - x)\right] \qquad (19\text{-}6.13)$$

Now k_1 and k_2 are related by the conservation of energy and momentum. The interaction transfers a momentum p in the x direction to a mass M, which, in

ISBN 0-8053-8061-7

the case of coherent regeneration, is the mass of a piece of the lattice. One has

$$k_2 = k_1 - p$$

$$\left(m_1^2 + k_1^2\right)^{1/2} + M = \left(m_2^2 + k_2^2\right)^{1/2} + \left(M^2 + p^2\right)^{1/2} \quad (19\text{-}6.14)$$

Because the mass M is very large compared to m_1 or m_2 we may neglect the recoil energy and write

$$\left(m_1^2 + k_1^2\right)^{1/2} - \left(m_2^2 + k_2^2\right)^{1/2} = p^2/2M = (k_1 - k_2)^2/2M \cong 0 \quad (19\text{-}6.15)$$

hence

$$m_1^2 + k_1^2 = m_2^2 + k_2^2 \qquad \text{or} \qquad -m\,\delta m = k\,\delta k \quad (19\text{-}6.16)$$

with $\delta m = m_2 - m_1$, $\delta k = k_2 - k_1$; that is

$$k_1 - k_2 = (m/k)\,\delta m \quad (19\text{-}6.17)$$

where m and k are the average mass and momentum of K_1 and K_2.

We can now sum the amplitudes originating through all layers of the slab and obtain from Eq. (19-6.13), for the amplitude at the right side of the slab,

$$A = \int_0^L N\,dx\,\exp(ik_2 x)\,f_{21}\,\exp\left[ik_1(L - x)\right]\exp\left[-\frac{m_1(L - x)}{k_1 2\tau}\right]$$

$$= \frac{Nf_{21}}{i(k_2 - k_1) + (m_1/2k_1\tau)}\left[\exp(ik_2 L) - \exp(ik_1 L)\exp\left(-\frac{m_1 L}{k_1 2\tau}\right)\right]$$

$$(19\text{-}6.18)$$

and by taking $|A|^2$ we obtain, for the probability of finding a K_1 at the edge of the slab,

$$W(l) = \left|\frac{Nf_{21}}{i(k_2 - k_1) + (m_1/2k_1\tau)}\right|^2 \left(1 - 2e^{-l}\cos 2l\delta^* + e^{-2l}\right) \quad (19\text{-}6.19)$$

with

$$l = m_1 L/2k_1\tau \qquad \delta^* = \tau(m_2 - m_1) = (k_1 - k_2)k_1\tau/m_1 \quad (19\text{-}6.20)$$

The magnitude of δ^* determines the behavior of $W(l)$. If $\delta^* \sim 1$, the function oscillates as shown in Fig. 19-7. The remarkable Eq. (19-6.19) has been verified experimentally.

ISBN 0-8053-8601-7

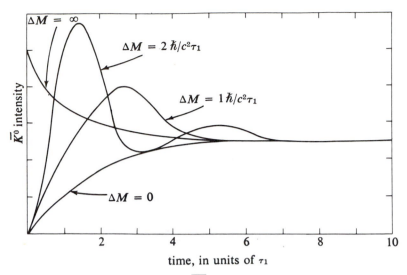

Figure 19-7 The intensity of the $\overline{K}^0$ component is shown as a function of time for three values of the mass difference in units of $\hbar/c^2\tau_1$. For $t = 0$ the beam is a pure K^0 beam. [From U. Camerini et al. (Wisconsin), R. W. Birge et al. (Berkeley); *Phys. Rev.*, **128**, 362 (1962).]

In addition to coherent regeneration proportional to N^2 there is an incoherent regeneration proportional to N, but we cannot go into more detail here. Other ingenious forms of regeneration experiments, for instance with two slabs separated by a gap, give the value of δm as well as its sign. (See Table 13-2.)

The tiny mass difference between K_1 and K_2 is analogous to that existing between other particles, for example, π^0, $\pi^\pm$; n, p, which differ in their electric charge, and as a consequence in their mass by an amount which can be crudely interpreted as being due to the electromagnetic energy $e^2/r = c^2\,\Delta m$, where r is the "radius" of the particle. Similarly, for weak interactions we expect a mass difference related to a difference in energy. We may try to guess the order of magnitude of this mass difference by a dimensional argument. By analogy with the electromagnetic case we guess that the mass difference is proportional to g^2 and we build the expression $(g^2/m'c^2)(\hbar/m'c)^{-6}$, which has the dimensions of an energy. For the mass m' we use the mass of the pion, although we cannot seriously justify doing so, and we obtain $\delta m \sim 6 \times 10^{-6}$ eV or $\delta m/m_K \sim 10^{-14}$! The experimental value is $m_2 - m_1 = 3.52 \times 10^{-6}$ eV.

We shall now discuss briefly the free decay of the K^0, which gives another way of revealing the mass difference between K_1 and K_2 (Treiman and Sachs, 1956). Suppose that we generate a beam of K^0 by strong interaction and write

ISBN 0-8053-8601-7

the $\psi(t)$ as

$$|\psi(t)\rangle = 2^{-1/2}|K_1\rangle \exp(-im_1 t) \exp(-\Gamma_1 t/2)$$

$$+ 2^{-1/2}|K_2\rangle \exp(-im_2 t) \exp(-\Gamma_2 t/2) \quad (19\text{-}6.21)$$

where $\Gamma_{1,2} = 1/\tau_{1,2}$, the reciprocal of the mean life of $K_{1,2}$. The ψ is such that for $t = 0$ it represents, as it should, K^0. At any time $|\psi(t)\rangle$ can be resolved into a K^0 and a $\overline{K^0}$ component. The former, for example, is

$$\langle K^0|\psi(t)\rangle = \tfrac{1}{2}\left\{ \exp\left[(-im_1 t) - \Gamma_1 t/2\right] + \exp\left[(-im_2 t) - \Gamma_2 t/2\right] \right\} \quad (19\text{-}6.22)$$

and its modulus square

$$\tfrac{1}{4}\left\{ \exp(-\Gamma_1 t) + \exp(-\Gamma_2 t) + 2\exp\left[-(\Gamma_1 + \Gamma_2)t/2\right] \cdot \cos(t\,\delta m) \right\} \quad (19\text{-}6.23)$$

gives the number of K^0 in the beam. The number of $\overline{K^0}$ is given by the same expression, except that the cosine term has a minus sign. If $m_1 = m_2$, Eq. (19-6.23) represents a combination of decreasing exponentials, but if $m_1 \neq m_2$, there are very peculiar oscillations in the number of K^0 or $\overline{K^0}$ in the beam (see Fig. 19-7). The quantitative aspects of the effect depend on the magnitude of the ratio

$$\frac{c^2\,\delta m/\hbar}{\Gamma_1} = \frac{\tau_1 c^2\,\delta m}{\hbar} \quad (\Gamma_2 \ll \Gamma_1) \quad (19\text{-}6.24)$$

The number of K^0 present in one section of the beam is shown by the semi-leptonic decays $K^0 \rightarrow e^+ + \nu_e + \pi^-$, $K^0 \rightarrow \mu^+ + \nu_\mu + \pi^-$ which are allowed by the $\Delta Q = \Delta S$ rule. Conversely, by the same rule, $\overline{K^0}$ has semileptonic decays containing π^+. This type of experiment gives only the magnitude, but not the sign, of δm. We have already mentioned that regeneration experiments with two regenerators give the sign. For best numerical values see Table 13-2.

19-7 *CP* VIOLATION IN *K* DECAY

In 1964 Christenson, Cronin, Fitch, and Turlay observed, to everybody's surprise, that a "stale" beam from which all K_1 should have disappeared gave beyond doubt a decay of K_2 into two pions in about 2 out of 1000 cases. This is a clear violation of *CP* conservation and as such of major importance. Is *CPT* conserved through a compensatory violation of *T* or is *CPT* also violated? The evidence favors the first hypothesis, preserving a very important tenet of the theory.

ISBN 0-8053-8061-7

Because of the obvious nonorthogonality of the states formerly called K_1 and K_2 demonstrated by their common decay into two pions, we introduce two new linear combinations of $|K^0\rangle$ and $|\overline{K^0}\rangle$, corresponding to the physical states having a long and a short decay constant, and we call them $|K_L\rangle$ and $|K_S\rangle$. Requiring *CPT* invariance limits the linear combination to the form

$$|K_S\rangle = \left[2(1 + |\epsilon|^2)\right]^{-1/2}\left[(1 + \epsilon)|K^0\rangle - (1 - \epsilon)|\overline{K^0}\rangle\right] \qquad (19\text{-}7.1)$$

$$|K_L\rangle = \left[2(1 + |\epsilon|^2)\right]^{-1/2}\left[(1 + \epsilon)|K^0\rangle + (1 - \epsilon)|\overline{K^0}\rangle\right] \qquad (19\text{-}7.2)$$

T invariance alone would give

$$|K_S\rangle = \left[2(1 + |\epsilon'|^2)\right]^{-1/2}\left[(1 + \epsilon')|K^0\rangle - (1 - \epsilon')|\overline{K^0}\rangle\right]$$

$$|K_L\rangle = \left[2(1 + |\epsilon'|^2)\right]^{-1/2}\left[(1 - \epsilon')|K^0\rangle + (1 + \epsilon')|\overline{K^0}\rangle\right]$$

It is thus clear that *CPT* and *T* invariance together require $\epsilon' = \epsilon = 0$. The description of the system by Eqs. (19-7.1) and (19-7.2) automatically guarantees *CPT* invariance.

The problem is to determine ϵ from experiment and to explain why it is different from zero. The phase of ϵ is determined only if one establishes a convention for the phase between K^0 and $\overline{K^0}$; the one usually adopted is that the transition amplitude of $K^0 \rightarrow 2\pi$ in the *i*-spin 0 state is real. An analysis of the decays gives the result

$$\arg \epsilon \cong \tan^{-1} \frac{2(m_L - m_S)}{(\Gamma_S - \Gamma_L)} \frac{c^2}{\hbar} \qquad (19\text{-}7.3)$$

where m_L, Γ_L and m_S, Γ_S refer to the mass and decay constants, respectively, of K_L and K_S. Other measurable quantities that can be used to determine ϵ are

$$\frac{\text{rate}(K_S \rightarrow \pi^- l^+ \nu)}{\text{rate}(K_S \rightarrow \pi^+ l^- \bar{\nu})} \cong \frac{\text{rate}(K_L \rightarrow \pi^- l^+ \nu)}{\text{rate}(K_L \rightarrow \pi^+ l^- \bar{\nu})} = 1 + 4 \operatorname{Re} \epsilon \qquad (19\text{-}7.4)$$

where l stands for lepton. Also measurable are the quantities

$$\frac{\text{ampl}(K_L \rightarrow \pi^a \pi^b)}{\text{ampl}(K_S \rightarrow \pi^a \pi^b)} = \eta_{ab}$$

with $a, b = +, -$ or $0, 0$.

If as it seems all *T* and *CP* violations are determined by the single small

ISBN 0-8053-8061-7

parameter ϵ, one has from theory

$$\eta_{00} \cong \eta_{+-} \cong \epsilon$$

Experiments, including some very difficult ones relative to η_{00}, give $|\eta_{00}| \cong |\eta_{+-}| \cong 2 \times 10^{-3}$. The final result for ϵ is that $|\epsilon| = 2 \times 10^{-3}$ and its phase is 43 deg.

Among the several hypotheses proposed to explain the *CP* violation, one of the most plausible (Wolfenstein, 1964) postulates the existence of a superweak interaction with a coupling constant of the order of 10^{-9} of that of the weak interaction, and obeying the selection rule $|\Delta S| = 2$. The decay of K_L into two pions would then be a two-step process. The superweak interaction would mix the K_L and K_S states and K_L would decay by the ordinary weak interaction into two pions. The mixing of K_L and K_S requires that the interaction have matrix elements between states differing by 2 units of strangeness. This hypothesis predicts $\epsilon = \eta_{+-} = \eta_{00}$, which is compatible with present observations.

It is noteworthy that the decay of the neutral K is the only known phenomenon showing a *CP* violation.

BIBLIOGRAPHY

Bég, M. A. B., and A. Sirlin, "Gauge Theories of Weak Interactions," *Ann. Rev. Nucl. Sci.*, **24**, 379 (1974).

Commins, E. D. (Com 73).

Gaillard, M. K., B. W. Lee, and J. L. Rosner, "Search for Charm," *Rev. Mod. Phys.*, **47**, 277 (1975).

Kleinknecht, K., "*CP* Violations and K^0 Decays," *Ann. Rev. Nucl. Sci.*, **26**, 1 (1976).

Lee, T. D., and C. S. Wu, "Weak Interactions," *Ann. Rev. Nucl. Sci.*, **15**, 381 (1965); **16**, 471 (1966).

Okun', L. B., *Weak Interactions of Elementary Particles*, Addison-Wesley, Reading, Mass., 1965.

Sachs, R. G., "Interference Phenomena of Neutral K Mesons," *Ann. Phys.*, **22**, 239 (1963).

Weinberg, S., "Recent Progress in Gauge Theories of the Weak, Electromagnetic, and Strong Interactions," *Rev. Mod. Phys.*, **46**, 255 (1974).

Williams, W. S. C. (W 71).

PROBLEMS

19-1 Why does Σ^0 decay into $\Lambda^0 + \gamma$ rather than directly into $n + \pi^0$ or $n + \gamma$?

19-2 Why do Ξ hyperons decay into $\Lambda + \pi$ rather than into a nucleon plus a pion?

19-3 What prevents K^0 from decaying into two gammas?

19-4 Show that the $|\Delta T| = \frac{1}{2}$ rule gives the branching ratio $1 : 2$ for $(K_S \rightarrow \pi^0 + \pi^0)/(K_S \rightarrow \pi^+ + \pi^-)$.

19-5 What prevents a $\bar{s}\gamma_\alpha(1 + \gamma_5)d$ term in the quark current?

19-6 Show that the ratio $R = (K^0 - \overline{K^0})/(K^0 + \overline{K^0})$ in a beam composed initially of K_2 is given by

$$R = \frac{2\{\exp[-(\Gamma_1 + \Gamma_2)(t/2)]\} \cos t\, \delta m}{\exp(-\Gamma_1 t) + \exp(-\Gamma_2 t)}$$

ISBN 0-8053-8601-7

How could one measure this ratio?

19-7   From the neutron mean life $\tau = 918$ sec, make a rough estimate of the decay rate (in sec^{-1}) you might expect for the decay $\Sigma^- \to \Sigma^0 + e^- + \bar{\nu}_e$. ($\Sigma^- - \Sigma^0$ mass difference $= 4.9$ MeV.)

19-8   (a) A subject of considerable interest is the beta decay of the Λ^0 particle $\Lambda^0 \to p + e + \bar{\nu}$. Assuming that the matrix element is one, the coupling constant is the same as in neutron decay, and neglecting (as in neutron decay) the Λ^0 recoil energy, calculate the expected beta-decay rate of the Λ^0. If the usual Λ^0 decay ($\Lambda^0 \to p + \pi^-$) has mean life 2.5×10^{-10} sec, what is the branching ratio for the beta decay? (b) What is the Λ^0 maximum recoil energy? (The mass of $\Lambda^0 = 1115$ MeV.)

19-9   The π^+ and π^- have precisely the same mass, whereas the Σ^+ and Σ^- do not. Why? Would you expect the K^+ and K^- to have the same mass? If so, why?

19-10   A K_2 monochromatic beam passes through a plate of thickness d_a and regeneration amplitude $f_{21}^a = |f_{21}^a| \exp(i\varphi_a)$, then through an empty interval of thickness L, and then through a second plate of thickness d_b and regenerating amplitude $f_{21}^b = |f_{21}^b| \exp(i\varphi_b)$. Show that the emerging K_1 amplitude is proportional to

$$1 + \alpha \exp[i(\varphi_a - \varphi_b)t + it\,\delta m - (t/2\tau\gamma)]$$

with

$$\alpha = \frac{|f_{21}^a|N_a d_a}{|f_{21}^b|N_b d_b}$$

and $N_{a,b}$ the number of nuclei per cubic centimeter in the absorbers; $t = L/\gamma v$; v is the velocity of K_2; and $\gamma = (1 - v^2)^{-1/2}$. This result gives a method of measuring δm. Try to develop it.

19-11   In Fig. 17-10, mark (a) the semileptonic decays with $\Delta S = \Delta C$ ($\sim \cos^2 \theta$), and (b) the nonleptonic decays with $\Delta S = \Delta C$ ($\sim \cos^4 \theta$) and the nonleptonic decays with $\Delta S \neq \Delta C$ ($\sim \cos^2 \theta \sin^2 \theta$) by joining the initial particle to the final hadron with a line.

19-12   Show that the decay $D^+ \to K^- \pi^+ \pi^+$ would lead to an *exotic* final state.

19-13   Analyze in terms of quarks and of the current–current interaction the decays: $K^- \to \mu^- + \bar{\nu}_\mu$; $\Lambda \to p + \pi^-$; $D^0 \to K^- + \pi^+$; $D^0 \to \bar{K}^0 + \pi^0$; $D^+ \to \bar{K}^0 + e^+ + \nu_e$; $D^0 \to \bar{K}^0 + \pi^0$.

ISBN 0-8053-8601-7

High-Energy Collisions of Hadrons

20-1 INTRODUCTION

Collisions of hadrons at energies very large compared with their rest mass, say above 5 GeV c.m., were once accessible only through cosmic rays. The new large accelerators now make available beams of several hundred GeV in the lab system; a typical figure is 300 GeV lab. Intersecting proton-proton storage rings reach c.m. energies of about 60 GeV, which is equivalent to 1918 GeV in the laboratory system. Cosmic rays remain unsurpassed for extremely high energies; in fact, events corresponding to a total energy of 10^{20} eV have been observed. As an example of an accelerator-produced high-energy event, Fig. 20-1 shows a bubble-chamber picture of a proton–proton collision.

Photographic emulsions and huge counter arrays are the most frequent observational tools for extremely high-energy events in cosmic rays. Figure 20-2 shows emulsions with high-energy events; the stars of low-energy physics are transformed into jets in the direction of the incoming particle. In photographic emulsion work the emerging tracks are usually divided into two types:

1. Shower particles which give a grain density <1.4 times the minimum grain density; these are relativistic particles, pions for the most part.

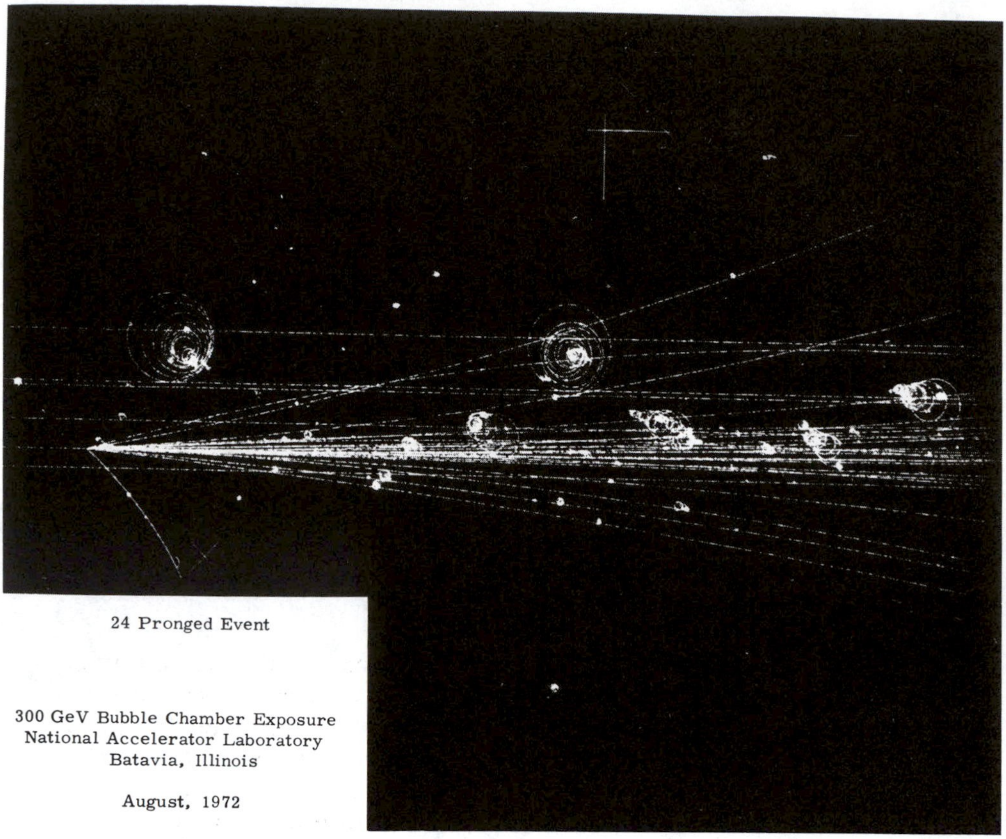

24 Pronged Event

300 GeV Bubble Chamber Exposure
National Accelerator Laboratory
Batavia, Illinois

August, 1972

Figure 20-1 Bubble-chamber picture of a 300-GeV proton interacting with a proton. [Courtesy of FNAL, Batavia, Ill.]

2.   Heavy ionizing particles with a grain density >1.4 times the minimum; these are subdivided into black tracks, with an ionization >10 times the minimum, and gray tracks, with a density of ionization between 1.4 and 10 times the minimum. A proton of 25 MeV is about at the limit of the black tracks.

It is obviously of great interest to determine the energy of the primary particle. An approximate method is based on the observation of the secondaries. Call γ_c and β_c the γ and β of the center of mass and γ_p, β_p the corresponding quantities of the incident particle, all relative to the lab system. Assume the primary to be a nucleus of mass M, colliding with a particle of mass N. One has ($c = 1$)

$$\gamma_c = \left[\frac{(M\gamma_p + N)^2}{(M^2 + N^2 + 2MN\gamma_p)} \right]^{1/2} \cong \left(\frac{M\gamma_p}{2N} \right)^{1/2} \tag{20-1.1}$$

ISBN 0-8053-8061-7

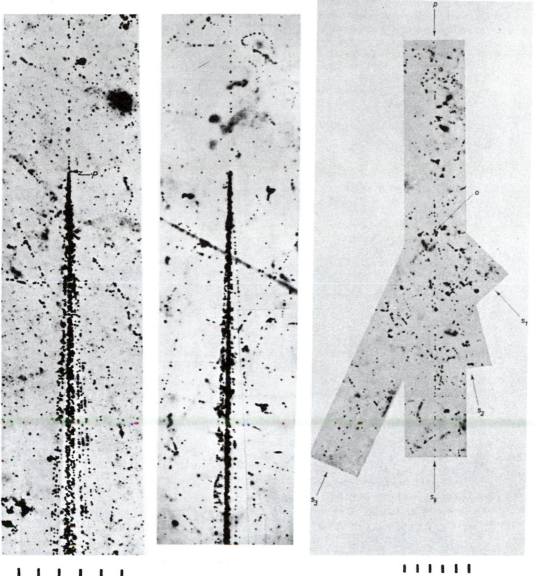

Figure   20-2   Jets of energy = 3000 GeV, 9000 GeV, and 40 000 GeV. For an analysis see (PFP 59), pp. 552 and 559.

where the approximation is for $\gamma_p \gg 1$. If one knows the nature of the target particle and γ_c, one then finds the total energy of the projectile $M\gamma_p$.

A determination of γ_c may be attempted by assuming that in the center-of-mass system the secondaries are emitted at angles $\bar{\theta}_s$ broadly distributed around the collision axis in the center of mass. Call $\bar{\gamma}_s$, $\bar{\beta}_s$, and $\bar{\theta}_s$ the quantities relative to a secondary and referred to the center-of-mass system. One has

$$\tan \theta_s = \frac{\sin \bar{\theta}_s}{\gamma_c \left(\cos \bar{\theta}_s + \beta_c / \bar{\beta}_s \right)} \tag{20-1.2}$$

where θ_s is in the laboratory system. If two secondaries 1 and 2 are emitted at angles $\bar{\theta}$ and $\pi - \bar{\theta}$ and with equal momenta, one has for the tracks in the laboratory

$$\tan \theta_2 \times \tan \theta_1 = \frac{\sin^2 \bar{\theta}}{\gamma_c^2 \left[-\cos^2 \bar{\theta} + \left(\beta_c / \bar{\beta}_s \right)^2 \right]} \tag{20-1.3}$$

Given many secondaries distributed in such a way that, for each θ_1, there is its corresponding θ_2, we have, since $\beta_c \approx \bar{\beta}_s \approx 1$,

$$\langle \tan \theta_1 \tan \theta_2 \rangle = \frac{1}{\gamma_c^2} \tag{20-1.4}$$

or

$$\log \gamma_c = -\frac{1}{n} \sum \log \tan \theta \tag{20-1.5}$$

For γ_c large and θ correspondingly small, Eq. (20-1.4) may be approximated by

$$\gamma_c \cong \frac{1}{\langle \theta \rangle} \tag{20-1.6}$$

where $\langle \theta \rangle$ is an average value of $|\theta|$ for the particles in the jet and corresponds roughly to particles escaping in the c.m. system at 90 deg to the line of flight.

Tests in emulsions with 200-GeV artificially accelerated protons showed that if one computes γ_c from Eq. (20-1.6) for a large number of events, the *average* γ_c is nearly correct, but the distribution of γ_c values ranges over at least one power of ten. The values computed from Eq. (20-1.6) are thus only qualitative and can be off by a factor of 2 or 3 in either direction.

Another important parameter of a collision is its inelasticity K, defined in a nucleon–nucleon collision as the fraction of the initially available energy radiated as newly created particles (pions, nucleon–antinucleon pairs). The inelasticity may be determined from photographic-emulsion experiments. For extremely high energies it proves to be small (0.20). Thus an incident particle of many thousands of GeV energy preserves about 80% of its energy after a collision.

ISBN 0-8053-8061-7

The highest-energy phenomena observed are colossal showers formed in air by incoming cosmic rays. These showers may contain as many as 10^9 particles and may cover areas of some square kilometers. The total energy has been found to reach 10^{20} eV. The jets noted in photographic emulsions may be the initial phase of such high-energy interactions, but the full development occurs outside the stack. The observations are made by covering an extended area with counters and measuring the location and time of arrival of the particles (Rossi, 1959).

Very high-energy collisions may give rise to very complicated events, with the creation of many new particles, and a detailed study of these phenomena has only begun. As an orientation we shall consider first a simple model.

20-2   STATISTICAL THEORY OF HIGH-ENERGY COLLISIONS

Fermi (1950) originated a crude but simple model that brings out some of the features that must of necessity be present in nuclear collisions. The idea is that when two nucleons collide, they suddenly release all their kinetic energy in a small volume of the order of magnitude of the cube of the range of nuclear forces. In spite of the small duration of the collision, it is assumed that the strong interactions are so strong as to allow the establishment of a statistical equilibrium in this small volume. The subsequent course of the reaction is then dominated by phase-space factors for the final states.

It is at once apparent that this type of consideration excludes all processes slower than those determined by strong interactions. For example, the very existence of electromagnetic radiation is ignored. On the other hand, the only parameter involved in the model is the volume Ω, in which equilibrium is established.

The distance at which nucleons interact is approximately the Compton wavelength of the pion $\hbar/m_\pi c$ (see Chap. 15); therefore the volume in which equilibrium is established must be reasonably near to

$$\Omega = \frac{4}{3}\pi\left(\frac{\hbar}{m_\pi c}\right)^3 = 1.18 \times 10^{-38} \text{ cm}^3 \tag{20-2.1}$$

or relativistically to

$$\Omega = \frac{4}{3}\pi\left(\frac{\hbar}{m_\pi c}\right)^3 \frac{2m_p c^2}{W} \tag{20-2.2}$$

The last factor represents the Lorentz contraction of the volume in which equilibrium is established in a nucleon–nucleon collision. W is the total energy in the center-of-mass system.

The model is especially suited to the treatment of multiple production of strongly interacting particles such as pions. An increase of Ω tends to increase the multiplicity in production phenomena.

ISBN 0-8053-8601-7

The conservation laws that are valid for strong interactions must be satis-fied, and this condition puts additional constraints on the statistical equilibrium. In practice, conservation of energy and momentum are always taken into account. However, in many calculations some of the other con-servation laws have been omitted for the sake of simplicity.

Statistical equilibrium means that all single states are equally probable. Hence the probability of obtaining a certain physical result is proportional to the number of single states (statistical weight) corresponding to it. For exam-ple, the statistical weight $S(n)$ of n independent spinless particles, with momenta $p_1 \cdots p_n$ and total energy W, in a volume Ω inside a large box of volume V, is computed as follows: Designate $Q(W)$ as the volume in momentum space inside the surface of constant energy W_0. Then the number of states per unit energy interval is

$$\frac{dN}{dW} = \frac{dQ}{dW} V^n \frac{1}{(2\pi\hbar)^{3n}} \tag{20-2.3}$$

The first factor comes from the momentum space; the second comes from the normalization volume V; the third is the volume of the unit cell in phase space.

The probability that all n independent particles will be in volume Ω is $(\Omega/V)^n$, and the statistical weight of the state where all n particles are in Ω with the specified momenta is

$$S(n) = \frac{\Omega^n}{(2\pi\hbar)^{3n}} \frac{dQ(W)}{dW} \tag{20-2.4}$$

However, the conservation of momentum effectively reduces the number of degrees of freedom of the system. The momentum space is $3(n - 1)$-dimensional, hence the exponent n is replaced by $n - 1$ in Eq. (20-2.4).

Other factors in $S(n)$ may come from the spin of the particles and from the identity of some of the particles. The isotopic spin conservation may also affect the number of accessible states. Finally the conservation of angular momentum should be taken into account. This last condition introduces con-siderable complications in the calculations but usually affects the results by relatively small numerical factors and is largely neglected.

Once the $S(n)$ are computed, the cross sections for different processes are obtained by multiplying a geometric cross section

$$\sigma_{geom} = \pi R^2 = \pi \left(\frac{\hbar}{m_\pi c}\right)^2 \cong 6 \times 10^{-26} \text{ cm}^2 \tag{20-2.5}$$

by $S(n)/\Sigma S(n)$.

ISBN 0-8053-8601-7

The model can be extended and simplified at extremely high energy when the volume Ω can be treated as a small volume heated at a temperature τ. This volume in very relativistic cases closely resembles the black body of radiation theory, except that the light quanta are replaced by pions or nucleon–antinucleon pairs. The particles formed must also be extremely relativistic. These conditions obtain for energies $> 10^{12}$ eV.

In extensive statistical studies Hagedorn and collaborators take into account some further experimental results not contained in the fundamental statistical hypotheses. Moreover, they have also varied the interaction volume for heavier mesons. The results agree qualitatively with the experimental data and are useful, at least in providing orientation for planning experiments.

Some of the ideas of the statistical model permeate all subsequent models. For instance, one may consider not one, but two excited volumes (fireballs), one in each of the colliding particles, with each fireball moving with half the total momentum of the original particle.

20-3 MAIN FEATURES OF HIGH-ENERGY COLLISIONS

The statistical model is the simplest, but, especially with the help of accelerators, one starts to see in high-energy collisions dynamic properties that go well beyond the statistical predictions. The most important are summarized here as simple empirical qualitative rules:

1. Rule of approximately constant cross sections. At high energy both total and elastic cross sections vary slowly and smoothly with energy. This variation is demonstrated by the samples shown in Fig. 20-3. Note, however, the considerable increase in the p–p cross section; similar increases are starting to appear in other collisions. The ratio of σ_e / σ_T has a value around 0.2.

2. Rule of small transverse momenta. The projections of the momenta of all outgoing hadrons on a plane perpendicular to the line of flight of the incident hadron are small, with an average value around 0.4 GeV/c. The energy and the nature of the incident particle does not influence the distribution. In elastic scattering this is clearly shown in Figs. 20-4 and 20-5; for inelastic processes compare Fig. 20-6. This rule is an excellent approximation, but in about 1 in 10^8 collisions one finds high (over 4 GeV/c) transverse momenta. This phenomenon recalls the wide-angle α scattering, discovered by Rutherford, that led to the conception of the nucleus. It is possible that such high transverse momentum collisions are due to the collisions of point-like grains contained in the proton (partons). (See Sec. 18-3 and Figs. 20-7 and 20-8.)

3. Rule of low multiplicities. The average number of outgoing hadrons increases slowly with energy, roughly as $\log s$ where s is the c.m. energy squared. (See Figs. 20-9 and 20-10.) In any case it is much smaller than if the energy were converted into particles. In other words, the energy of the incoming particle is transformed into the kinetic energy of the outgoing ones. Since

ISBN 0-8053-8061-7

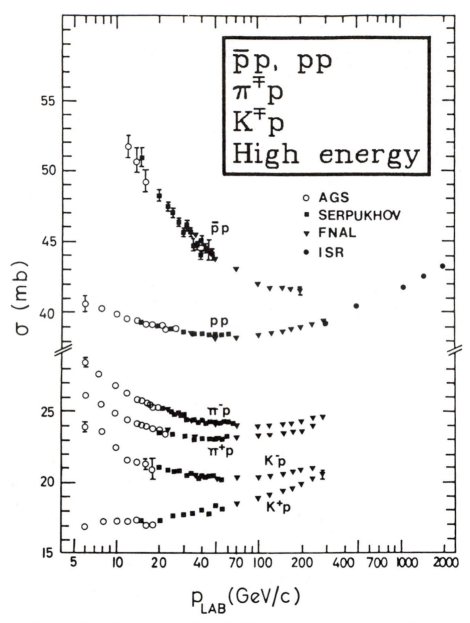

Figure   20-3   Total cross sections of different particles on protons. At higher energy some of the cross sections increase. [From Particle Data Group, "Review of Particle Properties" (1976)]

ISBN 0-8053-8061-7

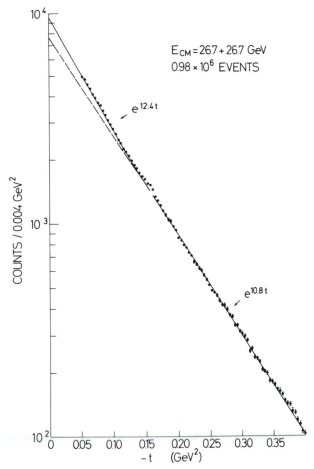

Figure 20-4 Elastic proton–proton scattering at small momentum transfer (ACGHT collaboration); CERN intersection rings. [*Phys. Letters*, **39B**, 663 (1972).]

the transverse momentum $p_\perp$ is much smaller than the longitudinal momentum $p_\parallel$, the direction of the incoming particle is an axis around which the other particles migrate. This is clearly seen in photographic emulsions or bubble-chamber pictures of "jets" (Figs 20-1 and 20-2).

Total cross sections confirm Pomeranchuk's theorem relating the total particle cross section $\sigma(x, A)$ with that of the antiparticle $\bar{\sigma} = \sigma(\bar{x}, A)$. According to Pomeranchuk's theorem, the difference $\Delta\sigma = \sigma - \bar{\sigma}$ tends to zero for large s. The relation is valid if the ratio Re $F/(\text{Im } F \cdot \log s)$ tends to zero for large s. (Here F is the forward scattering amplitude.) It is not clear, however, with what s dependence the difference between the cross sections

ISBN 0-8053-8601-7

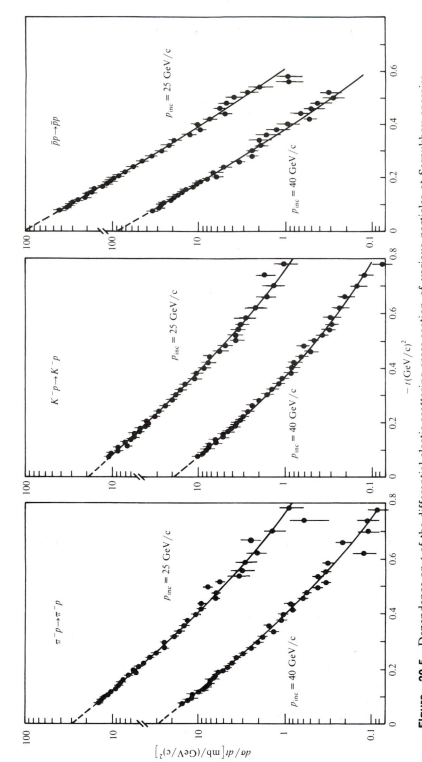

Figure 20-5 Dependence on t of the differential elastic scattering cross section of various particles at Serpukhov energies. [CERN-IHEP collaboration; see L. Foà, *Riv. Nuovo Cimento*, **3**, 283 (1973).]

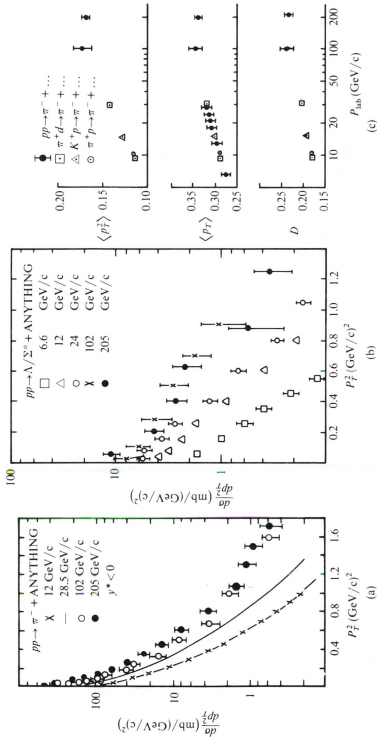

Figure 20-6 The p_T dependence of (a) π^- production and (b) the Λ^0/Σ^0 production cross section in p–p collisions. (c) Low-order moment of the p_T distribution for inclusive π^- production (integrated over all y). The dispersion D equals $[\langle p_T^2 \rangle - \langle p_T \rangle^2]^{1/2}$. [H. Bøggild and T. Ferbel, *Ann. Rev. Nucl. Sci.*, **24**, 451 (1974).]

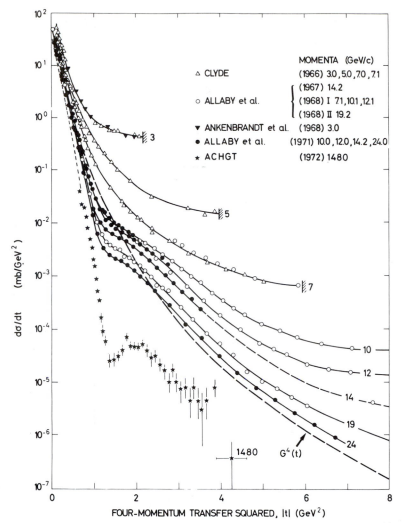

Figure 20-7 Elastic p–p scattering at high momentum transfer for different total energies. Note the curve of $G^4(t)$, the electronic form factor to the fourth power. See Sec. 18-2. [L. Foà, *Riv. Nuovo Cimento*, **3**, 283 (1973).]

should vanish. Experimentally, $\Delta\sigma$ decreases smoothly as $s^{-1/2}$ at energies now available. (See Figs. 20-11 and 20-12.)

Several total cross sections confirm the relations obtainable by simple quark models, as indicated in Sec. 17-4.

20-4   DIFFRACTION SCATTERING

We shall now consider some special cases in a little more detail. We begin with the simplest case, in which there are only two outgoing particles, and in

ISBN 0-8053-8061-7

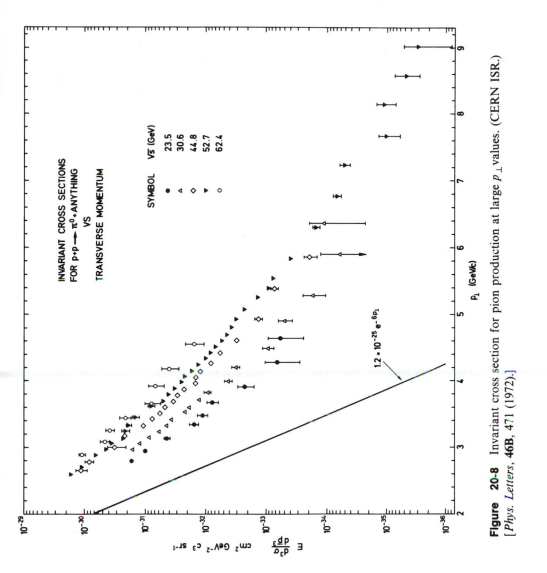

Figure 20-8 Invariant cross section for pion production at large $p_\perp$ values. (CERN ISR.) [*Phys. Letters*, **46B**, 471 (1972).]

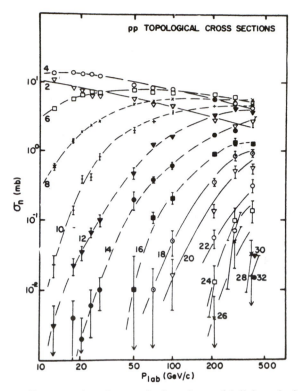

Figure 20-9   Cross section for producing the multiplicity of charged particles indicated on the curves versus incident momentum. The average number
according to this and other experiments is $\langle n \rangle = 0.69 + 1.93 \log s$. There are
about half as many neutral particles as there are charged ones. (CERN data.)
[H. Bøggild and T. Ferbel, *Ann. Rev. Nucl. Sci.*, **24**, 451 (1974).]

particular the elastic case in which $A + B = A' + B'$ with particle $A = A'$ and
$B = B'$. The momentum transfer squared is defined as

$$(p_A - p_{A'})^2 = (\mathbf{p}_A - \mathbf{p}_{A'})^2 - (E_A - E_{A'})^2 = -t \qquad (20\text{-}4.1)$$

and assume $\hbar = c = 1$. We then find that the differential cross section at
constant s and variable t is well represented by an exponential in t as long as
the momentum transfer is small (Fig. 20-4).

We can crudely represent the forward scattering peak by the equation

$$\frac{d\sigma}{dt} = a \exp(-bt) \qquad (20\text{-}4.2)$$

The optical theorem then gives $a = \sigma_T^2/16\pi$ provided the forward scattering
amplitude is purely imaginary. This is a good approximation even if the real
part is 0.2 of the imaginary one, and usually it is smaller. Of the two constants, a depends strongly on the nature of the hadrons; b varies from about 8

ISBN 0-8053-8601-7

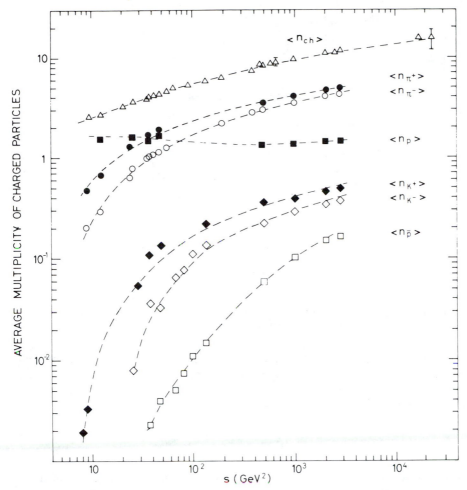

Figure 20-10 Average multiplicity of charged particles produced in p–p collisions (CERN data). [H. Bøggild and T. Ferbel, *Ann. Rev. Nucl. Sci.*, **24**, 451 (1974).]

to 12 $(\mathrm{GeV}/c)^{-2}$ on varying s and the particles involved. (See Fig. 20-5.) This forward peak is interpreted as shadow scattering or diffraction scattering, related to the total cross section by a well-known quantum-mechanical argument (see Sec. 11-5). At large s and large t Eq. (20-4.2) is generally not applicable. Structure strikingly reminiscent of a diffraction pattern is often observed, as in proton–proton scattering (Fig. 20-7).

One speaks of diffraction dissociation in the case in which, in the two-body reaction, $A \neq A'$, $B \neq B'$, provided the particles A, A' and B, B' have the same internal quantum numbers: baryon number, electric charge, strangeness, i-spin, and G parity (when defined). A' is then considered an excited

ISBN 0-8053-8601-7

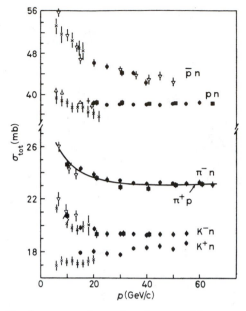

Figure 20-11 Total cross sections of various particles on neutrons (Serpukhov data). [L. Foà, *Riv. Nuovo Cimento*, **3**, 283 (1973).]

state of A and B' an excited state of B. In this case one has also forward peaking and approximate energy independence of the cross section. We can go one step further and replace one or both of the outgoing particles by groups of particles. As an example, a study of the reactions

$$\pi^- + p \Rightarrow \pi^- + \pi^- + \pi^+ + p \qquad (20\text{-}4.3)$$

has enabled us to divide the events observed into four groups. For all of them $A = \pi^-$ and $B = p$, but the outgoing particles are classified as follows.

$$
\begin{array}{lll}
\text{Group 1} & A' = \pi^-\pi^-\pi^+ & B' = p \\
\text{Group 2} & A' = \pi^- & B' = \pi^-\pi^+p \\
\text{Group 3} & A' = \pi^-\pi^- & B' = \pi^+p \\
\text{Group 4} & A' = \pi^-\pi^+ & B' = \pi^-p
\end{array} \qquad (20\text{-}4.4)
$$

Forward peaking is satisfied in all four cases. The internal quantum numbers are unchanged in cases 1 and 2; in the others there is charge exchange and hence the internal quantum number "charge" is changed. Only for groups 1 and 2 is the cross section approximately energy independent; for groups 3 and 4 it decreases with energy, until at very high energies processes 1

ISBN 0-8053-8061-7

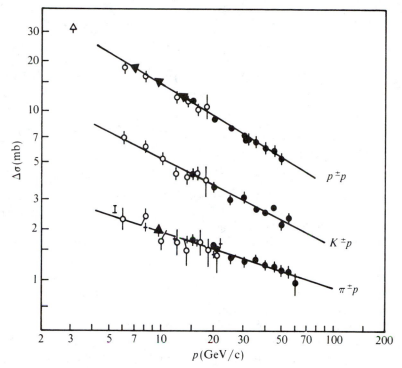

Figure 20-12 Energy dependence of particle–antiparticle cross-section differences (Serpukhov data). According to Pomeranchuk, the differences tend to zero for infinite energy. [L. Foà, *Riv. Nuovo Cimento*, **3**, 283 (1973).]

and 2 are dominant. Processes 1 and 2 are diffraction dissociation; the others not. In practice, one recognizes the hadrons forming A' because in the c.m. system they go in the same approximate direction and at the same speed as A; the same is true for B' and B.

20-5 EXCHANGE COLLISIONS; REGGE POLES

When the internal quantum numbers of A and A', B and B' are different and the momentum transfer is small, the reactions $A + B \Rightarrow A' + B'$ are described as one-particle exchange (OPE) collisions. Symbolically they are represented by the diagrams in Fig. 20-13. There are thus two ways to achieve the same result. The virtual particle exchanged C is different in the two channels, each characterized by the quantum numbers exchanged along it. These are determined by the conservation laws, as Table 20-1 shows. We distinguish the two channels experimentally from the angular distribution of the reaction. Call θ in the c.m. system the angle between the incoming and outgoing baryon; then the prevalence of small θ indicates a t-channel reaction

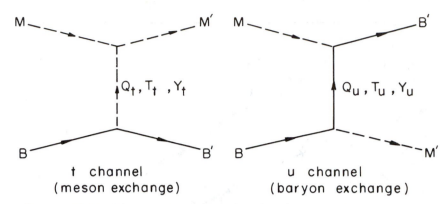

Figure   20-13   Diagrams for t-channel and u-channel exchanges in meson
(M)–baryon (B) interactions. Q, T, and Y are the charge, i spin, and hyper-
charge, respectively, of the exchanged system. [From J. D. Jackson, Proc. 13th
International Conference on High Energy Physics, Berkeley, (1967).]

and a prevalence of $\theta \cong 180$ deg indicates a u-channel reaction. Both
mechanisms may be present simultaneously, in which case there are forward
and backward peaks. Typical examples are demonstrated in Fig. 20-14. A
noteworthy empirical rule is that if there is no particle with the exchanged
quantum numbers that can be generated by a combination of quark-
–antiquark or by three quarks, the peaks are absent, showing that the OPE
reaction does not occur. If there is a $q\bar{q}$ or qqq particle corresponding to only
one channel, but not to the other, then only the channel along which a $q\bar{q}$ or
qqq particle can exchange gives a peak.

**TABLE 20-1   QUANTUM NUMBERS AND PARTICLES EXCHANGED IN SOME OPE
REACTIONS**

Reaction	t-Channel quantum number			Possible particles[a]	Forward peak	u-Channel quantum number			Possible particles	Backward peak
	Q_t	T_t	Y_t			Q_u	T_u	Y_u		
$\pi^- p \to \pi^0 n$	1	1	0	ρ	Yes	1	$\frac{1}{2}, \frac{3}{2}$	1	N, Δ	Yes
$\pi^+ p \to \pi^+ p$	0	0, 1	0	$\mathscr{P}, \rho$	Yes	0	$\frac{1}{2}, \frac{3}{2}$	1	N, Δ	Yes
$\pi^+ p \to \rho^+ p$	0	0, 1	0	$\mathscr{P}, \pi, \omega, \phi, A_2$	Yes	0	$\frac{1}{2}, \frac{3}{2}$	1	N, N'	Yes
$\pi^- p \to K^0 \Lambda$	1	$\frac{1}{2}$	1	K^*, K^{**}	Yes	1	1	0	Σ, Y_1^*	Yes
$\pi^- p \to K^+ \Sigma^-$	2	$\frac{3}{2}$	1	None	No	0	0, 1	0	Λ, Σ, Y_1^*	Yes
$\pi^- p \to \pi^- p$	0	0, 1	0	$\mathscr{P}, \rho$	Yes	2	$\frac{3}{2}$	1	Δ	Yes
$K^- p \to K^- p$	0	0, 1	0	$\mathscr{P}, \rho, \omega, \varphi$		2	1	2	None	No
$K^- p \to \overline{K^0} n$	1	1	0	ρ	Yes	1	0, 1	2	None	No

[a] $\mathscr{P}$ means a particle with quantum numbers of vacuum, called a *pomeron*.

ISBN 0-8053-8061-7

An example will show how to find the possible particles exchanged or, better, their quantum numbers. The first line of Table 20-1 presents the reaction $\pi^- + p \Rightarrow \pi^0 + n$. To find the particle exchanged in the t channel, put in Fig. 20-13 $B = p$, $B' = n$, $M = \pi^-$, and $M' = \pi^0$. We may also replace any particle by its antiparticle and reverse its momentum. In the t channel we may thus write $p + \bar{n} \Rightarrow X_t \Rightarrow \pi^+ + \pi^0$, and we see that X_t has $B_t = 0$, $Q_t = 1$, $T_t = 1$, $Y_t = 0$; candidates for X_t are therefore ρ, A_2, etc. At the (π^+, π^0) end we find the G-parity requirement for X_t, $G_t = 1$, and this singles out ρ. Similarly, for the u channel we find that $p + \pi^0 \Rightarrow X_u \Rightarrow n + \pi^+$ and $B_u = 1$, $Q_u = 1$, $T_u = \frac{1}{2}$ or $\frac{3}{2}$, and $Y = 1$. Possible particles are thus a proton or Δ. Similar arguments disclose the exchanged particle in other cases.

Calculations based on the OPE model [see, e.g., (W 71)] lead to an asymptotic cross section of the form

$$\frac{d\sigma}{dt} = F(t)s^{2J-2} \tag{20-5.1}$$

where J is the angular momentum of the particle exchanged. $F(t)$ is a function depending on the vertexes and gives the angular distribution, whereas the s dependence is contained in the other factor, s^{2J-2}. This formula, however, disagrees with experiment. For instance, $d\sigma/dt$ for the reaction $\pi^- + p$

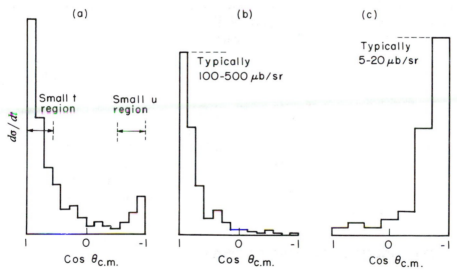

Figure 20-14 Representative production cross sections. In the most frequent type (a), the large forward peak at small t is caused by meson exchange; the small backward peak is caused by baryon exchange and diminishes on increasing s. (b) For this process only meson exchange is possible because of the values of the internal quantum numbers. (c) Only baryon exchange is possible. [From J. D. Jackson, Proc. 13th International Conference on High Energy Physics, Berkeley (1967).]

ISBN 0-8053-8601-7

$\Rightarrow \pi^0 + n$ varies approximately as s^{-1} (at constant t), whereas according to Eq. (20-5.1), with ρ^- exchange, it should be energy independent, since $J = 1$ for the ρ.

Moreover, exchange of higher-spin particles like the A_2 and f^0 $(J = 2)$ would lead, according to Eq. (20-5.1), to partial and therefore total cross sections eventually growing as s^2, but very general considerations by Froissart show that total cross sections cannot increase with energy faster than $(\log s)^2$. One way out of this apparent contradiction, still preserving the OPE model (there are others—e.g., Cheng and Wu's idea of s-channel unitarization, with an eikonal approximation, of the simple OPE diagram), is the application of Regge theory.

Regge's theory regards the J of the exchanged particle as a variable that is capable of assuming continuous values. This idea has a counterpart in the variable t. The scattering amplitude is a function of t, but t is different from the value $(t = m_C^2)$ it would have if the exchanged particle C were a real particle. However, we consider arbitrary values of t (off the mass shell). Could we, by analogy, consider the scattering amplitude a function of J, the angular momentum of C, for arbitrary, even complex, values of J? It will be necessary to interpolate the scattering amplitude by an analytic function that has poles for J integral or half integral, corresponding to real particles. The exchanged particle will then have an angular momentum $\alpha(t)$, in the same way as it has a mass, function of t. Only when $t = m_C^2$ does $\alpha(t)$ take on the J value corresponding to the real particle.

In Regge's theory Eq. (20-5.1) is generalized to

$$\frac{d\sigma}{dt} = F(t)s^{2\alpha(t) - 2} \qquad (20\text{-}5.2)$$

where $\alpha(t)$ coincides with J of a real particle only for special values of t. The α dependence on t is shown by the "Regge trajectories" of Fig. 17-9. If the particle exchanged is the ρ^-, we use the ρ^- trajectory and for t near zero we find $\alpha(0) \cong \frac{1}{2}$. This is the value to be inserted in Eq. (20-5.2), after which $d\sigma/dt$ becomes proportional to s^{-1}, in agreement with experiment.

We cannot go into a serious study of Regge's theory here; it is treated in, e.g., (W 71) and (Le 73).

20-6   INCLUSIVE REACTIONS; SCALING

In very high-energy processes, when many particles are produced, detailed study of the event is prohibitively complicated; even when only four or five particles have been produced it is almost impossible to keep track of all of them in a fruitful way. It proves more instructive to focus attention on only one particle and lump all the remaining ones together, paying no attention to them except for total energy and momentum. We saw an example of this

ISBN 0-8053-8061-7

procedure when we discussed the nucleon form factors as revealed by electron scattering.

A reaction $a + b \Rightarrow c +$ anything, when we focus attention on c irrespective of what is comprised in "anything," is an inclusive study. *The reaction may be described by giving the nature of a, b, and c; the total energy in the c.m. system; and* $\mathbf{p}_c$. For simplicity we omit any consideration of spins. For this study one could use s, t, and u, variables similar to the Mandelstam variables of Sec. 13-6, but it is common and useful to use different kinematic variables introduced by Feynman in 1969: $s, p_\perp$, and

$$x = 2p_\parallel^* / \sqrt{s} \tag{20-6.1}$$

where $\mathbf{p}^*$ is the momentum of c in the c.m. system.

Another set of variables is s, $p_\perp$, and y. This last quantity, called *rapidity*, has the following physical meaning: find a reference system moving with the z axis parallel to $p_\parallel$ in which c has only transverse momentum. This system moves with velocity β with respect to the lab. Then

$$y = \tanh^{-1} \beta \tag{20-6.2}$$

is the rapidity of c with respect to the lab. In the frame in which the momentum is only transverse, the energy of c is

$$\omega_c^2 = m_c^2 + p_\perp^2 \tag{20-6.3}$$

This defines the "transverse mass" ω_c (sometimes different letters are used).

We now establish some important relations between rapidity and other variables in an arbitrary frame K' moving with velocity β relative to the frame in which the momentum is only transverse. We have, with $y = \tanh^{-1} \beta$,

$$p_\perp' = p_\perp$$

$$p_\parallel' = \omega_c \sinh y \tag{20-6.4}$$

$$E' = \omega_c \cosh y$$

On changing from a frame K to another K', the rapidity changes in a simple way:

$$y' = y + u$$

where $u = \tanh^{-1} \beta'$ and β' is the velocity of the primed frame with respect to the unprimed one. The total energy of c in any frame is

$$E = \left(m_c^2 + p_\perp^2 + p_\parallel^2 \right)^{1/2} = \left(\omega_c^2 + p_\parallel^2 \right)^{1/2} = \omega_c \gamma = \omega_c \cosh y \tag{20-6.5}$$

ISBN 0-8053-8601-7

and its $p_\| = \omega_c \beta\gamma = \omega_c \sinh y$. The rapidity is connected to other kinematic variables by the relations

$$y = \tanh^{-1}\left(\frac{p_\|}{E}\right) = \frac{1}{2} \log \frac{E + p_\|}{E - p_\|} = \log \frac{E + p_\|}{\omega_c} \qquad (20\text{-}6.6)$$

which can be derived by using the fundamental properties of the hyperbolic functions and relativistic relations. A simple geometric meaning of rapidity is obtained by considering the angle θ between the trajectory of c and that of the incident particle. Obviously $p_\perp/p_\| = \tan\theta$. If $p_\perp^2 \gg m_c^2$, one obtains, from Eq. (20-6.6),

$$y \cong -\log \tan\left(\frac{\theta}{2}\right) \qquad (20\text{-}6.7)$$

and y in the laboratory is very easily measured in a photographic plate or in a bubble-chamber picture. In this context y has been familiar for a long time to cosmic-ray physicists, as discussed in Sec. 20-1.

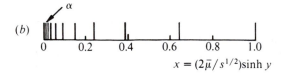

$$x = (2\bar{\mu}/s^{1/2})\sinh y$$

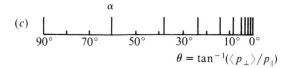

$$\theta = \tan^{-1}(\langle p_\perp \rangle/p_\|)$$

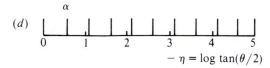

$$-\eta = \log \tan(\theta/2)$$

Figure 20-15 Relation among the variables y, x, θ and $\eta = -\log \tan(\theta/2)$. Center-of-mass energy, 53 GeV; $\bar{\mu} = (\langle p_\perp \rangle^2 + m_\pi^2)^{1/2}$. The first, second, etc. long vertical lines correspond to each other. Note the similarity between the η and y scales. [From L. Foà, *Riv. Nuovo Cimento*, **3**, 283 (1973).]

ISBN 0-8053-8601-7

Variables x and y are especially useful in the study of very high-energy phenomena. Figure 20-15 exhibits the relations among x, y, and θ and $\eta = -\log \tan(\theta/2)$ in collisions observed at the energy of the CERN ISR (p–p, 53 GeV c.m.).

In studying inclusive collisions producing a particle c one usually measures the quantity

$$E_c \frac{d^3\sigma}{dp^3} = \frac{d^3\sigma}{dp_\perp^2\, dy} = F_{ab}^c(p_\parallel, p_\perp, s) \tag{20-6.8}$$

or

$$\rho_{ab}^c = F_{ab}^c / \sigma_{ab} \tag{20-6.9}$$

where σ_{ab} is the total ab inelastic cross section. E_c is the energy of c, $\mathbf{p}$ its momentum of components $p_\perp$ and $p_\parallel$, and s the square of the total c.m. energy. The two forms are equivalent and both are relativistically invariant.

There are two related hypotheses on the asymptotic behavior of $F_{ab}^c(p_\parallel, p_\perp, s)$ that are reasonably verified by experiment. The first, or *limiting fragmentation* theory (Benecke, Chou, Yang, and Yen, 1969), assumes that at very high energy particles a and b fragment, each in its reference system, independently of the other. (Figure 20-16 shows the situation diagrammatically.) Limiting fragmentation therefore implies that at very high energies the distribution $F_{ab}^c(p_\parallel, p_\perp, s)$ of particles that are fragments of a depends on $p_\parallel$ and $p_\perp$ (measured in the rest frame of a), but not on s.

The other hypothesis, due to Feynman (1969), is that the invariant cross section in the c.m. system for $s \to \infty$ becomes

$$F_{ab}^c(p_\parallel^*, p_\perp, s) \to F_{ab}^c\left(p_\perp, \frac{2p_\parallel^*}{s^{1/2}}\right) = F_{ab}^c(p_\perp, x) \tag{20-6.10}$$

When this occurs we say that *scaling* obtains.

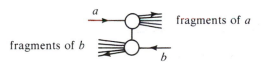

a

fragments of a

fragments of b

b

Figure 20-16 The collision of particles a and b seen in the c.m. system. The distribution of fragments in the reference frame of each particle does not depend on the energy of the other for $s \to \infty$. [From L. Foà, *Riv. Nuovo Cimento*, **3**, 283 (1973).]

ISBN 0-8053-8061-7

Limiting fragmentation and scaling are related, as can be seen by using the "one-dimensional gas" approximation. The phase space available to the fragments, written in the form $dy\, dp_\perp^2$, is limited to a small region because for high s the rapidity varies between y_a and y_b and $p_\perp$ is limited by the rule of small transverse momenta. It follows that the representative points of c are confined to a "bottle" of length $Y \cong y_a - y_b$ (see Problem 20-2) and radius $p_\perp^c$ (Fig. 20-17). One makes the further hypothesis that particles influence each other only if their rapidities differ by less than a certain amount (the correlation length in rapidity space). This hypothesis is supported by experimental evidence (see Fig. 20-18). It can be theoretically justified by the "multi-peripheral model" in which the reaction is described by a chain of peripheral collisions. One can then write

$$\rho_{ab}^c = \rho_{ab}^c(y_a - y_b, y_c - y_a, p_\perp) \tag{20-6.11}$$

where $y_a - y_b$ for $s \to \infty$ is $\log(s/m_a m_b)$ and $y_c - y_a$ gives $p_\parallel^c$ in the rest frame of a. If $y_c - y_a$ is finite while $s \to \infty$, particle c is within the correlation length with a, but not with b, and ρ_{ab}^c cannot depend on the nature or rapidity of b. The expression for ρ becomes a function of $y_c - y_a$ and $p_\perp$ only.

$$\rho_{ab}^c = f_a^c(y_c - y_a, p_\perp) = g_a^c(x, p_\perp) \tag{20-6.12}$$

in the limit as $Y \to \infty$ and with $y_c - y_a$ fixed, where x is the Feynman variable $2p_\parallel^*/s^{1/2}$ (see Problem 20-3). The first equality expresses the limiting fragmentation of a. A similar equation with $y_c - y_a$ replaced by $y_c - y_b$ gives the

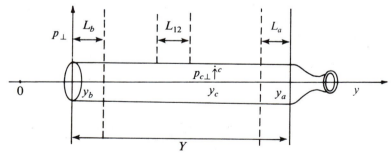

Figure 20-17 Feynman–Wilson gas picture. The "gas" of produced particles is confined in phase space to the general region of the "bottle." The length of the bottle is determined by the kinematics, while its radius is governed by the dynamic limitation of $p_\perp$. The finite correlation lengths at each end and in the center are denoted L_a, L_b, and L_{12}, respectively.

ISBN 0-8053-8061-7

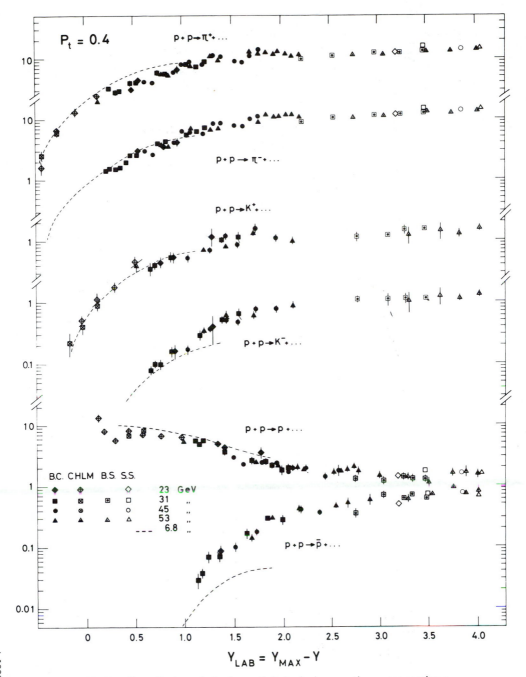

$\dfrac{E\,d\sigma}{d\bar{p}}$ (mb GeV⁻²c³)

$P_t = 0.4$

$p + p \rightarrow \pi^+ + \dots$

$p + p \rightarrow \pi^- + \dots$

$p + p \rightarrow K^+ + \dots$

$p + p \rightarrow K^- + \dots$

$p + p \rightarrow p + \dots$

$p + p \rightarrow \bar{p} + \dots$

B.C.	CHLM	B.S.	S.S.		
◆	⊕		◇	23	GeV
■	⊠	▣	□	31	"
●	⊛	◉	○	45	"
▲	▲	▲	△	53	"
		- - -		6.8	"

$Y_{LAB} = Y_{MAX} - Y$

Figure 20-18 Compilation of single-particle inclusive reaction cross sections in mb/(GeV⁻²c³) at $p_\perp = 0.4$ GeV/c for $p + p \rightarrow c +$ anything at various energies E as a function of $y_{lab} = y_c - y_a$. The center of the rapidity scale $(y_a + y_b)/2$ is at different values of y_{lab} depending on the total energy. (For $E = 23$ and 31 GeV it is at 3.2 and 3.5, respectively.) [CERN, J. V. Allaby et al., Proc. 4th International Conference on High Energy Collisions, Oxford (1972).]

ISBN 0-8053-8601-7

915

SINGLE PARTICLE DISTRIBUTIONS FROM EQUAL AND
UNEQUAL ENERGY RUNS

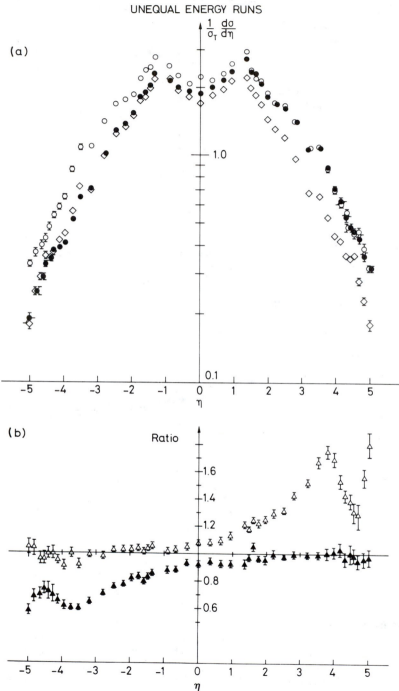

ISBN 0-8053-8061-7

limiting fragmentation of b. The second equality in Eq. (20-6.12) is an expression of Feynman scaling for $x > 0$, and by replacing a with b we obtain an expression of Feynman scaling for $x < 0$.

If particle c is produced many correlation lengths away from a and from b, in the central region of the bottle, ρ can depend only on $p_\perp$ and one has

$$\rho_{ab}^c = h^c(p_\perp) \qquad (20\text{-}6.13)$$

in the limit as $Y \to \infty$ and with $y_c - y_b \gg L_b, y_c - y_a \gg L_a$. The invariant cross section in this rapidity region is flat in rapidity and depends on a and b only through the normalization σ_{ab}. Experimental confirmation of this state of affairs is shown in Figs. 20-18 and 20-19.

Scaling may be related to a finer structure of the particles, such as is indicated by the quark hypothesis or parton hypothesis. The details of these relations are still not clear. It is interesting, however, to observe scaling phenomena at a level where the substructures are well known (i.e., in high-energy nuclear collisions). Thus, for instance, by hitting Be nuclei with p, d, or ^{4}He ions having 1–2 GeV per nucleon, one produces d, p, ^{3}H, ^{3}He, etc. In these collisions, even at relatively low momenta it is possible to recognize that the transverse momenta are small compared with longitudinal ones; one also observes several scaling phenomena. Figure 20-20 pertains to ^{4}He on Be and shows the rapidity distribution for the products p, d, ^{3}He, ^{3}H. All these rapidities show the same distribution and all peak at the incident alpha rapidity. This behavior corresponds to a diffractive dissociation. There is also a plateau at lower rapidity corresponding to the central region of the bottle (Fig. 20-17). The cross section for pion formation in $p + \text{Be} \to \pi +$ anything is shown in Fig. 20-20 as a function of the Feynman variable $x = p^*_\parallel / p^*_{max} \cong 2p^*_\parallel / \sqrt{s}$ and here we see the scaling on changing s, because each point in the figure contains results obtained with $\sqrt{s}$ varying from 10.28 to 12.60 GeV ($T_p = 1.05$–4.2 GeV).

Figure 20-19 Inclusive cross section of colliding proton beams as a function of $\eta = -\log \tan(\theta/2) \cong y$. Running conditions: $\diamondsuit$, 15 GeV/c on 15 GeV/c; ●, 15 GeV/c on 27 GeV/c; ○, 27 GeV/c on 27 GeV/c. The data obtained with asymmetric beams (●) coincide on the right with those obtained at 27 + 27 and at the left with those obtained at 15 + 15. This is shown in more detail in (b), where △ denotes the ratio of the ordinates of the ● points to the $\diamondsuit$ points and ▲ is the ratio of the ordinates of the ● points to those of the $\diamondsuit$ points. It follows that the distribution of particles downstream from a beam is independent of the energy of the other beam, in agreement with the limiting fragmentation theory. [CERN, H. Bøggild, AIP Conference–Particles and Fields, 1973, Am. Inst. of Physics, N.Y., 1973.]

ISBN 0-8053-8601-7

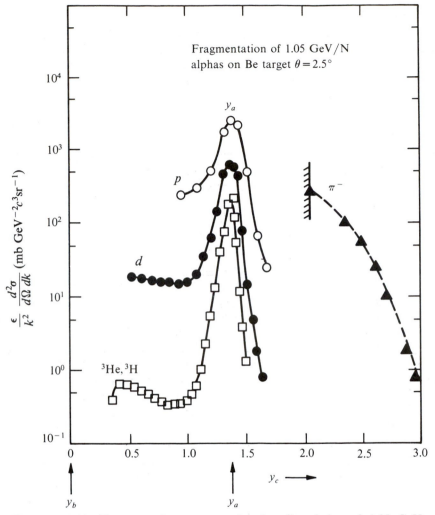

Figure 20-20 Fragmentation cross section for dissociation of 1.05 GeV nucleon^{-1} alpha particles on Be into p, d, ^{3}H, ^{3}He. $\theta = 2.5°$ lab; Be target. Ordinate, Lorentz-invariant cross section $(E/p^2)(d^2\sigma/d\Omega \, dp)$; abscissa, rapidity y_c. Arrows indicate the rapidity of the target and of the incident alphas. [From J. Jaros et al., Lawrence Berkeley Laboratory.]

BIBLIOGRAPHY

Amaldi, U., M. Jacob, and G. Matthiae, "Diffraction of Hadronic Waves," *Ann. Rev. Nucl. Sci.*, **26**, (1976).

Bøggild, H., and T. Ferbel, "Inclusive Reactions," *Ann. Rev. Nucl. Sci.*, **24**, 451 (1974).

Foà, L. "High Energy Hadron Physics," *Riv. Nuovo Cimento*, **3**, 283 (1973).

ISBN 0-8053-8601-7

Frazer, W. R., et al., "High Energy Multiparticle Reactions," *Rev. Mod. Phys.*, **44**, 284, (1972).

Hagedorn, R. and J. Ranft, "Statistical Thermodynamics of Strong Interactions at High Energies," *Suppl. Nuovo Cimento,* **6**, 169 (1968).

Jackson, J. D., "Models for High Energy Processes," *Rev. Mod. Phys.*, **42**, 12 (1970).

Jackson, J. D., "Introduction to Hadron Interaction at High Energies," Scottish Universities Summer School, 1974.

Kretzschmar, M., "Statistical Methods in High Energy Physics," *Ann. Rev. Nucl. Sci.*, **11**, 1 (1961).

Powell, C. F., P. H. Fowler, and D. H. Perkins (PFP 1959).

Schwitters, R. F., and K. Strauch, "The Physics of e^+–e^- Collisions," *Ann. Rev. Nucl. Sci.,* **26**, (1976).

Williams, W. S. C. (W 71).

PROBLEMS

20-1   Show that for two particles 1, 2

$$s = -(p_1 + p_2)^2 = m_1^2 + m_2^2 + 2\omega_1\omega_2 \cosh(y_1 - y_2) - 2\mathbf{p}_{1\perp} \cdot \mathbf{p}_{2\perp}$$

Hint: Use Eqs. (20-6.4) and the addition formula for cosh x.

20-2   Show that for $s^{1/2} \gg m_a$, m_b, in a two-body collision

$$y_a - y_b = Y \approx \log(s/m_a m_b)$$

In this same limit

$$Y^* \cong \tfrac{1}{2}(Y - \Delta)$$

$$y_a^* \cong \tfrac{1}{2}(Y + \Delta) \qquad -y_b^* \cong \tfrac{1}{2}(Y - \Delta)$$

where starred quantities refer to the center-of-mass system and $\Delta = \log(m_b/m_a)$.

20-3   Show that for finite $|x|$, fixed $p_\perp$, and large s

$$x = (\omega_c/m_a) \exp(y_c - y_a) \qquad x > 0$$

$$= -(\omega_c/m_b) \exp(y_b - y_c) \qquad x < 0$$

where $x = 2p_\parallel^* s^{-1/2}$; $s \to 2m_b(m_a^2 + p_a^2)^{1/2}$, and where $p_\parallel^*$ is the component of $\mathbf{p}_c$ parallel to the direction of the incident particle, a.

20-4   A pion of 10 GeV/c laboratory momentum impinges on a proton at rest. Find y for the pion and the proton in the lab and c.m. systems. In the elastic collision the pion is deflected by 1 deg in the laboratory. Find its new y and the y of the proton.

20-5   In a collision $a + b \Rightarrow c + X$, find the mass of X if you know s and $\mathbf{p}_c$. Note that the masses of a and b do not enter in the answer, but the mass m_c does.

20-6   Add entries to Table 20-1. Consider the reactions:

$$\pi^+ + p \Rightarrow \pi^0 + \Delta; \quad \omega + \Delta; \quad \rho + \Delta$$

$$K^- + p \Rightarrow \pi^+ + \Sigma^-; \quad K^+ + \Xi^-$$

ISBN 0-8053-8061-7

APPENDIX

Scattering from a Fixed
Center of Force

We assume that the potential U decreases faster than $1/r$ and is central. The scattering center is located at the origin of the coordinates. The Schrödinger equation is

$$\nabla^2\psi + \frac{2m}{\hbar^2}(E - U)\psi = 0 \tag{A-1}$$

We shall try to find a solution of Eq. (A-1) with an asymptotic behavior for ψ given by

$$\psi(\mathbf{r}) = e^{ikz} + f(\theta)\frac{e^{ikr}}{r} \tag{A-2}$$

$$k = p/\hbar \quad \text{or} \quad v = k\hbar/m \tag{A-3}$$

Physically Eq. (A-2) represents a plane wave e^{ikz} of amplitude 1 and a spherical outgoing wave of amplitude $f(\theta)$.

Emilio Segrè, Nuclei and Particles: An Introduction to Nuclear and Subnuclear Physics, Second Edition

ISBN 0-8053-8061-7

The outgoing flux through a large sphere centered at the origin is given by

$$\varphi = \frac{\hbar k}{m} \int |f(\theta)|^2 \, d\omega \tag{A-4}$$

as is seen from the expression for current in quantum mechanics. By definition the differential cross section is then

$$|f(\theta)|^2 = \frac{d\sigma}{d\omega} \tag{A-5}$$

We shall now calculate this quantity. We first need an identity

$$e^{ikz} = \frac{\pi\sqrt{2}}{(kr)^{1/2}} \sum_0^\infty i^l (2l+1)^{1/2} Y_{l,0}(\theta) J_{l+1/2}(kr) \tag{A-6}$$

with the spherical harmonic

$$Y_{l,0}(\theta) = \frac{(2l+1)^{1/2}}{(4\pi)^{1/2}} P_l(\cos\theta) \tag{A-7}$$

and $P_l(\cos\theta)$ a Legendre polynomial. The $Y_{l,0}(\theta)$ have the orthonormality property

$$\int Y_{l,0}(\theta) Y_{l',0}(\theta) \, d\omega = \delta_{l,l'} \tag{A-8}$$

The $J_{l+1/2}(x)$ are Bessel functions with the asymptotic expressions

$$J_{l+1/2}(x) = \frac{\sqrt{2}}{\sqrt{\pi}} \frac{x^{l+1/2}}{(2l+1)!!} \qquad x \ll 1 \tag{A-9}$$

$$J_{l+1/2}(x) = \left[\left(\frac{2}{\pi x} \right)^{1/2} \sin\left(x - \frac{\pi l}{2} \right) \right] \qquad x \gg 1 \tag{A-10}$$

The identity can be proved by developing e^{ikz} in a series of Legendre polynomials by the usual Fourier method. Figure A-1 shows the first few $j_l(x) = (\pi/2x)^{1/2} J_{l+1/2}(x)$.

We now also develop $f(\theta)$ in spherical harmonics,

$$f(\theta) = \sum_l a_l P_l(\cos\theta) = (4\pi)^{1/2} \sum_l \frac{a_l}{(2l+1)^{1/2}} Y_{l,0}(\theta) \tag{A-11}$$

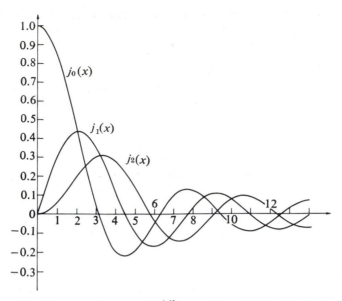

Figure A-1   The functions $(\pi/2x)^{1/2}J_{l+1/2}(x)=j_l(x)$ for $l=0$, 1, 2. They can also be expressed as

$$j_0(x) = \frac{\sin x}{x} \; ; \quad j_1(x) = \frac{\sin x}{x^2} - \frac{\cos x}{x} \; ; \quad j_2(x) = \left(\frac{3}{x^3} - \frac{1}{x} \right) \sin x - \frac{3 \cos x}{x^2}$$

and insert this expression into Eq. (A-2), obtaining asymptotically

$$\psi(r, \theta) = \frac{(4\pi)^{1/2}}{r} \sum_l \frac{Y_{l,0}}{(2l+1)^{1/2}} \left[e^{ikr} \left(+a_l - \frac{i}{2} \frac{2l+1}{k} \right) \right.$$

$$\left. + e^{-ikr} (-1)^l \frac{i}{2} \frac{2l+1}{k} \right] \tag{A-12}$$

The coefficient of e^{ikr} represents the amplitudes of outgoing waves of different angular momenta l, and similarly the coefficient of e^{-ikr} represents the amplitude of ingoing waves. In a stationary state they must be equal in modulus for each l separately in order to ensure conservation of matter; however, they may have different arguments. We thus have

$$- a_l + \frac{i}{2} \frac{2l+1}{k} = e^{2i\delta_l} \left(\frac{i}{2} \frac{2l+1}{k} \right) \tag{A-13}$$

or

$$a_l = \frac{2l+1}{2ik} (e^{2i\delta_l} - 1) \equiv (2l+1)A_l \tag{A-14}$$

ISBN 0-8053-8601-7

The real quantity δ_l is called the phase shift of the lth wave and A_l is called the scattering amplitude.

The δ_l are not determined by the conservation theorem, but by the asymptotic behavior of $\psi(\mathbf{r})$. The radial part $R(r)$ of $\psi(\mathbf{r})$ can be written $u(r)/r$. Insertion in Eq. (A-1) shows that the asymptotic behavior of $u(r)$ for $r \rightarrow \infty$ is

$$u_l(r) = \sin\left[kr - (\pi l/2) + \delta_l \right] \tag{A-15}$$

if, as assumed, $U(r)$ vanishes rapidly enough. On the other hand, for r small, we have

$$u_l''(r) - \frac{l(l+1)}{r^2} u_l + \frac{2m}{\hbar^2} \left[E - U(r) \right] u_l = 0 \tag{A-16}$$

Near the origin $u_l(r) = r^{l+1}/(2l+1)!!$, and if we can integrate Eq. (A-16), we find, by joining the solutions for r small and r large, the values of δ_l.

Once the δ_l are known, we find from Eqs. (A-14), (A-11), and (A-5),

$$\frac{d\sigma}{d\omega} = \frac{1}{4k^2} \left| \sum_l (2l+1) P_l (\cos\theta)(e^{2i\delta_l} - 1) \right|^2 \tag{A-17}$$

This can be integrated over the solid angle, and recalling Eq. (A-8), we have

$$\sigma = 4\pi \lambdabar^2 \sum_l (2l+1) \sin^2 \delta_l \tag{A-18}$$

We also have the important relation

$$\text{Im } f(0) = \sigma/4\pi\lambdabar \tag{A-19}$$

which is obtained from Eqs. (A-11), (A-14), and (A-18), remembering that $P_l(1) = 1$. Equation (A-19) is sometimes called the *optical theorem*.

At low energy δ_0 alone is important; the scattering is spherically symmetric, with a cross section

$$\sigma = 4\pi\lambdabar^2 \sin^2 \delta_0 = 4\pi a^2 \tag{A-20}$$

where a, called the *scattering length*, is susceptible to a simple and important geometric interpretation (see Sec. 10-2).

When the potential has a range r_0 the phase shifts different from 0 are only those for which $\hbar l < p r_0$. Semiclassically, consider the impact parameter b. For a collision to occur, b must be smaller than r_0; otherwise the particle passes outside the potential well. Now the angular momentum with respect to

ISBN 0-8053-8061-7

the center of scattering is $\hbar l = bp$, and this gives

$$\hbar l/p < r_0 \tag{A-21}$$

as a necessary condition for a collision. If

$$\hbar/p = \lambdabar \gg r_0 \tag{A-22}$$

only waves with $l = 0$ will be scattered.

The phase shifts δ_l are all important for the description of the collision. With modern computers the numerical integration of Eq. (A-16) pushed to r values where $u_l(r)$ becomes a sine function gives δ_l directly. We may, however, want an analytical approximation to δ_l. To obtain it, consider, along with $u_l(r)$, a function $v_l(r)$ obeying Eq. (A-16) when $U(r) = 0$, with the proper boundary conditions. The function $v_l(r)$ thus satisfies the equation

$$v_l''(r) - \frac{l(l+1)}{r^2} v_l + \frac{2mE}{\hbar^2} v_l = 0 \tag{A-23}$$

and vanishes at the origin. The function v_l explicitly is

$$v_l(r) = krj_l(kr) = (\pi kr/2)^{1/2}J_{l+1/2}(kr) \tag{A-24}$$

Multiply Eq. (A-16) by $v_l(r)$, Eq. (A-23) by $u_l(r)$, subtract, and integrate; these operations yield

$$\int_0^r (u_l'' v_l - v_l'' u_l)\, dr = u_l' v_l - v_l' u_l \Big|_0^r = \frac{2m}{\hbar^2} \int_0^r U(r)u_l(r)v_l(r)\, dr \tag{A-25}$$

At the lower limit $u_l(0) = v_l(0) = 0$. At large r from the asymptotic expressions of u_l and v_l, Eqs. (A-15), and Eq. (A-24) one has

$$-k \sin \delta_l = \frac{2m}{\hbar^2} \int_0^\infty U(r)u_l(r)v_l(r)\, dr \tag{A-26}$$

This integral cannot be calculated exactly unless one knows $u_l(r)$ and thus has already solved the problem, but if the phase shifts are small, we may replace $u_l(r)$ by $v_l(r)$ in the spirit of perturbation theory, obtaining

$$\sin \delta_l = -\frac{2mk}{\hbar^2} \int_0^\infty U(r)[j_l(kr)]^2 r^2\, dr \tag{A-27}$$

If all the phase shifts are small, we can replace $\sin \delta_l$ by δ_l and $\exp(2i\delta_l - 1)$

ISBN 0-8053-8061-7

by $2i\delta_l$, and we may write, from Eqs. (A-11) and (A-14),

$$f(\theta) = \sum_l \frac{(2l+1)}{k} \frac{1}{2i} \left[\exp(2i\delta_l) - 1\right] P_l(\cos\theta) \cong \sum_l \frac{(2l+1)}{k} \delta_l P_l(\cos\theta)$$

Replacing δ_l by its expressions, Eq. (A-27), and interchanging sum and integral, we have

$$-f(\theta) = \sum_l \frac{(2l+1)}{k} \frac{2mk}{\hbar^2} \int_0^\infty U(r)\left[j_l(kr)\right]^2 r^2 \, dr \, P_l(\cos\theta)$$

$$= \frac{2m}{\hbar^2} \int_0^\infty U(r)r^2 \, dr \sum_l (2l+1)P_l(\cos\theta)\left[j_l(kr)\right]^2 \qquad \text{(A-28)}$$

An identity proved in Whittaker and Watson and other mathematical treatises says that

$$\sum_l (2l+1)P_l(\cos\theta)\left[j_l(kr)\right]^2 = (\sin Kr)/Kr \qquad \text{(A-29)}$$

where $K = 2k\sin(\theta/2)$. Replacing this expression in Eq. (A-28), we obtain

$$-f(\theta) = \frac{2m}{\hbar^2 K} \int_0^\infty U(r)r \sin(Kr) \, dr \qquad \text{(A-30)}$$

the famous Born formula.

There are more direct ways to prove Born's formula. This one shows its relation to phase shifts, as well as the condition of its validity. If the potential is limited to magnitudes of the order of U_0 and extends on a radius a, the condition of validity may be written as $U_0 a/\hbar v \ll 1$, where v is the velocity of the particle.

Born's formula (A-30) may be rewritten in a form that is also valid relativistically:

$$\frac{d\sigma}{d\omega} = |f(\theta)|^2 = \frac{1}{4\pi^2\hbar^4} \frac{p^2}{v^2} \left| \int_0^\infty U(\mathbf{r}) \exp\left[i(\mathbf{k} - \mathbf{k}')\cdot\mathbf{r}\right] d\mathbf{r} \right|^2 \qquad \text{(A-31)}$$

where $\hbar\mathbf{k}$, $\hbar\mathbf{k}'$ are the momenta before and after collision.

ISBN 0-8053-8601-7

APPENDIX

Effective Range

In Chapter 10 we determined the phase shift of an s wave on being scattered by using a square-well potential. Here we shall discuss a more general method, which gives a very good approximation for this phase shift and at the same time does not require a specified form of the potential $V(r)$. Our final result will be Eq. (B-9).

We begin by writing Schrödinger's equation for an s wave and positive energy for the relative motion. Setting $r\psi = u$ as usual, we have

$$\frac{\hbar^2}{M}\frac{d^2u}{dr^2} + (E - V)u = 0 \tag{B-1}$$

or calling $MV/\hbar^2 = U$ and $ME/\hbar^2 = k^2$,

$$\frac{d^2u}{dr^2} + k^2u - U(r)u = 0 \tag{B-2}$$

Emilio Segrè, Nuclei and Particles: An Introduction to Nuclear and Subnuclear Physics, Second Edition

ISBN 0-8053-8061-7

Taking two different values of the energy E_a, E_b, we have

$$\frac{d^2u_a}{dr^2} + k_a^2 u_a - U(r)u_a = 0 \tag{B-3}$$

$$\frac{d^2u_b}{dr^2} + k_b^2 u_b - U(r)u_b = 0 \tag{B-4}$$

Multiplying the first equation [Eq. (B-3)] by u_b and the second by u_a, taking the difference, and integrating, we obtain

$$(u_b u_a' - u_b' u_a)\big|_0^\infty = (k_b^2 - k_a^2)\int u_a u_b \, dr \tag{B-5}$$

The functions u_a, u_b satisfy the boundary condition

$$u_a(0) = u_b(0) = 0 \tag{B-6}$$

Solutions of Eq. (B-2) for $U(r) = 0$ (free particle) are of the general form

$$v = A\,\sin(kr + \delta) \tag{B-7}$$

We look for solutions of this type which for large r are identical to u and for $r = 0$ reduce to unity. This last condition determines the constant $1/A = \sin \delta$. The identity of the asymptotic behavior determines δ.

For the two values of k, k_a and k_b, we then obtain in the same way as for u_a, u_a

$$(v_b v_a' - v_b' v_a)\big|_0^\infty = (k_b^2 - k_a^2)\int_0^\infty v_a v_b \, dr \tag{B-8}$$

We subtract Eq. (B-5) from Eq. (B-8). The u and v coincide at infinity and thus disappear from the difference; at 0, remembering that $u(0) = 0$ and $v(0) = 1$, we have

$$v_b'(0) - v_a'(0) = k_b \cot \delta_b - k_a \cot \delta_a = (k_b^2 - k_a^2)$$

$$\times \int_0^\infty (v_a v_b - u_a u_b)\, dr \tag{B-9}$$

This equation is exact and is the fundamental equation of the effective range theory.

Consider now the case when $k_a \to 0$, that is, the case of zero energy. By definition

$$k_a \cot \delta_a = -1/a \tag{B-10}$$
$$\lim k_a \to 0$$

ISBN 0-8053-8601-7

where a is the Fermi scattering length; substitution in Eq. (B-9) gives

$$k_b \cot \delta_b = -\frac{1}{a} + k_b^2 \int_0^\infty (v_0 v_b - u_0 u_b)\, dr \qquad \text{(B-11)}$$

Dropping the suffix b and calling

$$\tfrac{1}{2}\rho(E) = \int_0^\infty (v_0 v_b - u_0 u_b)\, dr \qquad \text{(B-12)}$$

we have

$$k \cot \delta = -\frac{1}{a} + \frac{1}{2} k^2 \rho(E) \qquad \text{(B-13)}$$

The integral $\tfrac{1}{2}\rho(E)$ has an integrand that vanishes for r greater than the range of nuclear forces. On the other hand, within this range u_a and u_0 are practically equal because the behavior of the function is dominated by the potential, with little influence from the energy as long as $|k^2| \ll |U|$. If u_a and u_0 are practically the same, v_b and v_0, which are determined by the asymptotic behavior of u_b and u_0, are also the same and we may replace, with good approximation, $\tfrac{1}{2}\rho(E)$ by

$$\tfrac{1}{2}\rho(0) = \int_0^\infty (v_0^2 - u_0^2)\, dr = \tfrac{1}{2} r_0 \qquad \text{(B-14)}$$

The constant r_0 is called the *effective range*. Substituting in Eq. (B-13), we have

$$k \cot \delta = -\frac{1}{a} + \frac{1}{2} k^2 r_0 + \cdots \qquad \text{(B-15)}$$

where the dots mean that this equation is only approximate.

ISBN 0-8053-8061-7

APPENDIX

Description of Polarized
Beams (Spin $\frac{1}{2}$)

A perfectly polarized beam of particles of spin $\frac{1}{2}$ may be represented as superposition with the proper phase relation of two beams having as spin eigenfunctions $\begin{vmatrix} 0 \\ 1 \end{vmatrix}$, $\begin{vmatrix} 1 \\ 0 \end{vmatrix}$, respectively.

$$\psi = a_1 \begin{vmatrix} 1 \\ 0 \end{vmatrix} + a_2 \begin{vmatrix} 0 \\ 1 \end{vmatrix} \tag{C-1}$$

Here the a's, which can be complex, contain all coordinates except spin. In the case of a free beam of momentum $\hbar\mathbf{k}$, for instance, we can write

$$a_1 = \frac{A_1}{v^{1/2}} e^{i\mathbf{k}\cdot\mathbf{r}} \qquad a_2 = \frac{A_2}{v^{1/2}} e^{i\mathbf{k}\cdot\mathbf{r}}$$

and the intensity of the beam is

$$|a_1|^2 + |a_2|^2 = I \tag{C-2}$$

Emilio Segrè, Nuclei and Particles: An Introduction to Nuclear and Subnuclear Physics, Second Edition

Copyright © 1977 by The Benjamin/Cummings Publishing Company, Inc., Advanced Book Program. All rights reserved. No part of this publication may be reproduced, stored in a retrieval system, or transmitted, in any form or by any means, electronic, mechanical photocopying, recording, or otherwise, without the prior permission of the publisher.

ISBN 0-8053-8601-7

We define as the degree of polarization in the z direction the difference between the probability of finding the spin parallel or antiparallel to z, that is,

$$P_z = \frac{|a_1|^2 - |a_2|^2}{|a_1|^2 + |a_2|^2} = \frac{|a_1|^2 - |a_2|^2}{I} \tag{C-3}$$

A beam represented by Eq. (C-1) always has, in a suitably chosen direction, the polarization 1, that is, the maximum obtainable value.

In order to treat the partially polarized beam, we introduce a *density matrix*; for the system described by Eq. (C-1) it has the form

$$\rho = \begin{vmatrix} a_1 a_1^* & a_1 a_2^* \\ a_2 a_1^* & a_2 a_2^* \end{vmatrix} \tag{C-4}$$

or

$$\rho_{ij} = a_i a_j^* \tag{C-4a}$$

or, more generally,

$$\rho_{ij} = \psi_i \psi_j^\dagger \tag{C-4b}$$

The density matrix affords the symbolism suitable for treating the case of unpolarized or partially polarized beams. In the first place, remembering that the sum of the diagonal terms of a matrix is called its *trace*, we have another way of writing Eq. (C-2), that is,

$$\mathrm{Tr}(\rho) = I \tag{C-5}$$

The degree of polarization P_z of Eq. (C-3) can also be expressed through the density matrix. Remembering that

$$\sigma_z = \begin{vmatrix} 1 & 0 \\ 0 & -1 \end{vmatrix}$$

we have

$$\frac{\mathrm{Tr}(\rho \sigma_z)}{\mathrm{Tr}(\rho)} = P_z \tag{C-6}$$

Thus far the density matrix has represented only a concise way of *writing* formulas. However, if we superimpose several beams *incoherently*, we may define a density matrix for this system by summing the density matrices of

ISBN 0-8053-8061-7

the components, which we call $\rho(\alpha)$,

$$\bar{\rho}_{ij} = \sum_{\alpha=1}^{N} \rho_{ij}(\alpha)w(\alpha) \tag{C-7}$$

where $w(\alpha)$ is the weight of each state, proportional to the intensity of the beam in the specific case. Note that the density matrix corresponding to the *coherent* superposition of two beams is Eq. (C-4) and not Eq. (C-7).

The intensity of the resulting beam is again

$$I = \text{Tr}(\bar{\rho}) = \sum_{\alpha} \sum_{i} w(\alpha)\rho_{ii}(\alpha) \tag{C-8}$$

and the expectation value of an operator $\mathcal{O}$ is also given by

$$\langle \mathcal{O} \rangle = \frac{\text{Tr}(\mathcal{O}\,\bar{\rho})}{\text{Tr}(\bar{\rho})} \tag{C-8a}$$

The 2×2 matrix $\bar{\rho}$ can always be written as

$$\bar{\rho} = a\begin{vmatrix} 1 & 0 \\ 0 & 1 \end{vmatrix} + b_x\begin{vmatrix} 0 & 1 \\ 1 & 0 \end{vmatrix} + b_y\begin{vmatrix} 0 & -i \\ i & 0 \end{vmatrix} + b_z\begin{vmatrix} 1 & 0 \\ 0 & -1 \end{vmatrix} \tag{C-9}$$

if suitable a, b_x, b_y, b_z are chosen. For instance,

$$\bar{\rho}_{11} = a + b_z \qquad \bar{\rho}_{12} = b_x - ib_y \qquad \text{etc.} \tag{C-10}$$

and we can rewrite Eq. (C-9) in a more compact form as

$$\bar{\rho} = a\mathbf{1} + \mathbf{b}\cdot\boldsymbol{\sigma} \tag{C-11}$$

where $\boldsymbol{\sigma}$ is the Pauli matrix vector of components σ_x, σ_y, σ_z. This way of writing has an especially simple physical interpretation.

We have

$$\text{Tr}(\bar{\rho}) = 2a = I \tag{C-12}$$

which gives the physical meaning of a in Eq. (C-11). The direct calculation of $\langle \sigma_x \rangle$, $\langle \sigma_y \rangle$, $\langle \sigma_z \rangle$ according to Eq. (C-8a) gives

$$\langle \sigma_x \rangle = \frac{b_x}{a} \qquad \langle \sigma_y \rangle = \frac{b_y}{a} \qquad \langle \sigma_z \rangle = \frac{b_z}{a} \tag{C-13}$$

The vector $\mathbf{b}$ is thus directed parallel to the expectation value of the spin, and

ISBN 0-8053-8601-7

the polarization P is simply

$$\mathbf{b}/a = \mathbf{P} \tag{C-14}$$

Hence the density matrix is, from Eqs. (C-9), (C-11), and (C-14), related to the directly observable quantities, intensity and polarization of the beam,

$$\bar{\rho} = \tfrac{1}{2}I(1 + \mathbf{P} \cdot \boldsymbol{\sigma}) \tag{C-15}$$

Note that an unpolarized beam has a density matrix multiple of the unit matrix.

How do we describe the effect of scattering on polarization? Suppose that the initial beam is in a pure state ψ_i [Eq. (C-1)], and consider the scattering from a spinless target. The final state will be ψ_f, also a two-row function. Now we define a 2×2 matrix operator S such that

$$S\psi^{(i)} = \psi^{(f)} \quad \text{and} \quad \psi^{(f)} = \psi^{(i)\dagger}S^{\dagger} \tag{C-16}$$

The matrix S is a special case of the "scattering matrix," and the dagger means "transposed" and "conjugate." The scattering matrix allows us to calculate the final density matrix $\rho^{(f)}$ from the initial density matrix $\rho^{(i)}$. By the definition, Eq. (C-4), and Eq. (C-16),

$$\rho_{lm}^{(i)} = \psi_l^{(i)}\psi_m^{(i)\dagger} \tag{C-17}$$

and

$$\rho_{lm}^{(f)} = \psi_l^{(f)}\psi_m^{(f)\dagger} \tag{C-18}$$

$$S\psi_l^{(i)}\psi_m^{(i)\dagger}S^{\dagger} = S\rho_{lm}^{(i)}S^{\dagger} \tag{C-19}$$

If we normalize $\psi^{(i)}$ in such a way as to represent a beam of unit intensity, the differential scattering cross section $d\sigma/d\omega$ is

$$\frac{\text{Tr}(\rho^{(f)})}{\text{Tr}(\rho^{(i)})} = \frac{d\sigma}{d\omega} \tag{C-20}$$

The matrix S will depend, for an unpolarized or spinless target, on the scattering angle and energy of the incident beam. Like any 2×2 matrix, S can be written

$$S = g\mathbf{1} + \mathbf{h} \cdot \boldsymbol{\sigma} \tag{C-21}$$

This way of writing is especially convenient. The vector quantity $\mathbf{h}$, which is a function of $\mathbf{k}^{(i)}$, $\mathbf{k}^{(f)}$, the initial and final momentum of the particles, must be

ISBN 0-8053-8601-7

an axial vector in order to give a scalar by multiplication with the axial vector $\boldsymbol{\sigma}$. Thus it must be expressed as

$$\mathbf{h} = \frac{\mathbf{k}^{(i)} \times \mathbf{k}^{(f)}}{|\mathbf{k}^{(i)} \times \mathbf{k}^{(f)}|} h(\theta, E) = \mathbf{n} h(\theta) \tag{C-22}$$

where θ is the scattering angle. The unit vector $\mathbf{n}$ perpendicular to the scattering plane represents the z axis, including sign. We then have

$$S = \begin{vmatrix} g + h & 0 \\ 0 & g - h \end{vmatrix} = g\mathbf{1} + h\mathbf{n} \cdot \boldsymbol{\sigma} \tag{C-23}$$

and $g + h, g - h$ are the scattering amplitudes for a perfectly polarized beam with spin parallel or antiparallel to z.

The functions $g(\theta, E)$ and $h(\theta, E)$ depend on energy and scattering angle. They can be calculated by a generalization of the phase-shift analysis given in Appendix A. The standard calculation is given in, for instance, (W 71). The results are

$$g(\theta) = \sum_{0}^{\infty} {}_l[(l + 1)A_{l+} + A_{l-}]P_l(\cos\theta) \tag{C-23a}$$

$$h(\theta) = -i\sum_{0}^{\infty} {}_l(A_{l+} - A_{l-}) \sin\theta \frac{dP_l(\cos\theta)}{d\cos\theta} \tag{C-23b}$$

where $A_{l\pm} = [\eta_{l\pm} \exp(2i\delta_{l\pm}) - 1]/2ik$ are partial wave amplitudes corresponding to the same l originating once as $j + \frac{1}{2}$ and once as $j - \frac{1}{2}$. They are functions of E. For instance, considering only s and p waves, A_{0+} corresponds to the $s_{1/2}$ wave, A_{1+} corresponds to the $p_{3/2}$ wave, and A_{1-} to the $p_{1/2}$ wave.

An unpolarized beam with the density matrix

$$\rho_0 = \tfrac{1}{2}I\begin{vmatrix} 1 & 0 \\ 0 & 1 \end{vmatrix} \tag{C-24}$$

becomes polarized after one scattering,

$$\rho_1 = S\rho_0 S^\dagger = \tfrac{1}{2}\begin{vmatrix} |g + h|^2 & 0 \\ 0 & |g - h|^2 \end{vmatrix} \tag{C-25}$$

and the amount of polarization is given, according to Eq. (C-6), by

$$P = \frac{1}{2}\frac{|g + h|^2 - |g - h|^2}{|g|^2 + |h|^2} = \frac{gh^* + g^*h}{|g|^2 + |h|^2} = \frac{2\operatorname{Re}(gh^*)}{|g|^2 + |h|^2} \tag{C-26}$$

ISBN 0-8053-8061-7

On the other hand, if the initial beam is totally polarized perpendicular to the plane of scattering,

$$\rho_0 = I \begin{vmatrix} 1 & 0 \\ 0 & 0 \end{vmatrix} \tag{C-27}$$

On scattering by an angle θ we obtain $\sigma(\theta)$,

$$\sigma(\theta) = \frac{\mathrm{Tr}(\rho_1)}{\mathrm{Tr}(\rho_0)} \tag{C-28}$$

where $\rho_1 = S\rho_0 S^\dagger$ [and S is given by Eq. (C-23)]. If we scatter by an angle $-\theta$, the corresponding S, according to Eq. (C-22), is

$$\begin{vmatrix} g - h & 0 \\ 0 & g + h \end{vmatrix} = S \tag{C-29}$$

The resulting cross sections are

$$\sigma(\pm\theta) = (|g|^2 + |h|^2)\left[1 \pm P(\theta)\right] \tag{C-30}$$

where P is given by Eq. (C-26). From these cross sections the asymmetry can be immediately calculated.

To characterize completely the scattering of a particle of spin $\frac{1}{2}$ from a spinless center, we need to know the two complex quantities g and h, that is, four real numbers, for instance $|g|$, $|h|$, the phase difference β between g and h, and the phase of h, α.

These numbers can be obtained, for example, by observing

$$\tfrac{1}{2}\left[\sigma(\theta) + \sigma(-\theta)\right] = |g|^2 + |h|^2 \tag{C-31}$$

Observation of P gives another relation. According to Eqs. (C-26) and (C-30)

$$|g + h| = \left[\sigma(\theta)(1 + P)\right]^{1/2} \tag{C-32}$$

$$|g - h| = \left[\sigma(\theta)(1 - P)\right]^{1/2} \tag{C-33}$$

The phase factor $e^{i\beta}$ can be determined (possibly with a sign ambiguity) from a triple scattering experiment; for instance, a measurement of R (Fig. 10-18) gives β through the relation

$$R = (1 - P^2)^{1/2} \cos(\beta - \theta) \tag{C-34}$$

where θ is, as usual, the scattering angle. Equations (C-32), (C-33), and (C-34)

ISBN 0-8053-8061-7

thus give g and h completely except for the phase α, which is not observable in scattering experiments from one center only.

These results can be extended, but the case of two nucleons (spin $\frac{1}{2}$) already gives a fairly complicated 4×4 scattering matrix. It is

$$M = \alpha\,1 + \beta\,(\boldsymbol{\sigma}_1 \cdot \mathbf{n})(\boldsymbol{\sigma}_2 \cdot \mathbf{n}) + \gamma(\boldsymbol{\sigma}_1 + \boldsymbol{\sigma}_2) \cdot \mathbf{n} + \delta\,(\boldsymbol{\sigma}_1 \cdot \mathbf{q})(\boldsymbol{\sigma}_2 \cdot \mathbf{q})$$

$$+\,\epsilon(\boldsymbol{\sigma}_1 \cdot \mathbf{p})(\boldsymbol{\sigma}_2 \cdot \mathbf{p}) \tag{C-35}$$

where $\mathbf{p}$ and $-\mathbf{q}$ are unit vectors in the direction of the momentum of the projectile (1) and of the target (2) after scattering. The α, β, γ, δ, ϵ are complex functions of E and θ.

PROBLEM

C-1   Considering only s and p waves, show that

$$g(\theta) = [A_{0+} + 2(A_{1+} + A_{1-}) \cos \theta]$$

$$h(\theta) = -i(A_{1+} - A_{1-}) \sin \theta$$

and from this calculate $d\sigma / d\omega$ and P.

ISBN 0-8053-8601-7

APPENDIX

Kinematics of Binary Collisions

In the relativistic binary collision

$$1 + 2 \rightarrow 3 + 4$$

it is convenient to use *Mandelstam variables* (Sec. 13-6). We denote by p_i the four momentum of particle i, of rest mass m_i, momentum $\mathbf{p}_i$, and total energy $E_i = m_i + T_i = (m_i^2 + \mathbf{p}_i^2)^{1/2}$. We use the convention $c = 1$, $p^2 = \mathbf{p}^2 - E^2$. (The convention $p^2 = E^2 - \mathbf{p}^2$ is also widely used, thus one has to be careful.) From these definitions $p_i^2 = -m_i^2$.

Conservation of energy and momentum gives

$$p_1 + p_2 = p_3 + p_4 \tag{D-1}$$

We now define the variables

$$-s = (p_1 + p_2)^2 = (p_3 + p_4)^2 \tag{D-2}$$

$$-t = (p_1 - p_3)^2 = (p_4 - p_2)^2 \tag{D-3}$$

$$-u = (p_1 - p_4)^2 = (p_3 - p_2)^2 \tag{D-4}$$

Emilio Segrè, Nuclei and Particles: An Introduction to Nuclear and Subnuclear Physics, Second Edition

Copyright © 1977 by The Benjamin/Cummings Publishing Company, Inc., Advanced Book Program. All rights reserved. No part of this publication may be reproduced, stored in a retrieval system, or transmitted, in any form or by any means, electronic, mechanical photocopying, recording, or otherwise, without the prior permission of the publisher.

ISBN 0-8053-8061-7

We know (Sec. 13-6) that $s + t + u = m_1^2 + m_2^2 + m_3^2 + m_4^2$.

In the c.m. system

$$s = (E_1^* + E_2^*)^2 = (E_3^* + E_4^*)^2 = (\text{total c.m. energy})^2 \tag{D-5}$$

$$t = m_1^2 + m_3^2 - 2(E_1^* E_3^* - \mathbf{p}^* \cdot \mathbf{q}^*) \tag{D-6}$$

$$u = m_1^2 + m_4^2 - 2(E_1^* E_4^* + \mathbf{p}^* \cdot \mathbf{q}^*) \tag{D-7}$$

where $\mathbf{p}^* = \mathbf{p}_1^* = -\mathbf{p}_2^*$ and $\mathbf{q}^* = \mathbf{p}_3^* = -\mathbf{p}_4^*$. Furthermore, calling

$$\lambda(a, b, c) = a^2 + b^2 + c^2 - 2(ab + ac + bc) \tag{D-8}$$

we have

$$\mathbf{p}^{*2} = \left[\lambda\left(s, m_1^2, m_2^2\right)\right]/4s \qquad \mathbf{q}^{*2} = \left[\lambda\left(s, m_3^2, m_4^2\right)\right]/4s \tag{D-9}$$

$$2\mathbf{p}^* \cdot \mathbf{q}^* = \left[(s + m_1^2 - m_2^2)(s + m_3^2 - m_4^2)/2s\right] - m_1^2 - m_3^2 + t \tag{D-10}$$

and the scattering angle between 1 and 3 is given by

$$\cos \theta^* = \frac{2s(t - m_1^2 - m_3^2) + (s + m_1^2 - m_2^2)(s + m_3^2 - m_4^2)}{\left[\lambda(s, m_1^2, m_2^2)\lambda(s, m_3^2, m_4^2)\right]^{1/2}} \tag{D-11}$$

In the laboratory system with particle 2 at rest

$$s = m_1^2 + m_2^2 + 2E_1^L m_2 \tag{D-12}$$

$$t = m_1^2 + m_3^2 - 2(E_1^L E_3^L - \mathbf{p}_1^L \cdot \mathbf{p}_3^L) \tag{D-13}$$

$$u = m_2^2 + m_3^2 - 2E_3^L m_2 \tag{D-14}$$

$$(\mathbf{p}_1^L)^2 = \left[\lambda(s, m_1^2, m_2^2)\right]/4m_2^2 \tag{D-15}$$

$$(\mathbf{p}_3^L)^2 = \left[\lambda(u, m_2^2, m_3^2)\right]/4m_2^2 \tag{D-16}$$

and the scattering angle is

$$\cos \theta_L = \frac{2m_2^2(t - m_1^2 - m_3^2) - (s - m_1^2 - m_2^2)(u - m_2^2 - m_3^2)}{\left[\lambda(s, m_1^2, m_2^2)\lambda(u, m_2^2, m_3^2)\right]^{1/2}} \tag{D-17}$$

ISBN 0-8053-8061-7

These and other formulas (see RPP 74), containing on one side directly observable quantities and on the other s, t, u, and other relativistic invariants, are often useful.

The passage from laboratory to c.m. system in the collision problem may be obtained graphically, according to Blaton (1950), as follows: first obtain

$$s^{1/2} = E^* = E_3^* + E_4^* = E_1^* + E_2^* = \left(m_1^2 + m_2^2 + 2E_1^L m_2 \right)^{1/2} \quad \text{(D-18)}$$

The $\bar{\beta}$ and $\bar{\gamma}$ of the c.m. with respect to the laboratory system are

$$\bar{\gamma} = E^L / s^{1/2} = \left(E_1^L + m_2 \right) / s^{1/2} \qquad \bar{\beta} = |\mathbf{p}_1^L| / \left(E_1^L + m_2 \right)$$

$$\bar{\eta} = \bar{\beta}\bar{\gamma} = |\mathbf{p}_1^L| / s^{1/2} \qquad \text{(D-19)}$$

We also have

$$E_3^* = E^* - E_4^* = E^* - \left(m_4^2 + E_3^{*2} - m_3^2 \right)^{1/2}$$

or

$$E_3^* = \left(m_3^2 - m_4^2 + s \right) / 2s^{1/2} \qquad \text{(D-20)}$$

In the nonrelativistic case the vector $\mathbf{p}_3^*$ starting from the origin with the center-of-mass angle θ^* is transformed to the laboratory system by adding to $\mathbf{p}^{*3}$ the component parallel to the direction of $\mathbf{p}_1$ a vector of magnitude $m_3\bar{\beta}$ (Fig. D-1). In the relativistic case the corresponding construction is more complicated. We must draw an ellipse of semiaxes $\bar{\gamma}|\mathbf{p}_3^*|$ and $|\mathbf{p}_3^*|$ concentric with the circle of radius $|\mathbf{p}_3^*|$. We then mark on the major axis of the ellipse the abscissa points $-\bar{\eta}E_3^* \equiv N_3$ and $+\bar{\eta}E_4^* \equiv N_4$. The vector $\mathbf{p}_3$ (laboratory) is obtained by joining point N_3 with point Q, where Q is the intercept of the ellipse with the parallel to its major axis passing through the end of $\mathbf{p}_3^*$; $\mathbf{p}_4$ is obtained by joining Q with N_4 (Fig. D-2). This construction is proved by the transformation rules of the energy–momentum four vector.

BIBLIOGRAPHY

Blaton, J., "On a Geometrical Interpretation of Energy and Momentum Conservation in Atomic Collisions and Disintegration Processes," *Kgl. Danske Vidensk. Selskab, Mat-fys. Medd.*, **24**, No. 20 (1950).

Dedrick, K. G., "Kinematics of High Energy Particles," *Rev. Mod. Phys.*, **34**, 429 (1962); **35**, 414 (1963).

Hagedorn, R., *Relativistic Kinematics*, Benjamin, New York, 1963.

Leon, M. (Le 73).

Particle Data Group (RPP 74).

ISBN 0-8053-8601-7

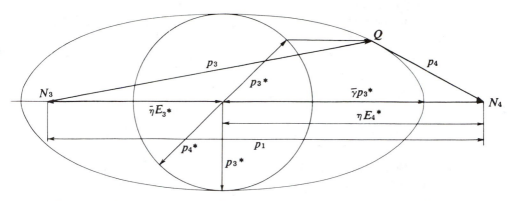

Figure D-1 Transformation from laboratory to center-of-mass reference system in a binary collision. Nonrelativistic case.

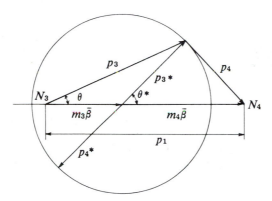

Figure D-2 Transformation from laboratory to center-of-mass system in a binary collision. Relativistic case. The three momenta are denoted by p_i.

PROBLEMS

D-1 Show that the laboratory threshold energy for producing new particles of mass $m_3 + m_4 + \cdots = M$ colliding a particle of mass m_1 on a particle of mass m_2 at rest is

$$T_1 = \frac{M}{m_2}\left(m_1 + m_2 + \frac{M}{2}\right)$$

D-2 Show that in a nucleon (M_1)–nucleus (M_2) elastic collision, one has the (nonrelativistic) formulas

$$\tan\theta_1 = \frac{M_2 \sin\theta_1^*}{M_1 + M_2 \cos\theta_1^*}$$

$$E_1^* = \left(\frac{M_2}{M_1 + M_2}\right)^2 E_1$$

Discuss possible values of θ_1 for different values of M_1/M_2. Calculate $d\cos\theta_1^*/d\cos\theta_1$.

ISBN 0-8053-8601-7

D-3   In a nucleon–nucleon elastic collision, one has the following relativistic formulas (m_2 at rest)

$$T_1 = 2m(\bar{\gamma}^2 - 1)$$

$$\tan\theta_1 = \bar{\gamma}^{-1}\tan(\theta_1^*/2)$$

$$\tan\theta_2 = \bar{\gamma}^{-1}\cot(\theta_1^*/2)$$

$$\bar{\gamma} = (T_1 + 2m)(4m^2 + 2T_1 m)^{-1/2}$$

$$\theta_1^* = \pi - \theta_2^*$$

Find the angle between $\mathbf{p}_1$, $\mathbf{p}_2$ after the collision.

Show that the kinetic energies after the collision are

$$T_1' = T_1 \cos^2(\theta_1^*/2)$$

$$T_2' = T_1 \sin^2(\theta_1^*/2)$$

and one has the relation

$$\left| \frac{d\cos\theta_1^*}{d\cos\theta_1} \right| = \gamma^{*2} \frac{\left[(g_1 + \cos\theta_1^*)^2 + (1/\gamma^{*2})\sin^2\theta_1^* \right]^{3/2}}{|1 + g_1 \cos\theta_1^*|}$$

where $g_i = \beta_i^*/\bar{\beta}$.

D-4   Justify the following mnemonic device, given by F. Crawford, for writing an approximate relativistic (R) formula if one remembers a nonrelativistic one (NR). Write the NR formula and add to the rest energy of each moving particle $\frac{1}{2}$ of the total kinetic energy in the center-of-mass system, $T/2$.

Examples:

  a. A particle m_1 incident on m_2 at rest has in the center-of-mass system the NR kinetic energy

$$T_1^* = T_1 \left(\frac{m_2}{m_1 + m_2} \right)^2$$

Relativistically,

$$T_1^* = T_1 \left(\frac{m_2}{m_2 + m_1 + T/2} \right)^2$$

  b. A particle at rest disintegrates into two with total kinetic energy T:

$$T = T_1 + T_2$$

NR:

$$T_1 = \frac{T m_2}{m_2 + m_1}$$

R:

$$T_1 = \frac{T(m_2 + T/2)}{m_1 + T/2 + m_2 + T/2} = \frac{T(m_2 + T/2)}{m_1 + m_2 + T}$$

ISBN 0-8053-8061-7

D-5 Show that $d\omega_{\text{c.m.}}/dt = 4\pi s/[\lambda(s_1, m_1^2, m_2^2)\lambda(s_1, m_3^2, m_4^2)]$.

D-6 Simplify Eqs. (D-1) to (D-17) for the elastic case $m_1 = m_3$, $m_2 = m_4$, and for the case in which $m_1 = m_2 = m_3 = m_4$.

D-7 Calculate Eqs. (D-1) to (D-17) for $m_1 = 0$ (neutrino).

D-8 Consider the reaction $K + p \Rightarrow (1) + (2) + (3)$. Define $m^2(1, 2)$ as the square of the energy in the c.m. system of particles 1 and 2. $m^2(1, 2) = -(p_1 + p_2)^2$ where the p_i are four momenta and $m^2(K, p) = s = E^{*2}$ is a relativistic invariant. We have also $p_i^2 = \mathbf{p}_i^2 - E_i^2 = -m_i^2$; $(p_1 + p_2 + p_3)^2 = -E^{*2}$ and $\mathbf{p}_1^* + \mathbf{p}_2^* + \mathbf{p}_3^* = 0$. We can then write $-E^{*2} = [p_1 + (p_2 + p_3)]^2 = -m^2(1) + m^2(2,3) - 2\mathbf{p}_1^{*2} - 2E_1^*(E^* - E_1^*)$ and from this obtain Eq. (15-8.4).

Composition of Angular Momenta

We give here, without proofs, a summary of the rules for obtaining the eigenfunctions of a system composed of two parts having angular momentum properties. These rules are the quantum-mechanical translation of the vector-model rules of composition of angular momenta. The components have eigenfunctions labeled according to their value of the total angular momentum j_1 and j_2 and of the z components of j_1 and j_2: m_1 and m_2. The compound system has eigenfunctions labeled according to the value of the total angular momentum of the compound system J and of its z component M. The eigenfunctions of the compound system are linear combinations of products of the component eigenfunctions. The coefficients of these linear combinations are called Clebsch–Gordan coefficients.

For the components one has the possible values of m_1, m_2:

$$m_1 = j_1, \quad j_1 - 1, \quad j_1 - 2, \ldots, \quad -j_1$$

$$m_2 = j_2, \quad j_2 - 1, \quad j_2 - 2, \ldots, \quad -j_2$$

and j_1, j_2 are positive integral or half-integral numbers. For the resultant

Emilio Segrè, Nuclei and Particles: An Introduction to Nuclear and Subnuclear Physics, Second Edition

ISBN 0-8053-8061-7

system one has

$$j_1 + j_2 \geqslant J \geqslant |j_1 - j_2| \qquad M = J, J - 1, \ldots, -J$$

All this is well known from the vector model and from quantum mechanics.

Now call $|j_1 m_1\rangle$ and $|j_2 m_2\rangle$ the eigenfunctions of the components corresponding to definite values of j_1, m_1 and j_2, m_2. The resultant eigenfunctions labeled according to J and M have the form (the sum must be extended over all values of m_1, m_2, such that $m_1 + m_2 = M$).

$$|JM\rangle = \sum_{m_1 + m_2 = M} \langle j_1 j_2 m_1 m_2 | JM \rangle |j_1 m_1\rangle |j_2 m_2\rangle$$

The coefficients $\langle j_1 j_2 m_1 m_2 | JM \rangle$ (Clebsch–Gordan coefficients) for some of the simplest cases are given in Table E-1, where each rectangle is labeled $j_1 \times j_2$. As an example, we see that if $j_1 = 1$, $j_2 = \frac{1}{2}$, we use the rectangle labeled $1 \times \frac{1}{2}$. J can have the values $\frac{1}{2}$ or $\frac{3}{2}$. For $J = \frac{3}{2}$, M can take the values $\frac{3}{2}$, $\frac{1}{2}$, $-\frac{1}{2}$, $-\frac{3}{2}$. The corresponding eigenfunctions are

$$|\tfrac{3}{2}\,\tfrac{3}{2}\rangle = |1\ 1\rangle|\tfrac{1}{2}\,\tfrac{1}{2}\rangle$$

$$|\tfrac{3}{2}\,\tfrac{1}{2}\rangle = \sqrt{\tfrac{2}{3}}\ |1\ 0\rangle|\tfrac{1}{2}\,\tfrac{1}{2}\rangle + \sqrt{\tfrac{1}{3}}\ |1\ 1\rangle|\tfrac{1}{2}\ -\tfrac{1}{2}\rangle$$

$$|\tfrac{3}{2}\ -\tfrac{1}{2}\rangle = \sqrt{\tfrac{2}{3}}\ |1\ 0\rangle|\tfrac{1}{2}\ -\tfrac{1}{2}\rangle + \sqrt{\tfrac{1}{3}}\ |1\ -1\rangle|\tfrac{1}{2}\,\tfrac{1}{2}\rangle$$

$$|\tfrac{3}{2}\ -\tfrac{3}{2}\rangle = |1\ -1\rangle|\tfrac{1}{2}\ -\tfrac{1}{2}\rangle$$

These mathematical formulas apply to any other set of operators (to i spin, for instance) obeying the same commutation relations as the angular momentum.

We also give here, for reference, the first spherical harmonics that are (as is well known) explicit forms of eigenfunctions of angular momentum labeled according to j and m for j integral number (Table E-2).

ISBN 0-8053-8601-7

TABLE E-1   CLEBSCH–GORDAN COEFFICIENTS[a]

$1/2 \times 1/2$

m_1	m_2	J M	1 +1	1 0	0 0	1 −1
+1/2	+1/2		1			
+1/2	−1/2			$\sqrt{1/2}$	$\sqrt{1/2}$	
−1/2	+1/2			$\sqrt{1/2}$	$-\sqrt{1/2}$	
−1/2	−1/2					1

$1 \times 1/2$

m_1	m_2	J M	3/2 +3/2	3/2 +1/2	1/2 +1/2	3/2 −1/2	1/2 −1/2	3/2 −3/2
+1	+1/2		1					
+1	−1/2			$\sqrt{1/3}$	$\sqrt{2/3}$			
0	+1/2			$\sqrt{2/3}$	$-\sqrt{1/3}$			
0	−1/2					$\sqrt{2/3}$	$\sqrt{1/3}$	
−1	+1/2					$\sqrt{1/3}$	$-\sqrt{2/3}$	
−1	−1/2							1

$3/2 \times 1/2$

m_1	m_2	J M	2 +2	2 +1	1 +1	2 0	1 0	2 −1	1 −1	2 −2
+3/2	+1/2		1							
+3/2	−1/2			$\sqrt{1/4}$	$\sqrt{3/4}$					
+1/2	+1/2			$\sqrt{3/4}$	$-\sqrt{1/4}$					
+1/2	−1/2					$\sqrt{1/2}$	$\sqrt{1/2}$			
−1/2	+1/2					$\sqrt{1/2}$	$-\sqrt{1/2}$			
−1/2	−1/2							$\sqrt{3/4}$	$\sqrt{1/4}$	
−3/2	+1/2							$\sqrt{1/4}$	$-\sqrt{3/4}$	
−3/2	−1/2									1

ISBN 0-8053-8601-7

2 × 1/2

m_1	m_2	J 5/2 M +5/2	5/2 +3/2	3/2 +3/2	5/2 +1/2	3/2 +1/2	5/2 −1/2	3/2 −1/2	5/2 −3/2	3/2 −3/2	5/2 −5/2
+2	1/2	1									
+2	−1/2		$\sqrt{1/5}$	$\sqrt{4/5}$							
+1	+1/2		$\sqrt{4/5}$	$-\sqrt{1/5}$							
+1	−1/2				$\sqrt{2/5}$	$\sqrt{3/5}$					
0	+1/2				$\sqrt{3/5}$	$-\sqrt{2/5}$					
0	−1/2						$\sqrt{3/5}$	$\sqrt{2/5}$			
−1	+1/2						$\sqrt{2/5}$	$-\sqrt{3/5}$			
−1	−1/2								$\sqrt{4/5}$	$\sqrt{1/5}$	
−2	+1/2								$\sqrt{1/5}$	$-\sqrt{4/5}$	
−2	−1/2										1

1 × 1

m_1	m_2	J 2 M +2	2 +1	1 +1	2 0	1 0	0 0	2 −1	1 −1	2 −2
+1	+1	1								
+1	0		$\sqrt{1/2}$	$\sqrt{1/2}$						
0	+1		$\sqrt{1/2}$	$-\sqrt{1/2}$						
+1	−1				$\sqrt{1/6}$	$\sqrt{1/2}$	$\sqrt{1/3}$			
0	0				$\sqrt{2/3}$	0	$-\sqrt{1/3}$			
−1	+1				$\sqrt{1/6}$	$-\sqrt{1/2}$	$\sqrt{1/3}$			
0	−1							$\sqrt{1/2}$	$\sqrt{1/2}$	
−1	0							$\sqrt{1/2}$	$-\sqrt{1/2}$	
−1	−1									1

TABLE E-1 (Continued)

3/2 × 1

m_1	m_2	J=5/2 M=+5/2	J=5/2 M=+3/2	J=3/2 M=+3/2	J=5/2 M=+1/2	J=3/2 M=+1/2	J=1/2 M=+1/2	J=5/2 M=-1/2	J=3/2 M=-1/2	J=1/2 M=-1/2	J=5/2 M=-3/2	J=3/2 M=-3/2	J=5/2 M=-5/2
+3/2	+1	1											
+3/2	0		$\sqrt{2/5}$	$\sqrt{3/5}$									
+1/2	+1		$\sqrt{3/5}$	$-\sqrt{2/5}$									
+3/2	-1				$\sqrt{1/10}$	$\sqrt{2/5}$	$\sqrt{1/2}$						
+1/2	0				$\sqrt{3/5}$	$\sqrt{1/15}$	$-\sqrt{1/3}$						
-1/2	+1				$\sqrt{3/10}$	$-\sqrt{8/15}$	$\sqrt{1/6}$						
+1/2	-1							$\sqrt{3/10}$	$\sqrt{8/15}$	$\sqrt{1/6}$			
-1/2	0							$\sqrt{3/5}$	$-\sqrt{1/15}$	$-\sqrt{1/3}$			
-3/2	+1							$\sqrt{1/10}$	$-\sqrt{2/5}$	$\sqrt{1/2}$			
-1/2	-1										$\sqrt{3/5}$	$\sqrt{2/5}$	
-3/2	+0										$\sqrt{2/5}$	$-\sqrt{3/5}$	
-3/2	-1												1

ISBN 0-8053-8061-7

ISBN 0-8053-8061-7

2 × 1

m_1 m_2	J = 3, M = +3	3, +2	2, +2	3, +1	2, +1	1, +1	3, 0	2, 0	1, 0	3, −1	2, −1	1, −1	3, −2	2, −2	3, −3
+2 +1	1														
+2 0		$\sqrt{1/3}$	$\sqrt{2/3}$												
+1 +1		$\sqrt{2/3}$	$-\sqrt{1/3}$												
+2 −1				$\sqrt{1/15}$	$\sqrt{1/3}$	$\sqrt{3/5}$									
+1 0				$\sqrt{8/15}$	$\sqrt{1/6}$	$-\sqrt{3/10}$									
0 +1				$\sqrt{6/15}$	$-\sqrt{1/2}$	$\sqrt{1/10}$									
+1 −1							$\sqrt{1/5}$	$\sqrt{1/2}$	$\sqrt{3/10}$						
0 0							$\sqrt{3/5}$	0	$-\sqrt{2/5}$						
−1 +1							$\sqrt{1/5}$	$-\sqrt{1/2}$	$\sqrt{3/10}$						
0 −1										$\sqrt{6/15}$	$\sqrt{1/2}$	$\sqrt{1/10}$			
−1 0										$\sqrt{8/15}$	$-\sqrt{1/6}$	$-\sqrt{3/10}$			
−2 +1										$\sqrt{1/15}$	$-\sqrt{1/3}$	$\sqrt{3/5}$			
−1 −1													$\sqrt{2/3}$	$\sqrt{1/3}$	
−2 0													$\sqrt{1/3}$	$-\sqrt{2/3}$	
−2 −1															1

[a]When calculating terms that are linear in these coefficients (e.g., interference, polarization), the sign convention becomes important. This table follows the one in Blatt and Weisskopf, Edmonds, Rose, Condon and Shortley, etc. Other authors (e.g., Schiff, Bethe, and de Hoffmann) use different conventions.

TABLE E-2   THE FIRST SPHERICAL HARMONICS

$$Y_0^0 = \sqrt{\frac{1}{4\pi}}$$

$$Y_1^0 = \sqrt{\frac{3}{4\pi}} \cos\theta; \qquad Y_1^1 = -\sqrt{\frac{3}{8\pi}} \sin\theta\, e^{i\phi}$$

$$Y_2^0 = \sqrt{\frac{5}{4\pi}} \left(\frac{3}{2}\cos^2\theta - \frac{1}{2}\right); \qquad Y_2^1 = -\sqrt{\frac{15}{8\pi}} \sin\theta \cos\theta\, e^{i\phi}$$

$$Y_2^2 = \frac{1}{4}\sqrt{\frac{15}{2\pi}} \sin^2\theta\, e^{2i\phi}$$

$$Y_3^0 = \sqrt{\frac{7}{4\pi}} \left(\frac{5}{2}\cos^3\theta - \frac{3}{2}\cos\theta\right); \qquad Y_3^1 = -\frac{1}{4}\sqrt{\frac{21}{4\pi}} \sin\theta(5\cos^2\theta - 1)e^{i\phi}$$

$$Y_3^2 = \frac{1}{4}\sqrt{\frac{105}{2\pi}} \sin^2\theta \cos\theta\, e^{2i\phi}; \qquad Y_3^3 = -\frac{1}{4}\sqrt{\frac{35}{4\pi}} \sin^3\theta\, e^{3i\phi}$$

$$(Y_l^m)^* = (-1)^m Y_l^{-m}$$

ISBN 0-8053-8061-7

Author Index

Abashian, A., 798
Abragam, A., 313
Abrams, G. S., 856
Adair, R. K., 809
Adler, S., 866
Ajzenberg-Selove, F., xviii, 339, 388, 522, 612
Albert, R. D., 416
Alburger, D. E., 388
Alder, K., 388
Aldrich, L. T., 202
Alff, C., 781, 794
Allaby, J. V., 915
Almqvist, E., 457, 600
Alper, T., 23
Alvarez, L. W., 112, 266, 702, 780
Amaldi, E., 6, 675
Amaldi, U., 918
Ambler, E., 8, 408, 445
Anderson, C. D., 6, 8, 732
Anderson, J. A., 796
Anderson, J. D., 567
Annis, M., 202
Arad, B., 388
Asaro, F., 202, 339
Ashkin, J., 36, 82
Astbury, A., 721
Aston, F. W., 5, 213
Atkinson, J. H., 83
Auerbach, N., 612

Auger, P., 57
Avogadro, A., 1
Ayres, D. S., 712

Bacon, G. E., 675
Backenstoss, G., 313
Badhuri, L. K., 498
Bainbridge, K. T., 313
Baldin, A. M., 612
Baldinger, E., 166
Ballam, J., 160
Baltay, C., 796
Bang, V. X., 796
Bardon, M., 735
Bargmann, V., 747
Barkas, W., 82, 106, 128, 717, 798
Barkla, C. G., 60
Barnes, S. W., 776
Barshay, S., 465
Bartlett, J. H., 474
Bassel, R. H., 565
Bastien, P. L., 794
Bayley, J., 743
Becker, H., 616
Becquerel, A. H., 2, 104
Bég, M. A. B., 889
Bell, P. R., 100
Bell, R. E., 199, 559
Ben David, G., 388

Benecke, J., 913
Berger, M. J., 82, 83
Bergkvist, E. K., 419
Bernardini, G., 610
Bernoulli, D., 1
Bethe, H. A., xvii, xviii, 31, 36, 82, 325, 329, 471, 480, 489, 498, 547, 653, 675, 725, 747, 947
Betz, H. D., 82
Beyer, R. T., 15
Bhaba, H. J., 80, 852
Bhaduri, R. K., 314
Bieri, R., 216
Birge, R. W., 886
Birkhoff, R. D., 45
Birks, J. B., 15
Birnbaum, W., 798
Bishop, G. R., 612
Bistirlich, J. A., 127
Bjork, C. W., 620
Bjørken, F. D., 847, 860
Bjorklund, R. F., 748
Blackett, P. M. S., 71, 111, 122, 457
Blair, J. S., 515, 595
Blaton, J., 938
Blatt, J. M., xviii, 388, 612, 658, 675, 947
Bleaney, B., 273
Blewett, J. P., xix, 166
Bloch, F., 30, 83, 266, 267
Bodansky, D., 512, 612
Bøggild, H., 901, 917, 918
Bohr, A., xviii, 239, 292, 293, 313, 388, 573, 584
Bohr, N., 5, 82, 519
Boltwood, R. B., 200
Boorse, H. A., 15
Booth, E. T., 610
Booth, N., 798
Born, M., 5, 925
Borst, L. B., 519
Bothe, W., xix, 6, 56, 83, 120, 128
Bowman, H. R., 588
Brabant, J. M., 104
Brack, M., 580, 612
Bradner, H., 128
Braga Marcazzan, G. M., 613
Bragg, W. H., 667
Bragg, W. L., 178, 665, 667
Breit, G., 130, 458, 477, 498, 501, 520
Brink, D. M., 388
Brode, R. B., 718
Brodsky, S. J., 860
Brolley, J. E., 593

Bromley, D. A., 457, 600
Brown, G. E., 313
Brown, R. H., 696
Bruck, H., 166
Brueckner, K. A., 471, 480, 774
Buck, B., 304, 310
Burcham, W. E., 612
Burgy, M. T., 438, 444, 673
Burhop, E. H., 82, 388
Burke, P. G., 796
Butler, C. C., 696
Butler, S. T., 612
Butterworth, I., 799

Cabibbo, N., 863, 868, 871
Camerini, U., 696, 886
Carlson, T. A., 396
Carmony, D. D., 796
Casimir, H. B. G., 253
Cassen, B., 459
Cavanagh, P. E., 80
Cerenkov, P. A., 31, 41
Chadwick, J., 6, 219, 237, 326, 391, 457, 616
Chamberlain, O., xvi, 7, 166
Champion, F. C., 457
Charpak, G., 128
Chase, R. L., 551
Cheston, W. B., 202, 814
Chew, G., 796, 799, 846
Chihiro, K., 15
Chinowsky, W., xiv
Chiu, C. B., 831, 839
Chou, T. T., 913
Christenson, J. H., 8, 887
Christofilos, N., 7, 131, 149
Church, E. L., 388
Cockcroft, J. D., 7, 130, 136
Cohen, B. L., xvii, 166, 290, 313, 612
Cole, T. E., 675
Combley, F., 747
Commins, E., xiv, xviii, 444, 747, 889
Compton, A. H., 60
Condon, E. U., xx, 321, 359, 459, 498, 947
Conversi, M., 8, 114, 732, 737
Cook, C. S., 417
Corben, S., 82
Corson, D., 718
Cosme, G., 854
Courant, E. D., 7, 131, 149, 150, 166
Courant, H., 796
Cowan, C. L., 397, 444
Cowan, E. W., 111

Cox, R. T., 408
Crandall, W. E., 748
Cranshaw, T. E., 114
Crewe, A. V., 766
Crispin, A., 82
Cronin, J. W., 8, 887
Crookes, W., 97
Crowe, K. M., 720, 798
Csikay, J., 395
Cumming, J. B., 609, 612
Curie, E., 15
Curie, I., 6, 15, 616
Curie, M., 2, 3, 15, 176
Curie, P., 176

Dalitz, R. H., 722, 784
Dalrymple, G. B., 202
Dalton, J., 1
Damgaard, J., 612
Danby, G., 444
Danos, M., 612
Danysz, M., 810, 814
Datz, S., 82
Davis, R., 401
Day, B. D., 498
De Beer, J. F., 114
De Benedetti, S., xviii, 82
de Broglie, L., 5
Dedrick, K. G., 938
De Franceschi, G., 799
Dehmelt, H. G., 253
de Hoffmann, F., xviii, 725, 747, 947
Demeur, M., 612
Dennison, D. M., 255
De Rújula, A., 747
De Shalit, A., 313
Deutsch, M., 77, 128, 388, 444
Devons, S., 388
Dewitt, C., xviii
Diambrini Palazzi, G., 83
Dirac, P. A. M., 5, 6, 228, 274, 394, 431, 694
Donnelly, T. W., 313
Drell, S. D., 860
Drijard, F. E., 202
Drisko, R. M., 565
Duckworth, H. E., 313
Du Mond, J., 127

Eadie, W. T., 202
Eberhard, P., 702
Eccles, S. F., 512

Eckart, C., 468
Edlund, M. C., xviii, 675
Edmonds, A. R., 551, 947
Einstein, A., 2
Eisberg, R. M., 612
Eisenbud, L., 470
Ellis, C. D., 326
Elster, J., 97
Elton, L. R. B., xvii
Endt, P. M., xviii, 612
Enke, C. G., 128
Erginsoy, C., 82
Erozolimsky, B. G., 438
Erwin, A. R., 796
Etherington, H., 675
Eve, A. S., 15
Everling, E., 216

Fabri, E., 784
Fabricand, B. P., 603
Fagg, L. W., 83
Fairstein, E., 128
Faissner, H., 498
Fajans, K., 4
Fano, U., xvi, 83
Faraday, M., 1
Farquhar, R. M., 202
Farwell, G., 512, 515
Feather, N., 41
Feenberg, E., 420
Feinberg, G., 444, 747
Feld, B. T., xviii, 675, 839
Feldman, L., 418
Feller, W., 202
Ferbel, T., 901, 918
Fermi, E., xv, xvii, xviii, xx, 6, 7, 15, 68, 393,
  404, 410, 444, 630, 669, 672, 675,
  725, 747, 799, 816, 895
Fernbach, S., 220, 221, 313, 531, 550
Ferrari, E., xix, 166
Ferro-Luzzi, M., 791
Feshbach, H., 512, 547
Feynman, R. P., xviii, 443, 725, 747, 753,
  862, 911, 913
Fickinger, W., 796
Fierz, M., 15
Firk, F. W. K., 612
Fisher, D. E., 597
Fisher, R. A., 202
Fitch, V. L., 8, 611, 887
Fleischer, R. L., 108
Flerov, G. N., 572

Flügge, S., xx, 339
Foà, L., 900, 902, 912, 918
Fowler, G. N., 82
Fowler, J. L., 675, 796
Fowler, P. H., xix, 128, 696, 699, 725, 919
Frank, I. M., 41
Fraser, J. S., 725, 798, 799
Frauenfelder, H., xvi, 272, 388, 429
Frazer, W. R., xvii, 846, 919
French, J. B., 613
Friedländer, G., xvii
Friedman, A. M., 741
Friedman, J. I., 848, 860
Friedrich, W., 5
Frisch, O. R., 15
Froissart, M., 799
Fry, W. F., 812
Fukui, S., 114
Fulco, J., 798, 846
Fuller, E. G., 602, 604, 612
Fulmer, R. H., 612

Gaillard, J. M., 444
Gaillard, M. K., 839, 860, 889
Gammel, J. L., 477
Gamow, G., 321, 422, 436
Gardner, E., 748
Garg, J. B., 557
Garvey, G. T., 248, 314
Garwin, E. L., 766
Garwin, R. L., 740
Gasiorowicz, S., xviii, 839
Gatto, R., 860
Geiger, H., 26, 328
Geitel, W., 97
Gell-Mann, M., 8, 443, 697, 699, 706, 799,
        817, 828, 839, 862, 882
Gentner, W., xix, 83, 128
Georgi, H., 747
Gerace, W. J., 314
Gershtein, S. S., 862
Geschwind, S. S., 314
Ghoshal, S. N., 521
Giacomelli, G., 776
Gibb, T. C., 388
Gilman, F. J., 860
Glaser, D., 112, 128
Glashow, S. L., 747, 832, 835, 872
Glasstone, S., xviii, 675
Glauber, R. J., 767, 799, 844
Glendenning, N. K., 612
Gloyna, E. F., 202

Goebel, C. J., 796
Goldanskii, V., 612
Goldberg, E., 596
Goldberg, M., 839
Goldberger, M. L., xiv, xviii, 675, 865, 875
Goldfarb, J. B., 388
Goldhaber, M., 361, 372, 388, 426, 444, 601
Goldstone, J., 471, 480
Goldwasser, E. L., 50, 52
Gomes, L. C., 498
Good, M. L., 702
Gordon, G. E., 599
Gordon, W., 750
Gorter, C. J., 273
Goudsmit, S., 255
Goulding, F. S., 92–94
Goulianos, K., 444
Gove, N. B., 218, 249, 314
Gozzini, A., 114
Graham, R. L., 199
Gray, L. H., 178
Graziano, W., 702
Green, A. E. S., 212, 314
Green, G. K., 150, 166
Greenberg, O. W., 836
Greenberg, W. M., 428
Greenwood, N. N., 388
Greinacher, H., 136
Greisen, K., 48, 70
Grill, A., 246
Grodzins, L., 426, 444
Gropp, A., 597
Gross, E. E., 466
Grouch, S. R., 128
Gugelot, P. C., 558
Gurney, R. W., 321

Haeberli, W., 166
Hagedorn, R., xviii, 897, 919
Hahn, J., 128
Hahn O., 372, 572
Haïssinsky, M., xviii
Halpern, I., 592
Hamilton, J. H., 306
Hanna, G. C., 320, 642
Hanna, S. S., 83, 385
Hansen, W. W., 267
Hanson, A. O., 50, 52, 97, 617
Hartree, D. R., 10
Harvey, J. A., 551
Hattersley, P. M., 721
Haxel, O., 281

Hayward, E., 602, 613
Hayward, R. W., 8, 408, 445
Heberle, J., 386
Heisenberg, W., 5, 237, 459, 474, 498
Hendricks, C. D., 613
Henkel, R. L., 592, 593
Henley, E. M., 444
Herb, R. G., 166
Hertz, H., 344
Hevesy, G., xix, 10
Hildebrand, R., 739, 765
Hintenberger, H., 202, 314
Hittmar, O. H., 612
Hodgson, P. E., 314, 613, 697
Hoffman, M. M. and D. C., 583
Hofstadter, R., 87, 101, 223, 314, 844, 860
Hollander, J. M., xix, 202
Holloway, M. G., 21
Hoppes, D. D., 8, 408, 445
Hudis, J., 610, 613
Hudson, R. P., 8, 408, 445
Hughes, D. J., xix, 541, 542, 613, 673, 675
Hughes, V. W., 745, 747
Huizenga, J. R., 613
Hulthén, L., 498
Humphrey, W. E., 114, 128
Hussain, M., 721
Huus, T., 388
Huwe, D. O., 808
Hyams, B. D., 740
Hyde, E. K., 599

Iizuka, J., 857
Iliopoulos, J., 832, 872
Inghram, M. G., 399, 444
Ivanenko, D., 237

Jackson, D. A., 263
Jackson, J. D., xiv, xx, 17, 83, 388, 908, 919
Jacob, G., 613
Jacob, M., xviii, 918
Jaffe, R. L., 314
James, M., 202
Jaros, J., 918
Jeffries, C. D., 314
Jensen, A. S., 612
Jensen, J. H. D., xix, 281, 314, 388
Joachain, C. J., 799
Joffe, B. L., 839
Johnson, C. H., 396
Joliot-Curie, F., 6, 15, 616

Jones, R. J., 132
Judd, D. L., xiv, xvi
Juric, M., 813, 814

Källén, G., xviii, 725
Kämpffer, F. A., xx, 725
Kapur, P. L., 512
Karplus, R., 814
Kaufmann, W., 2
Keepin, G. R., 589, 591, 675
Kellogg, J. M. B., 264, 266
Kelly, E., 559, 568
Kelly, F. M., 314
Kelson, I., 314
Kemp, A., 721
Kendall, H. W., 848, 860
Kennedy, J. W., xvii
Kerlee, D. D., 596
Kerman, A. K., 612
Kerst, D. W., 131, 166
Kinoshita, S., 104
Kinsey, K., 776
Klein, O., 61, 750
Kleinknecht, K., 889
Knipping, P., 5
Koch, H. W., 83
Koehler, W. C., 675
Kofoed-Hansen, O., 128, 227, 314, 388, 444
Kogut, J., 860
Konopinski, E. J., 444
Kopfermann, H., xix, 314
Korff, S. A., 43, 128
Kraybill, H., 796
Kretzschmar, M., 919
Krohn, V. E., 438, 444
Krueger, H., 253
Kuhn, H., 263
Kuhn, W., 605
Kühner, J. A., 457, 600
Kullander, S., 516
Kurie, F. N., 416, 417, 423

Lal, D., 202
Lamarsch, J. R., xix, 675
Lamb, W., 659
Landau, L. D., 394
Lander, R., 51
Lane, A. M., 552, 613
Langevin, P., 268
Lanphere, M. A., 202
Lapostolle, P. M., 166

Larsh, A. E., 599
Lattes, C. G., 8, 9, 748, 799
Laukien, G., 314
Lauritsen, C., 130
Lawrence, E. O., 7, 130, 131
Lawson, J. D., 657
Lazarus, D., 428
Leachman, R. B., 587
Ledbetter, J. O., 202
Lederer, C. M., xix, 202, 277
Lederman, L. M., 444, 740, 747
Ledley, B., 766
Lee, B. W., 839, 860, 889
Lee, T. D., 7, 394, 405, 444, 889
Lee, Y. K., 444
Leibfried, G., 82
Leon, M., xvii, 725, 799, 814, 938
Levinger, J. S., 613
Libby, W. F., 201, 202
Lillethun, E., 766
Lindenbaum, S. J., 610
Lindhard, J., 51
Lipkin, H. J., 429, 444, 839
Litt, J., 102
Littlejohn, C., 385
Livingston, M. S., xix, 7, 21, 131, 149, 162, 166
Livingston, R. S., 132
Low, F. A., 796
Lüders, G., 434, 712
Lum, G., xiv
Lutz, H. O., 82

McCarthy, A. L., 612
McClure, J. D., 567
McDaniel, D. K., 515
Macfarlane, M. H., 613
McGowan, F. K., 387, 388
MacGregor, M. H., 491
McKibben, J. L., 97
McMaster, W. H., 498
McMillan, E. M., 7, 10, 142, 166
Maglic, B. C., 780
Maiani, L., 832, 873
Maier-Leibnitz, H., xix, 83, 128
Majorana, E., 474, 498
Malmstadt, H. V., 128
Mandelstam, S., 713
Mandl, F., xx
Mang, H. J., 331, 339
March, R., 766, 796
Markowitz, S., 766

Marion, J. B., 675
Maris, T., 613
Mariscotti, M. A. J., 304, 310
Marney, M. C., 666
Marschall, H., 339
Marsden, E., 26
Marshak, R. E., 477, 725
Marshalek, E., 312, 314
Marshall, J., 484
Marshall, L., 669, 672
Mattauch, J., 216, 282, 283
Matthiae, G., 918
Mayer, M. G., xix, 281, 314, 388
Meadows, B. T., 839
Meitner, L., 364
Mencuccini, C., 794
Messiah, A., xx
Meunier, R., 102
Meyer, S. L., 611
Meyerhof, W. E., xvii
Michel, H. V., 202
Michel, L., 734, 747
Migneco, E., 585
Mihelich, J. W., 363
Milazzo Colli, L., 613
Milburn, R. H., 47, 53
Miller, D. G., 797
Miller, J. M., xvii, 610, 613
Millikan, R. A., 15,
Millman, S., 264
Mills, F. E., 50, 52
Milton, J. C. D., 588
Mistry, N., 444
Miyake, K., 776
Miyamoto, S., 114
Mo, L. W., 444
Møller, C., 80, 851
Mössbauer, R. L., 378–384, 388
Molière, G., 44
Moon, P. B., 378
Moorhouse, R. G., 799
Moravcsik, M. J., 491, 498
Moretto, L., xiv, 577, 613
Morita, M., 444, 747
Morpurgo, G., 839
Morrison, P., xvii
Morse, P. M., 554, 632
Moseley, H. G. J., 5, 208, 210
Moszkowski, S. A., xix, 209, 314, 357, 425, 445
Mott, N. F., 27, 80, 455
Mottelson, B. R., xviii, 239, 292, 293, 313, 388, 573, 584

Motz, J. W., 83
Motz, L., 15
Moyer, B. J., 104, 202, 748
Muirhead, H., 9, 696, 721
Mukherjee, P., 612
Murray, J. J., 808
Myers, W. D., 245, 247, 314, 574, 579

Nagel, H. H., 75
Nambu, Y., 699
Neddermeyer, S. H., 8, 732
Ne'eman, Y., 8, 817, 839
Nemirowsky, S. A., 314
Newman, E., 466
Nier, A. O., 215
Nierenberg, W. A., 314
Nigam, B. P., 44, 498
Nilsson, S. G., 296
Nishijima, K., xviii, 8, 697, 699, 725, 799, 839
Nishina, Y., 61
Nix, J. R., 299, 581, 582, 613
Nobles, R. A., 592
Nordheim, L., 285
Northcliffe, L. C., 35, 83
Novey, T. B., 438, 444
Noyes, H. P., 454
Nurmia, M., 326
Nutall, J. M., 328

Occhialini, G., 8, 9, 71, 111, 122, 172, 748, 799
Odishaw, H., xx, 359
Okubo, S., 828, 857
Okun, L. B., 839, 889
Omnès, R., xviii, 725, 799
Oppenheimer, R., 566
Orear, U., xv
Owen, G. E., 417
Owen, R. B., 99

Page, L. A., 83
Pais, A., 699, 882, 883
Palathingal, J. C., 428
Pancini, E., 8, 732, 737
Paneth, F., 10
Panofsky, W. K. H., 166, 719
Paschos, E. A., 847, 860
Pauli, H. C., 575, 612
Pauli, W., 5, 7, 234, 255, 393, 434, 712

Peierls, R., 512
Pellegrini, C., 166
Perkins, D. H., xvii, xix, 128, 725, 919
Perlman, I., xix, 202, 325, 339
Perlow, G. J., 385
Persico, E., xix, 166
Person, L. W., 312, 314
Petrjak, K. A., 572
Pevsner, A., 794
Pfeiffer, L., 388
Philipp, K., 329
Phillips, M., 566
Phillips, R. H., 720
Picasso, E., 747
Piccioni, O., 8, 732 737, 883
Piroué, P. A., 611
Planck, M., 2, 15
Pleasonton, F., 396
Pniewski, J., 810, 814
Poincaré, H., 2
Pomeranchuk, I. Y., 899
Pontecorvo, B., 6
Porter, C. E., 547, 612
Post, R. F., 656, 657, 675
Pound, R. V., 273
Powell, C. F., xix, 8, 9, 128, 172, 696, 725, 748, 799, 919
Present, R. D., 498
Preston, M. A., xviii, 314, 388, 444, 498
Preston, R. S., 385
Price, P. B., 108
Primakoff, H., 202, 747, 814
Prosperi, D., xiv
Prout, W., 5
Puppi, G., 866
Purcell, E. M., 267

Quigg, C., 799
Quinn, H. R., 747

Rabi, I. I., 264, 266
Radicati, L., 467
Raghavan, R. S., 388
Rainwater, J., 555
Ramachandran, G., 516
Raman, V. C., 236
Ramsey, N. F., xix, 250, 266, 314
Ranft, J., 919
Rasetti, F., xv, 6, 236, 239, 339
Rasmussen, J. O., xvi, 325, 339
Rayleigh, J. W. S., 60

Reale, A., 799
Regge, T., 829, 910
Reines, F., 397
Remsberg, L. P., 202
Reynolds, H. L., 596
Reynolds, J. H., 202, 216, 399, 444
Ribe, F. L., 675
Rich, A., 742, 747
Rickey, M. E., 512
Riesenfeld, W. B., 552, 613
Rindi, A., xiv
Ring, J., 776
Ringo, G. R., 438, 444, 673
Ritson, D. M., 128, 696
Roberts, A., 115
Roberts, W. J., 466
Robinson, H. P., 124
Robinson, P. C., 512
Rochester, G. D., xix, 696
Roos, M., 202
Rose, B., 776
Rose, H. J., 388
Rose, M. E., 273, 388, 947
Rosenblum, S., 328
Rosenbluth, M. N., 843
Rosenfeld, A. H., xv, xvi, 114, 128, 780
Rosner, J. L., 839, 860, 889
Rossi, A., 429
Rossi, B., xix, 8, 48, 70, 83, 119, 895
Rost, E., 565
Royds, T., 3
Rozental, S., 15
Ruderman, M., 814
Russell, A. S., 4
Rutherford, E., 2, 3, 5, 15, 25, 26, 83, 97,
　　120, 124, 168, 175, 219, 326, 616
Rutkowsky, R. W., 466

Sachs, R. G., 886, 889
Sadoulet, P., 202
Sailor, V. L., 519, 592
Sakata, S., 816, 817
Salam, A., 394, 872
Salpeter, E. E., 654, 655
Salvini, G., 799
Samios, N. P., 839
Sandweiss, J., 796
Sanford, J. R., 166, 796
Satchler, G. R., 565
Saxon, D. S., 222, 548
Sayres, E., 128
Schaefer, J. B., 808

Schardt, A. W., 363
Scharff Goldhaber, G., 304, 310
Schecter, L., 47, 53
Schiff, L. I., xx, 947
Schluter, R. A., xv
Schmidt, T., 257, 275, 278
Schmitz, N., 796
Schneider, J. M., 613
Schneps, J., 812
Schrödinger, E., 5
Schueler, H., 257
Schumacher, C. R., 489
Schwartz, M., 444
Schwarzschild, A. Z., 202
Schwinger, J., 434, 498, 675
Schwitters, R. F., 838, 860, 919
Scott, W. T., 44, 83
Seaborg, G. T., 10, 314
Segrè, E., xviii, 6, 7, 10, 15, 363
Segre, S. E., xix, 166
Seitz, F., 675
Seki, R., 314
Sellin, I. A., 306
Septier, A. L., 166
Serber, R., 550, 571
Shafer, R., xvi
Shakin, C. M., 612
Shapiro, G., xiv
Sheline, R. K., 312, 314
Shirley, D. A., 314
Shirley, V. S., 277
Shore, F. J., 592
Shortley, G. H., 947
Shugart, H., xiv
Shull, C. G., 666, 670
Siegbahn, K., xix, 125, 314, 388, 444
Signell, P., 477
Sikkeland, T., 599
Simievic, A., 246
Simon, F., 273
Sirlin, A., 889
Skarsgard, H. M., 559
Slàtis, S. H. E., 125
Sloan, D. H., 130
Smart, J. S., 670
Smith, F. M., 798
Smith, L., 166
Smith, P. B., xviii, 612
Smith, R. K., 592
Smyth, H. D., xix
Snell, A. H., 128
Snyder, H. S., 7, 44, 131, 149
Soddy, F., 2, 4, 168

Solmitz, F. T., 202
Sommerfeld, A., 10
Sorensen, R. A., 314
Specht, H. J., 585, 613
Spencer, L. V., 83
Stapp, H. P., 491
Staub, H. H., 523
Steffen, K. G., 166
Steinberg, A. I., 438
Steinberger, J., 444, 719
Steiner, H., xvi
Steller, J. S., 719
Stelson, P. H., 388
Stephens, F. S., 339
Stern, O., 264
Stevenson, M. L., 8, 780
Stone, Y., 92–94
Stonehill, D., 796
Strassmann, F., 572
Strauch, K., 838, 860, 919
Street, J. C., 8
Strutinsky, V. M., 575, 579, 612
Suess, H. E., 202, 281
Sundaresan, M. K., 44, 725
Sunyar, A. W., 426, 444
Susskind, D., 860
Suzuki, M., xiv
Swami, M. S., 812
Swiatecki, W. J., 245, 247, 314, 574, 579, 588
Szalay, A., 395
Szilard, L., 131

Taagepera, R., 326
Taft, H., 796
Talmi, I., 313, 314, 613
Tamm, I. E., 41
Taschek, R. F., 617
Tavendale, A. J., 128
Taylor, T. B., 550
Telegdi, V., xvi, 438, 444, 741, 747
Teller, E., 422, 436, 498, 601, 665, 675
Temmer, G. M., xiv, 309, 465, 518
Terrell, J., 587
Thaler, R. M., 477
Theissen, H., 607
Theobald, J. P., 585
Thibaud, J., 131
Thiele, W., 282, 283
Thomas, L. H., 142
Thomas, R. G., 552
Thomas, W., 605
Thomas, T. D., 613

Thompson, S. G., 588
Thomson, J. J., 2, 5, 15, 58, 213
Ticho, H. K., 702
Titterton, K. T., 590
Tobocman, W., 562, 613
Tolhoek, H. A., 81
Treiman, S. B., 796, 865, 875, 886
Trigg, G., 420
Trilling, G. H., xvi
Tripp, R. D., 799, 814
Turlay, R., 8, 887
Tuve, M. A., 130
Tyrén, H., 516

Uhlenbeck, G., 255
Ullman, J. D., 429
Unna, I., 613

Van De Graaff, R. J., 130
Vanderbosch, R., 613
Veksler, V., 7, 142
Vincent, D. H., 385
Visscher, W. M., 383
von Baeyer, H. J., 364
von Laue, M., 5, 321, 341
von Schweidler, E., 2
von Weizsaecker, C. F., 68, 243, 372

Walecka, J. D., 313, 498
Walker, R. M., 128
Walker, W. D., 796
Wallace, R., 104
Walton, E. T. S., 7, 130, 136
Wapstra, A. H., 218, 242, 249, 282, 283, 314
Warburton, E. K., 202
Watson, K. M., xviii, 552, 613, 799
Weinberg, A. M., xix, 675
Weinberg, S., 872, 874, 889
Weinrich, M., 740
Weisberger, W., 866
Weiss, M. S., 604
Weisskopf, V. F., xviii, 15, 388, 480, 498, 512, 514, 546, 547, 612, 658, 675, 947
Weneser, J., 361, 672, 388
Wenzel, W. A., 128
Wesley, J. C., 742, 747
West, E., 796
Wetherill, L. T., 202
Weyl, H., 394

White, H., 256
Whittaker, E. T., 15
Wick, G. C., 474, 725, 751
Wideroe, R., 130
Wiechert, E., 2
Wiegand, C. E., 7, 230, 314, 722
Wien, W., 166, 214
Wigner, E. P., xix, 452, 466, 468, 470, 473,
   498, 501, 512, 520, 526, 552, 675
Wilets, L., 314, 551, 747
Wilkinson, D. H., xix
Wilkinson, M. K., 675
Williams, W. S. C., xviii, 68, 617, 725, 799,
   814, 889, 919
Willis, B. H., 83
Wilson, C. T. R., 38, 110
Wilson, J. G., xix
Wilson, K. G., 914
Wilson, R., 498, 612
Wilson, R. R., 22, 34, 166
Winther, A., 388
Wojcicki, S. G., 702
Wolfenstein, L., 487, 489, 498, 613, 889
Wollan, E. O., 666, 675
Wong, C., 567
Wong, C. Y., 612
Woods, R. D., 548

Wu, C. S., xvi, xix, 8, 83, 128, 231, 314, 408,
   416, 418, 444, 445, 747, 889
Wu, T. Y., 44

Yamaguchi, Y., 699
Yang, C. N., 8, 394, 405, 444, 754, 796, 816,
   913
Yarnell, J. L., 677
Yen, E., 913
Yergin, P. J., 603
York, D., 202
York, H., 748
Yoshida, S., 388
Ypsilantis, T., 7
Yuan, L. C., 83, 128
Yukawa, H., 8, 15, 749, 799

Zacharias, J. R., 266
Zakharov, V. I., 839
Zamick, L., 388
Zeldes, N., 246
Zeldovich, Y. B., 862
Zemach, C., 787
Zucker, A., 466, 596, 597
Zweig, G., 817, 857

Subject Index

Absorption coefficient, for X rays, 64
Accelerators, classification, 130
  linear, 153
  voltage scale, 135
Adair method, 809
Age (neutron) 633, 636
Alpha decay, fine structure, 328
  hindrance, 327
  stability against, 320
  systematics, 330
Alpha particle, 2, 319
  long-range, 329
  spectra, 326
Amplifiers, 118
Angular momenta, composition, 942
Antiparticles, 6, 694, 695, 762, 803
Associated production, 697
Asymmetry (in scattering), 481
Atomic number, 5, 208
Atoms, exotic, 227, 717
Atoms, mesic, 227, 717
Attenuation, 14
Axial vector, 407

Barrier penetration, 319, 321, 524
Baryon, 682, 695, 801, 805
Beam transport, 165
Beta decay, double, 399
  energetics, 402
  Fermi constant, 405, 736

Fermi's theory, 7
Fermi transitions, 422, 425, 432, 436
Gamow-Teller transitions, 422, 425, 432, 436
main features, 391
parity conservation, 405
of pions, 864
rays, 4
Betatron, 136
  biased, 138
  focusing, 138
  oscillations, 138
Biological effects of radiation, 179
Blocking effect, 517
Boson, 234, 874
Bragg curve, 20, 33
  reflection for neutrons, 665
Bragg-Gray relation, 178
Breeding and converting, 652
Bremsstrahlung, 36, 69
Bubble chamber, 112
Buckling, material and geometric, 646
Bunching, 144, 158

Cabibbo's theory, 836, 866
Carbon cycle, 653
Carbon dating, 200
Cerenkov radiation, 31, 41
  counter, 101

Chain reaction, 638
  critical dimension, 647
  generation time, 638
  multiplication constant, 641
Charge conjugation, 708
Charge (electric) nuclear, 208, 223
Charge independence, 458
Charm, 818, 832, 869
Charmonium, 833
Clebsch-Gordan coefficients, 942
Cloud chamber, 109
  continuously operating, 111
Cloudy crystal ball, 548
Cockcroft-Walton accelerator, 136
Coincidence method, 120, 123, 197
  accidental, 121
  delayed, 197
Colliding beams, 157
Color quantum numbers, 836
Compound nucleus, 519
Compton scattering, 54, 58, 160
  angular distribution, 61, 81
  polarization, 81, 160
Compton wavelength, 12
Confidence level, 190, 196
Conservation of i spin, 464, 763
Conservation laws, 695, 704, 706
Conserved vector current, 862
Control rods, 649
Conversion, coefficient, 362, 368
  electrons, 362
Correlations, angular, 271, 373, 430, 439
  between momenta and helicities in beta
      decay, 432
  coefficients in nucleon-nucleon scatter-
      ing, 489
Coulomb excitation, 375
Counter, Cerenkov, 101
  Geiger-Müller, 95
  neutron, 97
  plateau, 96
  proportional, 94
  scintillation, 97
  solid-state, 90, 94
Counting loss, 190
Coupling constants, 405, 705
CPT theorem, 434, 712, 887
CP violation, 887
Creation operators, 411
Critical dimension, 647
Critical mass, 648
Criticality condition, 647
Cross section, 13, 505

Crossing relations, 713, 715
Current-current interaction, 865, 873
Cyclotron, 141

Dalitz pair, 722, 761
Dalitz plot, 784, 789, 806
Dead time, 87
Decay constant, 3, 169
  for alpha emission, 325
  for beta decay, 421, 436
  measurement, 195
  of muon, 736
  of pion, 721, 864, 875
  of strange particles, 866
Deformation, nuclear, 295
Delayed neutrons, 586, 589, 650
Delayed protons, 320
Delta ray, 21
Density of final states, 364, 411, 506, 738,
      784
Density matrix, 930
Destruction operators, 411
Detailed balance, 570, 756
Detector's parameters, 88
Deuterium cycle, 653
Deuteron radius, 450
  binding energy, 449
  electric quadrupole moment, 447, 471
  magnetic moment, and spin, 447, 471
  mass, 447
  photodisintegration, 496
  reactions, 570
Diatomic molecule, 232
Dichotomic variables, 459
Diffraction scattering, 532, 902
  dissociation, 905
Diffusion coefficient (of neutrons), 632
  of thermal neutrons, 636
Dipole (*see* Electric *or* Magnetic dipole)
Direct reactions, 514, 560
Displacement law, 4
Dual decay, 170
Dual fields, 350

Effective range, approximation, 453, 458,
      776, 926
Eightfold way, 817
Elasticity (in reactions), 542
Electric dipole (static), 250
Electric dipole radiation, 343, 346
  in n-p system, 497

in pion-nucleon system, 778
sum rule, 605
Electrometer fiber, 91
Electron, absorption, 42
  charge, 208
  classical radius, 12, 59, 852
  conversion, 362
  drift velocity, 42
  energy loss, 43, 62
  mass, 211
  mean free pass, 43
  mobility, 42
  spin, and magnetic moment, 254, 266, 743
Electronics, amplifiers, 118
  coincidence circuits, 119
  discriminators, 119
Epithermal neutrons, 627
Equilibrium, radioactive, 174
Error probable, 188
Eta-meson, 779, 781, 790, 794
Excitation functions, 504, 559, 568, 592, 609
Exotic atoms, 227, 717

Fano factor, 44, 91
Feather relation, 41
Fermi age, 632, 633, 635
Fermi energy, 238, 288
Fermi gas, 236, 554
Fermi temperature, 554
Fermion, 234
Fission, 6, 572
  alpha emitted on, 590
  asymmetry, 578, 583
  barrier, 573, 580
  competition, 591
  correlations, 593
  energy balance, 591
  fragments, stopping, 35, 594
  by heavy ions, 599
  neutron emission, 586, 650
  products activity, 578, 583
  spontaneous, 575, 586
  threshold, 576
  yield, 578
Flavor quantum numbers, 836
Fluctuations, Gauss formula, 184
  in ionization current, 191
  Poisson formula, 181, 184
  in radioactive decay, 181, 189
Fluorescence, nuclear, 377
Fluorescence radiation, 55, 57
Form factor, 224

neutron, 844
proton, 846
Formation, 780, 805
Four current, 431
Four-factors formula, 643
Four-vector, 431
Fragmentation, limited, 931
Ft value, 420, 425
  nomogram, 424
Fusion (*see* Reactions, nuclear)

Gamma rays, 4
Gamma emission, angular momentum, 354
  multipole, 344
  probability, 347, 355
  selection rules, 346, 354
  symmetry, 349
Gauge theories, 839, 871, 874
Geiger-Müller (*see* Counter)
Geiger-Nuttall law, 326
Gluons, 818, 839
Goldberger-Treiman relation, 866, 875
Golden rule No. 2, 364, 411, 505
Gray-Bragg relation, 178
Gyromagnetic ratio, 254

Hadrons, 622
Half-life (*see* Mean life)
Handedness, 394
Helicity, 79, 736
  of electron in beta decay, 429, 433
  of neutrino, 394, 427, 429, 433
Hindrance, in alpha decay, 331
Hypercharge, 699
Hyperfine structure, 257
  selection rules, 261
  Zeeman effect, 259, 261, 263
Hyperfragments, binding, 812
  decay, 810
Hyperon, 682

i spin (*see* Isotopic spin)
Identity, effects in scattering, 455
  of electrons, 211
  of particles, 232
Impact parameter, 24
Inhour, 651
Interactions, classification, 404, 704
  direct (*see also* Direct interactions), 560
  minimal electromagnetic, 706

Intermediate boson, 864
Ionization chamber, 88, 91
  for neutrons, 90
  saturation, 88
Ion mobility, 43, 89
  drift velocity, 42, 88
  sources, 132
Isobar, 13
  chains in fission, 579
Isomer, 13, 369
  islands, 372
Isospin (see Isotopic spin)
Isotone, 13
Isotope, abundance, 216, 218
  definition, 5, 13
  mass, 211
  shift, 384
  systematics, 218
Isotopic spin, 458, 683, 706, 763
  and beta decay, 469
  and gamma emission, 468
  conservation of, 464, 706, 764
  multiplets, 468
  rotation, 465

Jets, in nuclear reaction, 891

K particle, decay, 879, 880
  discovery, 698
  intrinsic parity, 762
  mass doublet, 886
  regeneration, 883
  strangeness, 702
Kinematics of nuclear reactions, 501, 621
  relativistic, 936
Kurie plot, 416, 423

Lambda particle decay, 875
  intrinsic parity, 762
  magnetic moment and spin, 878
  strangeness, 702
Lawson criterion, 657
Lepton, 682, 695, 730
  conservation, 440, 695
  heavy, 839
Lethargy, 624, 626
Level width, 539
Level spacing, 546, 552
Linear energy transfer, 179
Liquid drop model, 240

Long wave length approximation, 342
Luminosity, 157

Magic numbers, 281, 284
Magnet, deflecting, 164
Magnetic dipole moment (static), 254, 266, 274
  energy, 258
Magnetic dipole radiation, 350, 492
Magnetic resonance, 266
Magneton, nuclear, 254
Mandelstam variables, 713, 936
Mass, defect, 212, 240
  determination, 213
  doublets, 217
  excess, 212
  formula for nuclei, 240, 245
  of particles, 716, 824
Mass number, 210
Materials properties, 42, 65, 98, 103, 105
Matrix S, 532, 682
  unitarity, 534
Matrix element, in beta decay, 417, 421, 436
  in gamma emission, 355
Maximum likelihood, 193
Mean life, 169
  reduced, in alpha decay, 331
    in beta decay, 421, 424
    in gamma decay, 369
  of muons, 717
  of particles, 716, 721
  of pions, 719, 721, 723
Mesic atoms, 227, 717
Meson (see also Pion), η meson, 682
  K meson, 790, 795
  ρ meson, 779, 781
  ω meson, 779, 782
Michel parameters, 734
Minimal electromagnetic interaction, 706
Mirror nuclei, 226, 460
Moderator, 631
Mössbauer effect (see Recoilless emission), 344, 345, 348
Molecular-beam method, 264
Moment of inertia (nuclear), 302
  variable, 304
Monte Carlo method, 608
Multiple scattering, 44
  in photographic emulsion, 106
Multipole radiation, electric and magnetic, 345, 348
  in pion production, 778

selection rules, 353
symmetry properties, 350
Muon, capture, 737
  decay, 733
  magnetic moment, 741
  mass, 733
  mean life, 736
  production, 732
  spin, 740
Muonic number, 442, 731
Muonium, 745

Neutrino collisions, 853, 872
  detection, 393, 398, 730
  mass, 393, 731
  recoil, 393, 396
  species, 401, 731
  spin, 393, 731
Neutron, balance in reactor, 648
  beta decay, 436, 438
  capture by hydrogen, 491
  detection, 90, 97, 99
  diffusion, 631
  discovery, 6, 616
  emitted on fission, 586
  interference phenomena, 660
  magnetic moment, 266, 274
  mass, 211
  migration length, 648
  monochromator, 666, 669
  multiplication constant, 641, 648
  polarization, 673
  reactor, 619, 638, 639
  reflection, 672
  refraction, 670
  scattering (*see* Scattering)
  slowing down, 621
  source, 617, 626
  thermal, 626
  velocity selector, 505
Nilsson diagram, 297
Nomenclature of particles, 682, 683
Nordheim rules, 285
Nuclear forces, 446, 748, 753
  charge independence, 464
  exchange, 473, 476
  range, 446
  saturation, 446, 472
  spin dependence, 469
  tensor, 470
Nuclear matter, 223, 279, 480
Nuclear models, collective, 293

comparison, 209, 311
optical, 547
shell, 279
Nuclear reactions (*see* Reactions, nuclear)
Nucleon-antinucleon annihilation, 696, 726
Nucleon, 12, 753
Nucleonic cascade, 608

Occupation numbers, 287
Omega meson, 779, 782, 846
Omega minus baryon, 826
  strangeness, 704
OPE (one-particle exchange), 477, 908
Operator, charge conjugation, 708, 711
  creation, 411
  destruction, 411
  electric charge, 708, 711
  exchange, 473
  G-parity, 710, 711
  projection, 664
Optical model, 547
Optical theorem, 535
Orbits in accelerators, 126
  bunching, 143, 158
  equilibrium, 143
  instantaneous, 137
  synchronous, 143
Orientation, nuclear, 273

Packing fraction, 212, 240
Pair production, 54, 61, 63, 69
Pair spectrometer, 127
Pairing effect, 287, 289
Panofsky ratio, 719
Parity, conservation, 405, 430
  of electromagnetic radiation, 761
  intrinsic, 757, 762
  of states, 352
  operator, 708, 710
Particle tables, 684, 703
Partons, 847, 913
Pauli matrices, 266, 459
Peripheral collision, 599, 795
Phase oscillations, 147
Phase-shift analysis, in pion-nucleon system, 768
  in scattering, 772, 920
Phase stability, 143
Photodisintegration of deuteron, 492, 496
Photoelectric absorption, 54

Photographic emulsion, 104
  composition, 105
Photoproduction, of muons, 597
  of neutrons, 532
  of pions, 777
Pion, beams, 161, 721
  decay, 864, 875
  discovery, 8
  intrinsic parity, 761
  mass, 719
  pion-nucleon system, 763
  pion-pion interaction, 796
  production, 777
  spin, 754
Polarization, circular, of X rays, 81
  in electrons, 78
  of lambda particles, 676
  of neutrons, 673
  in nucleon scattering, 481
  of spin-1/2 particles, 929
Polarization, nuclear, 272
Pomeranchuk theorem, 899
Positron, 6
  annihilation, 74
  electron collision, 851
Positronium, 76
Potential barrier, nuclear, 321, 334
  in nuclear reactions, 508, 524
  transparency, 324
Preequilibrium emission, 517
Pregnant nuclei, 650
Production (of particles), 780, 805
Propagation of errors, 188
Proportional counter, 94
Proton, magnetic moment, 254, 274
  mass, 211
  spin, 255
Pseudoscalars and scalars, 407, 410
Pseudovectors, 407
Psi and *J* particles, 854

Q in reactions, 502
Quadrupole electric (radiation), 344
Quadrupole electric (static), 250, 257, 294
  energy, 253
  intrinsic, 307, 310
Quadrupole lens, 161
  matrix method for calculating, 163
Quantum number, additive and multiplica-
      tive, 706
  G, 710
  of particles, 684, 703
  strangeness, 699

Quantum numbers for 2 pions, 783
  for nucleon antinucleon, 726
Quarks, 816, 818, 847
Quark combination, 821
  and cross sections, 829
  and magnetic moments, 828
  and masses, 826
Quasi particles, 289

Radiation dose, 176
Radiation length, 71
Radioactive branching, 170, 175
Radioactive decay, data, 2, 171
  chronological and geological applica-
      tions, 200
Radioactive families, 170
  equilibrium, 174
Radius, deuteron, 450
  electron (classical), 12, 59, 852
  neutron, 845
  nuclear, 26, 219
  proton, 225, 845
Random walk, 629
Range energy relation, 32, 34, 39
Range of particles, 20
Rapidity, 911
Reactions, nuclear, chain reaction (*see*
      Chain reaction)
  channel, 525, 526
  competition, 520
  endothermic, 509
  excitation function, 504
  exothermic, 508
  fusion, 652, 654
  by heavy ions, 595
  high energy, 897
  inclusive, 911
  intermediate energy, 608
  inverse, 496, 510, 723, 756
  kinematics, 503, 621
  knock-on, 564
  photonuclear, 601
  pick-up, 571
  resonance (*see* Resonance)
  statistical theory, 896
  stripping, 566, 571
  table, 513
Reactor, nuclear, 638
  control of, 649
  criticality condition, 647
  relaxation time of, 649
Recoilless emission, 271, 378, 381
  Doppler effect, 382

Zeeman effect, 386
Recurrence of particles, 829
Reference system, change, 938
Regge poles, 907, 910
  trajectories, 830
Resonances, as elementary particles, 780
Resonance in nuclear reactions, 514, 535, 538, 585
  Breit-Wigner formula, 520, 526, 537, 542, 591
  capture escape, 774
  integral, 643
  in pion-nucleon scattering, 774
Resonance giant, 550, 601, 607
Rho meson, 779, 781, 846
Roentgen unit, 176
Rotational levels (nuclear), 300
Rutherford scattering, 22, 222

Scalars and pseudoscalars, 407
Scaling, 847, 910, 913
Scattering amplitude, 529
  Bhabha, 80, 852
  coherent and incoherent, 661
  diffraction, 902
  elastic, 508, 527, 530, 621
  inelastic, 509
  magnetic, of neutrons, 674
  matrix, 491
  Møller, 80, 851
  Mott, 80, 843
  multiple, 44
  of neutrons in parahydrogen, 664
  optical theorem, 923
  phase-shift analysis, 920
  plural, 44
  polarization, 491
  Rayleigh, 54
  Rutherford, 22, 222
  spin-flip, 663
  Thomson, 54
  triple, 489
Scattering length, 458, 659, 673
  determination, 660
  and refraction index, 671
Scattering, neutron-proton, 450, 621, 673
  coherent and incoherent, 664
  length, 451, 453, 458, 673
  polarization, 484
  spin dependence, 452, 664
  proton-proton, 455
  length, 458
Schmidt lines, 278

Scintillation counters, 97
Scintillator, characteristics, 98, 101
Selection rules, in beta decay, 421, 432
  in charmed particles decay, 871
  in strange particles decay, 866, 871
  for gamma emission, 347, 351, 354
  parity, 352
Semiconductor, as detector, 90, 94
Separation energy, 282
Shell model, 226, 279
Showers, 71, 895
Sigma particle (*see* Strange particles *and* Lambda particle)
Solar cycles, 652
Spallation, nuclear, 612
Spark chamber, 114
Spectator, 560
Spectrometer, alpha, 124
  beta, 124
  Doppler, 382
  gamma, 126
  mass, 213
  pair, 128
Spherical harmonics, table, 948
Spin and magnetic moment, 254, 265, 271, 274, 809
Spin-orbit coupling, 280, 485
Spin and statistics, 236
Spinthariscope, 97
Standard deviation, 182, 186, 188
Statistics and spin, 232
Stigmatic system, 147
Stopping power, 20, 32, 38
  scaling relations, 33
Straggling, 20, 48
Strangeness, 8, 697, 699
Strange particles, associated production, 697
  quantum numbers, 702
Streamer chamber, 117
Strong focusing, 149
SU(2) and (3), 817
Sum rule, 602, 606
Synchrocyclotron, 148
Synchrotron, 149, 152
  radiation, 140

Temperature, nuclear, 554
Tensor forces, 470
Thermal column, 637
  utilization factor, 642
Thomson scattering, 58
  polarization, 60

Time reversal, 511, 534
Tolerance dose, 180
Tracks in emulsions, 104
  in solids, 107
Transition radiation, 41
Treiman-Yang angle, 796
Triangular inequality, 853

Unitary symmetry, 817
Units, atomic, 10
  of nuclear mass, 211
  radioactive, and radiation, 177

(V-A) interaction, 444, 882
Van De Graaff accelerator, 135
Velocity selector, 165
Vibrational levels (nuclear), 505

Virtual binding, 334
  emission, 752
  states, 338
Voltage multiplier, 136

Width, 539, 548
Wien filter, 166, 214

Xi particle, intrinsic parity, 762
  strangeness, 702

Yukawa interaction, 748
  fields, 750

Zitterbewegung, 429
Zweig rule, 857

Emilio Segrè

received his doctorate in 1928 from the University of Rome with Professor Fermi. In 1932 he was appointed Assistant Professor at the University of Rome.

In 1936 Professor Segrè became Director of the Physics Laboratory at the University of Palermo. He remained there until 1938, when he emigrated to the United States and joined the University of California at Berkeley. Later he was a group leader in the Los Alamos Laboratory and returned to Berkeley in 1946 as a Professor of Physics.

Dr. Segrè's work has been mainly in atomic and nuclear physics. In atomic spectroscopy he studied forbidden lines and the Zeeman effect. Since 1934 he has worked in nuclear physics, first collaborating with Professor Fermi on neutron research and the discovery of slow neutrons. Following his research in this field of nuclear physics, he worked in radiochemistry and participated in the discovery of the elements technetium, astatine, and plutonium. In particle physics, Professor Segrè has worked on the interaction between nucleons and related polarization phenomena, and on antinucleons. The 1959 Nobel Prize in physics was awarded jointly to the author and to O. Chamberlain for discovery of the antiproton.

Professor Segrè is the editor of *Experimental Nuclear Physics* and the author of *Enrico Fermi, Physicist.* He is a member of the National Academy of Sciences (U.S.A.) and of the Accademia Nazionale dei Lincei, Rome, and many other scientific and professional societies.

Dr. Segrè is currently an active participant in the field of nuclear physics at Lawrence Radiation Laboratory, University of California at Berkeley.

Nuclei and Particles

Second Edition

From the *Preface to the Second Edition*

"The second edition preserves the goals, level, and spirit of the first. In the last decade nuclear and particle physics have become increasingly technical, and both theory and experiment have grown more complicated. This is an unavoidable trend that tends to increase the gap between textbooks and original research literature in the journals, which often is intelligible to only a very restricted circle of initiates. However, the student should somehow be given a glimpse of what the specialists are doing before joining them and should acquire an idea of the forest before concentrating on the tree. One of the aims of this book is to convey such a general, but not superficial, view of the subject.

"I have revised the whole text, updating numerical data and improving derivations and writing style. I have added or replaced many figures, and a study of the illustrations should be rewarding for the inquisitive student. The bibliographies have been modernized, especially by the addition of easily accessible review articles that are known to me for their quality and clarity. Problems have been added and changed. I have tried not to increase the length of the book and, whenever possible, subjects that are common knowledge or are usually taught in other courses have been omitted."

The Benjamin/Cummings Publishing Company, Inc.
Advanced Book Program
Reading, Massachusetts 01867